Intermediate Filaments

Subcellular Biochemistry
Volume 31

SUBCELLULAR BIOCHEMISTRY

Recent Volumes in This Series:

Volume 22 **Membrane Biogenesis**
Edited by A. H. Maddy and J. R. Harris

Volume 23 **Physicochemical Methods in the Study of Biomembranes**
Edited by Herwig J. Hilderson and Gregory B. Ralston

Volume 24 **Proteins: Structure, Function, and Engineering**
Edited by B. B. Biswas and Siddhartha Roy

Volume 25 **Ascorbic Acid: Biochemistry and Biomedical Cell Biology**
Edited by J. Robin Harris

Volume 26 ***myo*-Inositol Phosphates, Phosphoinositides, and Signal Transduction**
Edited by B. B. Biswas and Susweta Biswas

Volume 27 **Biology of the Lysosome**
Edited by John B. Lloyd and Robert W. Mason

Volume 28 **Cholesterol: Its Functions and Metabolism in Biology and Medicine**
Edited by Robert Bittman

Volume 29 **Plant–Microbe Interactions**
Edited by B. B. Biswas and H. K. Das

Volume 30 **Fat-Soluble Vitamins**
Edited by Peter J. Quinn and Valerian E. Kagan

Volume 31 **Intermediate Filaments**
Edited by Harald Herrmann and J. Robin Harris

Intermediate Filaments

Subcellular Biochemistry
Volume 31

Edited by

Harald Herrmann
German Cancer Research Center
Heidelberg, Germany

and

J. Robin Harris
University of Mainz
Mainz, Germany

PLENUM PRESS • NEW YORK AND LONDON

The Library of Congress cataloged the first volume of this title as follows:

Sub-cellular biochemistry.
London, New York, Plenum Press.
v. illus. 23 cm. quarterly.
Began with Sept. 1971 issue. Cf. New serial titles.
1. Cytochemistry—Periodicals. 2. Cell organelles—Periodicals.
QH611.S84 574.8'76 73-643479

ISSN 0306-0225

ISBN 0-306-45854-3

This series is a continuation of the journal *Sub-Cellular Biochemistry,* Volumes 1 to 4 of which were published quarterly from 1972 to 1975

A Division of Plenum Publishing Corporation
233 Spring Street, New York, N.Y. 10013

http://www.plenum.com

10 9 8 7 6 5 4 3 2 1

Printed in the United States of America

Contributors

Ueli Aebi Maurice E. Müller Institute, Biozentrum, University of Basel, CH-4056 Basel, Switzerland

Ichiro Azuma Institute of Immunological Science, Hokkaido University, Sapporo 060, Japan

Racine M. Barton Departments of Medicine and of Anatomy and Cell Biology, College of Physicians and Surgeons, Columbia University, New York, New York 10032

Michael Bloemendal Department of Physics and Astronomy, and Institute for Molecular Biological Sciences, Vrije Universiteit, 1081 HV Amsterdam, The Netherlands

Yassemi Capetanaki Department of Cell Biology, Baylor College of Medicine, Houston, Texas 77030

Pierre A. Coulombe Departments of Biological Chemistry and Dermatology, The Johns Hopkins University School of Medicine, Baltimore, Maryland 21205

Robert M. Evans Department of Pathology, University of Colorado Health Sciences Center, Denver, Colorado 80262

Itzhak Fischer Department of Neurobiology and Anatomy, Allegheny Medical College, Philadelphia, Pennsylvania 19129

Roberto L. Flores Departments of Pathology and of Anatomy and Cell Biology, Columbia University College of Physicians and Surgeons, New York, New York 10032

David L. Gard Department of Biology, University of Utah, Salt Lake City, Utah 84112-0140

Harald Herrmann Division of Cell Biology, German Cancer Research Center, D-69120 Heidelberg, Germany

Ilse Hofmann Division of Cell Biology, German Cancer Research Center, D-69120 Heidelberg, Germany

Masaki Inagaki Laboratory of Biochemistry, Aichi Cancer Center Research Institute, Chikusa-ku, Nagoya, Aichi 464, Japan

Naoyuki Inagaki Laboratory of Biochemistry, Aichi Cancer Center Research Institute, Chikusa-ku, Nagoya, Aichi 464, Japan

Paul A. Janmey Division of Experimental Medicine, Brigham and Women's Hospital, and Program in Biological and Biomedical Sciences, Harvard Medical School, Boston, Massachusetts 02115

Klaus-Peter Janssen Adolf-Butenandt-Institut für Zellbiologie, Ludwig-Maximilians-Universität, D-80336 Munich, Germany

Cheolwha Jung Center for Cellular Neurobiology and Neurodegeneration Research, Department of Biological Sciences, University of Massachusetts at Lowell, Lowell, Massachusetts 01854

Michael W. Klymkowsky Molecular, Cellular, and Developmental Biology, University of Colorado at Boulder, Boulder, Colorado 80309-0347

Martin Kooijman Department of Physics and Astronomy, and Institute for Molecular Biological Sciences, Vrije Universiteit, 1081 HV Amsterdam, The Netherlands; *present address*: Centre for Protein Technology TNO-WAU, 6700 EV Wageningen, The Netherlands

Hidetaka Kosako Laboratory of Biochemistry, Aichi Cancer Center Research Institute, Chikusa-ku, Nagoya, Aichi 464, Japan

Georg Krohne Division of Electron Microscopy, Biocenter of the University of Würzburg, D-97074 Würzburg, Germany

Nam-On Ku Department of Medicine, VA Palo Alto Health Care System, Palo Alto, California 94304; and Stanford University School of Medicine, Stanford, California 94305-5487

Rudolf E. Leube Department of Anatomy, Johannes Gutenberg University, D-55099 Mainz, Germany

Conrad L. Leung Department of Biochemistry and Molecular Biophysics, Columbia University College of Physicians and Surgeons, New York, New York 10032

Jian Liao CLONTECH Laboratories, Palo Alto, California 94303

Ronald K. H. Liem Departments of Pathology and of Anatomy and Cell Biology, Columbia University College of Physicians and Surgeons, New York, New York 10032

DongLing Ma Department of Neurobiology and Anatomy, Allegheny Medical College, Philadelphia, Pennsylvania 19129

Thomas M. Magin Department of Molecular Genetics, Institute of Genetics, University of Bonn, and Bonner Forum Biomedizin, D-53117 Bonn, Germany

Jürgen Markl Institute of Zoology, Johannes Gutenberg University of Mainz, D-55099 Mainz, Germany

Shigeo Masaki Department of Biochemistry, Institute for Developmental Research, Aichi Human Service Center, Kasugai, Aichi 480-03, Japan

Kaori Matsuzawa Laboratory of Biochemistry, Aichi Cancer Center Research Institute, Chikusa-ku, Nagoya, Aichi 464, Japan; and Institute of Immunological Science, Hokkaido University, Sapporo 060, Japan

Kevin McGowan Departments of Biological Chemistry and Dermatology, The Johns Hopkins University School of Medicine, Baltimore, Maryland 21205

Derek J. Milner Department of Cell Biology, Baylor College of Medicine, Houston, Texas 77030

Dietmar Mischke Institut für Immungenetik, Charité Medizinische Fakultät der Humboldt-Universität zu Berlin, D-14050 Berlin, Germany

Roland Moll Department of Pathology, Philipps University of Marburg Medical School, D-35033 Marburg, Germany

M. Bishr Omary Department of Medicine, VA Palo Alto Health Care System, Palo Alto, California 94304; and Stanford University School of Medicine, Stanford, California 94305-5487

Daniel Price Department of Medicine, VA Palo Alto Health Care System, Palo Alto, California 94304; and Stanford University School of Medicine, Stanford, California 94305-5487

Roy A. Quinlan Department of Biochemistry, University of Dundee, Dundee DD1 4HN, Scotland

Aileen Sandilands Department of Biochemistry, University of Dundee, Dundee DD1 4HN, Scotland

Nisson Schechter Department of Psychiatry and Behavioral Science, and Department of Biochemistry and Cell Biology, Institute for Cell and Developmental Biology, Health Science Center, State University of New York at Stony Brook, Stony Brook, New York 11794

Manfred Schliwa Adolf-Butenandt-Institut für Zellbiologie, Ludwig-Maximilians-Universität, D-80336 Munich, Germany

Susan C. Schweitzer Department of Pathology, University of Colorado Health Sciences Center, Denver, Colorado 80262

Jagesh V. Shah Division of Experimental Medicine, Brigham and Women's Hospital, Boston, Massachusetts 02115

Thomas B. Shea Center for Cellular Neurobiology and Neurodegeneration Research, Department of Biological Sciences, University of Massachusetts at Lowell, Lowell, Massachusetts 01854

Sergey M. Troyanovsky Department of Dermatology, Washington University Medical School, St. Louis, Missouri 63110

Herbert van Amerongen Department of Physics and Astronomy, and Institute for Molecular Biological Sciences, Vrije Universiteit, 1081 HV Amsterdam, The Netherlands

Howard J. Worman Departments of Medicine and of Anatomy and Cell Biology, College of Physicians and Surgeons, Columbia University, New York, New York 10032

Jason Yabe Center for Cellular Neurobiology and Neurodegeneration Research, Department of Biological Sciences, University of Massachusetts at Lowell, Lowell, Massachusetts 01854

Qian Ye Departments of Medicine and of Anatomy and Cell Biology, College of Physicians and Surgeons, Columbia University, New York, New York 10032; *present address*: Laboratory of Developmental Hematopoiesis, Sloan-Kettering Institute for Cancer Research, New York, New York 10021

Foreword

The more we learn about intermediate filament (IF) proteins, the more puzzling the structures they form appear to be. When in the beginning of the main wave of IF reseach, i.e., 1978–1985, most of the emphasis was laid on the elucidation of the complexity of this multigene family and their specific expression patterns during mammalian—in particular, human embryonic—development and in diseases, the interest soon spread to other vertebrates, such as amphibia and fish, which are at least as well equipped with a diversity of IF proteins. Moreover, it seems rather safe to speculate that the ongoing sequencing of the human genome will bring more IF proteins to light. As with other IF proteins, such as cytokeratin 20, specific antibodies to certain IF proteins may then spot and characterize individual cells and pathways of differentiation.

Recent IF "gene knockout" data with mice and gene analyses in human hereditary diseases have already taught us that the consequences of them missing or present in mutated forms can be as diverse as their presumed functions in the individual cell types. IF proteins have evolved in such a versatile way that they are not only engaged in the thermoregulation of the higher vertebrate as major elements of the epidermis, the fur and the feather coat, but we find them also forming, from insects to man, the hallmark structure of the nuclear envelope, the nuclear lamina. Accordingly, their function will differ quite distinctly when either taking part in the general organization of nuclear architecture or in the specific constitution of a body barrier like epidermis. Very little is known about the role of IF proteins in the life of a white blood cell, yet vimentin is found in lymphocytes from fish to man. Thus, while we have no idea of its raison d'être, evolution must know.

We also know from experiments with transgenic mice that the ectopic expression of IF proteins may have a most pronounced effect: For example, epidermal cytokeratins, synthesized in the islet cells of the pancreas, interfere with cell function and are a death sentence to these cells. Also, moderate modifications of neuronal IF proteins may cause drastic changes of cellular performance leading ultimately to motor neuron death. We know an increasing number of examples where the presence of the right IF proteins in the right cell is important.

This book contains articles by a number of "IF aficionados" and leaders in this field of research, and I am particularly pleased to see several of my former and present co-workers and students among them. The scope of the book is remarkably broad, covering the field from tumor typing to gene "knockouts," from IF assembly mechanism to the subgroup of "beaded chain" filaments, from nuclear lamins to the organization of the desmosomal complex. I am sure that this book will provide a comprehensive overview of the essentials of IF research as well as of very recent developments, including also a wealth of unpublished data and concepts. I am convinced this will stimulate new experiments and again broaden the interest in these "tough filaments."

Dr. Werner W. Franke

German Cancer Research Center
Heidelberg, Germany

Preface

Among the first proteins to be analyzed by X-ray crystallography, in the 1930s by William Astbury, were those that make up our hair and nails, the keratins. In the same year when the first protein sequence was reported by Fred Sanger, an important structural feature of keratins was proposed by Francis Crick based on these early results (F. Crick, 1953, The packing of α-helices: Simple coiled-coils, *Acta Crystallogr.* **6**:689–697). However, nearly 20 years later, a biology textbook (H. R. Mahler and E. H. Cordes, 1971, *Biological Chemistry*, 2nd ed., Harper and Row, New York) referred only briefly to keratins as "major constituents of skin, hair, and other epidermal structures." Since the 1970s an explosion of information regarding the structurally significant proteins within the complex mixture referred to as keratins, the intermediate filament (IF) proteins, has occurred. We are now witnessing the functional exploration of the members of a growing multigene family, employing all the latest methods of biochemistry, molecular biology, and cell biology as well as biophysical chemistry. Crystals of these extended coiled-coil-forming proteins have not been obtained yet, so that high-resolution structural information, besides the assumed coiled-coil formation of the central, mostly alpha-helical rod domain, is lacking.

A general introduction in the concept that cellular ultrastructure is organized and dominated by fibrous proteins is presented by Manfred Schliwa (*The Cytoskeleton*, Springer-Verlag, Vienna, 1986) and Alexander D. Bershadsky and Juri M. Vasiliev (*Cytoskeleton*, Plenum Press, New York, 1988). Much of the progress occurring in the last 20 years is reviewed in books edited by Jerry W. Shay (*Cell and Molecular Biology of the Cytoskeleton*, Plenum Press, New York, 1986) and by Robert D. Goldman and Peter M. Steinert (*Cellular and Molecular Biology of Intermediate Filaments*, Plenum Press, New York, 1990). Other valuable sources of information on intermediate filament biology have been edited by George E. Rogers *et al.* (*Biology of Wool and Hair*, Chapman and Hall, London, 1989) and Eugenia Wang, Donald Fischman, Ronald K. H. Liem, and Tung-Tien Sun (*Annals of the New York Academy of Sciences*, Vol. 455, "Intermediate Filaments," 1985).

Relevant monographs have been published by Peter Traub (*Intermediate Filaments*, Springer-Verlag, Berlin, 1985) and David A. D. Parry and Peter M. Steinert (*Intermediate Filament Structure*, Springer-Verlag, Berlin, 1995). Finally, concise and lucid overviews can be found in volumes of the *Annual Review of Biochemistry* (1982, 1988, and 1994) and the *Annual Review of Cell Biology* (1985).

The need for another book on intermediate filaments arose following an exciting Gordon Research Conference in 1996, where many of the participants reported new data and concepts that were apparently not available in the literature. Along these lines, we have encouraged our contributors to report data and ideas that were in progress or that were, for one reason or another, "still in the top drawer." Thus the material presented in this book aims to build upon the literature mentioned above and to draw attention to new and exciting research directions. This book also can be read and understood by the nonspecialist, since all the necessary basic knowledge is presented. We wish to thank all the authors for their efficiency and enthusiasm. We also gratefully acknowledge the expert secretarial assistance of Eva Ouis. Without her help this book project would have taken considerably longer.

Intermediate filament (IF) proteins are grouped into five sequence homology classes (see Chapter 10). Therefore, it was only logical to proceed from classes I and II, whose members always go together, to class V. According to their biological significance, cytokeratins (class I and II IF proteins) form the first section of this book, including Chapters 1–8. The second part of the book deals with the assembly of IF proteins, and introduces several important biophysical methods. This section mostly describes vimentin, a class III IF protein that is found in species from trout to humans. In the third section, the complexity of neuronal IF proteins (class IV) as well as of the nuclear IF proteins (class V) and their interacting non-IF partners is presented. Whereas the latter IF proteins, the lamins, were detected and numbered last, they very probably were the founding members of the IF protein family, and one might reasonably have started with them.

The evolution of the animal kingdom proceeded to a high degree of complexity even before some descendants of bony fish decided to explore dry land. Although this important step may have led to the development of new structural elements of the body plan in these pioneering species, the layout for cellular intermediate filament proteins must have been established long before, since bony fish are known to contain complex sets of IF proteins, and in addition cartilagenous fish such as sharks and cyclostomes such as lampreys also have IF proteins very similar to our own. In Chapter 1 the fish biologists Jürgen Markl and Nisson Schechter provide detailed information on this topic, reminding us of the fact that nothing in biology makes sense except in the light of evolution.

Amphibia, the classical subjects of embryology and developmental biology, also express IF proteins in very complex patterns. Some of them, such as the clawed toad, *Xenopus laevis*, build huge, intricately structured oocytes that allow early development to occur in an explosive way, leading to a neurulating organism 20 hr after fertilization. These oocytes contain a complex IF system that responds to maturation and fertilization in a very distinct manner, as reported by David L. Gard and Michael W. Klymkowsky in Chapter 2.

Behind such structural complexity lies a multigene family with a peculiar genomic organization. Dietmar Mischke in Chapter 3 renders an account of the epidermal differentiation complex, showing how the generation of structural diversity could be laid down in the DNA blueprint of our cells and how far we are from a clear and total understanding of the emergence and maintenance of cellular structure.

In Chapter 4 M. Bishr Omary and co-workers discuss the various types of modifications that cytokeratins may go through, especially during mitosis; in particular, they point out that glycosylation is a frequent posttranslational modification of these cytoskeletal structures, previously assumed to be relatively inert. Obviously, these biochemical reactions modulate protein–protein interactions and help make IF proteins very dynamic.

In Chapter 5 Thomas M. Magin gives a comprehensive overview of what happens when IF proteins, which are apparently central to developmental processes, as they are expressed early in the blastocyst, are taken away in the mouse. This "knockout" procedure—the targeted inactivation of genes—is assumed to be a crucial test for the functional importance of a gene product. Quite a number of researchers were surprised, however, when they realized that "important" molecules were not being missed that much, although one could argue that really important molecules are always "backed up" by other components in case they happen to be lost.

After taking part in the pioneering work of Elaine Fuchs' group on the importance of cytokeratin mutations for the instability of the epidermis in various pathological situations, Pierre A. Coulombe continued to work on cytokeratins that are expressed in physiologically relevant situations. In Chapter 6 he and Kevin McGowan report on IF proteins that are engaged in such mysterious events as those involved in wound healing, which not only needs a very complex biochemistry, but, on a higher level of organization, depends on a multitude of cellular interactions that employ dynamic changes in various cellular components and a fine-tuned cascade of signaling activities.

Human expression patterns of cytokeratins reflect the fact that our many cell types are highly selective with regard to the type of IF protein synthesized. In Chapter 7 Roland Moll describes how this selectivity provides a tool for identifying cells that have lost control of their growth and may have migrated through the body to places where they are physiologically simply not wanted. The phenomenon that most tumor cells retain the expression of their *original* IF proteins still has to be explained in terms of molecular pathology. However, modulations of cytokeratin patterns also need to be considered when applying such markers in histopathological tumor diagnosis.

The integration of IFs into the cellular architecture is mediated by complex junctional structures centered around transmembrane proteins that bind in a calcium-dependent way with their extracellular domain to similar molecules on neighboring cells (the cadherins), thereby generating a truly transcellular cytoskeleton. The intracellular part of such a cell–cell junction is a tight proteinaceous structure referred to as the desmosomal *plaque*, since it binds much heavy metal stain during processing for electron microscopy. This does not mean, however, that the proteins in this structure are not interacting dynamically. Plakoglobin, a central player in the establishment of desmosomal organization, carries 12 unique repeat domains, presumed folding units called "arm" repeats, after the *Drosophila* sister molecule Armadillo, in which they were first identified. (Recent X-ray analyses of the arm repeat domain of a cadherin-binding protein confirm that these repeats do indeed fold in a very distinctive way; see Huber, A. H., Nelson, W. J., and Weiss, W. I., 1997, Three-dimensional structure of the armadillo repeat region of beta-catenin, *Cell* **90**:871–882). Sergey M. Troyanovsky and Rudolf E. Leube in Chapter 8 describe how the dynamic interaction of desmoglein and desmocollin (both desmosomal cadherins) may take place with plakoglobin.

Differentiation of the eye lens tissue into a transparent entity involves a program of reorganizational events, including the removal of the nucleus and other cellular organelles, to convert an epithelial cell layer into fiber cells that guarantee the elasticity and translucency required for vision. Two new IF proteins involved in the establishment of the lens cytoskeleton have been recently cloned and characterized. Aileen Sandilands, Shigeo Masaki, and Roy A. Quinlan describe in Chapter 9 how the network of these unusual IF proteins is actively remodeled during lens fiber cell differentiation, by employing posttranslational modifications, such as proteolysis and phosphorylation, and the chaperone-like action of α-crystallins.

IF proteins are unique in that they can be renatured from high concentrations of urea very easily. Filament assembly can thereby be studied *in vitro*, and this, in combination with recombinant techniques, allows evaluation of the influence of sequence motifs as well as complete subdomains on assembly. Electron microscopy is a powerful tool, especially when used in the scanning transmission mode to quantitatively determine the molecular mass in filaments. Harald Herrmann and Ueli Aebi open the section of the book dealing with IF assembly in Chapter 10, in which these authors give a new view of how IFs assemble, which is completely different from what has been presented in textbooks and reviews.

This is followed by the presentation of a viscometric assay that is very convenient for monitoring the dynamics of IF formation (as required to assay the influence of various parameters, including temperature, ions, and mutations), by Ilse Hofmann in Chapter 11. Moreover, a short overview of how to use surface plasmon resonance biosensors for the determination of binding affinities of various IF proteins for one another is also included.

A new and surprising intrinsic property of IFs, the increase of their viscoelasticity upon application of stress, can be measured with the help of a laser-operated torsion pendulum that can be constructed simply from Lego bricks. Paul A. Janmey, Manfred Schliwa, and their colleagues from Boston and Munich elegantly describe in Chapter 12 how this is achieved.

The study of transient electric birefringence goes back to a phenomenon described by Kerr in 1875. It can be used to measure molecular properties, in particular of rod-shaped particles, with great precision. However, this topic requires a thorough and detailed understanding of biophysical methods and theory. This powerful method is presented in Chapter 13 by Herbert van Amerongen and co-workers.

IF proteins are subject to phosphorylation by various kinases. In this area one group has performed eminent work in the identification of the sites that are indeed modified, giving us the molecular information to understand the black "phosphorylation" spots on X-ray films. In Chapter 14 Masaki Inagaki and his group report on an exciting new connection of IFs to the Rho system of small GTP-binding proteins, the Rho-kinases.

The interaction of IFs containing vimentin with lipid droplets is described by Susan C. Schweitzer and Robert M. Evans in Chapter 15. These authors explore the potential role of this cytoskeletal component within lipid metabolism. Significantly, this concept goes back to early observations made by Peter Traub on the tenacious and selective binding of lipids to vimentin.

In Chapter 16 Yassemi Capetanaki and Derek J. Milner report on the drastic consequences that "knockout" of the muscle-specific IF protein desmin has for the architecture of the heart. In the absence of desmin there is extensive damage of the myocardium, and complete loss of coordination and alignment between neighboring myocytes. It will be interesting to determine from future studies how the organization of desmin IFs with other major players like titin, nebulin, plectin, and desmoplakin is coordinated.

Ronald K. H. Liem first published on intermediate filaments in nervous tissues some 20 years ago. In Chapter 17, he and his colleagues develop a clear picture of the complexity of the cytoskeleton in the nervous system. Given the number of neuronal and nonneuronal cell types, as well as the multitude of interactions these cells elaborate, the nervous system is by far the most complex part of the body.

The dynamics of neurofilaments and the intriguing questions of neurofilament transport are described in Chapter 18 by Thomas B. Shea and co-workers. They show that posttranslational modification, assembly, and transport through the axon of neurofilament subunits are distinct events, and that multiple populations of neurofilaments and neurofilament subunits coexist in neuronal cells.

It was a big surprise to the IF community when we learned that cytoplasmic IFs have a counterpart inside the nucleus, constituting the nuclear lamina, which stabilizes and helps to structurally integrate the nuclear envelope with its integral pore complexes. How these nuclear IFs, the lamins, do this job is discussed by Georg Krohne in Chapter 19. The other supporting nuclear IF elements, the lamin B receptor and lamina-associated proteins, are presented by Howard J. Worman and colleagues in Chapter 20.

Harald Herrmann
J. Robin Harris

Heidelberg, Germany
Mainz, Germany

Contents

Chapter 1
Fish Intermediate Filament Proteins in Structure, Evolution, and Function
Jürgen Markl and Nisson Schechter

Chapter 3

The Complexity of Gene Families Involved in Epithelial Differentiation: Keratin Genes and the Epidermal Differentiation Complex

Dietmar Mischke

Chapter 4

Keratin Modifications and Solubility Properties in Epithelial Cells and *in Vitro*

M. Bishr Omary, Nam-On Ku, Jian Liao, and Daniel Price

Chapter 5
Lessons from Keratin Transgenic and Knockout Mice
Thomas M. Magin

Chapter 6
The Wound Repair-Associated Keratins 6, 16, and 17: Insights into the Role of Intermediate Filaments in Specifying Keratinocyte Cytoarchitecture
Kevin McGowan and Pierre A. Coulombe

Chapter 9
Lens Intermediate Filament Proteins
Aileen Sandilands, Shigeo Masaki, and Roy A. Quinlan

Chapter 12

Viscoelasticity of Intermediate Filament Networks

Paul A. Janmey, Jagesh V. Shah, Klaus-Peter Janssen, and Manfred Schliwa

Chapter 13

Transient Electric Birefringence in the Study of Intermediate Filament Assembly

Herbert van Amerongen, Martin Kooijman, and Michael Bloemendal

Chapter 14

Possible Regulation of Intermediate Filament Proteins by Rho-Binding Kinases

Kaori Matsuzawa, Hidetaka Kosako, Ichiro Azuma, Naoyuki Inagaki, and Masaki Inagaki

Chapter 15

Vimentin and Lipid Metabolism

Susan C. Schweitzer and Robert M. Evans

Chapter 16
Desmin Cytoskeleton in Muscle Integrity and Function
Yassemi Capetanaki and Derek J. Milner

Chapter 17
The Complexity of Intermediate Filaments in the Nervous System
Conrad L. Leung, Roberto L. Flores, and Ronald K. H. Liem

Chapter 18

Extensive Phosphorylation and Axonal Transport of Triton-Soluble Neurofilament Subunits

Thomas B. Shea, Cheolwha Jung, Jason Yabe, DongLing Ma, and Itzhak Fischer

Chapter 19
Lamin Assembly *in Vivo*
Georg Krohne

Chapter 20
Nuclear Lamin-Binding Proteins
Qian Ye, Racine M. Barton, and Howard J. Worman

Chapter 1

Fish Intermediate Filament Proteins in Structure, Evolution, and Function

Jürgen Markl and Nisson Schechter

1. INTRODUCTION

Intermediate filaments (IFs) represent one of the three major cytoskeletal protein filament systems of most vertebrate cells. IFs form networks that extend from the nucleus to the plasma membrane, attach to desmosomes, and interact with a variety of cell structures. They support the tensile strength and shape of cells and probably even transfer information. In addition, the nuclear lamins are recognized as IF proteins, providing the cell nucleus with a putative skeleton for chromatin attachment. An intriguing molecular diversity of the IF proteins leads to tissue-specific expression patterns, which makes them ideal molecular markers in developmental biology and pathology. [For reviews, see Steinert and Roop (1988), Carmo-Fonseca and David-Ferreira (1990), and Fuchs and Weber (1994)].

Jürgen Markl Institute of Zoology, Johannes Gutenberg University of Mainz, D-55099 Mainz, Germany. **Nisson Schechter** Department of Psychiatry and Behavioral Science, and Department of Biochemistry and Cell Biology, Institute for Cell and Developmental Biology, Health Science Center, State University of New York at Stony Brook, Stony Brook, New York 11794.

Subcellular Biochemistry, Volume 31: Intermediate Filaments, edited by Herrmann and Harris. Plenum Press, New York, 1998.

1.1. Higher Vertebrate IF Proteins Are Expressed in Tissue-Specific Programs

From frogs to humans, the different cytoplasmic IF proteins are almost always specifically expressed: keratins in epithelia, vimentin in mesenchymally derived cells and tissues, desmin in muscle cells, glial fibrillary acidic protein (GFAP) in astrocytes, and neurofilament proteins in neurons [for exceptions, see, for example, Quinlan *et al.* (1995)]. According to their primary structure and gene organization, keratins are separated into type I and type II proteins; keratin filaments are characteristically assembled from type I/II heterodimers. Desmin, vimentin, and GFAP all represent type III proteins; they are more closely related to each other than to keratins, and form homopolymers. Distinct from type I to type III IF proteins, as suggested from several structural features, are the neurofilament proteins (type IV proteins; some type III neuronal proteins also exist). The type I to type IV IF proteins from terrestrial (tetrapod) vertebrates (amphibians, reptiles, birds, and mammals) are more closely related to each other than to the nuclear lamins (type V proteins), which also belong to the IF multigene family.

1.2. Invertebrate IF Proteins Are Closely Related to the Nuclear Lamins

Representative species from nearly all invertebrate phyla and from lower chordates have been screened for cytoplasmic IF proteins by electron microscopy and immunofluorescence, and in most cases IFs have been found, with the exception of the arthropods and their closest relatives (Bartnik and Weber, 1989). The absence of cytoplasmic IFs might be correlated with the supporting strength of a rigid exoskeleton, which probably reduces the need for a cytoskeleton; however, the fruit fly, *Drosophila*, has lamins (Riemer and Weber, 1994). Neuronal and nonneuronal IF proteins from annelids, mollusks, and nematodes have been sequenced, but the primary structures could not be correlated with any of the mammalian cytoplasmic IF protein classes: they contain, in the 1b domain of their α-helical rod, six additional heptads with 42 amino acids, which is typical for the nuclear lamins (e.g., Weber *et al.*, 1989; Dodemont *et al.*, 1990, 1994; Szaro *et al.*, 1991; Tomarev *et al.*, 1993; Adjaye *et al.*, 1995; Bovenschulte *et al.*, 1995). This, together with other key features, such as a lamin tail similarity, indicates that the neuronal and nonneuronal IF proteins of protostomic invertebrates are phylogenetically more closely related to the ubiquitous nuclear lamins (type V proteins) than to the type I to type IV IF proteins of terrestrial vertebrates.

In deuterostomic invertebrates, IF proteins have been more closely studied in echinoderms and ascidians (e.g., Terakado, 1988; Schroeder and Otto, 1991; Mencarelli *et al.*, 1991), although no sequence information is available from these species. However, Riemer *et al.* (1992) sequenced a gene coding for a cytoplasmic, nonneuronal IF protein (termed IF1) from the closest existing relative of the vertebrates, the cephalochordate *Branchiostoma lanceolatum*. The predicted primary structure of this protein lacks the extended coil 1b version and the lamin tail homology found in protostomic IF proteins. However, although its gene organization closely resembles vertebrate type III genes, it could not be clearly assigned to any of the vertebrate type I to type III proteins (Riemer *et al.*, 1992). This suggests that the differentiation into type I to type III proteins appeared later in evolution, probably when primordial vertebrates (with a nonstratified epidermis like their invertebrate ancestors) gradually evolved from sluggish filter feeders into rapid swimmers and predators. Such a lifestyle required a highly rigid, but flexible skin, which led to the evolution of a stratified epidermis with its tensile strength reinforced by a novel cytoskeletal element, α-keratin. The diversification of primordial IF proteins into type I and II keratins on one hand and the various type III proteins (notably vimentin, desmin, and GFAP) on the other is therefore hidden somewhere in the evolution of the most ancient and diverse of all vertebrate groups, the fishes.

1.3. Fish IF Proteins—Bewildering Chaos or Scientific Goldmine?

One good reason to investigate IF proteins in fish is to trace the phylogenetic pathways which have led to the different IF protein types and to their tissue-specific expression programs, as found today in mammals and other terrestrial vertebrates. Moreover, such studies could provide hints for possible biological functions of the surprising molecular diversity of IF proteins. They also might reveal, in specialized organs or tissues of this enormously diverse group of aquatic animals, totally novel IF proteins. Useful markers for cell and tissue typing in fish differentiation and pathology could ultimately be obtained, notably in genetic model organisms such as the zebrafish.

In teleost fishes, equivalents to the various IF protein types known from mammals have been noted. The occurrence of keratinlike tonofilaments has been reported in electron microscopic studies on fish epidermal keratinocytes (e.g., Schliwa, 1975; Bereiter-Hahn *et al.*, 1979; Cooper and Schliwa, 1986); interestingly, cold treatment induces a structural transformation of fish epidermal tonofilaments into dense amorphous aggregates that is readily reversed upon rewarming the cells (Schliwa and Euteneuer, 1979). The presence of two distinct subtypes of keratin polypeptides (type I and type II) in teleosts has been demonstrated by nucleic acid cross-hybridization using cDNA probes coding for human keratins (Fuchs and Marchuk, 1983). In the cichlid *Tilapia mossambica*, the presence of distinct genes of desmin and vimentin has been demonstrated by hybridization with hamster cDNA probes (Quax *et al.*, 1984). RNA coding for GFAP was detected in optic nerve astrocytes from the carp, *Cyprinus carpio* (Cohen *et al.*, 1993). A survey of different neurofilament proteins in fish based on immunoblot analyses can be found in Mencarelli *et al.* (1991).

With respect to biochemical detail of the IF protein system, the most thoroughly studied fish tissue is the goldfish optic nerve (see Section 3 of this chapter), whereas the most comprehensively studied fish with respect to the number of investigated cell types, tissues and organs is the rainbow trout (see Section 2). In both animals, IF protein expression patterns have been observed which differ fundamentally from those published for terrestrial vertebrates. The goldfish optic nerve, for example, contains keratins as major IF proteins (e.g., Giordano *et al.*, 1989). Many mesenchymal cells and tissues of the rainbow trout express keratins (Markl and Franke, 1988; Markl *et al.*, 1989); in this fish, vimentin is restricted to a small number of cell types (Herrmann *et al.*, 1996). It is difficult to integrate such fundamental interspecies differences alongside functional concepts of the diversity of intermediate filament proteins, a topic which has already been discussed in detail elsewhere (Markl, 1991; see also Section 2.6 of this chapter).

In addition to the rainbow trout studies, immunocytochemical keratin and/or vimentin surveys on several other teleost fish species are available, notably on the Northern pike (Thompson *et al.*, 1987), the striped bass and the medaka (Bunton, 1993), the Indian freshwater catfish (Ainis *et al.*, 1995), and the common carp (Groff *et al.*, 1997). However, these authors used in part different antibodies and also different tissue fixation procedures compared to the trout studies, and a comprehensive keratin catalog based on two-dimensional polyacrylamide gel electrophoresis (2D-PAGE) as in trout (Markl *et al.*, 1989) has still not been published for other fish species. Many fish IF protein studies focus on a single organ, tissue, or cell type, as for example, epidermis (Tsai, 1996), liver (Pankov *et al.*, 1986), testis (Arenas *et al.*, 1995), electrocytes (Frail *et al.*, 1990; Cordeiro *et al.*, 1996), melanophores (Gyoeva *et al.*, 1987), slime gland thread cells (Koch *et al.*, 1994), olfactory system (Byrd and Brunjes, 1995), retina (De Guevara *et al.*, 1994), spinal cord (Bodega *et al.*, 1995), or optic nerve and brain (see Section 3). These authors collected a wealth of useful information on their respective topic, but they did not attempt to reveal the entire IF protein system of their studied fish species.

As a result it is still difficult to present, on the basis of published work, a comprehensive view of the intermediate filament protein patterns in fish, although the rainbow trout and the goldfish could provide reference models for the biochemistry and immunocytochemistry of teleost IF proteins. In Section 2, which describes fish type I to type III proteins and biochemical and immunocytochemical information, we include a broad comparative survey. This compilation has taken several years and contains data which are not yet fully published. Overall, this project (J.M.) aims to identify biochemically, characterize, and sequence individual fish keratins and type III proteins, to analyze their expression patterns, and to link this knowledge to the data available from higher vertebrates and from invertebrates. This survey includes the rainbow trout, *Oncorhynchus mykiss*, the zebrafish, *Danio rerio*, the carp, *Cyprinus carpio*, the shark, *Scyliorhinus stellaris*, and the lamprey, *Lampetra fluviatilis*. Section 3 reviews the current knowledge on the goldfish (*Carassius auratus*) neurofilament protein system and provides new data on the zebrafish (N.S.).

2. FISH KERATINS AND TYPE III PROTEINS: STRUCTURE AND EVOLUTION

In mammals, a variety of distinct keratins have been characterized and, according to their expression patterns, classified into several subsets. One is typical for epidermal keratinocytes, another, predominantly consisting of equivalents of human keratins K8 and K18, is typical for simple epithelia and some nonepithelial cell types (Moll *et al.*, 1982). A comparable situation is found in the clawed toad, *Xenopus laevis* (e.g., Fouquet *et al.*, 1988), which is significant in view of the fact that the last common ancestor of amphibia and mammals lived in the Devonian, some 380 million years ago [for the geological time scale used in this chapter, see Gould (1993)].

2.1. Salmonid Teleosts: The Rainbow Trout as a Reference Model

It was 450 million years ago in the Ordovician when the bony fishes separated into two different phylogenetic lines: the sarcopterygians and the actinopterygians. The one gave rise, in the Devonian, to terrestrial vertebrates, the other, in the Permian, to teleost fishes. In *Oncorhynchus mykiss*, the rainbow trout, two distinct subsets of type I and type II keratins have been defined (Figure 1A; Markl *et al.*, 1989), termed E keratins (from "epidermal") and S keratins (from "simple epithelial"), respectively. In trout, E keratins are typical for stratified epithelia, with the entire subset being expressed in epidermal keratinocytes; several of the E keratins also occur in the simple epithelium of the intestinal mucosa. The entire subset of S keratins, on the other hand, is expressed in the fibroblastoid trout cell line RTG2; S keratins occur in all simple epithelia including those of the digestive tract. In this respect, the trout differs from terrestrial vertebrates in that simple epithelia in the trout digestive tract express a mixture of E and S keratins, whereas in mammals and *Xenopus* such tissues exclusively contain S keratins. A more fundamental difference was observed, however, in that many mesenchymal cell types of the trout do not express vimentin, as in the equivalent cells in terrestrial vertebrates, but S keratins (Markl and Franke, 1988; Markl *et al.*, 1989). Among these

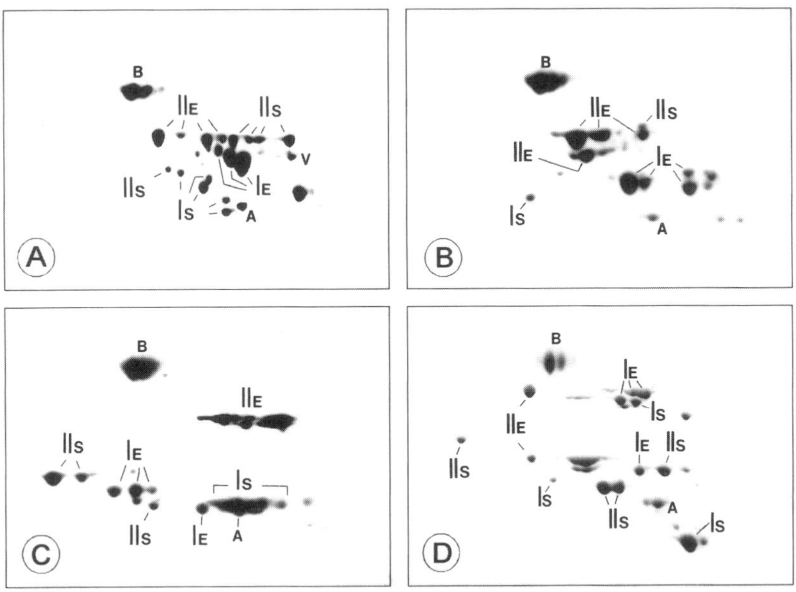

FIGURE 1. Two-dimensional coelectrophoreses of cytoskeletal proteins from (A) rainbow trout (*Oncorhynchus mykiss*) epidermis and RTG-2 cells, (B) carp (*Cyprinus carpio*) epidermis and eye, (C) shark (*Scyliorhinus stellaris*) epidermis and intestine, and (D) lamprey (*Lampetra fluviatilis*) epidermis and gill. I_E, II_E, I_S, II_S, type I and type II E (epidermal) and S (nonepidermal, *i.e.*, simple epithelial, mesenchymal, glial) keratins, respectively, as deduced from CKBB assays; V, vimentin; A, actin marker; B, bovine serum albumin marker. According to Markl *et al.* (1989), Brauer (1993), Bauer (1996), and Schaffeld and Markl (1997).

keratin-containing cell types are, for example, fibroblasts and other interstitial cells, chondrocytes, endothelial and mesothelial cells, and gill pillar cells (Figure 2A). Another puzzling result was the presence of S keratins, together with GFAP and vimentin in trout optic nerve astrocytes (Markl and Franke, 1988; de Guevara *et al.*, 1994; Herrmann *et al.*, 1996). Consequently, the mammalian view that "keratins are typical for epithelia" is obsolete.

Recently, a cDNA coding for a trout type II S keratin (sK_{II}) has been sequenced, and revealed as an orthologue of human K8 (Münick and Markl, 1995). Another sequenced trout cDNA codes for a type I S keratin (sK_{I}) which corresponds to human K18 (Conrad *et al.*, 1997). The structural relationships of K8 and K18 orthologues from various fishes and other vertebrates are illustrated in Figure 3. From this phylogenetic diagram it is suggested that the observed interspecies differences in expression patterns are indeed based on equivalent keratin subtypes.

Trout vimentin was also identified, and its expression was found to be mainly restricted to white blood cells, chromatophores, ocular lens tissue, and cells of iris and retina; moreover, it also occurs in eggs (Herrmann *et al.*, 1996). Several other trout cell types were found to coexpress vimentin and keratins, for instance, the eye lens epithelium, chondrocytes, and (rather surprisingly) the basal germinal cell layer of the epidermis (Figures 2B′, 2B). Sequence analysis of trout vimentin and comparison with vimentins from other fishes and from terrestrial vertebrates revealed a highly conserved primary structure (Figure 3; Herrmann *et al.*, 1996). In spite of this, the vimentins from ectotherms like trout and *Xenopus*, and endotherms like birds and mammals, differ with respect to a temperature dependence of their assembly properties. In trout vimentin, for example, the temperature optimum for filament assembly *in vitro* is between 18 and 24°C, and above 27°C filament assembly is apruptly abolished. This can be interpreted as an adaptation to the animal's cold-water habitat and may have a function in itself, probably with respect to mechanisms constituting a "cellular thermometer" (Herrmann *et al.*, 1996).

The peculiar expression patterns of keratins and type III proteins in the rainbow trout raised the crucial question as to whether this situation is a unique curiosity, or whether it is typical for fish in general.

2.2. Cyprinid Teleosts: Goldfish Optic Nerve, Carp, and Zebrafish Tissues

Cyprinid fishes belong to the Cypriniformes, the ancestors of which separated from the ancestors of the Salmoniformes in the Permian period, about 260 million years ago. This means that in terms of phylogenetic distance, trout and goldfish are about as divergent as, for example, alligator and penguin. In cyprinids, a carefully studied biochemical example is the goldfish optic nerve, which, as in the rainbow trout, coexpresses several keratins (see Figure 4, in Section 3) as well as vimentin and GFAP. A type

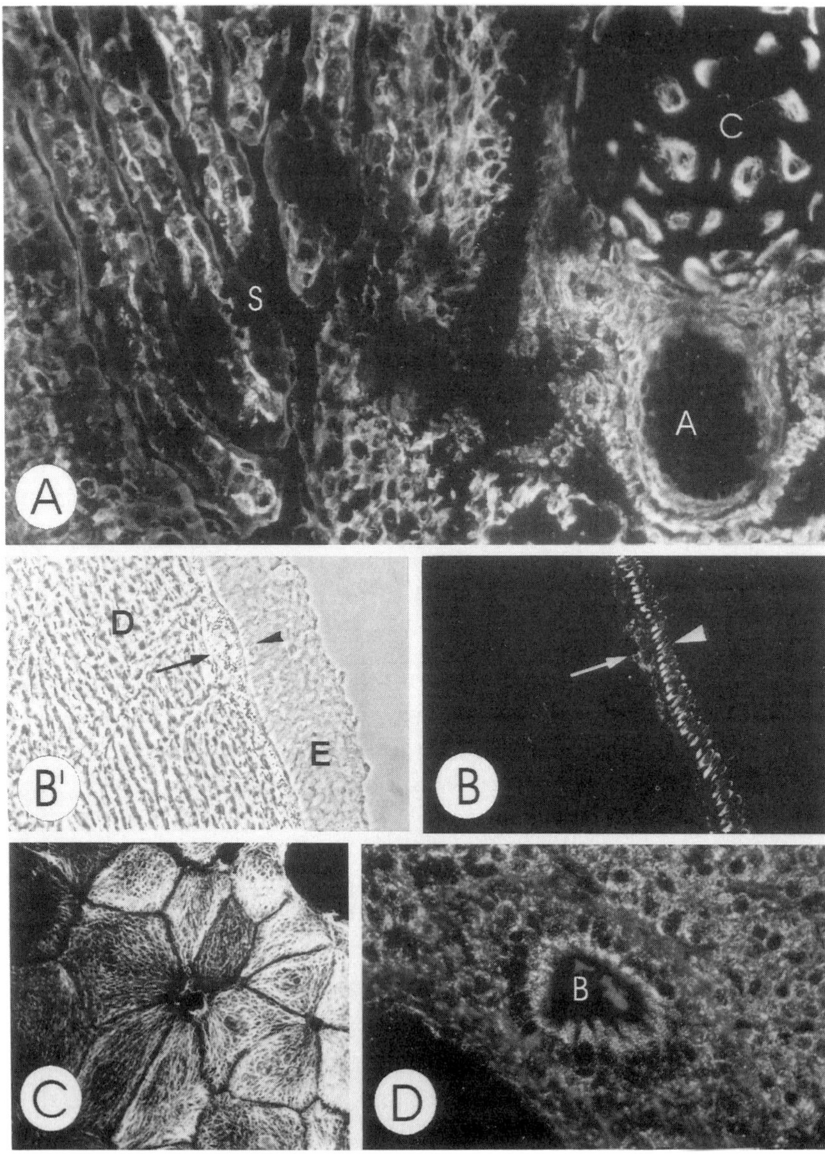

II keratin (ON_3) and three type I keratins (GK48, GK49, GK50) have been sequenced (Giordano *et al.*, 1989, 1990; Druger *et al.*, 1992, 1994). ON_3 and GK48 are goldfish correlates of human K8 and K18, respectively, and also correspond to the two sequenced S keratins from the rainbow trout (Figure 3). GK49 and GK50, on the other hand, are related to human E keratins, as illustrated by their position relative to human epidermal keratin K10 in the phylogenetic tree (Figure 3). The primary structures of a goldfish glial fibrillary acidic protein ($GFAP_1$) and two goldfish vimentins (termed α and β) are also known (Figure 3; Glasgow and Schechter, 1993; Glasgow *et al.*, 1994a), and the picture is completed by sequence data on neurofilament proteins (see Section 3; for the type III neuronal protein plasticin, see also Figure 3). In contrast to such detailed structural knowledge on goldfish neuronal tissues, the nonneuronal body parts of this fish have not been investigated in detail with respect to their IF proteins. However in a closely related cyprinid, the common carp, *Cyprinus carpio*, differential expression of keratins in various nonneuronal tissues has been documented by immunocytochemistry (Groff *et al.*, 1997); moreover, these authors collected evidence for keratin expression in many mesenchymally derived cells and tissues, including scale-associated cells, fibroblasts, mesothelia, and adipocytes. A carp vimentin and two carp GFAPs (termed 1 and 2) have been sequenced (Cohen *et al.*, 1991, 1993), and their primary structures fit well into the phylogenetic tree (Figure 3). A number of carp E and S keratins could be identified (Figure 1B; Bauer, 1996), and their expression patterns will be documented (J. Markl and colleagues, unpublished). In summary, the results from carp closely resemble those from another cyprinid, the zebrafish (see below). This similarity has an advantage, because some experiments on the IF protein system for which

FIGURE 2. Immunofluorescence microscopy of (A, B, D) rainbow trout frozen tissue sections and (C) *Xenopus* cultured cells, using various monoclonal antibodies. (B′) Phase contrast optics. (A) Trout gill tissue immunostained with anti-keratin antibody lu-5, showing extensive labeling of epithelial as well as mesenchymal cell types (Markl *et al.*, 1989); A, artery; C, cartilage ray; S, secondary gill lamellae. (B′, B) Trout skin immunostained with anti-vimentin antibody Vim-14.13, showing positive reaction with dermal chromatophores (arrow) and with the basal germinal cell layer of the epidermis (arrowhead); note that dermal fibroblasts are not significantly labelled (Herrmann *et al.*, 1996); D, dermis; E, epidermis. (C) *Xenopus* cultured cell line XLKE-A6 immunostained with antibody 1.3, showing labeling of a typical keratin filament network; this antibody specifically recognized a keratin (XK2/18) in immunoblots (from Fouquet, 1991). (D) Trout liver tissue immunostained with antibody 1.3, which here specifically stains mitochondria visible as countless small dots; in immunoblots, a 60-kDa protein enriched in extracts from purified liver mitochondria was specifically labeled (M. Schaffeld and J. Markl, unpublished); B, bile duct.

the zebrafish is simply too small can be performed more easily in the much larger carp.

In the zebrafish, *Danio rerio*, a panel of type I and type II E and S keratins has recently been identified by 2D-PAGE, complementary keratin blot-binding (CKBB) assay, and immunoblotting (Conrad *et al.*, 1998; see also Figure 5 in Section 3 of this chapter). These keratins range from 56 to 46 kDa in molecular mass and from pH 6.6 to 5.2 in isoelectric point. The type II zebrafish keratins exhibit significantly higher molecular masses (56–52 kDa) compared to the type I keratins (50–48 kDa), but the isoelectric points show no significant difference between the two keratin subclasses (type II: pH 6.0–5.5; type I: pH 6.7–5.2). The keratin catalog of zebrafish almost exactly corresponds to that of the carp (for the latter, see Figure 1B). A panel of monoclonal anti-keratin antibodies was used for immunoblotting of zebrafish cytoskeletal preparations and immunofluorescence microscopy of frozen tissue sections. These antibodies showed differential

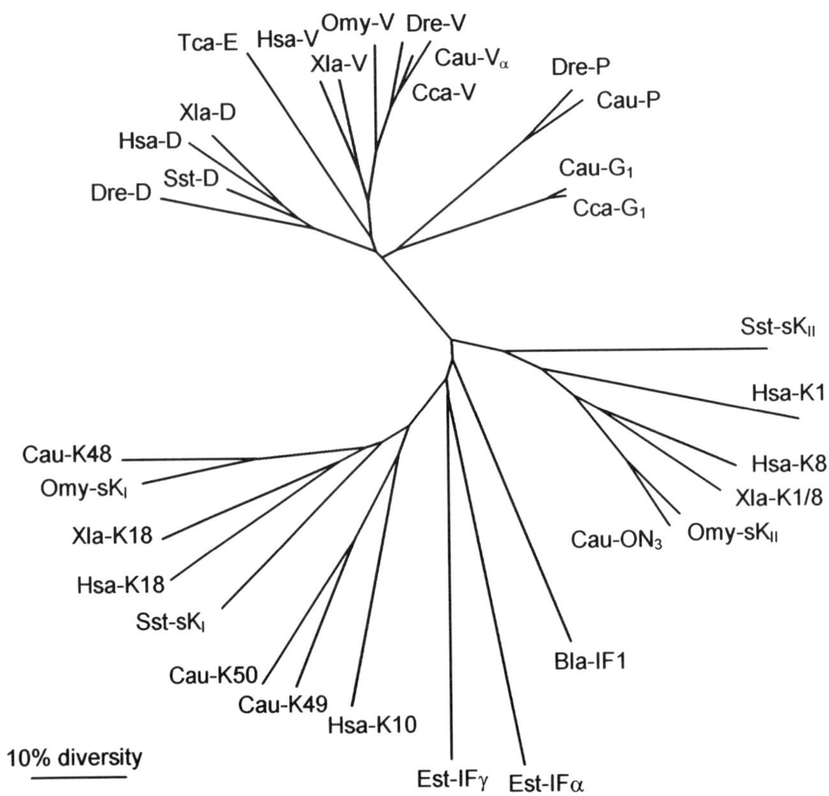

cytoplasmic expression of keratins, which not only included epithelia, but also a variety of mesenchymally derived cells and tissues (Conrad *et al.*, 1998). The primary structure of zebrafish vimentin was recently determined (Figure 3; J. Cerda and H. Herrmann, unpublished). The sequence of zebrafish desmin is also now available (Lim *et al.*, 1996), which, together with desmins from terrestrial vertebrates, forms a separate branch in the IF protein phylogenetic tree (Figure 3).

Insofar as the experiments on goldfish, carp, and zebrafish can be compared, the results are rather similar, which underlines the close phylogenetic relationship of the three cyprinids. The observed IF protein expression patterns are fundamentally different from those found in terrestrial vertebrates, notably in mesenchymal and optic nerve tissues. In spite of many differing details, in principle the cyprinid IF protein system resembles the rainbow trout system, and therefore probably holds true for teleost fishes in general. But are these features—notably the expression of keratins in a variety of mesenchymal cells—restricted to teleosts, or do they also exist in other fishes, for example, in elasmobranchs?

FIGURE 3. Radial representation of a phylogenetic tree of type I and II keratins (lower part) and type III IF proteins (upper part) of fish and other chordates. The tree is based on CLUSTAL multiple alignment of the various sequences [for source references, see Section 2 of this chapter and Quinlan *et al.* (1995)]. The phylogenetic tree (which appears here more like a *bush*) has the following major characteristics: Type I keratins, type II keratins, and the different subtypes of type III IF proteins (D, V, P, G) constitute discrete branches; GFAP (G) and plasticin (P) have a common root, which is also true for vimentin (V) and desmin (D); electric ray "electrin" (Tca-E) inserts between desmin and vimentin; the mammalian/amphibian keratin subtype K18 also exists in teleosts and shark; the goldfish optic nerve keratins K49 and K50 form a branch together with human K10, but are more closely related to each other; the sequenced shark type II protein from simple epithelia (Sst-sK_{II}) is most probably the direct K8 equivalent, although human K1 is more closely related to the K8 branch; the two hagfish thread cell IF proteins (Est-IFα and γ) go back to a common root which branches off from the tree between type I and type II keratins; the cephalochordate nonneuronal IF protein (Bla-IF1) branches off in the same region. [Note that in human, K8 and K18 are typical for simple epithelia, whereas K1 and K10 characterize cornifying stratified squamous epithelia (Moll *et al.* (1982).] Hsa, *Homo sapiens* (human); Xla, *Xenopus laevis* (clawed toad); Omy, *Oncorhynchus mykiss* (rainbow trout); Cau, *Carassius auratus* (goldfish); Cca, *Cyprinus carpio* (common carp); Dre, *Danio rerio* (zebrafish); Sst, *Scyliorhinus stellaris* (lesser-spotted dogfish); Tca, *Torpedo californica* (electric ray); Est, *Eptatretus stouti* (Pacific hagfish); Bla, *Branchiostoma lanceolatum* (cephalochordate). The following IF proteins have been included: D, desmin; V, vimentin; E, "electrin" (ray electrocyte IF protein); P, plasticin (see also Figs. 6 and 7); G, glial fibrillary acidic protein (form 1 of two different sequences); K, keratin. Cau-ON_3 and Cau-K48 are identical to goldfish K8 and K18, respectively (see Figure 4). With respect to the trout catalog in Markl *et al.* (1989), Omy-sK_I is identical to keratin S5, and Omy-sK_{II} corresponds to keratin S2.

2.3. Elasmobranchs: Shark Tissues, Ray Electrocytes

The last common ancestor of teleosts and elasmobranchs lived in the Ordovician, some 450 million years ago, which means that trout and shark are about as distant phylogenetically as are trout and human or shark and human. This makes sharks and other cartilaginous fishes such as rays key organisms in the study of IF protein evolution. In various tissues of a shark, the lesser-spotted dogfish, *Scyliorhinus stellaris*, a number of type II and type I keratins have been identified by CKBB assays, and have been separated into E and S keratins (Figure 1C; Schaffeld and Markl, 1997). The entire panel of these proteins was specifically labeled, in immunoblots, with guinea pig antibodies raised against trout keratins. On the other hand, in immunofluorescence microscopy of shark frozen tissue sections, the same antibodies exclusively stained epithelial tissues (Schaffeld and Markl, 1997). From these data it is suggested that in this shark, and probably in elasmobranchs in general, keratins are not expressed in mesenchymally derived cells and tissues as in teleosts. Moreover, it was demonstrated that the simple epithelium of the intestinal mucosa exclusively contains S keratins as in mammals and not a mixture of E and S keratins as in trout. In summary, the IF protein expression patterns documented in shark are rather complex, and in many details differ from those found in trout, frog, or mammals. However, the apparent absence of mesenchymal keratins, and the exclusive presence of S keratins in the intestinal mucosa, both resemble the situation in terrestrial vertebrates rather than that in teleost fishes.

Two shark S keratins have been sequenced (Schaffeld *et al.*, 1997a, b). One is clearly a type II keratin (sK_{II}) and related to human K8, and most probably represents its direct equivalent in spite of the position of sK_{II} relative to human epidermal keratin K1 in the phylogenetic tree (Figure 3). The other keratin (sK_{I}) is obviously equivalent to human K18, which expands the K18 subfamily from mammals via frog and teleosts to shark; the K18 sequences strikingly trace the current scheme of the subsequent branching points of vertebrate evolution (Figure 3).

A highly specialized elasmobranch organ without any equivalent in mammals is the intriguing muscle tissue derivative, the electric organ of electric rays. In the characteristic cells of this organ, the electrocytes, large amounts of an IF protein have been observed, which initially appeared to be a desmin (Walker *et al.*, 1985; Sealock *et al.*, 1989). However, sequence analysis of this protein from the electric ray, *Torpedo californica*, revealed that although it is related to desmin, it could not be clearly distinguished from vimentin (Frail *et al.*, 1990). Apparently, this protein branched off from the phylogenetic tree at a point which was very close to the separation of the various type III proteins, and probably before or soon after the separation of desmin and vimentin (Figure 3). However, is this electrocyte type III protein (for which "electrin" would be an appropriate term) a specialized cytoskeletal element of the ray electric organ, or do there exist in rays and other cartilaginous fishes individual vimentins and desmins in addition? [One should keep in mind that Quax *et al.* (1984) detected distinct genes for both proteins in all vertebrate classes.]

This question can be answered at least for the shark *Scyliorhinus stellaris*. In CKBB assays using trout vimentin, a protein can be identified in this shark which is apparently vimentin. Several other proteins—the putative shark desmin and its variants—were also labeled in this assay, reflecting the close structural relationship of both type III proteins, but this reaction was significantly weaker. Moreover, a shark cDNA clone coding for the complete head domain, central rod domain, and tail domain of a protein has been sequenced which is a typical desmin (Figure 3; Löbbecke *et al.*, 1997).

2.4. Cyclostomes: Lamprey Tissues, Hagfish Gland Thread Cells

According to fossil records, about 480 million years ago in the Ordovician the Vertebrata already existed as two distinct phylogenetic lines: the Gnathostomata, which include all vertebrates discussed so far—from elasmobranchs to mammals—and the Agnatha, with a single recent group, the cyclostomes. The latter are of particular interest with respect to IF protein phylogeny, because they evolved separately for almost the entire existence of the vertebrates. The cyclostomes include two different families, the lampreys and the hagfishes, and information on IF proteins is available from both groups.

Using a panel of monoclonal antibodies, Zaccone *et al.* (1995) observed differential expression of keratins in the epidermis and horny teeth of the adult sea lamprey, *Petromyzon marinus*. The epidermis of the larval (ammocoete) sea lamprey also showed cell type-specific keratin expression patterns, as deduced from immunolocalization (Alarcon *et al.*, 1993). Several keratins were detected, by immunolocalization, 1D-PAGE, and immunoblotting in the liver of the larval and adult sea lamprey (Alcaron *et al.*, 1997). In the adult freshwater lamprey, *Lampetra fluviatilis*, a keratin catalog based on 2D-PAGE, immunoblotting, and CKBB assays was elaborated (Figure 1D; Brauer, 1993); it contains various type I and type II keratins which have been found to be differentially expressed in various lamprey tissues. With a few exceptions, e.g., endomysia, blood vessel endothelia, and astrocytes, keratins in lamprey seem to be restricted to epithelia.

From the lamprey data, together with the results from cartilaginous fishes such as the shark, it appears that the abundance of mesenchymal keratins observed in trout and cyprinids is indeed restricted to teleosts. However "only in teleosts" would be an inadequate statement in view of the fact that they represent the most diverse vertebrate group on earth, comprising about 30,000 species. It should also kept in mind that nonteleost bony fishes, such as sturgeons, gars, and lungfishes, have not been investigated so far.

The slime gland thread cell of the Pacific hagfish, *Eptatretus stouti*, provides a unique cellular system for the study of intermediate filaments. This epidermal cell type produces enormous amounts of an extracellularly released cytoplasmic thread, which consists of a massive, conical, IF-rich biopolymer of ca. 60 cm length and 3 μm width (Downing *et al.*, 1984). It modulates the properties of mucus released by the hagfish into seawater (Koch *et al.*, 1991a, b). The two major IF proteins (termed α and γ) of this thread, which coassemble *in vitro* to form 10-nm filaments, have been isolated and characterized by Spitzer *et al.* (1984, 1988). Protein IFα has been sequenced by Koch *et al.* (1994), and shows a unique central rod that bears no preferred identity to any other currently sequenced IF subunit; however, it also has several characteristics of a conventional type II keratin. When the various fish IF proteins sequenced more recently are included in the comparison, hagfish IFα branches off from somewhere between type I and type II keratins (Figure 3). Interestingly, hagfish IFγ, which has also been sequenced (Koch *et al.*, 1995) is somewhat more closely related to its heterologous partner than to other keratins (Figure 3). It therefore appears that the two hagfish thread keratins branched off from the phylogenetic tree soon after the separation of a primordial IF protein into type I and type II keratins.

Using the available sequence information on type I to type III IF proteins from fish, it is fascinating to see that the nonneuronal protein IF1 from the lower chordate *Branchiostoma lanceolatum*, which has a gene organization resembling more closely type III genes than type I and II genes (Riemer *et al.*, 1992), branches off from the phylogenetic tree between type I and type II keratins, and this rather close to the branching point which leads to the peculiar hagfish thread keratins (Figure 3). It is an interesting question whether or not the *Branchiostoma* IF protein represents a primordial keratin version which still forms homopolymers. It does indeed appear that we are here very close to the phylogenetic roots of the type I to type III IF proteins.

2.5. Keratinlike Epitopes in Fish Mitochondria?

In the course of our comprehensive survey of fish IF proteins, a puzzling observation was made with two monoclonal antibodies raised by

Bernadette Fouquet in the laboratory of Werner W. Franke against *Xenopus* cytoskeletal preparations. Antibody 1.3, which in *Xenopus* specifically reacts with keratin XK 2/18 in immunoblots and labels keratin filaments in immunohistology (Figure 2C; Fouquet, 1991), in the rainbow trout selectively stains mitochondria (Figure 2D); the latter was particularly obvious in gill chloride cells and in liver tissues. Conversely, antibodies 69.19 and 145.2 stain mitochondria in *Xenopus* (Fouquet, 1991), but keratins in shark and lamprey, respectively. In low-salt extracts from isolated rainbow trout liver mitochondria, a 60-kDa protein of pI 5.4 was detected which, in immunoblots, specifically reacted with antibody 1.3; this protein is presently undergoing further study (J. Markl, unpublished). Since three different monoclonal antibodies are involved in this keratin/mitochondrial relationship, an explanation on the basis of random cross-reactivities seems to be unlikely. At least in mammals, a close association of IFs with mitochondria has been frequently reported from electron microscopic examination of a variety of tissues and cultured cells, and short filamentous cross-bridges between individual IFs and mitochondria have been observed (for review, see Carmo-Fonseca and David-Ferreira, 1990). In rat hepatocytes, a type II keratin (K8) is apparently anchored to the inner mitochondrial membrane, and this interaction is probably of regulatory relevance (Görlach *et al.*, 1995, and literature cited therein). Therefore, our observations could result from antibody binding to a cytoplasmic keratin filament, with the relevant epitope only accessible close to the mitochondrial attachment site. Alternatively, we may have observed a novel keratinlike mitochondrial protein, possibly a specialized linker between the IF network and the mitochondrion.

2.6. IF Protein Diversity: Product of Neutral Evolution or Functional Demands?

The fundamental differences observed in IF protein expression patterns between teleost fishes and terrestrial vertebrates provide a caveat for functional concepts of the IF protein diversity (Markl, 1991). For example, how can one explain in functional terms why mammalian fibroblasts exclusively contain vimentin, whereas trout fibroblasts lack vimentin, but express a whole panel of keratins? From this it appears that the observed IF protein diversity could result from neutral evolution—a mere historical divergence without any adaptive value. According to such a view, the different cell types would need specific amounts of IF proteins, but it would be irrelevant as to which subtype. On the other hand, the observed interspecies variations could result from different demands on the cytoskeleton in different anatomical constructions and animal habitats. Moreover, the striking conservation of the primary structure of vimentin from trout to human (Figure 3), its invariant expression in certain cell types such as immune cells, glial cells, and chromatophores, and the ubiquitous dominance of keratins in epithelia are good arguments that different IF proteins should have different functions.

More specifically, the "exotic" expression of keratins in fish optic nerve could correlate with the unique regeneration capability of the latter (see Section 3). Merrick *et al.* (1995) were unable to find a homologue of GFAP in larval sea lamprey, but using anti-keratin monoclonal antibodies they documented strong expression of several S keratins in glial cells of the *Petromyzon* ammocoete. In spinal cord and brain they were able to identify, by 2D-PAGE and immunoblotting, three different glial keratins. The expression of keratins in glial cells of this larval cyclostome may have significance for the extraordinary ability of the lamprey spinal cord axons to regenerate. It should be noted that in adult goldfish and trout, central nervous glial cells lack keratins, but express GFAP, whereas in glial elements of the optic nerve, which readily regenerates, keratins are expressed. Glial keratins have also been described in amphibia with axonal regeneration (for references, see Lurie *et al.*, 1994). Apparently a correlation exists between the regenerative capacity of axons and the presence of keratins in neighboring glia. Presumably, the network of desmosomal junctions that interconnect keratin-containing cells stabilizes the glial architecture following injury, thereby preserving guidance cues during axon regeneration (Lurie *et al.*, 1994), and providing convincing support from the *fish world* for the concept that different IF proteins have different functions.

2.7. Neurofilament Proteins Originated Twice in Evolution

Not only the supporting glial cells, but also the central player of this regenerative system, the growing fish axons, differentially express IF proteins. Among these neuronal IF proteins is a type III protein termed plasticin (Figure 3), but they comprise predominantly type IV proteins. From specific structural features, notably an extended and charged tailpiece and a peculiar gene structure, type IV neurofilament proteins are believed to represent a branch of the phylogenetic tree of vertebrate IF proteins which is distinct from type I to type III proteins (Lewis and Cowan, 1986), and exists separately already in lower chordates such as the ascidian *Ciona intestinalis* (Mencarelli *et al.*, 1991). Type IV neurofilament proteins evolved independently of the neurofilament proteins of protostomic invertebrates (Szaro *et al.*, 1991), and originated at an as-yet-unidentified point of the deuterostomic lineage (Riemer *et al.*, 1992).

3. FISH NEUROFILAMENT PROTEINS IN DEVELOPMENT AND REGENERATION

3.1. General Aspects of Fish Neurofilaments

Fish neurofilament proteins have the same structural organization as almost all cytoplasmic intermediate filament proteins. The primary structures of all sequenced fish neurofilament proteins consist of a highly conserved α-helical central domain (rod) of approximately 310 amino acids, flanked by variable amino-terminal (head) and carboxy-terminal (tail) domains. Despite similarities in the structures of these proteins, the diversity of fish neurofilament proteins and their temporal and spatial patterns of expression suggest that their structures support the developmental program of nerve growth (Giordano *et al.*, 1995).

Among fish, the goldfish optic nerve has been one of the most extensively examined tissues for its intermediate filament protein composition. This analysis has included neuronal as well as nonneuronal intermediate filament proteins (Quitschke and Schechter, 1984). These studies show that the intermediate filament protein compositions of goldfish optic nerve differ and are more complex than that of adult mammalian nerve tissue (Quitschke and Schechter, 1983, 1984; Quitschke *et al.*, 1985; Giordano *et al.*, 1989). This was most strikingly observed when intermediate filament proteins from fish and mammals were compared by resolving them by two-dimensional gel electrophoresis (Quitschke and Schechter, 1983, 1984). In fish nerve tissue, there were greater numbers of intermediate filament proteins and their molecular weights varied over a wide range. This complexity arises from the expression of some of the typical intermediate filament proteins such as vimentin (Glasgow *et al.*, 1994a) and GFAP (Cohen *et al.*, 1993), the unexpected expression of several different keratin pairs in glial-type cells (Giordano *et al.*, 1989, 1990; Druger *et al.*, 1992, 1994), and a greater number of different neurofilament proteins (Quitschke and Schechter, 1984; Quitschke *et al.*, 1985). Some of the fish neurofilament proteins are particularly noteworthy because they are structurally similar to higher vertebrate neurofilament proteins, which are expressed, for the most part, during development (Escurat *et al.*, 1990; Gorham *et al.*, 1990; Glasgow *et al.*, 1992, 1994b; Fliegner *et al.*, 1994). Thus, it was suggested that these neurofilament proteins have structural attributes that support the fish's capacity for continuous neuronal growth, plasticity, and regeneration.

Unlike mammals, the central nervous systems of anamniotic vertebrates, such as frogs and fish, display a remarkable capacity for continued growth and functional regeneration. This regenerative capacity may be linked to the retention of embryonic attributes throughout life. A good example of this is the goldfish visual pathway. It develops continuously throughout life (Meyer, 1978; Stuermer and Easter, 1984) and its optic axons exhibit robust functional regeneration after optic nerve injury (Attardy and Sperry, 1963). Since optic nerve injury induces a sequelae of molecular events which regulate and support the growth process, this pathway has emerged as a powerful model system for determining the molecular underpinnings of nerve growth and development (Grafstein, 1986). These molecular responses to injury arise from the soma of the injured neuron itself as well as from surrounding nonneuronal cells. For example, the expression of growth factors (Schwalb *et al.*, 1995), extracellular matrix proteins (Ford-Holevinski *et al.*, 1986), and structural proteins (Heacock and Agranoff, 1976) are induced after optic nerve injury. One class of well-characterized structural proteins in this system is made up of the neurofilament proteins (Quitschke and Schechter, 1984, 1986).

3.2. Neurofilament Proteins in Goldfish Optic Nerve

Goldfish optic nerve neurofilament proteins have been extensively analyzed by 2D-PAGE (Quitschke and Schechter, 1984, 1986; Quitschke *et al.*, 1985). Their synthesis, transport, and phosphorylation have been studied in the normal optic nerve and after injury (Quitschke and Schechter, 1983, 1984; Tesser *et al.*, 1986). Goldfish optic axons contain the typical NF-L and NF-M proteins, but lack the NF-H protein (Quitschke *et al.*, 1985). However, the predominant intermediate filament proteins of the goldfish optic nerve can be resolved into at least four isoelectric variants (Figure 4; originally designated ON_1–ON_4) having a molecular mass of 58 kDa, and an additional IF protein of 48 kDa (Quitschke and Schechter, 1984). Peptide mapping studies demonstrated that ON_1 and ON_2 differ from ON_3 and ON_4. In addition, ON_1 is similar to ON_2, and ON_3 is similar to ON_4 (Quitschke and Schechter, 1986). Proteins ON_2 and ON_4 are most likely phosphorylated forms of ON_1 and ON_3, respectively.

A comparison of goldfish optic nerve IF proteins before and after optic nerve crush provided information on which proteins were of neuronal and which were of nonneuronal origin (Quitschke and Schechter, 1984). The concentrations of ON_1 and ON_2 dramatically decreased after optic nerve crush and were restored to normal levels with regeneration of the nerve. In contrast, the concentrations of ON_3, ON_4, and the 48-kDa protein remained stable (Quitschke and Schechter, 1983, 1984). Thus, the results of the biochemical experiments showed that ON_1 and ON_2 are neurofilament proteins, synthesized in retinal ganglion cells and transported into the optic nerve. In contrast, ON_3, ON_4, and the 48-kDa protein are of nonneuronal

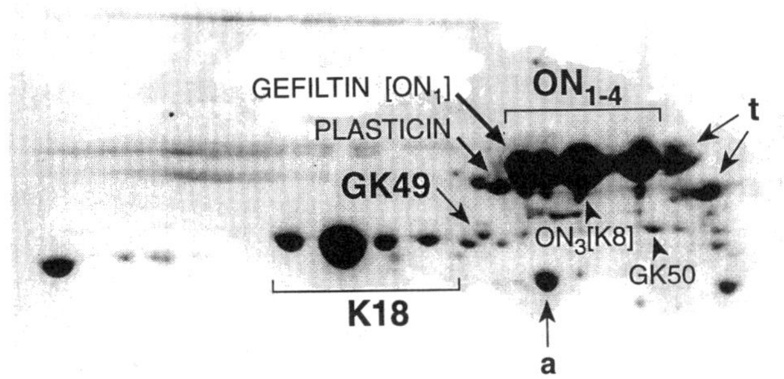

FIGURE 4. Two-dimensional PAGE of goldfish optic nerve cytoskeletal proteins. The four ON proteins are indicated by a bracket. ON_3 (K8), K18 (GK48), GK49, and GK50 are keratins. The positions of tubulin (t) and actin (a) are indicated for reference.

origin and are synthesized in glial-type cells (Quitschke and Schechter, 1984; Giordano *et al.*, 1990; Druger *et al.*, 1992, 1994). A histological analysis using polyclonal and monoclonal antibodies to the ON_1/ON_2 and ON_3/ON_4 proteins confirmed these findings (Jones *et al.*, 1986, 1989; Jones and Schechter, 1987).

3.3. Molecular Cloning of Goldfish Neurofilament Proteins

The primary structure is now known for many of the goldfish visual pathway IF proteins. For the most part, these structures were determined by screening regenerating retina or optic nerve cDNA libraries. Very often, the transcribed and ^{35}S-methionine-labeled translated products from an isolated cDNA were mixed with intermediate filament protein preparations from retina or optic nerve and subsequently resolved by two-dimensional gel electrophoresis. Consequently, the predicted amino acid sequence of the cDNA and its translation product could be correlated with a specific intermediate filament protein resolved on the two-dimensional gel. In this way, it was determined that ON_3, ON_4, and the 48-kDa protein, along with several proteins of low concentration, were keratins (Giordano *et al.*, 1989, 1990; Druger *et al.*, 1992, 1994). The significance of this is discussed in Section 2. Other, more typical intermediate filament proteins whose primary structures have been determined include multiple forms of vimentin (Glasgow *et al.*, 1994a), GFAP (Cohen and Schwartz, 1993; Cohen *et al.*, 1993), and NF-M protein (Glasgow *et al.*, 1994b).

Other structurally novel intermediate filament proteins have also been discovered and characterized. The entire nucleotide and predicted amino acid sequences for ON_1 were determined by Glasgow *et al.* (1994c). Comparison of the predicted amino acid sequence for ON_1 with sequences for other intermediate filament proteins indicated that it was a novel type IV intermediate filament protein. Because it represented a new goldfish eye intermediate filament protein, it was named gefiltin (goldfish eye filament protein). Furthermore, on screening a regenerating retinal cDNA library, another neuronal protein was isolated; its predicted amino acid sequence indicated that it was a new type III intermediate filament protein. This protein was named plasticin because it was discovered and characterized in a neuronal pathway known for its plasticity (Glasgow *et al.*, 1992). As deduced from *2D-PAGE*, plasticin levels are much lower than those of gefiltin (Figure 4).

3.4. Plasticin and Gefiltin Expression in Normal and Regenerating Goldfish Optic Nerve

In normal goldfish retina, plasticin expression is restricted to very young retinal ganglion cells at the retinal margin and to the youngest axons as the nerve exits the eye (Fuchs *et al.*, 1994). Following optic nerve crush, there is a dramatic increase in plasticin immunoreactivity in all the retinal ganglion cells. In contrast, gefiltin is constitutively expressed in the older retinal ganglion cells and is the predominant neurofilament protein in the goldfish optic nerve. These findings indicated that the two proteins have different functions during nerve development. This observation was further supported by a quantitative analysis of the plasticin and gefiltin mRNA from retinal ganglion cells after optic nerve crush and during regeneration (Glasgow *et al.*, 1994c). In the normal retina, plasticin mRNA is expressed at very low levels, whereas gefiltin mRNA is moderately abundant. After optic nerve crush, plasticin mRNA levels increase dramatically, preceding gefiltin mRNA expression. Gefiltin mRNA expression lags behind plasticin and peaks at approximately 40 days. It is noteworthy that the initial increase in gefiltin mRNA occurs at about the time the first ganglion cell axons reach their tectal targets. Maximal levels of expression occur at about 30 days when the optic axons have invaded the entire tectum, with a return to precrush levels once the retinotectal projection is restored. These results indicate that the upregulation of plasticin mRNA is independent of tectal target connections, whereas the upregulation of gefiltin mRNA depends on tectal connections. These histological and biochemical results show that plasticin and gefiltin are differentially expressed in ganglion cell subpopulations of the normal retina. They also vary temporally in their induction after optic nerve crush (Fuchs *et al.*, 1994; Glasgow *et al.*, 1994c). These studies support the idea that different developmentally regulated neurofilament proteins have structural attributes that support the program of axonal growth.

3.5. Plasticin and Gefiltin Expression in Zebrafish

Zebrafish (*Danio rerio*) and goldfish are both teleosts belonging to the cyprinid family which are closely related phylogenetically. Furthermore, they share the capacity for nerve regeneration (Attardy and Sperry, 1963; Bernhardt *et al.*, 1996). However, the goldfish is not well suited for determining the function of specific proteins during nerve development. In addition, goldfish are tetraploid, making genetic studies impossible. On the other hand, the diploid zebrafish is emerging as an important model system for studying early vertebrate neurogenesis. Zebrafish are amenable to genetic and transgenic manipulations permitting visualization of phenotypes resulting from altered gene expression. Other advantages of this system have been well reviewed elsewhere (Eisen, 1991, 1996).

It is perhaps not surprising that the two-dimensional gel patterns of goldfish and zebrafish optic nerve proteins are strikingly similar (Figures 4 and 5). This indicates that goldfish cDNA probes representing the variable domains of plasticin and gefiltin would be suitable for screening a zebrafish retinal cDNA library. With this approach, full-length plasticin and gefiltin cDNAs were isolated and sequenced. Only small differences in the predicted amino acid sequences were found between goldfish plasticin and gefiltin and their zebrafish orthologues (Figures 6 and 7). Both zebrafish and goldfish plasticin share with the other type III intermediate filament proteins the highly conserved consensus sequence IKTVETRDG in the tail

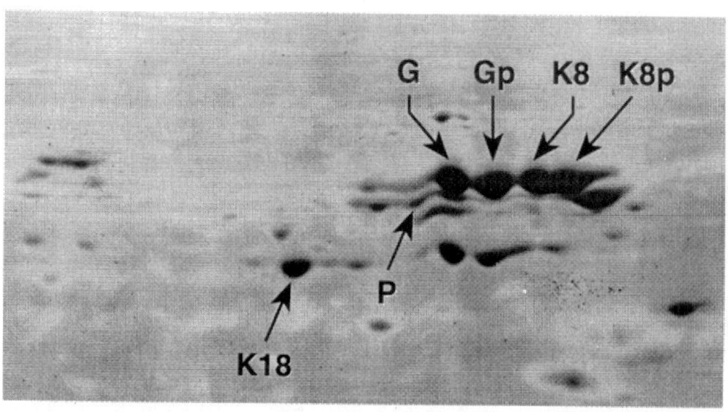

FIGURE 5. Two-dimensional PAGE of zebrafish optic nerve proteins. G (ON_1), gefiltin; Gp, phosphorylated form of gefiltin; P, plasticin; K8 (ON_3), $K8_p$ (phosphorylated form of K8), and K18 are keratins.

Table I
Percent Amino Acid Identities between Plasticin and Selected Type III Neurofilament Proteins[a]

Zebrafish	gfPlasticin	XIF3	Peripherin
Head	70	28	38
Coil	92	74	69
Tail	87	40	35
Overall	89	65	61

[a]The amino acid sequence identities between zebrafish plasticin and the selected intermediate filament proteins were calculated manually using the algorithm of the GAP program from the University of Wisconsin's Genetics Computer Group package (Devereux *et al.*, 1984). For a more detailed comparison, the identities were also calculated individually for the N-terminal "head," central coil, and C-terminal "tail" domains.

domain (Leonard *et al.*, 1988). On the other hand, the SSYRR(T/M)FG motif which is in the head domain of peripherin, vimentin, desmin, and XIF III is not in goldfish plasticin and is only partially preserved in zebrafish plasticin with the SYR sequence in the head domain.

3.6. Relationships between Plasticin and Gefiltin and the Developmentally Related Mammalian Neurofilaments

When compared to mammalian neurofilament proteins, plasticin and gefiltin have unique amino acid sequences particularly in their variable domains. However, databank search shows that the amino acid sequence of plasticin is most homologous to that of peripherin (human, rat, and mouse; Figure 6, Table I), whereas gefiltin is most homologous to α-internexin (mouse and rat; Figure 7, Table II). In addition, a gefiltin homologue was

FIGURE 6. The predicted amino acid sequences of the type III neurofilament proteins zebrafish plasticin (Asch *et al.*, 1998), goldfish plasticin (Glasgow *et al.*, 1992), *Xenopus* XIF3 (Sharpe *et al.*, 1989), and rat peripherin (Thompson and Ziff, 1989). The sequences were manually adjusted following computer alignment using PILEUP from the University of Wisconsin's Genetics Computer Group package (Devereux *et al.*, 1984). The alignments were calculated using a gap weight of 3.00 and a gap length weight of 0.10. Periods are inserted to represent gaps in the optimized alignment.

```
               1                                                    50
 zfPlasticin MSHY...... .....SYR.. .....T.SHH ..RSYGTPS. ..VSSR.IGG
 gfPlasticin MSHS...... .......... .....TFSHL FSPHFGAPVY SPVSSR.IGG
        XIF3 MSHS..GLRS T..STSYRRT LGSSPVPS.. ..SYSSSSRL ..STSRHFGS
  Peripherin MSHHSSGLRS SISSTSYRRT FGPPPSLSPG AFSYSSSSRF ..SSSRLLGS

             51                                                    100
 zfPlasticin RY.......T SSIPSRPVDF R.SRSSAPAP RL........ SYDKVDFSLA
 gfPlasticin RYV......S SSVPTRSVDF R.SRSSAPAP RL........ SYDKVDFSSA
        XIF3 PSPGPSSRSS SS........ ........AF RVRSSTPVRV SLDRVDFSVA
  Peripherin GSP......S SSARLGS..F RAPR..AGAL RLPS...... ..ERLDFSMA

             101                                                   150
 zfPlasticin EAVNQEFLAT RSNEKRELQE LNDRFASFIE KVRYLEQQNS RLTQELGQFK
 gfPlasticin EAINQEFFAT RSNEKRELQE LNDRFASFIE KVRHLEQQNS KLILELGQYK
        XIF3 EAVNQEFLTT RSNEKAELQE LNDRFASFIE KVRYLEQQNA VLVTEINQAR
  Peripherin EALNQEFLAT RSNEKQELQE LNDRFANFIE KVRFLEQQNA ALRGELSQAR

             151                                                   200
 zfPlasticin EQQQGPSGRL GDLCQQEMRE LRRQLELMGK DRDQMQVERD NLAEDVALLK
 gfPlasticin DQHQGSTGRI NELCQQEMRE LRRQLELMAK DRDQMQVERD NLAEDVALLN
        XIF3 SKE...PTRA SDLCQQELRE LRKQLELLGK DRDHIQVERD NFAEDLAFLK
  Peripherin GQE...PARA DQLCQQELRE LRRELELLGR ERDRVQVERD GLAEDLGALK

             201                                                   250
 zfPlasticin QRLNEEMQKR EEAENNLTLF RKDVDDATLA RLELEKKIES LMDEIEFLKK
 gfPlasticin QRLNEEMGKR QEAENNLTLF RKDVDDATLA RLELERKIES LMDEIEFLKK
        XIF3 QRLDEEVHKR EDAENNLVLF RKDVDDATLS TLELERKIES LMDEIEFLKK
  Peripherin QRLEEETRKR EDAEHNLVLF RKDVDDATLS RLELERKIES LMDEIEFLKK

             251                                                   300
 zfPlasticin MHDEEIQDVQ VSVQSQQM.K MEVMETSTRP DLTAALRDIR SQYESIASKN
 gfPlasticin MHDEEIQDVQ VSVQSQQM.K MEVMETSSRP DLTGALRDIR AQYESIATKN
        XIF3 LHEEELNDVQ VSVQAQPV.H MEI.EAAKQP DLTSALRDIR SQYETIAAKN
  Peripherin LHEEELRDLQ VSVESQQVQQ VEV.EATVKP ELTAALRDIR AQYENIAAKN

             301                                                   350
 zfPlasticin MQESEEWYKS KFADLTDSAK RNAEAMRQSK QDNNDLRRQI QAQNCDIEVL
 gfPlasticin MQESEEWYKS KFADLTDSAK RNAEAMRQGK QENNDLRRQI QAQNCDIDSL
        XIF3 VQESEDWYKS KFADLSDAAN RNSEALRQAK QDMNESRRQI QSLTCEVDGL
  Peripherin LQEAEEWYKS KYADLSDAAN RNHEALRQAK QEMNESRRQI QSLTCEVDGL

             351                                                   400
 zfPlasticin KSTNEALLRQ MREMEEQFTL EAGNYQDAVG RLEEEIRHLK EEMSRHLREY
 gfPlasticin KRTNEALLRQ MREMEEQFAA EARNYQDTVS RLEDEIRNLK EEMSRHLREY
        XIF3 KGTNEALLRQ MKNMEEQFGM EAANYQDTIG GLEQEVQHMK EEMSRHLREY
  Peripherin RGTNEALLRQ LRELEEQFAL EAGGYQAGAA RLEEELRQLK EEMARHLREY

             401                                                   450
 zfPlasticin QDLLNVKMAL DIEIATYRKL LEGEESRIAV PIMKMPSMSG HSGDYGQFSD
 gfPlasticin QDLLNVKMAL DIEIATYRKL LEGEENRIVV PIMKMPSMSG YSGDYGQFSD
        XIF3 QDLLNVKMAL DIEIATYRKL LEGEESRIAV PIHSLTSLSI KSPAAPEI.D
  Peripherin QELLNVKMAL DIEIATYRKL LEGEESRISV PVHSFASLSL K.TTVPEV.E

             451                                              497
 zfPlasticin PAPG....KKV VIKTVEIRD GE.VVKESTK EKARD.EKKD
 gfPlasticin TRAG....QKV VIKTVETRD GE.VVKESTK EKGRD.EKKD SHGQGKD
        XIF3 PSTETHTRKTV AIKTIETRD GEQVVTESRK EQSSEGEK
  Peripherin PPQDSHSRKMV LIRTIETRD GEKVVTESQK EQHSELDKSS IHSY
```

```
                  1                                                   50
      zfGefiltin  MSYGSDIY.. SASSYRKIFG DSTRFSASPS RLSSSRS... ....GFKSQS
      gfGefiltin  MSYGSDIY.. SASSYRKIFG DSTRYSASPP RLSSSRS... ....GFKSQS
        Xefiltin  MSFGSDHY.. LSSSYRKIFG DPPR..ASSA RLGPSGSSRT TVAGGYRSHS
Alpha-Internexin  MSFGSEHYLC SASSYRKVFG DGSRLSARLS GPGASGS... .....FRSQS

                  51                                                 100
      zfGefiltin  VNRPNIP.GS YKRSTRS... GFPSSSLNL. DSFDFTQSTV LNNEFKIIRT
      gfGefiltin  TTRSSIP.SS YKRGTRS... AFPSSSLTL. ESFDFTQSTA LNNEFKIIRT
        Xefiltin  QSRSNVPSSS YRRAPRG... A...GYLAA. DTVDLSQTSA VNNEYKIIRT
Alpha-Internexin  LSRSNVASTA ACSSASSLGL GLAYRRLPAS DGLDLSQAAA RTNEYKIIRT

                  101                                                150
      zfGefiltin  NEKEQMQGLN DRFAMFIEKV RNLEQHNKVL ETELVTLRQR QTEPSRLAEL
      gfGefiltin  NEKEQMQGLN DRFAMFIDKV RNLEQHNKVL EAELVTLRQR QTEPSRLAEL
        Xefiltin  NEKEQLQGLN DRFAMFIEKV HNLEQQNKVL ETELTALRQR QSEPSRLGEL
Alpha-Internexin  NEKEQLQGLN DRFAVFIEKV HQLETQNRAL EAELAALRQR HAEPSRVGEL

                  151                                                200
      zfGefiltin  YQQEIRELRS QLEELNAEKN QMMFERDSIE EDLQKLHEKF EEEMRMREEA
      gfGefiltin  YQQEIRELRS QLEELNAEKN QMMFERDNIE EDLQKLQEKF EEEMRIREEA
        Xefiltin  YQQEMKELRA QVEDLNAEKA QIIIERDNLE DDLEKLKGKY EDEMRMREET
Alpha-Internexin  FQRELRELRA QLEEASSARA QALLERDGLA EEVQRLRARC EEESRGREGA

                  201                                                250
      zfGefiltin  EQTLRAFKKD VDNASMVRLD LEKKVESLLD EINFLRKVHE EEVAELMNMI
      gfGefiltin  EQTLKAFKKD VDNATMVRLD LEKKVEALLD EINFIRKVHE EEVIELMNMI
        Xefiltin  EYALKSHKKD VDDATLARLD LEKKVESLLD EISFLRKVHD EEVTELMSML
Alpha-Internexin  ERALKAQQRD VDGATLARLD LEKKVESLLD ELAFVRQVHD EEVAELLATL

                  251                                                300
      zfGefiltin  QAS...QVSV EMEMAKPDLT SALKDIRGQY EAMASKNLQS AEEWYKSKFA
      gfGefiltin  QAA...QVSV EMEVAKPDLT SALKEIRGQY EAMANKNLHS AEEWYKSKFT
        Xefiltin  QAS...QVSV EMEMPKPDLT SALKEIRSQY ESLAAKNLHS ADEWYKSKFA
Alpha-Internexin  QASSQAAAEV DVAVAKPDLT SALREIRAQY ESLAAKNLQS AEEWYKSKFA

                  301                                                350
      zfGefiltin  DLSEQANKSN EVIRASREEL NEFRRQLQSK TIEIESLRGT NESLERQIHE
      gfGefiltin  DLSEQANKSN EVIRASREEL NEFRRQLQSK TIEIESLRGT NESLERQIHE
        Xefiltin  NLTEQASRST ELIRANREEM NDYRRQLQSK TIETESLRGT NESLERQLQE
Alpha-Internexin  NLNEQAARST EAIRASREEI HEYRRQLQAR TIEIEGLRGA NESLERQILE

                  351                                                400
      zfGefiltin  MEDRHNAEIM GYQDSIGELE NDLRTTKSEM ARHLREYQDL LNVKMALDIE
      gfGefiltin  MEDTHNAEVM GYQDTIGQLD NELRTTKSEM ARHLREYQDL LNVKMALDIE
        Xefiltin  MEDRHMAETA GLQDTINQLE NELRNTKGEM ARHLHEYQDL LNVKMALDIE
Alpha-Internexin  LEERHSAEVA GYQDSIGQLE SDLRNTKSEM ARHLREYQDL LNVKMALDIE

                  401                                                450
      zfGefiltin  IAAYRKLLEG EETRISTG.. .ITYPTPASV SSYSYQSRLY SSTSMSSKKE
      gfGefiltin  IAAYRKLLEG EETRISTG.. .ITYPTPTSG SSYSYQSRMY SSSSVSGKKE
        Xefiltin  IAAYRKLLEG EETRFITAGI SMMPSNPHPP PSYTYQSRVL SSSSSKVTPA
Alpha-Internexin  IAAYRKLLEG EETRFSTSGL SISGLNPLPN PSYLLPPRIL SSTTSKVSSA

                  451                                                500
      zfGefiltin  V.....KDDD DK.QQSGKPG KGSSQSDDSK KN.DKIDSGD LNSTNQKN..
      gfGefiltin  V.....KDDD DKHQQSSKPG KGSSQSDDYK KS.DKIDSGD VNPTNQKN..
        Xefiltin  V......DSR KKEETDGVSK IGSKQSARIG ETYEEIIEET VVSTKKKDKP
Alpha-Internexin  GLSLKKEEEE EEEEEEGASK EVTKKTSKVG ESFEETLEET VVSTKKTEKS

                  501        513
      zfGefiltin  .......... ...
      gfGefiltin  .......... ...
        Xefiltin  DQNDEAMVKP ...
Alpha-Internexin  TIEEITTSSS QKM
```

Table II
Percent Amino Acid Identities between Gefiltin and Selected Type IV Neurofilament Proteins[a]

Zebrafish	gfGefiltin	Xefiltin	α-Internexin
Head	87	50	40
Coil	91	75	68
Tail	84	32	28
Overall	89	65	58

[a]The amino acid sequence identities between zebrafish gefiltin and the selected type IV intermediate filament proteins were calculated manually using the algorithm of the GAP program from the University of Wisconsin's Genetics Computer Group package (Devereux *et al.*, 1984). For a more detailed comparison, the identities were also calculated individually for the N-terminal "head," central coil, and C-terminal "tail" domains.

cloned from *Xenopus* and named xefiltin (Zhao and Szaro, 1997). This protein was highly expressed in the frog retina and other regions of the central nervous system. It appears that gefiltin (Glasgow *et al.*, 1994), xefiltin (Zhao and Szaro, 1997), α-internexin (Ching and Liem, 1991), and an additional *Xenopus* neurofilament protein, XNIF (Charnas *et al.*, 1992), all belong to a subgroup of neurofilament type IV proteins distinct from NF-L (Zhao and Szaro, 1997). In general, this subgroup of neurofilament proteins has analogous patterns of expression during neuronal development.

Plasticin, gefiltin, and xefiltin proteins are expressed in animals whose neurons have an extensive capacity for continued growth plasticity and regeneration. The expression of peripherin (Escurat *et al.*, 1990; Gorham *et al.*, 1990; Troy *et al.*, 1990) and α-internexin (Escurat *et al.*, 1990; Fliegner *et al.*, 1994; Chien and Liem, 1995) in earlier phases of mammalian development and axonal remodeling in the peripheral nervous system and central nervous system is somewhat analogous to the neurofilament expression in anamniotic vertebrates. In adult mammals, where the capacity for con-

FIGURE 7. The predicted amino acid sequences of the type IV neurofilament proteins zebrafish gefiltin (Asch *et al.*, 1998), goldfish gefiltin (Glasgow *et al.*, 1994a), *Xenopus* xefiltin (Zhao and Szaro, 1997), and a-internexin (Ching and Liem, 1991). The sequences were computer aligned using PILEUP from the University of Wisconsin's Genetics Computer Group package (Devereux *et al.*, 1984). The alignments were calculated using a gap weight of 3.00 and a gap length weight of 0.10. Periods are inserted to represent gaps in the optimized alignment.

tinued neurogenesis and functional regeneration is limited, peripherin and α-internexin are much less abundant and, to a large extent, are replaced by the neurofilament triplet proteins which support a more rigid neuronal structure.

4. CONCLUSIONS

The phylogenetic pathways which have led from invertebrate IFs to the different amphibian/mammalian IF protein subtypes, with their specific expression programs, have been partially elucidated by carefully surveying a limited number of fish reference species such as trout, goldfish, shark, and lamprey. At least 40 different primary structures of fish IF proteins are now available and contribute to our understanding of IF protein classification and phylogeny. With more and more additional sequences, this picture will gradually clarify. Ultimately however, it will be necessary to sequence the entire IF protein complement of several key species of fish as well as lower chordates, to trace successfully the origin of the different IF protein subtypes found today in terrestrial vertebrates. A surprising complexity of fish IF proteins has been revealed, and in part the observed tissue-specific expression patterns are fundamentally different from those found in mammals. The most "exotic" expression mode has been detected in salmonid and cyprinid teleosts, whereas in shark and lamprey the situation appears to be more "mammalian-like." In highly specialized fish cells and tissues such as hagfish gland thread cells, ray electrocytes, and goldfish optic nerve, novel IF protein subtypes have been detected and sequenced. Arguments for possible biological functions of the molecular diversity of IF proteins have been collected, notably from studies on the visual pathway of cyprinids. These data suggest that different developmentally regulated neuronal IF proteins have structural attributes that support the program of axonal growth. Finally, with respect to cell and tissue typing in fish, the information collected from the reference species described in this chapter (*i.e.*, compilation of IF expression patterns and antibody specificities, elaboration of keratin catalogs based on *2D-PAGE*, and analysis of primary structures) provides a firm basis for the future application of IF proteins as molecular markers in fish differentiation and pathology.

NOTE ADDED IN PROOFS. After having submitted this review it was detected in the group of J.M. that the shark keratins designated as IIE, IIs, and Is in Figure 1C all are variants of only three genuine keratins. The amino acid sequences of the IIs and the Is shark keratin (termed here Sst-sKII and Sst-sKI, respectively) are used in Figure 3. Moreover, the primary structure of the shark IIE keratin is now available; it sheds new light on the evolution of IIE keratins in general (Schaffeld *et al.*, 1998).

ACKNOWLEDGMENTS. We thank Dr. Laura Fochtmann and Dr. J. Robin Harris for a critical reading of portions of this manuscript. A part of this work was supported by a grant from the Stiftung Rheinland-Pfalz für Innovation (836-386261/138) to J.M., another part by a grant from the National Institutes of Health (EY05212) to N.S.

5. REFERENCES

Adjaye, J., Plessmann, U., Weber, K., and Dodemont, H., 1995, Characterization of neurofilament NF70 mRNA from the gastropod *Helix aspersa* reveals that neuronal and non-neuronal intermediate filament proteins of cerebral ganglia arise from separate lamin-related genes, *J. Cell Sci.* **108:**3581–3590.

Ainis, L., Tagliafierro, G., Mauceri, A., Licata, A., Ricca, M. B., and Fasulo, S., 1995, Cytokeratin type distribution in the skin and gill epithelia of the Indian freshwater catfish, *Heteropneustes fossilis* as detected by immunohistochemistry, *Folia Histochem. Cytobiol.* **33:**77–81.

Alarcon, V. B., Filosa, M. F., and Youson, J. H., 1993, Keratin polypeptides in the epidermis of the larval (ammocoete) sea lamprey, *Petromyzon marinus* L., show a cell type-specific immunolocalization, *Can. J. Zool.* **72:**190–194.

Alcaron, V. B., Filosa, M. F., and Youson, J. H., 1997, Cytokeratins in the liver of the sea lamprey (*Petromyzon marinus*) before and after metamorphosis, *Cell Tiss. Res.* **287:**365–374.

Arenas, M. I., Fraile, B., De Miguel, M., and Paniagua, R., 1995, Intermediate filaments in the testis of the teleost mosquito fish *Gambusia affinis holbrooki*: A light and electron microscope immunocytochemical study and Western blotting analysis, *Histochem. J.* **27:**329–337.

Asch, W. S., Leake, D., Canger, A. K., Passini, M. A., Argenton, F., and Schechter, N., 1998, Cloning of zebrafish neurofilament cDNAs for plasticin and gefiltin: Increased mRNA expression in ganglion cells after optic nerve injury, *J. Neurochem.* **71**: in press.

Attardy, D. G., and Sperry, R. W., 1963, Preferential selection of central pathways by regenerating optic fibers, *Exp. Neurol.* **7:**46–64.

Bartnik, E., and Weber, K., 1989, Widespread occurence of intermediate filaments in invertebrates; common principles and aspects of diversion, *Eur. J. Cell Biol.* **50:**17–33.

Bauer, H., 1996, Biochemische und immunhistologische Charakterisierung von Intermediärfilamentproteinen des Karpfens *Cyprinus carpio*, Diploma thesis, Johannes Gutenberg University of Mainz, Mainz, Germany.

Bereiter-Hahn, J., Osborn, M., Weber, K., and Voth, M., 1979, Filament organization and formation of microridges at the surface of fish epidermis, *J. Ultrastruct. Res.* **69:**316–330.

Bernhardt, R. R., Tongiorgi, E., Anzini, P., and Schachner, M., 1996, Increased expression of specific recognition molecules by retinal ganglion cells and by optic pathway glia accompanies the successful regeneration of retinal axons in adult zebrafish, *J. Comp. Neurol.* **376:**253–264.

Bodega, G., Suarez, I., Rubio, M., and Fernandez, B., 1995, Type II cytokeratin expression in adult vertebrate spinal cord, *Tissue Cell* **27:**555–559.

Bovenschulte, M., Riemer, D., and Weber, K., 1995, The sequence of a cytoplasmic intermediate filament (IF) protein from the annelid *Lumbricus terrestris* emphasizes a distinctive feature of protostomic IF proteins, *FEBS Lett.* **360:**223–226.

Brauer, J., 1993, Biochemische und immunologische Charakterisierung der Cytokeratine des Neunauges *Lampetra fluviatilis*, Diploma thesis, Johannes Gutenberg University of Mainz, Mainz, Germany.

Bunton, T. E., 1993, The immunocytochemistry of cytokeratin in fish tissues, *Veterin. Pathol.* **30:**418–425.

Byrd, C. A., and Brunjes, P. C., 1995, Organization of the olfactory system in adult zebrafish: Histological, immunohistological, and quantitative analysis, *J. Comp. Neurol.* **358:**247–259.

Carmo-Fonseca, M., and David-Ferreira, J. F., 1990, Interactions of intermediate filaments with cell structures, *Electron Microsc. Rev.* **3:**115–141.

Charnas, L. R., Szaro, B. G., and Gainer, H., 1992, Identification and developmental expression of a novel low molecular weight neuronal intermediate filament protein expressed in *Xenopus laevis*, *J. Neurosci.* **12:**3010–3024.

Chien, C. L., and Liem, R. K. H., 1995, The neuronal intermediate filament, alpha-internexin is transiently expressed in amacrine cells in the developing mouse retina, *Exp. Eye Res.* **61:**749–756.

Ching, G. Y., and Liem, R. K. H., 1991, Structure of the gene for the neuronal intermediate filament protein α-internexin and functional analysis of its promoter, *J. Biol. Chem.* **266:**19459–19468.

Cohen, I., and Schwartz, M., 1993, cDNA clones from fish optic nerve, *Comp. Biochem. Physiol.* **104**[B]**:**439–447.

Cohen, I., Shani, Y., Blaugrund, E., and Schwartz, M., 1991, Isolation and sequence analysis of two intermediate filament cDNA clones from fish optic nerve, *Mol. Brain Res* **11:**181–195.

Cohen, I., Shani, Y., and Schwartz, M., 1993, Cloning and characteristics of fish glial fibrillary acidic protein: Implications for optic nerve regeneration, *J. Comp. Neurol.* **334:**431–443.

Conrad, M., Lemb, K., Schubert, T., and Markl, J., 1998, Biochemical identification and tissue-specific expression patterns of keratins in the zebrafish Danio rerio, *Cell Tiss. Res.*, in press.

Conrad, M., Lieb, B., and Markl, M., 1997, Rainbow trout *Oncorhynchus mykiss* type I keratin "S5" mRNA (complete cds), EMBL/GenBank/DDBJ Databases, Accession Number Y14289.

Cooper, M. S., and Schliwa, M., 1986, Motility of cultured fish epidermal cells in the presence and absence of direct current electric fields, *J. Cell Biol.* **102:**1384–1399.

Cordeiro, M. C. R., Benchimol, M., Mermelsteil, C. S., Sá, L. A., Faria, M. V. C., Chargas, C., and Moura Neto, V., 1996, Desmin filaments in the electrocytes of the electric organ of the electric eel *Electrophorus electricus*, *Cell Tiss. Res.* **285:**387–393.

De Guevara, R., Pairault, C., and Pinganaud, G., 1994, Expression of vimentin and GFAP during the development of the trout retina, *C. R. Acad. Sci. Paris Life Sci.* **317:**737–741.

Devereux, J., Haeberli, P., and Smithies, O., 1984, A comprehensive set of sequence analysis programs for the VAX, *Nucleic Acids Res.* **12:**387–395.

Dodemont, H., Riemer, D., and Weber, K., 1990, Structure of an invertebrate gene encoding cytoplasmic intermediate filament (IF) proteins: Implications for the origin and diversification of IF proteins, *EMBO J.* **9:**4083–4094.

Dodemont, H., Riemer, D., Ledger, N., and Weber, K., 1994, Eight genes and alternative RNA processing pathways generate an unexpectedly large diversity of cytoplasmic intermediate filament proteins in the nematode *Caenorhabditis elegans*, *EMBO J.* **13:**2625–2638.

Downing, S. W., Spitzer, R. H., Koch, E. A., and Salo, W. L., 1984, The hagfish slime gland thread cell. I. A unique cellular system for the study of intermediate filament–microtubule interactions, *J. Cell Biol.* **98:**653–669.

Druger, R. K., Levine, E. M., Glasgow, E., Jones, P. S., and Schechter, N., 1992, Cloning of a type I keratin from goldfish optic nerve: Differential expression of keratins during regeneration, *Differentiation* **52:**33–43.

Druger, R. K., Glasgow, E., Fuchs, C., Levine, E. M., Matthews, J. P., Park, C. Y., and Schechter, N., 1994, Complex expression of keratins in goldfish optic nerve, *J. Comp. Neurol.* **340:**269–280.

Eisen, J. S., 1991, Developmental neurobiology of the zebrafish, *J. Neurosci.* **11:**311–317.

Eisen, J. S., 1996, Zebrafish make a big splash, *Cell* **87:**969–977.

Escurat, M., Djabali, K., Gumpel, M., Gros, F., and Portier, M.-M., 1990, Differential expression of two neuronal intermediate-filament proteins, peripherin and the low-molecular-mass neurofilament protein (NF-L), during development in rat, *J. Neurosci.* **10:**764–784.

Fliegner, K. H., Kaplan, M. P., Wood, T. L., Pintar, J. E., and Liem, R. K. H., 1994, Expression of the gene for the neuronal intermediate filament protein alpha-internexin coincides with the onset of neuronal differentiation in the developing rat nervous system, *J. Comp. Neurol.* **342:**161–173.

Ford-Holevinski, T. S., Hopkins, J. M., McCoy, J. P., and Agranoff, B. W., 1986, Laminin supports neurite outgrowth from explants of axotomized adult rat retinal neurons, *Brain Res.* **393:**121–126.

Fouquet, B., 1991, Expression von Intermediärfilament-Proteinen in *Xenopus laevis*, Ph.d. thesis, Faculty of Biology, University of Heidelberg, Germany.

Fouquet, B., Herrmann, H., Franz, J. K., and Franke, W. W., 1988, Expression of intermediate filament proteins in *Xenopus laevis* III. Identification of mRNAs encoding cytokeratins typical for complex epithelia, *Development* **104:**533–548.

Frail, D. E., Mudd, J., and Merlie, J. P., 1990, Nucleotide sequence of an intermediate filament cDNA from *Torpedo californica*, *Nucleic Acids Res.* **8:**7.

Fuchs, C., Glasgow, E., Hitchcock, P. F., and Schechter, N., 1994, Plasticin, a newly identified neurofilament protein, is preferentially expressed in young retinal ganglion cells of adult goldfish, *J. Comp. Neurol.* **350:**452–462.

Fuchs, E. V., and Marchuk, D., 1983, Type I and type II keratins have evolved from lower eukaryotes to form the epidermal intermediate filaments in mammalian skin, *Proc. Natl. Acad. Sci. USA* **80:**5857–5861.

Fuchs, E., and Weber, K., 1994, Intermediate filaments: Structure, dynamics, function, and disease, *Annu. Rev. Biochem.* **63:**345–382.

Giordano, S., Glasgow, E., Tesser, P., and Schechter, N., 1989, A type II keratin is expressed in glial cells of the goldfish visual pathway, *Neuron* **2:**1507–1516.

Giordano, S., Hall, C., Quitschke, W., Glasgow, E., and Schechter, N., 1990, Keratin 8 of simple epithelia is expressed in glia of the goldfish nervous system, *Differentiation* **44:**163–172.

Giordano, S., Glasgow, E., Druger, R., and Schechter, N., 1995, Intermediate filaments: A molecular link to nerve development and regeneration in the goldfish visual pathway, in *Neuron–Glia Interactions during Phylogeny. Plasticity and Regeneration* (A. Vernadakis and B. I. Roots, eds.), Vol. 2, p. 616, Humana Press, Totowa, New Jersey.

Glasgow, E., and Schechter, H., 1993, Goldfish *Carassius auratus* vimentin mRNA (complete cds), EMBL/GenBank/DDBJ Databases, Accession Number L23876.

Glasgow, E., Druger, R. K., Levine, E. M., Fuchs, C., and Schechter, N., 1992, Plasticin, a novel type III neurofilament protein from goldfish retina: Increased expression during optic nerve regeneration, *Neuron* **9:**373–381.

Glasgow, E., Druger, R. K., Fuchs, C., Levine, E. M., Giordano, S., and Schechter, N., 1994a, Cloning of multiple forms of goldfish vimentin: Differential expression in CNS, *J. Neurochem.* **63:**470–481.

Glasgow, E., Hall, C. M., and Schechter, N., 1994b, Organization, sequence, and expression of a gene encoding goldfish neurofilament medium protein, *J. Neurochem.* **63:**52–61.

Glasgow, E., Druger, R. K., Fuchs, C., Lane, W. S., and Schechter, N., 1994c, Molecular cloning of gefiltin (ON1): Serial expression of two new neurofilament mRNAs during optic nerve regeneration, *EMBO J.* **13:**297–305.

Gorham, J. D., Baker, H., Kegler, D., and Ziff, E. B., 1990, The expression of the neuronal intermediate filament protein peripherin in the rat embryo, *Dev. Brain Res.* **57:**235–248.

Görlach, M., Meyer, H. E., Eisermann, B., and Soboll, S., 1995, cAMP-dependent phosphorylation of cytokeratin in hepatic inner mitochondrial membrane, *Biol. Chem. Hoppe-Seyler* **376:**51–55.

Gould, S. J., ed., 1993, *The Book of Life*, Ebury Hutchinson/Random House UK, London.

Grafstein, B., 1986, The retina as a regenerating organ, in *The Retina: A Model for Cell Biology Studies* (R. Adler and D. Farber, eds.), pp. 275–335, Academic Press, New York.

Groff, J. M., Naydan, D. K., Higgins, R. J., and Zinkl, J. G., 1997, Cytokeratin-filament expression in epithelial and non-epithelial tissues of the common carp (*Cyprinus carpio*), *Cell Tiss. Res.* **287:**375–384

Gyoeva, F. K., Leonova, E. V., Rodionov, V. I., and Gelfand, V. I., 1987, Vimentin intermediate filaments in fish melanophores, *J. Cell Sci.* **88:**649–655.

Heacock, A. M., and Agranoff, B. W., 1976, Enhanced labeling of a retinal protein during regeneration of optic nerve in goldfish, *Proc. Natl. Acad. Sci. USA* **73:**828–832.

Herrmann, H., Münick, M. D., Brettel, M., Fouquet, B., and Markl, J., 1996, Vimentin in a cold-water fish, the rainbow trout: Highly conserved primary structure but unique assembly properties, *J. Cell Sci.* **109:**569–578.

Jones, P. S., and Schechter, N., 1987, Distribution of specific intermediate-filament proteins in the goldfish retina, *J. Comp. Neurol.* **266:**112–121.

Jones, P. S., Tesser, P., Keyser, K. T., Quitschke, W., Samadi, R., Karten, H. J., and Schechter, N., 1986, Immunohistochemical localization of intermediate filament proteins of neuronal and nonneuronal origin in the goldfish optic nerve: Specific molecular markers for optic nerve structures, *J. Neurochem.* **47:**1226–1234.

Jones, P. S., Tesser, P., Borchert, J., and Schechter, N., 1989, Monoclonal antibodies differentiate neurofilament and glial filament proteins in the goldfish visual pathway: Probes for monitoring neurite outgrowth from retinal explants, *J. Neurosci.* **9:**454–465.

Koch, E. A., Spitzer, R. H., and Pithawalla, R. B., 1991a, Structural forms and possible roles of aligned cytoskeletal biopolymers in hagfish (slime eel) mucus, *J. Struct. Biol.* **106:**205–210.

Koch, E. A., Spitzer, R. H., Pithawalla, R. B., and Downing, S. W., 1991b, Keratin-like components of gland thread cells modulate the properties of mucus from hagfish (*Eptatretus stouti*), *Cell Tiss. Res.* **264:**79–86.

Koch, E. A., Spitzer, R. H., Pithawalla, R. B., and Parry, D. A. D., 1994, An unusual intermediate filament subunit from the cytoskeletal biopolymer released extracellularly into seawater by the primitive hagfish (*Eptatretus stouti*), *J. Cell Sci.* **107:**3133–3144.

Koch, E. A., Spitzer, R. H., Pithawalla, R. B., Castillos, F. A., and Parry, D. A., 1995, Hagfish biopolymer: A type I/type II homologue of epidermal keratin intermediate filaments, *Int. J. Biol. Macromol.* **17:**283–292.

Leonard, D. G. B., Gorham, J. D., Cole, P., Greene, L. A., and Ziff, E. B., 1988, A nerve growth factor-regulated messenger RNA encodes a new intermediate filament protein, *J. Cell Biol.* **106:**181–193.

Lewis, S. A., and Cowan, N. J., 1986, Anomalous placement of introns in a member of the intermediate filament multigene family: An evolutionary conundrum, *Mol. Cell. Biol.* **6:**1529–1534.

Lim, T. M., Chan, W. T., Gong, Z., and Chua, K. L., 1996, Zebrafish *Danio rerio* desmin mRNA (complete cds), EMBL/GenBank/DDBJ Databases, Accession Number Q90441.

Löbbecke, A., Schultess, J., Schaffeld, M., Lieb, B., and Markl, J., 1997, Shark *Scyliorhinus stellaris* desmin mRNA (incomplete cds), EMBL/GenBank/DDBJ Databases, Accession Number Y15064.

Lurie, D. I., Pijak, D. S., and Selzer, M. E., 1994, Structure of reticulospinal axon growth cones and their cellular environment during regeneration in the lamprey spinal cord, *J. Comp. Neurol.* **344:**559–580.

Markl, J., 1991, Cytokeratins in mesenchymal cells: Impact on functional concepts of the diversity of intermediate filament proteins, *J. Cell Sci.* **98:**261–264.

Markl, J., and Franke, W. W., 1988, Localization of cytokeratins in tissues of the rainbow trout: Fundamental differences in expression pattern between fish and higher vertebrates, *Differentiation* **39:**97–122.

Markl, J., Winter, S., and Franke, W. W., 1989, The catalog and the expression complexity of cytokeratins in a lower vertebrate: Biochemical identification of cytokeratins in a teleost fish, the rainbow trout, *Eur. J. Cell Biol.* **50:**1–16.

Mencarelli, C., Magi, B., Marzocchi, B., Contorni, M., and Pallini, V., 1991, Evolutionary trends in neurofilament proteins in fish, *Comp. Biochem. Physiol.* **100**[B]**:**733–740.

Merrick, S. E., Pleasuer, S. J., Lurie, D. I., Pijak, D. S., Selzer, M. E., and Lee, V. M.-Y., 1995, Glial cells of the lamprey nervous system contain keratin-like proteins, *J. Comp. Neurol.* **355:**199–210.

Meyer, R. L., 1978, Evidence from thymidine labeling for continuing growth of retina and tectum in juvenile goldfish, *Exp. Neurol.* **59:**99–111.

Moll, R., Franke, W. W., Schiller, D. L., Geiger, B., and Krepler, R., 1982, The catalog of human cytokeratin polypeptides: Patterns of expression in normal epithelia, tumors and cultured cells, *Cell* **31:**11–24.

Münick, M., and Markl, J., 1995, Rainbow trout *Oncorhynchus mykiss* type II "S" keratin mRNA (complete cds), EMBL/GenBank/DDBJ Databases, Accession Number X92522.

Pankov, R. G., Uschewa A. A., Tasheva, B. T., and Markov, G. G., 1986, Vertebrate liver cytokeratins: A comparative study, *Biochem. Cell Biol.* **65:**547–557.

Quax, W., van der Heuvel, R., Egberts, W. V., Quax-Jeuken, Y., and Bloemendal, H., 1984, Intermediate filament cDNAs from BHK-21 cells: Demonstration of distinct genes for desmin and vimentin in all vertebrate classes, *Proc. Natl. Acad. Sci. USA* **81:**5970–5974.

Quinlan, R., Hutchinson, C., and Lane, B., 1995, Intermediate filament proteins, in: *Protein Profile*, Vol. 2 (P. Sheterline, ed.), pp. 801–952, Academic Press, London.

Quitschke, W., and Schechter, N., 1983, *In vitro* protein synthesis in the goldfish retinotectal pathway during regeneration: Evidence for specific axonal proteins of retinal origin in the optic nerve, *J. Neurochem.* **41:**1137–1142.

Quitschke, W., and Schechter, N., 1984, 58,000 dalton intermediate filament proteins of neuronal and nonneuronal origin in the goldfish visual pathway, *J. Neurochem.* **42:**569–576.

Quitschke, W., and Schechter, N., 1986, Homology and diversity between intermediate filament proteins of neuronal and nonneuronal origin in goldfish optic nerve, *J. Neurochem.* **46:**545–555.

Quitschke, W., Jones, P. S., and Schechter, N., 1985, Survey of intermediate filament proteins in optic nerve and spinal cord: Evidence for differential expression, *J. Neurochem.* **44:**1465–1476.

Riemer, D., and Weber, K., 1994, The organization of the gene for *Drosophila* lamin C: Limited homology with vertebrate lamin genes and lack of homology versus the *Drosophila* lamin Dm^0 gene, *Eur. J. Cell Biol.* **63:**299–306.

Riemer, D., Dodemont, H., and Weber, K., 1992, Analysis of the cDNA and gene encoding a cytoplasmic intermediate filament (IF) protein from the cephalochordate *Branchiostoma lanceolatum*; implications for the evolution of the IF protein family, *Eur. J. Cell Biol.* **58:**128–135.

Schaffeld, M., and Markl, J., 1997, Expression pattern and biochemical properties of cytokeratins in the lesser spotted dogfish *Scyliorhinus stellaris*, *Verh. Dtsch. Zool. Ges.* **90:**213.

Schaffeld, M., Lieb, B., and Markl, J., 1997a, Shark *Scyliorhinus stellaris* type I keratin sK18 mRNA (complete cds), EMBL/GenBank/DDBJ Databases, Accession Number Y14647.

Schaffeld, M., Lieb, B., and Markl, J., 1997b, Shark *Scyliorhinus stellaris* type II keratin sK8 mRNA (complete cds), EMBL/GenBank/DDBJ Databases, Accession Number Y14648.

Schaffeld, M., Löbbecke, A., Bernhard Lieb, B., and Markl, J., 1998, Tracing keratin evolution: Catalog, expression patterns and primary structure of shark (*Scyliorhinus stellaris*) keratins, *Eur. J. Cell Biol.*, in press.

Schliwa, M., 1975, Cytoarchitecture of surface layer cells of teleost epidermis, *J. Ultrastruct. Res.* **52:**377–386.

Schliwa, M., and Euteneuer, U., 1979, Structural transformation of epidermal tonofilaments upon cold treatment, *Exp. Cell Res.* **122:**93–101.

Schroeder, T. E., and Otto, J. J., 1991, Snoods: A periodic network containing cytokeratin in the cortex of starfish oocytes, *Dev. Biol.* **144:**240–247.

Schwalb, J. M., Boulis, N. M., Gu, M., Winickoff, J., Jackson, P. S., Irwin, N., and Benowitz, L. I., 1995, Two factors secreted by the goldfish optic nerve induce retinal ganglion cells to regenerate axons in culture, *J. Neurosci.* **15:**5514–5525.

Sealock, R., Murnane, A. A., Paulin, D., and Froehner, S. C., 1989, Immunochemical identification of desmin in *Torpedo* postsynaptic membranes and at the rat neuromuscular junction, *Synapse* **3:**315–324.

Sharpe, C. R., Pluck, A., and Gurdon, J. B., 1989, XIF3, a *Xenopus* peripherin gene, requires an inductive signal for enhanced expression in anterior neural tissue, *Development* **107:**701–714.

Spitzer, R. H., Downing, S. W., Koch, E. A., Salo, W. L., and Saidel, L. J., 1984, Hagfish slime gland thread cells. II. Isolation and characterization of intermediate filament components associated with the thread, *J. Cell Biol.* **98:**670–677.

Spitzer, R. H., Koch, E. A., and Downing, S. W., 1988, Maturation of hagfish gland thread cells: Composition and characterization of intermediate filament polypeptides, *Cell Motil. Cytoskel.* **11:**31–45.

Steinert, P. M., and Roop, D. R., 1988, Molecular and cellular biology of intermediate filaments, *Annu. Rev. Biochem.* **57:**593–625.

Stuermer, C. A., and Easter, S. S., Jr., 1984, A comparison of the normal and regenerated retinotectal pathways of goldfish, *J. Comp. Neurol.* **223:**57–76.

Szaro, B. G., Pant, H. C., Way, J., and Battey, J., 1991, Squid low molecular weight neurofilament proteins are a novel class of neurofilament proteins, *J. Biol. Chem.* **266:**15035–15041.

Terakado, K., 1988, The pattern of organization of intermediate filaments and their asymmetrical association with dense bodies in smooth muscle of an ascidian *Halocynthia roretzi*, *Cell Tiss. Res.* **252:**23–32.

Tesser, P., Jones, P. S., and Schechter, N., 1986, Elevated levels of retinal neurofilament mRNA accompany optic nerve regeneration, *J. Neurochem.* **47:**1235–1243.

Thompson, J. S., Virtanen, I., and Lehto, V.-P., 1987, Intermediate filaments in normal tissues and lymphomas of Northern pike, *Esox lucius* L., from the Aland islands of Finland, *J. Comp. Pathol.* **97:**257–266.

Thompson, M. S., and Ziff, E. B., 1989, Structure of the gene encoding peripherin, an NGF-regulated neuronal-specific type III intermediate filament protein, *Neuron* **2:**1043–1053.

Tomarev, S. I., Zinovieva, R. D., and Piatigorsky, J., 1993, Primary structure and lens-specific expression of genes for an intermediate filament protein and a β-tubulin in cephalopods, *Biochim. Biophys. Acta* **1216:**245–254.

Troy, C. M., Brown, K., Greene, L. A., and Shelanski, M. L., 1990, Ontogeny of the neuronal intermediate filament protein, peripherin, in the mouse embryo, *Neuroscience* **36:**217–237.

Tsai, J.-C., 1996, Characterization of the pattern of cytokeratin proteins in the epidermal cells of loach, *Misgurnus anguillicaudatus* (Cypriniformes), *Fish Physiol. Biochem.* **15:**307–316.

Walker, J. H., Boustead, C. M., Witzemann, V., Shaw, G., Weber, K., and Osborn, M., 1985, Cytoskeletal proteins at the cholinergic synapse: Distribution of desmin, actin, fodrin, neurofilaments, and tubulin in *Torpedo* electric organ, *Eur. J. Cell Biol.* **38:**123–133.

Weber, K., Plessmann, U., and Ulrich, W., 1989, Cytoplasmic intermediate filament proteins of invertebrates are closer related to nuclear lamins than are vertebrate intermediate filament proteins; sequence characterization of two muscle proteins of a nematode, *EMBO J.* **8:**3221–3227.

Zaccone, G., Howie, A. J., Mauceri, A., Fasulo, S., Lo Casio, P., and Youson, J. H., 1995, Distribution patterns of cytokeratins in epidermis and horny teeth of the adult sea lamprey, *Petromyzon marinus*, *Folia Histochem. Cytobiol.* **33:**69–75.

Zhao, Y., and Szaro, B. G., 1997, Xefiltin, a new low molecular weight neuronal intermediate filament protein of *Xenopus laevis*, shares sequence features with goldfish gefiltin and mammalian alpha-internexin and differs in expression from XNIF and NF-L, *J. Comp. Neurol.* **377:**351–364.

Chapter 2

Intermediate Filament Organization during Oogenesis and Early Development in the Clawed Frog, *Xenopus laevis*

David L. Gard and Michael W. Klymkowsky

1. INTRODUCTION

Amphibian oocytes are remarkable cells, whose distinctive features have been exploited repeatedly in the history of experimental embryology and developmental biology. For a number of reasons, the African clawed frog, *Xenopus laevis*, has become the model system for the study of oogenesis and early development in amphibians (Kay and Peng, 1991). First, unlike other commonly used amphibians, the ovaries of a female *Xenopus* contain oocytes at all stages of development (Dumont, 1972). In addition, *Xenopus* oocytes can be cultured *in vitro*, allowing investigation of many developmental processes that are not accessible in other species. For example, much of our understanding of the mechanisms for targeting

---

David L. Gard Department of Biology, University of Utah, Salt Lake City, Utah 84112-0140. **Michael W. Klymkowsky** Molecular, Cellular, and Developmental Biology, University of Colorado at Boulder, Boulder, Colorado 80309-0347.

Subcellular Biochemistry, Volume 31: Intermediate Filaments, edited by Herrmann and Harris. Plenum Press, New York, 1998.

and localizing developmentally important maternal mRNA has resulted from studies of cultured *Xenopus* oocytes (Kloc and Etkin, 1995; Kloc *et al.*, 1996). In addition, late-stage oocytes can be manipulated *in vitro* and then surgically implanted in a female to generate fertilizable eggs (Heasman *et al.*, 1991; Torpey *et al.*, 1992a). Because zygotic transcription does not begin until embryonic stage 8.5 (the midblastula transition; Newport and Kirschner, 1982), this approach can be used to analyze the functions of maternally encoded (or experimentally expressed) proteins during early embryogenesis (Heasman *et al.*, 1994a, b; Kofron *et al.*, 1997). Finally, the recent development of transgenic technology for *Xenopus* (Kroll and Amaya, 1996) opens the possiblity of extending experimental studies from the oocyte and early embryo to later developmental stages.

In addition to its role as a model system for studying development, the *Xenopus* oocyte and embryo have also been an important system for studying basic cellular structure and processes. For example, current ideas of chromosome structure owe much to the analysis of "lampbrush" chromosomes present in the amphibian oocyte. Likewise, our understanding of many diverse cellular processes, such as nuclear transport, the regulation of the eukaryotic cell cycle, spindle assembly, and intracellular signaling, has benefitted greatly from pioneering studies carried out in amphibian oocytes and embryos.

Xenopus oocytes also provide a unique model for studying the specification and generation of cellular and embryonic polarity. Fully grown *Xenopus* oocytes and eggs are polarized along an animal–vegetal (A–V)* axis, made visible by the differential localization of pigment to the animal cortex. A–V polarization also extends to many other features of the oocyte, including the position of the germinal vesicle (GV; the oocyte nucleus), the distribution of yolk, localization of developmentally important maternal mRNAs, and cytoskeletal organization (Dent and Klymkowsky, 1989; Klymkowsky and Karnovsky, 1994; Gard, 1995; Gard *et al.*, 1995c). Many of these A–V asymmetries are retained during oocyte maturation and, after fertilization, contribute to the determination of embryonic axes. Although the mechanisms of axis specification and establishment remain unclear, A–V polarization may be an elaboration of structural asymmetries present in oogonia and postmitotic stage 0 oocytes (Gerhart, 1980; Klymkowsky and Karnovsky, 1994; Gard, 1995).

*Abbreviations: A-V, animal–vegetal; BFAG, bullous pemphoid antigen; GV, germinal vesicle; GVBD, germinal vesicle breakdown; IF, intermediate filament; IFAP, IF-associated protein; MAP, MT-associated protein; MC, mitochondrial cloud; MF, microfilament; MPF, maturaton-promoting factor; MT, microtubule; MTOC–TMA, MT-organizing center–transient MT array.

Research in our laboratories has focused on the organization and regulation of the cytoskeleton during oogenesis and early development in *X. laevis*, and its role(s) in the specification and establishment of cellular and embryonic polarity. Like most metazoan cells, *Xenopus* oocytes contain a complex, interconnected cytoskeletal network consisting of the three major cytoplasmic filaments: microtubules (MTs), composed of tubulin and its associated proteins (MAPs); microfilaments (MFs), composed of actin and its associated proteins; intermediate filaments (IFs), composed of simple epithelial keratins; and the cross-linking proteins, such as plectin (Herrmann and Wiche, 1987; Foisner and Wiche, 1991; Foisner *et al.*, 1994; Svitkina *et al.*, 1996), that interconnect them. Recent studies have demonstrated that MTs and MFs play important roles in the formation and maintenance of the A–V axis of stage VI *Xenopus* oocytes. For example, both are required for the proper localization of maternal mRNAs to the vegetal cortex during axis formation (Yisraeli *et al.*, 1990), and disassembly of either MTs or MFs results in displacement of the GV from its normal position within the animal hemisphere cytoplasm (Gard, 1993; Roeder and Gard, 1994). Less is known, however, regarding the role of IFs during oogenesis and early development in vertebrates. In fact, recent analysis of mice that are homozygous for null mutations in the IF subunit proteins vimentin (Colucci *et al.*, 1994), glial fibrillary acidic protein (Gomi *et al.*, 1995; Pekny *et al.*, 1995; Liedtke *et al.*, 1996; McCall *et al.*, 1996), and the simple epithelial keratin K8 (Baribault and Oshima, 1991; Baribault *et al.*, 1993, 1994) suggests that the defects associated with the absence of specific types of IFs can be remarkably subtle. In this chapter, we summarize recent results that may provide new insights into the assembly and organization of IFs during oogenesis in the African frog, *Xenopus laevis*, with emphasis placed upon the keratin filament system. In addition, we discuss the possible functions of keratin filaments during oogenesis and early development in *Xenopus*.

2. IF PROTEINS EXPRESSED IN THE *XENOPUS* OOCYTE

2.1. *Xenopus* Oocytes Contain a Nuclear Lamina Composed of Multiple Lamins

Like most metazoan cells, *Xenopus* oocytes contain both nuclear and cytoplasmic IF proteins. Although originally described as containing only a single lamin, L_{III} (Benavente *et al.*, 1985; Stick and Hausen, 1985), it is now clear that the oocyte contains three lamins: L_I, L_{II}, and L_{III}. L_{III} is the major lamin species found in oocytes, being ten-fold more abundant than L_{II}, and

100-fold more abundant than L_I (Lourim *et al.*, 1996). Using the whole-mount staining method developed by Dent (Dent *et al.*, 1989), it is possible to view the changing morphology of the GV during oogenesis (Figure 1). In early oocytes (stages I–II), the oocyte nuclear membrane appears uniformly stained with the anti-lamin antibody 14a9 (available from the Developmental Studies Hybridoma Bank; University of Iowa, Iowa City, Iowa). As oogenesis proceeds, the surface of the nuclear envelope becomes increasingly convoluted, and discrete domains of folded nuclear envelope (which can also be seen in phase-contrast images of freshly isolated GVs) are apparent.

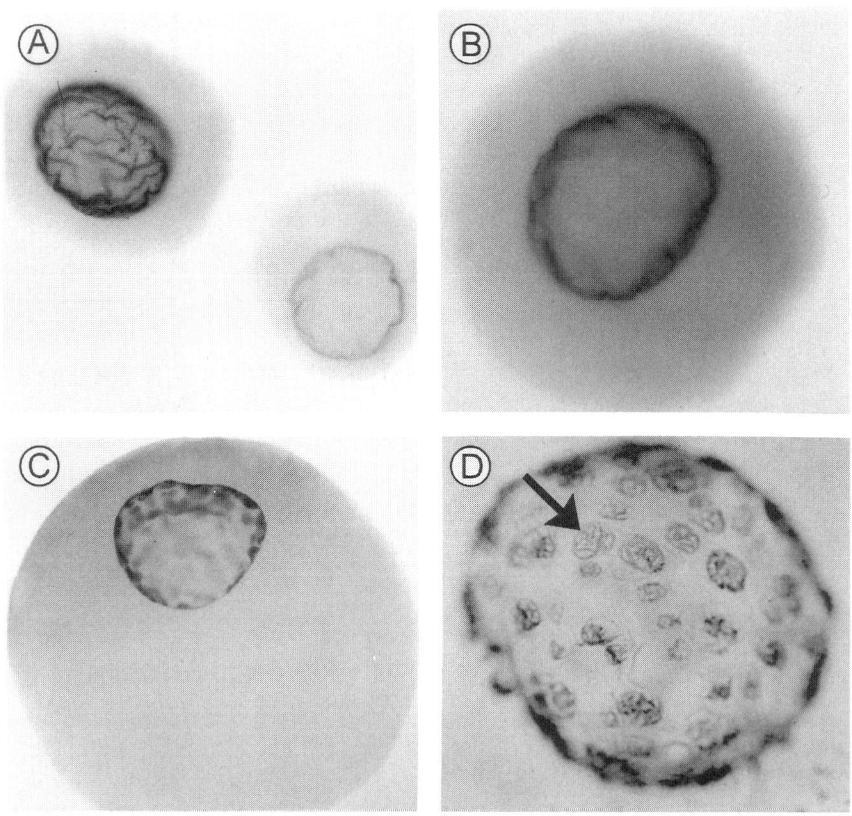

FIGURE 1. Lamins in the oocyte. Xenopus oocytes were isolated from adult females, fixed, bleached, and stained with the antilamin antibody 14a9. In early stage oocytes (A, B) the nucleus is symmetrically positioned and relatively smooth; lamin staining is uniform. In later stage V oocytes (C, D) the asymmetric position of the nucleus is apparent and the presences of folded domains (arrow in part D) can easily be seen. How these domains form is unknown.

2.2. Do *Xenopus* Oocytes Contain IFs Composed of Vimentin?

Stage VI *Xenopus* oocytes have been reported to contain two biochemically and spatially distinct systems of cytoplasmic IFs, one composed of keratins (Franz *et al.*, 1983; Gall *et al.*, 1983; Godsave *et al.*, 1984a) and another composed of vimentin (Godsave *et al.*, 1984b; Torpey *et al.*, 1990, 1992b). Reports of vimentin filaments in *Xenopus* oocytes, however, have proven controversial. Franz *et al.* (1983) did not observe vimentin immunoreactivity in oocytes, even though vimentin could be clearly visualized in the surrounding thecal cells of the follicle (Godsave *et al.*, 1986; Dent and Klymkowsky, 1989). Several laboratories, including our own, have been unable to detect vimentin or vimentin-containing filaments in oocytes during stages I–VI of oogenesis (see Dent *et al.*, 1992, and references therein). Moreover, significant transcription of vimentin mRNA was not detected until after completion of gastrulation and beginning of neurulation (stage 14) (Herrmann *et al.*, 1989) and vimentin was first detectable in the late gastrula/early neurula-stage embryos (Herrmann *et al.*, 1989; Dent *et al.*, 1989). Finally the incorporation of *Xenopus* vimentin, expressed from mRNA microinjected into stage VI oocytes, into filaments was relatively slow compared to the rapid incorporation of a *Xenopus* keratin (encoded by synthetic mRNA for *Xenopus* keratin-8) into the preexisting keratin filament network (Bachant, 1993; Klymkowsky, 1995; Bachant and Klymkowsky, 1996). Such behavior is consistent with the *de novo* assembly of a vimentin filament network within the oocyte, rather than incorporation into a preexisting filament network. Taken together, these observations lead us to conclude that reports of vimentin in late-stage *Xenopus* oocytes are due to the presence of an unrelated polypeptide that cross-reacts with some vimentin antisera.

2.3. Late-Stage *Xenopus* Oocytes Contain an Extensive Keratin Filament Network

In contrast to the case with vimentin, the presence of keratins and keratin filaments in the *Xenopus* oocyte has been well established by molecular, biochemical, and cytological analyses. Keratin filament networks have also been found in the oocytes of starfish and sea urchins (Boyle and Ernst, 1989; Schroeder and Otto, 1991). The composition of keratin filaments of stage VI *Xenopus* oocytes is most similar to that of simple epithelia, consisting of two type I (acidic) keratins (probably similar to mammalian keratins K18/Endo B and K19) and a single type II (neutral/basic) keratin. Molecular analysis indicates that the type II keratin is the

Xenopus homologue of the mammalian K8 (Endo A) found in simple epithelia (Franz and Franke, 1986).*

Initial reports (Franz *et al.,* 1983; Godsave *et al.,* 1984a) indicated that these keratins were assembled into a complex array consisting of three interlinked networks: (1) an extensive network of cortical filaments; (2) a meshwork or cage of filaments surrounding the GV; and (3) a network of radially oriented cytoplasmic filaments linking the cortex and perinuclear arrays. Subsequent examination of cortical keratin filament network in stage VI *Xenopus* oocytes using whole-mount immunocytochemistry revealed a striking A–V polarization in the organization of keratin filaments (Klymkowsky *et al.*, 1987): a relatively sparse, disconnected network of keratin filaments or short fragments was observed in the animal cortex, while a complex, anastomosing network of thick keratin filament bundles was apparent in the vegetal cortex.

More recently, confocal immunofluorescence microscopy has revealed that an extensive network of keratin filaments extends throughout the cortex of stage VI oocytes (Figure 2), and has confirmed that the distribution of keratin filaments in the cortex is polarized along the A–V axis of the oocyte (Gard *et al.*, 1997). The increased contrast and depth of field afforded by confocal microscopy revealed that the sparse or incomplete network of keratin filaments seen by conventional immunocytochemical techniques was but a portion of an extensive three-dimensional network of keratin filaments present in the animal cortex. The A–V polarity of the cortical network of keratin filaments is apparent in both tangential and cross sections of the oocyte cortex: the keratin filament network in the animal cortex exhibits a finer apparent mesh size than that of the vegetal cortex (compare Figures 2A and 2B), and extends two to three times deeper into the subcortical cytoplasm than the coarser, nearly planar network of keratin filaments observed in the vegetal cortex (compare Figures 2C and 2D). The differing views of the keratin filament network in the animal cortex obtained by conventional versus confocal microscopy likely result from the three-dimensional nature of cortical keratin network, which is not readily apparent using the former technique.

Optical cross sections obtained by confocal microscopy also revealed a network of keratin filaments extending throughout the cytoplasm of stage VI oocytes. In the animal hemisphere, keratin filaments surrounded the GV and a roughly radial network of keratin filaments extended to

*It should be remembered, however, the *X. laevis* is pseudotetraploid, and that the polypeptides encoded by the two pseudoalleles are not identical (Klymkowsky, 1995). It is therefore likely that two forms of each of the keratins exist within the oocyte.

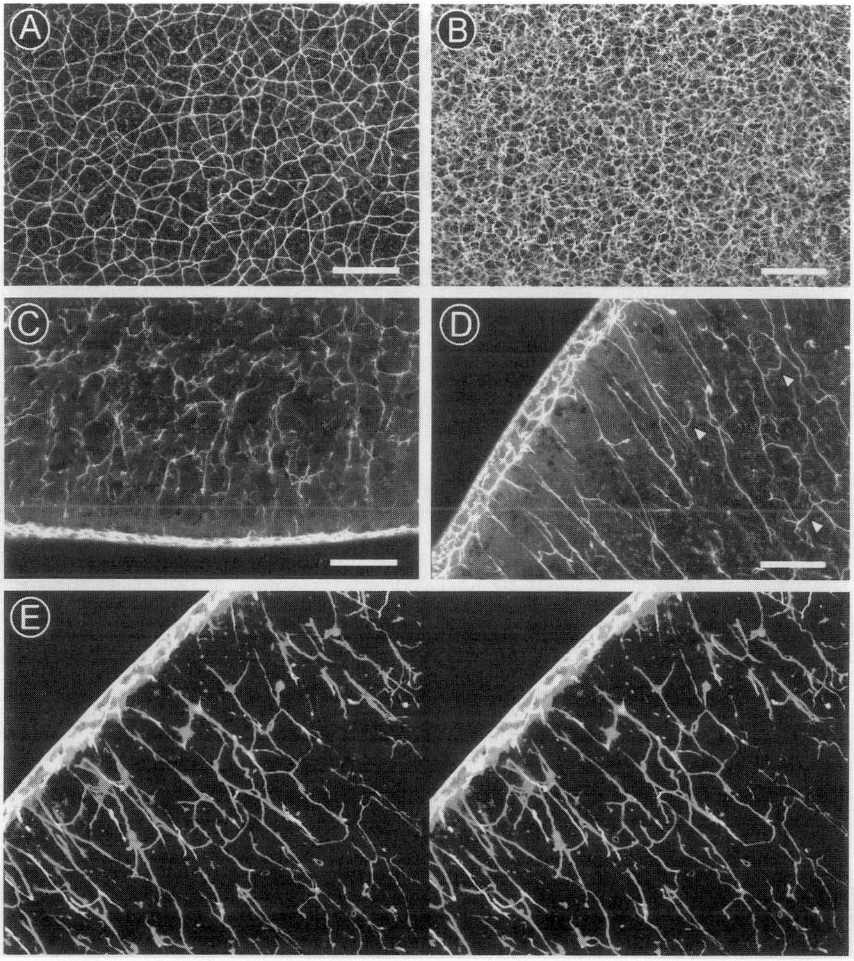

FIGURE 2. Stage VI *Xenopus* oocytes contain a complex cortical network of keratin filaments. (A) and (B): Grazing views (projections of four and five sections, respectively) of the vegetal (A) and animal cortex (B) of stage VI oocytes stained with antikeratin antibodies reveal complex networks of filaments in the cortex of both hemispheres. Note the characteristic difference in complexity and apparent mesh size of keratin networks in the vegetal and animal cortex. (C) and (D): Optical cross-sections (projections of five serial sections collected at 1 μm intervals) of the vegetal (C) and animal (D) hemispheres revealed the cortical keratin filaments, as well as a substantial network of keratin filaments in the underlying subcortical cytoplasm. Note the difference in thickness of the cortical network in the vegetal and animal cortex, and the radial organization of keratin filaments in the animal hemisphere. Arrowheads in D denote transverse keratin filaments (see text). (E): A stereo pair showing filament organization in the animal cytoplasm of a stage VI oocyte. Scale bars in A–D are 25 μm.

the animal cortex (Figures 2D and 2E). Three-dimensional reconstructions from serial optical sections (Figure 2E) also revealed a population of transversely oriented keratin filaments linking adjacent radial filaments. In the vegetal cytoplasm, the radial keratin filaments appear less well ordered, presumably because they must wind their way between the large, densely packed yolk platelets characteristic of the vegetal hemisphere.

To further study the regulation of IF assembly in the late-stage oocyte, we have used the injection of capped RNAs encoding epitope-tagged forms of *Xenopus* vimentin (Dent *et al.*, 1992) and keratin K8 (Bachant and Klymkowsky, 1996). When RNA encoding either vimentin or keratin was injected into either the animal or vegetal pole of the oocyte the newly synthesized proteins assembled locally into a network. In the case of keratin, the exogenous protein rapidly integrates into the endogenous keratin filament network (Bachant and Klymkowsky, 1996). However, since the oocyte contains no preexisting vimentin network (see above) and vimentin does not copolymerize with the keratins (Aubin *et al.*, 1980; Henderson and Weber, 1981), the network of vimentin filaments assembled from the exogenous vimentin forms *de novo*. The increasing insolubility of exogenous vimentin in the oocyte with time after injection (Klymkowsky, 1995) supports this conclusion. Interestingly, the organization of vimentin filaments assembled from the exogenous vimentin mirrored that of the endogenous keratin filament network as viewed by conventional microscopy: in the vegetal cortex, vimentin IFs formed an extensive, anastomosing network, while the vimentin rarely formed discrete filaments in the animal cortex (Dent *et al.*, 1992; see below). While these results have not been confirmed by confocal microscopy, they suggest that the assembly and distribution of IFs from exogenously expressed vimentin is regulated by the factors/mechanisms that are asymmetrically distributed within the oocyte. The nature of these factors remains unclear. However, analysis by two-dimensional gel electrophoresis reveals no obvious difference in the level of posttranslational modification of vimentin expressed in the animal or vegetal hemispheres (Klymkowsky, 1995).

3. KERATIN FILAMENT ORGANIZATION IN THE OOCYTE

3.1. Keratin Organization and Animal–Vegetal Polarity Depend upon Microfilaments and Microtubules

Loss of the normal organization of MF or MTs in stage VI *Xenopus* oocytes treated with cytochalasin B, an inhibitor of microfilament assembly, or nocodazole, an inhibitor of MT assembly, leads to dramatic changes in keratin network organization, suggesting that the organization and polarization of the cortical keratin network in stage VI oocytes is dependent upon both MFs and MTs (Figure 3) (Klymkowsky, 1995; Gard *et al.,* 1997). Treatment of stage VI oocytes with cytochalasin B causes much of the keratin filament network to be released from the cortex, stretching the keratin filaments into the subcortical cytoplasm (Figures 3A and 3B). Cytochalasin B also induces formation of numerous keratin filament aggregates in deeper regions of the cytoplasm (Figure 3B) (Gard *et al.*, 1997), which are reminiscent of vimentin-containing IF aggregates in somatic cells treated with drugs that depolymerize MTs (Knapp *et al.*, 1983a, b). Interestingly, treatment with cytochalasin B also leads to the partial solubilization of the vegetally localized Vg1 mRNA (Yisraeli *et al.*, 1990) and to the disorganization of other cortical components, including γ-tubulin (Gard, 1994), indicating that the organization and integrity of the oocyte cortex is highly dependent upon an intact MF system.

In contrast, treatment of stage VI oocytes with nocodazole or other MT inhibitors results in the loss of the characteristic A–V polarity of keratin filament distribution in the oocyte cortex. In the absence of MTs, keratin filament networks of both the animal and vegetal cortices exhibited a finer mesh size (compare Figures 3C and 3C with each other, and with Figures 2A and 2B) (Gard *et al.,* 1997). In addition, keratin filaments appeared to be more closely apposed to the animal cortex in nocodazole-treated oocytes, explaining the increased prominence of the keratin filament network (Klymkowsky, 1995). These observations suggested that MTs play an important role in the maintenance of the polarized distribution of keratin filaments in the cortex of stage VI *Xenopus* oocytes.

Despite the dependence of keratin filament organization upon MFs and MTs, neither MF nor MT organization (Figures 2G–2I) is affected by nearly complete disruption of the cytoplasmic keratin filament network induced by the microinjection of monoclonal anti-keratin antibodies (Gard *et al.,* 1997). This result is similar to that observed previously in cultured cells: whereas disruption of MF or MT organization can lead to a profound reorganization of the IF network (Knapp *et al.,* 1983a, b), the disruption of IF organization has no apparent effect on either MT or MF organization (Klymkowsky *et al.*, 1983, 1989). These observations suggest that cytoskeletal organization in *Xenopus* oocytes, as in other cell types, is dependent upon a hierarchy of interactions in which IF organization is dependent upon interactions with MTs and MFs, while MT/MF organization is largely independent of the presence of IFs (Figure 4A; Klymkowsky *et al.,* 1989).

A model depicting the organization of the keratin filament cytoskeleton of stage VI oocytes based upon these interactions is shown in

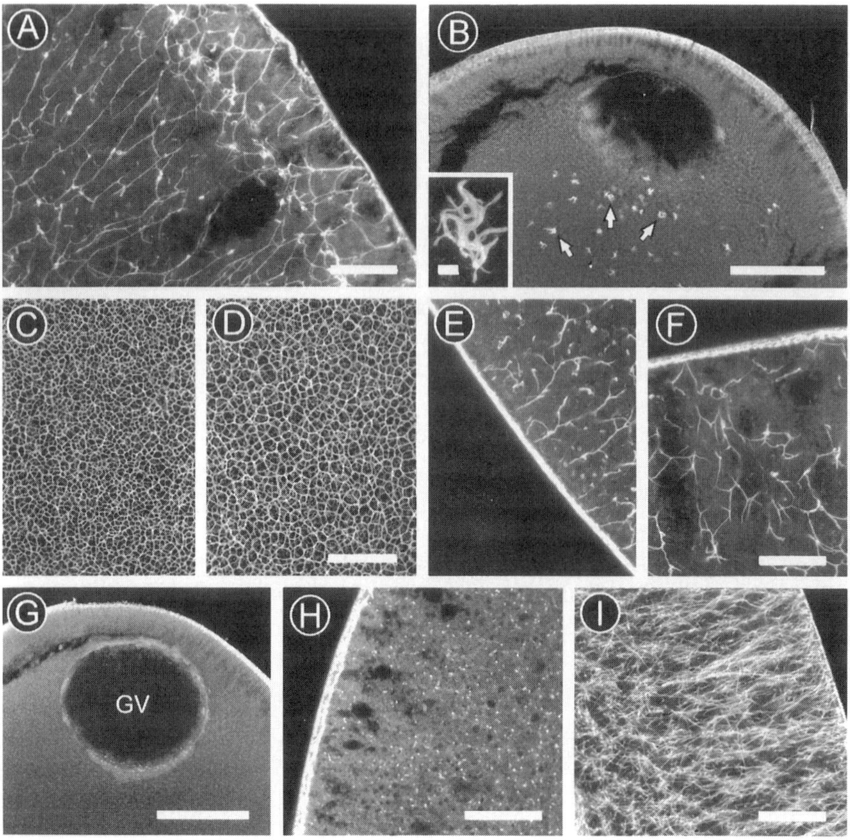

FIGURE 3. Cytoskeletal inhibitors disrupt the organization of the keratin filament network in stage VI *Xenopus* oocytes. (A): Stretching keratin filaments into the subcortical cytoplasm is apparent in the animal hemisphere after treatment of this oocyte with cytochalasin B (20 μg/ml for 20 hours; a projection of five optical sections). (B): Numerous brightly stained keratin aggregates (arrows) are apparent below the GV of this cytochalasin-treated oocyte (20 μg/ml cytochalasin B for 20 hours). The inset shows one aggregate at higher magnification. Note the thickened cortical keratin filament network in the animal hemisphere. (C) and (D): Note the similarity and fineness of the keratin meshwork in these grazing views (projections of two and three optical sections, respectively) of the vegetal (C) and animal (D) cortex of oocytes treated with nocodazole (20 μg/ml for 48 hours). (E) and (F): Note the nearly equal thickness of the keratin network in these cross-sectional views (projections of five optical sections) of the vegetal (E) and animal (F) hemispheres of nocodazole-treated oocytes (20 μg/ml nocodazole for 48 hours). (G): The GV is normally positioned in the animal hemisphere 18 hours after injection of antikeratin IgG, despite nearly complete elimination of cytoplasmic keratin filaments. (H): Cytoplasmic keratin filaments are nearly eliminated 18 hours after injection of antikeratin IgG. (I): MT organization (stained with antitubulin) appears normal 24 hours after injection of antikeratin IgG. Scale bars are 25 μm in A, D, F, and I, 250 μm in B (5 μm in inset) and G, and 50 μm in H.

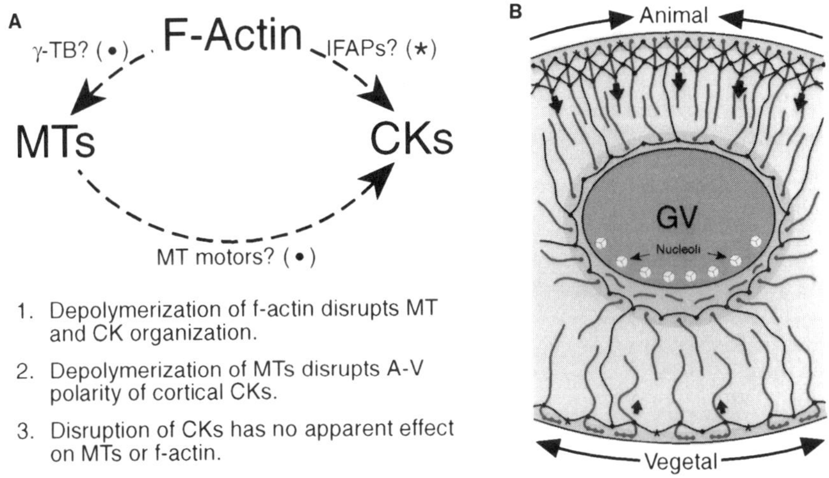

FIGURE 4. (A) Organization and polarization of the cytoskeleton in stage VI *Xenopus* oocytes is dependent upon a hierarchy of interactions between F-actin, MTS, and keratin filaments. (B) A model for the organization of the cytoskeleton of stage VI *Xenopus* oocytes. MTs (grey lines) are anchored to cortical and perinuclear actin via the centrosomal protein γ-tubulin (dots). Keratin filaments (black lines) are anchored to F-actin in the cortical and perinuclear cytoplasm by an unidentified protein complex (*). Keratins are also linked to MTs, possibly by an MT-dependent motor protein. In this model, the A-V polarity of the MT cytoskeleton is specified by the asymmetric distribution of the centrosomal protein γ-tubulin in the oocyte cortex, and the A–V polarity of the cortical keratin network is dependent upon the asymmetric organization of MTs in the animal and vegetal hemispheres (see text).

Figure 4B. In this view, both MTs and keratin filaments are anchored, either directly or indirectly, to cortical MFs. Keratin filaments also interact, directly or indirectly, with radially oriented MTs in both the animal and vegetal hemispheres. In this model, the A–V asymmetry of the cortical keratin filament network does not result from the polarized assembly of keratin filaments, but rather from the polarized distribution of MTs. Previous studies have shown that MTs are more numerous and exhibit a greater degree of radial organization in the animal hemisphere (Gard, 1991). The abundance of organized MTs in the animal cytoplasm exerts two "forces" on the meshwork of keratin filaments in the oocyte cortex. The first, a lateral force, causes the keratin filaments to become concentrated or "bunch up" in the animal cortex. The second is an inward force or tension that stretches the keratin filament network into the subcortical cytoplasm of the animal hemisphere. A simple way to visualize this process is to imagine the keratin filament network of the oocyte as a uniform elastic network, much like a hairnet or "snood," on the surface of a sphere. Now imagine grabbing that network on one side of the sphere and constricting it—this would lead to a thickened, denser filament network on one side of the sphere and a more open, "stretched" network on the opposite side. We have proposed that the MT network of the animal hemisphere exerts just such an asymmetric force on the keratin filament network, leading to a concentration of keratin filaments in the animal cortex, and the stretching of the keratin filament network in the vegetal cortex, forming the characteristic, open keratin meshwork observed by both conventional and confocal microscopy.

The nature of the proposed linkages between keratin filaments, MTs, and MFs remain unknown. Results from several studies suggest that microtubule-associated proteins (MAPs) or kinesin, a MT-dependent motor protein, may mediate interactions between IFs and MTs in somatic cells (Bloom and Vallee, 1983; Hirokawa *et al.,* 1988; Falconer *et al.,* 1989; Rodionov *et al.,* 1985; Draberova and Draber, 1993; Gyoeva and Gelfand, 1991). Recently, a family of IF-associated proteins (IFAPs), which includes plectin, desmoplakin, and bullous pemphigoid antigen (BPAG), have been proposed to link IFs to both MFs and MTs in mammalian cells (Foisner and Wiche, 1991; Foisner *et al.*, 1994; Svitkina *et al.*, 1996; Fuchs *et al.*, 1997, and references therein). *Xenopus* oocytes contain at least two MAPs (XMAP215 and XMAP230; Gard and Kirschner, 1987; Andersen *et al.,* 1995), and probably multiple members of the kinesin (Houliston *et al.*, 1994; Vernos *et al.*, 1993, 1995; Niclas *et al.*, 1996; Walczak *et al.*, 1996, 1997; Boleta *et al.,* 1996; Robb *et al.,* 1996) and cytoplasmic dynein (Heald *et al.,* 1996) families. However, their complement of IFAPs remains unknown. Thus, the determination of which, if any, of these proteins mediate the interactions between keratin filaments, MFs, and MTs in *Xenopus* oocytes awaits further analysis.

3.2. Assembly of the Keratin Filament Network during Early Oogenesis

During their extended arrest in meiotic prophase, *Xenopus* oocytes increase nearly one million-fold in volume and specify, and then establish, the A–V polarity characteristic of stage VI oocytes and unfertilized eggs. We have found that the MT and MF cytoskeletons of *Xenopus* oocytes are remodeled on several occasions during oocyte growth and axis formation (Gard *et al.,* 1995b; Roeder and Gard, 1994; see Gard, 1995, and Gard *et al.,* 1995c, for reviews). Previous reports indicated that keratin filaments are present throughout stages I–VI of oogenesis (Franz *et al.,* 1983; Godsave *et al.,* 1984a). In these studies, keratin filaments were first observed during stage I of oogenesis, where they form a sparse network in the cortex and perinuclear cytoplasm. Keratin filaments also formed a "cage" surrounding the mitochondrial cloud, a peculiar aggregate of thousands of mitochondria, cytoplasmic organelles, granular-fibrillar germplasm, and maternal mRNA formed during mid–late stage I of oogenesis. Later, keratin filaments formed a radial cytoplasmic array and a dense cortical network. However, it was unclear from these studies how the characteristically polarized keratin filament network of stage VI oocytes became established.

To better understand the biogenesis of the keratin filament array and the possible role of keratin filaments during oogenesis and axis formation, we used confocal immunofluorescence microscopy to examine the assembly and organization of keratin filaments throughout oogenesis, from oogonia through stage VI (Gard *et al.*, 1997). The results from these studies are summarized in Figure 5 and are discussed below.

Confocal immunofluorescence microscopy reveals that stage 0 oocytes contain prominent MT and MF networks. In contrast, the cytoplasm of mitotic oogonia and early stage 0 oocytes are devoid of keratin filaments (Figure 6). The follicle cells that surround the oocytes contain dense, intensely stained networks of keratin filaments (Figure 6A). Although they lack keratin filaments, we cannot exclude the possibility that proliferating oogonia and stage 0 oocytes contain IFs composed of vimentin or some other cytoplasmic IF protein. By stage I, however, vimentin appears to be absent from the oocyte (Franz *et al.*, 1983; Dent, 1992; see above).

By early–mid stage I, oocytes contain extensive networks of cytoplasmic MTs and cortical MFs, and are assembling networks of MFs in the nucleus and actin cables in the cytoplasm (Gard, 1991; Gard *et al.*, 1995a; Roeder and Gard, 1994; see Gard, 1995; Gard *et al.*, 1995c). Assembly of keratin filaments, on the other hand, does not begin until mid–late stage I of oogenesis (Gard *et al.*, 1997). Keratin filaments first appear in association with the GV and early mitochondrial cloud (MC), which itself appears during mid–stage I (Kloc and Etkin, 1995; Kloc *et al.*, 1996) (Figure 6C). By late stage I, a network of keratin filaments extends throughout the oocyte cortex. Keratin filaments surround the GV and MC, and a sparse network of cytoplasmic keratin filaments link these organelles to the cortex (Figures 6D and 6E). The network of cytoplasmic keratin filaments is further elaborated during stage II of oogenesis to form an extensive, but irregularly ordered, three-dimensional meshwork linking the perinuclear and cortical keratin networks (Figure 7A).

Aside from their association with the GV, MC, and cortex, no discrete foci of keratin filament assembly was apparent in these studies, suggesting that a discrete IF-organizing center, analogous to a microtubule-organizing center (MTOC), does not exist in the *Xenopus* oocyte. This conclusion is also supported by the assembly of dispersed networks of vimentin

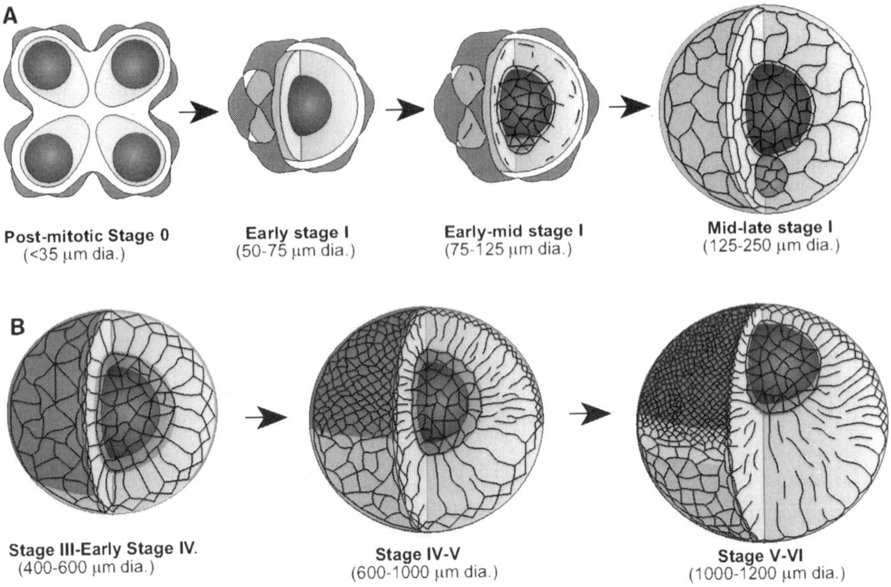

FIGURE 5. A summary cf keratin filament and organization and polarization during oogenesis in *Xenopus laevis.* (A) During stage 0 and early stage I, oocytes lack keratin filaments, which are found in the surrounding follicle cells. By early-mid–stage I, keratin filaments form a network surrounding the GV, surround and penetrate the mitochondrial mass, and begin to extend into the cytoplasm. By mid-late–stage I, and extensive network of cytoplasmic keratin filaments links the perinuclear network to the mitochondrial mass and cortex. (B) Keratin filaments are symmetrically distributed during early stage IV. Polarization of the cortical and cytoplasmic keratin filament networks is apparent during mid-stage IV-stage V, as first the animal (during mid-stage IV), then vegetal (during late stage IV-stage V) keratin filament networks adopt the organization characteristic of stage VI. Summarized from Franz *et al.* (1983); Godsave *et al.* (1984a), Klymkowsky *et al.* (1987); Torpey *et al.* (1992b), Ryabova *et al.* (1993), and Gard *et al.* (1997).

filaments in oocytes microinjected with vimentin mRNA (Dent *et al.,* 1992). Similarly, the *de novo* assembly of IF networks in cultured cells fails to provide any evidence for a discrete IF organizing center (Klymkowsky *et al.*, 1989).

The A–V axis of *Xenopus* oocytes becomes externally apparent as pigment is differentially localized to the animal cortex during early stage IV of oogenesis. During stage III–early stage IV, the poorly ordered network of cytoplasmic keratin filaments is reorganized to form a more radially oriented, but still apparently unpolarized, array of keratin filaments linking the cage of keratin filaments surrounding the GV to the extensive cortical

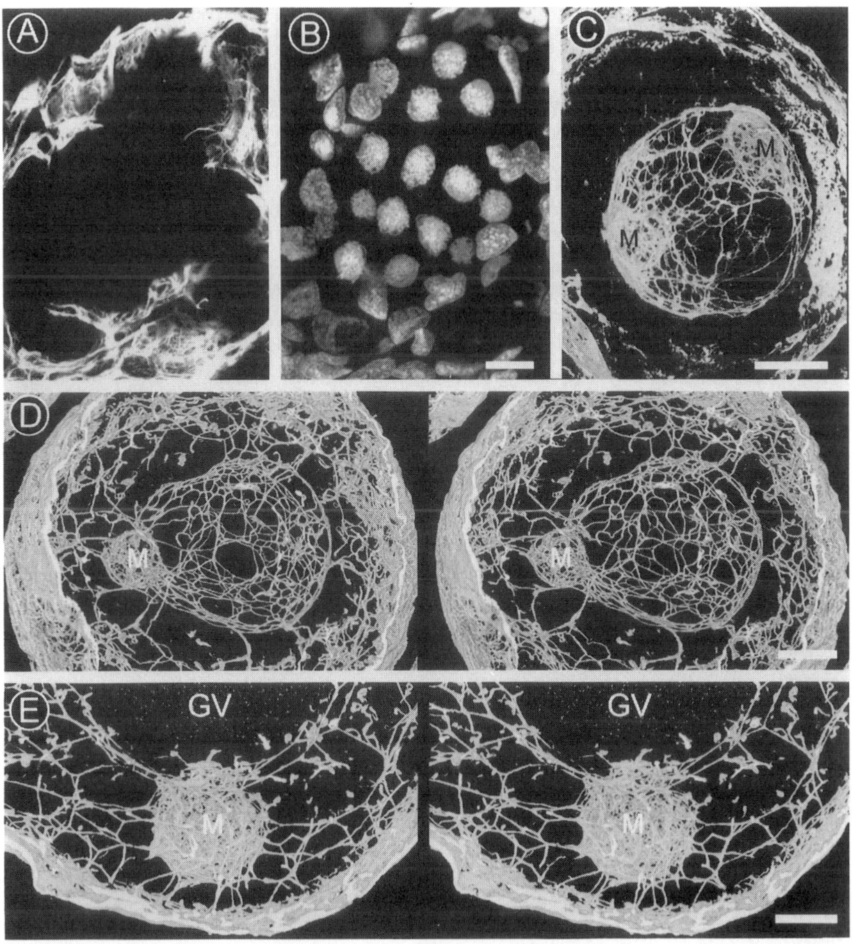

FIGURE 6. Keratin filaments are first apparent during mid-stage I of oogenesis. (A, B) A nest of early stage 0 oocytes isolated from a juvenile frog and stained with (A) antikeratin antibodies and (B) ethidium homodimer to reveal chromatin. Keratin staining is only apparent in the follicle cells surrounding the oocytes. (C) A reconstruction (from 53 optical sections) of a mid-stage I (125 μm diameter) oocyte from a juvenile frog stained with antikeratin, showing the assembly of a perinuclear keratin filament network and the association of keratin filaments with two early mitochondrial aggregates (M; forerunners of the mitochondrial cloud). (D) A stereo view (reconstructed from 76 optical sections) of a mid–late-stage I (185 μm) oocyte from an adult frog stained with antikeratin, showing a meshwork of keratin filaments surrounding the GV and linking the GV and mitochondrial cloud (M) to the oocyte cortex. (E) A stereo view (reconstructed from 25 optical sections) of the mitochondrial cloud of a stage I oocyte (approximately 200 μm diameter) isolated from an adult frog and stained with antikeratin. Note the extensive network of keratin filaments surrounding and penetrating the mitochondrial cloud. Scale bars are 10 μm in planels A–B, and 25 μm in panels C–E.

keratin network (Figures 7B and 7C). Despite the visible polarization of the oocyte cortex during early stage IV, the cortical and cytoplasmic keratin filament networks appear symmetric until mid–late stage IV of oogenesis, as oocytes reach diameters of 600–800 μm. At this time, keratin filaments

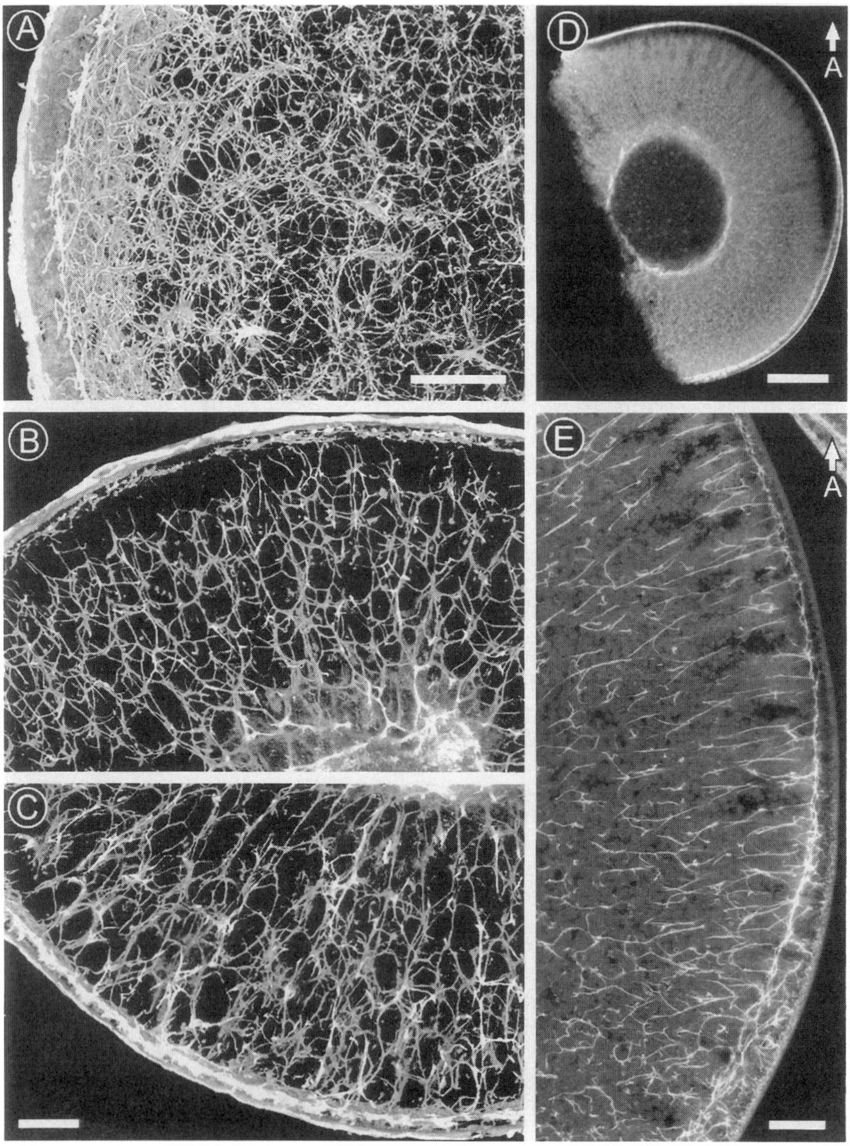

in the animal cortex and cytoplasm begin to adopt the organization characteristic of stage VI oocytes (Figures 7D and 7E); a radial array of cytoplasmic keratin filaments links the perinuclear keratin network to a 10- to 12-μm-thick, fine-meshed net of keratin filaments in the animal cortex. In contrast, final "maturation" of the keratin filament network in the vegetal cortex to form the coarse, open meshwork seen in stage VI oocytes does not occur until late stage IV or stage V, as oocytes reach diameters of 1.0–1.1 mm.

3.3. Keratin Filaments Are Disassembled during Oocyte Maturation

Most somatic cells undergo a dramatic morphological transformation as they enter M phase, which is accompanied by extensive reorganization of their cytoskeleton, including IFs. The nuclear lamina, for instance, is completely disassembled during M phase, apparently in direct response to phosphorylation of the lamins by cdc2 kinase (Peter *et al.*, 1990; Ward and Kirschner, 1990).

The fate of cytoplasmic IFs is rather more variable. In many types of somatic cells, the IF networks remain intact, but "collapse," forming a "cage" surrounding the mitotic spindle (Blose and Chako, 1976; Blose, 1979; Aubin *et al.*, 1980; Blose and Bushnell, 1982; Zieve *et al.*, 1980). As discussed above, IF organization is dependent upon interactions with MTs and MFs, and the M-phase collapse of IF organization may be due simply to the cell cycle-dependent reorganization of MF/MT systems. However, in

FIGURE 7. Polarization of the keratin filament network occurs during mid–late stage IV of oogenesis. (A) A reconstruction (from 39 optical sections) of a late stage II or early stage III (350–400 μm diameter) oocyte. Little radial order is seen in the cytoplasmic keratin network. The dense, subcortical keratin network lies approximately 10 μm below the oocyte surface. (B, C) Reconstructions (both from 15 optical sections) of the (B) animal and (C) vegetal regions of an early stage IV oocyte (450–500 μm diameter). No evidence of polarity is seen in the organization of cytoplasmic or cortical keratin filaments in these views of the same oocyte. Note the numerous transverse connections between the radially oriented keratin filaments in both views. (D) A mid-stage IV (600–650 μm diameter) oocyte. Note the difference in thickness of the cortical keratin network in the animal and vegetal hemispheres (arrow denotes the direction of the animal pole). (E) A composite of five images (each a projection of five optical sections) spanning the equatorial region of a mid-stage IV (800–850 μm diameter) oocyte. Note the pronounced difference in cytoplasmic and cortical keratin organization in the animal and vegetal regions (arrow denotes the direction of the animal pole). Keratin organization in the animal cortex and cytoplasm is similar to that of fully grown stage VI oocytes. The keratin network of the vegetal cortex remains much thicker than that of fully grown stage VI oocytes (compare to Fig. 2). Scale bars are 25 μm panels in panels A–C and E, and 100 μm in panel D.

other cell types, the IF network is fragmented during mitosis (Horwitz *et al.*, 1981; Franke *et al.*, 1982, 1984; Lane *et al.*, 1982; Rosevear *et al.*, 1990). In these cases, IF proteins remain insoluble, but electron microscopy reveals only "amorphous aggregates" in place of the typical 8- to 10-nm IFs (Franke *et al.*, 1982; Lane *et al.*, 1982). In yet other cell types, there appears to be localized disassembly of IFs, particularly in the region of the contractile ring (Blose, 1979; Blose and Bushnell, 1982; Nishizawa *et al.*, 1991). The reorganization of IFs during M phase is reversible, and typical 8- to 10-nm IFs reappear rapidly as cells reenter interphase. Interestingly, whether a particular cell type undergoes IF network collapse or fragmentation during M phase can be influenced by relatively subtle changes in culture conditions (Tölle *et al.*, 1987). In all cases examined, the reorganization of IFs during mitosis is correlated with the hyperphosphorylation of IF subunit proteins (Ku *et al.*, 1996).

In response to stimulation by progesterone, stage VI *Xenopus* oocytes reenter the meiotic cell cycle, completing the first meiotic division, and arrest in second meiotic metaphase. This process, usually referred to as "oocyte maturation," is accompanied by substantial reorganization of the oocyte microtubule cytoskeleton, culminating in assembly of the meiotic spindles (Gard, 1991, 1992). The organization of the keratin filament network of *Xenopus* oocytes is also dramatically modified during the meiotic maturation of *Xenopus* oocytes (Klymkowsky *et al.*, 1987, 1991; Klymkowsky and Maynell, 1989). However, in contrast to the formation of insoluble keratin filament aggregates during mitosis in somatic cells (see above), during the late stages of meiosis in *Xenopus* oocytes the keratin network is nearly completely disassembled into soluble oligomers (Klymkowsky *et al.*, 1991; Bachant and Klymkowsky, 1996). These oligomers have an apparent molecular weight of ~750kDa (as determined by size exclusion chromatography) (Bachant and Klymkowsky, 1996). Although keratin filament disassembly often appears complete, the large size of the resulting keratin complexes relative to the tetrameric keratin filament subunit (~750kDa vs. the calculated ~200kDa of the tetramer) suggests that some oligomeric structure is retained.

Conventional and confocal microscopy reveal that disassembly of the cortical keratin filament is first apparent in the animal hemisphere, beginning soon after germinal vesicle breakdown (GVBD) (Figure 8A). Somewhat paradoxically, the keratin filament network in the animal cortex immediately overlying the GV often becomes transiently more apparent during or soon after GVBD, appearing more robust and more highly interconnected (Klymkowsky *et al.*, 1991; Gard *et al.*, 1997). This probably results as a secondary effect of the extensive disassembly and reorganization of cytoplasmic MTs that accompanies oocyte maturation, resulting

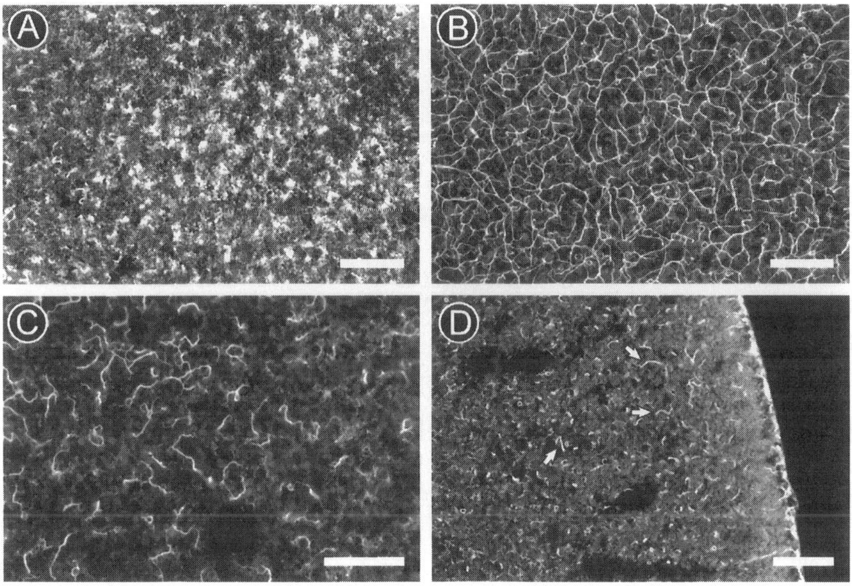

FIGURE 8. The cortical and cytoplasmic keratin filament arrays are disassembled during progesterone-induced maturation. (A) Keratin filaments in the animal cortex of this oocyte have disassembled nearly completely within 0–10 min after formation of the white maturation spot, leaving only brightly staining keratin aggregates. (B) In contrast, the keratin filament network in the vegetal cortex remains intact 0–10 min after white spot formation. (C) By 20–30 min after white spot formation, the vegetal keratin network exhibits substantial fragmentation. (D) The radial network of cytoplasmic keratin filaments in the animal hemisphere has disassembled nearly completely by 0–10 min after white spot formation. Arrows denote fragments of keratin filaments, consistent with filament severing (see text). Panels A–C are projections of six serial optical sections. Panel D is a single optical section. All scale bars are 25 μm.

in an effect and appearance similar to that observed in stage VI oocytes treated with nocodazole or other MT inhibitors (see above). Disassembly of the vegetal keratin filament network in the vegetal cortex occurs somewhat later, but is generally complete within 60 min after GVBD (Figures 8B and 8C).

The radially oriented network of cytoplasmic keratin filaments is also rapidly disassembled during meiotic maturation. Disassembly of the cytoplasmic keratin filament network occurs more rapidly than that of the overlying cortical network, and the cytoplasm is nearly devoid of long keratin filaments within 10 min after GVBD. Confocal immunofluorescence microscopy reveals numerous short keratin filament fragments scattered throughout the animal cytoplasm immediately after GVBD (Figure 8D), consistent with a severing mechanism for keratin filament disassembly (Klymkowsky *et al.*, 1991). As with the keratin filaments in the cortex, disassembly of cytoplasmic keratin filaments proceeds in an animal–vegetal direction (Klymkowsky and Maynell, 1989).

Interestingly, as maturation proceeds, short keratin filament fragments or keratin oligomers are concentrated in the novel MT-organizing center–transient MT array complex (MTOC–TMA complex) that is assembled at GVBD (Figures 9A–9D; Figure 10A shows an MTOC–TMA complex stained with anti-tubulin) (Gard, 1992; Gard *et al.*, 1995c, and references therein). After forming near the basal surface of the GV, the MTOC–TMA complex migrates to the animal pole, where it serves as the immediate precursor of the first meiotic spindle (Gard, 1992). As this migration occurs, concentrations of short keratin filaments or oligomers are observed in association with base of the MTOC–TMA complex (Figures 9E and 9F). During the early stages of M1 spindle assembly, these keratin filaments or oligomers are often observed displaced to one side of the nascent spindle. At later stages of spindle assembly, no association of keratin filaments with either the M1 or M2 spindles is apparent.

The association of these short keratin filaments or oligomers with the MTOC–TMA and nascent M1 spindle suggested that keratin filaments might play a structural role in the assembly or organization of this complex, or in subsequent assembly of the M1 spindle. However, nearly complete elimination of these short keratin filaments by the injection of anti-keratin monoclonal antibodies had no apparent affect on formation or migration of the MTOC–TMA complex, or on the assembly of the first meiotic spindle (Figure 10). Similarly, the down regulation of keratin mRNA by antisense oligonucleotide injection did not impair the completion of meiosis, fertilization, or early development (Torpey *et al.*, 1992a). The association of keratin oligomers with the MTOC–TMA complex thus appears to have little functional importance, and the role of the short keratin filaments during oocyte maturation, if any, remains uncertain.

Disassembly of the oocyte's keratin filament network is induced by the injection of maturation-promoting factor (MPF), but unlike events directly regulated by MPF, keratin network disassembly requires ongoing protein synthesis to proceed; in fact, if protein synthesis is blocked, the keratin filament network will reassemble (Klymkowsky *et al.*, 1987; Klymkowsky and Maynell, 1989). During maturation, the type II keratin of the oocyte is hyperphosphorylated (Klymkowsky *et al.*, 1991). Whether this is the direct cause of keratin filament disassembly has yet to be determined. Although phosphorylation of the oocyte keratin is triggered by MPF, it is unlikely that the keratin proteins are directly phosphorylated by the $p34^{cdc2}$/MPF kinase,

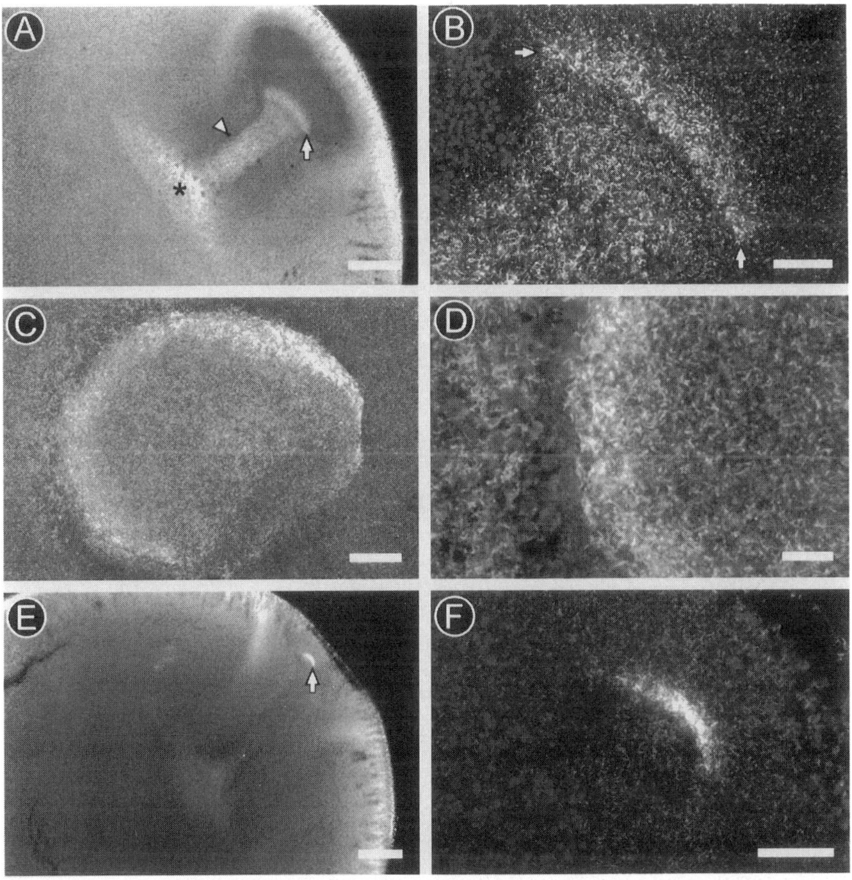

FIGURE 9. Keratin filaments or oligomers are associated with the transient MTOC and MT array assembled during maturation. (A, B) Short fragments of keratin filaments or keratin oligomers are concentrated in the base of the transient MTOC–MT array (the MTOC–TMA complex) 0–10 min after white spot formation (arrows in A and B). Keratin filaments are also concentrated in the yolk-free cytoplasm beneath the MTOC–TMA complex, forming a cylindrical "stalk" (arrowhead in A), and in a broad "base" corresponding to the initial site of MTOC–TMA assembly (asterisk in A). (C, D) Polar views (looking down on the animal pole) of MTOC–TMA at 0–10 min after white spot formation, showing concentrations of keratin filaments or oligomers in the base of the. At higher magnification (D), short keratin filaments can be seen. (E, F) Keratins are concentrated in the base of the MTOC–TMA as it nears the animal cortex 10–20 min after white spot formation. Scale bars are 100 μm in panels A and E, 25 μm in panels B, C, and F, and 10 μm in panel D.

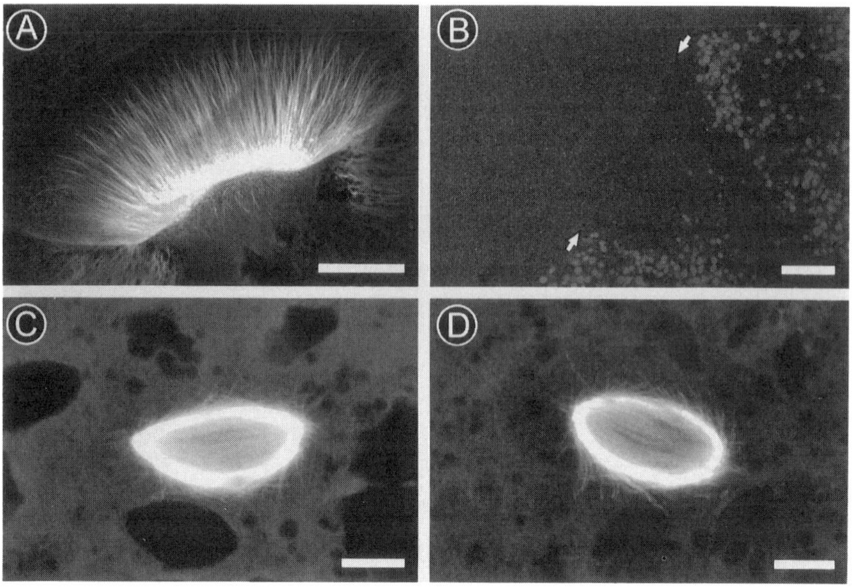

FIGURE 10. Keratin filaments are not required for assembly of the MTOC–TMA complex or meiotic spindles. (A) A normal MTOC–TMA complex formed after depletion of keratin filaments by microinjection of antikeratin antibodies (fixed 0–10 min after white spot formation and stained with anti-α-tubulin). (B) A similar MTOC–TMA in an oocyte injected with antikeratin and then stained with antikeratin antibodies shows marked depletion of keratin filaments in the MTOC–TMA complex (between arrows; compare to Figures 9B and 9F). (C) A normal M1 spindle after depletion of keratin by microinjection of antikeratin IgG (fixed 50–60 min after white spot formation and stained with antitubulin). (D) A normal M1 spindle in a control oocyte (fixed 30–40 min after white spot formation and stained with antitubulin). Scale bars are 50 μm in panel A, 25 μm in panel B, and 10 μm in panels C and D.

since sequence analysis reveals that the keratin proteins lack consensus sites for MPF kinase-mediated phosphorylation.

That host factors specifically influence the nature of the M-phase reorganization of IFs is dramatically illustrated by the behavior of vimentin in the oocyte. As mentioned above, the oocyte does not contain an endogenous vimentin network. If oocytes are injected in the animal hemisphere with vimentin RNA, vimentin accumulates, but an interconnected vimentin filament network does not form (Figure 11A; Dent *et al.*, 1992). Upon maturation, however, a vimentin network appears and grows more connected and robust with time (Figures 11B and 11C), even as the endogenous keratin filament network is *disassembling*. Vimentin does contain an MPF kinase phosphorylation site and is hyperphosphorylated during oocyte maturation (and during mitosis in somatic cells; Chou *et al.*, 1990), but MPF-mediated phosphorylation does not appear to control vimentin network

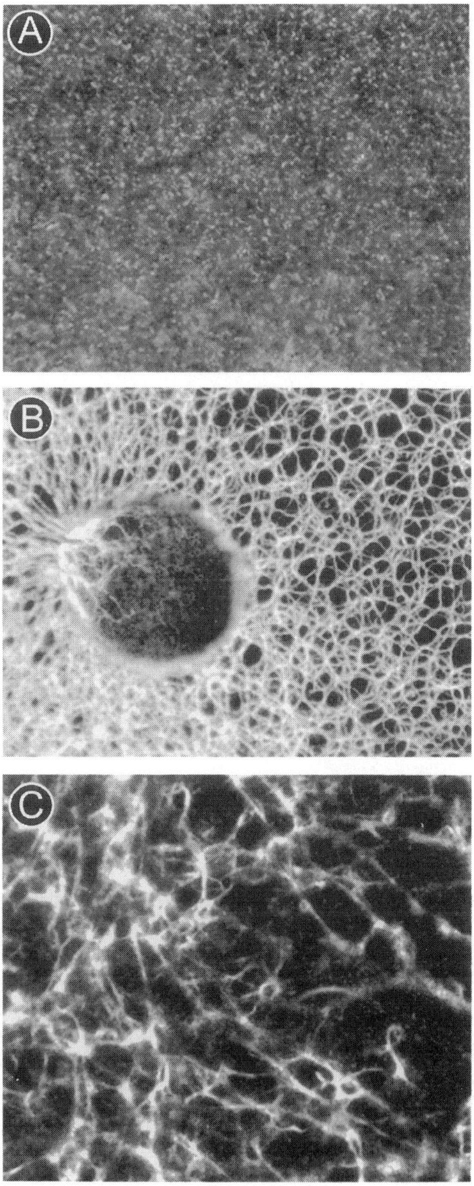

FIGURE 11. Vimentin reorganization during oocyte maturation. Stage VI oocytes were injected in the animal hemisphere with capped RNA encoding myc-tagged *Xenopus* vimentin-1; after 24–48 hr in culture, the oocytes were fixed and stained with the anti-myc antibody 9E10 and fluorescein-conjugated secondary antibody. In panel A, the punctated vimentin staining seen in prophase oocytes is apparent. Upon oocyte maturation (B, C), this punctate system is transformed into an interconnected network of vimentin filaments. In panel B, the circular hole in the network corresponds to the site of injection; typically during maturation the oocyte swells and pushes a cytoplasmic projection through the hole in the vitelline envelope formed during injection. This "injection bleb" is often lost during fixation and staining. The experiments from which these pictures were taken were performed by Joe Dent (Dent, 1992; Dent *et al.*, 1992).

assembly, since mutation of the putative target site for MPF kinase does not significantly change the behavior of vimentin (Dent *et al.*, 1992). Thus, the host cell factors that regulate the assembly and disassembly of filaments composed of exogenously expressed vimentin remain unknown.

3.4. Cytoplasmic Keratin Filaments Are Reassembled after Fertilization

After completion of maturation, unfertilized *Xenopus* eggs lack an organized keratin filament network, and generally contain only a few short keratin filament fragments or punctate keratin aggregates (Klymkowsky *et al.*, 1987). However, beginning ~30–45 min after fertilization, the keratin filament network begins to reappear in the vegetal hemisphere, as disorganized filaments which become increasing organized with time (Figure 12; Klymkowsky *et al.*, 1987). As the microtubule-based rotation of the egg cortex, which specifies the dorsal–ventral axis of the developing embryo, is completed, the keratin network is elaborated to form the anastomosing network of cortical filament bundles and the fine radial array characteristic of the early embryo (Figure 12; Klymkowsky *et al.*, 1987). Dorsal–ventral asymmetries in the levels of β-catenin (Larabell *et al.*, 1997) and mitochondria (Yost *et al.*, 1995) have been reported, but the keratin system appears, superficially at least, to be symmetric.

Analysis of the process of embryonic keratin filament reassembly (Figure 12) indicates that there is no apparent interaction between the reassembled keratin network and germ plasm located in the vegetal cortex and visualized by staining with propidium iodide (Savage, 1993).

4. DO KERATIN FILAMENTS PLAY A ROLE DURING OOGENESIS AND AXIS FORMATION?

Unlike MTs and MFs, for which polymer assembly and/or organization can be manipulated by treating cells with well-characterized and (relatively) specific drugs, pharmacological agents for disrupting IF organization are not available (Klymkowsky *et al.*, 1989). Rather, the cellular and developmental roles played by IFs have been studied using either the intracellu-

FIGURE 12. The embryonic keratin network. Eggs were fixed at 30, 45, or 120 min after fertiliation and stained with either the antikeratin antibody 1h5 (A, B, C) or propidium iodide (A*, B*), which specifically labels the germ plasm under these conditions (Savage and Danilchik, 1993). By 30 min after fertilization (A, A*) punctate aggregates of keratin have appeared; however, these appear to bear no relationship to propidium-stained germ plasm. By 45 min after fertilization, filamentous structures can be seen emerging from the keratin aggregates (B). Again, the distribution of germ plasm appears to be completely independent of keratin (B*). By the two-cell stage, a robust and oriented keratin filament network has formed in the vegetal hemisphere (C). The parallel orientation of these keratin filaments is likely to be due to interactions between the newly forming keratin network and the cortical microtubule system (Klymkowsky *et al.*, 1989). The experiments from which these pictures were taken were performed by Robert Savage, working with Mike Danilchik, Univesity of Oregon Health Sciences Center (unpublished results, used with permission).

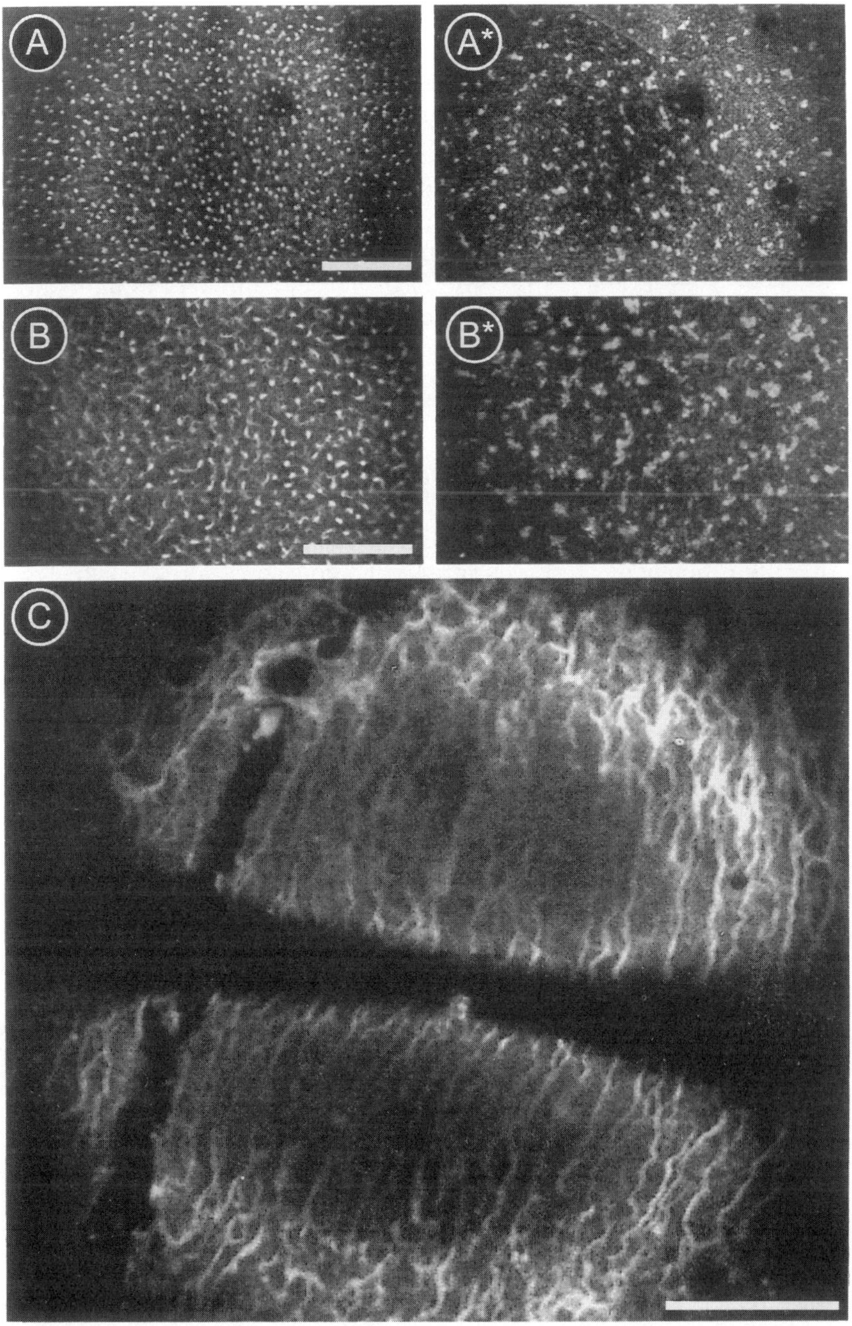
A
A*
B
B*
C

lar injection of specific antibodies, the expression of "dominant negative" forms of IF proteins, the use of "antisense" reagents, or targeted "null" mutations (in the mouse) (Klymkowsky *et al.*, 1989; Klymkowsky and Evans, 1995).

In cultured cells, the absence or disruption of IF networks generally produces little if any effect on cellular morphology or behavior (Klymkowsky *et al.,* 1989; Klymkowsky and Evans, 1995). In order to test whether the oocyte's rather elaborate cortical, radial, and perinuclear keratin filament networks play a critical role in oocyte differentiation or organization, we have examined the effects of injected anti-keratin antibodies (Gard *et al.,* 1997) or the expression of mutated keratin proteins (Bachant, 1993) on various aspects of oocyte structure. While results from these experiments have failed to reveal a definitive role for keratin filaments during oogenesis or very early development, when combined with the time course of keratin filament assembly and distribution of keratin filaments during oogenesis, they place constraints on the roles that keratin filaments might play.

For example, it is unlikely that keratin filaments contribute directly to the specification of the A–V axis during early oogenesis. Although polarization of *Xenopus* oocytes is not externally apparent until early stage IV of oogenesis, substantial evidence suggests that specification of the A–V axis occurs as early as stage I, with formation of the mitochondrial cloud. During stages II–III of oogenesis, components of the MC, including mitochondria and cytoplasmic organelles, the granular-fibrillar nuage material of the germplasm, and maternal RNAs such as Xlsirts and XCAT–1, disperse to the cortex of what will become the vegetal pole of the oocyte (Kloc *et al.*, 1996). The MC of stage I oocytes may thus represent the earliest definitive marker for the A–V axis that will be elaborated later in oogenesis. Going back even further, postmitotic stage 0 oocytes are highly polarized along a nuclear–centrosomal axis established after their last mitotic division, and it has been suggested that this axis may be the forerunner of the A–V axis [for views and reviews see Gerhart (1980), Dent and Klymkowsky (1989), Klymkowsky and Karnovsky (1994), and Gard (1995)]. However, stage 0 oocytes lack significant numbers of keratin filaments. Assembly of the cortical and cytoplasmic arrays of keratin filaments does not begin until mid–late stage I, often after the position of the mitochondrial cloud has been established. Moreover, polarization of cortical and cytoplasmic keratin filament networks is not evident until quite late in oogenesis (late stage IV). Thus, keratin filaments are either absent, or present in an unpolarized array, during the stages of oogenesis when axis specification and establishment are presumed to occur.

The concentration of keratin filaments within and surrounding the mitochondrial mass of stage I *Xenopus* oocytes raises some interesting questions about their role in the organization of this novel aggregate of organelles and its role in RNA targeting and transport. Recent studies have shown that a class of maternal RNAs localize specifically to the distal (cortical) side of the MC during late stage I, from which they disperse to the (future) vegetal cortex during stage II of oogenesis (Kloc and Etkin, 1995; Kloc *et al.*, 1996). This has led to a suggestion that the MC serves as an organizer or "way station" for the targeting and transport of developmentally important maternal RNAs to the cortex. All three of the major cytoskeletal elements found in oocytes, including keratin filaments, MTs, and MF cables, are concentrated in the MC, both surrounding and penetrating the mass and linking it to the GV and cortex. Disruption of either MTs or MFs has no apparent effect on the cohesion of the MC itself (Heasman *et al.*, 1984; Gard, 1992; Roeder and Gard, 1994), suggesting that keratin filaments or other factors may play a role in its formation. However, three-dimensional reconstructions from serial confocal sections reveal no evidence that any of the cytoskeletal elements within the MC are regionally specialized or polarized in a manner that might shed light on their potential role(s) in RNA localization or transport (Gard, 1992; Roeder and Gard, 1994; Gard *et al.*, 1997; D. L. Gard, unpublished results).

The dramatic A–V asymmetry of the keratin filament distribution in the cortex of stage VI oocytes, combined with the isolation of specific maternal mRNAs from detergent-insoluble, keratin-rich cytoskeletons and isolated vegetal cortices (Pondel and King, 1988; Yisreali *et al.*, 1990; Elinson *et al.*, 1993; Forristall *et al.*, 1995), led to the suggestion that the network of keratin filaments in the vegetal cortex plays an important role in the localization and/or anchoring of developmentally important mRNAs during oogenesis. However, direct attachment of maternal mRNAs to cortical keratin filaments is inconsistent with observations that Vg1 mRNA is not released from the oocyte cortex after disruption of the normal keratin filament organization by the expression of dominant negative keratins (Bachant, 1993), and that disassembly of the cortical keratin network precedes release of Vg1 mRNA from the cortex during oocyte maturation (Klymkowsky *et al.*, 1991). In addition, confocal immunofluorescence microscopy reveals that the animal cortex of stage VI oocytes contains a cortical keratin network that is as elaborate as (or even more so than) that in the vegetal cortex (Elinson *et al.*, 1993; Gard *et al.*, 1997). Moreover, analysis of keratin filament organization during oogenesis indicates that polarization of the cortical keratin filament network does not occur until mid–late stage IV of oogenesis, subsequent to the localization of most maternal mRNA to the cortex during stages II–early IV of oogenesis. It is unlikely, then, that the asymmetric distribution of cortical keratin filaments plays a direct role in targeting maternal mRNA to the vegetal cortex. However, the stretching of the keratin filament network in the vegetal cortex implicit in our model of cytoskeletal organization and polarization (see above) may provide a cytoskeletal basis for the lateral spreading of components, including maternal RNAs, anchored to the vegetal cortex.

Finally, keratin filaments could serve a purely structural role, similar to that proposed for keratin filaments in epithelial cells (Fuchs, 1994). In any case, despite the presence of numerous radially oriented keratin filaments linking the GV to the cortex, keratin filaments do not appear to play any major role in positioning of the oocyte nucleus or maintaining gross cytoplasmic organization. This conclusion stems from two observations. First, the number of radially oriented keratin filaments (estimated at 9000–15,000 per oocyte; Gard *et al.,* 1997) is much smaller than the number of radially oriented MTs (estimated to be 500,000–800,000; Gard, 1991). In addition, in contrast to the dramatic displacement of the GV observed after disruption of either MTs or MFs (Gard, 1993; Roeder and Gard, 1994), nearly complete elimination of the cytoplasmic network of keratin filaments has no apparent effect on the position of the GV in the animal hemisphere or on the distribution of pigment or yolk (Gard *et al.,* 1997).

The extensive network of cortical keratin filaments may contribute to the elasticity and integrity of the oocyte cortex. Indeed, the cortex of stage VI oocytes is much more resistant to mechanical disruption than is that of unfertilized eggs, which lack cortical keratin networks (D. L. Gard, unpublished observations). Although cortical stiffness and elasticity of the oocyte have not yet been quantified, oocytes do not appear to be significantly more fragile after disruption of keratin filaments by the microinjection of anti-keratin antibodies (D. L. Gard, unpublished observations). In addition, such a structural role would not explain, or predict, the characteristic and dramatic A–V polarization of keratin filament distribution in the cortex of stage VI oocytes.

5. KERATIN FUNCTIONS WITHIN THE EMBRYO

After the fertilization-induced reassembly of keratin filaments, embyronic cells contain both a cortical and a subcortical keratin network, in a number of ways similar to that found in the oocyte. The keratin network associated with the embryonic surface is quite robust and is formed by large filament bundles (Klymkowsky *et al.*, 1987; Klymkowsky and Karnovsky, 1994). Underlying this cortical system is a finer network of radial fibers. Interior cells of the embryo lack the dense cortical keratin network, but still contain both a fine cortical and radial keratin filmaent system.

In contrast to the rather unsatisfying lack of defined functions for keratin filaments during oogenesis, results from two approaches suggest that keratin filaments do play important roles in early embryonic development in *Xenopus*. Disruption of keratin filaments in early embryos, either by antisense-mediated depletion of keratin K8 mRNA (Torpey *et al.*, 1992a) or by the injection of anti-keratin antibodies (Klymkowsky *et al.*, 1992), resulted in gastrulation defects. In the case of antisense-depletion studies, much of the cortical keratin network of blastula stage embryos was disrupted (see Figure 2 of Torpey *et al.*, 1992a) and associated with this disruption was the loss of a "compacted" surface and defects in wound healing and gastrulation. In the case of our antibody injection studies, we found that different anti-keratin antibodies differentially affected the embryonic keratin filament networks. Injection of the monoclonal anti-keratin antibody 1h5 disrupted the cortical system, but appeared to leave the radial system intact. Disruption of the cortical keratin network had little effect on gastrulation. In contrast, injection of the monoclonal anti-keratin antibodies AE1 and AE3 disrupted the radial keratin filament network, but left the cortical system intact. AE1 and AE3 disrupted gastrulation (Klymkowsky *et al.*, 1992). Neither 1h5, AE1, nor AE3 injection produced defects in wound healing (A. Karnovsky and M. W. Klymkowsky, unpulished observations), nor have wound healing defects been observed in epithelia from K8 −/− mice (Brock *et al.*, 1996).

In analogy to other systems, we propose that the cortical keratin filament networks present in superficial embryonic cells function primarily to strengthen the embyronic surface, much like the keratin network of epidermal cells. The radial keratin filament networks (and the keratin networks of nonsuperficial cells) appear to influence the morphogenetic properties of blastomeres, perhaps by cross-bridging cytoskeletal systems into a mechanically integrated whole. We presume that it is the much larger size of the embryo, and the concomitantly increased mechanical stringency associated with gastrulation in *Xenopus*, that makes the presence of an embryonic keratin filament network critical. In other organisms, such as the mouse, the presence of a keratin filament network is clearly not necessary for early development (Baribault and Oshima, 1991; Baribault *et al.*, 1993), although the sterility of female K8 −/− mice suggests that keratin-based defects in maternal components play a critical role in early mouse development.

6. SUMMARY AND PERSPECTIVES

In *Xenopus*, as in other sytems, the functions of IFs can be subtle and difficult to define. In the oocyte, an intricate network of keratin filaments forms early in oogenesis, and is remodeled, disassembled during oocyte maturation, and reassembled following fertilization. It is clear that keratin organization is dependent upon interactions with MTs and MFs, although the exact molecular "linker" and regulatory mechanisms involved are unknown. Disruption of the oocyte's keratin network produces little overt effect on oocyte organization, and its functions remain obscure. In the embryo, antibody injection and antisense depletion suggest that keratin filaments play an important role in morphogenesis. Based on the absence of early morphogenetic defects in K8 $-/-$ mouse embryos, it seems likely that these morphogenetic functions of keratins are associated with idiosyncratic properties of the *Xenopus* embryo, rather than defects in more generalized cellular processes. The recent development of transgenic methodologies for *Xenopus* open up new avenues for investigating IF function, combining methods heretofore available only in mammals with the significant advantages of experimental accessibility and speed afforded by the *Xenopus* system.

ACKNOWLEDGMENTS. The authors would like to thank their collaborators and members of their laboratories, who have contributed their time and insight to the work presented in this chapter. D.L.G. was supported by a grant from the National Science Foundation (MCB 9506051), M.W.K. by grants from the National Institutes of Health (GM54001) and the National Science Foundation (MCB 94-18715).

7. REFERENCES

Andersen, S., Buendia, B., Dominguez, J., Sawyer, A., and Karsenti, E., 1995, Effect on microtubules dynamics of XMAP230, a microtubule asscoiated protein present in *Xenopus laevis* eggs and dividing cells, *J. Cell Biol.* **127:**1289–1299.

Aubin, J. E., Osborn, M., Franke, W. W., and Weber, K., 1980, Intermediate filaments of the vimentin-type and the cytokeratin-type are distributed differently during mitosis, *Exp. Cell Res.* **129:**149–165.

Bachant, J. B., 1993, Assembly and dynamics of keratin intermediate filaments in *Xenopus laevis* oocytes, Ph.D. thesis, University of Colorado, Boulder.

Bachant, J. B., and Klymkowsky, M. W., 1996, The identification of a supratetrameric species as the major soluble form of keratin in *Xenopus* oocytes and rabbit reticulocyte lysates, *J. Cell Biol.* **132:**153–165.

Baribault, H., and Oshima, R. G., 1991, Polarized and functional epithelia can form after the targeted inactivation of both mouse keratin 8 alleles, *J. Cell Biol.* **115:**1675–1684.

Baribault, H., Price, J., Miyai, K., and Oshima, R. G., 1993, Mid-gestational lethality in mice lacking keratin 8, *Genes Dev.* **7:**1191–1202.

Baribault, H., Penner, J., Iozzo, R. V., and Wilson, H. M., 1994, Colorectal hyperplasia and inflammation in keratin 8-deficient FVB/N mice, *Genes Dev.* **8:**2964–2973.

Benavente, R., Krohne, G., and Franke, W. W., 1985, Cell type-specific expression of nuclear lamina proteins during development of *Xenopus laevis*, *Cell* **41:**177–190.

Bloom, G. S., and Vallee, R. B., 1983, Association of microtubule-associated protein 2 (MAP 2) with microtubules and intermediate filaments in cultured brain cells, *J. Cell Biol.* **96:**1523–1531.

Blose, S. H., 1979, Ten-nanometer filaments and mitosis: Maintenance of structural continuity in dividing endoethelial cells, *Proc. Natl. Acad. Sci. USA* **76:**3372–3376.

Blose, S., and Bushnell, A., 1982, Observations on the vimentin-10-NM filaments during mitosis in BHK21 cells, *Exp. Cell Res.* **142:**57–62.

Blose, S. H., and Chako, S., 1976, Rings of intermediate (100Å) filament bundles in the perinuclear region of vascular endothelial cells. Their mobilization by colcemid and mitosis, *J. Cell Biol.* **70:**459–466.

Boyle, J. A., and Ernst, S. G., 1989, Sea urchin oocytes possess cortical arrays of microfilaments, microtubules and intermediate filaments, *Dev. Biol.* **134:**72–84.

Boleta, H., Karsenti, E., and Vernos, I., 1996, Xklp2, a novel *Xenopus* centrosomal kinesin-like protein required for centrosome separation during mitosis, *Cell* **84:**49–59.

Brock, J., McCluskey, J., Baribault, H., and Martin, P., 1996, Perfect wound healing in the keratin 8 deficient mouse embryo, *Cell Motil. Cytoskel.* **35:**358–366.

Chou, Y. H., Bischoff, J. R., Beach, D., and Goldman, R. D., 1990, Intermediate filament reorganization during mitosis is mediated by p34cdc2 phosphorylation of vimentin, *Cell* **62:**1063–1071.

Colucci, G. E., Portier, M. M., Dunia, I., Paulin, D., Pournin, S., and Babinet, C., 1994, Mice lacking vimentin develop and reproduce without an obvious phenotype, *Cell* **79:**679–694.

Dent, J. A., 1992, Expression and organization of vimentin filaments in oocytes and embryos of *Xenopus laevis*, Ph.d. thesis, University of Colorado, Boulder.

Dent, J. A., and Klymkowsky, M. W., 1989, Whole-mount analyses of cytoskeletal reorganization and function during oogenesis and early embryogenesis in *Xenopus*, in *The Cell Biology of Fertilization* (G. Shatten and H. Shatten, eds.), pp. 63–103, Academic Press, New York.

Dent, J. A., Polson, A. G., and Klymkowsky, M. W., 1989, A whole-mount immunocytochemical analysis of the expression of the intermediate filament protein vimentin in *Xenopus*, *Development* **105:**61–74.

Dent, J. A., Cary, R. B., Bachant, J. B., Domingo, A., and Klymkowsky, M. W., 1992, Host cell factors controlling vimentin organization in the *Xenopus* oocyte, *J. Cell Biol.* **119:** 855–866.

Draberova, E., and Draber, P., 1993, A microtubule-interacting protein involved in coalignment of vimentin intermediate filaments with microtubules, *J. Cell Sci.* **106:**1263–1273.

Dumont, J. N., 1972, Oogenesis in *Xenopus laevis* (Daudin) I. Stages of oocyte development in laboratory maintained animals, *J. Morphol.* **136:**153–180.

Elinson, R. P., King, M. L., and Forristall, C., 1993, Isolated vegetal cortex from *Xenopus* oocytes selectively retains localized mRNAs, *Dev. Biol.* **160:**554–562.

Falconer, M. M., Vielkind, U., and Brown, D. L., 1989, Association of acetylated microtubules, vimentin intermediate filaments, and MAP 2 during early neural differentiation in EC cell culture, *Biochem. Cell Biol.* **67:**537–544.

Foisner, R., and Wiche, G., 1991, Intermediate filament-associated proteins, *Curr. Opin. Cell Biol.* **3:**75–81.

Foisner, R., Feldman, B., Sander, L., Seifert, G., Artlieb, U., and Wiche, G., 1994, A panel of monoclonal antibodies to rat plectin: Distinction by epitope mapping and immunoreactivity with different tissues and cell lines, *Acta Histochem.* **96:**421–438.

Forristall, C., Pondel, M., Chen, L., and King, M. L., 1995, Patterns of localization and cytoskeletal association of two vegetally localized RNAs, Vg1 and Xcat-2, *Development.* **121:**201–208.

Franke, W. W., Schmid, E., Grund, C., and Geiger, B., 1982, Intermediate filament proteins in nonfilamentous structures: Transient disintegration and inclusion of subunit proteins in granular aggregates, *Cell* **30:**103–113.

Franke, W. W., Grund, C., Kuhn, C., Lehto, V., and Virtanen, I., 1984, Transient change of organization of vimentin filaments during mitosis as demonstrated by a monoclonal antibody, *Exp. Cell Res.* **154:**567–580.

Franz, J. K., and Franke, W. W., 1986, Cloning of cDNA and amino acid sequence of a keratin expressed in oocytes of *Xenopus* laevis, *Proc. Natl. Acad. Sci. USA* **83:**6475–6479.

Franz, J. K., Gall, L., Williams, M. A., Picheral, B., and Franke, W. W., 1983, Intermediate-size filaments in a germ cell: Expression of keratins inoocytes and eggs of the frog *Xenopus*, *Proc. Natl. Acad. Sci. USA* **80:**6254–6258.

Fuchs, E., 1994, Intermediate filaments and disease: Mutations that cripple cell strength, *J. Cell Biol.* **125:**511–516.

Fuchs, E. V., Yang, Y., Dowling, J., Yu, Q.-C., and Guo, L., 1997, Intermediate filament cytoarchitecture and BPAG1: A gene encoding two different types of intermediate filament linker protein, in *Cytoskeletal–Membrane Interactions and Signal Transduction* (P. Cowin and M. W. Klymkowsky, eds.), pp. 167–179, Landes, Bioscience, Austin, Texas.

Gall, L., Picheral, B., and Gounon, P., 1983, Cytochemical evidence for the presence of intermediate filaments and microfilaments in the egg of *Xenopus laevis*, *Biol. Cell* **47:**331–342.

Gard, D. L., 1991, Organization, nucleation, and acetylation of microtubules in *Xenopus laevis* oocytes: A study by confocal immunofluorescence microscopy, *Dev. Biol.* **143:**346–362.

Gard, D. L., 1992, Microtubule organization during maturation of *Xenopus* oocytes: Assembly and rotation of the meiotic spindles, *Dev. Biol.* **151:**516–530.

Gard, D. L., 1993, Ectopic spindle assembly during maturation of *Xenopus* oocytes: Evidence for functional polarization of the oocyte cortex, *Dev. Biol.* **159:**298–310.

Gard, D. L., 1994, λ-Tubulin is asymmetrically distributed in the cortex of *Xenopus* oocytes, *Dev. Biol.* **161:**131–140.

Gard, D. L., 1995, Axis formation during amphibian oogenesis: Reevaluating the role of the cytoskeleton, *Curr. Top. Dev. Biol.* **30:**215–252.

Gard, D. L., and Kirschner, M. W., 1987, A microtubule-associated protein from *Xenopus* eggs that specifically promotes assembly at the plus-end, *J. Cell Biol.* **105:**2203–2215.

Gard, D. L., Affleck, D., and Error, B. M., 1995a, Microtubule organization, acetylation, and nucleation in *Xenopus laevis* oocytes: II. A developmental transition in microtubule organization during early diplotene, *Dev. Biol.* **168:**189–201.

Gard, D. L., Cha, B.-J., and Roeder, A. D., 1995b, F-actin is required for spindle anchoring and rotation in *Xenopus* oocytes: A re-examination of the effects of cytochalasin B on oocyte maturation, *Zygote* **3:**17–26.

Gard, D. L., Cha, B, J., and Schroeder, M. M., 1995c, Confocal immunofluorescence microscopy of microtubules, microtubule-associated proteins, and microtubule-organizing centers during amphibian oogenesis and early development, *Curr. Top. Dev. Biol.* **31:**383–431.

Gard, D. L., Cha, B, J., and King, E., 1997, The organization and animal-vegetal asymmetry of cytokeratin filaments in stage VI *Xenopus* oocytes is dependent upon F-actin and microtubules, *Dev. Biol.* **184:**95–114.

Gerhart, J. C., 1980, Mechanisms regulating pattern formation in the amphibian egg and early embryo, in *Biological Regulation and Development*, vol. 2. (R. F. Goldberger, ed.), pp. 133–316, Plenum Press, New York.

Godsave, S. F., Wylie, C. C., Lane, E. B., and Anderton, B. H., 1984a, Intermediate filaments in the *Xenopus* oocyte: The appearance and distribution of keratin-containing filaments, *J. Embryol. Exp. Morphol.* **83:**157–167.

Godsave, S. F., Anderton, B. H., Heasman, J., and Wylie, C. C., 1984b, Oocytes and early embryos of *Xenopus laevis* contain intermediate filaments which react with anti-mammalian vimentin antibodies, *J. Embryol. Exp. Morphol.* **83:**169–187.

Godsave, S. F., Anderton, B. H., and Wylie, C. C., 1986, The appearance and distribution of intermediate filament proteins during differentiation of the central nervous system, skin and notochord of *Xenopus laevis*, *J. Embryol. Exp. Morphol.* **97:**201–223.

Gomi, H., Yokoyama, T., Fujimoto, K., Ikeda, T., Katoh, A., Itoh, T., and Itohara, S., 1995, Mice devoid of the glial fibrillary acidic protein develop normally and are susceptible to scrapie prions, *Neuron* **14:**29–41.

Gyoeva, F. K., and Gelfand, V. I., 1991, Coalignment of vimentin intermediate filaments with microtubules depends on kinesin, *Nature* **353:**445–448.

Heald, R., Tournebize, R., Blank, T., Sandaltzopoulos, R., Becker, P., Hyman, A., and Karsenti, E., 1996, Self-organization of microtubules into bipolar spindles around artificial chromosomes in *Xenopus* egg extracts, *Nature* **382:**420–425.

Heasman, J., Wylie, C. C., Hausen, P., and Smith, J. C., 1984, Fates and states of determination of single vegetal pole blastomeRes. of *X. laevis*, *Cell* **37:**185–194.

Heasman, J., Crawford, A., Goldstone, K., Garner-Hamrick, P., Gumbiner, B., McCrea, P., Kintner, P., Noro, C. Y., and Wylie, C., 1994a, Overexpression of cadherins and underexpression of β-catenin inhibit dorsal mesoderm induction in early *Xenopus* embryos, *Cell* **79:**791–803.

Heasman, J., Ginsberg, D., Geiger, B., Goldstone, K., Pratt, T., Yoshida, N. C., and Wylie, C., 1994b, A functional test for maternally inherited cadherin in *Xenopus* shows its importance in cell adhesion at the blastula stage, *Development* **120:**49–57.

Heasman, J., Holwill, S., and Wylie, C. C., 1991, Fertilization of cultured *Xenopus* oocytes and use in studies of maternally inherited molecules, *Meth. Cell Biol.* **36:**213–230.

Henderson, D., and Weber, K., 1981, Immuno-electron microscopical identification of the two types of intermediate filaments in established epithelial cells, *Exp. Cell Res.* **132:**297–311.

Herrmann, H., and Wiche, G., 1987, Plectin and IFAP-300K are homologous proteins binding to microtubule-associated proteins 1 and 2 and to the 240-kilodalton subunit of spectrin, *J. Biol. Chem.* **262:**1320–1325.

Herrmann, H., Fouquet, B., and Franke, W. W., 1989, Expression of intermediate filament proteins during development of *Xenopus laevis*, I, cDNA clones encoding different forms of vimentin, *Development* **105:**279–98.

Hirokawa, N., Hisanaga, S., and Shiomura, Y., 1988, MAP2 is a component of crossbridges between microtubules and neurofilaments in the neuronal cytoskeleton: Quick-freeze, deep-etch immunoelectron microscopy and reconstitution studies, *J. Neurosci.* **8:**2769–2779.

Horwitz, B., Kupfer, H., Eshhar, Z., and Geiger, B., 1981, Reorganization of arrays of prekeratin filaments during mitosis: Immunofluorescence microscopy with multiclonal and monoclonal prekeratin antibodies, *Exp. Cell Res.* **134:**281–290.

Houliston, E., Le, G. R., Kress, M., Philippe, M., and Le, G. K., 1994, The kinesin-related protein Eg5 associates with both interphase and spindle microtubules during *Xenopus* early development, *Dev. Biol.* **164:**147–159.

Kay, B. H., and Peng, H. P., 1991, *Xenopus laevis*: *Practical Uses in Cell and Molecular Biology*, Academic Press, New York.

Kloc, M., and Etkin, L. D., 1995, Two distinct pathways for the localization of RNAs at the vegetal cortex in *Xenopus* oocytes, *Development* **121:**287–297.

Kloc, M., Larabell, C., and Etkin, L. D., 1996, Elaboration of the messenger transport organizer pathway for localization of RNA to the vegetal cortex of *Xenopus* oocytes, *Dev. Biol.* **180:**119–130.

Klymkowsky, M. W., 1995, Intermediate filament organization, reorganization and function in the clawed frog *Xenopus. Curr, Top. Dev. Biol.* **31:**455–486.

Klymkowsky, M. W., and Evans, R. M., 1995, Intermediate filaments: A medical overview, in *Princip of Medical Biology* (E. Bittar and N. Bittar, eds.), Vol. 2, pp. 147–188, JAI Press, Greenwich, Connecticut.

Klymkowsky, M. W., and Karnovsky, A., 1994, Morphogenesis and the cytoskeleton: Studies of the *Xenopus* embryo, *Dev. Biol.* **165:**372–384.

Klymkowsky, M. W., and Maynell, L. A., 1989, MPF-induced breakdown of cytokeratin filament organization in the maturing *Xenopus* oocyte depends upon the translation of maternal mRNAs, *Dev. Biol.* **134:**479–485.

Klymkowsky, M. W., Miller, R. H., and Lane, E. B., 1983, Morphology, behavior, and interaction of cultured epithelial cells after the antibody-induced disruption of keratin filament organization, *J. Cell Biol.* **96:**494–509.

Klymkowsky, M. W., Maynell, L. A., and Polson, A. G., 1987, Polar asymmetry in the organization of the cortical keratin system of *Xenopus laevis* oocytes and embryos, *Development* **100:**543–557.

Klymkowsky, M. W., Bachant, J. B., and Domingo, A., 1989, Functions of intermediate filaments, *Cell Motil. Cytoskel.* **14:**309–331.

Klymkowsky, M. W., Maynell, L. A., and Nislow, C., 1991, Cytokeratin phosphorylation, cytokeratin filament severing and the solubilization of the maternal mRNA Vg1, *J. Cell Biol.* **114:**787–797.

Klymkowsky, M. W., Shook, D. R., and Maynell, L. A., 1992, Evidence that the deep keratin filament system of the *Xenopus* embryo acts to ensure normal gastrulation, *Proc. Natl. Acad. Sci. USA* **89:**8726–8740.

Knapp, L. W., O'Guin, W. M., and Sawyer, R. H., 1983a, Drug-induced alterations of cytokeratin organization in cultured epithelial cells, *Science* **219:**501–503.

Knapp, L. W., O'Guin, W. M., and Sawyer, R. H., 1983b, Rearrangement of the keratin cytoskeleton after combined treatment with microtubule and microfilament inhibitors, *J. Cell Biol.* **97:**1788–1794.

Kofron, M., Spagnuolo, A., Klymkowsky, M. W., Wylie, C., and Heasman, J., 1997, The roles of maternal α-catenin and plakoglobin in the early *Xenopus* embryo, *Development* **124:**1553–1560.

Kroll, K. L., and Amaya, E., 1996, Transgenic *Xenopus* embryos from sperm nuclear transplantations reveal FGF signaling requirements during gastrulation, *Development* **122:**3171–3183.

Ku, N. O., Liao, J., Chou, C. F., and Omary, M. B., 1996, Implications of intermediate filament protein phosphorylation, *Cancer Metast. Rev.* **15:**429–444.

Lane, E. B., Goodman, S. L., and Trejdosiewicz, L. K., 1982, Disruption of the keratin filament network during epithelial cell division, *EMBO J* **1:**1365–1372.

Larabell, C. A., Torres, M., Rowning, B. A., Yost, C., Miller, J. R., Wu, M., Kimelman, D., and Moon, R. T., 1997, Establishment of the dorso-ventral axis in *Xenopus* embryos is presaged by early asymmetries in β-catenin that are modulated by the Wnt signaling pathway, *J. Cell Biol.* **136:**1123–1136.

Liedtke, W., Edelmann, W., Bieri, P. L., Chiu, F. C., Cowan, N. J., Kucherlapati, R., and Raine, C. S., 1996, GFAP is necessary for the integrity of CNS white matter architecture and long-term maintenance of myelination, *Neuron* **17:**607–615.

Lourim, D., Kempf, A., and Krohne, G., 1996, Characterization and quantitation of three B-type lamins in *Xenopus* oocytes and eggs: Increase of lamin LI protein synthesis during meiotic maturation, *J. Cell Sci.* **109:**1775–1185.

McCall, M. A., *et al.*, 1996, Targeted deletion in astrocyte intermediate filament (GFAP) alters neuronal physiology, *Proc. Natl. Acad. Sci. USA* **93:**6361–6366.

Newport, J., and Kirschner, M., 1982, A major developmental transition in early *Xenopus laevis* embryos: 1. Characterization and timing ofcellular changes at the midblastula stage, *Cell* **30:**675–686.

Niclas, J., Allan, V, J., and Vale, R. D., 1996, Cell cycle regulation of dynein association with membranes modulates microtubule-based organelle transport, *J. Cell Biol.* **133:**585–593.

Nishizawa, K., Yano, T., Shibata, M., Ando, S., Saga, S., Takahashi, T., and Inagaki, M., 1991, Specific localization of phosphointermediate filament protein in the constricted area of dividing cells, *J. Biol. Chem.* **266:**3074–3079.

Pekny, M., Leveen, P., Pekna, M., Eliasson, C., Berthold, C. H., Westermark, B., and Betsholtz, C., [illegible], Mice lacking glial fibrillary acidic protein display astrocytes devoid of intermediate filaments but develop and reproduce normally, *EMBO J.* **14:**1590–1598.

Peter, M., Nakagawa, J., Doree, M., Labbe, J., and Nigg, E., 1990, *In vitro* disassembly of the nuclear lamina and M phase-specific phosphorylation of lamins by cdc2 kinase, *Cell* **61:**591–602.

Pondel, M. D., and King, M. L., 1988, Localized maternal mRNA related to transforming growth factor beta mRNA is concentrated in a cytokeratin-enriched fraction from *Xenopus oocytes*, *Proc. Natl. Acad. Sci. USA* **85:**7612–7616.

Robb, D., Heasman, J., Raats, J., and Wylie, C., 1996, A kinesin-like protein is required for germ plam aggregation in *Xenopus*, *Cell* **87:**823–831.

Rodionov, V. I., Nadezhdina, E. S., Leonova, E. V., Vaisberg, E. A., Kuznetsov, S. A., and Gelfand, V. I., 1985, Identification of a 100 kD protein associated with microtubules, intermediate filaments and coated vesicles in cultured cells, *Exp. Cell Res.* **159:**377–387.

Roeder, A. D., and Gard, D. L., 1994, Confocal microscopy of F-actin distribution in *Xenopus* oocytes, *Zygote* **2:**111–124.

Rosevear, E. R., McReynolds, M., and Goldman, R. D., 1990, Dynamic properties of intermediate filaments—Disassembly and reassembly during mitosis in baby hamster kidney cells, *Cell Motil. Cytoskel.* **17:**150–166.

Ryabova, L. V., Virtanen, I., Lehtonen, E., Wartiovaara, J., and Vassetzky, S. G., 1993, Morphology of the keratin cytoskeleton of *Xenopus* oocytes as studied using heterologous monoclonal antibodies, *Russ. J. Dev. Biol.* **24:**364–372.

Savage, R., 1993, Germ plasm dynamics in *Xenopus*, Ph.D. thesis, Wesleyan University, Middletown, Connecticut.

Savage, R. M., and Danilchik, M. V., 1993, Dynamics of germ plasm localization and its inhibition by ultraviolet irradiation in early cleavage *Xenopus* embryos, *Dev. Biol.* **157:**371–382.

Schroeder, T. E., and Otto, J. J., 1991, Snoods: A periodic network containing cytokeratin in the cortex of starfish oocytes, *Dev. Biol.* **144:**240–247.

Stick, R., and Hausen, P., 1985, Changes in the nuclear lamina composition during early development of *Xenopus laevis*, *Cell* **41:**191–200.

Svitkina, T. M., Verkhovsky, A. B., and Borisy, G. G., 1996, Plectin sidearms mediate interaction of intermediate filaments with microtubules and other components of the cytoskeleton, *J. Cell Biol.* **135:**991–1007.

Tölle, H. G., Weber, K., and Osborn, M., 1987, Keratin filament disruption in interphase and mitotic cells–how is it induced? *Eur. J. Cell Biol.* **43:**35–47.

Torpey, N. P., Heasman, J., and Wylie, C. C., 1990, Identification of vimentin and novel vimentin-related proteins in *Xenopus* oocytes and early embryos, *Development* **110:**1185–1195.

Torpey, N., Wylie, C. C., and Heasman, J., 1992a, Function of maternal keratin in *Xenopus* development, *Nature* **357:**413–415.

Torpey, N. P., Heasman, J., and Wylie, C. C., 1992b, Distinct distribution of vimentin and keratin in *Xenopus* oocytes and early embryos, *J. Cell Sci.* **101:**151–160.

Vernos, I, Heasman, J., and Wylie, C., 1993, Multiple kinesin-like transcripts in *Xenopus* oocytes, *Dev. Biol.* **157:**232–239.

Vernos, I., Raats, J., Hirano, T., Heasman, J., Karsenti, E., and Wylie, C., 1995, Xklp1, a chromosomal *Xenopus* kinesin-like protein essential for spindle organization and chromosome positioning, *Cell* **81:**117–127.

Walczak, C. E., Mitchison, T. J., and Desai, A., 1996, XKCM1: A *Xenopus* kinesin-related protein that regulates microtubule dynamics during mitotic spindle assembly, *Cell* **84:**37–47.

Walczak, C. E., Verma, S., and Mitchison, T. J., 1997, XCTK2: A kinesin-related protein that promotes mitotic spindle assembly in *Xenopus laevis* egg extracts, *J Cell Biol.* **136:**859–870.

Ward, G., and Kirschner, M., 1990, Identification of cell cycle-regulated phosphorylation sites on nuclear lamin C, *Cell* **61:**561–577.

Yisraeli, J. K., Sokol, S., and Melton, D. A., 1990, A two-step model for the localization of maternal mRNA in *Xenopus* oocytes: Involvement of microtubules and microfilaments in the translocation and anchoring of Vg1 mRNA, *Development* **108:**289–298.

Yost, H. J., Phillips, C. R., Boore, J. L., Bertman, J., Whalon, B., and Danilchik, M. V., 1995, Relocation of mitochondria to the prospective dorsal marginal zone during *Xenopus* embryogenesis, *Dev. Biol.* **170:**83–90.

Zieve, G. W., Heidemann, S. R., and McIntosh, J. R., 1980, Isolation and partial characterization of a cage of filaments that surround the mammalian mitotic spindle, J. *Cell Biol.* **87:**160–169.

Chapter 3

The Complexity of Gene Families Involved in Epithelial Differentiation

Keratin Genes and the Epidermal Differentiation Complex

Dietmar Mischke

1. INTRODUCTION

The majority of the cells in the human body are of epithelial origin. They form specialized, dynamic tissues that are part of organ systems as diverse as the skin, the digestive tract and its associated glands (e.g., liver, pancreas, gall bladder), the lung, and the eye, to name only some.

Among these, the human skin is the largest organ of the body and its uppermost layer, the epidermis, together with its appendages such as hair and nails provides a protective barrier against desiccation and such external noxious agents as mechanical, chemical, thermal, or bacterial assaults. Accordingly, the epidermis is a highly adapted and complex tissue that presents as a continually self-renewing stratified squamous epithelium composed of several functionally different layers (Figure 1). To promote

Dietmar Mischke Institut für Immungenetik, Charité, Medizinische Fakultät der Humboldt-Universität zu Berlin, D-14050 Berlin, Germany.

Subcellular Biochemistry, Volume 31: Intermediate Filaments, edited by Herrmann and Harris. Plenum Press, New York, 1998.

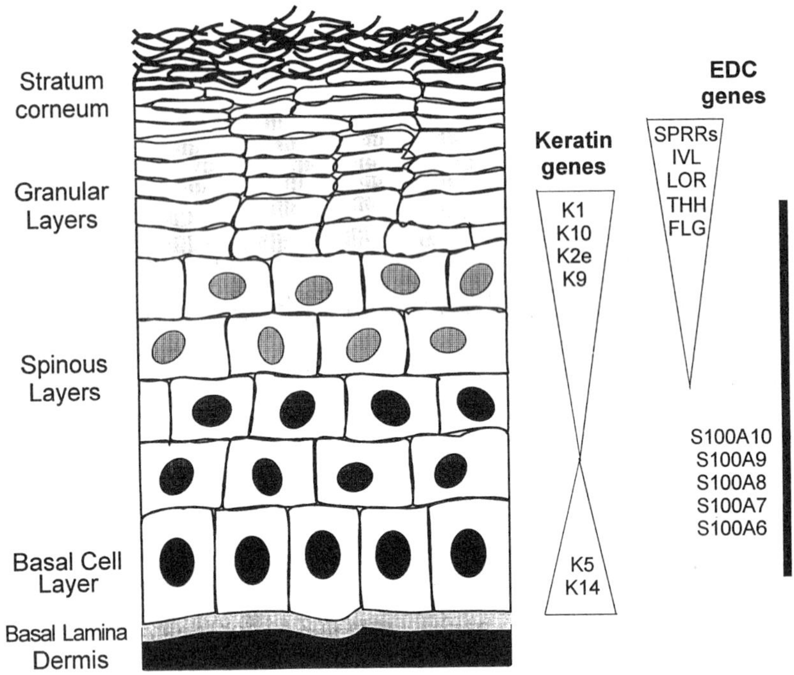

FIGURE 1. Schematic representation of the histology of the human epidermis. Four distinct layers of epidermal cells arise above a connective tissue compartment, the dermis (left). The expression patterns of genes involved in maturation of the epidermis are indicated on the right (see text).

and maintain its remarkable architecture, an orchestrated program of differentiation is executed during which undifferentiated proliferative keratinocytes of the basal cell layer are converted into nondividing differentiating cells of the upper, spinous, and granular cell layers. The series of morphological and biochemical changes eventually culminates in the production of the stratum corneum from which dead, flattened, and enucleated squames are exfoliated.

This extensively analyzed process, which occurs every 2–4 weeks and continues throughout life, is brought about by the temporally and spatially regulated expression of many genes. The keratins, forming the intermediate filaments of all epithelial cells, provide a rigid cytoskeletal scaffold that contributes fundamentally to the functional integrity of the entire tissue, as has been learned from the analysis of several inherited disorders of epidermal differentiation (see below). Cornification, the final step of keratinocyte differentiation, shapes the outermost protective structure, the almost completely insoluble cornified cell envelope, that is assembled on the inside of the plasma membrane by transglutaminase-catalyzed cross-linking of loricrin, involucrin, trichohyalin, small proline-rich proteins (e.g., Steinert and Marekov, 1995), and possibly other, not-yet-identified precursors. At the same time, profilaggrin is processed into filaggrin, which leads to keratin filaments becoming densely packed inside cornified cells. In parallel, multilamellar lipid structures are also formed.

This vectorial program of differentiation is under the control of a variety of intrinsic and extrinsic factors, among which calcium plays an important role, for example, in stratification per se, desmosome formation, the activation of enzymes, and gene regulation. Hence, calcium-binding proteins like those belonging to the S100A family are present to aid in sequestering calcium and/or to directly interact with the structural constituents of the epidermis or their genes.

2. KERATIN GENES

2.1. Introduction

Considering the diversity of epithelial cells and tissues of the body, it comes as no surprise that the family of intermediate filament proteins expressed exclusively in epithelial cells, i.e., the keratins, is exceptionally large. The catalog of human cytokeratins presented by Moll *et al.* (1982) has provided a systematic framework for this plentiful family that has found wide acceptance. Based on the meticulous analysis of cytoskeletal protein fractions from many different epithelial tissues, the polypeptides are subdivided into two classes according to their relatedness and corresponding migration behavior after two-dimensional gel electrophoretic separations: the neutral-to-basic (cyto)keratins (K1–K8) and the acidic (cyto)keratins (K9–K19). With the advent of a still-growing number of cloned cDNAs and characterized genes of intermediate filament proteins, these two families were accordingly classified as type I (acidic) and type II (neutral–basic) keratins. Like other intermediate filament proteins, the keratins share a central α-helical domain of approximately 310 amino acids which is flanked by nonhelical amino-terminal head and carboxyl-terminal tail domains that vary considerably in size and sequence (for reviews see Fuchs and Weber, 1994; Parry and Steinert, 1992; Steinert and Roop, 1988).

The 1982 catalog still holds true and provides a useful framework despite some additions and one deletion. Changes come from the detection of new or additional genes due to the finer distinction of family members by DNA sequence data, like the two discrete K2 genes K2e and K2p (Collin *et al.*, 1992a,b), the at least four isoforms of K6 genes (Takahashi *et al.*, 1995), and the biochemical and immunological characterization of spot IT by Moll *et al.* (1982) as a type I protein, K20 (Moll *et al.*, 1990) and its subsequent cloning (Moll *et al.*, 1993). Moreover, the steadily expanding number of identified human hair keratins (M. A. Rogers *et al.*, 1997, 1996, 1995, 1994; Bowden *et al.*, 1994; Fink *et al.*, 1994; Heid *et al.*, 1986) have not been integrated into the catalog despite their unequivocal lineage with keratin intermediate filaments, but are described as hHa1–hHa5 (type I) and hHb1–hHb6 (type II). One item of the catalog, K11, may be considered deleted because it does not seem to distinguish a product of an independent gene, but rather represents an allele of the highly polymorphic K10 gene (Mischke and Wild, 1987; and below).

The term "cytokeratin," intended to distinguish the intracellular "keratin-like" proteins from the hard keratins of hair and nail presenting on the outside of cells, has not been endorsed. In the OMIM (Online Mendelian Inheritance in Man) database these intermediate filament proteins are listed as keratins (K) and the abbreviation KRT has been chosen by the nomenclature committee of the Human Genome Organization (HUGO) to designate the corresponding genes. A more unifying nomenclature for all keratins (and keratin-associated proteins) proposed by G. E. Rogers and Powell (1993) has not found general acceptance for human keratins. In this scheme, the abbreviations *K1.m* or *K2.m* are used to designate type I or type II keratins (*KRT1.m* or *KRT2.m* for their corresponding genes), respectively, with (*m*) being the number of the Moll catalog or a new number. Alleles, variants, or pseudogenes may then be simply indicated by adding an appropriate extension [see Powell and Rogers (1997) for a detailed discussion]. For example, allele #4 of K10 or allele #1 of K1 (see below) would then be denoted as K1.10.4 (KRT1.10.4) or K2.1.1 (KRT2.1.1), respectively.

The functional significance of the multiplicity of keratins has not been resolved completely, but the interactions of keratin filaments with intermediate-filament-associated proteins and cellular organelles (Chou *et al.*, 1997) differ, as do the assembly properties of keratins. Keratin intermediate filaments are obligate heteropolymers requiring at least one type I and one type II chain for filament assembly *in vivo* and *in vitro*. While *in vitro* reassembly experiments have demonstrated that, with few restrictions, virtually any type I keratin can copolymerize with any type II keratin (e.g., Hofmann and Franke, 1997; Hatzfeld and Franke, 1985), *in vivo* only specific pairs of type I and type II keratins are expressed according to a developmentally controlled and differentiation-specific program (Oshima, 1992; Steinert and Roop, 1988; Sun *et al.*, 1984; Moll *et al.*, 1982). For example, in basal cells of stratified squamous epithelia only K5 and K14 are present, but their synthesis is downregulated (by shutting off transcription of the corresponding genes) when keratinocytes begin to differentiate and new sets of differentiation-specific keratins are produced. In non-keratinizing squamous epithelia, maturing suprabasal cells express K4 and K13, whereas in the keratinizing epidermis K1 and K10 or in the corneal epithelium K3 and K12 are expressed suprabasally. Likewise, the expression of the pair K6 and K16 is induced in the epidermis only during wound healing (see Chapter 6 by K. McGowan and P. A. Coulombe in this book), but is invariably expressed in suprabasal cells of upper digestive tract stratified squamous epithelia. Except for K2e, which is synthesized as a regular constituent of the epidermis when the cells have entered the spinous layer, and for K9, the corresponding type I or type II *in vivo* partners have not yet been identified. Because mutations in K1 may give rise to a similar disease phenotype as mutations in K9 (palmoplantar keratosis; Reis *et al.*, 1994; Kimonis *et al.*, 1994), by inference, K9 may assemble with K1 in the suprabasal epidermal layers of distinct body regions like palmar and plantar epidermis to endow specific function (most likely strength) to these especially exposed areas of the skin.

2.2. Gene Structure

Almost all keratins have now been cloned from mRNA, and genomic DNA sequence information of numerous genes is available as well [see the respective entries in Genbank (http://www.ncbi.nlm.nih.gov) or consult GeneCards, an electronic encyclopedia integrating information about the functions of human genes and their products (http://bioinfo.weizmann.ac.il/cards) for the keratins]. Comparison of the intron/exon distribution reveals a very similar gene organization within each keratin family, but a markedly different one between the two families: type I genes contain eight exons separated by seven introns and type II genes possess nine exons and eight introns (shown schematically in relation to the proteins in Figure 2). Yet some of the genes deviate from that general pattern. For example, in KRT10 the intron 7 is shifted toward the C-terminus, with exon 8 comprising only four coding nucleotides (one amino acid), the translational stop codon, and the untranslated nucleotides at the 3′ end (Rieger and Franke, 1988). Likewise, intron 7 of KRT9 and hHa5 are located together within the 3′ noncoding region (Reis *et al.*, 1994, and M. A. Rogers *et al.*, 1996, respectively). In KRT19, which is an exceptional tailless keratin, only five

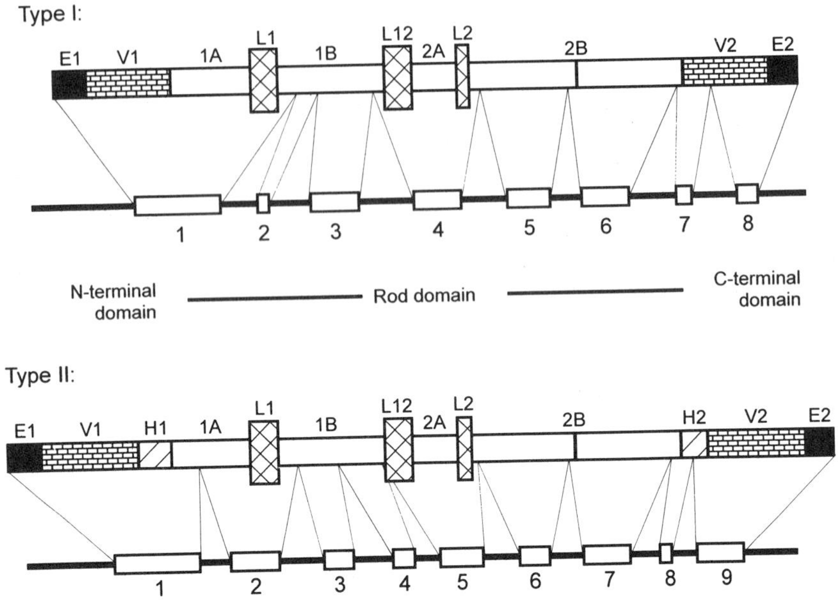

FIGURE 2. Intron/exon organization of human keratin genes. The subdomains of both types of keratins (Steinert and Roop, 1988) are indicated above, and the distribution of exons (filled boxes) and introns (lines) are aligned below.

introns have been detected, introns 6 and 7 not being present (Bader *et al.*, 1988) [in addition to the genomic human KRT19 sequence, Bader *et al.* (1988) also give the genomic KRT15 sequence, neither of which I could find in the databases]. Among the type II genes, KRT8 diverges from the schematic presentation shown in Figure 2 because the gene lacks intron 5, and intron 8 is shifted by approximately 20 nucleotides toward the C-terminal end (Krauss and Franke, 1990).

Only one copy per haploid genome appears to exist for all keratin genes. Although a seeming exception, the at least four "isoformic" human keratin 6 genes (KRT6A–KRT6D) that are highly related in gene structure, nucleotide sequence, and predicted amino acid sequence (Takahashi *et al.*, 1995) nevertheless appear to be individual type II keratin genes (rather then polymorphic variants; see below) because they are differentially expressed in many epithelial tissue types. Among them, K6A is the dominant isoform in skin tissue samples and cultured epithelial cells and was among the first human keratins to be cloned (Tyner *et al.*, 1985). Similarly, K2e and K2p also represent discrete keratins despite their cognate designation because they differ considerably in their head and tail domains and share only 72% sequence identity in the rod domain (Collin et al., 1992b).

2.3. Chromosomal Localization of Intermediate Filament Genes

The genes of intermediate filament proteins are scattered within the human genome (Figure 3). For example, the genes of the type III intermediate filament proteins glial fibrillary acidic protein (GFAP), desmin, peripherin, and vimentin are localized on chromosomes 17q21 (Bongcam-Rudloff *et al.*, 1991), 2q35 (Viegas-Pequignot *et al.*, 1989), 12q11–q13 (Moncla *et al.*, 1992), and 10p13 (Ferrari *et al.*, 1987), respectively. The genes of the type IV neurofilament proteins have been mapped to 8p21 (neurofilament light chain, NEFL; Hurst *et al.*, 1987) and 22q12–q13 (neurofilament heavy chain, NEFH; Mattei *et al.*, 1988), and the genes of the type V nuclear intermediate filament proteins have been localized on chromosomes 1q21.3 (lamin A/C), 5q23 (lamin B1), and 19p13.3 (lamin B2), respectively (Wydner *et al.*, 1996).

At the same time, using miscellaneous mapping techniques like *in situ* hybridization, analysis of somatic cell hybrids, and contig building of cloned sequences, the work of many groups has established that the keratin genes are clustered in the genome: all type I keratins (except KRT18) are located on chromosome 17q11–q21 and all type II keratins (plus KRT18) are located on chromosome 12q11–q13 (e.g., Ceratto *et al.*, 1997; Milisavljevic *et al.*, 1996; Yoon *et al.*, 1994; Reis *et al.*, 1994, 1992; Rosenberg *et al.*, 1991; Waseem *et al.*, 1990a,b; Savtchenko *et al.*, 1990; Barletta *et al.*, 1990; Popescu *et al.*, 1989; Lessin *et al.*, 1988; Bader *et al.*, 1988; Romano *et al.*, 1988, 1987). The clusters of keratin genes most likely also harbor the respective types of hair keratin genes (M. A. Rogers *et al.*, 1996, 1995; Healy *et al.*, 1995).

For the type II cluster of keratin genes on human chromosome 12q11–q13 (most likely 12q13) an almost complete and integrated high-resolution physical and genetic map is available that is based on an extended analysis of genetic markers and clones of yeast artificial chromosomes (YACs) in that region (Yoon *et al.*, 1994). Besides the nine keratin genes already positively identified (Figure 4a), a total of 15 keratin or keratin-like sequences may be present within this contig of 630kb, including possibly the genes of KRT3, KRT2p, the isoforms of KRT6, and the type II hair keratins.

A particular point relating to this map and an associated controversy is the exception to the rule, i.e., the localization of the type I keratin KRT18 not only within the type II cluster, but also next to its type II partner, KRT8. In apparent accord with its type I nature, Heath *et al.* (1990) had localized

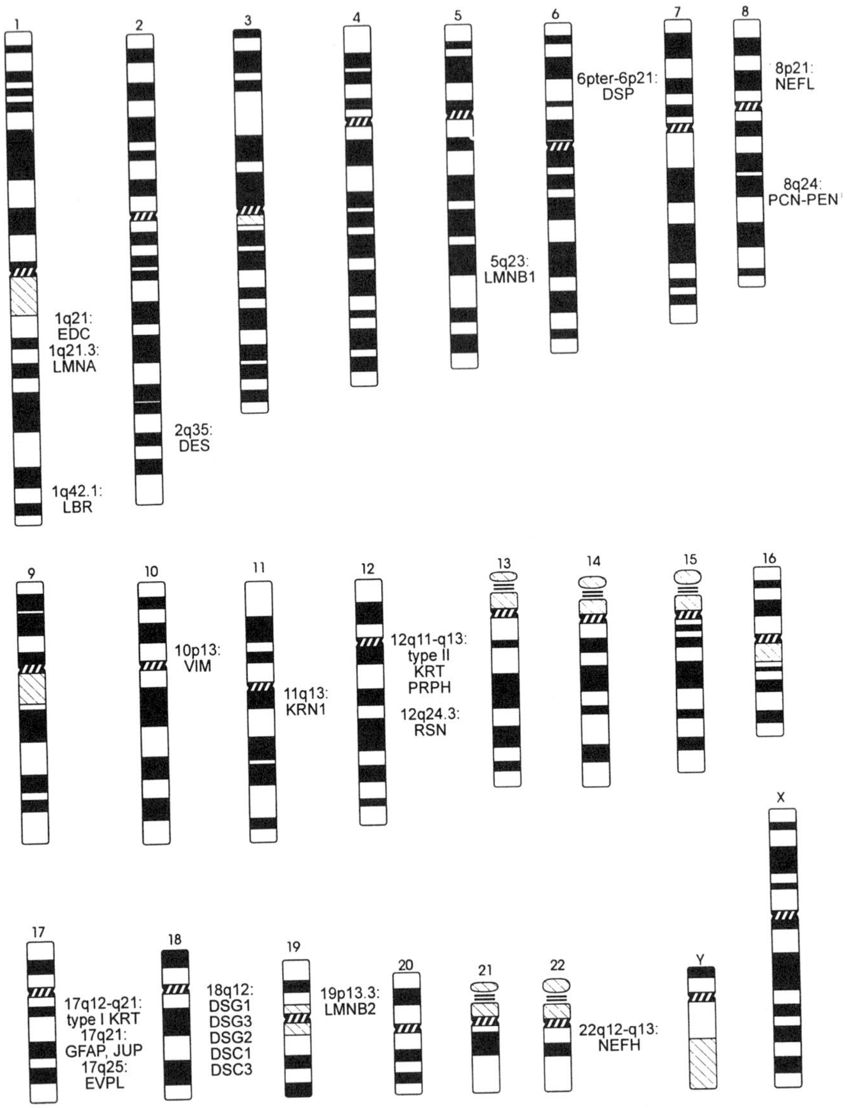

FIGURE 3. Chromosomal localization of human genes encoding intermediate filament and intermediate filament-associated proteins as well as other selected proteins contributing to the integrity of epithelial tissues. The most likely position of the genes is indicated next to the idiogram of the respective human chromosome. DES, desmin; DSP, desmoplakin; DSG, desmogleins; DSC, desmocollins; EDC, epidermal differentiation complex; EVPL, envoplakin; GFAP, glial fibrillary acidic protein; JUP, plakoglobin; KRT, keratin; KRN1, ultrahigh-sulfur keratin 1; LMNA, lamin A/C; LBR, lamin B receptor; LMNB1, lamin B1; LMNB2, lamin B2; NEFL, neurofilament light chain; NEFH, neurofilament heavy chain; PCN-PEN, plectin; PRPH, peripherin; RSN, restin; VIM, vimentin. The Idiograms are courtesy of David Adler, Department of Pathology, University of Washington, Seattle, Washington (http://www.pathology.washington. edu/IdiogramAlbums/idiograms/Human).

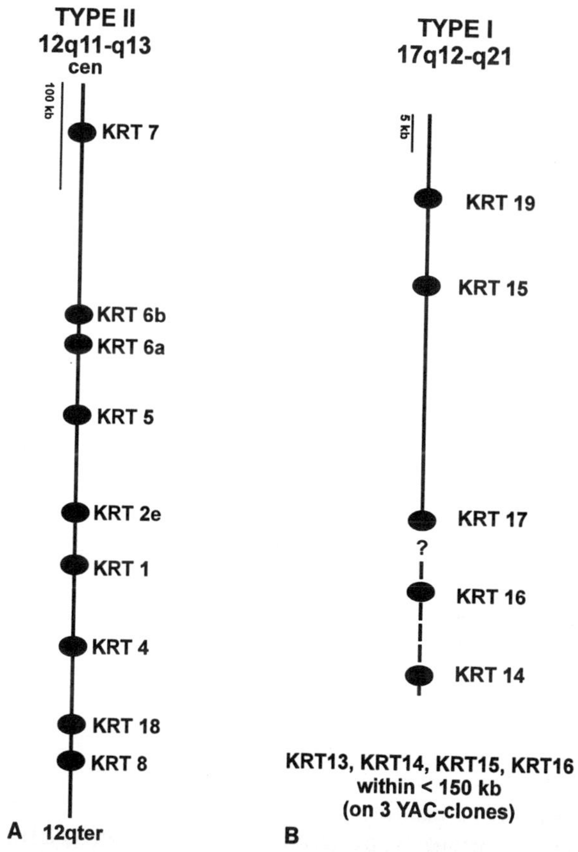

FIGURE 4. Maps of the two keratin gene clusters in 12q11–q13 and 17q12–q21. (a) Map of the keratin gene cluster in 12q11–q13 according to Yoon *et al.* (1994). Eight of the type II genes and the type I gene KRT18 are shown. A total of 15 keratin/keratin-like genes are present within this 630-kbp contig. The orientation of the cluster is indicated. (b) Collection of mapping information for the type I keratin gene cluster in 17q12–q21. KRT19, KRT15, and KRT17 have been mapped within a clone of 55 kb that may even harbor at least one additional keratin gene between KRT15 and KRT17 (Milisavljevic *et al.*, 1996), confirming the linkage of KRT19 and KRT15 within 17 kb (Bader *et al.*, 1988). KRT16 and KRT14 may extend the map (Milisavljevic *et al.*, 1996; Savtchenko *et al.*, 1990; Rosenberg *et al.*, 1988). The arrangement of the type I genes on three YAC clones, however, appears to be KRT 16–KRT13–KRT14–KRT 15 (Ceratto *et al.*, 1997). The KRT9, KRT10, KRT12, and KRT20 as well as the type I hair keratin genes may also reside in this cluster, whose orientation is still unknown.

a human keratin 18 cDNA sequence by *in situ* hybridization to 17p12–p11 (a region that also harbors pseudogenes of KRT16 and KRT14; Rosenberg *et al.*, 1988, see below) and close to the type I cluster at 17q, while Waseem *et al.* (1990b) assigned the KRT18 gene by detection of intron-specific polymerase chain reaction (PCR) products in somatic cell hybrids correctly, but in disagreement with the emerging dictum, to chromosome 12q. The location of these two genes, which are coexpressed in embryonic and simple epithelia, on a single chromosome and in close proximity is consistent with the hypothesis that keratins 8 and 18 may be closer to an ancestral gene than the keratins of more highly differentiated epithelia, and hence sheds some light on the evolutionary relationship among the keratin families (Waseem *et al.*, 1990b).

Our knowledge of the physical organization of keratin genes within the type I cluster on chromosome 17q12–q21 is unfortunately still quite rudimentary (Figure 4b). Only recently Milisavljevic *et al.* (1996) mapped the KRT19, KRT15, and KRT17 genes within approximately 50kb confirming an earlier report of Bader *et al.* (1988), who found KRT19 and KRT15 to be linked within about 17kb. To these genes, KRT16 and KRT14 may be closely linked as well (Milisavljevic *et al.*, 1996; Savtchenko *et al.*, 1990; Rosenberg *et al.*, 1988). The KRT10, KRT13, and KRT15 genes have also been reported to reside in that region and to map to the same restriction fragment of 140kb and to be present within one YAC clone (Romano *et al.*, 1991). This account, however, seems not to be endorsed in a recent paper from the same group (Ceratto *et al.*, 1997), which now rather claims KRT13, KRT14, KRT15, and KRT16 to be contained within less than 150kb of DNA, based on the analysis of three YAC clones. Accordingly, the awaited complete physical map of the type I keratin gene cluster will need to confirm these results, establish the decisive order of genes with respect to each other and the centromere, and integrate the remaining type I keratins such as KRT9 and KRT12 that have already been localized to this chromosomal segment (Reis *et al.*, 1992, and Irvine *et al.*, 1997a, respectively), as well as KRT10, KRT20, and the type I hair keratins, some of which have been shown to be contiguous, too (M. A. Rogers *et al.*, 1996, 1995).

Another point of interest is the physical distance between the type I cluster of keratins and two other genes that have been mapped to 17q21, i.e., the type III intermediate filament gene GFAP (Bongcam-Rudloff *et al.*, 1991) and the gene for a protein shown to localize to two different types of cellular junctions, plakoglobin (JUP, junctional plakoglobin; Franke *et al.*, 1989). By genetic means, the latter appears to map between KRT10 and D17S858, an interval that also contains the BRCA1 gene for familial breast cancer (Aberle *et al.*, 1995).

In other mammalian species the keratin genes are also clustered, notably on chromosomes that show homology of synteny to human chromosomes 12 and 17, respectively (Table I). For example, in the mouse, the type I keratin genes are localized on chromosome 11 in a segment known to harbor genes affecting the development of hair (Nadeau *et al.*, 1989), and in the sheep, a 100-kb segment of DNA containing six type II keratin genes, three hair and three hair-related, has been cloned and characterized (Powell *et al.*, 1994; Powell and Beltrame, 1994).

To summarize this section, the count of individual keratin genes identified in the human genome totals 34 at the time of writing: 16 type I keratin genes (KRT9, KRT10, KRT12, KRT13, KRT14, KRT15, KRT16, KRT17, KRT18, KRT19, KRT20, plus five hair keratin genes) and 18 type II keratin genes (KRT1, KRT2e, KRT2p, KRT3, KRT4, KRT5, KRT6A, KRT6B, KRT6C, KRT6D, KRT7, KRT8, plus six hair keratin genes). It is likely that some additional isoformic KRT6 genes also need to be considered, and the curators of the human genome database (GDB; http://gdbwww.gdb.org) have reserved the symbol KRT22 for an entirely new keratin gene still to be detected. A haze over this reasonably simple picture, however, comes from the presence of pseudogenes in the genome and the genetic polymorphism of many keratins.

2.4. Pseudogenes

The two clusters of keratin genes located on human chromosome 17 and chromosome 12 apparently comprise the transcriptionally active genes of type I and type II keratins that are important for the developmental and differentiation-specific expression of keratins in epithelial tissues.

In addition, keratin-like or more distantly related sequences are also found scattered elsewhere in the genome, many of which represent pseudogenes. For example, using *in situ* hybridization of a keratin 15 sequence, Barletta *et al.* (1990) detected small clusters of grains on the long

Table I
Chromosomal Localization of Keratin Genes in Different Species

Species	Type I keratin genes	Type II keratin genes	Reference
Human	17q12–q21	12q11–q13	See text
Mouse	11, band D	15	Correll *et al.* (1992), Nadeau *et al.* (1989)
Cow	19q16–qter	5q14–q23	Fries *et al.* (1991)
Sheep	11q25–q29	3q14–q22	Hediger *et al.* (1991)

arm of chromosome 14 at bands q31–q32 and on the long arm of chromosome 21 in addition to the major hybridization site at chromosome 17q21–q23. Chromosomal regions containing nonfunctional genes for KRT16 and KRT14 have also been found at 17p12–p11 (Rosenberg *et al.*, 1988) and the curators of GDB have assigned the symbols KRT16L1 and KRT16L2 and KRT14L1, KRT14L2, and KRT14L3 for these keratin-like pseudogenes. Pseudogenes corresponding to keratin 17 and keratin 19 have been found in the human genome as well (Troyanovsky and Leube, 1994, and Savtchenko *et al.*, 1988, respectively). The two keratins coexpressed in very early embryonic stem cells and simple epithelia, K8 and K18, present with a particularly large number of pseudogenes: 9–24 for K8 (Waseem *et al.*, 1990a) and 15–20 for K18 (Kulesh and Oshima, 1988). It has been speculated that this multiplicity may be due to the early embryonic gene activity making K8 and K18 mRNAs specifically prone to virus-mediated reverse transcription and therefore generation and integration of pseudogenes into the germ line.

Among the other keratin-like sequences detected so far, a human hair cuticle ultrahigh-sulfur keratin gene (KRN1) and an ultrahigh-sulfur-like gene (KRN1L) are located on chromosome 11q13.5 and 11p15.5, respectively (MacKinnon *et al.*, 1991) and a keratin-like gene S that shows significant similarities with the amino acid sequences of loricrin, K1, and K10 has been identified in the class I region of the human major histocompatibility complex on chromosome 6 (Zhou and Chaplin, 1993).

2.5. Polymorphism and Mutation

In the course of the extensive study of keratin genes and their expression patterns, additional levels of heterogeneity and complexity soon became apparent. One such level relates to naturally occurring protein variants of keratins that are caused by genetic polymorphism of the respective genes and the other to mutations in keratin genes that cause diseases of various phenotypes and reflect the underlying pattern of tissue- and differentiation-specific expression of the affected genes.

While the term polymorphism has been used in its general meaning quite frequently in intermediate filament research to describe the many variations in appearance of structures and patterns, like the variable mass per length, width, or composition of reconstituted keratin filaments (e.g., Steinert, 1990; Engel *et al.*, 1985; Steven *et al.*, 1983), the variation in unit cell patterns of paracrystals of GFAP (Stewart *et al.*, 1989), or the alternative splicing of the unique peripherin gene leading to three peripherin polypeptides in nerve cells (Landon *et al.*, 1989), in the context of this review the term polymorphism will be used strictly in its genetic sense.

Accordingly, it will describe the appearance of alternative forms of a gene (i.e., alleles) in a population that occur at a certain frequency (generally >1%) and cannot be attributed solely to continuous recurrence of new mutations. Other well-known examples of genetic polymorphisms are the alleles of the genes controlling human blood groups or the alleles of genes residing in the major histocompatibility complex.

The earliest reports of genetic polymorphism of intermediate filament proteins went largely unnoticed, as they appeared well before the explosion of interest in intermediate filaments. For example, already in 1976 the heavy chain of the three neurofilament proteins of rabbits was shown to be polymorphic (Willard, 1976) and in 1982 a polymorphic variant of human glial fibrillary acidic protein was described (Comings, 1982).

The genetic polymorphism of keratins was also first observed at the protein level (Wild and Mischke, 1986; Mischke and Wild, 1987). Analyses of keratin-enriched cytoskeletal residues of upper digestive tract squamous epithelium and the epidermis that had been obtained from several hundred probands and analyzed separately by high-resolution sodium dodecyl sulfate (SDS) gel electrophoresis showed interindividual variations within the corresponding polypeptide patterns of the respective tissue for four keratin polypeptides, namely the type II keratins K1, K4, and K5 and the type I keratin K10. Such variations were most easily explained by postulating a genetic polymorphism of the underlying keratin genes. Accordingly, the allelic proteins were shown to be codominantly expressed and transmitted as true Mendelian traits (Wild and Mischke, 1986; Mischke and Wild, 1987; Mischke *et al.*, 1990).

To determine the molecular basis of the genetic polymorphism of these keratins, the N- and C-terminal domains of their genes were analyzed in detail, because these domains particularly specify the members of each class by their characteristic size, composition, and sequence. By PCR amplification, single-strand conformation gel electrophoresis, and sequencing, the alleles of KRT1 and KRT10 revealed size differences in their carboxyl-terminal V2 subdomains, whereas the alleles of KRT4 and KRT5 differed in their amino-terminal domains (Korge *et al.*, 1992a,b; Wanner *et al.*, 1993).

For example, the difference between alleles KRT4.1 (K4a1 in the original report of Wanner *et al.*, 1993) and KRT4.2 (K4a2 originally) is due to a nucleotide transition from T to C causing a neutral amino acid substitution from valine to alanine. Alleles KRT4.2 and KRT4.3 (K4b originally), however, are at variance by a much larger alteration, i.e., the in-frame deletion of 42 bp equaling the loss of 14 amino acids in the V1 domain. The alleles KRT5.1 (K5a originally) and KRT5.2 (K5b1 originally) are distinguished by a transversion from A to G replacing a glycine by a glutamic acid in the H1 subdomain, whereas alleles KRT5.2 and KRT5.3 (K5b2 originally) reveal a third base C-to-T transition having no effect at the amino acid level of the V1 subdomain (Wanner *et al.*, 1993).

In contrast, the allelic differences among the alleles of the KRT1 and the KRT10 genes do not reside in the N-terminal, but in the C-terminal domain. They are generally represented by alleles of different size that directly reflect the estimated molecular weight differences of the corresponding polypeptides (Korge *et al.*, 1992a,b; Mischke and Wild, 1987). While the difference between the largest and the smallest KRT10 allele is as large as 114bp or 38 amino acids (Korge *et al.*, 1992a), more subtle size differences among the alleles and point mutations are also present. Correspondingly, the true number of KRT10 alleles significantly exceeds the number of variants detectable by protein analysis (Mischke and Wild, 1987). Within a Caucasian population of European origin, 12 alleles of KRT10 have been identified so far and their appropriately combined allele frequencies correspond well to those obtained for the polypeptide variants in a different sample (Table II; Mischke, 1993; Korge *et al.*, 1992b; Mischke and Wild, 1987). Two alleles, KRT10.4 and KRT10.5, encode polypeptide variants that are not resolved on two-dimensional gels and furnish protein spot K10a, while the alleles KRT10.2 and KRT10.3 yield spot K10b in such analyses (Mischke and Wild, 1987). Of these, variant K10a most likely corresponds to K10 and variant K10b to K11 of the Moll catalog. This

Table II
Polymorphic Human Keratins in a Population of Caucasian Origin[a]

	KRT1	KRT2e	KRT2p	KRT3	KRT4	KRT5	KRT10
Gene	KRT2.1	KRT2.2e	KRT2.2p	KRT2.3	KRT2.4	KRT2.5	KRT1.10
Allele number	2	3	2	2	3	3	12
Allele frequency	0.1(a): 0.470	0.1: 0.167	0.1: 0.250	0.1: 0.393	0.1(a1): 0.158	0.1(a): 0. 082	0.1: 0.014
	0.2(b): 0.530	0.2: 0.816	0.2: 0.750	0.2: 0.607	0.2(a2): 0.682	0.2(b1): 0.783	0.2: 0.397
		0.3: 0.017			0.3(b): 0.160	0.3(b2): 0.135	0.3: 0.308
							0.4: 0.123
							0.5: 0.096
							0.6: 0.007
							0.7: 0.014
							0.8: 0.020
							0.9: 0.007
							0.10: 0.007
							0.11: 0.007
							0.12: 0.007
PIC	0.37	0.27	0.30	0.36	0.44	0.33	0.68

[a]The designation of alleles follows the rule KRTx.1/ KRTx.2, etc. Previous allele assignments for KRT1, KRT4, and KRT5 (Wanner *et al.*, 1993; Korge *et al.*, 1992a) are given in parentheses. Additional details may also be found in the references cited in the text. PIC, Polymorphism information content, a value that can be calculated from the frequencies of all alleles.

would also be in accordance with the very similar tryptic peptide maps of K10 and K11 (Moll *et al.*, 1982) and the inconsistent occurrence of K11 in epidermal specimens reported in the literature. Accordingly, K11 can be considered as an allele of the K10 gene and removed as an individual entity from the catalog.

Extending this type of analysis, we (M. Zirra and D. Mischke, unpublished results) have now also identified polymorphic variants of the KRT2e, KRT2p, and KRT3 keratin genes at the genomic DNA sequence level, though not at the protein level. For example, the differences among the KRT2e alleles (Figure 5) reside in the N-terminal domain, with allele KRT2e.2 showing an in-frame deletion of 18 bp, equivalent to the loss of six amino acids as compared to KRT2e.1. Alleles KRT2e.2 and KRT2e.3 differ by a nucleotide transition from G to A leading to an amino acid substitution from glycine to serine, thereby generating an *Alu*I restriction fragment length polymorphism (RFLP).

KRT2p and KRT3 are also polymorphic. The two alleles of KRT2p differ by a C-to-T transition in the 5′ noncoding region, generating a *Hpa*II RFLP. The two alleles of KRT3 again differ by a 30-bp deletion in the N-terminal V1 subdomain (unpublished results). Two polymorphic loci have also been described in the cuticular hHa2 gene that correspond to amino acid substitutions in the short helix termination region (Winter *et al.*, 1996).

The above genetic polymorphisms of KRT1, KRT2e, KRT3, KRT4, KRT5, and KRT10 occur within the translated sequence of the respective genes, resulting in polypeptides with amino acid substitutions or, more frequently, with deletions of many amino acids within the N- or C-terminal domains (Figure 5; Wanner *et al.*, 1993; Korge *et al.*, 1992a,b). These subdomains of the keratin chain are characterized by a particular structural protein motif termed the *glycine loop* that consists of glycine-rich quasipeptide repeats flanked by aromatic residues (Steinert *et al.*, 1991). Such glycine loops are thought to be involved in the structural organization of keratin filaments in epithelial cells, thereby contributing to the flexibility and elasticity of the entire tissue. The size differences among the alleles, most interestingly, invariably correspond to the deletion of one (as shown in Figure 5 for KRT2e) or two complete glycine loops including the aromatic residue or, in some cases, in a size reduction of the loop. Therefore, the functional, physiological, and hitherto possibly undetected pathophysiological consequences of allelic variations on intermediate filament formation and, consequently, the "quality" of epithelial tissues will merit further investigations.

Correspondingly, and in contrast to the point mutations occurring in the conserved helix-initiation or helix-termination peptides at the beginning or end of the rod domains of keratins or other hot spots of mutation which cause severe diseases of the skin (see below), none of the described

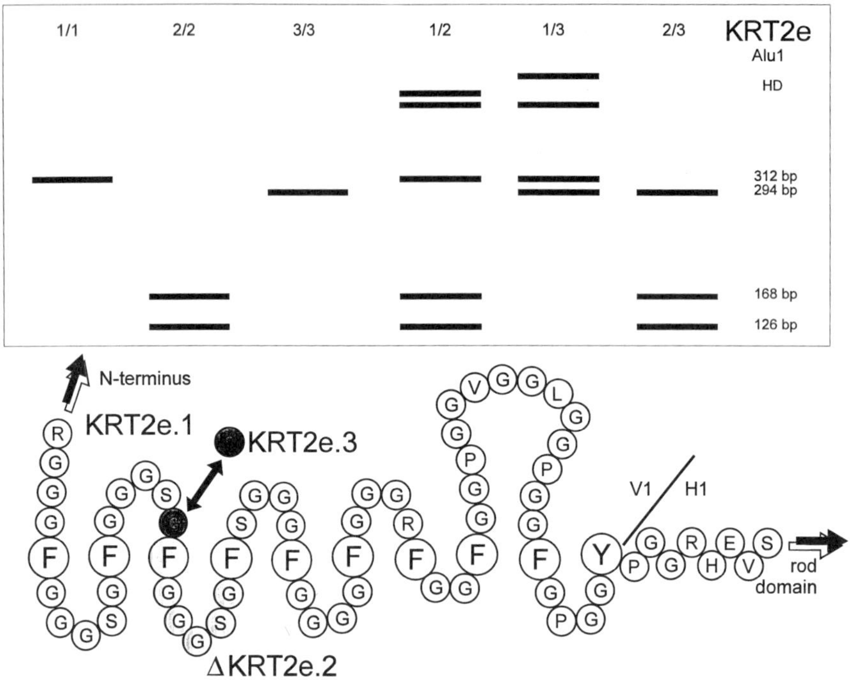

FIGURE 5. The alleles of KRT2e. (Top) Schematic representation of a gel electrophoretic separation of allele-specific PCR products generated from the N-terminal domain of KRT2e with the primer pair K2e–3a: TTGGCAGTCGGAGTCTTGTTGG/K2e–3b: TGACAGAGACTTCGTGGATGCC. All amplicons have been digested with the restriction enzyme *Alu*I before application to the gel as double-stranded DNAs. The KRT2e.1 allele yields a 312-bp product, alleles KRT2e.2 and KRT2e.3 present with products of 294 bp, of which allele KRT2e.2 is cut by *Alu*I yielding restriction fragments of 168 and 126 bp. Heterozygous individuals carrying alleles KRT2e.1/KRT2e.2 and KRT2e.1/KRT2e.3 may be detected by their particular heteroduplices (HD); discrimination between heterozygotes with alleles KRT2e.2 and KRT2e.3 requires *Alu*I digestion. (Bottom) Schematic representation of the deduced amino acid sequence of the N-terminal domain of K2e. The difference between alleles K2e.1 and K2e.2/K2e.3 is due to an in-frame deletion of 18 bp equivalent to the loss of six amino acids or one glycine loop (shaded). K2e.2 and K2e.3 differ by a nucleotide transition from G to A (Gly to Ser) generating the *Alu*I RFLP.

alleles and their underlying sequence differences seem to be associated with an apparent impairment of epithelial development and differentiation. For example, the mutation at codon 137 in the H1 subdomain of K5 that alters the glycine of allele K5.1 to the glutamic acid of allele K5.2 (Wanner *et al.*, 1993) appears to be well tolerated, whereas a leucine-to-proline mutation located only 11 amino acids further down in the same subdomain of the KRT1 causes defective keratin filament formation in one family with epidermolytic hyperkeratosis (Chipev *et al.*, 1992), and a proline-to-leucine mutation at codon 24 of K5 gives rise to a rare autosomal dominant disorder, epidermolysis bullosa simplex with mottled pigmentation (Irvine *et al.*, 1997b; Uttam *et al.*, 1996).

Allelic variations in noncoding regions will be tolerated even more easily and have been described for several keratin genes. For example, KRT9 harbors a microsatellite sequence (K9CAi4) in intron 4 (Reis *et al.*, 1994), KRT10 displays an *Ava*II RFLP (McLean and Lane, 1992), KRT14 an *Alu* polymorphism in intron 3 (Humphries *et al.*, 1992), KRT8 two RFLPs that can be detected by PCR (McLean *et al.*, 1994), and hHa2 a third polymorphic site in intron 6 (Winter *et al.*, 1996). In addition, polymorphisms have also been described for several other genes encoding intermediate filament proteins.

In view of the understanding that intermediate filament proteins and, even more so, the members of the keratin families are closely related and have evolved through common, presumably lamin-like ancestors (Dodemont *et al.*, 1990; Blumenberg, 1989; Steinert and Roop, 1988), genetic polymorphism may not be surprising, but rather mirror ongoing evolutionary processes leading to diversification and specialization. In particular, repetition of tandem sequence motifs followed by mutational changes has been suggested to advance the diversity within the variable N- and C-terminal domains of keratin proteins (Klinge *et al.*, 1987; Sun *et al.*, 1984). Furthermore, since the keratin genes are clustered, they may also be particularly prone to recombination events like unequal crossing-over or gene conversion. Successive gene duplication events have also been considered by Takahashi *et al.* (1995) to explain the multiplicity of human K6 genes. In fact, the N- and C-terminal sequences of polymorphic keratin genes reveal parallel and antiparallel repeats that are reminiscent of tandem duplication and insertion. Slipped-strand mispairing followed by illegal crossing-over can generate the interstitial deletions that are characteristic for the alleles presenting with deleted glycine loops like KRT1, KRT2e, KRT3, KRT4, and KRT10 (see Wanner *et al.*, 1993, for an example).

Sequence variability is constrained by functional requirements, however, as many mutations will be lost during evolution due to deleterious effects on filament formation and function. With respect to keratins, the requirements for heteropolymerization of type I and type II proteins are also likely to be less tolerant to alterations in the polypeptides than the requirements of homopolymerization as in most other intermediate filament proteins (Myers *et al.*, 1987). Correspondingly, it has been shown that the end domains have subtle roles in intermediate filament network formation and/or 10-nm filament assembly and stability (Bader *et al.*, 1991), with recent findings suggesting that specific sequence motifs in the end domains and their spatial arrangement are particularly important (Herrmann *et al.*, 1996, 1992; K. R. Rogers *et al.*, 1995; Hatzfeld and Burba, 1994; Heins and Aebi, 1994; Parry and Steinert, 1992; Lu and Lane, 1990).

The question at what time in evolution the alleles arose remains open. However, despite their close physical linkage within 200–300 kb of the type II keratin gene cluster on chromosome 12q11–q13 (Yoon *et al.*, 1994), the apparent lack of linkage disequilibrium among the KRT4 and KRT5 alleles (Mischke *et al.*, 1990) may definitely argue in favor of very long histories for these alleles.

To elucidate the potential involvement of keratin genes in hereditary disorders of squamous cell differentiation, the polymorphic keratin genes will be much more informative markers in linkage analyses than anonymous ones residing outside of the two keratin gene clusters because of the lower likelihood of recombination due to their close physical association and or direct involvement. Indeed, the allele frequency (as determined in a Caucasian population of European origin) for the highly polymorphic KRT10 reveals an observed heterozygosity of 74% resulting in a calculated polymorphism information content (PIC) value of 0.68 (Mischke, 1993) that is almost high enough to qualify for a polymorphism of DNA micro-satellites (Table II). Hence, the KRT10 alleles are highly informative intragenic markers when type I genes are suspected to cause a disease, such as epidermolytic palmoplantar keratoderma (Reis *et al.*, 1994) or epidermolytic hyperkeratosis (Rothnagel *et al.*, 1992).

The type II keratin genes appear to be less polymorphic, displaying only two or three alleles each, and accordingly yield lower heterozygosity indices and PIC values (Table II). Combined, however, the genetic polymorphism of KRT1, KRT4, and KRT5 together with the allelic variations of the KRT2e, KRT2p, and KRT3 genes will also improve haplotype analyses for the type II keratin gene cluster in families affected with inherited disorders of epithelial differentiation. Accordingly, the polypeptide polymorphism of K5 has already been helpful in unraveling the case of one epidermolysis bullosa family (Lane *et al.*, 1992), as has the K1 polymorphism in a family with epidermolytic hyperkeratosis (Compton *et al.*, 1992; Chipev *et al.*, 1992). The polymorphisms of keratins have likewise been used to definitely exclude the association of genes in the two keratin clusters as causing the striated form of palmoplantar keratoderma (Hennies *et al.*, 1995). In conclusion, genetic polymorphism of the keratin genes is rendering an already complex multigene family even more complex. The identification of alterations in keratin genes leading to disease phenotypes will, however, continue to benefit from the improved possibilities for genetic analyses in affected families.

Since the first accounts of keratin gene abnormalities in the blistering disease epidermis bullosa simplex (Bonifas *et al.*, 1991; Coulombe *et al.*, 1991), many more reports have appeared and the number is growing. Only very few keratins have not been implicated with a disorder of epithelial differentiation. Accordingly, a summary of the association of mutations in keratin genes with diseases of epithelial differentiation deserves a review of its own. Since many such reviews have been published in recent years, this topic is restricted here to Table III, listing only some of the recent additions

Table III
Disease Association of Keratin and Epidermal Differentiation Complex Genes[a]

Keratin gene	Disease association
KRT1	Epidermolytic hyperkeratosis (MIM no. 113800)
	Epidermolytic palmo-plantar-keratoderma (MIM no. 144200)
KRT2e	Ichthyosis bullosa Siemens (MIM no. 146800)
KRT3	Meesmann corneal dystrophy (MIM no. 122100) (Irvine *et al.*, 1997a)
KRT4	White sponge nevus (MIM no. 193900)
KRT5	Epidermolysis bullosa simplex (MIM nos. 131900, 131800, 131760, 131960)
KRT6A	Pachyonychia congenita (MIM no. 167200)
KRT9	Epidermolytic palmo-plantar-keratoderma (MIM no. 144200)
KRT10	Epidermolytic hyperkeratosis (MIM nos. 113800, 600648)
KRT12	Meesmann corneal dystrophy (MIM no. 122100) (Irvine *et al.*, 1997a)
KRT13	White sponge nevus (MIM no. 193900)
KRT14	Epidermolysis bullosa simplex (MIM nos. 131900, 131800, 131760)
KRT16	Pachyonychia congenita (MIM no. 167200)
	Nonepidermolytic palmo-plantar-keratoderma (MIM no. 600962)
KRT17	Pachyonychia congenita (MIM no. 167210)
	Steatocystoma multiplex (MIM no. 184500; Smith *et al.*, 1997)
KRT18	Cryptogenic cirrhosis (Ku *et al.*, 1997)
hHb1–6	Monilethrix (MIM no. 158000)
	(Healy *et al.*, 1995; Korge *et al.*, 1996; Winter *et al.*, 1997)
EDC gene	
LOR	Vohwinkel syndrome (MIM no. 124500) (Maestrini *et al.*, 1996)

[a] Except for KRT2p, KRT7, KRT8, KRT15, KRT19, KRT20, and the additional isoforms of KRT6, all other keratins, i.e., 14 genes, have been shown to be associated with diseases of epithelial differentiation. The human type II hair keratin genes have also been associated with a disease. For references to most disease associations of keratins see Fuchs (1996), Korge and Krieg (1996), and McLean and Lane (1995), or consult the website of Online Mendelian Inheritance in Man (to which the MIM numbers refer) at http://www3.ncbi.nlm.nih.gov/omim/.

to the list. The interested reader should consult such excellent overviews as those of Fuchs (1996), Korge and Krieg (1996), Corden and McLean (1996), and McLean and Lane (1995) for more details.

All these studies have demonstrated the fundamental importance of keratins in maintaining the integrity of the epithelium: In the absence of an effective keratin filament network, epithelial cells lose their resistance to normal mechanical stress or physical pressure, resulting, for example, in blistering skin disorders of various phenotypes, hyperkeratotic diseases, or other ailments of epithelial tissues. In almost all of these cases, the disease-causing defect can be traced to a dominant-negative point mutation in a particular keratin gene occurring either at the beginning or end of the central α-helical domain, or in other well-restricted sections of the keratin chain, suggesting that these regions are of special functional importance either for intermediate filament assembly, network formation, and stability, or the interaction of filaments with other cellular organelles that contribute to the cytoarchitecture of epithelial cells (e.g., Chou *et al.*, 1997; Herrmann *et al.*, 1996, 1992; K. R. Rogers *et al.*, 1995; Heins and Aebi, 1994; Parry and Steinert, 1992; Hatzfeld and Weber, 1992; Lu and Lane, 1990).

All of these diseases are generally inherited in an autosomal dominant fashion because a mutation in one keratin polypeptide will already impair the assembly of obligate heteropolymeric filaments or their normal interactions with other intracellular components. In accord with the concept of pairwise expression of keratins (see above), a mutation in either a type I or a type II keratin of a pair coexpressed in a particular tissue layer or epithelial cell will therefore give rise to the same disease or one with a similar phenotype.

3. THE EPIDERMAL DIFFERENTIATION COMPLEX

In addition to keratins, many other proteins are required to generate a tissue as intricately composed as the epidermis. Interestingly enough, the genes for some of them are also physically linked in the genome. In particular, region 1q21 of human chromosome 1 unites several structurally, functionally, and evolutionarily related genes that play an important role in the final steps of terminal differentiation of the human epidermis (Mischke *et al.*, 1996, and references therein; Marenholz *et al.*, 1996; Lee *et al.*, 1993a; Fietz *et al.*, 1992). These genes encode proteins that are indispensable for the cornification of epidermal cells, such as the cornified cell envelope precursors loricrin (LOR), involucrin (IVL), and small proline-rich proteins (SPRR1, SPRR2, and SPRR3), the intermediate filament-associated proteins profilaggrin (FLG) and trichohyalin (THH), as well as several

S100A calcium-binding proteins involved in signal transduction, cell cycle progression, and cytoskeletal membrane interaction.

Like keratins, these proteins also share a number of structural features, as shown schematically in Figure 6. Apart from a very similar gene structure in terms of exon–intron distribution, the cornified cell envelope precursor proteins contain a central region of short peptide repeats. The number of repeats varies from 39 (involucrin) and 20 (loricrin) to 3–14 for the different small proline-rich proteins (Backendorf and Hohl, 1992). The S100A calcium-binding proteins all carry two calcium-binding domains of the helix-turn-helix EF-hand motif and share both a high degree of homology at the amino acid level and a strikingly similar gene organization. Since these genes used to carry an excess of different names, a new logical nomenclature (S100A1–S100A13) has recently been suggested and approved by GDB (Schäfer *et al.*, 1995). The genes of the intermediate filament-associated proteins in 1q21 appear to be "fused genes" encoding at the 5′-end two EF-hand calcium-binding motifs followed by tandem peptide repeats that are characteristic for the cornified cell envelope precursors, thus rendering the proteins multifunctional (Presland *et al.*, 1995, 1992; Markova *et al.*, 1993; Fietz *et al.*, 1993; Lee *et al.*, 1993b).

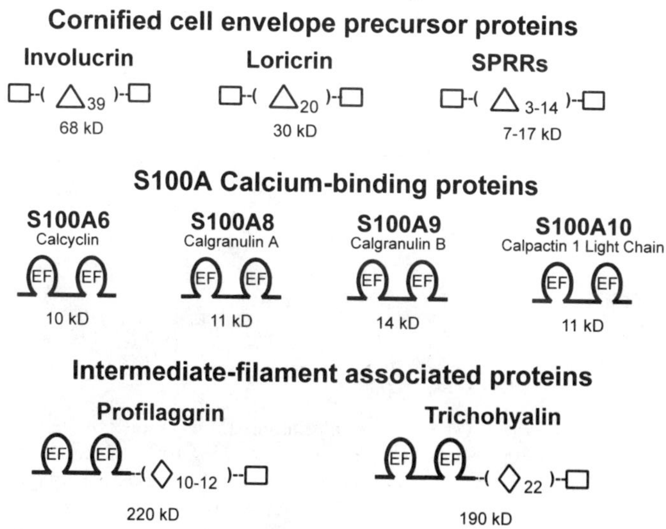

FIGURE 6. Similarities among the gene families of the epidermal differentiation complex in 1q21. Schematic representation of the main structural features of the three polypeptide families encoded in the EDC. Boxes: N- and C-terminal polypeptide-specific domains; triangles, diamonds: internal peptide repeats, EF: calcium binding helix-turn-helix motifs.

As shown in Figure 7, a total of 28 genes have now been physically mapped within about 2 Mb of 1q21 using both high-resolution rotating-field gel electrophoresis (Ziegler and Volz, 1992) of large restriction fragments of genomic DNA and YAC contig building (Mischke *et al.*, 1996; Marenholz

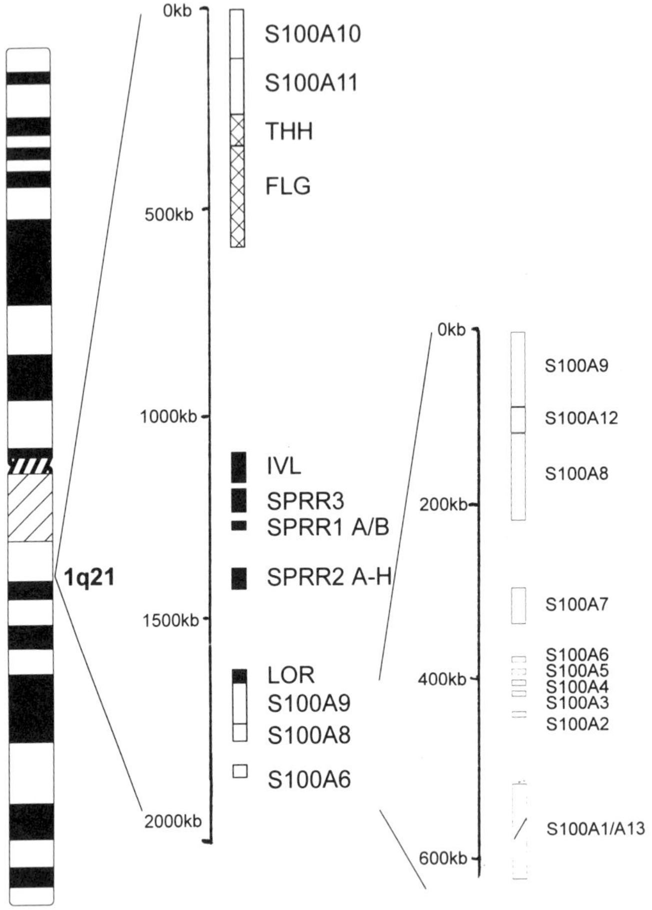

FIGURE 7. Integrated map of the epidermal differentiation complex in 1q21. The physical map of the EDC is shown in the correct orientation on the long arm of chromosome 1. Twenty-eight genes have been mapped so far (Wicki *et al.*, 1996a, b; Marenholz *et al.*, 1996; Mischke *et al.*, 1996; Schäfer *et al.*, 1995; Gibbs *et al.*, 1993; Engelkamp *et al.*, 1993; Volz *et al.*, 1993). The boxes indicate the smallest distance to which the respective genes have been mapped, but not the size of the actual gene. Black boxes represent genes encoding cornified cell envelope precursors, open boxes genes of S100A calcium-binding proteins, and hatched boxes the apparently fused genes of the multifunctional proteins. The idiogram is courtesy of David Adler, University of Washington, Seattle, Washington.

et al., 1996). Thirteen genes encode cornified cell envelope precursor proteins [LOR, IVL, one SPRR3, two SPRR1, and 8 SPRR2; see Gibbs *et al.* (1993) for the multiplicity of the SPRR family], 2 genes (THH, FLG) encode intermediate filament associated proteins, and 13 genes encode S100A proteins (Schäfer *et al.*, 1995; Wicki *et al.*, 1996a, b). The latter flank this array of genes, bringing the trichohyalin and profilaggrin genes in close proximity to both the centromeric S100A genes and the cornified cell envelope precursor genes. From this, interesting questions about the evolution of the "fused" genes arise.

Colocalization of functionally interacting genes within the human genome appears to be rather frequent. Two different situations may be distinguished. One is the case of gene clusters, i.e., the linkage of genes that share a high degree of homology and therefore belong to the same gene family, as exemplified by the type I and type II keratin clusters on chromosomes 17q and 12q (see above), or the immunoglobulin heavy chain cluster on chromosome 14q (Cook *et al.*, 1994). The other concept regards colocalization as an avenue by which a gene complex is created to unite functionally interacting genes that belong to different gene families. A well-known example for a gene complex is the major histocompatibility (HLA) complcx on chromosome 6p (Milner and Campbell, 1992) bringing together many genes that are crucial for the normal function of the immune system. Some encode the highly polymorphic HLA class I and class II molecules involved in antigen presentation, while other, unrelated genes encode either proteins responsible for the intracellular processing of antigens or the transport of peptides to be presented by HLA class I molecules.

Correspondingly, the close physical linkage of genes in 1q21 belonging to different gene families and their functional interdependence during the final steps of terminal epidermal differentiation thus merits the designation "epidermal differentiation complex (EDC)" to characterize their pivotal involvement in cornification and maturation of the epidermis (Mischke *et al.*, 1996).

The functional similarities between the S100A proteins and the calcium-binding intermediate filament-associated proteins as well as many other gene- and protein-structural relationships among all three gene families within the EDC may be more than coincidental. This raises questions that have likewise been addressed in the context of the MHC alluded to above, namely on the biological role and evolutionary history of the EDC: e.g., are the genes kept together simply by accident, or are there functional constraints in keeping genes in close physical distance over millions of years and perhaps across species barriers? In this regard, analysis of the human α-globin and the mouse class II MHC loci has shown that colocalization of genes can signify the existence of a locus control region, whose dominant presence is necessary for optimal gene expression and perhaps is also of regulatory and transcriptional advantage to the linked genes (Dillon and Grosveld, 1993; Carson and Wiles, 1993). In analogy, some genes of the EDC may also share superior locus control regulatory elements coordinating their tissue- and differentiation-specific transcription. Accordingly, analysis of the EDC throughout vertebrate evolution should shed light on the functional importance of linkage within the epidermal differentiation complex. In this respect it is noteworthy that the mouse loricrin and profilaggrin genes are also linked (Rothnagel *et al.*, 1994) and that the pericentromeric region of both arms of human chromosome 1 has been found to be syntenic with a continuous segment of mouse chromosome 3 defining a putative homologous linkage group (Mosely and Seldin, 1989).

Furthermore, the colocalization of epidermal structural genes together with calcium-binding genes within the EDC is even more intriguing because calcium levels are critically involved in coordinating the process of epidermal differentiation and tightly control the *in vitro* expression of at least the profilaggrin and loricrin genes (Hohl *et al.*, 1991; Yuspa *et al.*, 1989). Some S100A proteins (S100A10 and S100A11) have even been found to be directly incorporated into the cornified cell envelope (Robinson *et al.*, 1997).

Several diseases have been associated with the genes of the EDC and the 1q21 region as well. Among these are the nonrandom association with several types of epithelial tumors (reviewed in Solomon *et al.*, 1991) and inherited disorders of keratinization. Notably, Vohwinkel keratoderma, a mutilating palmoplantar hyperkeratosis, has recently been shown to be due to a mutation in an EDC gene, that for loricrin (Maestrini *et al.*, 1996). Highly aberrant expression of profilaggrin and abnormal cornified cell envelopes have also been described in a number of ichthyotic disorders (Resing and Dale, 1991; Michels and Juhlin, 1990) and S100A7, S100A8, and S100A9 (formerly psoriasin, PSOR1; calgranulin A, MRP8; and calgranulin B, MRP 14) have been shown to be highly upregulated in psoriatic epidermis and primary human keratinocytes undergoing abnormal differentiation (Hoffmann *et al.*, 1994; Saintigny *et al.*, 1992). Similarly, a defect in the gene encoding cathepsin K which is also located in 1q21 has been identified in pycnodysostosis (Gelb *et al.*, 1996; Johnson *et al.*, 1996).

The knowledge of the chromosomal organization of the EDC genes in 1q21 and the identification and mapping of easily screenable markers will then also be of great value for genetic linkage analyses in affected families and for unraveling the underlying defects.

4. CONCLUDING REMARKS

In the past two decades considerable advances toward understanding the fundamental questions relating to the complexity, evolution, and differential expression of genes encoding intermediate filament proteins and particularly the keratins have been achieved. The multitude of these genes and those located within the epidermal differentiation complex has been dealt with primarily in this review, while other topics have only be considered in passing. But it is already clear that the answers to the questions of gene regulation and differential expression will be as complex and interesting as the gene families themselves.

NOTE ADDED IN PROOFS. Since submission of this review, the number of keratin genes (in particular of hair keratins) has continued to grow (see the forthcoming papers by M. Rogers, J. Schweizer, H. Winter *et al.*). Likewise, the number of genes located within the EDC in 1q21 has almost doubled (see papers from the European 1q21 consortium: C. Backendorf, P. Djian, C. W. Heizmann, D. Hohl, A. Forus, D. Mischke, D. Nizetic, J. Ragoussis, and F. Watt).

ACKNOWLEDGMENTS. Generous funding by grants from the Deutsche Forschungsgemeinschaft (Mi 210/6, Mi210/9), the European Commission (BMH4-CT96-0319), the Berliner Krebsgesellschaft, and the Maria Sonnenfeld Stiftung is gratefully acknowledged.

Over the years I have enjoyed collaborations (the benefit is documented in the papers published) and stimulating discussions with many colleagues providing essential encouragement and succor. Some of them deserve special mention: Dr. Hans-Heinrich Förster, Dr. Hans-Christian Hennies, Dr. Bernhard Korge, Ingo Marenholz, Dr. André Reis, Martina Richter-Freund, Irmgard Tillmans, Dr. Armin Volz, Dr. Reinhard Wanner, Dr. Gerd Wild, Gabriele Wille, Dr. Andreas Ziegler, and Maja Zirra.

I especially acknowledge Drs. M. Blumenberg, J. Jorcano, G. E. Rogers, and T. T. Sun for sharing their unpublished data.

5. REFERENCES

Aberle, H., Bierkamp, C., Torchard, D., Serova, O., Wagner, T., Natt, E., Wirsching, J., Heidkämper, C., Montagna, M., Lynch, H. T., Lenoir, G. M., Scherer, G., Feunteun, J., and Kemler, R., 1995, The human plakoglobin gene localizes on chromosome 17q21 and is subjected to loss of heterozygosity in breast and ovarian cancers, *Proc. Natl. Acad. Sci. USA* **92:**6384–6388.

Backendorf, C., and Hohl, D., 1992, A common origin for cornified envelope proteins, *Nat. Genet.* **2:**91.

Bader, B. L, Jahn, L., and Franke, W. W., 1988, Low level expression of cytokeratins 8, 18 and 19 in vascular smooth muscle cells of human umbilical cord and in cultured cells derived therefrom, with an analysis of the chromosomal locus containing the cytokeratin 19 gene, *Eur. J. Cell Biol.* **47:**300–319.

Bader, B. L., Magin, T. M., Freudenmann, M., Stumpp, S., and Franke, W. W., 1991, Intermediate filaments formed *de novo* from tail-less cytokeratins in the cytoplasm and in the nucleus, *J. Cell Biol.* **115:**1293–1307.

Barletta, C., Batticane, N., Ragusa, R. M, Leube, R., Peschle, C., and Romano, V., 1990, Subchromosomal localization of two human cytokeratin genes (KRT4 and KRT15) by *in situ hybridization, Cytogenet. Cell Genet.* **54:**148–150.

Blumenberg, M., 1989, Evolution of homologous domains of cytoplasmic intermediate filament proteins and lamins, *Mol. Biol. Evol.* **6:**53–65.

Bongcam-Rudloff, E., Nister, M., Betsholtz, C., Wang, J.-L., Stenman, G., Huebner, K., Croce, C. M., and Westermark, B., 1991, Human glial fibrillary acidic protein: Complementary DNA cloning, chromosome localization, and messenger RNA expression in human glioma cell lines of various phenotypes, *Cancer Res.* **51:**1553–1560.

Bonifas, J. M., Rothman, A. L., and Epstein, E. H., Jr., 1991, Epidermolysis bullosa simplex: Evidence in two families for keratin gene abnormalities, *Science* **254:**1202–1205.

Bowden, P. E., Hainey, S., Parker, G., and Hodgins, M. B., 1994, Sequence and expression of human hair keratin genes, *J. Dermatol. Sci.* **7**(Suppl.)**:**S152–S163.

Carson, S., and Wiles, M. V., 1993, Far upstream regions of class II MHC Ea are necessary for position-independent, copy-dependent expression of Ea transgene, *Nucleic Acids Res.* **21:**2065–2072.

Ceratto, N., Dobkin, C., Carter, M., Jenkins, E., Yao, X.-L., Cassiman, J.-J., Aly, M. S., Bosco, P., Leube, R., Langbein, L., Feo, S., and Romano, V., 1997, Human type I cytokeratin genes are a compact cluster, *Cytogenet. Cell Genet.* **77:**169–174.

Chipev, C. C., Korge, B. P., Markova, N., Bale, S. J., DiGiovanna, J. J., Compton, J. G., and Steinert, P. M., 1992, A leucine > proline mutation in the H1 subdomain of keratin 1 causes epidermolytic hyperkeratosis, *Cell* **70:**821–828.

Chou, Y. H., Skalli, O., and Goldman, R. D., 1997, Intermediate filaments and cytoplasmic networking: New connections and more functions, *Curr. Opin. Cell Biol.* **9:**49–53.

Collin, C., Moll, R., Kubicka, S., Ouhayoun, J. P., and Franke, W. W., 1992a, Characterization of human cytokeratin 2, an epidermal cytoskeletal protein synthesized late during differentiation, *Exp. Cell Res.* **202:**132–141.

Collin, C., Ouhayoun, J. P., Grund, C., and Franke, W. W., 1992b, Suprabasal marker proteins distinguishing keratinizing squamous epithelia: Cytokeratin 2 polypeptides of oral masticatory epithelium and epidermis are different, *Differentiation* **51:**137–148.

Comings, D. E., 1982, Two-dimensional gel electrophoresis of human brain proteins. IV. Disorders of glial proliferation and a polymorphism of glial fibrillary acidic protein—GFAP Duarte, *Clin. Chem.* **28:**805–812.

Compton, J. G., DiGiovanna, J. J., Santucci, S. K., Kearns, K. S., Amos, C. I., Abangan, D. L., Korge, B. P., McBride, O. W., Steinert, P. M., and Bale, S. J., 1992, Linkage of epidermolytic hyperkeratosis to the type II keratin gene cluster on chromosome 12q, *Nat. Genet.* **1:**301–305.

Cook, G. P., Tomlinson, I. M., Walter, G., Riethman, H., Carter, N. P., Buluwela, L., Winter, G., and Rabbitts, T. H., 1994, A map of the human immunoglobulin VH locus completed by analysis of the telomeric region of chromosome 14q, *Nat. Genet.* **7:**162–168.

Corden, L. D., and McLean, W. H., 1996, Human keratin diseases: Hereditary fragility of specific epithelial tissues, *Exp. Dermatol.* **5:**297–307.

Correll, A. T., Webb, G. C., Ford, J. H., Rogers, G. E., and Powell, B. C., 1992, Localization by *in situ* hybridization of a type I keratin intermediate filament gene (Krt-1.14) to band D of mouse chromosome 11, *Cytogenet. Cell Genet.* **60:**245–246.

Coulombe, P. A., Hutton, M. E., Letai, A., Hebert, A., Paller, A. S., and Fuchs, E., 1991, Point mutations in human keratin 14 genes of epidermolysis bullosa simplex patients: Genetic and functional analyses, *Cell* **66:**1301–1311.

Dillon, N., and Grosveld, F., 1993, Transcriptional regulation of multigene loci: Multilevel control, *Trends Genet.* **9:**134–137.

Dodemont, H., Riemer, D., and Weber, K., 1990, Structure of an invertebrate gene encoding cytoplasmic intermediate filament (IF) proteins: Implications for the origin and the diversification of IF proteins, *EMBO J.* **9:**4083–4094.

Engel, A., Eichner, R., and Aebi, U., 1985, Polymorphism of reconstituted human epidermal keratin filaments: Determination of their mass-per-length and width by scanning transmission electron microscopy (STEM), *J. Ultrastruct. Res.* **90:**323–335.

Engelkamp, D., Schäfer, B. W., Mattei, M. G., Erne, P., and Heizmann, C. W., 1993, Six S100 genes are clustered on human chromosome 1q21: Identification of two genes coding for the two previously unreported calcium-binding proteins S100D and S100E, *Proc. Natl. Acad. Sci. USA* **90:**6547–6551.

Ferrari, S., Cannizzaro, L. A., Battini, R., Huebner, K., and Baserga, R., 1987, The gene encoding human vimentin is located on the short arm of chromosome 10, *Am. J. Hum. Genet.* **41:**616–626.

Fietz, M. J., Rogers, G. E., Eyre, H. J., Baker, E., Callen, D. F., and Sutherland, G. R., 1992, Mapping of the trichohyalin gene: Co-localization with profilaggrin, involucrin, and loricrin genes, *J. Invest. Dermatol.* **99:**542–544.

Fietz, M. J., McLaughlan, C. J., Campbell, M. T., and Rogers, G. E., 1993, Analysis of the sheep trichohyalin gene: Potential structural and calcium-binding roles of trichohyalin in the hair follicle, *J. Cell Biol.* **121:**855–865.

Fink, P., Rogers, M. A., Korge, B., Winter, H., and Schweizer, J., 1995, A cDNA encoding the human type I hair keratin hHa1, *Biochim. Biophys. Acta* **1264:**12–14.

Franke, W. W., Goldschmidt, M. D., Zimbelmann, R., Mueller, H. M., Schiller, D. L., and Cowin, P., 1989, Molecular cloning and amino acid sequence of human plakoglobin, the common junctional plaque protein, *Proc. Natl. Acad. Sci. USA* **86:**4027–4031.

Fries, R., Threadgill, D. W., Hediger, R., Gunawardana, A., Blessing, M., Jorcano, J. L., Stranzinger, G., and Womack, J. E., 1991, Mapping of bovine cytokeratin sequences to four different sites on three chromosomes, *Cytogenet. Cell Genet.* **57:**135–141.

Fuchs, E., 1996, The cytoskeleton and disease: Genetic disorders of intermediate filaments, *Annu. Rev. Genet.* **30:**197–231.

Fuchs, E., and Weber, K., 1994, Intermediate filaments: Structure, dynamics, function, and disease, *Annu. Rev. Biochem.* **63:**345–382.

Gelb, B. D., Shi, G. P., Chapman, H. A, and Desnick, R. J., 1996, Pycnodysostosis, a lysosomal disease caused by cathepsin K deficiency, *Science* **273:**1236–1238.

Gibbs, S., Fijnemann, R., Wiegant, J., van Kessel, A. G., van de Putte, P., and Backendorf, C., 1993, Molecular characterization and evolution of the SPRR family of keratinocyte differentiation markers encoding small proline-rich proteins, *Genomics* **16:**630–637.

Hatzfeld, M., and Burba, M., 1994, Function of type I and type II keratin head domains: Their role in dimer, tetramer and filament formation, *J. Cell Sci.* **107:**1959–1972.

Hatzfeld, M., and Franke, W. W., 1985, Pair formation and promiscuity of cytokeratins: Formation *in vitro* of heterotypic complexes and intermediate-sized filaments by homologous and heterologous recombinations of purified polypeptides, *J. Cell Biol.* **101:**1826–1841.

Hatzfeld, M., and Weber, K., 1992, A synthetic peptide representing the consensus sequence motif at the carboxy-terminal end of the rod domain inhibits intermediate filament assembly and disassembles preformed filaments, *J. Cell Biol.* **116:**157–166.

Healy, E., Holmes, S. C., Belgaid, C. E., Stephenson, A. M., McLean, W. H. I., Rees, J. L., and Munro, C. S., 1995, A gene for monilethrix is closely linked to the type II keratin gene cluster at 12q13, *Hum. Mol. Genet.* **4:**2399–2402.

Heath, P., Elvin, P., Jenner, D., Gammack, A., Morten, J., and Markham, A., 1990, Localisation of a cDNA clone for human cytokeratin 18 to chromosome 17p11–p12 by *in situ* hybridisation, *Hum. Genet.* **85:**669–670.

Hediger, R., Ansari, H. A., and Stranzinger, G. F., 1991, Chromosome banding and gene localizations support extensive conservation of chromosome structure between cattle and sheep, *Cytogenet. Cell Genet.* **57:**127–134.

Heid, H. W., Werner, E., and Franke, W. W., 1986, The complement of native alpha-keratin polypeptides of hair-forming cells: A subset of eight polypeptides that differ from epithelial cytokeratins, *Differentiation* **32:**101–119.

Heins, S., and Aebi, U., 1994, Making heads and tails of intermediate filament assembly, dynamics and networks, *Curr. Opin. Cell Biol.* **6:**25–33.

Hennies, H. C., Küster, W., Mischke, D., and Reis, A., 1995, Localization of a locus for the striated form of palmoplantar keratoderma to chromosome 18q near the desmosomal cadherin gene cluster, *Hum. Mol. Genet.* **4:**1015–1020.

Herrmann, H., Hofmann, I., and Franke, W. W., 1992, Identification of a nonapeptide motif in the vimentin head domain involved in intermediate filament assembly, *J. Mol. Biol.* 22**3:**637–650.

Herrmann, H., Häner, M., Brettel, M., Müller, S. A., Goldie, K. N., Fedtke, B., Lustig, A., Franke, W. W., and Aebi, U., 1996, Structure and assembly properties of the intermediate filament protein vimentin: The role of its head, rod and tail domains, *J. Mol. Biol.* **264:**933–953.

Hoffmann, H. J., Olsen, E., Etzerodt, M., Madsen, P., Thogersen, H. C., Kruse, T., and Celis, J. E., 1994, Psoriasin binds calcium and is upregulated by calcium to levels that resemble those observed in normal skin, *J. Invest. Dermatol.* **103:**370–375.

Hofmann, I., and Franke, W. W., 1997, Heterotypic interactions and filament assembly of type I and type II cytokeratins *in vitro*: Viscometry and determinations of relative affinities, *Eur. J. Cell Biol.* **72:**122–132.

Hohl, D., Lichti, U., Breitkreutz, D., Steinert, P. M., and Roop, D. R., 1991, Transcription of the human loricrin gene *in vitro* is induced by calcium and cell density and suppressed by retinoic acid, *J. Invest. Dermatol.* **96:**414–418.

Humphries, M. M., Sheils, D. M., Jordan, S. A., Farrar, G. J., Kumar-Singh, R., Humphries, P., 1992, Alu polymorphism in the human type I keratin (KRT14) gene, *Hum. Mol. Genet.* **1:**453.

Hurst, J., Flavell, D., Julien, J.-P., Meijer, D., Mushynski, W., and Grosveld, F., 1987, The human neurofilament gene (NEFL) is located on the short arm of chromosome 8, *Cytogenet. Cell Genet.* **45:**30–32.

Irvine, A. D., Corden, L. D., Swensson, O., Swensson, B., Moore, J. E., Frazer, D. G., Smith, F. J., Knowlton, R. G., Christophers, E., Rochels, R., Uitto, J., and McLean, W. H., 1997a,

Mutations in cornea-specific keratin K3 or K12 genes cause Meesmann's corneal dystrophy, *Nat. Genet.* **16:**184–187.

Irvine, A. D., McKenna, K. E., Jenkinson, H., and Hughes, A. E., 1997b, A mutation in the V1 domain of keratin 5 causes epidermolysis bullosa simplex with mottled pigmentation, *J. Invest. Dermatol.* **108:**809–810.

Johnson, M. R., Polymeropoulos, M. H., Vos, H. L., Ortiz-de-Luna, R. I., and Francomano, C. A., 1996, A nonsense mutation in the cathepsin K gene observed in a family with pycnodysostosis, *Genome Res.* **6:**1050–1055.

Kimonis, V., DiGiovanna, J. J., Yang, J. M., Doyle, S. Z., Bale, S. J., and Compton, J. G., 1994, A mutation in the V1 end domain of keratin 1 in non-epidermolytic palmar-plantar keratoderma, *J. Invest. Dermatol.* **103:**764–769.

Klinge, E. M., Sylvestre, Y. R., Freedberg, I. M., and Blumenberg, M., 1987, Evolution of keratin genes: Different domains evolve by different pathways, *J. Mol. Evol.* **24:**319–329.

Korge, B. P., and Krieg, T., 1996, The molecular basis for inherited bullous diseases, *J. Mol. Med.* **74:**59–70.

Korge, B. P., Compton, J. G., Steinert, P. M., and Mischke, D., 1992a, The two size alleles of human keratin 1 are due to a deletion in the glycine-rich carboxyl-terminal V2 subdomain, *J. Invest. Dermatol.* **99:**697–702.

Korge, B. P., Gan, S. Q., McBride, O. W., Mischke, D., and Steinert, P. M., 1992b, Extensive size polymorphism of the human keratin 10 chain resides in the C-terminal V2 subdomain due to variable numbers and sizes of glycine loops, *Proc. Natl. Acad. Sci. USA* **89:**910–914.

Korge, B. P., Richards, G., Punter, C., Mauch, C., Junginger, H., Lin, J.-P., Rogers, M., Winter, H., Hamm, H., Schweizer, J., Krieg, T., and Traupe, H., 1996, Monilethrix links to the keratin type II cluster at 12q13 and cloning of a possible candidate gene, *J. Invest. Dermatol.* **106:**843.

Krauss, S., and Franke, W. W., 1990, Organization and sequence of the human gene encoding cytockeratin 8, *Gene* **86:**241–249.

Ku, N.-O., Wright, T. L., Terrault, N. A., Gish, R., and Omary, M. B., 1997, Mutation of human keratin 18 in association with cryptogenic cirrhosis, *J. Clin. Invest.* **99:**19–23.

Kulesh, D. A., and Oshima, R. G., 1988, Cloning of the human keratin 18 gene and its expression in nonepithelial mouse cells, *Mol. Cell Biol.* **8:**1540–1550.

Landon, F., Lemonnier, M., Benarous, R., Huc, C., Fiszman, M., Gros, F., and Portier, M. M., 1989, Multiple mRNAs encode peripherin, a neuronal intermediate filament protein, *EMBO J.* **8:**1719–1726.

Lane, E. B., Rugg, E. L., Navsaria, H., Leigh, I. M., Heagerty, A. H. M., Ishida-Yamamoto, A., and Eady, R. A. J., 1992, A mutation in the conserved helix termination peptide of keratin 5 in hereditary skin blistering, *Nature* **356:**244–246.

Lee, S. C., Wang, M., McBride, O. W., O'Keefe, E. J., Kim, I. G., and Steinert, P. M., 1993a, Human trichohyalin gene is clustered with the genes for other epidermal structural proteins and calcium-binding proteins at chromosomal locus 1q21, *J. Invest. Dermatol.* **100:**65–68.

Lee, S. C., Kim, I. G., Marekov, L. N., O'Keefe, E. J., Parry, D. A., and Steinert, P. M., 1993b, The structure of human trichohyalin. Potential multiple roles as a functional EF-hand-like calcium-binding protein, a cornified cell envelope precursor, and an intermediate filament-associated (cross- linking) protein, *J. Biol. Chem.* **268:**12164–12176.

Lessin, R. S, Hübner, K., Isobe, M., Croce, C. M., and Steinert, P. M., 1988, Chromosomal mapping of human keratin genes: Evidence of non-linkage, *J. Invest. Dermatol.* **91:**572–578.

Lu, X., and Lane, E. B., 1990, Retrovirus-mediated transgenic keratin expression in cultured fibroblasts: Specific domain functions in keratin stabilization and filament formation, *Cell* **62:**681–696.

MacKinnon, P. J., Powell, B. C., Rogers, G. E., Baker, E. G., MacKinnon, R. N., Hyland, V. J., Callen, D. F., and Sutherland, G. R., 1991, An ultrahigh-sulphur keratin gene of the human hair cuticle is located at 11q13 and cross-hybridizes with sequences at 11p15, *Mammalian Genome* **1:**53–56.

Maestrini, E., Monaco, A. P., McGrath, J. A., Ishida-Yamamoto, A., Camisa, C., Hovnanian, A., Weeks, D. E., Lathrop, M., Uitto, J., and Christiano, A. M., 1996, A molecular defect in loricrin, the major component of the cornified cell envelope, underlies Vohwinkel's syndrome, *Nat. Genet.* **13:**70–77.

Marenholz, I., Volz, A., Davies, A., Ragoussis, I., Ziegler, A., Korge, B. P., and Mischke, D., 1996, Genetic analysis of the epidermal differentiation complex (EDC) on human chromosome 1q21: Chromosomal orientation, new markers, and a 6 Mbp YAC contig, *Genomics* **37:**295–302.

Markova, N. G., Marekov, L. N., Chipev, C. C., Gan, S. Q., Idler, W. W., and Steinert, P. M., 1993, Profilaggrin is a major epidermal calcium-binding protein, *Mol. Cell. Biol.* **13:**613–625.

Mattei, M.-G., Dautigny, A., Pham-Dinh, D., Passage, E., Mattei, J.-F., and Jolles, P., 1988, The gene encoding the large human neurofilament subunit (NF-H) maps to the q121–q131 region on human chromosome 22, *Hum. Genet.* **80:**293–295.

McLean, W. H., and Lane, E. B., 1992, AvaII RFLP of human keratin 10 (KRT-10) detected by PCR, *Hum. Mol. Genet.* **1:**659.

McLean, W. H., and Lane, E. B., 1995, Intermediate filaments in disease, *Curr. Opin. Cell Biol.* **7:**118–125.

McLean, W. H., Sultan, N., Parfitt, E., and Lane, E. B., 1994, Polymorphisms in the keratin 8 gene detected by PCR, *Hum. Mol. Genet.* **3:**1031.

Michels, S., and Juhlin, L., 1990, Cornified envelopes in congenital disorders of keratinization, *Br. J. Dermatol.* **122:**15–21.

Milisavljevic, V., Freedberg, I. M., and Blumenberg, M., 1996, Close linkage of the two keratin gene clusters in the human genome, *Genomics* **34:**134–138.

Milner, C. M., and Campbell, R. D., 1992, Genes, genes and more genes in the human major histocompatibility complex, *Bioessays* **14:**565–571.

Mischke, D., 1993, Frequencies of human keratin 10 alleles, *Hum. Mol. Genet.* **2:**618.

Mischke, D., and Wild, G., 1987, Polymorphic keratins in human epidermis, *J. Invest. Dermatol.* **88:**191–197.

Mischke, D., Wille, G., and Wild, A. G., 1990, Allele frequencies and segregation of human polymorphic keratins K4 and K5, *Am. J. Hum. Genet.* **46:**548–552.

Mischke, D., Korge, B. P., Marenholz, I., Volz, A., and Ziegler, A., 1996, Genes encoding structural proteins of epidermal cornification and S100 calcium-binding proteins form a gene complex ("epidermal differentiation complex") on human chromosome 1q21, *J. Invest. Dermatol.* **106:**989–992.

Moll, R., Franke, W. W., Schiller, D. L., Geiger, B., and Krepler, R., 1982, The catalog of human cytokeratins: Patterns of expression in normal epithelia, tumors and cultured cells, *Cell* **31:**11–24.

Moll, R., Schiller, D. L., and Franke, W. W., 1990, Identification of protein IT of the intestinal cytoskeleton as a novel type I cytokeratin with unusual properties and expression patterns, *J. Cell Biol.* **111:**567–580.

Moll, R., Zimbelmann, R., Goldschmidt, M. D., Keith, M., Laufer, J., Kasper, M., Koch, P. J., and Franke, W. W., 1993, The human gene encoding cytokeratin 20 and its expression during fetal development and in gastrointestinal carcinomas, *Differentiation* **53:**75–93.

Moncla, A., Landon, F., Mattei, M.-G., and Portier, M.-M., 1992, Chromosomal localisation of the mouse and human peripherin genes, *Genet. Res.* **59:**125–129.

Moseley, W. S., and Seldin, M. F., 1989, Definition of mouse chromosome 1 and 3 gene linkage groups that are conserved on human chromosome 1: Evidence that a conserved linkage group spans the centromere of human chromosome 1, *Genomics* **5:**899–905.

Myers, M. W., Lazzarini, R. A., Lee, V. M.-L., Schlaepfer, W. W., and Nelson, D. L., 1987, The human mid-size neurofilament subunit: A repeated protein sequence and the relationship of its gene to the intermediate filament gene family, *EMBO J.* **6:**1617–1626.

Nadeau, J. H., Berger, F. G., Cox, D. R., Crosby, J. L., Davisson, M. T., Ferrara, D., Fuchs, E., Hart, C., Hunihan, L., Lalley, P. A., Langley, S. H., Martin, G. R., Nichols, L., Phillips, S. J., Roderick, T. H., Roop, D. R., Ruddle, F. H., Skow, L. C., and Compton, J. G., 1989, A family of type I keratin gene and the homeobox-2 gene complex are closely linked to the rex locus on mouse chromosome 11, *Genomics* **5:**454–462.

Oshima, R. G., 1992, Intermediate filament molecular biology, *Curr. Opin. Cell Biol.* **4:**110–116.

Parry, D. A. D., and Steinert, P. M., 1992, Intermediate filament structure, *Curr. Opin. Cell Biol.* **4:**94–98.

Popescu, N. C., Bowden, P. E., and DiPaolo, J. A., 1989, Two type II keratin genes are localized on human chromosome 12, *Hum. Genet.* **82:**109–112.

Powell, B. C., and Beltrame, J. S., 1994, Characterization of a hair (wool) keratin intermediate filament gene domain, *J. Invest. Dermatol.* **102:**171–177.

Powell, B. C., and Rogers, G. E., 1997, The role of keratin proteins and their genes in the growth, structure and properties of hair, in *Formation and Structure of Human Hair* (P. Jollés, H. Zahn, and H. Höcker, eds.), pp. 59–148, Birkhäuser, Basel.

Powell, B., Crocker, L., and Rogers, G., 1994, Hair follicle differentiation: Expression, structure and evolutionary conservation of the hair type II keratin intermediate filament gene family, *Development* **114:**417–433.

Presland, R. B., Haydock, P. V., Fleckman, P., Nirunsuksiri, W., and Dale, B. A., 1992, Characterization of the human epidermal profilaggrin gene. Genomic organization and identification of an S-100-like calcium binding domain at the amino terminus, *J. Biol. Chem.* **267:**23772–23781.

Presland, R. B., Bassuk, J. A., Kimball, J. R., and Dale, B. A., 1995, Characterization of two distinct calcium-binding sites in the amino-terminus of human profilaggrin, *J. Invest. Dermatol.* **104:**218–223.

Reis, A., Küster, W., Eckardt, R., and Sperling, K., 1992, Mapping of a gene for epidermolytic palmoplantar keratoderma to the region of the acidic keratin gene cluster at 17q12–q21, *Hum. Genet.* **90:**113–116.

Reis, A., Hennies, H. C., Langbein, L., Digweed, M., Mischke, D., Drechsler, M., Schrock, E., Royer-Pokora, B., Franke, W. W., Sperling, K., and Küster, W., 1994, Keratin 9 gene mutations in epidermolytic palmoplantar keratoderma (EPPK), *Nat. Genet.* **6:**174–179.

Resing, K. A., and Dale, B. A., 1991, Proteins of keratohyalin, in *Physiology, Biochemistry, and Molecular Biology of the Skin* (L. A. Goldsmith, ed.), pp. 148–167, Oxford University Press, Oxford.

Rieger, M., and Franke, W. W., 1988, Identification of an orthologous mammalian cytokeratin gene. High degree of intron sequence conservation during evolution of human cytokeratin 10, *J. Mol. Biol.* **204:**841–856

Robinson, N. A., Lapic, S., Welter, J. F., and Eckert, R. L., 1997, S100A11, S100A10, annexin I, desmosomal proteins, small proline rich proteins, plasminogen activator inhibitor 2, and involucrin are components of the cornified envelope of cultured human epidermal keratinocytes, *J. Biol. Chem.* **272:**12035–12046.

Rogers, G. E., and Powell, B. C., 1993, Organization and expression of hair follicle genes, *J. Invest. Dermatol.* **101:**50S–55S.

Rogers, K. R., Eckelt, A., Nimmrich, V., Janssen, K. P., Schliwa, M., Herrmann, H., and Franke, W. W., 1995, Truncation mutagenesis of the non-alpha-helical carboxyterminal tail domain of vimentin reveals contributions to cellular localization but not to filament assembly, *Eur. J. Cell Biol.* **66:**136–150.

Rogers, M. A., Schweizer, J., Kreig, T., and Winter, H., 1994, A novel human type I hair keratin gene: Evidence for two keratin hHa3 isoforms, *Mol. Biol. Rep.* **20:**155–161.

Rogers, M. A., Nischt, R., Korge, B., Krieg, T., Fink, T. M., Lichter, P., Winter, H., and Schweizer, J., 1995, Sequence data and chromosomal localization of human type I and type II hair keratin genes, *Exp. Cell Res.* **220:**357–362.

Rogers, M. A., Winter, H., Langbein, L., Krieg, T., and Schweizer, J., 1996, Genomic characterization of the human type I cuticular hair keratin hHa2 and identification of an adjacent novel type I hair keratin gene hHa5, *J. Invest. Dermatol.* **107:**633–638.

Rogers, M. A., Langbein, L., Praetzel, S., Moll, I., Krieg, T., Winter, H., and Schweizer, J., 1997, Sequences and differential expression of three novel human type-II hair keratins, *Differentiation* **61:**187–194.

Romano, V., Bosco, P., Costa, G., Leube, R., Franke, W. W., Rocchi, M., and Romeo, G., 1987, Chromosomal assignment of cytokeratin genes [Abstract], *Cytogenet. Cell Genet.* **46:**683.

Romano, V., Bosco, P., Rocchi, M., Costa, G., Leube, R. E., Franke, W. W., and Romeo, G., 1988, Chromosomal assignments of human type I and type II cytokeratin genes to different chromosomes, *Cytogenet. Cell Genet.* **48:**148–151.

Romano, V., Bosco, P., Raimondi, E., Feo, S., Leube, R., Franke, W., and Ceratto, N., 1991, Chromosomal mapping and physical linkage analysis of human acidic cytokeratin genes [Abstract], *Cytogenet. Cell Genet.* **58:**2009–2010.

Rosenberg, M., RayChaudhury, A., Shows, T. B., Le Beau, M. M., and Fuchs, E., 1988, A group of type I keratin genes on human chromosome 17: Characterization and expression, *Mol. Cell. Biol.* **8:**722–736.

Rosenberg, M., Fuchs, E., Le Beau, M. M., Eddy, R. L., and Shows, T. B., 1991, Three epidermal and one simple epithelial type II keratin genes map to human chromosome 12, *Cytogenet. Cell Genet.* **57:**33–38.

Rothnagel, J. A., Dominey, A. M., Dempsey, L. D., Longley, M. A., Greenhalgh, D. A., Gagne, T. A., Huber, M., Frenk, E., Hohl, D., and Roop, D. R., 1992, Mutations in the rod domains of keratins 1 and 10 in epidermolytic hyperkeratosis, *Science* **257:**1128–1139.

Rothnagel, J. A., Longley, M. A., Bundman, D. S., Naylor, S. L., Lalley, P. A., Jenkins, N. A., Gilbert, D. J., Copeland, N. G., and Roop, D. R., 1994, Characterization of the mouse loricrin gene: Linkage with profilaggrin and the flaky tail and soft coat mutant loci on chromosome 3, *Genomics* **23:**450–456.

Saintigny, G., Schmidt, R., Shroot, B., Juhlin, L., Reichert, U., and Michels, S., 1992, Differential expression of calgranulin A and B in various epithelial cell lines and reconstructed epidermis, *J. Invest. Dermatol.* **99:**639–644.

Savtchenko, E. S., Freedberg, I. M., Choi, I. Y., and Blumenberg, M., 1988, Inactivation of human keratin genes: The spectrum of mutations in the sequence of an acidic keratin pseudogene, *Mol. Biol. Evol.* **5:**97–108.

Savtchenko, E. S., Tomic, M., Fuker, R., and Blummenberg, M., 1990, Three parallel linkage groups of human acidic keratin genes, *Genomics*, **7:**394–407.

Schäfer, B. W., Wicki, R., Engelkamp, D., Mattei, M. G., and Heizmann, C. W., 1995, Isolation of a YAC clone covering a cluster of nine S100 genes on human chromosome 1q21:

Rationale for a new nomenclature of the S100 calcium-binding protein family, *Genomics* **25:**638–643.

Smith, F. J. D., Corden, L. D., Rugg, E. L., Ratnavel, R., Leigh, I. M., Moss, C., Tidman, M. J., Hohl, D., Huber, M., Kunkeler, L., Munro, C. S., Lane, E. B., and McLean, W. H. I., 1997, Missense mutations in keratin 17 cause either pachyonychia congenita type 2 or a phenotype resembling steatocystoma multiplex, *J. Invest. Dermatol.* **108:**220–223.

Solomon, E., Borrow, J., and Goddard, A. D., 1991, Chromosome aberrations and cancer, *Science* **254:**1153–1160.

Steinert, P. M., 1990, The two-chain coiled-coil molecule of native epidermal keratin intermediate filaments is a type I-type II heterodimer, *J. Biol. Chem.* **265:**8766–8774.

Steinert, P. M., and Marekov, L. N., 1995, The proteins elafin, filaggrin, keratin intermediate filaments, loricrin, and small proline-rich proteins 1 and 2 are isodipeptide cross-linked components of the human epidermal cornified cell envelope, *J. Biol. Chem.* **270:**17702–17711.

Steinert, P. M., and Roop, D. R., 1988, Molecular and cellular biology of intermediate filaments, *Annu. Rev. Biochem.* **57:**593–625.

Steinert, P. M., Mack, J. W., Korge, B. P., Gan, S.-Q., Haynes, S. R., and Steven, A. C., 1991, Glycine loops in proteins: Their occurrence in certain intermediate filament chains, loricrins and single-stranded RNA binding proteins, *Int. J. Biol. Macromol.* **13:**130–139.

Steven, A. C., Hainfeld, J. F., Trus, B. L., Wall, J. S., and Steinert, P. M., 1983, Epidermal keratin filaments assembled *in vitro* have masses-per-unit-length that scale according to average subunit mass: Structural basis for homologous packing of subunits in intermediate filaments, *J. Cell Biol.* **97:**1939–1944.

Stewart, M., Quinlan, R. A., and Moir, R. D., 1989, Molecular interactions in paracrystals of a fragment corresponding to the alpha-helical coiled-coil rod portion of glial fibrillary acidic protein: Evidence for an antiparallel packing of molecules and polymorphism related to intermediate filament structure, *J. Cell Biol.* **109:**225–234.

Sun, T.-T., Eichner, R., Schermer, A., Cooper, D., Nelson, W. G., and Weiss, R. A., 1984, Classification, expression, and possible mechanisms of evolution of mammalian epithelial keratins: A unifying model, in *Cancer Cells 1, The Transformed Phenotype* (A. Levine, W. Topp, G. vande Woude, and J. D. Watson, eds.), pp. 169–176, Cold Spring Harbor Laboratory, Cold Spring Harbor New York.

Takahashi, K., Paladini, R. D., and Coulombe, P. A., 1995, Cloning and characterization of multiple human genes and cDNAs encoding highly related type II keratin 6 isoforms, *J. Biol. Chem.* **270:**18581–18592.

Troyanovsky, S. M., and Leube, R. E., 1994, Activation of the silent human cytokeratin 17 pseudogene-promoter region by cryptic enhancer elements of the cytokeratin 17 gene, *Eur. J. Biochem.* **225:**61–69.

Tyner, A. L., Eichman, M. J., and Fuchs, E., 1985, The sequence of a type II keratin gene expressed in human skin: Conservation of structure among all intermediate filament genes, *Proc. Natl. Acad. Sci. USA* **82:**4683–4687.

Uttam, J., Hutton, E., Coulombe, P. A., Anton-Lamprecht, I., Yu, Q. C., Gedde-Dahl, T., Jr., Fine, J. D., and Fuchs, E., 1996, The genetic basis of epidermolysis bullosa simplex with mottled pigmentation, *Proc. Natl. Acad. Sci. USA* **93:**9079–9084.

Viegas-Pequignot, E., Lin, L. Z., Dutrillaux, B., Apiou, F., and Paulin, D., 1989, Assignment of human desmin gene to band 2q35 by nonradioactive *in situ* hybridization, *Hum. Genet.* **83:**33–36.

Volz, A., Korge, B. P., Compton, J. G., Ziegler, A., Steinert, P. M., and Mischke, D., 1993, Physical mapping of a functional cluster of epidermal differentiation genes on chromosome 1q21, *Genomics* **18:**92–99.

Wanner, R., Förster, H. H., Tilmans, I., and Mischke, D., 1993, Allelic variations of human keratins K4 and K5 provide polymorphic markers within the type II keratin gene cluster on chromosome 12, *J. Invest. Dermatol.* **100:**735–741.

Waseem, A., Alexander, C. M., Steel, J. B., and Lane, E. B., 1990a, Embryonic simple epithelial keratins 8 and 18: Chromosomal location emphasizes difference from other keratin pairs, *New Biol.* **2:**464–478.

Waseem, A., Gough, A. C., Spurr, N. K., and Lane, E. B., 1990b, Localization of the gene for human simple epithelial keratin 18 to chromosome 12 using polymerase chain reaction, *Genomics* **7:**188–194.

Wicki, R., Marenholz, I., Mischke, D., Schäfer, B. W., and Heizmann, C. W., 1996a, Characterization of the human S100A12 (calgranulin C, p6, CAAF1, CGRP) gene, a new member of the S100 gene cluster on chromosome 1q21, *Cell Calcium* **20:**459–464.

Wicki, R., Schäfer, B. W., Erne, P., and Heizmann, C. W., 1996b, Characterization of the human and mouse cDNAs coding for S100A13, a new member of the S100 protein family, *Biochem. Biophys. Res. Commun.* **227:**594–599.

Wild, G. A., and Mischke, D., 1986, Variation and frequency of cytokeratin polypeptide patterns in human squamous non-keratinizing epithelium, *Exp. Cell Res.* **162:**114–126.

Willard, M. B., 1976, Genetically determined protein polymorphism in the rabbit nervous system, *Proc. Natl. Acad. Sci. USA* **73:**3641–3645.

Winter, H., Rogers, M. A., Mahr, B., Cremer, M., Krieg, T., Navsaria, H. A., and Leigh, I. M., 1996, The region coding for the helix termination motif, and the adjacent intron 6 of the human type I hair keratin gene hHa2 contains three natural, closely spaced polymorphic sites, *J. Invest. Dermatol.* **106:**544–548.

Winter, H., Rogers, M. A., Langbein, L., Stevens, H. P., Leigh, I. M., Labrèze, C., Roul, S. Taieb, A., Krieg, T., and Schweizer, J., 1997, Mutations in the hair cortex keratin hHb6 cause the inherited hair disease monilethrix, *Nat. Genet.* **16:**372–374.

Wydner, K. L., McNeil, J. A., Lin, F., Worman, H. J., and Lawrence, J. B., 1996, Chromosomal assignment of human nuclear envelope protein genes LMNA, LMNB1, and LBR by fluorescence in situ hybridization, *Genomics* **32:**474–478.

Yoon, S. J., LeBlanc-Straceski, J., Ward, D., Krauter, K., and Kucherlapati, R., 1994, Organization of the human keratin type II gene cluster at 12q13, *Genomics* **24:**502–508.

Yuspa, S. H., Kilkenny, A. E., Steinert, P. M., and Roop, D. R., 1989, Expression of murine epidermal differentiation markers is tightly regulated by restricted extracellular calcium concentrations *in vitro*, *J. Cell Biol.* **109:**1207–1217.

Zhou, Y., and Chaplin, D. D., 1993, Identification in the HLA class I region of a gene expressed late in keratinocyte differentiation, *Proc. Natl. Acad. Sci. USA* **90:**9470–9474.

Ziegler, A., and Volz, A., 1992, Rotating field gel electrophoresis (ROFE), in *Methods in Molecular Biology, Vol. 12: Pulsed Field Gel Electrophoresis* (M. Burmeister and L. Ulanovsky, eds.), pp. 63–72, Humana Press, Totowa, New Jersey.

Chapter 4

Keratin Modifications and Solubility Properties in Epithelial Cells and *in Vitro*

M. Bishr Omary, Nam-On Ku, Jian Liao, and Daniel Price

1. GENERAL OVERVIEW OF KERATINS AND THEIR POSTTRANSLATIONAL MODIFICATIONS

Keratins make up the largest subfamily of intermediate filament (IF)* proteins (reviewed by Lazarides, 1980; Moll *et al.*, 1982; Steinert and Roop, 1988; Coulombe, 1993; Fuchs and Weber, 1994). This chapter will focus on the posttranslational modifications and solubility of the "soft" keratins (also termed cytokeratins) that are found in the cytoplasm of epithelial

*Abbreviations: Emp, Empigen BB; GlcNAc, N-acetylglucosamine; hsp, heat shock proteins; IF, intermediate filament(s); IFAP, intermediate filament-associated protein; K, keratin; NP40, Nonidet P40; PAGE, polyacrylamide gel electrophoresis; PKA, cAMP-dependent protein kinase; PKC, protein kinase C.

M. Bishr Omary, Nam-On Ku, and Daniel Price Department of Medicine, VA Palo Alto Health Care System, Palo Alto, California 94304; and Stanford University School of Medicine, Stanford, California 94305-5487. **Jian Liao** CLONTECH Laboratories, Palo Alto, California 94303.

Subcellular Biochemistry, Volume 31: Intermediate Filaments, edited by Herrmann and Harris. Plenum Press, New York, 1998.

cells. It will not address the "hard," essentially insoluble keratins that are found in epidermal appendages such as hair, nails, and animal horns. The IF keratin family is divided into two subtypes (Table I) based on their isoelectric point, molecular mass, and sequence (Moll *et al.*, 1982). Keratins and the cells that express them manifest several distinguishing features that include the following: (i) All epithelial cells that express keratins express at least one type I and one type II keratin in an epithelial cell-type specific manner. For example, simple single-layered (i.e., glandular/secretory) epithelia, as found in the gastrointestinal tract, liver, and exocrine pancreas, express keratin polypeptides 8 (K8) and 18 (K18) with variable levels of K19 (Eckert, 1988; Stasiak *et al.*, 1989) and K20 (Moll *et al.*, 1990; Calnek and Quaroni, 1993) depending on the cell type (Table I). (ii) Keratin mutations are associated with a growing list of skin, oral, and corneal diseases (Fuchs and Coulombe, 1992; Steinert and Bale, 1993; McLean and Lane, 1995; Fuchs, 1996; Irvine *et al.*, 1997). In addition, a K18 mutation was identified in a patient with liver disease of "unknown etiology" (Ku *et al.*, 1997). This and a liver-targeted phenotype in mice that express a mutant K18 (Ku *et al.*, 1995, 1996b) or have a null mutation in K8 (Baribault *et al.*, 1993) or K18 (see Chapter 5 by T. Magin in this book) have led to the hypothesis that K18 or K8 mutations may either predispose to, or in some circumstances cause, human liver disease (reviewed by Omary and Ku, 1997).

Keratins undergo several posttranslational modifications (Table II), with the best-studied modification being phosphorylation (reviewed by Ku *et al.*, 1996a). The study of these modifications is important from regulatory/mechanistic and functional perspectives. In terms of regulation, keratin

Table I
Keratins and Their Tissue-Specific Distribution

A. Keratins	
Type I (acidic)	Keratins 9–20 (K9–K20)
Type II (relatively basic)	Keratins 1–8 (K1–K8)
B. Tissue distribution	
Simple epithelia[a]	K8, K18, K19, K20
Basal keratinocytes	K5, K14
Suprabasal keratinocytes	K1, K10
Esophageal epithelia	K4, K13
Corneal epithelia	K3, K12
Wound-healing skin	K6, K16

[a]Examples include cells in the gastrointestinal tract, liver, exocrine pancreas, and sweat glands.

Table II
Posttranslational Modifications of Keratins

Modification	Keratin	Nature or site of modification
Phosphorylation[a]	K1, K6, K8, K18	Serine/threonine
Glycosylation	K8, K18, K13, K19	O-GlcNAc-serine
Acetylation[b]	Epidermal keratins	N-terminal serine
Lipid	K8, K14, K18, K19	Palmitate, retinoylation
Proteolysis	K18, K19	Caspase mediated
Transglutamination	K8/18, K1/10	Cross-linking

[a]Based on isoelectric focusing and SDS–PAGE analysis, it is highly likely that all keratins are phosphoproteins. However, the indicated keratins are the only ones with already identified *in vivo* phosphorylation sites.
[b]The initial description involved analysis of bovine skin keratins (Steinert and Idler, 1975). It is likely that all keratins are N-blocked by acetylation.

phosphorylation and glycosylation are dynamic modifications (Chou *et al.*, 1992; Liao *et al.*, 1995a) that are located in the head and tail domains of the keratin molecule (Figure 1). These are the domains that are responsible for the structural heterogeneity and hence the tissue-specific expression among the different keratins. The dynamic nature of these two modifications implies that they play at least some role in modulating keratin function, which likely includes keratin subcellular localization and interaction with other cytoplasmic proteins. Although still poorly understood, regulation may also potentially occur by keratin-associated proteins such as the heat shock protein (hsp) 70 family (Liao *et al.*, 1995c), the 14-3-3 protein family (Liao and Omary, 1996), and desmosomal-associated proteins such as desmoplakin (Stappenbeck *et al.*, 1993), IFAP 300 (Skalli *et al.*, 1994), and Band 6 (Hatzfeld *et al.*, 1994). In addition, keratin regulation may also take place jointly by a keratin posttranslational modification and an associated protein. For example, K18 Ser33 phosphorylation is a modification that regulates the interaction of K8/18 with 14-3-3 proteins (Liao and Omary, 1996; Ku *et al.*, 1998).

Although it is well accepted that modifications such as phosphorylation and glycosylation are likely to regulate keratin function, one important stumbling block has been that keratin function, and in fact the function of all other cytoplasmic IF proteins, is poorly understood. One function that has clearly emerged, based on the keratin transgenic animal models and the human skin diseases that have been identified, is the maintenance of cellular mechanical integrity particularly upon cell or tissue stress (Fuchs and Weber, 1994; Morley *et al.*, 1995; Omary and Ku, 1997). An important aspect of studying the posttranslational modifications of keratins is their use

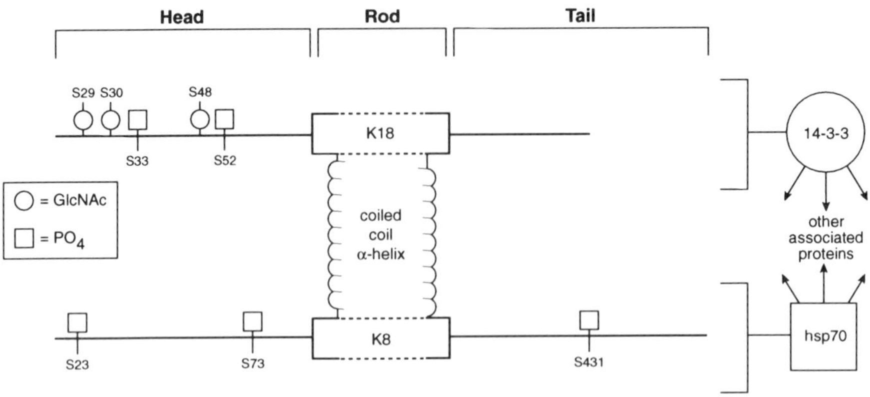

FIGURE 1. Phosphorylation and glycosylation of K8/18. Glycosylation and phosphorylation of K18 are located in the head domain. In the case of K8, glycosylation sites have not been identified, but phosphorylation occurs in the head and tail domains. Since both modifications are regulatory and occur in the domains that impart most of the structural heterogeneity among different IF proteins, it is hypothesized that these modifications regulate the tissue-specific function(s) of keratin IF proteins. The sizes of the head, rod, and tail domains are not drawn to scale but consist for K8 of 87, 315, and 80 amino acids, respectively, and for K18 of 82, 308, and 39 amino acids, respectively. The schematic also highlights the association of 14-3-3 with K18, and heat shock protein 70 (hsp 70) with K8 (see text). In addition, both 14-3-3 and hsp70 interact with other proteins.

as a handle for studying the function of these proteins. Due to the lack of a better assay, a presumed function-related parameter and correlate that investigators have measured is filament assembly and IF protein solubility. The basis for these measurements is that perturbations in assembly/solubility are likely to reflect changes in IF protein functions, whatever these functions may be. This chapter will review our current understanding of keratin posttranslational modifications and solubility, and the role that they are playing in improving our understanding of the function and regulation of these proteins.

2. KERATIN PHOSPHORYLATION

2.1. General Features

Phosphorylation of keratins was initially described in epidermal cells and tissues (Decker, 1968; Sun and Green, 1978; Gilmartin *et al.*, 1980), then in glandular keratins (Bravo *et al.*, 1982; Oshima, 1982; Celis *et al.*, 1983). Although significant gains still need to be made in understanding keratin phosphorylation, this modification is by far the most studied and best understood. The complexity in functional characterization of keratin phosphorylation is in part related to the large number of potential phosphorylation sites. For example, intact K8 and K18 contain 61 and 37 serines, respectively (Ku and Omary, 1997; Oshima *et al.*, 1986). The lack of a keratin function also makes assaying putative roles for keratin phosphorylation difficult. Serine is the major keratin phosphorylation site (Gilmartin *et al.*, 1980; Oshima, 1982), with small amounts of threonine phosphorylation in epidermal keratins (Steinert *et al.*, 1982; Steinert, 1988). The phosphotyrosine content of keratins is small, if any (Steinert *et al.*, 1982), and no *in vivo* tyrosine phosphorylation site has been reported for any IF protein (Ku *et al.*, 1996a). Although some IF proteins may be phosphorylated by pp^{60src} *in vitro*, including desmin (Collett *et al.*, 1980) and K8/18 (unpublished observations), the biological significance of such phosphorylation is unclear. The *in vivo* stoichiometry of phosphorylation (phosphate molecules/keratin molecule) varies between 0.25 and 0.3 for rat K8 and K18 (Yano *et al.*, 1991), and 3 and 5.45 for mouse and bovine epidermal keratins (Steinert *et al.*, 1982). Phosphate turnover occurs very rapidly in cells. For example, one half of the $^{32}PO_4$ incorporated into mitotic amnion cells (after a 10-min pulse) is lost within 13 min (Fey *et al.*, 1983).

2.2. Characterization of Keratin Phosphorylation

The methods for identifying *in vivo* $^{32}PO_4$-labeled keratin phosphorylation sites have relied on microsequencing of CNBr and/or protease-generated phosphopeptides (Steinert, 1988). Alternatively, tryptic or chymotryptic peptides were generated from keratins that were isolated from $^{32}PO_4$-labeled cells, followed by two-dimensional peptide mapping to separate the peptides and manual Edman degradation to identify the position of the radiolabeled phosphorylated residue(s) within the peptide (Ku and Omary, 1994b, 1997). The potential phosphorylation sites are then predicted based on known keratin sequences, and are mutated followed by transfection of the wild-type and point-mutated cDNAs, *in vivo* $^{32}PO_4$ labeling, and peptide mapping to confirm the position of the phosphorylated residues (Ku and Omary, 1994b, 1997). This direct determination of the phosphorylation sites within a known phosphopeptide or phosphoprotein can be envisioned as a "forward" approach, in contrast to an alternate "reverse immunologic" approach that is outlined in Figure 2. This latter approach has been successfully used to identify phospho-Ser73 of K8, whose identification could otherwise have been more difficult using a standard "forward" approach (Liao *et al.*, 1997). The approach outlined in Figure 2 also generated two independent antibodies that recognized the

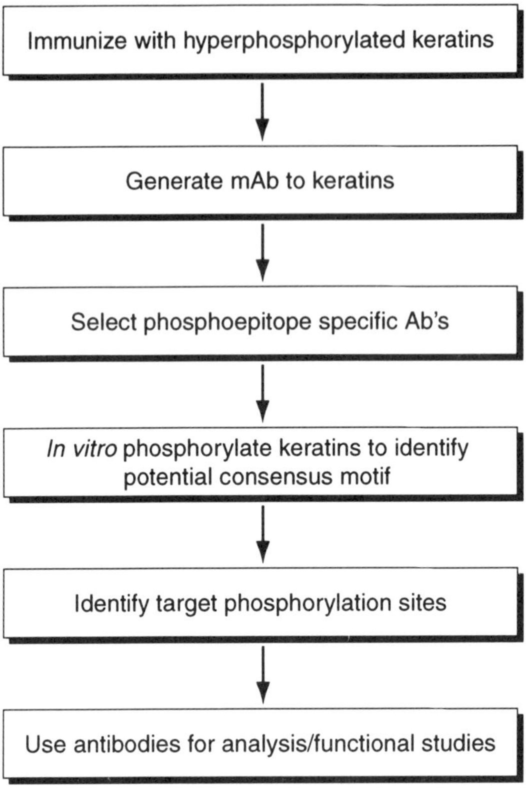

FIGURE 2. "Reverse immunologic" approach to generate antiphosphoepitope-specific antibodies and identify keratin phosphorylation sites. This strategy was used to generate several mAb that recognize unique K8 and K18 phosphoepitopes. Hyperphosphorylated keratins were generated by treating cultured cells with okadaic acid, followed by immunoprecipitation of the keratins, release of the immunoprecipitated keratins using urea (the antibody was covalently linked to agarose and is not released), then dialysis to remove the urea and renature the keratins. Antibodies were then chosen for their selective binding to phosphorylated keratins, by immunoblotting against keratins that were isolated from cells which were cultured in the presence or absence of okadaic acid. Identification of the phosphorylation site that is recognized by the antibody is carried out by *in vitro* phosphorylation of keratins, with a variety of kinases, then immunoblotting with the antibodies to determine those kinases that best phosphorylate a site that is recognized by an antibody in question. Based on the consensus sequence of the kinase(s) that provides a strong antibody-binding signal and the known sequences of K8 and K18, potential target phosphorylation sites are evaluated using biochemical and mutational approaches. Antibodies are then used for characterization and functional studies. This approach was successfully used to identify phospho-Ser73 of K8 (Liao *et al.*, 1997).

already identified phospho-Ser431 of K8 (Ku and Omary, 1997) and other antibodies that remain to be characterized. Hence, this approach complements the directed generation of antiphosphoepitope antibodies (reviewed by Inagaki *et al.*, 1997), whereby antibodies are generated to phosphopeptides that correspond to known phosphorylation sites.

2.3. Dynamics of Keratin Phosphorylation

The characterized *in vivo* phosphorylation sites of keratins are summarized in Figure 3. All the currently known keratin phosphorylation sites reside in the tail and/or head domains. In the case of K1, most of the sites have the *basic-X-S/T*, motif which suggests phosphorylation by cAMP-dependent protein kinase (PKA). The function of these phosphorylation sites is not known, but interestingly, phosphorylation of K1/10 in human foreskin appears more dynamic (i.e., has faster *in vivo* $^{32}PO_4$ incorporation) as compared with K5/14 (Steinert, 1988). Assuming that this does not reflect less accessibility of the $^{32}PO_4$ to proliferating basal (preferentially K5/14-expressing) versus more differentiated suprabasal (preferentially K1/10-expressing) keratinocytes, this result is somewhat surprising in that the more differentiated cells are still quite active in terms of their keratin phosphorylation. Significant differences in phosphate turnover among keratin partners also occur. For example, $^{32}PO_4$ pulse-labeling (30 min) of K8/18 in colonic cultured cells followed by chase shows that the K18-incorporated radioactivity is lost by 6 h, while 30% remains in K8 (Liao *et al.*, 1995a). In addition, the dynamic nature of K18 versus K8 phosphorylation can also be evident after treatment of cells with phosphatase inhibitors as noted in okadaic acid-treated human HT29 cells (Ku and Omary, 1994b) or in microcystin-LR-treated rat hepatocytes (Toivola *et al.*, 1997), which result in dramatic and preferential hyperphosphorylation of K18 as compared with K8. Furthermore, drug-induced hepatotoxicity (Ku *et al.*, 1996b) and mitosis (Liao and Omary, 1996) result in preferential K18 hyperphosphorylation. Variability in turnover at specific phosphorylation sites also occurs, but is dependent on the state of the cells analyzed. For example, turnover of phospho-Ser52 K18 during interphase is faster than that of phospho-Ser33 K18, while phosphorylation of Ser33 during mitosis increases to a greater extent as compared with Ser52 (Ku *et al.*, 1998).

2.4. Significance of Keratin Phosphorylation

Several *in vivo* K8 and K18 phosphorylation sites have been characterized (Figure 1), and specific functional roles and/or correlations with unique

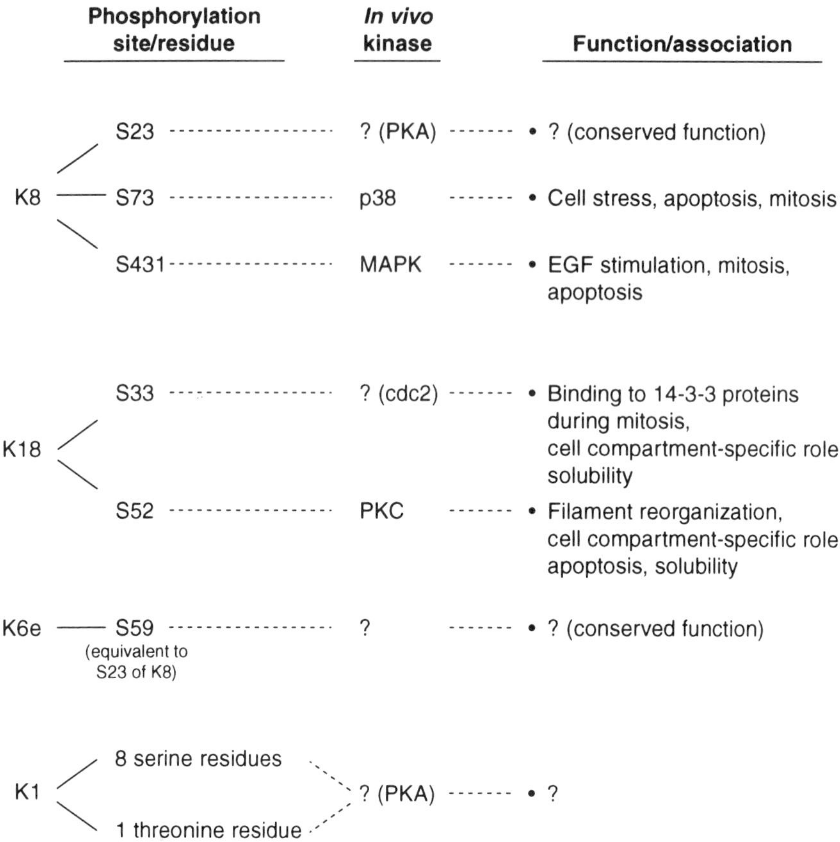

FIGURE 3. Kinases and functions/associations of keratin *in vivo* phosphorylation sites. The *in vivo* phosphorylation sites (S = serine) of the listed keratins are shown together with what is known regarding their function and kinases. The kinases indicated in parentheses are putative, and their assignment is based on *in vitro* phosphorylation data or consensus sequence information. Evidence for the remaining kinases is as follows: p38 stress-activated protein kinase (unpublished data), mitogen-activated protein kinase (MAPK) (Ku and Omary, 1997), and PKC (Omary *et al.*, 1992; Ku and Omary, 1994b). The assignment of a "conserved function" is a hypothesis that is based on the conserved nature of the indicated sites (S23 in K8 and S59 in K6e) in all type II keratins (i.e., K1–K8) (Ku and Omary, 1997).

cellular events have been defined for these sites (Figure 3 and Table III). For example, K8 Ser23, which is a major physiologic phosphorylation site, is highly conserved among all type II keratins, and its equivalent (i.e., Ser59 of K6e) is also phosphorylated (Ku and Omary, 1997). The *in vivo* kinase that phosphorylates K8 Ser23 (or K6e Ser59) is not known, but this site can be phosphorylated *in vitro* by PKA (Ando *et al.*, 1996). The conserved

Table III
Functions or Associated Cellular Events of Keratin Phosphorylation Sites

Function or associated cell event	Keratins and sites	Evidence	Reference
Conserved (to be determined) type II keratin function	K8 Ser23 K6e Ser59	Sequence conservation Identified phosphorylation in more than one keratin	Ku and Omary (1997)
Filament reorganization and solubility	K18 Ser52	Mutational analysis in cultured cells Staining of cultured cells with phosphoepitope-specific antibodies	Ku and Omary (1994b), Liao *et al.* (1995a)
Growth factor stimulation/mitosis	K8 Ser431, 73 K18 Ser52, 33	Analysis in cultured cells Partial hepatectomy in mice	Liao *et al.* (1995a, 1997), Ku *et al.* (1997), Ku *et al.* (1998)
Interaction with 14-3-3 proteins	K18 Ser33	Mutational analysis in cultured cells	Liao and Omary (1996), Ku *et al.* (1998)
Cell compartment specific	K18 Ser52, 33	Staining of tissues using phosphoepitope-specific antibodies	Liao *et al.* (1995a), Ku *et al.* (1998)
Protection against cell stress	Generalized increase in phosphorylation K8 Ser73	Association of K8/18 hyperphosphorylation with a variety of stress models in cultured cells and mice	Liao *et al.* (1995b, 1997), Ku *et al.* (1996b)
Apoptosis	K8 Ser73, 431 K18 Ser52	Induction of apoptosis in cultured cells	Liao *et al.* (1997), Ku *et al.* (1997), Caulin *et al.* (1997)

phosphorylation of K8 Ser23 in more than one keratin suggests a conserved function for this phosphorylation site (Ku *et al.*, 1996a).

K8/18 phosphorylation can also play an important role in filament reorganization and solubility (see Section 5 page 124). For example, K18 Ser52 phosphorylation increases, from a basal level during interphase, nearly threefold during mitosis in association with filament reorganization. Also, phosphoepitope-specific antibodies that recognize phospho-Ser52 bind preferentially to the reorganized filaments rather than to the normal-appearing filaments (Liao *et al.*, 1995a). Transfection of a Ser52 → Ala K18 mutant interfered with the ability of K8/18 filaments to reorganize upon mitotic arrest or treatment with okadaic acid (Ku and Omary, 1994b). K18 Ser52 is a likely PKCε-related kinase phosphorylation site *in vivo*, based on its increased phosphorylation in a calcium-independent fashion upon PKC activation (Chou and Omary, 1991), K8/18 colocalization with PKCε by immune staining, significant *in vitro* phosphorylation by this kinase, association of K8/18 immunoprecipitates with a PKCε-related kinase, and inhibition of its phosphorylation in cells and *in vitro* by a PKCε-specific pseudosubstrate (Omary *et al.*, 1992). The role of PKC in keratin phosphorylation was also implicated in several other cell systems including HeLa (Tolle *et al.*, 1987), mouse hepatocytes (Cadrin *et al.*, 1992), and thyroid follicle cells (Deery, 1993) after culturing in the presence of phorbol ester.

Another K18 physiologic phosphorylation site is Ser33, which regulates the binding of the 14-3-3 family of proteins to K18 (Ku *et al.*, 1998). The 14-3-3 protein family was first identified in rabbit brain (Moore and Perez, 1967), and consists of several members (α, β, ζ, τ, ε, δ, σ, η) that are highly conserved between plants and mammals (Aitken, 1995). The α and δ isoforms are phosphorylated forms of the β and ζ isoforms (Aitken *et al.*, 1995). The 14-3-3 proteins exist as homo- and heterodimers (Jones *et al.*, 1995), and have the unique feature of binding to a wide range of phosphoproteins including kinases, phosphatases, oncogene proteins, and other enzymes such as tryptophan hydroxylase (Aitken, 1995, 1996). Phosphorylation of the 14-3-3-binding proteins appears to be required for binding to 14-3-3 as determined for tryptophan hydroxylase (Furukawa *et al.*, 1993), Raf and Bcr kinases (Michaud *et al.*, 1995; Muslin *et al.*, 1996), K18 (Liao and Omary, 1996), and the apoptosis-promoting protein BAD (Zha *et al.*, 1996). One role for keratin-14-3-3 binding, based on *in vitro* studies, is to sequester keratins from the cytoskeletal to the soluble compartments during mitosis and upon other stimuli (Figure 4A and Section 5.2). It remains to be investigated if keratin–14-3-3 binding serves to displace other 14-3-3-associated proteins or if it acts as a "surface" or "adapter" to bring together proteins that bind either keratins and/or 14-3-3.

Several hormones and growth factors induce a rapid (within 30 min) increase in keratin phosphorylation. Examples include treatment of rat hepatocytes (Baribault *et al.*, 1989) and human colonic HT29 cells (Ku and Omary, 1997) with epidermal growth factor, rat vaginal cells with estradiol-17β (Gupta *et al.*, 1990), and human amnionic WISH cells with pro-urokinase (Busso *et al.*, 1994). The rapid phosphorylation response suggests a potential role in downstream events. The increase in K8/18 phosphorylation during mitosis and growth factor stimulation (Figure 3 and Table III) result in a general phosphorylation increase in several K8 and K18 sites. The interplay among these different hyperphosphorylated sites is not entirely clear, but several already characterized distinct phosphorylation roles for K18 Ser52 (i.e., filament reorganization) and K18 Ser33 (i.e., 14-3-3 binding) indicate that what appears to be a "global" increase in K8/18 phosphorylation is likely to reflect unique functions for each of the involved sites. In addition, filament reorganization is likely to represent distinct rearrangements that depend on the stimulus (e.g., epidermal growth factor or other mitogens) or event (e.g., mitosis or apoptosis). However, the functional significance of filament reorganization in various contexts remains to be defined.

Several common features of the K18 Ser33 and Ser52 phosphorylation sites deserve mention. For example, both sites are phosphorylated during interphase and both sites increase their phosphorylation during mitosis (Liao *et al.*, 1995a; Ku *et al.*, 1998). However, K18 Ser33 phosphorylation is required for 14-3-3–K18 binding during mitosis based on the dramatic increase in K18 phosphorylation (Liao and Omary, 1996), particularly of Ser33 phosphorylation, and the finding that a K18 Ser33 → Ala mutant completely inhibits 14-3-3 binding, while a Ser52 → Ala K18 or a K18 glycosylation mutant (Ser 29, 30, 48 → Ala) does not alter the binding (Ku *et al.*, 1998). This, coupled with the basal K18 Ser33 phosphorylation during interphase in the absence of detectable 14-3-3 binding, indicates that K18 Ser33 phosphorylation is necessary, but not sufficient for 14-3-3 binding, and that a second event is needed for this binding to occur *in vivo* (Ku *et al.*, 1998). The nature of this second event remains to be investigated. Another similarity between phospho-Ser52 and phospho-Ser33 K18 is that they are both preferentially, but not exclusively, localized in the apical domain of hepatocytes, which suggests a compartment-specific functional role. This colocalization is based on double staining with K18 Ser52 or Ser33 antiphosphoepitope antibodies, and by double staining of anti-phospho-Ser33 or phospho-Ser52 K18 with antiaminooligopeptidase (an apical-specific marker) or the CE9 antigen (a basolateral-specific marker; Bartles *et al.*, 1985) (Ku *et al.*, 1998). The assignment of K18 phospho-Ser52 to the liver apical compartment differs from our previous assignment to the basolateral compartment (Liao *et al.*, 1995a), which was based on single-staining morphology without performing double staining. Despite similar colocalization of the phospho-Ser33 and Ser52 K18 species, as determined by immunofluorescence staining, biochemical fractionation, then characterization of the cytosolic/membrane and cytoskeletal liver compartments, indicated that mouse liver K18 molecules are preferentially phosphorylated on either Ser52 or Ser33, with minimal phosphorylation on both sites of the same molecules. This observation lends further support to the functional segregation of these two K18 phosphorylation sites.

Another role for keratin phosphorylation may involve protection against cell stress (Table III). This hypothesis is supported by the association of K8/18 hyperphosphorylation with a variety of stresses, including heat and rotavirus infection of cultured cells (Liao *et al.*, 1995b) and hepatotoxicity in mice (Ku *et al.*, 1996b). Stress-induced hyperphosphorylation was also noted in other IF proteins including neurofilaments (Giasson and Mushynski, 1996) and vimentin (Cheng and Lai, 1994), and as such appears to be a general phenomenon. The stress-associated hyperphosphorylation of K8/18 is dramatic, occurs at a relatively early stage when cells are still able to recover from the stress-induced damage, and involves the formation of a distinct K8 species (termed HK8) that migrates on one-dimensional gels slower than the remaining phosphorylated and nonphosphorylated K8 (Liao *et al.*, 1995b). HK8 corresponds to K8 Ser73 phosphorylation, and represents a marker that accumulates during cell stress and apoptosis, in contrast with the low HK8 levels that are found during mitosis (Liao *et al.*, 1997). Of note, transgenic mice that express Arg89 → Cys human K18 are far more susceptible to drug-induced liver injury as compared to mice that express normal human K18, and are significantly less able to hyperphosphorylate their keratins (Ku *et al.*, 1996b). This is despite the fact that mutant K8/18 (i.e., wild-type K8 and Arg89 mutant K18) are basally more phosphorylated than wild-type K8/18 in transfected cells and in transgenic mice (Ku *et al.*, 1995). The reason for this basal hyperphosphorylation is unclear, but it may reflect the stress induced by the increase in hepatocyte fragility, or the fact that the keratins in the mutation-induced rearranged filaments have altered accessibility to relevant kinases and/or phosphatases. The mechanism of any potential stress-protective effect of IF protein hyperphosphorylation remains to be determined. Possibilities include storage of phosphates as reserve and/or to slow down normal energy-requiring cell processes (Liao *et al.*, 1995b), acting as a "phosphate sink" to protect cells from untoward effects of activated kinases (Lai *et al.*, 1993), or facilitating other putative protective signals that remain to be defined (Ku *et al.*, 1996a).

Keratin hyperphosphorylation was also recently noted in association with apoptosis in cultured cells. For example, increased phosphorylation of K8 Ser73/Ser431, and K18 Ser52, but not Ser33, occurs during apoptosis (Liao *et al.*, 1997; Ku *et al.*, 1997). The lack of increased K18 Ser33 phosphorylation during apoptosis correlates with absence of 14-3-3 binding (Figure 4A). The apoptosis-induced increase in K8/18 phosphorylation occurs much earlier (within 60 min of exposure to the apoptosis-inducing agent) than the observed K18 degradation into stable fragments after ~12 h (see Section 4.2). The significance of this hyperphosphorylation remains to be determined. To that end, hyperphosphorylated keratins that are purified from apoptotic cells are no more sensitive to *in vitro* proteolysis by the caspase CPP32 or to digestion by lysates that were obtained from apoptotic cells than are basally phosphorylated keratins (Ku *et al.*, 1997). This, coupled with the distinct time separation of K18 hyperphosphorylation from fragmentation, suggests that the increase in K8/18 phosphorylation does not have a direct effect on the caspase-mediated proteolysis.

3. KERATIN GLYCOSYLATION

3.1. Introduction

Glycosylation of keratins was first implicated in human K13 that was isolated from cultured newborn human foreskin keratinocytes, based on its metabolic labeling with ^{3}H-glucosamine, but not with ^{3}H-mannose or ^{3}H-fucose (Roberts and Jenner, 1983; Roberts and Brunt, 1986). The incorporated sugar was assumed to be O-linked (i.e., to Ser/Thr), rather than asparagine N-linked, since it was not altered by tunicamycin pretreatment of the cells (Roberts and Jenner, 1983). A similar keratin labeling by ^{3}H-glucosamine was observed in cultured pig epidermis (King, 1986), with subsequent conclusive identification of the incorporated sugar into K13 as single O-linked N-acetylglucosamines (GlcNAc) (King and Hounsell, 1989). This type of cytoplasmic protein glycosylation was first identified in lymphocyte cytoplasmic proteins (Torres and Hart, 1984), and subsequently shown to be a very common modification that is found in a growing list of broad-range proteins. These O-GlcNAcylated proteins include transcription factors, kinases and phosphatases, nuclear pore proteins, oncogene and tumor suppressor products, and viral and cytoskeletal proteins (reviewed by Hart *et al.*, 1996). The other characterized IF proteins with the O-GlcNAc modification are K8 and K18 (Chou *et al.*, 1992; Chou and Omary, 1993; Ku and Omary, 1995) and neurofilaments (Dong *et al.*, 1993, 1996). Keratin glycosylation (particularly for K18 as compared with

K8) is conserved in mammalian and insect cells (Ku and Omary, 1994a). It is likely that most, if not all IF proteins have this modification. For example, vimentin can be metabolically labeled by ^{3}H-glucosamine, but the precise nature of the attached sugar has not been verified (unpublished observations). Other forms of keratin/IF glycosylation may also occur, although the evidence has been indirect. For example, nuclear lamin binding to concanavalin A (Ferraro *et al.*, 1989) may suggest direct or indirect attachment of mannose, which has not been further confirmed. However, a cautionary note is the potential of immunologic mimicry between GlcNAc and the peptide backbone of IF proteins (Shikhman *et al.*, 1994). Another potential mode of keratin glycosylation is modification by keratan sulfate based on binding of epidermal keratins to antikeratan antibodies and abolition of the binding upon treatment with endo-β-D-galactosidase and keratinase (Schafer and Sorrell, 1993).

3.2. Characterization of Keratin Glycosylation

Several O-GlcNAcylation sites have been identified in IF and non-IF proteins. In keratins, three K18 glycosylation sites have been defined (Ser29, 30, 48; Ku and Omary, 1995). The location of the O-GlcNAc sites in IF proteins so far has been restricted to the head (i.e., K18; Chou and Omary, 1993; Ku and Omary, 1995) and head/tail domains (i.e., neurofilaments; Dong *et al.*, 1993, 1996). No uniform consensus sequence has been associated with this modification, but it involves Ser/Thr residues that are proximal to proline and/or valine, and tends to be present in domains that are rich in Ser/Thr (Ku and Omary, 1995; Hart *et al.*, 1996). One potential consensus sequence that fits several IF and non-IF identified glycosylation sites is the sequence Pro-X_1-X_2-X_3, where one X_{1-3} is a hydrophobic residue and two X_{1-3} correspond to Ser/Thr residues or Ser/Thr and any other amino acid (Ku and Omary, 1995). This motif is found in the head domain of several IF proteins, including desmin, lamins, and K16 (Ku and Omary, 1995), but glycosylation of these IF proteins remains to be established. The stoichiometry of glycosylation varies among the proteins that have this modification. For example, the number of O-GlcNAc's per protein molecule was 0.1 and 0.15 for neurofilament-L and -M, respectively (Dong *et al.*, 1993), 1.5–2 for K8/18 (Chou *et al.*, 1992), 2–10 for nuclear pore proteins (Holt *et al.*, 1987), and 8 for transcription factor Sp1 (Jackson and Tjian, 1988). Recently, a unique O-GlcNAc transferase that is localized in the cytosol and nucleus, and is involved in the above-described glycosylation, was cloned independently by the Hart and Hanover groups (Kreppel *et al.*, 1997; Lubas *et al.*, 1997).

Detection and confirmation of keratin glycosylation is performed using a combination of techniques, including *in vitro* galactosylation, *in vivo* labeling, and lectin binding (Roquemore *et al.*, 1994). *In vitro* galactosylation is carried out using galactosyltransferase, which adds galactose to accessible terminal GlcNAc's, and UDP-^{3}H-galactose. This can be done using immunopurified keratins, and if solubility is a problem as for epidermal keratins, then galactosylation may be performed on protease-generated soluble keratin fragments. Labeling of the protein or peptide suggests the presence of a terminal GlcNAc. This can be further examined using β-elimination, which generates β-D-^{3}H-Gal[1–4]-D-GlcNAc if the terminal GlcNAc is single O-linked rather than a compound sugar. Metabolic labeling of cells is performed using ^{3}H-glucosamine, and the nature of the attached saccharides is assessed after acid hydrolysis of the purified protein. Examination of the *in vitro* and *in vivo* labeled keratins by two-dimensional gel electrophoresis (first dimension: isoelectric focusing; second dimension: SDS–polyacrylamide gel electrophoresis) provides further information regarding the relationship of the phosphorylated (i.e., more-acidic isoforms) and glycosylated species (Liao *et al.*, 1996). The use of lectins such as wheat germ agglutinin can also provide supportive information, although binding of wheat germ agglutinin to terminal O-GlcNAc is variable and appears to depend on the density of the exposed sugars (Roquemore *et al.*, 1994).

3.3. Features of O-GlcNAcylated Proteins Including Keratins

The functional significance of the IF O-GlcNAcylation remains unclear, but several properties of this glycosylation as well as recent findings in other proteins with this modification deserve mention. The first important property of this modification is that it occurs on phosphorylated proteins that are involved in forming multimeric complexes. Presence of O-GlcNAc on Ser/Thr residues, which are also potential phosphorylation sites, raises the attractive possiblity that phosphorylation and the O-GlcNAc modification may regulate each other (Hart *et al.*, 1989). In fact in some cases, phosphorylation and O-GlcNAcylation are almost mutually exclusive in that they occur on different populations of the same protein as found in the 67-kDa initiation factor-2-associated protein (Datta *et al.*, 1989), K18 (Chou *et al.*, 1992), and RNA polymerase II (Kelly *et al.*, 1993). Similarly, phosphorylation and glycosylation of c-Myc occurs at the same site (i.e., Thr58) within its transactivation domain (Chou *et al.*, 1995). However, in the case of K8/18 it appears that phosphorylation and glycosylation are for the most part independent, since there are several conditions that increase both modifications, as found after mitotic arrest (Chou and Omary, 1993), heat stress (Liao *et al.*, 1995b), and K8/18 filament disruption in cultured cells or in transgenic mice that express an Arg89 → Cys K18 mutant (Ku *et al.*, 1995). Furthermore, none of the identified K8 (Ser23, 73, 431) and K18 (Ser33, 52) phosphorylation sites appear to be glycosylated (Ku *et al.*, 1996b, and unpublished observations). A second important property of the O-GlcNAc modification is that it is highly dynamic. For example, turnover of O-GlcNAc in K8/18 and αB-crystallin is far greater than protein turnover (Chou *et al.*, 1992; Roquemore *et al.*, 1996), and O-GlcNAcylation can be induced in lymphocytes upon mitogen stimulation (Kearse and Hart, 1991). Third, O-GlcNAcylation of a protein may, in some cases, be tissue specific. For example, porcine smooth muscle and chicken gizzard talin are modified by O-GlcNAc, while talin from human activated or inactivated platelets is not O-GlcNAcylated (Hagmann *et al.*, 1992).

The increase in K8/18 glycosylation in association with mitotic arrest (Chou and Omary, 1993) does not occur during normal progression of cells through mitosis as assayed after release of an aphidicolin-induced G1/S block or enrichment of G2/M cells using mechanical shaking (Chou and Omary, 1993, 1994). The nocodazole-induced increase in K8/18 glycosylation appears to be unique to keratins. For example, other proteins with the single O-GlcNAc type modification, such as nuclear pore protein p62 and transcription factor Sp1, are not affected (Haltiwanger and Philipsberg, 1997). Furthermore, mitotic arrest does not affect UDP-GlcNAc, O-GlcNAc transferase, or O-GlcNAc-β-N-acetyl-glucosaminidase levels (Haltiwanger and Philipsberg, 1997), although it is not known if the distribution of the transferase is altered. Due to the independent effect of increased keratin phosphorylation, the mitotic arrest-induced increase in K8/18 glycosylation is more likely to be related the filament reorganization that accompanies such arrest, rather than to be a cause of the reorganization.

Although phosphorylation in a site-specific manner is a known contributor to K8/18 filament reorganization (see Section 2), events that result in increased K8/18 phosphorylation and filament reorganization are not uniformaly associated with increased K8/18 glycosylation. For example, rotavirus infection of HT29 cells (Liao *et al.*, 1995b) and feeding mice the hepatotoxin griseofulvin (Ku *et al.*, 1996b) are both associated with increased K8/18 phosphorylation, but with either no change or a decrease in K8/18 glycosylation, respectively. In addition, filament reorganization takes place during mitosis coupled with increased K8/18 phosphorylation and without any change in K8/18 glycosylation (Chou and Omary, 1994). Also, mutation of the K18 glycosylation sites does not have any obvious effect on K8/18 filament assembly as determined by immunofluorescence staining (Ku and Omary, 1995). Therefore, it appears that K8/18 glycosylation is unlikely to have any significant effect on filament assembly or organization.

3.4. Potential Functions to Consider for Keratin Glycosylation

Given our current limited understanding of K8/18 glycosylation and the O-GlcNAcylation of other proteins, what "to-be-tested" roles can we hypothesize for keratin glycosylation? In the case of Sp1, hypoglycosylation as occurs during nutritional depletion correlates with increased degradation by the proteosome pathway (Han and Kudlow, 1997). This raises the possiblity that O-GlcNAc may serve a protective role against degradation. However, this potential will need further testing since nutritional depletion may also be associated with other compounding variables that have not been assessd, including changes in other posttranslational modifications. Alternatively, O-GlcNAc may play a role in subcellular targeting of the protein it modifies or in complex formation with other cellular proteins. For example, O-GlcNAcylated ankyrin_G, which is hypophosphorylated, is preferentially concentrated at the node of Ranvier and not in adjacent areas of myelinated axons (Zhang and Bennett, 1996). In addition, ankyrin_G that is expressed in a glycosylation-competent rabbit reticulocyte lysate has a larger Stokes radius than when expressed in wheat germ lysates which are O-GlcNAc glycosylation-incompetent (Zhang and Bennett, 1996), although the phosphorylation states of ankyrin_G in the two systems was not examined. Therefore, functional candidates for the O-GlcNAcylation of keratins include one or more of the following: protein–protein interaction at the keratin–keratin (i.e., an assembly or filament reorganizaton role) or keratin-associated protein levels, cell- or tissue-specific functions if tissue-specific differences in glycosylation occur in keratins as found for talin, subcellular localization, regulation of protein turnover, and modulation of phosphorylation, which would imply regulation of phosphorylation-specific functions.

4. OTHER POSTTRANSLATIONAL MODIFICATIONS

4.1. Acetylation and Lipid Modifications

In addition to phosphorylation and glycosylation, keratins undergo several other posttranslational modifications (Table II). For example, most if not all of the keratins contain an N-terminal serine that is N-acetylated, which precludes N-terminal sequencing using standard techniques (Steinert and Idler, 1975; Ku *et al.*, 1997). Aside from acetylation, keratins may undergo lipid modification based on metabolic labeling of human foreskin with ^{3}H-retinoic acid (N. Takahashi *et al.*, 1991; Breitman and Takahashi, 1996), and labeling of mouse mammary epithelial cells with ^{3}H-palmitate (Asch *et al.*, 1993). However, technical problems make further validation of these findings difficult due to the extremely low (~1%) overall labeling of the keratins by these two fatty acids. The presumed sites of modification are not known, and the only evidence for covalent attachment is the inability to remove the attached radiolabeled lipid by exhaustive organic solvent extraction. Hence, further analysis will be needed to confirm and characterize these modifications.

4.2. Proteolysis of Keratins

Proteolysis is another keratin posttranslational modification that will likely become "mainstream," particularly as related to apoptosis. One of the earliest descriptions of keratin proteolysis was in mouse and human skin (Bowden *et al.*, 1984). The degradation appears to involve cleavage of 2- to 5-kDa fragments of the upper granular layer keratins, although the sites and significance of proteolysis and the enzymes involved are not known. Keratin degradation and filament collapse was also noted in adenovirus-infected HeLa and 293 cells (Chen *et al.*, 1993; Zhang and Schneider, 1994). In this case, degradation involved K7 and K18, and occured at methionine-73 of K18 which is a consensus sequence of the L3 adenovirus proteinase. Although the site of K7 degradation was not determined, the head and tail domains of K7 each contain one potential L3 cleavage site (Chen *et al.*, 1993). Keratin cleavage occurs at a late stage of adenovirus infection and can be inhibited by 2-aminopurine, which prevents the adenovirus shutoff of host protein synthesis (Zhang and Schneider, 1994). Therefore, it appears that keratin degradation and disassembly and the inability to synthesize new keratins (and other proteins) may play a role in disrupting cell integrity and virus-induced cell lysis.

Another type of recently identified keratin proteolysis was noted in K18 and K19 upon induction of cells to undergo apoptosis (Caulin *et al.*, 1997; Ku *et al.*, 1997). For example, induction of apoptosis in HT29 cells generates two stable and dominant K18 fragments, termed p29 (29kDa) and p23 (23kDa), and variable levels of smaller fragments that have not been characterized (Ku *et al.*, 1997). Similar findings in other K8/18-expressing systems were recently described (Caulin *et al.*, 1997). Proteolysis occurs at the carboxy-terminal side of K18 Asp237, which has the caspase-specific consensus sequence ^{234}Val-Glu-Val-Asp that is within the linker L1-2 region of the keratin backbone. Identical sequences are also found in K19,

K20, and lamin B, and related sequences are found in other IF proteins. Similar K19 fragments are generated during apoptosis, and the cut site occurs at asp237 of K19 (Ku *et al.*, 1997 and unpublished observations). Of note, caspase-specific cleavage of lamins also occurs (Takahashi *et al.*, 1996), and mutation of the relevant asp site in lamin B delayed apoptosis progression (Rao *et al.*, 1996). The caspase-mediated K18 proteolysis can be prevented by using caspase-specific inhibitors (not shown). The p29 and p23 K18 fragments correspond to the N-terminal and C-terminal portions of the K18 molecule, respectively, as determined by microsequencing and immunoblotting with epitope-specific antibodies. An easily detectable portion of these two fragments remains associated with the K8/18 complex that is immunoprecipitated from NP40-solubilized apoptotic cells, while the majority of these fragments are in the cytoskeletal (insoluble) fractions. K8 is highly resistant to caspase-mediated proteolysis, and the post-Emp-solubilized pellet contains very high levels of K8 with minimal intact K18 (Ku *et al.*, 1997).

The significance of K18 and K19 proteolysis during apoptosis remains to be evaluated in terms of whether keratin degradation is obligate, important, or simply a bystander phenomenon. Involvement of two type I keratins (K18 and K19) suggests that this is likely to be a general keratin, if not IF protein, feature and is supported by the presence of similar potential caspase sequence targets in other keratins. It is likely that keratin fragmentation and network collapse would facilitate the progression of apoptosis, but less likely that keratin caspase degradation is an obligate event, since mutation of the lamin caspase-cleavage site delayed, but did not inhibit apoptosis (Rao *et al.*, 1996). The role of keratin fragmentation versus hyperphosphorylation in filament disruption will be an important question to address. As mentioned above, hyperphosphorylation of K8 and K18 occurs early in apoptosis, while proteolysis is a late event which suggests lack of direct relatedness of these two modifications (see also last paragraph of Section 2.4.).

Keratin fragments are also noted in patients with cancer and have been tested in a variety of settings as tumor markers (reviewed by van Dalen, 1996). The earlier descriptions referred to "tissue polypeptide antigen" (Bjorklund, 1980), which was subsequently identified as being related to K8, K18, and K19 (Weber *et al.*, 1984). The sizes of these tumor-associated fragments are similar to some of the apoptosis-generated fragments described above. For example, SDS–PAGE analysis of tissue-polypeptide-specific antigen shows 13-kDa and 22-kDa species and a doublet at 42 kDa, and molecular characterization of the 13-kDa species indicated that it corresponds to residues 284–396 of K18 (Rydlander *et al.*, 1996). The abundance of keratins in cells (e.g., 5% of the total cellular protein of cultured simple epithelial cells is K8/18; Chou *et al.*, 1993) makes them potential useful targets for fragment detection in sera obtained from patients with disease states that involve apoptosis and/or necrosis.

4.3. Keratin Transglutamination

Keratins are also modified by transglutamination under normal physiologic and pathologic conditions (Osborn and Weber, 1989). This modification involves transglutaminases that mediate peptide cross-linking by forming a covalent bond between a glutamine γ- carboxyl group and a lysine ε-amino group. Such cross-links have been identified under physiologic conditions in the normal epidermis (Buxman and Wuepper, 1975; Abernethy *et al.*, 1977; Steinert and Idler, 1979). For example, in the cornified cell envelope several such cross-links have been identified including K1–loricrin, K2e–loricrin, and K10–loricrin cross-links (Steinert and Marekov, 1995). Interestingly, a transglutaminase-related antigen was found to be associated with vimentin in cultured mouse dermal fibroblasts (Trejo-Skalli *et al.*, 1995).

In contrast with epidermal keratins, simple epithelial keratins do not appear to contain ε-(γ-glutamyl) lysine cross-links under normal physiologic conditions. However, these cross-links are found in livers of mice fed the hepatotoxin griseofulvin, and generated upon accumulation a cytoplasmic hyalin termed "Mallory bodies" (Zatloukal *et al.*, 1992). The griseofulvin mouse Mallory body model (Denk *et al.*, 1975) mimics a human histologic marker that is found in a number of liver diseases, particulary alcoholic liver disease (Mallory, 1911). The trigger for the formation of these cross-links and their pathogenic role are not known, and it is unclear if the observed hyperphosphorylation of Mallory body keratins contributes to their formation (Salmhofer *et al.*, 1994; Cadrin *et al.*, 1995).

5. KERATIN SOLUBILITY

5.1. Definition and Nature of the Soluble Keratin Fraction

The soluble keratin pool may be defined as the keratin fraction that remains in the supernatant of cell or tissue homogenates that have been subjected to high-speed centrifugation. Strictly speaking, the homogenization should be performed in an aqueous, nondetergent-containing buffer. If written a few years ago, this section would have been entitled "keratin insolubility," since the dogma had been that keratins and other IF proteins are essentially insoluble polymers. The basis of this dogma is that epidermal keratins are indeed <1% soluble in aqueous or nonionic detergents, and measurement of the soluble fraction of other IF proteins such as vimentin afforded a soluble pool of 0.18–0.39% (Soellner *et al.*, 1985). The relative insolubility of keratins led to what is now a routine protocol and a very useful technique of partially purifying keratins, by sequential extraction with a buffer that contains Triton X-100, then with a high-salt buffer, thereby leaving a highly enriched keratin fraction pellet (Achtstaetter *et al.*, 1986). This keratin fraction can be solubilized in buffers that contain 2% SDS or ≥6 M urea. However, as keratins differ in their tissue distribution, they also differ significantly in their solubility. For example, K5/14 can be solubilized in 2 M urea, whereas K1/10 require 4–6 M urea-containing buffers for solubilization (Eichner and Kahn, 1990). In addition, several studies hinted that K8/18 may be more soluble than had been assumed for keratins in general. These studies include the identification of a soluble K8/18 fraction in rat liver (Sahyoun *et al.*, 1982) and in the human PLC hepatoma cell line (Franke *et al.*, 1987), and the estimation that 1–10% of *Xenopus laevis* oocyte and egg keratins are soluble (Gall and Karsenti, 1987). Quantitation of the soluble K8/18 fraction in cultured human colonic cells that are asynchronously growing and near confluency (i.e., primarily G0/G1 stage of the cell cycle) showed that 5% of the keratin pool was soluble (Figure 4; Chou *et al.*, 1993). Detailed studies of the soluble pool of *Xenopus* keratins also showed that 5% of the keratin pool is soluble (Bachant and Klymkowsky, 1996). The cell-cycle stage and the particular cell stimulus play very important roles in dictating the levels of the highly dynamic soluble keratin fraction (Figure 4).

The soluble keratins that are isolated from cultured human HT29 cells (Chou *et al.*, 1993), cultured rat liver cells (Quinlan *et al.*, 1985; Franke *et al.*, 1987), and *Xenopus* oocytes (Klymkowsky *et al.*, 1991) are tetrameric (i.e., two K8 and two K18 molecules) based on cross-linking and velocity sedimentation analysis. However, it appears that the molecular form of the soluble keratin pool depends on the method of isolation. For example, velocity gradient sedimentation yields 7 S tetrameric species, while size-exclusion chromatography yields larger complexes (~750 kDa) that may represent octamers or tetramers that are associated with other proteins (Bachant and Klymkowsky, 1996). This larger species is unstable and incubation at 4°C for 16 hr (i.e., approximate duration of sedimentation) causes most of it to revert back to a smaller (~150 kDa) form after reanalysis by size-exclusion chromatography.

Although the soluble keratin pool is a significant biosynthetic intermediate (Liao *et al.*, 1995b; Bachant and Klymkowsky, 1996), it does not simply represent a posttranslational intermediate of the cytoskeletal pool. This is supported by the low turnover of keratins (Denk *et al.*, 1987; Chou

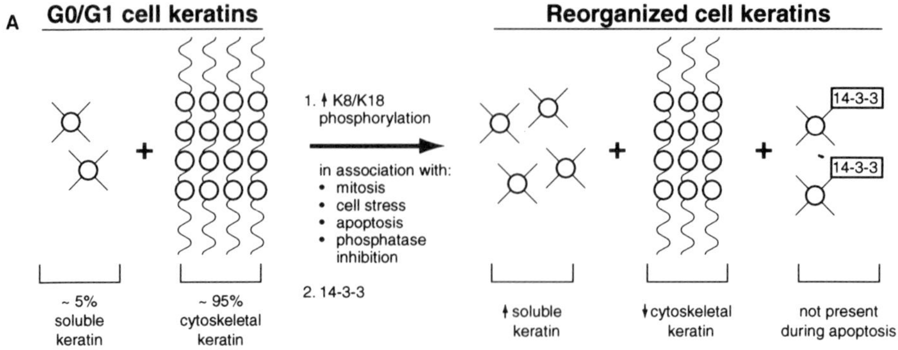

B

Cell fraction:	cytosolic	membrane	cytoskeletal-insoluble	
% of total keratin:	(5%)	(10%)	(45%)	(40%)
Solubilization:	No detergent	Nonidet P40	Empigen	Remaining
Distribution				
overall keratin phosphorylation:	++++	+++	+	+
14-3-3-keratin binding:	++	++	–	–
phospho-ser33 K18:	+++	+++	+/–	+/–
phospho-ser52 K18	+++	+++	+	–
phospho-ser73 K8:	+++	+++	+	+++
apoptotic K18 fragments:	+	+	++	+++

FIGURE 4. Modulation of keratin solubility and subcellular distribution of various K8/18 species. (A) Modulation of keratin solubility depending on the cell physiologic state: During the G0/G1 stage of the cell cycle, ~5% of the total cellular keratin in cultured colonic HT29 is soluble after cell disruption in the absence of detergent followed by high-speed ultracentrifugation. This soluble fraction consists of K8/18 tetramers (i.e., two K18 and two K8 molecules; Chou *et al.*, 1993). Keratin solubility increases at least twofold (e.g., during mitosis) or considerably more (e.g., after phosphatase inhibition) in association with several stimuli/treatments. In all cases, except apoptosis, the increase in K8/18 phosphorylation is associated with binding of the soluble fraction to 14-3-3 proteins. Binding of 14-3-3 and keratin phosphorylation result in sequestration of additional soluble K8/18. (B) Distribution of overall and site-specific K8/18 phosphorylated species, 14-3-3-bound K8/18, and K18 apoptotic fragments in different subcellular compartments. The indicated subcellular compartments are generated after cell disruption in a detergent-free buffer followed by centrifugation, then sequential solubilization with the nonionic detergent Nonidet P40 (NP40), removal of the NP40 by centrifugation, solubilization of the pellet with the zwitterionic detergent Empigen BB (Emp), removal of the Emp fraction after spinning, then solubilization of the residual pellet using sodium dodecyl sulfate sample buffer (Liao *et al.*, 1996). The NP40 fraction contains primarily membrane solubilized components plus a fraction of "loosely" attached filamentous K8/18, while the Emp fraction contains a significant portion of the usually "insoluble" filamentous component (Lowthert *et al.*, 1995) plus any NP40-resistant, but Emp-solubilized membrane fraction. The numbers in parentheses correspond to the approximate portion of the total cellular keratin that is in each indicated compartment. These numbers can vary significantly, with an increase in the cytosolic/membrane fractions and a corresponding decrease in the cytoskeletal-insoluble fractions (Chou *et al.*, 1993; Lowthert *et al.*, 1995) in association with the conditions outlined in panel A.

et al., 1992), and yet the relatively high soluble fraction. In addition, the exchange between the soluble and cytoskeletal (filamentous) keratin compartments occurs within minutes (Miller *et al.*, 1993). Therefore, the equilibrium between the soluble and cytoskeletal pools is highly dynamic. Also, a 90-fold increase in the biosynthetic rate of oocyte keratins that was induced by microinjection of keratin types I and II mRNA did not significantly alter the chase kinetics or fraction of the soluble keratin pool (Bachant and Klymkowsky, 1996).

5.2. Mechanisms of Generating the Soluble Keratin Fraction

The mechanisms that contribute to generating and modulating the soluble keratin fraction are partially understood. Two factors that clearly play a role in promoting keratin solubility are phosphorylation and association with an IF-associated protein (IFAP). In the case of phosphorylation, site-specific and a generalized hyperphosphorylation both contribute to the observed increase in solubility. With regard to site-specific phosphorylation, K18 Ser33 hyperphosphorylation, which occurs in association with mitosis, cell stress, or treatment of cells with phosphatase inhibitors, results in binding with 14-3-3 proteins and sequestration of additional soluble K8/18 (Liao and Omary, 1996). With regard to the role of IFAPs in regulating keratin solubility, keratin–IFAP interaction may occur in a phosphorylation-dependent or -independent manner (Ku *et al.*, 1996a). To that end, there are several reports of IFAPs that "solubilize" IF proteins independent of phosphorylation. For example, α-crystallin increases the solubility of vimentin and glial fibrillary acidic protein *in vitro* (Nicholl and Quinlan, 1994), and S100 proteins bind and sequester desmin (Garbuglia *et al.*, 1996) and glial fibrillary acidic protein (Bianchi *et al.*, 1993) into a soluble pool in vitro. However, there are no examples to date of a keratin solubility cofactor that binds in a phosphorylation-independent fashion. Hence although hsp70 associates with K8/18, this association does not appear to be phosphorylation dependent and does not sequester keratins into the soluble pool *in vitro* (Liao *et al.*, 1995c).

In contrast to hsp70, 14-3-3 proteins act as a solubility cofactor in a phosphorylation-dependent manner. This was demonstrated by *in vitro* reconstitution studies which showed that 14-3-3 bound to the NP40-solubilized fraction of K8/18 that was isolated from mitotic cells, but not to the keratins that were isolated from the remaining Emp fraction, nor to keratins that were isolated from NP40-solubilized nonmitotic cells (Liao and Omary, 1996). This phosphokeratin-dependent binding results in a significant increase in the soluble keratin fraction and in disassembly of the keratin filament network *in vitro* (Liao and Omary, 1996). It is not known if 14-3-3 proteins serve a similar solubility-cofactor role with other keratins, but the sequence context of K18 phospho-Ser33 (i.e., the 14-3-3-binding site of K18) has similar, but not identical features with other keratins (not shown). Additional potential keratin posttranslational modifications, such as the *in vitro* ADP-ribosylation of desmin and its effect on filament assembly (Zhou *et al.*, 1996), may also contribute to the solubility properties of keratins. The possible role of keratin-associated proteins or modifications other than phosphorylation as solubility cofactors may be highlighted by *Xenopus* oocyte keratins. For example, *Xenopus* vegetal and animal hemisphere keratins show extended and fragmented filaments, respectively, yet there no evidence that their overall phosphorylation states are different (reviewed by Klymkowsky, 1995). However, site-specific phosphorylation could play a significant role in modulating the soluble keratin fraction in *Xenopus* oocytes.

There is compelling evidence that keratin phosphorylation plays a crucial role in regulating keratin solubility. First, many *in vitro* studies of all IF proteins that have been examined (reviewed by Inagaki *et al.*, 1996) showed that *in vitro* phosphorylation with any one of several kinases results in an increase in the soluble IF fraction (including keratins; Yano *et al.*, 1991). Second, *in vitro* filament assembly of *in vivo* phosphorylated keratins that were purified from rotavirus-infected or heat-stressed cells (Liao *et al.*, 1995b), or from mitotic cells (Liao and Omary, 1996) (i.e., hyperphosphorylation states), generated soluble keratin fractions that were significantly higher than keratins that were purified from cells grown under basal conditions. Third, several hyperphosphorylation states in cultured cells (e.g., mitosis, mitotic arrest, stress conditions such as heat and virus infection) are all associated with increased keratin phosphorylation (Chou *et al.*, 1993; Chou and Omary, 1993, 1994; Liao *et al.*, 1995c). In addition, treatment of cultured mouse keratinocytes (Kasahara *et al.*, 1993) and rat hepatocytes (Toivola *et al.*, 1997) with phosphatase inhibitors resulted in a dramatic increase in keratin solubility and phosphorylation. Fourth, simple epithelial cells exhibit a keratin phosphorylation gradient that parallels keratin solubility (see Section 5.3). However, it is important to note that keratin phosphorylation per se does not necessarily always imply increased solubility. For example, keratin species that are phosphorylated on K8 Ser73 show a significant segregation with the insoluble keratin fraction (Figure 4B). Hence, hyperphosphorylation can also correlate with protein insolubility as noted for the hyperphosphorylated tau protein that is found in paired helical filaments of Alzheimer disease (reviewed by Billingsley and Kincaid, 1997).

The parallel increase in keratin solubility with generalized hyperphosphorylation of several K8 and K18 sites, in association with mitosis and cell stress, raises the question of the relative contribution to keratin solubility of single versus multiple phosphorylation sites. It may be more biologically relevant to consider the increase in keratin solubility as a collective readout of multiple intracellular events, with each event being modulated by one or more keratin phosphorylation site. If so, then keratin (or other IF protein) solubility becomes simply a biophysical parameter that is reflected in the solubility equilibrium constant of a given keratin pair depending on its phosphorylation state. This concept does not take into account the potential role of keratin-associated proteins and other keratin posttranslational modifications.

5.3. Simple Epithelial Cells Have Parallel Keratin Solubility and Phosphorylation Gradients

The polar nature of phosphate groups and the need to have phosphorylation sites be exposed to cytosolic kinases raise the reasonable prediction that phosphoproteins that have soluble and insoluble fractions (e.g., IF proteins) are likely to have a higher specific activity of the phosphorylated species in the soluble pool. Interestingly, this did not appear to be the case for vimentin (Soellner *et al.*, 1985) or in our initial studies with K8/18 (Chou *et al.*, 1993). However, subsequent studies with K8/18 in cultured cells and in intact tissues showed that the original prediction is for the most part true and that one important factor that resulted in underestimating the phosphorylation density of the soluble keratin fraction was the lack of potent phosphatase inhibition (i.e., okadaic acid) in our isolation buffers. Hence, two-dimensional gel analysis of keratins that were immunopurified from cytosolic, NP40, Emp, and the remaining pellet fractions of $^{32}PO_4$-labeled cells showed a gradual decrease in the level of keratin phosphorylation (Liao and Omary, 1996). In addition, immunoblotting of NP40 versus Emp fractions that were obtained from cultured colonic human cells and from intact mouse liver, with several anti-keratin phosphoepitope-specific antibodies showed preferential distribution of most of the phosphorylated species in the cytosolic/NP40 fractions (Ku *et al.*, 1998; Figure 4B). The functional significance of such a gradient is likely reflected in the function of the individual phosphokeratin species. Of note, a similar underestimate of tau protein phosphorylation was also observed if tau is isolated in the absence of okadaic acid (Mawal-Dewan *et al.*, 1994). This also underscores the importance of cellular phosphatases in maintaining the desired degree of cell protein phosphorylation. It remains to be determined if this gradient applies to other keratin and nonkeratin IF proteins, and if cell-specific differences occur.

6. CONCLUDING REMARKS

Significant and exciting gains have been made in linking human keratin mutations to a variety of diseases that reflect the tissue-specific expression of the various keratins. It is likely that the biochemistry and cell biology of these proteins will catch up with the molecular medicine gains that have been made. To that end, systematic and detailed biochemical and cell biological studies dealing with their posttranslational modifications will encompass one needed approach to better our understanding of the regulation and function of these proteins. Another emerging approach is to identify and characterize keratin-associated proteins, and even for this approach posttranslational modifications are likely to be important participant regulatory players. A common function that has arisen from the human keratin diseases and the animal disease models is a role in maintaining the mechanical integrity of cells in several tissues (reviewed by Steinert and Bale, 1993; McLean and Lane, 1995; Fuchs, 1996; Omary and Ku, 1997). In addition, the study of keratin posttranslational modifications, particularly phosphorylation, is beginning to unfold other related and unrelated functions, such as protection during cell stress and a possible role in cell signaling (Ku *et al.*, 1996a). The use of animal models and tools such as phosphoepitope-specific antibodies should provide important adjuncts to other biochemical and molecular methods for characterizing the *in vivo* dynamics, functions, and complexities of keratin and other IF posttranslational modifications. This is particularly relevant to IF proteins, whereby unraveling some of their functions may be feasible at the tissue rather than single-cell levels, as was the case in demonstrating the importance of these proteins in maintaining cell integrity in the context of a tissue in the human and animal-model keratin diseases.

The importance and concurrent complexity of keratin (and likely other IF protein) phosphorylation is exemplified by using a transgenic animal model that expresses a human K18 with a Ser52 → Ala mutation (N.-O. Ku and M. B. Omary, unpublished observations). As expected, this mutation obliterates K18 phosphorylation at that site. However, and unexpectedly, the level of K18 phosphorylation did not appear to change dramatically as determined by two-dimensional gel analysis (not shown), which suggests that compensatory phosphorylation at other sites has taken place. This raises the question of whether there is functional redundancy of different keratin phosphorylation sites, and if cells need to maintain a certain level of keratin phosphorylation to optimally sustain their normal physiology. Further analysis of this transgenic animal model and analyses of similar models that represent other posttranslational modifications should help clarify the biological significance of these modifications and the keratin functions that underlie them.

In studying the biology of keratin posttranslational modifications, K8/18 have been highly useful models for investigation, primarily because of their relative solubility as compared with other keratins. Another helpful aspect has been the availability of many tissue-cultured cell lines that express K8 and K18, and the abundance and relative accessibility of the organs that express them. This is making the difficult task of studying keratin function and regulation more attainable, and allowing for the generation and design of specific reagents (e.g., phosphoepitope-specific antibodies) and model systems for study (e.g., tissue culture mitosis/stress/apoptosis models, modification-targeted transgenic mice).

7. SUMMARY

The gains that have been made in characterizing some of the keratin posttranslational modifications have helped answer some questions regarding these modifications and have generated an information base for asking additional refined questions in furture studies. Highlights of where we believe we currently stand with regard to keratin posttranslational modifications are as follows:

1. Keratin glycosylation, via O-GlcNAc, is a dynamic modification that has been conclusively identified in K13, K8, and K18. Three serine glycosylation sites in the head domain of K18 have been identified, and it is possible that all keratins are glycosylated. The function of this modification remains to be defined, but is likely to be different from phosphorylation, since the two modifications are generally segregated on different molecules and several examples exist whereby both modifications increase simultaneously.

2. Keratin phosphorylation occurs within the tail and/or head domains of all keratins that have been examined. Several serine phosphorylation sites and some of the relevant kinases have been characterized in K8, K6, and K18, and serine/threonine sites have been identified in K1. Functions of keratin phosphorylation that have significant experimental support include a role in filament solubility and reorganization and a role in regulating keratin binding with other cytoplasmic proteins. The significance of filament reorganization and increased solubility under a variety of physiologic conditions such as mitosis and cell stress are important areas of future and ongoing investigation. Other associations with keratin phosphorylation include protection against cell stress, cell signaling, apoptosis, and

cell compartment-specific roles. At this stage, however, it is not known if these associations play direct or indirect roles.

3. Keratin transglutamination occurs in epidermal and simple epithelial keratins under physiologic and pathologic states, respectively. In the physiological context, the role of this modification is clear in terms of providing a compact protective structure, while in the pathologic context of liver disease the role remains ambiguous.

4. Proteolysis of K18 and K19 by caspases occurs during apoptosis, and generates stable keratin fragments that are highly enriched within the cytoskeletal compartment. Proteolysis of the type II keratins appears to be spared for reasons that remain to be defined. It is likely that this apoptosis-associated degradation involves all type I keratins. Keratin fragments are also noted in sera of patients in association with a variety of epithelial tumors. If a signal does exist for the apoptosis-associated fragmentation, aside from caspase activation, then it appears that the overall increase in keratin phosphorylation during apoptosis does not account for this signal.

5. Keratins undergo several other posttranslational modifications including disulfide bond formation (not found in K8/18 due to lack of cysteines) and acetylation of their N-terminal serines. Modification by lipids is also possible, but this modification requires further confirmation.

6. Keratin solublility is highly dynamic and varies profoundly depending on the keratin pair and the physiologic state of the cell. Within the keratin family, simple epithelial keratins are among the most soluble (~5% of K8/18 is soluble at basal conditions). Phosphorylation plays an important role in modulating keratin solubility, and distinct differences occur in site-specific phosphorylation depending on the soluble versus cytoskeletal partitioning of the keratin. Keratin solubility (at least for K8/18) also appears to be regulated by 14-3-3 proteins via K18 Ser33 phosphorylation.

ACKNOWLEDGMENTS. We are very grateful to Kris Morrow for preparing the figures and Ruthie Stevens for assistance with the word processing. Our work is supported by a Veterans Affairs Merit and Career Development Awards and NIH Grants DK47918 and DK52951.

8. REFERENCES

Abernethy, J. L., Hill, R. L., and Goldsmith, S. A., 1977, ε-(γ-glutamyl) lysine crosslinks in human stratum corneum, *J. Biol. Chem.* **252:**1837–1839.

Achtstaetter, T., Hatzfeld, M., Quinlan, R. A., Parmelee, D. C., and Franke, W. W., 1986, Separation of cytokeratin polypeptides by gel electrophoretic and chromatographic techniques and their identification by immunoblotting, *Meth. Enzymol.* **134:**355–371.

Aitken, A., 1995, 14-3-3 proteins on the MAP, *Trends Biochem. Sci.* **20:**95–97.

Aitken, A., 1996, 14-3-3 and its possible role in co-ordinating multiple signalling pathway, *Trends Cell Biol.* **6:**341–347.

Aitken, A., Howell, S., Jones, D., Madrazos, J., and Patel, Y., 1995, 14-3-3 α and δ are the phosphorylated forms of Raf-activating 14-3-3 β and ζ, *J. Biol. Chem.* **270:**5706–5709.

Ando, S., Tokui, T., Yano, T., and Inagaki, M., 1996, Keratin 8 phosphorylation *in vitro* by cAMP-dependent protein kinase occurs within the amino- and carboxl-terminal end domains, *Biochem. Biophys. Res. Commun.* **221:**67–71.

Asch, H. L., Mayhew, E., Lazo, R. O., and Asch, B. B., 1993, Lipids covalently bound to keratins of mouse mammary epithelial cells, *Biochem. Mol. Biol. Int.* **29:**1161–1169.

Bachant, J. B., and Klymkowsky, M. W., 1996, A nontetrameric species is the major soluble form of keratin in *Xenopus* oocytes and rabbit reticulocyte lysates, *J. Cell Biol.* **132:**153–165.

Baribault, H., Blouin, R., Bourgon, L., and Marceau, N., 1989, Epidermal growth factor-induced selective phosphorylation of cultured rat hepatocyte 55-kD cytokeratin before filament reorganization and DNA synthesis, *J. Cell Biol.* **109:**1665–1676.

Baribault, H., Price, J., Miyai, K., and Oshima, R. G., 1993, Mid-gestational lethality in mice lacking keratin 8, *Genes Dev.* **7:**1191–1202.

Bartles, J. R., Braiterman, L. T., and Hubbard, A. L., 1985, Endogenous and exogenous domain markers of the rat hepatocyte plasma membrane, *J. Cell Biol.* **100:**1126–1138.

Bianchi, R., Giambanco, I., and Donato, R., 1993, S-100 protein, but not calmodulin, binds to the glial fibrillary acidic protein and inhibits its polymerization in a Ca^{2+}-dependent manner, *J. Biol. Chem.* **268:**12669–12674.

Billingsley, M. L., and Kincaid, R. L., 1997, Regulated phosphorylation and dephosphorylation of tau protein: Effects on microtubule interaction, intracellular trafficking and neurodegeneration, *Biochem. J.* **323:**577–591.

Bjorklund, B., 1980, On the nature and clinical use of tissue polypeptide antigen (TPA), *TumorDiagnostik* **1:**9–20.

Bowden, P. E., Quinlan, R. A., Breitkreutz, D., and Fusenig, N. E., 1984, Proteolytic modification of acidic and basic keratins during terminal differentiation of mouse and human epidermis, *Eur. J. Biochem.* **142:**29–36.

Bravo, R., Small, J. V., Fey, S. J., Larsen, P. M., and Celis, J. E., 1982, Architecture and polypeptide composition of HeLa cytoskeletons: Modification of cytoarchitectural polypeptides during mitosis, *J. Mol. Biol.* **154:**121–143.

Breitman, T. R., and Takahashi, N., 1996, Retinoylation of proteins in mammalian cells, *Biochem. Soc. Trans.* **24:**723–727.

Busso, N., Masur, S. K., Lazega, D., Waxman, S., and Ossowski, L., 1994, Induction of cell migration by pro-urokinase binding to its receptor: Possible mechanism for signal transduction in human epithelial cells, *J. Cell Biol.* **126:**259–270.

Buxman, M. M., and Wuepper, K. D., 1975, Keratin cross-linking and epidermal transglutaminase. A review with observations on the histochemical and immunochemical localization of the enzyme, *J. Invest. Dermatol.* **65:**107–112.

Cadrin, M., McFarlane Anderson, N., Assheim, L. H., Kawahara, H., Franks, D. J., Marceau, N., and French, S. W., 1992, Differential phosphorylation of CK8 and CK18 by 12-O-tetradecanoyl-phorbol-13-acetate in primary cultures of mouse hepatocytes, *Cell. Signal.* **4:**715–722.

Cadrin, M., McFarlane Anderson, N., Aasheim, L. H., Kawahara, H., Franks, D. J., and French, S. W., 1995, Modifications in cytokeratin and actin in cultured liver cells derived from griseofulvin-fed mice, *Lab. Invest.* **72:**453–460.

Calnek, D., and Quaroni, A., 1993, Differential localization by *in situ* hybridization of distinct keratin mRNA species during intestinal epithelial cell development and differentiation, *Differentiation* **53:**95–104.

Caulin, C., Salvesen, G. S., and Oshima, R. G., 1997, Caspase cleavage of keratin 18 and reorganization of intermediate filaments during epithelial cell apoptosis, *J. Cell Biol.* **138:**1379–1394.

Celis, J. E., Larsen, P. M., Fey, S. J., and Celis, A., 1983, Phosphorylation of keratin and vimentin polypeptides in normal and transformed mitotic human epithelial amnion cells: Behavior of keratin and vimentin filaments during mitosis, *J. Cell Biol.* **97:**1429–1434.

Chen, P. H., Ornelles, D. A., and Shenk, T., 1993, The adenovirus L3 23-kilodalton proteinase cleaves the amino-terminal head domain from cytokeratin 18 and disrupts the cytokeratin network of HeLa cells, *J. Virol.* **67:**3507–3514.

Cheng, T.-J., and Lai, Y.-K., 1994, Transient increase in vimentin phosphorylation and vimentin-HSC70 association in 9L rat brain tumor cells experiencing heat-shock, *J. Cell. Biochem.* **54:**100–109.

Chou, C.-F., and Omary, M. B., 1991, Phorbol acetate enhances the phosphorylation of cytokeratins 8 and 18 in human colonic epithelial cells, *FEBS Lett.* **282:**200–204.

Chou, C.-F., and Omary, M. B., 1993, Mitotic-arrest associated enhancement of O-linked glycosylation and phosphorylation of human keratins 8 and 18, *J. Biol. Chem.* **268:**4465–4472.

Chou, C.-F., and Omary, M. B., 1994, Mitotic arrest with anti-microtubule agents or okadaic acid is associated with increased glycoprotein terminal GlcNAc's, *J. Cell Sci.* **107:**1833–1843.

Chou, C.-F., Smith, A. J., and Omary, M. B., 1992, Characterization and dynamics of O-linked glycosylation of human cytokeratin 8 and 18, *J. Biol. Chem.* **267:**3901–3906.

Chou, C.-F., Riopel, C. L., Rott, L. S., and Omary, M. B., 1993, A significant soluble keratin fraction in "simple" epithelial cells: Lack of an apparent phosphorylation and glycosylation role in keratin solubility, *J. Cell Sci.* **105:**433–445.

Chou, T.-Y., Hart, G. W., and Dang, C. V., 1995, c-Myc is glycosylated at threonine 58, a known phosphorylation site and a mutational hot spot in lymphomas, *J. Biol. Chem.* **270:**18961–18965.

Collett, M. S., Purchio, A. F., and Erikson, R. L., 1980, Avian sarcoma virus-transforming protein, $pp60^{src}$ shows protein kinase activity specific for tyrosine, *Nature* **285:**167–169.

Coulombe, P. A., 1993, The cellular and molecular biology of keratins: Beginning a new era, *Curr. Opin. Cell Biol.* **5:**17–29.

Datta, B., Ray, M. K., Chakrabarti, D., Wylie, D. E., and Gupta, N. K., 1989, Glycosylation of eukaryotic peptide chain initiation factor 2 (eIF-2)-associated 67-kDa polypeptide (p^{67}) and its possible role in the inhibition of eIF-2 kinase-catalyzed phosphorylation of the eIF-2 α-subunit, *J. Biol. Chem.* **264:**20620–20624.

Decker, R. H., 1968, The identification of a phosphoprotein in acantholytic epidermis, *J. Invest. Dermatol.* **51:**141–146.

Deery, W. J., 1993, Role of phosphorylation in keratin and vimentin filament integrity in cultured thyroid epithelial cells, *Cell Motil. Cytoskel.* **26:**325–339.

Denk, H., Gschnait, F., and Wolff, K., 1975, Hepatocellular hyalin (Mallory bodies) in long term griseofulvin-treated mice: A new experimental model for the study of hyalin formation, *Lab. Invest.* **32:**773–776.

Denk, H., Lackinger, E., Zatloukal, K., and Franke, W. W., 1987, Turnover of cytokeratin polypeptides in mouse hepatocytes, *Exp. Cell Res.* **173:**137–143.

Dong, D. L.-Y., Xu, Z.-S., Chevrier, M. R., Cotter, R. J., Cleveland, D. W., and Hart, G. W., 1993, Glycosylation of mammalian neurofilaments, *J. Biol. Chem.* **268:**16679–16687.

Dong, D. L.-Y., Xu, Z.-S., Hart, G. W., and Cleveland, D. W., 1996, Cytoplasmic O-GlcNAc modification of the head domain and the KSP repeat motif of the neurofilament protein neurofilament-H, *J. Biol. Chem.* **271:**20845–20852.

Eckert, R. L., 1988, Sequence of the human 40-kDa keratin reveals an unusual structure with very high sequence identity to the corresponding bovine keratin, *Proc. Natl. Acad. Sci. USA* **85:**1114–1118.

Eichner, R., and Kahn, M., 1990, Differential extraction of keratin subunits and filaments from normal human epidermis, *J. Cell Biol.* **110:**1149–1158.

Ferraro, A., Eufemi, M., Cervoni, L., Marinetti, R., and Turano, C., 1989, Glycosylated forms of nuclear lamins, *FEBS Lett.* **257:**241–246.

Fey, S. J., Larsen, P. M., and Celis, J. E., 1983, Evidence for coordinated phosphorylation of keratins and vimentin during mitosis in transformed human amnion cell. Phosphate turnover of modified proteins, *FEBS Lett.* **157:**165–169.

Franke, W. W., Winter, S., Schmid, E., Sollner, P., Hammerling, G., and Achtstatter, T., 1987, Monoclonal cytokeratin antibody recognizing a heterotypic complex: Immunological probing of conformational states of cytoskeletal proteins in filaments and in solution, *Exp. Cell Res.* **173:**17–37.

Fuchs, E., 1996, The cytoskeleton and disease: Genetic disorders of intermediate filaments, *Annu. Rev. Genet.* **30:**197–231.

Fuchs, E., and Coulombe, P. A., 1992, Of mice and men: Genetic skin diseases of keratin, *Cell* **69:**899–902.

Fuchs, E., and Weber, K., 1994, Intermediate filaments: Structure, dynamics, function and disease, *Annu. Rev. Biochem.* **63:**345–382.

Furukawa, Y., Ikuta, N., Omata, S., Yamauchi, T., Isobe, T., and Ichimura, T., 1993, Demonstration of the phosphorylation-dependent interaction of tryptophan hydroxylase with the 14-3-3 protein, *Biochem. Biophys. Res. Commun.* **194:**144–149.

Gall, L., and Karsenti, E., 1987, Soluble cytokeratins in *Xenopus laevis* oocytes and eggs, *Biol. Cell* **61:**33–38.

Garbuglia, M., Verzini, M., Giambanco, I., Spreca, A., and Donato, R., 1996, Effects of calcium-binding proteins (S-100a$_o$, S-100a, S-100b) on desmin assembly *in vitro*, *FASEB J.* **10:**317–324.

Giasson, B. I., and Mushynski, W. E., 1996, Aberrant stess-induced phosphorylation of perikaryal neurofilaments, *J. Biol. Chem.* **271:**30404–30409.

Gilmartin, M. E., Culbertson, V. B., and Freedberg, I. M., 1980, Phosphorylation of epidermal keratins, *J. Invest. Dermatol.* **75:**211–216.

Gupta, P. D., Swarup, G., and Reddy, A. G., 1990, Modulation of phosphorylation and dephosphorylation of keratin and other polypeptides by estradiol-17β in rat vaginal epithelium, *FEBS Lett.* **273:**135–138.

Hagmann, J., Grob, M., and Burger, M. M., 1992, The cytoskeletal protein talin is O-glycosylated, *J. Biol. Chem.* **267:**14424–14428.

Haltiwanger, R. S., and Philipsberg, G. A., 1997, Mitotic arrest with nocodazole induces selective changes in the level of O-linked N-acetylglucosamine and accumulation of incompletely processed N-glycans on proteins from HT29 cells, *J. Biol. Chem.* **272:**8752–8758.

Han, I., and Kudlow, J. E., 1997, Reduced O glycosylation of Sp1 is associated with increased proteasome susceptibility, *Mol. Cell Biol.* **17:**2550–2558.

Hart, G. W., Haltiwanger, R. S., Holt, G. D., and Kelly, W. G., 1989, Glycosylation in the nucleus and cytoplasm, *Annu. Rev. Biochem.* **58:**841–874.

Hart, G. W., Kreppel, L. K., Comer, F. I., Arnold, C. S., Snow, D. M., Ye, Z., Cheng, X., DellaManna, D., Caine, D. S., Earles, B. J., Akimoto, Y., Cole, R. N., and Hayes, B. K., 1996, O-GlcNAcylation of key nuclear and cytoskeletal proteins: Reciprocity with O-phosphorylation and putative roles in protein multimerization, *Glycobiology* **6:**711–716.

Hatzfeld, M., Kristjansson, G. I., Plessmann, U., and Weber, K., 1994, Band 6 protein, a major constituent of desmosomes from stratified epithelia, is a novel member of the *armadillo* multigene family, *J. Cell Sci.* **107:**2259–2270.

Holt, G. D., Snow, C. M., Senior, A., Haltiwanger, R. S., Gerace, L., and Hart, G. W., 1987, Nuclear pore complex glycoproteins contain cytoplasmically disposed O-linked N-acetylglucosamine, *J. Cell Biol.* **104:**1157–1164.

Inagaki, M., Matsuoka, Y., Tsujimura, K., Ando, S., Tokui, T., Takahashi, T., and Inagaki, N., 1996, Dynamic property of intermediate filaments: Regulation by phosphorylation, *BioEssays* **18:**481–487.

Inagaki, M., Inagaki, N., Takahashi, T., and Takai, Y., 1997, Phosphorylation-dependent control of structures of intermediate filaments: A novel approach using site- and phosphorylation state-specific antibodies, *J. Biochem.* **121:**407–414.

Irvine, A. D., Corden, L. D., Swensson, O., Swensson, B., Moore, J. E., Frazer, D. G., Smith, F. J. D., Knowlton, R. G., Christophers, E., Rochels, R., Uitto, J., and McLean, W. H. I., 1997, Mutations in cornea-specific keratin K3 or K12 genes cause Meesmann's corneal dystrophy, *Nat. Genet.* **16:**184–187.

Jackson, S. P., and Tjian, R., 1988, O-glycosylation of eukaryotic transcription factors: Implications for mechanisms of transcriptional regulation, *Cell* **55:**125–133.

Jones, D. H., Ley, S., and Aitken, A., 1995, Isoforms of 14-3-3 protein can form homo- and heterodimers *in vivo* and *in vitro*: Implications for function as adapter proteins, *FEBS Lett.* **368:**55–58.

Kasahara, K., Kartasova, T., Ren, X.-Q., Ikuta, T., Chida, K., and Kuroki, T., 1993, Hyperphosphorylation of keratins by treatment with okadaic acid of BALB/MK-2 mouse keratinocytes, *J. Biol. Chem.* **268:**23531–23537.

Kearse, K. P., and Hart, G. W., 1991, Lymphocyte activation induces rapid changes in nuclear and cytoplasmic glycoproteins, *Proc. Natl. Acad. Sci. USA* **88:**1701–1705.

Kelly, W. G., Dahmus, M. E., and Hart, G. W., 1993, RNA polymerase II is a glycoprotein: Modification of the COOH-terminal domain by O-GlcNAc, *J. Biol. Chem.* **268:**10416–10424.

King, I. A., 1986, Incorporation of [^{3}H] glucosamine into keratin-related polypeptides in pig epidermis, *FEBS* **201:**114–118.

King, I. A., and Hounsell, E. F., 1989, Cytokeratin 13 contains O-glycosidically linked N-acetylglucosamine residues, *J. Biol. Chem.* **264:**14022–14028.

Klymkowsky, M. W., 1995, Intermediate filament organization, reorganization, and function in the clawed frog *Xenopus*, *Curr. Top. Dev. Biol.* **31:**455–486.

Klymkowsky, M. W., Maynell, L. A., and Nislow, C., 1991, Cytokeratin phosphorylation, cytokeratin filament severing, and the solubilization of the maternal mRNA Vg1, *J. Cell Biol.* **114:**787–797.

Kreppel, L. K., Blomberg, M. A., and Hart, G. W., 1997, Dynamic glycosylation of nuclear and cytosolic proteins, *J. Biol. Chem.* **272:**9308–9315.

Ku, N.-O., and Omary, M. B., 1994a, Expression, glycosylation, and phosphorylation of human keratins 8 and 18 in insect cells, *Exp. Cell Res.* **211:**24–35.

Ku, N.-O., and Omary, M. B., 1994b, Identification of the major physiologic phosphorylation site of human keratin 18: Potential kinases and a role in filament reorganization, *J. Cell Biol.* **127:**161–171.

Ku, N.-O., and Omary, M. B., 1995, Identification and mutational analysis of the glycosylation sites of human keratin 18, *J. Biol. Chem.* **270:**11820–11827.

Ku, N.-O., and Omary, M. B., 1997, Phosphorylation of human keratin 8 *in vivo* at conserved head domain serine 23 and at epidermal growth factor-stimulated tail domain serine 431, *J. Biol. Chem.* **272:**7556–7564.

Ku, N.-O., Michie, S., Oshima, R. G., and Omary, M. B., 1995, Chronic hepatitis, hepatocyte fragility, and increased soluble phosphoglycokeratins in transgenic mice expressing a keratin 18 conserved arginine mutant, *J. Cell Biol.* **131:**1303–1314.

Ku, N.-O., Liao, J., Chou, C.-F., and Omary, M. B., 1996a, Implications of intermediate filament protein phosphorylation, *Cancer Metast. Rev.* **15:**429–444.

Ku, N.-O., Michie, S. A., Soetikno, R. M., Resurreccion, E. Z., Broome, R. L., Oshima, R. G., and Omary, M. B., 1996b, Susceptibility to hepatotoxicity in transgenic mice that express a dominant-negative human keratin 18 mutant, *J. Clin. Invest.* **98:**1034–1046.

Ku, N.-O., Liao, J., and Omary, M. B., 1997, Apoptosis generates stable fragments of type I keratins, *J. Biol. Chem.* **273:**33197–33203.

Ku, N.-O., Liao, J., and Omary, M. B., 1998, Phosphorylation of human Keratin 18 serine-33 regulates binding to 14-3-3 proteins, *EMBO J.* **17:**1892–1906.

Ku, N.-O., Wright, T. L., Terrault, N. A., Gish, R., and Omary, M. B., 1997, Mutation of human keratin 18 in association with cryptogenic cirrhosis, *J. Clin. Invest.* **99:**19–23.

Lai, Y.-K., Lee, W.-C., and Chen, K.-D., 1993, Vimentin serves as a phosphate sink during the apparent activation of protein kinases by okadaic acid in mammalian cells, *J. Cell. Biochem.* **53:**161–168.

Lazarides, E., 1980, Intermediate filaments as mechanical integrators of cellular space, *Nature* **283:**249–256.

Liao, J., and Omary, M. B., 1996, 14-3-3 proteins associate with phosphorylated simple epithelial keratins during cell cycle progression and act as a solubility cofactor, *J. Cell Biol.* **133:**345–358.

Liao, J., Ku, N.-O., and Omary, M. B., 1996, Two-dimensional gel analysis of glandular keratin intermediate filament phosphorylation, *Electrophoresis* **17:**1671–1676.

Liao, J., Ku, N.-O., and Omary, M. B., 1997, Stress, apoptosis, and mitosis induce phosphorylation of human keratin 8 at ser73 in tissues and cultured cells, *J. Biol. Chem.* **272:**17565–17573.

Liao, J., Lowthert, L. A., Ku, N.-O., Fernandez, R., and Omary, M. B. 1995a, Dynamics of human keratin 18 phosphorylation: Polarized distribution of phosphorylated keratins in simple epithelial tissues, *J. Cell Biol.* **131:**1291–1301.

Liao, J., Lowthert, L. A., and Omary, M. B., 1995b, Heat stress or rotavirus infection of human epithelial cells generates a distinct hyperphosphorylated form of keratin 8, *Exp. Cell Res.* **219:**348–357.

Liao, J., Lowthert, L. A., Ghori, N., and Omary, M. B., 1995c, The 70-kDa heat shock proteins associate with glandular intermediate filaments in an ATP-dependent manner, *J. Biol. Chem.* **270:**915–922.

Lowthert, L. A., Ku, N.-O., Liao, J., Coulombe, P. A., and Omary, M. B., 1995, Empigen BB: A useful detergent for solubilization and biochemical analysis of keratins, *Biochem. Biophys. Res. Commun.* **206:**370–379.

Lubas, W. A., Frank, D. W., Krause, M., and Hanover, J. A., 1997, O-linked GlcNAc transferase is a conserved nucleocytoplasmic protein containing tetratricopeptide repeats, *J. Biol. Chem.* **272:**9316–9324.

Mallory, F. B., 1911, Cirrhosis of the liver. Five different types of lesions from which it may arise, *Bull. Johns Hopkins Hosp.* **22:**69–75.

Mawal-Dewan, M., Henley, J., Van de Voorde, A., Trojanowski, J. Q., and Lee, V. M.-Y., 1994, The phosphorylation state of tau in the developing rat brain is regulated by phosphoprotein phosphatases, *J. Biol. Chem.* **269:**30981–30987.

McLean, W. H. I., and Lane, E. B., 1995, Intermediate filaments in disease, *Curr. Opin. Cell Biol.* **7:**118–125.

Michaud, N. R., Fabian, J. R., Mathes, K. D., and Morrison, D. K., 1995, 14-3-3 is not essential for Raf-1 function: Identification of Raf-1 proteins that are biologically activated in a 14-3-3- and Ras-independent manner, *Mol. Cell. Biol.* **15:**3390–3397.

Miller, R. K., Khuon, S., and Goldman, R. D., 1993, Dynamics of keratin assembly: Exogenous type I keratin rapidly associates with type II keratin *in vivo*, *J. Cell. Biol.* **122:**123–135.

Moll, R., Franke, W. W., Schiller, D. L., Geiger, B., and Krepler, R., 1982, The catalog of human cytokeratins: Patterns of expression in normal epithelia, tumors and cultured cells, *Cell* **31:**11–24.

Moll, R., Schiller, D. L., and Franke, W. W., 1990, Identification of protein IT of the intestinal cytoskeleton as a novel type I cytokeratin with unusual properties and expression patterns, *J. Cell Biol.* **111:**567–580.

Moore, B. W., and Perez, V. J., 1967, Specific acidic proteins of the nervous system, in *Physiological and Biochemical Aspects of Nervous Integration* (F. D. Carlson, ed.), pp. 343–359, Prentice-Hall, Englewood Cliffs, New Jersey.

Morley, S. M., Dundas, S. R., James, J. L., Gupta, T., Brown, R. A., Sexton, C. J., Navsaria, H. A., Leigh, I. M., and Lane, E. B., 1995, Temperature sensitivity of the keratin cytoskeleton and delayed spreading of keratinocyte lines derived from EBS patients, *J. Cell Sci.* **108:**3463–3471.

Muslin, A. J., Tanner, J. W., Allen, P. M., and Shaw, A. S., 1996, Interaction of 14-3-3 with signaling proteins is mediated by the recognition of phosphoserine, *Cell* **84:**889–897.

Nicholl, I. D., and Quinlan, R. A., 1994, Chaperone activity of α-crystallins modulates intermediate filament assembly, *EMBO J.* **13:**945–953.

Omary, M. B., and Ku, N.-O., 1997, Intermediate filament proteins of the liver: Emerging disease association and functions, *Hepatology* **25:**1043–1048.

Omary, M. B., Baxter, G. T., Chou, C.-F., Riopel, C. L., Lin, W. Y., and Strulovici, B., 1992, PKCε-related kinase associates with and phosphorylates cytokeratin 8 and 18, *J. Cell Biol.* **117:**583–593.

Osborn, M., and Weber, K., 1989, Keratins, transglutaminase, and Mallory bodies—The really insoluble stuff, *Lab. Invest.* **61:**585–587.

Oshima, R. G., 1982, Developmental expression of murine extra-embryonic endodermal cytoskeletal proteins, *J. Biol. Chem.* **257:**3414–3421.

Oshima, R. G., Millan, J. L., and Cecena, G., 1986, Comparison of mouse and human keratin 18: A component of intermediate filaments expressed prior to implantation, *Differentiation* **33:**61–68.

Quinlan, R. A., Cohlberg, J. A., Schiller, D. L., Hatzfeld, M., and Franke, W. W., 1984, Heterotypic tetramer (A_2D_2) complexes of non-epidermal keratins isolated from cytoskeletons of rat hepatocytes and hepatoma cells, *J. Mol. Biol.* **178:**365–388.

Rao, L., Perez, D., and White, E., 1996, Lamin proteolysis facilitates nuclear events during apoptosis, *J. Cell Biol.* **135:**1441–1455.

Roberts, G. P., and Brunt, J., 1986, Biosynthesis of a glycosylated keratin by human keratinocytes, *Biochim. Biophys. Acta* **883:**413–419.

Roberts, G. P., and Jenner, L., 1983, Glycoproteins and glycosaminoglycans synthesized by human keratinocytes in culture, *Biochem. J.* **212:**355–363.

Roquemore, E. P., Chou, T.-Y., and Hart, G. W., 1994, Detection of O-linked N-acetylglucosamine (O-GlcNAc) on cytoplasmic and nuclear proteins, *Meth. Enzymol.* **230:**443–460.

Roquemore, E. P., Chevrier, M. R., Cotter, R. J., and Hart, G. W., 1996, Dynamic O-GlcNAcylation of the small heat shock protein αB-crystallin, *Biochemistry* **35:**3578–3586.

Rydlander, L., Ziegler, E., Bergman, T., Schoberl, E., Steiner, G., Bergman, A.-C., Zetterberg, A., Marberger, M., Bjorklund, P., Skern, T., Einarsson, R., and Jornvall, H., 1996, Molecular characterization of a tissue-polypeptide-specific-antigen epitope and its relationship to human cytokeratin 18, *Eur. J. Biochem.* **241:**309–314.

Sahyoun, N., Stenbuck, P., Levine III, H., Bronson, D., Moncharmont, B., Henderson, C., and Cuatrecasas, P., 1982, Formation and identification of cytoskeletal components from liver cytosolic precursors, *Proc. Natl. Acad. Sci. USA* **79:**7341–7345.

Salmhofer, H., Rainer, I., Zatloukal, K., and Denk, H., 1994, Post-translational events involved in griseofulvin-induced keratin cytoskeleton alterations, *Hepatology* **20:**731–740.

Schafer, I. A., and Sorrell, J. M., 1993, Human keratinocytes contain keratin filaments that are glycosylated with keratan sulfate, *Exp. Cell Res.* **207:**213–219.

Shikhman, A. R., Greenspan, N. S., and Cunningham, M. W., 1994, Cytokeratin peptide SFGSGFGGGY mimics N-acetyl-beta-D-glucosamine in reaction with antibodies and lectins, and induces *in vivo* anti-carbohydrate antibody response, *J. Immunol.* **153:**5593–5606.

Skalli, O., Jones, J. C. R., Gagescu, R., and Goldman, R. D., 1994, IFAP 300 is common to desmosomes and hemidesmosomes and is a possible linker of intermediate filaments to these junctions, *J. Cell. Biol.* **125:**159–170.

Soellner, P., Quinlan, R. A., and Franke, W. W., 1985, Identification of a distinct soluble subunit of an intermediate filament protein: Tetrameric vimentin from living cells, *Proc. Natl. Acad. Sci. USA* **82:**7929–7933.

Stappenbeck, T. S., Bornslaeger, E. A., Corcoran, C. M., Luu, H. H., Virata, M. L. A., and Green, K. J., 1993, Functional analysis of desmoplakin domains: Specification of the interaction with keratin versus vimentin intermediate filament networks, *J. Cell Biol.* **123:**691–705.

Stasiak, P. C., Purkis, P. E., Leigh, I. M., and Lane, E. B., 1989, Keratin 19: Predicted amino acid sequence and broad tissue distribution suggest it evolved from keratinocyte keratins, *J. Invest. Dermatol.* **92:**707–716.

Steinert, P. M., 1988, The dynamic phosphorylation of the human intermediate filament keratin 1 chain, *J. Biol. Chem.* **263:**13333–13339.

Steinert, P. M., and Bale, S. J., 1993, Genetic skin diseases caused by mutations in keratin intermediate filaments, *Trends Genet.* **9:**280–284.

Steinert, P. M., and Idler, W. W., 1975, The polypeptide composition of bovine epidermal α-keratin, *Biochem. J.* **151:**603–614.

Steinert, P. M., and Idler, W. W., 1979, Post-synthetic modification of mammalian epidermal α-keratin, *Biochemistry* **18:**5664–5669.

Steinert, P. M., and Marekov, L. N., 1995, The proteins elafin, filaggrin, keratin intermediate filaments, loricrin, and small proline-rich proteins 1 and 2 are isodipeptide cross-linked components of the human epidermal cornified cell envelope, *J. Biol. Chem.* **270:**17702–17711.

Steinert, P. M., and Roop, D. R., 1988, Molecular and cellular biology of intermediate filaments, *Annu. Rev. Biochem.* **57:**593–625.

Steinert, P. M., Wantz, M. L., and Idler, W. W., 1982, O-Phosphoserine content of intermediate filament subunits, *Biochemistry* **21:**177–183.

Sun, T. T., and Green, H., 1978, The keratin filaments of cultured human epidermal cells, *J. Biol. Chem.* **253:**2053–2060.

Takahashi, A., Alnemri, E. S., Lazebnik, Y. A., Fernandes-Alnemri, T., Litwack, G., and Moir, R. D., 1996, Cleavage of lamin A by Mch2 alpha but not CPP32: Multiple interleukin 1 beta-converting enzyme-related proteases with distinct substrate recognition properties are active in apoptosis, *Proc. Natl. Acad. Sci. USA* **93:**8395–8400.

Takahashi, N., Jetten, A. M., and Breitman, T. R., 1991, Retinoylation of cytokeratins in normal human epidermal keratinocytes, *Biochem. Biophys. Res. Commun.* **180:**393–400.

Toivola, D. M., Goldman, R. D., Garrod, D. R., and Eriksson, J. E., 1997, Protein phosphatases maintain the organization and structural interactions of hepatic keratin intermediate filaments, *J. Cell Sci.* **110:**23–33.

Tolle, H.-G., Weber, K., and Osborn, M., 1987, Keratin filament disruption in interphase and mitotic cells—How is it induced? *Eur. J. Cell Biol.* **43:**35–47.

Torres, C.-R., and Hart, G. W., 1984, Topography and polypeptide distribution of terminal N-acetylglucosamine residues on the surfaces of intact lymphocytes, *J. Biol. Chem.* **259:**3308–3317.

Trejo-Skalli, A. V., Velasco, P. T., Prasanna Murthy, S. N., Lorand, L., and Goldman, R. D., 1995, Association of a transglutaminase-related antigen with intermediate filaments, *Proc. Natl. Acad. Sci. USA* **92:**8940–8944.

van Dalen, A., 1996, Significance of cytokeratin markers TPA, TPA (cyk), TPS and CYFRA 21.1 in metastatic disease, *Anticancer Res.* **16:**2345–2350.

Weber, K., Osborn, M., Moll, R., Wiklund, B., and Luning, B., 1984, Tissue polypeptide antigen (TPA) is related to the non-epidermal keratins 8, 18, 19 typical of simple and non-squamous epithelia: Re-evaluation of a human tumor marker, *EMBO J.* **3:**2707–2714.

Yano, T., Tokui, T., Nishi, Y., Nishizawa, K., Shibata, M., Kikuchi, K., Tsuiki, S., Yamauchi, T., and Inagaki, M., 1991, Phosphorylation of keratin intermediate filaments by protein kinase C, by calmodulin-dependent protein kinase and by cAMP-dependent protein kinase, *Eur. J. Biochem.* **197:**281–290.

Zatloukal, K., Fesus, L., Denk, H., Tarcsa, E., Spurej, G., and Böck, G., 1992, High amount of ε-(γ-glutamyl) lysine cross-links in Mallory bodies, *Lab. Invest.* **66:**774–777.

Zha, J., Harada, H., Yang, E., Jockel, J., and Korsmeyer, S. J., 1996, Serine phosphorylation of death agonist BAD in response to survival factor results in binding to 14-3-3 not BCL-X_L, *Cell* **87:**619–628.

Zhang, X., and Bennett, V., 1996, Identification of O-linked N-acetylglucosamine modification of $ankyrin_G$ isoforms targeted to nodes of Ranvier, *J. Biol. Chem.* **271:**31391–31398.

Zhang, Y., and Schneider, R. J., 1994, Adenovirus inhibition of cell translation facilitates release of virus particles and enhances degradation of the cytokeratin network, *J. Virol.* **68:**2544–2555.

Zhou, H., Huiatt, T. W., Robson, R. M., Sernett, S. W., and Graves, D. J., 1996, Characterization of ADP-ribosylation sites on desmin and restoration of desmin intermediate filament assembly by de-ADP-ribosylation, *Arch. Biochem. Biophys.* **334:**214–222.

Chapter 5

Lessons from Keratin Transgenic and Knockout Mice

Thomas M. Magin

1. INTRODUCTION

Keratins are the most diverse group of all intermediate filament (IF) proteins, with more than 30 different genes including those encoding hair keratins (Fuchs and Weber, 1994). Corroborative studies based on assembly experiments with purified individual keratins and the characterization of cloned genes (Hanukoglu and Fuchs, 1982, 1983; Quinlan *et al.*, 1985; Schiller *et al.*, 1982; Steinert *et al.*, 1976) documented that keratin genes are organized in two families, coding for the more acidic type I and the more basic type II proteins. At least one member of each family is required to form heterotypic IF. While much progress was made to isolate keratin genes and to describe their expression patterns in normal and malignant tissues (Moll, 1994; Moll *et al.*, 1982; Quinlan *et al.*, 1985), the functional role of keratins as well as their pathogenic relevance in human disorders remained obscure in the early days of keratin research.

Thomas M. Magin Department of Molecular Genetics, Institute of Genetics, University of Bonn, and Bonner Forum Biomedizin, D-53117 Bonn, Germany.

Subcellular Biochemistry, Volume 31: Intermediate Filaments, edited by Herrmann and Harris. Plenum Press, New York, 1998.

In the early 1980s two seminal discoveries were made: It was shown that cultured cells retained their typical morphology and growth properties in the absence of cytoplasmic IF, thus reflecting the methodical limitations of classical cell culture experiments (Venetianer *et al.*, 1983). Furthermore, altered keratin arrangements were demonstrated in a group of human bullous and hyperkeratotic skin disorders, suggesting a putative rolė in their underlying pathology (Anton-Lamprecht, 1983). In due course it was recognized that cDNAs coding for truncated keratin polypeptides behaved as dominant-negative subunits upon transfection in cultured cells (Albers and Fuchs, 1987, 1989). The consecutive expression of such constructs in transgenic mice (Vassar *et al.*, 1989) had a major impact on our understanding of the consequences of altered keratin expression *in vivo*. Those transgenic mice expressing a mutant human K14 gene in basal epidermis developed a cutaneous phenotype bearing resemblance to the human disorder epidermolysis bullosa simplex (EBS). This prompted the mutation analysis of keratin genes in EBS patients and other disorders with a suspected keratin involvement (for a review, see Corden and McLean, 1996).

This review will briefly summarize the current knowledge of keratin regulatory elements. The main focus, however, will be on the analysis of keratin function by overexpression of normal or mutant genes in conventional transgenic and gene knockout mice.

1.1. Regulation of Keratin Expression in Transgenic Mice

1.1.1. Regulation of Simple Epithelial Keratins

Keratins 8 and 18 are coexpressed in all embryonic and many adult simple epithelia. While liver hepatocytes only express K8 and K18, other simple epithelia like intestinal and colon epithelia are known for their coexpression of keratins 7, 19, and 20 (Moll *et al.*, 1982, 1990). The precise regulatory sequences and essential transcription factors involved in tissue-specific regulation are only partially characterized. The following section will review the key findings in K8, 18, and 19 gene regulation.

1.1.1a. Keratin 19. For human Kl9, it was shown shown that 3.6 kb of 5′ in addition to a 1.2-kb fragment ranging from −8.6 to −6.4 of 5′ upstream sequences and 0.7 kb of 3′ sequences were sufficient to provide tissue-specific expression in several lines of transgenic mice (Bader and Franke, 1990). While the expression was clearly independent of integration effects, it is unclear whether the tested gene fragment contained all elements required to direct copy-number independence. Furthermore, the presence of regulatory sequences in introns remains to be investigated.

1.1.1b. Keratin 18. It was shown that the transcription control of the human K18 gene resided in a 10-kb genomic fragment containing K18. Flanked by 2.5 kb of 5′ and 3.5 kb of 3′ regulatory sequences, it reproduced quite faithfully the expression pattern of mouse K18 in a copy-number-dependent and position-independent manner (Abe and Oshima, 1990). The amount of K18 mRNA was directly proportional to the number of K18 gene copies (Oshima *et al.*, 1996). This was also found with a smaller construct with only 1 kb of 5′ sequences including the 3′ flanking region. Furthermore, both constructs showed ectopic expression in transgenic lines carrying the highest copy numbers (Neznanov *et al.*, 1993; Neznanov and Oshima, 1993), including K18 mRNA expression even in tissues like muscle, which do not express keratins.

A similar observation was made in K8 transgenic mice Casanova *et al.*, 1995). It is unclear whether this indicates the relief of repression or the necessity of more extensive flanking sequences. However, it could also reflect an alteration of chromatin structure as a consequence of the head-to-tail array of the transgenes. A closer analysis of the K18 5′ region revealed the presence of two *Alu* elements immediately upstream of the proximal promoter (Neznanov and Oshima, 1993; Thorey *et al.*, 1993b). Pol III-mediated transcription of the conserved *Alu* sequences in the opposite direction relative to the K18 promoter seemed positively correlated with K18 expression. By mutating the Pol III promoter in the *Alu* element, it was shown that integration-site independence and copy-number dependence are separable entities: without 3′ flanking sequences copy-number dependence is lost and a general decrease in tissue-specific expression was observed. The integration-site independence was maintained at least in some tissues (Thorey *et al.*, 1993b). The deletion of the 3.5 kb of 5′ flanking DNA reduced liver-specific expression by about 90% with minor effects in other tissues (Neznanov *et al.*, 1993; Neznanov and Oshima, 1993).

The search for DNase I-hypersensitive and transcription factor-binding sites in the 10 kb encompassing the K18 genes revealed three regions of interest: (a) The proximal promoter of 250 bp contains a TATA box and four potential SP1-binding sites. (b) An enhancer element occurs in the first intron, which is conserved among human and mouse K18 genes. This enhancer contains an AP-1 sequence, which is known to bind members of the Jun and Fos transcription factor families (Oshima *et al.*, 1990). (c) A regulatory element occurs in exon 6 with an additional AP-1 consensus site. This element seems to cooperate with the intron 1 sites (Oshima *et al.*, 1996). Its functional relevance *in vivo* has to be reevaluated, given that K18 remains expressed in c-Jun knockout mice (Hilberg *et al.*, 1993).

In addition to AP-1 sites, binding sites for protoencogenes of the avian erythroblastosis family (ETS) were identified in the intron 1 enhancer

(Oshima *et al.*, 1990; Pankov *et al.*, 1994). As the ETS family comprises positive- and negative-acting proteins, their effect on K18 expression in mice is unclear. Similar to c-Jun knockout mice, it has been mentioned that K18 expression is maintained in double-knockout ETS-ES cells (Oshima *et al.*, 1996). AP-1-binding proteins and ETS family members are also mediators of those oncogenes which use the Ras signal transduction cascade. After the deletion of cognate AP-1- or ETS-binding sites in the first intron, the expression of K18 was drastically reduced in embryonal cells (Oshima *et al.*, 1996). Finally, there are three silencer elements in the first intron of human K18, two of which are also present in the murine orthologue. It has been suggested that these elements are involved in repressing K18 transcription in undifferentiated embryonal carcinoma cells, but not in tissue-specific regulation in mice (Oshima *et al.*, 1996).

The significant contribution of transgenic mice became apparent when the methylation state of the K18 gene was analyzed. As in K8 (Ichinose *et al.*, 1988; Krauss and Franke, 1990; Kulesh and Oshima, 1988; Oshima *et al.*, 1988; Tamai *et al.*, 1991), the promoter of K18 is embedded in a CpG island which otherwise is typical of housekeeping genes (Bird, 1986). In most lines of cultured cells the lack of K18 was in perfect agreement with the methylation of sequences in the first exon (Oshima *et al.*, 1988). Moreover, treatment of certain cultured cells with 5-azacytidine, which inhibits DNA methyltransferase, led to the expression of K8 and 18 (Semat *et al.*, 1986). This suggested that differential methylation of promoter sequences are critical for tissue-specific expression. Further studies revealed that methylation of the CpG island is a common cell culture artifact not acting *in vivo* (Antequera *et al.*, 1990). In fact the K18 promoter remained unmethylated in transgenic mice (Neznanov *et al.*, 1993; Neznanov and Oshima, 1993). A recent and refined analysis revealed the existence of two differentially methylated CpGs in the K18 first intron enhancer (Oshima *et al.*, 1996). One of them was part of an ETS-binding site which could be involved in K18 repression in nonepithelial tissues. However, this methylation pattern was only analyzed in mice carrying a deletion construct of the K18 gene (Oshima *et al.*, 1996). On the basis of current data on DNA methylation, it seems a reasonable assumption that methylation plays a modulatory, but not a decisive role in tissue-specific K18 expression.

1.1.1c. Keratin 8. The analysis of the human K8 gene in transgenic mice has been less extensive than that of K18. A genomic fragment containing about 1.1 kb of upstream and 3.2 kb of downstream sequences showed specific and faithful expression in most transgenic lines. Similar to those K18 mice which carried a high transgene copy number, the line of K8 mice expressing the highest amount of K8 mRNA showed ectopic transgene expression, e.g., in striated muscle or in basal epidermis (Casanova *et al.*, 1995). When the properties of flanking sequences were tested by ligating them to a *lacZ* marker gene, no transgene expression was found at all. This indicated that additional flanking and possibly intragenic sequences are necessary for K8 expression. Although some K18 transgenic mice displayed *lacZ* expression, this was restricted to embryonic development in a mosaic pattern (Thorey *et al.*, 1993b). Generally, the study of human K8, 18, and 19 genes in transgenic mice has shown that about 5 kb of upstream and 3.5 kb of downstream sequences provide copy-number-dependent and integration-site-independent expression of these genes. For tissue-specific expression of K18 at least, there is an additional necessity for an intron 1 enhancer with AP-1 and ETS, as well as binding sites for silencing factors. In addition, a sequence element in exon 6 is necessary. Removal of downstream flanking sequences reduced keratin expression in some, but not all tissues.

1.1.2. Regulation of Epidermal Keratin Genes

Keratins 5 and 14 are the prototype keratins of stratified epithelia predominantly expressed in basal cells (Fuchs and Green, 1980, 1978). In terminal skin differentiation, when basal keratinocytes move upward, K5 and 14 are sequentially replaced by K1, 2e, and 10 (Collin *et al.*, 1992; Fuchs and Green, 1980, 1978). During wound healing or in hyperproliferative disorders, the keratin triplet 6/16/17 is transiently expressed instead of suprabasal keratins 1/2e/10 (Mansbridge and Knapp, 1987; Stoler *et al.*, 1988; Weiss *et al.*, 1984).

1.1.2a. Keratins 5 and 14. The first keratin transgenic mouse experiment was set up to examine regulatory sequences of the human K14 gene (Vassar *et al.*, 1989). A K14 cDNA with a neuropeptide P epitope tag, controlled by 2.5 kb of 5′ and 0.7 kb of 3′ sequences, was faithfully expressed in basal epidermis and its appendages. This indicated that no intron sequences are required for appropriate K14 expression. However, this study was confined to founder mice. In this setting, those with a high copy number showed mosaic K14 expression. The lack of a relationship between transgene copy numbers and the amount of K14 mRNA indicated that 2.5 kb of up- and 0.7 kb of downstream human K14 sequences are required for the appropriate tissue specificity. In this context, it is reasonable to assume that yet-undefined elements responsible for integration independence and copy-number dependence play a role in the overall K14 regulation.

More recently, the sequences involved in human keratin 5 expression were assayed in transgenic mice (Byrne and Fuchs, 1993). In combination with a *lacZ* reporter gene carrying SV40 splice donor/acceptor sites and an SV40 polyA site, a 6-kb upstream promoter fragment directed tissue-specific expression of *lacZ*. Apart from expression in a few neuronal cells, this construct reproduced the typical K5 pattern faithfully. The data allowed no quantification of expression levels between endogenous mouse K5 and *lacZ*. Deletion of all but 90 bp of promoter flanking sequences resulted in a novel *lacZ* expression pattern: *lacZ* was found in epidermis and hair follicles, but was absent from other stratified epithelia. In epidermis, the staining was no longer basal, but, reminiscent of keratins 1/10 or K6/16, was detected in suprabasal layers. In addition, ectopic *lacZ* activity was found in striated and smooth muscle cells, kidney, and a few other cells (Byrne and Fuchs, 1993).

The search for transcription factor-binding sites in this 90-bp promoter fragment of the K5 gene revealed two Sp1 and a variety of other transcription factor-binding sites, but no AP-2 site. Additional AP-2 sites reside further upstream with a presumptive role in epidermal keratin gene regulation. The importance of AP-2 for keratin regulation may have to be reconsidered, given that AP-2 knockout mice showed continued presence of keratins (Schorle *et al.*, 1996). The same might be true for Sp1 in the light of a recent Sp1 knockout (Marin *et al.*, 1997), which has placed into question the presumed participation of Sp1 in the transcription of many genes. Another set of regulatory proteins for K5 and K14 had binding sites around the translation initiation sites of these genes. The role of these recently discovered factors for the specific expression of K5 and K14 *in vivo* remains to be shown.

Data on the regulation of the bovine keratin 5 gene correspond nicely with those discussed before (Byrne and Fuchs, 1993). The largest construct used had the *lacZ* gene inserted in the first exon with 5.3 kb of upstream and about 11 kb of downstream sequences (Ramirez *et al.*, 1994). Like the human counterpart, the bovine K5 gene reproduced the pattern of the mouse gene in epidermis, but also showed ectopic and altered expression. Remarkably, reticular cells of the thymus, epithelium of the renal pelvis, and the basal layer of the trachea were devoid of bovine K5 expression. A construct containing only 5.3 kb of upstream sequences had an expression pattern very similar to the larger construct, but displayed a higher degree of variation among different lines of mice. Also, a quantitative analysis of *lacZ* expression revealed a lower expression level of the shorter construct.

Deleting more 5′-sequences down to 1.3 kb flanking the promoter abolished K5-mediated *lacZ* expression completely. This is different from the residual activity of the human K5 promoter, which retained some activity and partial tissue selectivity even with a 90-bp promoter fragment (Byrne and Fuchs, 1993).

In general, it seems that about 5 kb of upstream sequences can direct the expression of K5 and K14 to developing and adult epidermis. The different residual activity of 5′ deletion constructs observed in various experiments can be explained best by assuming an array of positive and negative regulatory elements present in these genes. Evidence for such an organization derived from a recent analysis of the human K6a gene (Takahashi and Coulombe, 1997) (see below). Given that all reports described either the lack or the ectopic expression of these genes in certain tissues or groups of cells, one might predict further control elements for the specific expression of K5 and K14. These could reside further away in flanking regions or in introns. The recently discovered protein factor Skn-1 (Andersen *et al.*, 1993) with its postulated involvement in keratin gene regulation is probably not an essential factor, given that knockout mice for Skn-1 have an identical pattern of keratin expression in epidermis when compared to normal mice (Andersen *et al.*, 1997).

1.1.2b. Keratins 6 and 16. K6 and K16, unlike most other keratins, are encoded by several functional and highly related genes, the significance of which is not yet understood. For bovine K6, an 8.8-kb promoter fragment was able to support *lacZ* expression in a pattern similar to the mouse K6 gene (Ramirez *et al.*, 1995). As for other keratin transgenes, mosaic or an overexpression of *lacZ*, beginning in the stratum granulosum instead of the stratum spinosum, was observed. While the latter most likely reflected the lack of control elements not present in the transgene construct, the former could be the result of the clonal organization of epidermal stem cells (Carroll *et al.*, 1993; Clausen and Potten, 1990; Mackenzie, 1997; Vassar *et al.*, 1991). The K6 *lacZ* transgene was inducible by retinoic acid and by the tumor promoter phorbol 12-myristate 13-acetate (PMA) (Ramirez *et al.*, 1995), indicating the presence of control elements responsible for basal and inducible gene expression. A transgene with only 0.8 kb of promoter sequences showed no *lacZ* activity. Therefore, it remained unclear whether there are common or separate control elements for basal and inducible K6 expression. The same authors also reported that a K6 transgene which carried *lacZ* inserted in the first exon was nearly completely inactive. This could indicate the presence of transcription factor-binding sites in the first intron, which need close association with promoter-proximal sequences. Alternatively, sequences in the first exon might be required.

Recently, a similar analysis was carried out for the human K6a gene, which is the predominant keratin in human epidermis inducible by injury or a variety of other stimuli (Takahashi and Coulombe, 1996). Unexpectedly, a 5.3-kb upstream promoter fragment failed to express *lacZ* in transgenic mice in a manner expected from its murine orthologue. The only reported expression was found in vibrissae. Here, however, the presence of K6 may represent its basal expression rather than its induced state. The same study showed that *lacZ* expression in transgenes carrying either 1, 2.5, or 5.2 kb of promoter sequences was inducible either by injury, retinoic acid, the phorbol-ester PMA, or a classical two-step carcinogenesis program (Takahashi and Coulombe, 1997). Quantitative analysis showed a higher degree of induction in those transgenic lines with a larger 5′-promoter region. In contrast, a 550-bp promoter fragment showed no transgene expression. Therefore, regulatory sequences necessary for K6a induction by various stimuli are located between −960 and −550 bp. It follows that those binding sites mediating basal K6a expression are at least partially located at different sites, either further upstream, in introns, or downstream from the 3′-untranslated region.

It is difficult to reconcile the differences between the human K6a and the bovine K6 gene promoters. A possibility is that they are not orthologues and have different expression domains in human and cow. Alternatively, the human K6a gene might only be short of a few hundred base pairs, which could provide the elements necessary for those sites where it is normally expressed (Takahashi and Coulombe, 1996). Moreover, both studies have corrected findings from cell culture studies: These had concluded (e.g., Jiang *et al.*, 1993) that 390 bp of 5′ regulatory sequences are sufficient to drive K6 expression. As transgenic mice have shown, this is not the case. Furthermore, transgenic mouse studies have shown that the expression of K6 *in vivo* can be clearly separated from cell proliferation occurring either long before or after onset of mitosis (Paladini *et al.*, 1996).

1.1.2c. Keratin 1. The analysis of human K1, which together with K10 forms intermediate filaments (IFs) in the upper strata of stratified squamous epithelia, revealed an epidermis-specific expression of human K1 from a 10.8-kb fragment. This construct contained about 1.2 kb of 5′ and 4.4 kb of 3′ sequences. All three lines of mice established showed a copy-number-independent expression (Chung *et al.*, 1994). While the pattern of human K1 closely mirrored that of murine K1 during embryonic development, its expression was different in two aspects: (a) In forestomach, which in the mouse does express K1, no human K1 was found. (b) In epidermis, where K1 is expressed strictly in suprabasal cells, about 20–30% of basal cells showed human K1 expression. Moreover, a mosaic human K1 expression was observed in upper epidermis (Chung *et al.*, 1994).

Whether these deviations from the typical K1 expression resulted from a lack of regulatory elements absent in the 10.8-kb fragment or from a clonal variability of epidermal stem cells remained open (Clausen and Potten, 1990; Mackenzie, 1997, and references therein; Mackenzie and Bickenbach, 1985). An additional transgene, which carried 9.3 kb of upstream and the same 4.4 kb of downstream DNA, did show human K1 expression indistinguishable from the mouse K1 (Bickenbach *et al.*, 1996; Chung *et al.*, 1994), strongly suggesting that the mosaic expression patterns seen with the shorter 10.8-kb fragment was due to position effects. The fine mapping of *cis*-regulatory elements of the human K1 gene in mice remains to be done. Indications from cell transfection experiments pointed to a 250-bp Calzium-responsive element in the 4.4-kb fragment located 3′ downstream of the gene (Rothnagel *et al.*, 1993). Important elements must also reside in the 8.1-kb upstream fragment between −1.2 and −9.3 kb.

2. FUNCTIONAL STUDIES

2.1. Expression of Mutant Keratins

2.1.1. Mutant Keratin 14 Mice as Models for Epidermolysis Bullosa Simplex

The observation that substochiometric amounts of neurofilaments or keratin subunits (Albers and Fuchs, 1987, 1989; Gill *et al.*, 1990; Monteiro *et al.*, 1990; Wong and Cleveland, 1990) interfered with IF assembly opened a new avenue to explore IF function. The transfection of deleted keratin subunits into epithelial cells caused rearrangement or collapse of existing keratin IFs, depending on the level of deleted subunits present in cells and the extent of the deletion. In particular, deletion of the head domain or of the conserved rod end domains caused a more severe phenotype than tail or internal rod deletions (Wilson *et al.*, 1992).

In order to explore the consequences of such deletions for IF organization *in vivo*, a construct was prepared with 6 kb of the human Kl4 promoter sequences controlling the expression of a mutant K14 cDNA with a neuropeptide P tag (Vassar *et al.*, 1991). This mutant lacked 135 amino acids (aa) of its C-terminus. Transgenic mice expressing this truncated K14P cDNA suffered from a high perinatal lethality due an extremely fragile skin. Therefore, only mice derived from a mosaic founder could be analyzed. Western blotting of epidermal extracts from these mice showed expression levels of mutant K14 ranging from 5% to 90% of the total amount of K14. Histology revealed cytolysis in the basal and spinous layers of epidermis, while the granular and cornified layers remained intact. In agreement with these histological findings, immunocytochemistry revealed the expression of mutant K14 in basal and spinous layers, suggesting that its expression was responsible for the changes in epidermis. The overall program of terminal differentiation was unaltered, as judged by the normal expression of Kl/10 and filaggrin and the presence of a cornified envelope. In groups of cells, K6, one of the keratins typically found in wound healing and altered differentiation, was expressed (Kopan and Fuchs, 1989; Mansbridge and Knapp, 1987; Stoler *et al.*, 1988; Weiss *et al.*, 1984). Electron microscopy revealed the presence of keratin clumps in basal and suprabasal cells, often coinciding with cytolysis. These keratin aggregates stained positive with a neuropeptide P antibody, indicating they contained mutant keratin. In contrast, filamentous keratin showed only sparse anti-P labeling. Hemidesmosomes and desmosomes appeared normal, even in the presence of large keratin aggregates.

This study strongly suggested a close link between the existence of mutations in epidermal keratin genes and a group of inherited human skin disorders, namely epidermolysis bullosa simplex (EBS). It was shown independently by three groups (Bonifas *et al.*, 1991; Coulombe *et al.*, 1991a; Lane *et al.*, 1992) that single amino acid changes as caused by point mutations in K5 or K14 were involved in EBS. Patients suffering from the most severe, Dowling–Meara variant of EBS are clinically characterized by a generalized blistering skin phenotype leading to an increased lethality in the first years of life (Anton-Lamprecht and Schnyder, 1982; Fine *et al.*, 1991). Ultrastructural hallmarks of this disorder are keratin aggregates in basal cells, broken keratin IFs between the basement membrane and the nucleus, as well as basal cell cytolysis (Anton-Lamprecht and Schnyder, 1982; Fine *et al.*, 1991). Although K14 mutant mice mirrored the phenotype of Dowling–Meara EBS, there were also some differences: In the K14 mutant mice, many more basal cells displayed cytolysis, had few intact IFs, and numerous, large keratin aggregates. Cytolysis and K6 expression were noted in the spinous layer (Vassar *et al.*, 1991), while no K6/16 expression has been observed in EBS patients (Corden and McLean, 1996; Fine *et al.*, 1991; McLean and Lane, 1995). These differences may be due to the following: (a) A higher level of mutant K14 may be present in the mouse compared to human. (b) The truncated K14 mutant expressed in mouse could cause a more severe phenotype than point mutations in human. (c) The persistence of larger amounts of truncated K14 in the upper epidermis most likely led to cytolysis. The suprabasal expression of mutant K14 could either be due to an increased stability of the human K14 in the mouse or to the suprabasal activity of the K14 promoter fragment used in the experiment. These data, in conjunction with previous electron microscopic findings (Anton-Lamprecht, 1983), convincingly demonstrated that keratins do have an important function as a cytoskeleton in epidermis.

The same authors generated a series of transgenic mice expressing various Kl4 mutants, using a 2.3-kb fragment of the human K14 promoter (Coulombe *et al.*, 1991b). Comparing the phenotypes of mice expressing K14 mutants lacking 50 aa of the C-terminus, 117 aa of the N-terminus plus 42 aa of the C-terminus, or 135 aa of the C-terminus showed a positive correlation between the degree of filament dispruption and the severity of disease. The phenotype was more severe in animals expressing higher levels of the transgene, whereas mice with less disruptive mutants showed a tendency of local instead of generalized blistering. This was reminiscent of the clinical variablity of EBS, suggesting that mutations in the same gene can cause disorders with distinct phenotypes. However, patients harboring the same mutation often show considerable differences in the severity of their clinical phenotype, suggesting that other factors modulate disease severity (Corden and McLean, 1996).

The analysis of these transgenic mice by electron microscopy suggested that cytolysis occurred also in the absence of large keratin aggregates, thus indicating a prominent pathogenetic role of less stable keratin IF. The analysis of recovering areas of previously blistered skin in mice gave a first insight into why EBS patients healed their wounds without scarring: Regenerating keratinocytes were flat instead of having their typical columnar shape. This might have rendered them more resistent to mechanical strain. The correlation between cell shape and the composition and intracellular distribution of keratin IF is unknown, but will be a challenging area of future research. Certainly, other factors contribute to the lack of scarring, such as the strength of the suprabasal epidermis including the cornified envelope, where keratins make up more than 70% of total protein (Fuchs and Green, 1980, 1978; Lane, 1993), and the dermis. Their supportive role is emphasized in rare EBS patients completely lacking K14 (Chan *et al.*, 1994; Jonkman *et al.*, 1996; Rugg *et al.*, 1994). Despite the absence of a keratin cytoskeleton in basal epidermis, they develop a relatively mild phenotype compared to those with point mutations in the conserved rod end domains (Corden and McLean, 1996; Fuchs, 1994; McLean and Lane, 1995).

2.1.2. Mutant Keratin 10 Mice as Models for Bullous Congenital Ichthyosiform Erythroderma

Transgenic mice expressing a truncated K10/14 hybrid in their suprabasal epidermis provided the first indication that another group of skin disorders, bullous congenital ichthyosiform erythroderma (BCIE), might be caused by mutated K1 or K10. BCIE is characterized by cytolysis of suprabasal keratinocytes, local or generalized blistering in early infancy, and hyperkeratosis (Traupe, 1989). Making maximal use of their previously established K14 truncation construct, mice were generated which expressed a hybrid minigene made of 7.5 kb of human K10 promoter, the amino end of K10 including coil 1A, and parts of 1B (Fuchs *et al.*, 1992). To the amino end of K10, K14 sequences were fused in frame. In effect, this hybrid had the same 135-aa C-terminal deletion previously shown to disrupt keratin IF (Vassar *et al.*, 1991). The analysis of mice transgenic for this K10/14 hybrid was restricted to newborn founder animals with a range of phenotypes probably dependent on the transgene copy number. These mice showed some resemblance to the phenotype encountered in BCIE patients (Traupe, 1989; Wilgram and Caulfield, 1966), but were generally far more severely affected. In animals 1 day of age, cytolysis and skin blistering were observed, accompanied by a thickening of the granular (acanthosis) and the cornified layer (hyperkeratosis). Also, K6 expression was induced in postmitotic suprabasal cells and the labeling index of basal keratinocytes was increased three- to fivefold. In suprabasal cells, keratin IFs were rearranged to large aggregates, which sometimes formed perinulear shells, although less frequently than seen in BCIE patients (Ishida-Yamamoto *et al.*, 1994). The overall amount of suprabasal keratin seemed increased in these mice compared to human patients. Another interesting feature in the K10/14 transgenic mice was the observation of binucleate cells, a phenomenon previously described in BCIE patients (Wilgram and Caulfield, 1966). A possible explanation would be a mechanical obstruction imposed by keratin shells interfering with nuclear division.

Within 1 year, several reports verified the occurence of point mutations in keratins 1 and 10 among BCIE patients (Bonifas *et al.*, 1992; Chipev *et al.*, 1992; Cheng *et al.*, 1992; for a review, see Corden and McLean, 1996).

In due course, a mouse model expressing a truncated human K1 gene lacking 60 amino acids of the coil 2B region and the tail was established (Bickenbach *et al.*, 1994, 1996). In contrast to the K10/14 mice, the founder pups expressing the truncated K1 were born without a disease phenotype. Breeding to homozygosity provided a sufficiently high level of truncated K1 (although not stated how much) to cause neonatal appearance of a fragile epidermis with numerous blisters. The condition of mice improved with onset of fur growth, and, after the first months of life, blisters were only noted on foot pads. No nuclear shape changes or binucleate cells were noted, in agreement with our data (Porter *et al.*, 1996; Reichelt *et al.*, 1997). Unlike in BCIE, where keratohyalin granules were reported to increase in size and number (Ishida-Yamamoto *et al.*, 1994), they were smaller in the truncated K1-expressing mice (Bickenbach *et al.*, 1996). In the adult mice, collapsed keratin IF were no longer seen; however, acanthosis and hyperkeratosis persisted. As in the previous K10/14 mouse, K6 was expressed, although at a limited extent in the adult, which also presented only a few skin lesions. The labeling index of basal keratinocytes in back skin, which was only mildly affected, was increased fivefold. An unexpected result was that it was increased only by the same degree in the severely affected foot pad; for the relationship between epidermal thickening and proliferation, see Porter *et al.* (1998).

2.1.3. Expression of Mutant Keratin 6 in Epidermis

The human K6a promoter was used to overexpress a mutant K6a gene in mice (Takahashi and Coulombe, 1996). The mutant consisted of an internal 52-amino acid deletion spanning the head–coil 1A domain. After induction of the transgene by the phorbol ester PMA, the mutant K6a protein was expressed in the granular layer, accompanied by hyperkeratosis. Depending on the expression level and after mechanical trauma, e.g., tape stripping, cytolysis was seen. While this mouse model shared only a few characteristics of the human hyperkeratotic skin disorder ichthyosis bullosa of Siemens, which is caused by K2e mutations (McLean and Lane, 1995), it certainly has its value because of its inducible features.

2.1.4. Expression of Mutant Keratin 18

In analogy to the known point mutations in epidermal keratin genes, which have been linked to a defined group of skin disorders (Corden and McLean, 1996), analogous mutations in simple epithelial keratins could cause a clinical phenotype by acting in the same dominant-negative fashion as most their epidermal counterparts. Therefore, a point mutation was introduced into human K18 (Ku *et al.*, 1995) at an arginine position (R89), which converted it to cysteine or histidine, analogous to those at an arginine residue, which is the most frequently mutated residue in epidermal keratins (Corden and McLean, 1996). In K18 mutant mice, an increase of phosphorylation of soluble K8/18 was noted. In this setting, only liver tissue showed a recognizable phenotype, resulting from a widespread collapse of keratin IF, leaving mostly small aggregates throughout hepatocytes. Further hallmarks were liver necrosis and inflammation and a twofold increase in serum transglutaminase (Ku *et al.*, 1995). Among other organs, only pancreas and colon showed cytoskeletal abnormalities which were not associated with a clinical phenotype.

This study underlined the importance of a conserved arginine residue in the integrity of simple epithelial and epidermal keratins. In terms of keratin biology, it is notable that the effect of the mutations on cytoskeletal integrity was manifest only in liver and very mildly in pancreas and colon, which in addition to K18 express the type I keratins 19 and 20 in some or many cells (Moll *et al.*, 1982, 1990). This is different from epidermal keratin mutations which lead to a disease phenotype irrespective of the complexity of the number of keratins present. It is unknown whether this is due to distinct properties of various keratins or of their associated proteins. Another explanation would be that the cytoskeletal role of keratins is more pronounced in epidermis than in simple epithelia, which is supported by findings in keratin knockout mice (Baribault *et al.*, 1994, 1993; Magin *et al.*, 1998). The results from mutant K18 mice (Ku *et al.*, 1995) further raised the issue of whether chronic liver inflammation in humans could be a consequence of mutations in K8 or K18. At least one patient suffering from liver disease of unknown etiology has been described with a K18 mutation (Ku *et al.*, 1997).

Extending their initial study (Ku *et al.*, 1996), these authors provided good evidence that expression of mutant K18 at appropriate levels led to an increased susceptibility to liver toxins, namely griseofulvin and acetaminophen (Ku *et al.*, 1996). The administration of these drugs resulted in an increased mortality, associated with histopathological and biochemical changes in the liver, when compared to normal or transgenic mice overexpressing human K18. In these mice, the keratin IF appeared diffuse, but no Mallory bodies (Hazan *et al.*, 1986), which are the hallmark of long-term griseofulvin treatment in mice or of alcoholic hepatitis in humans, were detected (Denk and Lackinger, 1986). A comparison of keratin phosphorylation before and after griseofulvin treatment revealed significant changes. While the overall phosphorylation rate was increased, the mutant K18 mice showed a much lower degree of keratin phosphorylation than normal mice or those overexpressing human K18.

Whether the changes in keratin phosphorylation and glycosylation (Ku *et al.*, 1996) are related to the increased susceptibility to drug-induced hepatotoxicity remains to be proven. Certainly, hepatocytes expressing the mutant K18 are more fragile than normal cells (Ku *et al.*, 1996). However, additional factors, for example, infectious ot toxic agents, are likely to be involved in the development of extensive liver damage.

2.2. Ectopic and Overexpression of Keratins

2.2.1. Keratin 14 Expression in Liver

Recent studies used the well-established, epitope-tagged human K14 cDNA, which was linked to the transthyretin promoter for expression in the liver of transgenic mice (Albers *et al.*, 1995). These mice developed liver degeneration, consisting of infiltration of inflammatory cells, accumulation of lipids and glycogen, and a decrease in bile acid flow. The observed changes accumulated with age, similar to those findings in old K18 knockout mice (T. M. Magin *et al.*, unpublished). K14P-mutant mice shared several common features with those expressing mutant K18, including the lack of Mallory bodies (Ku *et al.*, 1996), but had a less severe phenotype. This could be due to the low level of K14P, which accumulated to less than 1% of total protein (Albers *et al.*, 1995), while mutant K18 was present at similar or even higher levels than its murine wild-type orthologue (Ku *et al.*, 1995). In addition, cell transfection experiments indicated that K14P formed IFs with K8 (Albers *et al.*, 1995; Lu and Lane, 1990). In contrast, mutant K18 seemed unable to form extended IFs *de novo* and disrupted preexisting filaments (Ku *et al.*, 1995). The reason why K14P caused a disease phenotype at all remained unclear. It could be speculated that it is a consequence of K14P incorporation into IFs, leading to a decreased stability of K8/14/18 IFs compared to K8/18 filaments. The same study showed an altered interaction of K14P-containing filaments with desmosomes leading to a widened intercellular gap (Albers *et al.*, 1995). This could be a consequence of altered interactions between keratin IFs and desmosomal proteins, which have been proposed to mediate force transduction between cells (Arnn and Staehelin, 1981).

On the other hand, a number of transgenic mouse lines have been established to overexpress normal human keratins at their normal sites, including K8 (Casanova *et al.*, 1995), K18 (Abe and Oshima, 1990), K19 (Bader and Franke, 1990), K14 (Vassar *et al.*, 1989), and K1 (Bickenbach *et al.*, 1996, 1994). Possibly, a subtle phenotype in these mice has gone unreported, or they have not been kept long enough to develop any clinical abnormalities. The data allow the following predictions: Expression of an orthologous keratin in its "natural" tissue should have no consequences as long as the amount of transgene protein is balanced by the corresponding type I or II keratin partner. Beyond a certain degree, overexpression of a single keratin in its original cell type may cause a clinical phenotype for various reasons. Balanced overexpression of a keratin pair in orthotypic tissues, though not yet reported, might be tolerated up to a certain amount, beyond which a clinical phenotype develops. If this balanced state is disrupted by ectopic expression of a keratin subunit, this might lead to a clinical manifestation (Blessing *et al.*, 1993; Powell and Rogers, 1990).

2.2.2. Expression of Sheep Wool Keratin in Hair Cells

The first overexpression of a keratin used a type II sheep wool keratin gene for expression in hair cells of transgenic mice (Powell and Rogers, 1990). Those mice carrying 250 transgene copies developed a recurring pattern of near-complete hair loss and regrowth, starting at day 19 after birth. This cycle followed the natural progression of the mouse hair cycle. The overexpression of the sheep wool type II keratin was accompanied by a decrease of keratin-associated high-sulfur proteins typical of hair, possibly due to a mechanical obstruction in the cytoplasm of hair cells. An explanation could be a delayed translation of mRNA coding for IF-associated proteins, which are expressed late in hair development. Trangenic hair was wavy and brittle, causing hair to break off close to the skin surface (Powell and Rogers, 1990). This study strongly indicated that a proper balance among keratins and their associated proteins is required for normal growth, maintenance, and structure of hair.

2.2.3. Expression of Keratin 16 in Epidermis

For the overexpression of human Kl6 in transgenic mice, a genomic fragment which included 6 kb of 5′ and 1.5 kb of 3′ noncoding sequences was used (Takahashi *et al.*, 1994). With increasing transgene copy number, mice developed acanthotic and hyperkeratotic skin lesions with erythemae, starting a few days after birth. Mice expressing large amounts of human K16 had a strikingly increased mortality rate in the postnatal period, most likely due to their cutaneous phenotype. Immunohistochemistry confirmed the expression of human K16 mostly in suprabasal epidermis, where a variety of changes were noted. K1 and 10 were found to be reduced, while endogenous mouse K6, and probably also K16 and K17, were increased. The expression of K14 persisted in suprabasal layers, as had been found in transgenic mice expressing either wild-type K14 or a K14 mutant (Vassar *et al.*, 1991, 1989). The expression of filaggrin, an epidermal keratin-associated protein, was drastically decreased. The latter is in contrast to all other keratin transgenic or knockout mice developed as animal models for BCIE (Bickenbach *et al.*, 1996, 1994; Fuchs *et al.*, 1992; Porter *et al.*, 1996; Reichelt *et al.*, 1997). Among morphological changes, a thickening of the outer root sheath of hair follicles, acanthosis, and hyperkeratosis of interfollicular epidermis were noted. In addition to its suprabasal expression, human K16 was also found in a number of basal keratinocytes, where keratin aggregates were formed. The consequences of this inappropriate expression have not been studied in further detail, but may be highly significant in the context of the encountered phenotype as well as the observed reduction of K1/10 and filaggrin expression in the upper epidermis. Besides these changes, a split of desmosomes was observed in some cells, surprisingly not accompanied by cytolysis. In order to explain the described phenotype, the authors speculated that human K16 with its intrinsic sequence properties could behave as a mutant in mice. Alternatively, the expression of human K16 may interfere with the balance of type I keratins. They suggested that K16 in general had unusual assembly properties. This was further supported in a subsequent study (Paladini *et al.*, 1995) and agreed with ultrastructural findings in K10 knockout mice (Reichelt *et al.*, 1997). However, along with the increase of K16 in transgenic mice, there was also an increase of endogenous mouse K6 and 16. Their contribution is not separable from that of human K16 in "classic" transgenes, but would require a combination of K6/16 knockout mice which could be crosed with conventional transgenic mice expressing human K16.

2.2.4. Expression of Keratins 1 and 10 in Pancreas

A well-documented and sometimes overlooked study on ectopic keratin expression was carried out by directing the expression of human K1 and K10, either separately or together, to beta cells of pancreas, using the rat insulin promoter (Blessing *et al.*, 1993). While mice expressing K10, which coassembled with endogenous K8 and 18, showed no clinical phenotype, those expressing K1 alone or in combination with K10 developed a specific form of insulin-dependent diabetes mellitus, starting between 2 and 8 weeks of age. This was accompanied by a reduction of insulin/K1-producing cells and of insulin-secretory vesicles. The underlying reasons for the selective development of diabetes mellitus in K1- but not K10-expressing mice are not known. With respect to the balance of keratins in a given cell or tissue, one can conclude that the type of individual keratin expressed is of central importance. Beyond the manifestation of diabetes mellitus, two important conclusions can be drawn from this study: (a) As revealed by the expression of a single keratin in pancreatic islet cells of transgenic mice, either of the epidermal keratins 1 or 10 coassembled with the endogenous K8/18 keratins to a cytoskeleton which appeared normal by ultrastructural terms. (b) In double-transgenic mice, K1/10 formed densely packed filament arrays closely resembling those seen in the uppper strata of epidermis. It has been assumed by the majority of researchers that this so-called "keratin pattern" is a result of IF bundling through filaggrin (Dale *et al.*, 1994). Although the presence of filaggrin analogues in pancreatic islet cells could not be excluded, it seemed likely that the typical IF bundling resulted from the intrinsic structural properties of K1/10, in agreement with previous *in vitro* data (Eichner *et al.*, 1986).

Considering the limited number of mouse models, it is difficult to predict whether only some or all keratin subunits carry a "disease potential." Deletion analysis and binding studies, possibly including the yeast-two-hybrid system (Fields and Sternglanz, 1994), might help to unravel the mechanisms involved. One final aspect to consider here is the use of keratin genes from other species, mostly human, in transgenic mice. Due to the high degree of sequence similarity of orthologues, it seems safe to assume that basic keratin properties are interchangeable among different species. However, if one doubts that "a mouse is a mouse is a mouse," one might argue that minor sequence differences between keratins lead to subtle changes which can go unnoted in transgenic animals.

From the data discussed so far, it is apparent that transgenic mice expressing dominant-negative keratin mutants had a major impact in elucidating keratin function. Although imperfect animal models for the human skin disorders EBS and BCIE (Fuchs *et al.*, 1992; Vassar *et al.*, 1991), they provided the first evidence that epidermal keratins function as esssential cytoskeletal proteins. Most importantly, these results initiated the investigation of functionally important mutations of human keratin genes as compared to a "neutral" polymorphisms.

3. KERATIN KNOCKOUT MICE

Knockout mice have provided exciting and sometimes unexpected findings in keratin biology. So far, three groups of keratin genes have been knocked out: epidermal keratin genes 14 and 10, the corneal-specific gene K12, and the simple epithelial keratin genes 8, 18, and 19.

3.1. Knockout Mice for Epidermal Keratins

3.1.1. Keratin 14 Knockout Mice

K14 knockout mice were generated by removing sequences around and including the first exon to produce a complete knockout (Lloyd *et al.*, 1995). Homozygotes displayed a generalized blistering skin disease associated with an increased mortality, whereas heterozygotes showed no clinical abnormalities. Survivers improved and showed blisters only at hair-free body sites. While the skin condition improved with age, the overall condition of mice deteriorated, and by 3 months of age all homozygotes were reported dead. Closer analysis of several tissues including epidermis and esophagus revealed a significant, however incomplete reduction of basal keratinocyte IFs. This might have been expected if K5 and 14 were the only keratins present in the basal layer of epidermis. Although K15 had previously been described as a minor keratin (Leube *et al.*, 1988; Moll *et al.*, 1982), its functional significance has been convincingly highlighted by K14 knockout mice. In basal keratinocytes, the overall number of keratin filaments was reduced when compared to wild-type animals. The subcellular distribution of K5/15 filaments was altered to a predominant localization in areas close to desmosomes and hemidesmosomes. Filaments also appeared shorter (Lloyd *et al.*, 1995). In addition, cytolysis of basal cells was noted. This could have been a consequence of either the reduced number of keratin IFs or of the different properties of K5/15, which might not be functional equivalents for the normal K5/14/15 filaments. In agreement with the former hypothesis, the analysis of esophagus demonstrated a high expression of K15 at birth, when no tissue damage was present. On the other hand, the reduction of K15 expression in adults was paralleled by an increased fragility of the esophagus. However, external factors, such as the change of diet from liquid to solid, might be a contributing factor (Lloyd *et al.*, 1995). As in human EBS, no K6/16 induction in suprabasal epidermis was detected (in contrast to the opposite observation in mutant K14 mice; Vassar *et al.*, 1991). The analysis of other tissues in K14 knockout mice showed no clear correlation between the integrity of a given tissue and the number of keratin IF present. This suggested that the relationships between cell shape, cytoskeleton, and its associated proteins are distinctive features of individual cells and tissues. In a genetic context, K14-null mice closely resembled those patients with K14-null mutations (Chan *et al.*, 1994; Jonkman *et al.*, 1996; Rugg *et al.*, 1994). While the latter are only mildly affected, leading to the diagnosis of EBS-Koebner (Fine *et al.*, 1991), the mice developed a far more severe phenotype, closely resembling the human Dowling–Meara variant of EBS caused by K5 or 14 point mutations (Corden and McLean, 1996; Coulombe *et al.*, 1991b). These striking differences seen in mice and humans could be related to the contribution of genetic modifiers (Baribault *et al.*, 1994) or to different biological properties of human and murine skin. The rediscovered importance of K15 in K14 knockout mice merits its evaluation in subforms of EBS with involvement of vagina, esophagus, and a few other tissues.

3.1.2. Keratin 10 Knockout Mice

Our group has described keratin knockout mice which turned out to be a very good model for BCIE (Porter *et al.*, 1996, 1997; Reichelt *et al.*, 1997). Homozygous mice had extremely fragile skin with generalized blistering and died in the first hours after birth. Heterozygotes appeared inconspicuous at birth, but developed scaley, hyperkeratotic skin before the onset of fur growth. In adults, hyperkeratotic lesions and erosions were restricted to tail, paw, and the ear in hairy mice, while in hairless mice, a mild generalized hyperkeratosis was noted. The unexpected phenotype in heterozygotes was then shown to be the consequence of a truncated keratin 10 fragment expressed in epidermis as a result of the design of the targeting vector. This fragment comprised the head, coil 1A, and 27 amino acids of coil 1B and was present in detergent-insoluble cytoskeletal preparations (Porter *et al.*, 1996; Reichelt *et al.*, 1997). Similar to BCIE, acanthosis and hyperkeratosis in both genotypes of mice was observed, but far more severely in homozygotes. There was an inverse relationship between the number of intact keratin filaments in suprabasal epidermis and the degree of cytolysis. In conjunction with the reduced number of keratin aggregates in homo- compared to heterozygotes, it became evident that cell fragility in BCIE, as in EBS (Coulombe *et al.*, 1991b; Lloyd *et al.*, 1995), is a consequence of reduced mechanical stability resulting from the reduced number of normal IFs and not of an increase in keratin aggregates as previously suggested (Vassar *et al.*, 1991). In line with BCIE, an increase in K6/16/17 expression was found in the upper epidermis of K10 knockout mice (Porter *et al.*, 1996, 1997; Reichelt *et al.*, 1997). A striking observation in K10 knockout or conventional transgenic animals (Bickenbach *et al.*, 1994, 1996; Fuchs *et al.*, 1992) as well as in BCIE patients (Corden and McLean, 1996) was that the upregulation of K6/16/17 did not improve the stability of their skin. Based on the assumption that K6/16 are unable to form stable, extended IFs (Coulombe *et al.*, 1995; Paladini *et al.*, 1995), it is tempting to speculate that the phenotype of K10 mice and BCIE patients is not only a consequence of absent or mutant K1/10 IF, but may also result from the poor IF-forming capacity of K6/16 (Coulombe *et al.*, 1995; Paladini *et al.*, 1995). In fact, cytolysis was prevalent in upper spinous and granular cells, where K6/16 expression was upregulated in mice. The absence of cytolysis in lower spinous cells coincided with the lack of K6/16 expression (Reichelt *et al.*, 1997). In addition, the expression of K5/14 in lower spinous cells could be a contributing factor to cell integrity.

Another compelling feature in K10 knockout mice was the increased expression of filaggrin in the granular layer, similar to that noted in BCIE (Ishida-Yamamoto *et al.*, 1994). Electron microscopy revealed a perturburbation of keratin–filaggrin interactions. Instead of the normal K1/10 association, predominant K6/16–filaggrin interactions were observed. The altered composition and ultrastructure of keratins probably resulted in an altered barrier function of epidermis, to which filaggrin might contribute (Steinert and Marekov, 1995) as indicated by an increased water loss in homozygotes (Reichelt *et al.*, 1997; and unpublished data) and an altered lipid composition of the cornified layer in K10 knockout mice (T. Doering, K. Sandhoff, J. Reichelt, and T. M. Magin, unpublished). Finally, a loss of K2e protein in paw sole epidermis was noted in homozygous mice, while its mRNA level remained unchanged. This would argue that in normal paw epidermis, both K1 and K2e form IFs with K10. In homozygotes, residual K1 preferentially assembled with truncated K10. This resulted in the degradation of K2e.

Epidermal thickening is a phenomenon common to many genodermatoses, but little is known about the underlying causes. In K10 knockout mice, a highly variable degree of epidermal thickening has been observed, offering an exciting opportunity to investigate the relationship between keratin mutations and epidermal homeostasis. Therefore, we performed bromodeoxyuridine, proliferating cell nuclear antigen (PCNA), and Ki-67 labeling studies and compared the extent of labeling at various body sites to the degree of skin thickening (Porter *et al.*, 1998). These data indicated that a significant increase of basal cell proliferation was limited to a few sites such as ear and eyelid epidermis. Other sites, such as trunk or foot pad epidermis, showed no increase in their labeling index. This strongly suggested that an increase in cell proliferation is only partially responsible for epidermal thickening. Other mechanisms, such as decreased desquamation, are likely to play a major role. No positive correlation between cell proliferation and K6/16 expression was detectable. The latter seemed to result from the extent of tissue fragility caused by the K10 knockout (Reichelt *et al.*, 1997).

3.1.3. Keratin 12 Knockout Mice

Keratins 3 and 12 are the corneal-specific equivalents of epidermal keratins 1 and 10 (Lane, 1993; Moll *et al.*, 1982). Knocking out the K12 gene led to a mild corneal dystrophy in homozygotes without a noticeable clinical phenotype in heterozygotes (Kao *et al.*, 1996). Upon gentle mechanical stress, corneal cells appeared fragile and could easily be removed from the underlying tissue. In contrast to skin disorders with keratin mutations, the number of cell layers was decreased. This was also seen in corneas of K14-deficient mice, and these authors suggested that cytolysis reduced the number of migrating cells, which in cornea have to move centripetally due to the location of stem cells (Lloyd *et al.*, 1995). In analogy to epidermis of K14-null animals, basal cell IFs retained their normal appearance, but their overall amount was shown to be reduced. Upper cells had even less keratin IF, and superficial cells were devoid of keratin IFs (Kao *et al.*, 1996). The importance of keratins for the integrity of cornea was also confirmed by the recent discovery of point mutations in the genes coding for K3 or K12, which are associated with inherited, autosomal-dominant Meesmann corneal dystrophy (Irvine *et al.*, 1997).

3.2. Knockout Mice for Simple Epithelial Keratins

In early embryonic development and in several simple epithelia of the adult, K8 and 18 are the only keratins coexpressed. Based on this observation, one might have predicted that K8 or K18 knockouts would manifest in early embryonic death.

3.2.1. Keratin 8 Knockout Mice

In was quite surprising that ES cells in which both K8 alleles were targeted formed embryoid bodies indistinguishable from wild-type ES cells, despite the absence of keratin IF (Baribault and Oshima, 1991). Residual amounts of K18 and K19 were colocalized in a nonfilamentous form along desmosomes. In functional terms, it was documented that K8-depleted embryoid bodies secreted alpha-fetoprotein and laminin in a normal manner.

This study was extended by the generation of K8 knockout mice (Baribault *et al.*, 1994, 1993), which, depending on the genetic background, had different phenotypes due to an incomplete penetrance of the targeted K8 allele. In the C57/Bl6 strain of mice, the majority of animals died around day 12, after major organ formation had taken place, indicating that embryonic development up to this stage can proceed without a keratin cytoskeleton. Analysis of the K8 litter around day 12 showed that homozygous K8 pups appeared growth-retarded and suffered from internal bleeding, possibly due to the rupture of fetal hepatocytes. If mechanical fragility at this time of development occurs as a consequence of the K8 knockout, it is reminiscent of plakoglobin null mice dying of a "broken heart" around embryonic day 11 (Ruiz *et al.*, 1996).

By crossing the K8 allele into the FVB/N strain of mice, about 55% of homozygotes survived to adulthood (Baribault *et al.*, 1994). These developed hyperplastic lesions in all parts of the colon, but not in the small intestine. Typically, villi were extended 5–10 times, accompanied by submucosal inflammation. No major changes were reported in other organs, although a few older animals showed liver necrosis and fibrosis. It was observed, however, that K8-null females were sterile for yet-unknown reasons. Recently, the capacity of K8 embryos for wound healing was analyzed (Brock *et al.*, 1996). This showed that after amputation of the hindlimb of 11.5-day embryos, wound healing was normal in K8-null mice, possibly because cell migration relies predominantly on the actin cytoskeleton. The reason for the incomplete penetrance in different mouse strains is not understood. Evidence suggests that unknown genetic factors are more likely to be involved rather than the expression of other type II keratins which could complement the missing K8.

3.2.2. Keratin 18 Knockout Mice

Since K8 and 18 represent subunits of the same 10-nm filament (Lane, 1993), a similar phenotype for K18 knockout mice mice might have been expected. Surprisingly, K18-null mice were viable and fertile and did not suffer from midgestational death or colorectal hyperplasia (Magin *et al.*, 1998). As in K8-null mice, in K18-null mice hepatocytes were completely devoid of keratin IFs as shown by Western blotting and light and electron microscopy. Even in the same mouse strain as K8 knockout mice (C57/B16), K18 animals showed no phenotypic abnormalities in embryonic or young adult states. In particular, hepatocytes appeared to be normal and growth retardation was not observed as with K8-null mice (Baribault *et al.*, 1993). This strongly indicates that even for keratins expressed in the same tissue a gene knockout can have different consequences.

There are at least four more interesting observations in K18 knockout mice. (a) It was shown by light and electron microscopy that in hepatocytes desmosomes are formed and maintained even in the absence of IFs. Although not formally shown in adults, the same was found in K8-negative embryoid bodies (Baribault and Oshima, 1991). (b) K7, a type II keratin highly similar to K8, was absent or drastically reduced in several tissues of K18-null mice, despite continued transcription of its mRNA. This was most obvious in uterus, which normally expresses K7, 8, 18, and 19. In this tissue of K18 knockout mice we detected only K8 and 19, but no K7 in the cytoskeletal fraction, but small amounts of K7 in the soluble pool. It is highly suggestive that in contrast to previous *in vitro* obeservations (Hatzfeld and Franke, 1985), keratins *in vivo* are much more selective and form IFs preferentially from K8/18, K8/19, and K7/18, but not from K7 and K19. The molecular basis for selective assembly properties is not known. (c) The deletion of K18 left the tailless K19 as the only type I keratin in some tissues. This enabled us to analyze structures formed between K8 and K19. By immune electron microscopy it was shown that K8/19 formed bona fide IFs *in vivo*, with normal length and diameter. These interacted in a typical manner with desmosomes. In the rapidly dividing intestinal and colon epithelia, K8/19 IFs seemed to be able to functionally compensate for the normally expressed K8/18/19 IFs. (d) Old K18 mice (about 18 months old) were found to reexpress K8 in hepatocytes. In the absence of other keratins, K8 was deposited in small dots in the cell periphery and in large, irregular aggregates in the cytoplasm. These were identified as Mallory bodies by several criteria (T. M. Magin, C. Stumptner, H. Denk, and K. Zatloukal, unpublished). In conjunction with the absence of Mallory bodies in keratin-negative hepatocytes of young K18- (and of K8-) null mice, these findings strongly suggest that the presence of K8 is a minimal prerequisite for Mallory body formation. If the development of these specific cell inclusions is an indicator of chronic liver disease, our findings point toward a protective role for keratins in hepatocytes.

4. CONCLUSIONS AND PERSPECTIVES

The analysis of transgenic mice has been extremely useful in demonstrating the importance of keratins as cytoskeletal proteins and in correcting the simplified picture of gene regulation suggested by cell transfection experiments. While certainly valuable at their time, it seems now that "classic" transgenic mice generated by pronuclear injection are superceded by those carrying targeted mutations introduced by gene targeting in ES cells. One drawback in conventional transgenic animals is the copy number effect, which is difficult to interpret when dealing with heterooligomeric protein complexes like keratins.

The effect of dominant-negative mutations is incompletely understood. Is it solely a qualitative or quantitative effect of the mutant relative to the wild type? Or are there additional secondary effects like the induction of K6 and K16 in suprabasal epidermis as a consequence of genetic manipulation? Can we really assume that the use of orthologues in transgenic mice provides the same answers as one would get from analyzing endogenous homologues? In their review of keratin-related disorders, Corden and McLean (1996) pointed out that the same mutation in the same gene can give rise to a wide range of phenotypes, even in the same family. This would strongly argue in favor of using endogenous genes, as highlighted in experiments expressing neurofilaments in transgenic mice (Lee and Cleveland, 1996).

Over the past few years, refined gene targeting methods have become available to establish better models for human diseases or to analyze regulatory regions of genes in their natural environment. Such methods include double-replacement gene targeting (Askew *et al.*, 1993; Stacey *et al.*, 1994) to introduce precise alterations into the germline. The development of the cre–lox system (Jiang and Gridley, 1997) allows tissue-specific gene alterations and, in combination with regulatable promoters (Gossen *et al.*, 1995; Kistner *et al.*, 1996), should permit the analysis of stage-specific gene alterations. It might be useful to standardize mouse strains in order to minimize phenotypic variation and, as they become established, to make cre-expressing mice freely available.

Transgenics will need complementation by improving methods to culture cells in organotypic fashion. This could help to evaluate gene therapy protocols. Finally, the beginning application of the yeast-two-hybrid system (Fields and Sternglanz, 1994; Meng *et al.*, 1996) will be a useful extension of current methods studying protein interactions.

ACKNOWLEDGMENTS. I am most grateful to Dr. Rolf Schröder for stimulating discussion and active help in finishing the manuscript. Work in my lab is supported by the Deutsche Forschungsgemeinschaft (SFB 284) and the Bonner Forum Biomedizin.

5. REFERENCES

Abe, M., and Oshima, R. G., 1990, A single human keratin 18 gene is expressed in diverse epithelial cells of transgenic mice, *J. Cell Biol.* **111:**1197–1206.

Albers, K., and Fuchs, E., 1987, The expression of mutant epidermal keratin cDNAs transfected in simple epithelial and squamous cell carcinoma lines, *J. Cell Biol.* **105:**791–806.

Albers, K., and Fuchs, E., 1989, Expression of mutant keratin cDNAs in epithelial cells reveals possible mechanisms for initiation and assembly of intermediate filaments, *J. Cell Biol.* **108:**1477–1493.

Albers, K. M., Davis, F. E., Perrone, T. N., Lee, E. Y., Liu, Y., and Vore, M., 1995, Expression of an epidermal keratin protein in liver of transgenic mice causes structural and functional abnormalities, *J. Cell Biol.* **128:**157–169.

Andersen, B., Schonemann, M. D., Flynn, S. E., Pearse, R. V. d., Singh, H., and Rosenfeld, M. G., 1993, Skn-1a and Skn-1i: Two functionally distinct Oct-2-related factors expressed in epidermis, *Science* **260:**78–82; Erratum, *Science* **262:**78–82 (1993).

Andersen, B., Weinberg, W. C., Rennekampff, O., McEvilly, R. J., Bermingham, Jr., J., Hooshmand, F., Vasilyev, V., Hansbrough, J. F., Pittelkow, M. R., Yuspa, S. H., and Rosenfeld, M. G., 1997, Functions of the POU domain genes Skn-1a/i and Tst-1/Oct-6/SCIP in epidermal differentiation, *Genes Dev.* **11:**1873–1884.

Antequera, F., Boyes, J., and Bird, A., 1990, High levels of *de novo* methylation and altered chromatin structure at CpG islands in cell lines, *Cell* **62:**503–514.

Anton-Lamprecht, I., 1983, Genetically induced abnormalities of epidermal differentiation and ultrastructure in ichthyoses and epidermolyses: Pathogenesis, heterogeneity, fetal manifestation, and prenatal diagnosis, *J. Invest. Dermatol.* **81:**149s–156s.

Anton-Lamprecht, I., and Schnyder, U. W., 1982, Epidermolysis bullosa herpetiformis Dowling-Meara, *Dermatology* **164:**221–235.

Arnn, J., and Staehelin, L. A., 1981, The structure and function of spot desmosomes, *Int. J. Dermatol.* **20:**330–339.

Askew, G. R., Doetschman, T., and Lingrel, J. B., 1993, Site-directed point mutations in embryonic stem cells: A gene-targeting tag-and-exchange strategy, *Mol. Cell. Biol.* **13:**4115–4124.

Bader, B. L., and Franke, W. W., 1990, Cell type-specific and efficient synthesis of human cytokeratin 19 in transgenic mice, *Differentiation* **45:**109–118.

Baribault, H., and Oshima, R. G., 1991, Polarized and functional epithelia can form after the targeted inactivation of both mouse keratin 8 alleles, *J. Cell Biol.* **115:**1675–1684.

Baribault, H., Price, J., Miyai, K., and Oshima, R. G., 1993, Mid-gestational lethality in mice lacking keratin 8, *Genes Dev.* **7:**1191–1202.

Baribault, H., Penner, J., Iozzo, R. V., and Wilson-Heiner, M., 1994, Colorectal hyperplasia and inflammation in keratin 8-deficient FVB/N mice, *Genes Dev.* **8:**2964–2973.

Bickenbach, J. R., Rothnagel, J. R., Longley, M. A., Bundman, D. S., Dominey, A. M., and Roop, D. R., 1994, An animal model for the dominantly inherited skin disease epidermolytic hyperkeratosis (EHK), *J. Invest. Dermatol.* **102:**545–552.

Bickenbach, J. R., Longley, M. A., Bundman, D. S., Dominey, A. M., Bowden, P. E., Rothnagel, J. A., and Roop, D. R., 1996, A transgenic mouse model that recapitulates the clinical features of both neonatal and adult forms of the skin disease epidermolytic hyperkeratosis, *Differentiation* **61:**129–139.

Bird, A. P., 1986, CpG-rich islands and the function of DNA methylation, *Nature* **321:**209–213.

Blessing, M., Ruther, U., and Franke, W. W., 1993, Ectopic synthesis of epidermal cytokeratins in pancreatic islet cells of transgenic mice interferes with cytoskeletal order and insulin production, *J. Cell Biol.* **120:**743–755.

Bonifas, J. M., Rothman, A. L., and Epstein, E. H., 1991, Epidermolysis bullosa simplex: Evidence in two families for keratin gene abnormalities, *Science* **2554:**1202–1205.

Bonifas, J. M., Bare, J. W., Chen, M. A., Lee, M. K., Slater, C. A., Goldsmith, L. A., and Epstein, E. H., 1992, Linkage of the epidermolytic hyperkeratosis phenotype and the region of the type II keratin cluster on chromosome 12, *J. Invest. Dermatol.* **99:**524–527.

Brock, J., McCluskey, J., Baribault, H., and Martin, P., 1996, Perfect wound healing in the keratin 8 deficient mouse embryo, *Cell Motil. Cytoskel.* **35:**358–366.

Byrne, C., and Fuchs, E., 1993, Probing keratinocyte and differentiation specificity of the human K5 promoter *in vitro* and in transgenic mice, *Mol. Cell. Biol.* **13:**3176–3190.

Carroll, J. M., Albers, K. M., Garlick, J. A., Harrington, R., and Taichman, L. B., 1993, Tissue- and stratum-specific expression of the human involucrin promoter in transgenic mice, *Proc. Natl. Acad. Sci. USA* **90:**10270–10274.

Casanova, L., Bravo, A., Were, F., Ramirez, A., Jorcano, J. J., and Vidal, M., 1995, Tissue-specific and efficient expression of the human simple epithelial keratin 8 gene in transgenic mice, *J. Cell Sci.* **108:**811–820.

Chan, Y.-M., Anton-Lamprecht, I., Yu, Q.-C., Jäckel, A., Zabel, B., Ernst, J.-P., and Fuchs, E., 1994, A human keratin 14 "knockout": The absence of K14 leads to severe epidermolysis bullosa simplex and a function for an intermediate filament protein, *Genes Dev.* **8:**2574–2587.

Cheng, J., Syder, A. J., Yu, Y. C., Letai, A., Paller, A. S., and Fuchs, E., 1992, The genetic basis of epidermolytic hyperkeratosis: A disorder of differentiation-specific epidernal keratin genes, *Cell* **70:**811–819.

Chiper, C. C., Korge, B. P., Markova, N., Bale, S. J., Di Giovanna, J. J., Compton, J. G., and Steinert, P. M., 1992, A leucin–proline mutation in the H1 subdomain of keratin 1 causes epidermolytic hypakeratosis, *Cell* **70:**821–828.

Chung, S. Y., Cheng, C. K., Rothnagel, J. A., Yu, S. H., Nakazawa, H., Mehrel, T., Hohl, D., Rosenthal, D. S., Steinert, P. M., Yuspa, S. H., and Roop, D. R., 1994, Expression of the human keratin gene (K1) in transgenic mice is tissue- and developmental-specific but altered with respect to differentiation state, *Mol. Cell. Differentiation* **2:**61–81.

Clausen, O. P., and Potten, C. S., 1990, Heterogeneity of keratinocytes in the epidermal basal cell layer, *J. Cutan. Pathol.* **17:**129–143.

Collin, C., Ouhayoun, J.-P., Grund, C., and Franke, W. W., 1992, Suprabasal marker proteins distinguishing keratinizing squamous epithelia: Cytokeratin 2 polypeptides of oral masticatory epithelium and epidermis are different, *Differentiation* **51:**137–148.

Corden, L. D., and McLean, W. H., 1996, Human keratin diseases: Hereditary fragility of specific epithelial tissues, *Exp. Dermatol.* **5:**297–307.

Coulombe, P. A., Hutton, M. E., Letai, A., Hebert, A., Paller, A. S., and Fuchs, E., 1991a, Point mutations in human keratin 14 genes of epidermolysis bullosa simplex patients: Genetic and functional analyses, *Cell* **66:**1301–1311.

Coulombe, P. A., Hutton, M. E., Vassar, R., and Fuchs, E., 1991b, A function for keratins and a common thread among different types of epidermolysis bullosa simplex diseases, *J. Cell Biol.* **115:**1661–1674.

Coulombe, P. A., Bravo, N. S., Paladini, R. D., Nguyen, D., and Takahashi, K., 1995, Overexpression of human keratin 16 produces a distinct skin phenotype in transgenic mouse skin, *Biochem. Cell Biol.* **73:**611–618.

Dale, B. A., Resing, K. A., and Presland, R. B., 1994, Keratohyalin granule proteins, in *The Keratinocyte Handbook* (I. M. Leigh, E. B. Lane, and F. M. Watt, eds.), pp. 323–350, Cambridge University Press, Cambridge.

Denk, H., and Lackinger, E., 1986, Cytoskeleton in liver diseases, *Semin. Liver Dis.* **6:**199–211.

Eichner, R., Sun, T. T., and Aebi, U., 1986, The role of keratin subfamilies and keratin pairs in the formation of human epidermal intermediate filaments, *J. Cell Biol.* **102:**1767–1777.

Fields, S., and Sternglanz, R., 1994, The two-hybrid system: An assay for protein–protein interactions, *Trends Genet.* **10:**286–292.

Fine, J. D., Bauer, E. A., Briggaman, R. A., Carter, D. M., Eady, R. A., Esterly, N. B., Holbrook, K. A., Hurwitz, S., Johnson, L., Lin, A., *et al.*, 1991, Revised clinical and laboratory criteria for subtypes of inherited epidermolysis bullosa. A consensus report by the Subcommittee on Diagnosis and Classification of the National Epidermolysis Bullosa Registry, *J. Am. Acad. Dermatol.* **24:**119–135.

Fuchs, E., 1994, Intermediate filaments and disease: Mutations that cripple cell strength, *J. Cell Biol.* **125:**511–516.

Fuchs, E., and Green, H., 1980, Changes in keratin gene expression during terminal differentiation of the keratinocyte, *Cell* **19:**1033–1042.

Fuchs, E., and Green, H., 1978, The expression of keratin genes in epidermis and cultured epidermal cells, *Cell* **15:**887–897.

Fuchs, E., and Weber, K., 1994, Intermediate filaments: Structure, dynamics, function and disease, *Annu. Rev. Biochem.* **63:**345–382.

Fuchs, E., Esteves, R. A., and Coulombe, P. A., 1992, Transgenic mice expressing a mutant keratin 10 gene reveal the likely genetic basis for epidermolytic hyperkeratosis, *Proc. Natl. Acad. Sci. USA* **89:**6906–6910.

Gill, S. R., Wong, P. C., Monteiro, M. J., and Cleveland, D. W., 1990, Assembly properties of dominant and recessive mutations in the small mouse neurofilament (NF-L) subunit, *J. Cell Biol.* **111:**2005–2019.

Gossen, M., Freundlieb, S., Bender, G., Muller, G., Hillen, W., and Bujard, H., 1995, Transcriptional activation by tetracyclines in mammalian cells, *Science* **268:**1766–1769.

Hanukoglu, I., and Fuchs, E., 1982, The cDNA sequence of a human epidermal keratin: Divergence of sequence but conservation of structure among intermediate filament proteins, *Cell* **31:**243–252.

Hanukoglu, I., and Fuchs, E., 1983, The cDNA sequence of a type II cytoskeletal keratin reveals constant and variable structural domains among keratins, *Cell* **33:**915–924.

Hatzfeld, M., and Franke, W. W., 1985, Pair formation and promiscuity of cytokeratins: Formation *in vitro* of heterotypic complexes and intermediate-sized filaments by homologous and heterologous recombinations of purified polypeptides, *J. Cell Biol.* **101:**1826–1841.

Hazan, R., Denk, H., Franke, W. W., Lackinger, E., and Schiller, D. L., 1986, Change of cytokeratin organization during development of Mallory bodies as revealed by a monoclonal antibody, *Lab. Invest.* **54:**543–553.

Hilberg, F., Aguzzi, A., Howells, N., and Wagner, E. F., 1993, C-*jun* is essential for normal mouse development and hepatogenesis, *Nature* **365,** 179–181; Erratum, *Nature* **366,** 368.

Ichinose, Y., Morita, T., Zhang, F. Y., Srimahasongcram, S., Tondella, M. L., Matsumoto, M., Nozaki, M., and Matsushiro, A., 1988, Nucleotide sequence and structure of the mouse cytokeratin *endoB* gene, *Gene* **70:**85–95.

Irvine, A. D., Corden, L. D., Swensson, O., Swensson, B., Moore, J. E., Frazer, D. G., Smith, F. J., Knowlton, R. G., Christophers, E., Rochels, R., Uitto, J., and McLean, W. H., 1997, Mutations in cornea-specific keratin K3 or K12 genes cause Meesmann's corneal dystrophy, *Nature Genet.* **16:**184–187.

Ishida-Yamamoto, A., Eady, R. A. J., Underwood, R. A., Dale, B. A., and Holbrook, K. A., 1994, Filaggrin expression in epidermolytic ichthyosis (epidermolytic hyperkeratosis), *Br. J. Dermatol.* **131:**767–779.

Jiang, R., and Gridley, T., 1997, Gene targeting: Things go better with Cre, *Curr. Biol.* **7:**321–323.

Jiang, C. K., Magnaldo, T., Ohtsuki, M., Freedberg, I. M., Bernard, F., and Blumcenberg, M., 1993, Epidemal growth factor and transforming growth factor alpha specifically induce the activation- and hyperproliferation-associated keratins 6 and 16, *Proc. Natl. Acad. Sci. USA* **90:**6786–6790.

Jonkman, M. F., Heeres, K., Pas, H. H., Luyn, M. J. A. v., Elema, J. D., Corden, L. D., Smith, F. J. D., McLean, W. H. I., Ramaekers, F. C. S., Burton, M., and Scheffer, H., 1996, Effects of keratin 14 ablation on the clinical and cellular phenotype in a kindred with recessive epidermolysis bullosa simplex, *J. Invest. Dermatol.* **107:**764–769.

Kao, W. W., Liu, C. Y., Converse, R. L., Shiraishi, A., Kao, C. W., Ishizaki, M., Doetschman, T., and Duffy, J., 1996, Keratin 12-deficient mice have fragile corneal epithelia, *Invest. Ophthalmol. Vis. Sci.* **37:**2572–2584.

Kistner, A., Gossen, M., Zimmermann, F., Jerecic, J., Ullmer, C., Lubbert, H., and Bujard, H., 1996, Doxycycline-mediated quantitative and tissue-specific control of gene expression in transgenic mice, *Proc. Natl. Acad. Sci. USA* **93:**10933–10938.

Kopan, R., and Fuchs, E., 1989, The use of retinoic acid to probe the relation between hyperproliferation-associated keratins and cell proliferation in normal and malignant epidermal cells, *J. Cell Biol.* **109:**295–307.

Krauss, S., and Franke, W. W., 1990, Organization and sequence of the human gene encoding cytokeratin 8, *Gene* **86:**241–249.

Ku, N. O., Michie, S., Oshima, R. G., and Omary, M. B., 1995, Chronic hepatitis, hepatocyte fragility, and increased soluble phosphoglycokeratins in transgenic mice expressing a keratin 18 conserved arginine mutant, *J. Cell Biol.* **131:**1303–1314.

Ku, N. O., Michie, S. A., Soetikno, R. M., Resurreccion, E. Z., Broome, R. L., Oshima, R. G., and Omary, M. B., 1996, Susceptibility to hepatotoxicity in transgenic mice that express a dominant-negative human keratin 18 mutant, *J. Clin. Invest.* **98:**1034–1046.

Ku, N. O., Wright, T. L., Terrault, N. A., Gish, R., and Omary, M. B., 1997, Mutation of human keratin 18 in association with cryptogenic cirrhosis, *J. Clin. Invest.* **99:**19–23.

Kulesh, D. A., and Oshima, R. G., 1988, Cloning of the human keratin 18 gene and its expression in nonepithelial mouse cells, *Mol. Cell. Biol.* **8:**1540–1550.

Lane, E. B., 1993, Keratins, in *Connective Tissue and Its Heritable Disorders: Molecular, Genetic, and Medical Aspects* (P. M. Royce and B. Steinmann, eds.), pp. 237–247, Wiley-Liss, New York.

Lane, E. B., Rugg, E. L., Navsaria, H., Leigh, I. M., Heagerty, A. H. M., Ishida-Yamamoto, A., and Eady, R. A. J., 1992, A mutation in the conserved helix termination peptide of keratin 5 in hereditary skin blistering, *Nature* **356:**244–246.

Lee, M. K., and Cleveland, D. W., 1996, Neuronal intermediate filaments, *Annu. Rev. Neurosci.* **19:**187–217.

Leube, R. E., Bader, B. L., Bosch, F. X., Zimbelmann, R., Achtstaetter, T., and Franke, W. W., 1988, Molecular characterization and expression of the stratification-related cytokeratins 4 and 15, *J. Cell Biol.* **106:**1249–1261.

Lloyd, C., Yu, Q.-C., Cheng, J., Turksen, K., Degenstein, L., Hutton, E., and Fuchs, E., 1995, The basal keratin network of stratified squamous epithelia: Defining K15 function in the absence of K14, *J. Cell Biol.* **129:**1329–1344.

Lu, X., and Lane, E. B., 1990, Retrovirus-mediated transgenic keratin expression in cultured fibroblasts: Specific domain functions in keratin stabilization and filament formation, *Cell* **62:**681–696.

Mackenzie, I. C., 1997, Retroviral transduction of murine epidermal stem cells demonstrates clonal units of epidermal structure, *J. Invest. Dermatol.* **109:**377–383.

Mackenzie, I. C., and Bickenbach, J. R., 1985, Label-retaining keratinocytes and Langerhans cells in mouse epithelia, *Cell Tiss. Res.* **242:**551–556.

Magin, T. M., Schröder, R., Leitgeb, S., Wanninger, F., Zatloukal, K., Grund, C., and Melton, D. W., 1998, Lessons from keratin 18 Knockout mice: Formation of novel keratin filaments, secondary loss of keratin 7 and accumulation of liver-specific K8-positive aggregates, *J. Cell Biol.* **140:**1441–1451.

Mansbridge, J. N., and Knapp, A. M., 1987, Changes in keratinocyte maturation during wound healing, *J. Invest. Dermatol.* **89:**253–263.

Marin, M., Karis, A., Visser, P., Grosveld, F., and Philipsen, S., 1997, Transcription factor Sp1 is essential for early embryonic development but dispensable for cell growth and differentiation, *Cell* **89:**619–628.

McLean, W. H., and Lane, E. B., 1995, Intermediate filaments in disease, *Curr. Opin. Cell Biol.* **7:**118–125.

Meng, J. J., Khan, S., and Ip, W., 1996, Intermediate filament protein domain interactions as revealed by two-hybrid screens, *J. Biol. Chem.* **271:**1599–1604.

Moll, R., 1994, Cytokeratins in the histological diagnosis of malignant tumors, *Int. J. Biol. Markers* **9:**63–69.

Moll, R., Franke, W. W., Schiller, D. L., Geiger, B., and Krepler, R., 1982, The catalog of human cytokeratins: Patterns of expression in normal epithelia, tumors and cultured cells, *Cell* **31:**11–24.

Moll, R., Schiller, D. L., and Franke, W. W., 1990, Identification of protein IT of the intestinal cytoskeleton as a novel type I cytokeratin with unusual properties and expression patterns, *J. Cell Biol.* **111:**567–580.

Monteiro, M. J., Hoffman, P. N., Gearhart, J. D., and Cleveland, D. W., 1990, Expression of NF-L in both neuronal and nonneuronal cells of transgenic mice: Increased neurofilament density in axons without affecting caliber, *J. Cell Biol.* **111:**1543–1557.

Neznanov, N. S., and Oshima, R. G., 1993, *Cis* regulation of the keratin 18 gene in transgenic mice, *Mol. Cell. Biol.* **13:**1815–1823.

Neznanov, N., Thorey, I. S., Cecena, G., and Oshima, R. G., 1993, Transcriptional insulation of the human keratin 18 gene in transgenic mice, *Mol. Cell. Biol.* **13:**2214–2223.

Oshima, R. G., Trevor, K., Shevinsky, L. H., Ryder, O. A., and Cecena, G., 1988, Identification of the gene coding for the Endo B murine cytokeratin and its methylated, stable inactive state in mouse nonepithelial cells, *Genes Dev.* **2:**505–516.

Oshima, R. G., Abrams, L., and Kulesh, D., 1990, Activation of an intron enhancer within the keratin 18 gene by expression of c-*fos* and c-*jun* in undifferentiated F9 embryonal carcinoma cells, *Genes Dev.* **4:**835–848.

Oshima, R. G., Baribault, H., and Caulin, C., 1996, Oncogenic regulation and function of keratins 8 and 18, *Cancer Metast. Rev.* **15:**445–471.

Paladini, R. D., Takahashi, K., Gant, T. M., and Coulombe, P. A., 1995, cDNA cloning and bacterial expression of the human type I keratin 16, *Biochem. Biophys. Res. Commun.* **215:**517–523.

Paladini, R. D., Takahashi, K., Bravo, N. S., and Coulombe, P. A., 1996, Onset of reepithelialization after skin injury correlates with a reorganization of keratin filaments in wound edge keratinocytes: Defining a potential role for keratin 16, *J. Cell Biol.* **132:**381–397.

Pankov, R., Neznanov, N., Umezawa, A., and Oshima, R. G., 1994, AP-1, ETS, and transcriptional silencers regulate retinoic acid-dependent induction of keratin 18 in embryonic cells, *Mol. Cell. Biol.* **14:**7744–7757.

Porter, R. M., Leitgeb, S., Melton, D. W., Swensson, O., Eady, R. A. J., and Magin, T. M., 1996, Gene targeting at the mouse cytokeratin 10 locus: Severe skin fragility and changes of cytokeratin expression in the epidermis, *J. Cell Biol.* **132:**925–936.

Porter, R. M., Reichelt, J., Lunny, D. P., Magin, T. M., and Lane, E. B., 1998, The relationship between hyperproliferation and epidermal thickening in a mouse model for BCIE, *J. Invest. Dermatol.* **110:**101–107.

Powell, B. C., and Rogers, G. E., 1990, Cyclic hair-loss and regrowth in transgenic mice overexpressing an intermediate filament gene, *EMBO J.* **9:**1485–1493.

Quinlan, R. A., Schiller, D. L., Hatzfeld, M., Achtstaetter, T., Moll, R., Jorcano, J. L., Magin, T. M., and Franke, W. W., 1985, Patterns of expression and organization of cytokeratin intermediate filaments, *Ann. N. Y. Acad. Sci.* **455:**282–306.

Ramirez, A., Bravo, A., Jorcano, J. L., and Vidal, M., 1994, Sequences 5′ of the bovine keratin 5 gene direct tissue- and cell-type-specific expression of a *lacZ* gene in the adult and during development, *Differentiation* **58:**53–64.

Ramirez, A., Vidal, M., Bravo, A., Larcher, F., and Jorcano, J. L., 1995, A 5'-upstream region of a bovine keratin 6 gene confers tissue-specific expression and hyperproliferation-related induction in transgenic mice, *Proc. Natl. Acad. Sci. USA* **92:**4783–4787.

Reichelt, J., Bauer, C., Porter, R. M., Lane, E. B., Herzog, V., and Magin, T. M., 1997, Out of balance, *J. Cell Sci.* **110:**2175–2186.

Rothnagel, J. A., Greenhalgh, D. A., Gagne, T. A., Longley, M. A., and Roop, D. R., 1993, Identification of a calcium-inducible, epidermal-specific regulatory element in the 3'-flanking region of the human keratin 1 gene, *J. Invest. Dermatol.* **101:**506–513.

Rugg, E. L., McLean, W. H. I., Lane, E. B., Pitera, R., McMillan, J. R., Dopping-Hepenstal, P. J. C., Navsaria, H. A., Leigh, I. M., and Eady, R. A. J., 1994, A functional “knockout” for human keratin 14, *Genes Dev.* **8:**2563–2573.

Ruiz, P., Brinkmann, V., Ledermann, B., Behrend, M., Grund, C., Thalhammer, C., Vogel, F., Birchmeier, C., GŸnthert, U., Franke, W. W., and Birchmeier, W., 1996, Targeted mutation of plakoglobin in mice reveals essential functions of desmosomes in the embryonic heart, *J. Cell Biol.* **135:**215–225.

Schiller, D. L., Franke, W. W., and Geiger, B., 1982, A subfamily of relatively large and basic cytokeratin polypeptides as defined by peptide mapping is represented by one or several polypeptides in epithelial cells, *EMBO J.* **1:**761–769.

Schorle, H., Meier, P., Buchert, M., Jaenisch, R., and Mitchell, P. J., 1996, Transcription factor AP-2 essential for cranial closure and craniofacial development, *Nature* **381:**235–238.

Semat, A., Duprey, P., Vasseur, M., and Darmon, M., 1986, Mesenchymal–epithelial conversions induced by 5-azacytidine: Appearance of cytokeratin Endo-A messenger RNA, *Differentiation* **31:**61–66.

Stacey, A., Schnieke, A., McWhir, J., Colman, A., and Melton, D. W., 1994, Use of double-replacement gene targeting to replace the murine alpha-lactalbumin gene with its human counterpart in embryonic stem cells and mice., *Mol. Cell. Biol.* **14:**1009–1016.

Steinert, P. M., and Marekov, L. N., 1995, The proteins elafin, filaggrin, keratin intermediate filaments, loricrin, and small proline-rich proteins 1 and 2 are isodipeptide cross-linked components of the human epidermal cornified cell envelope, *J. Biol. Chem.* **270:**17702–17711.

Steinert, P. M., Idler, W. W., and Zimmerman, S. B., 1976, Self-assembly of bovine epidermal keratin filaments *in vitro*, *J. Mol. Biol.* **108:**547–567.

Stoler, A., Kopan, R., Duvic, M., and Fuchs, E., 1988, Use of monospecific antisera and cRNA probes to localize the major changes in keratin expression during normal and abnormal epidermal differentiation, *J. Cell Biol.* **107:**427–446.

Takahashi, K., and Coulombe, P. A., 1996, A transgenic mouse model with an inducible skin blistering disease phenotype, *Proc. Natl. Acad. Sci. USA* **93:**14776–14781.

Takahashi, K., and Coulombe, P. A., 1997, Defining a region of the human keratin 6a gene that confers inducible expression in stratified epithelia of transgenic mice, *J. Biol. Chem.* **272:**11979–11985.

Takahashi, K., Folmer, J., and Coulombe, P. A., 1994, Increased expression of keratin 16 causes anomalies in cytoarchitecture and keratinization in transgenic mouse skin, *J. Cell Biol.* **127:**505–520.

Tamai, Y., Takemoto, Y., Matsumoto, M., Morita, T., Matsushiro, A., and Nozaki, M., 1991, Sequence of EndoA gene encoding mouse cytokeratin and its methylation state in the CpG-rich region, *Gene* **104:**169–176.

Thorey, I. S., Cecena, G., Reynolds, W., and Oshima, R. G., 1993a, Alu sequence involvement in transcriptional insulation of the keratin 18 gene in transgenic mice, *Mol. Cell. Biol.* **13:**6742–6751.

Thorey, I. S., Meneses, J. J., Neznanov, N., Kulesh, D. A., Pedersen, R. A., and Oshima, R. G., 1993b, Embryonic expression of human keratin 18 and K18-beta-galactosidase fusion genes in transgenic mice, *Dev. Biol.* **160:**519–534.

Traupe, H., 1989, The epidermolytic (acanthokeratolytic) ichthyoses, in *The Ichthyoses* (H. Traupe, ed.), pp. 139–153, Springer-Verlag, Berlin.

Vassar, R., Rosenberg, M., Ross, S., Tyner, A., and Fuchs, E., 1989, Tissue-specific and differentiation-specific expression of a human K14 keratin gene in transgenic mice, *Proc. Natl. Acad. Sci. USA* **86:**1563–1567.

Vassar, R., Coulombe, P. A., Degenstein, L., Albers, K., and Fuchs, E., 1991, Mutant keratin expression in transgenic mice causes marked abnormalities resembling a human genetic skin disease, *Cell* **64:**365–380.

Venetianer, A., Schiller, D. L., Magin, T., and Franke, W. W., 1983, Cessation of cytokeratin expression in a rat hepatoma cell line lacking differentiated functions, *Nature* **305:**730–733.

Weiss, R. A., Eichner, R., and Sun, T.-T., 1984, Monoclonal antibody analysis of keratin expression in epidermal diseases: A 48- and 56-kdalton keratin as molecular markers for hyperproliferative keratinocytes, *J. Cell Biol.* **98:**1397–1406.

Wilgram, G. F., and Caulfield, J. B., 1966, An electron microscopy study of epidermolytic hyperkeratosis, *Arch. Dermatol.* **94:**127–143.

Wilson, A. K., Coulombe, P. A., and Fuchs, E., 1992, The roles of K5 and K14 head, tail, and R/K L L E G E domains in keratin filament assembly *in vitro*, *J. Cell Biol.* **119:**401–414.

Wong, P. C., and Cleveland, D. W., 1990, Characterization of dominant and recessive assembly-defective mutations in mouse neurofilament NF-M, *J. Cell Biol.* **111:**1987–2003.

Chapter 6

The Wound Repair-Associated Keratins 6, 16, and 17

Insights into the Role of Intermediate Filaments in Specifying Keratinocyte Cytoarchitecture

Kevin McGowan and Pierre A. Coulombe

1. INTRODUCTION TO KERATIN EXPRESSION AND FUNCTION

The major structural proteins of epithelial cells are the keratins, which are members of the superfamily of intermediate filament proteins. There are in excess of 30 keratin genes, and these can be clearly partitioned into two sequence types, I and II, based on operational criteria such as genomic structure and nucleotide sequence homology (Fuchs and Weber, 1994). Most keratin genes, including the type II K1–K8 and the type I K9–K20, are expressed in "soft" epithelia, ranging from glandular tissues such as pancreas and salivary glands to cornifying stratified epithelia such as epidermis

---

Kevin McGowan and Pierre A. Coulombe Departments of Biological Chemistry and Dermatology, The Johns Hopkins University School of Medicine, Baltimore, Maryland 21205.

Subcellular Biochemistry, Volume 31: Intermediate Filaments, edited by Herrmann and Harris. Plenum Press, New York, 1998.

(Moll *et al.*, 1982a; O'Guin *et al.*, 1990). In addition, a group of five type I (Ha1–Ha5) and six type II (Hb1–Hb6) keratins are expressed in hard epithelia such as hair and nail (Heid *et al.*, 1988; O'Guin *et al.*, 1990). As the polymerization of keratin intermediate filaments (KIFs) begins obligatorily with the formation of a type I–type II coiled-coil heterodimer (reviewed by Coulombe, 1993), an epithelial cell must coordinate the expression of at least one type I and one type II gene in order to produce a keratin filament network in its cytoplasm. The expression of several of these type I and type II genes is in fact regulated in a pairwise and differentiation-specific fashion. This is best illustrated in stratified epithelia, where expression of the type II K5 and the type I K14 and K15 occurs in the progenitor basal layer, while differentiating suprabasal keratinocytes express distinct keratin pairings depending upon the differentiation program being executed. For instance, the K1–K10 pair is characteristic of cornifying epithelia such as epidermis, the K4–K13 pair is expressed in the wet epithelia lining the oral mucosa, tongue, and esophagus, while the K3–K12 pair is specific to the cornea (O'Guin *et al.*, 1990). The diversity and pattern of expression of keratin genes are highly conserved among higher vertebrate species, and studies are in progress in many laboratories to decipher the underlying significance.

In epithelial cells, where they predominate among cytoskeletal elements, KIFs are typically organized in a pancytoplasmic network involving their noncovalent linkage at the surface of the nucleus and at membrane-spanning adhesion complexes at the cell periphery. These adhesion sites consist of desmosomes, which mediate adhesion between cells, and hemidesmosomes, mediating the adhesion of epithelial cells to the underlying basal lamina (Schwarz *et al.*, 1990; Jones and Green, 1991; Coulombe and Fuchs, 1994). In stratified epithelia, where desmosomes are especially abundant, the organization of KIFs literally transcends that of the cell—it is in that supracellular context that the functional significance of KIFs takes its full meaning. Indeed, KIFs are major determinants of the structural resilience of surface epithelia, such that even a partial loss of function leads to the fragilization of the epithelium involved. This role of mechanical support is entirely consistent with the properties, abundancy, and intracellular organization of KIFs. This intuitively logical function has recently been convincingly demonstrated in the context of studies involving transgenic mice. Disruption of the KIF network in epidermal keratinocytes via the targeted expression of powerful dominant-negative keratin mutants (Vassar *et al.*, 1991; Coulombe *et al.*, 1991b; Fuchs *et al.*, 1992; Rothnagel *et al.*, 1993b; Porter *et al.*, 1996; Takahashi and Coulombe, 1996) or the introduction of null mutations (Lloyd *et al.*, 1995) results in lysis of the targeted cell population whenever the skin of such mice is subjected to trivial mechanical trauma. These studies collectively established that an intact KIF network is essential to maintain the structural integrity of the epidermis in the face of the mechanical stress to which the skin is routinely subjected (Coulombe and Fuchs, 1994).

Such transgenic mouse findings led to the obvious realization that a number of inherited blistering diseases that feature autosomal dominant inheritance at the genetic level, trauma-induced epithelial cell lysis at the pathological level, and an altered KIF network at the ultrastructural level could in fact be keratin gene disorders (Fuchs and Coulombe, 1992). This was rapidly confirmed by genetic linkage (Bonifas *et al.*, 1991) and DNA sequencing studies (Coulombe *et al.*, 1991a; Bonifas *et al.*, 1991), initially focused on the keratin 14 gene of patients with epidermolysis bullosa simplex. Since then, mutations have been found in as many as 13 additional keratin genes, producing a heterogeneous array of blistering disease phenotypes (Table I). The identity of the keratin gene affected and the character of the mutation are the main determinants of the clini-

Table I
List of Keratin Genes Implicated in Diseases

Keratin gene	Disease	Distribution[a]	Reference[b]
K5, K14	Epidermolysis bullosa simplex	Epidermis (basal layer)	Coulombe *et al.* (1991a) Bonifas *et al.* (1991) Lane *et al.* (1992)
K1, K10	Epidermolytic hyperkeratosis	Epidermis (low suprabasal)	Rothnagel *et al.* (1992) Cheng *et al.* (1992) Chipev *et al.* (1992)
K2e	Ichtyosis bullosa of Siemens	Epidermis (high suprabasal)	Rothnagel *et al.* (1994) McLean *et al.* (1994)
K9	Palmoplantar keratoderma, *epidermolytic* variant	Palm/sole epidermis only (suprabasal layers)	Reis *et al.* (1994) Bonifas *et al.* (1994)
K16, K1	Palmoplantar keratoderma, *nonepidermolytic* variant	Palm/sole epidermis only (suprabasal layers)	Kimonis *et al.* (1994) Shamser *et al.* (1995)
K6a, K16, K17	Pachyonychia congenita	Nail and other stratified epithelia	McLean *et al.* (1995) Bowden *et al.* (1995) Smith *et al.* (1997)
K17	Steatocystoma multiplex	Skin	Smith *et al.* (1997)
K4, K13	Oral white sponge nevus	Oral mucosa	Rugg *et al.* (1995) Richard *et al.* (1995)
K3, K12	Meesman corneal dystrophy	Cornea	Irvine *et al.* (1997)
K18	Predisposition to chronic liver disease	Simple epithelia	Ku *et al.* (1997)

[a]The distribution of the mutated keratin protein generally correlates with the tissue(s) affected by the disease process.
[b]Only the original discoveries are tabulated for the sake of brevity.

cal presentation of the disease, since they specify the target epithelial cell population and the extent of its sensitivity to various forms of trauma (see reviews by Coulombe and Fuchs, 1994; Fuchs *et al.*, 1994; Steinert and Bale, 1993; McLean and Lane, 1995). From a biological standpoint, the discovery of naturally occuring keratin mutations in a variety of blistering diseases affecting the skin and its appendages, the oral mucosa, and the cornea, together with the above-mentioned transgenic mouse studies, unequivocally established the critical importance of keratin filaments to the resilience and function of stratified epithelia.

2. KERATIN EXPRESSION IS ALTERED DURING THE REEPITHELIALIZATION OF SKIN WOUNDS

Disruption of the integrity of stratified epithelia results in a loss of their barrier properties, which rapidly triggers a multistep homeostatic response primarily designed to restore epithelial continuity over the wound site. This response has been particularly well studied in skin, where the superficial epidermis provides a vital barrier function (Bereiter-Hahn, 1986; Clark, 1993). Wound closure is brought about by the combined action of keratinocyte migration originating from the surrounding healthy tissue along with a contraction of the connective wound bed, itself mediated by migratory fibroblasts that display contractile properties. In all the mammalian species surveyed, tangible signs of keratinocyte migration at the cut edges of the skin wound are first manifested at ~18–24 hr after injury. Whenever the dimensions of the wound site warrant it, the process of migration can be sustained over several days, and is supported by an enhanced rate of epithelial cell proliferation taking place in the basal layer, somewhat distally from the cut edge (Bereiter-Hahn, 1986; Clark, 1993). In stratified epithelia including epidermis, migration into the wound site occurs in the form of a stratified sheet, and it is widely believed that suprabasal keratinocytes, which at the time of injury were already engaged in the differentiation pathway, participate to reepithelialization (Garlick and Taichman, 1994; Coulombe and Takahashi, 1996). To meet the new demands placed upon them, keratinocytes at the wound edge must initially undergo an activation phase (Grinnell, 1992) during which they reorganize their available "resources" and generate the missing ones needed for migration as part of an epithelial sheet. Accordingly, significant changes in gene expression occur in wound-edge keratinocytes during the initial 24–48 hr postinjury (Coulombe, 1997). Of particular interest, many of the genes that are induced in wound-edge epidermis are also characteristic of diseases states featuring hyperproliferation or abnormal differentiation (Coulombe, 1997).

The activated keratinocyte at the edge of an acute skin wound shows a dramatically altered ultrastructure (Paladini *et al.*, 1996, and references therein). In addition to cellular hypertrophy and a shift from a polygonal to an elongated and polarized shape, the KIF network fragments and retracts from the cell periphery, while the number of desmosomes at the cell surface decreases. This keratin reorganization is of interest, since one could reasonably argue that a keratinocyte that displays thick bundles of K1–K10-rich filaments radially extending from the cell center to nearly all the membrane-spanning desmosomes would not be capable of the plasticity required during cell migration (Coulombe and Takahashi, 1996). Based on our current understanding of KIF dynamics and organization we can formulate several potential mechanisms to account for this reorganization. A most simple one, however, would consist of a rapid change in keratin gene expression. As a matter of fact, it has been known for several years that keratin expression undergoes significant changes after damage to the epidermis, as the so-called "hyperproliferation-associated" keratins K6 and K16 appear at the wound edge (Mansbridge and Knapp, 1987). Recent studies have indicated that the induction of K6, K16, and yet another type I keratin, K17, occurs within 6–12 hr specifically in suprabasal keratinocytes after full epidermal thickness injury to human and mouse skin (Paladini *et al.*, 1996; K. McGowan and P. A. Coulombe, unpublished data), and that their accumulation correlates with the observed changes in KIF organization and desmosomes in this keratinocyte subpopulation. This accumulation of new keratins occurs at the expense of those normally produced in abundance in early differentiating keratinocytes, K1 and K10 (Mansbridge and Knapp, 1987; Paladini *et al.*, 1996; see Figure 1). Such rapid changes in the keratin pattern represent a conserved response to skin injury in mammalian species, insofar as it has been described in human, bovine, and mouse (Roop *et al.*, 1984; Schweizer, 1993; Ramirez *et al.*, 1995). Moreover, enhanced expression of K6, K16, and K17 is not restricted to the wound-edge epidermis, as it also occurs in proximal appendages; in mouse, it has also been described following injury to footpads, oral mucosa, and cornea (Takahashi and Coulombe, 1996, 1997). The "wound-induced" keratins 6, 16, and 17 may thus serve a specific functional need during reepithelialization that cannot be supported by the normal complement of keratins expressed in stratified epithelia under resting conditions.

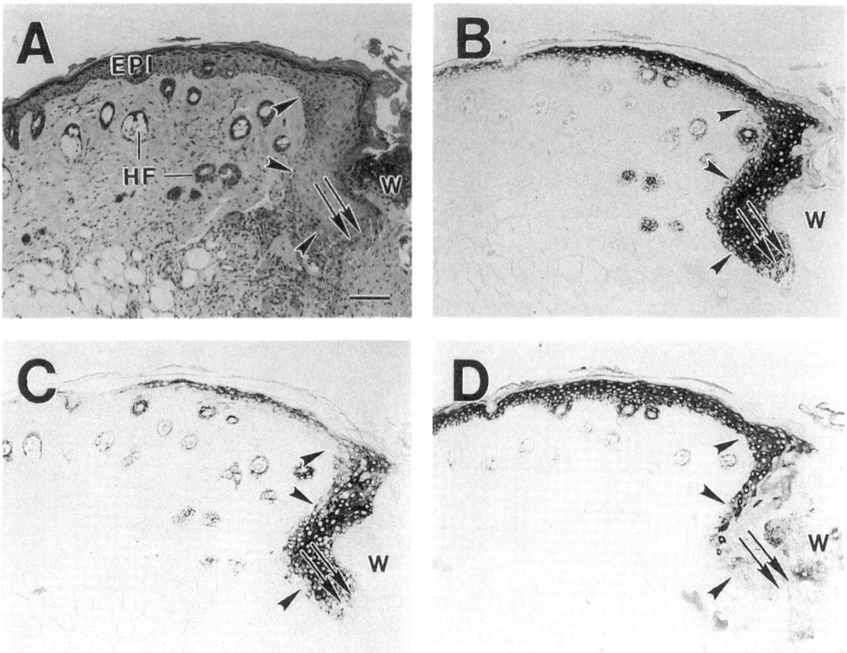

FIGURE 1. Altered keratin expression during the reepithelialization of skin wounds. A mouse was anesthesized and subjected to full-skin-thickness injury as described (Takahashi and Coulombe, 1997). Four days later, the wound site was surgically harvested, fixed in Bouin's, and embedded in paraffin. Cross sections of the wound site were either stained with hematoxylin-eosin (A) or immunostained using antisera to mouse K6 (B), mouse K17 (C), or human/mouse K10 (D) followed by peroxidase-based detection. Migration of an epithelial sheet originating from the edges of the wound is well established at 4 days (see double arrows). The migrating epithelium and proximal epidermis at the wound edge are strongly stained for K6 and K17, while the epidermis distal from the wound edge is negative. In contrast, K10 is strongly expressed in the suprabasal layers of normal epidermis, but is noticeably absent from the migrating epithelium. The arrowheads depict the dermoepidermal interface. EPI, epidermis; HF, hair follicle; W, wound site. Bar in panel A equals 200 μm.

3. ANALYSIS OF K6, K16, AND K17 SEQUENCES AND THEIR EXPRESSION IN NORMAL, WOUNDED, AND DISEASED EPITHELIA

3.1. Cloning and Sequence Analysis

Human K6 was initially cloned as a partial cDNA by Hanukoglu and Fuchs (1983). Later, using this cDNA as a probe, Tyner *et al.* (1985) isolated a human genomic clone showing very high but imperfect homology with the

K6 sequence reported, notably at the level of the 3′ noncoding sequence. Further studies by Tyner and Fuchs (1986) evidenced two types of K6 mRNA transcripts and corresponding genes, designated K6a (corresponding to the partial cDNA clone) and K6b (full-length genomic clone). While attempting to obtain a genomic clone for the human K6a gene many years later, we discovered the existence of additional K6 isoform genes—as many as four additional ones—in the human genome. These K6 isoforms (K6a–K6f) are highly related at the gene structure, nucleotide sequence (including introns, and 5′ and 3′ noncoding segments), and predicted amino acid sequence levels (Takahashi *et al.*, 1995). As a result, it is impossible to discriminate among them using conventional hybridization-based strategies, providing an explanation for the detection of only two distinct transcripts and genes by Tyner and Fuchs (1986). Together with the findings of Tyner *et al.* (1985) and Rosenberg *et al.* (1991), our data led us to conclude that the human K6 isoform genes are organized in tandem as part of the large cluster of type II keratin genes on the long arm of chromosome 12 (12q11–13) (see Chapter 3 of this book). A survey of K6 mRNA expression using a reverse transcriptase–polymerase chain reaction (RT-PCR)-based approach (Takahashi *et al.*, 1995) indicated that the dominant isoform in human skin from various body sites indeed corresponds to the partial K6a cDNA clone isolated by Hanukoglu and Fuchs (1983; see Tyner and Fuchs, 1986). Compared to K6b and especially K6a, the K6c–e isoform mRNAs are poorly expressed in human scalp and footsole, in a squamous cell carcinoma *in situ*, and in epidermal cell lines (Takahashi *et al.*, 1995). One should emphasize that there is no evidence demonstrating expression of specific human K6 isoform genes at the protein level, as the K6 isoform proteins cannot be distinguished using conventional separation based on size, charge, or immunological properties.

The coding sequences of human K6 isoforms are clearly related to that of K5 at a nucleotide sequence level, such that it appears likely that the K5 gene and the primordial K6 gene arose by duplication of a common ancestor gene (Blumenberg, 1988). The predicted amino acid sequence of human K5 and K6a is 66%, 90%, and 56% identical for the amino-terminal head, the central α-helical rod, and the carboxy-terminal tail domains, respectively (analysis not shown). Subsequent duplication events would have given rise to the current group of K6 human isoform genes [see Takahashi *et al.* (1995) for discussion]. Of interest, at least three functional K6 genes have been characterized in the bovine (Blessing *et al.*, 1987; Ramirez *et al.*, 1995; Navarro *et al.*, 1995), while two functional K6 genes have been uncovered in the mouse (Rothnagel *et al.*, 1993; K. Takahashi and P. A. Coulombe, unpublished data). Careful sequence comparison fails to reveal a common pattern in the diversification of K6 sequences in the human, bovine, and mouse, suggesting that the duplication of K6 genes occurred independently, subsequent to the divergence of these species during evolution (Takahashi *et al.*, 1995; K. Takahashi and P. A. Coulombe, unpublished data).

There is considerable confusion in the literature with respect to the K16 sequence, due in part to difficulties arising from the very high homology between K14, K16, and K17. A human genomic clone initially identified as encoding K17 (RayChaudhury *et al.*, 1986) was later corrected and reinterpreted to encode K16 (Rosenberg *et al.*, 1988). The cDNA clone that we generated by RT-PCR from cultured human epidermal keratinocyte mRNA (GenBank Accession No. 579867; Paladini *et al.*, 1995) is virtually identical to the corrected GK-3 sequence reported by Rosenberg *et al.* (1988) and to a partial K16 cDNA sequence obtained from tracheal epithelium by Huang *et al.* (1994). On the other hand, the amino acid sequence predicted from a K17 genomic clone isolated by Troyanovsky *et al.* (1992) is remarkably similar to that reported by RayChaudury *et al.* (1986). An obvious explanation for this strange occurrence would be the existence of multiple K16 isoforms in human (McLean *et al.*, 1995); however, we have been unable to find evidence for this while applying RT-PCR on mRNA samples prepared from human foot sole and scalp skin (K. Takahashi and P. A. Coulombe, unpublished data).

The existence of a mouse orthologue gene for K16 has been questioned (Schweizer, 1993). If true, this would represent the first example of a human keratin gene with no orthologue in mouse. Schweizer and colleagues have reported that they were unable to isolate a sequence corresponding to K16 while screening relevant mouse cDNA libraries with a probe derived from the 3′ end of the human K16 cDNA (obtained from Rosenberg *et al.*, 1988), even though this strategy has been successful when applied to other keratin genes. In our hands a similar 3′ noncoding probe cross-hybridizes with related type I keratin genes even when applied under stringent conditions [K. McGowan, unpublished observations; for a different outcome see Porter *et al.* (1996)]. On the other hand, we found that a polyclonal antiserum raised against a 22-mer peptide corresponding to the carboxy-terminus of K16 (Takahashi *et al.*, 1994) reacts with a ~49- to 51-kDa antigen that is induced in an intermediate filament extract prepared from wounded mouse skin (Coulombe *et al.*, 1995). This antigen can be electrophoretically and immunologically resolved from K17 and K14 proteins in such mouse skin extracts (K. McGowan and P. A. Coulombe, unpublished data). When applied on frozen sections prepared from unfixed mouse tissues, this antiserum stains the hair follicle outer root sheath in normal skin, the wound-edge keratinocytes after acute injury to the skin, and the filiform papillae of the dorsal tongue epithelium (Figure 2). As this distribution is reminescent of

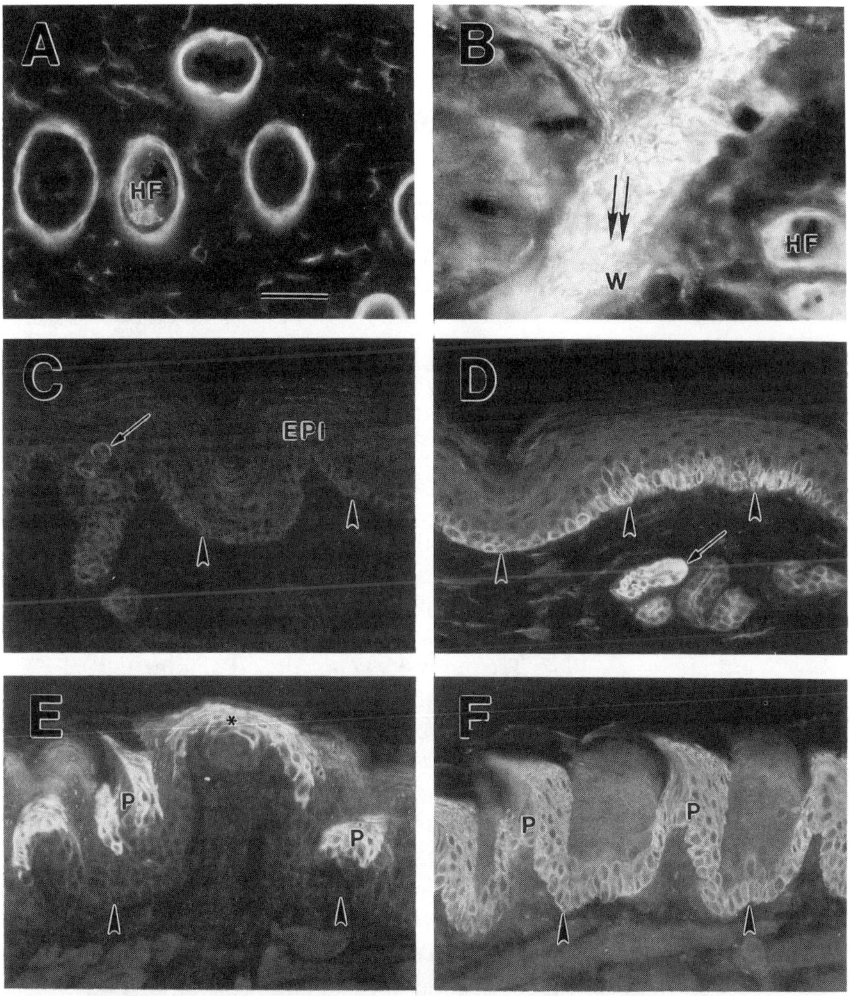

FIGURE 2. An anti-human K16 antibody cross-reacts with a mouse epithelial antigen. Mouse tissues were snap-frozen and sections were obtained using a cryostat. The sections were incubated with rabbit polyclonal antisera directed against human K16 (Takahashi *et al.*, 1994) or mouse K6 (Roop *et al.*, 1984) followed by a FITC-conjugated goat anti-rabbit secondary. (A) Normal skin. The anti-human K16 antiserum stains the outer root sheath of hair follicles (HF), but not epidermis (not shown). (B) Wounded skin. Anti-human K16 stains the migrating epithelium (see double arrow), indicating that the antigen recognized is induced by injury. (C, D) Foot sole. The epidermis (EPI) is negative with the anti-K16 antiserum, while sweat gland ducts are faintly positive (see arrow in panel C). In contrast, the basal epidermal layer and the sweat gland ducts (arrow) are strongly positive for K6 (D), using the anti-mouse K6 produced by Roop *et al.* (E, F) Dorsal tongue. The anterior and posterior regions of filiform papillae (P) and the fongiform papillae (asterisk) are strongly positive for anti-human K16. In addition to these, the anti-K6 antiserum stains the basal layer in the interpapillar epithelium. Arrowheads depict the interface between epithelium and connective tissue. Bar equals 50 μm.

that of K16 in human epithelia (see below), the mouse antigen recognized by this polyclonal antiserum is likely to correspond to the K16 orthologue. If true, the information currently available would suggest that mouse and human K16 have diverged more significantly than other type I keratins genes such as K14 and K17.

The isolation and characterization of a full-length genomic clone for human K17 was described in 1992 by Troyanovsky *et al.* The information derived from this effort enabled the unequivocal identification of a partial cDNA clone previously isolated from a mouse skin cDNA library, pkSCC50 (Knapp *et al.*, 1987), as encoding mouse K17 (Schweizer, 1993). We recently isolated and characterized a full-length genomic clone for mouse K17 using a probe derived from this partial cDNA clone. Taken together these data establish that human and mouse K17 are very highly related at the nucleotide and especially the predicted amino acid sequence level (K. McGowan, and P. A. Coulombe, manuscript in preparation).

Sequence comparisons reveal that the human K14, K16, and K17 genes show the structure typical of type I keratin genes. In fact, the extent of identity of their nucleotide and amino acid sequences is so pronounced that they unequivocally make up a subgroup among type I keratins (RayChaudhury *et al.*, 1986; Troyanovsky *et al.*, 1992; Paladini *et al.*, 1995). The central α-helical rod domains of these three keratin proteins are nearly 90% identical, and in some cases the analogy extends to the usually less conserved end domain sequences. Thus, human K14 (472 amino acids [aa], 50 kDa; Marchuk *et al.*, 1984) and K16 (473 aa, 48 kDa; Paladini *et al.*, 1995) differ significantly only at the level of their carboxy-terminal 30 amino acids. On the other hand, the amino-terminal head domain of K17 (432 aa, 47 kDa; Troyanovsky *et al.*, 1992) is relatively distinct within this group, while its carboxy-terminal domain is clearly related to that of K14 (Troyanovsky *et al.*, 1992). Again, it appears very likely that these three genes arose by duplication of a common ancestral gene during evolution.

3.2. Distribution within Normal Epithelia

Although they are better known for their much enhanced expression during wound repair and especially in a variety of disease states, K6, K16, and K17 proteins are constitutively expressed in a number of complex epithelia.

In skin, these keratins are expressed in all the appendages and in the specialized epidermis of foot and sole. In the pilosebaceous unit, they occur in the inner differentiating layers of the outer root sheath of the hair and in the apocrine sebaceous glands (Moll *et al.*, 1983; Yoshikawa *et al.*, 1995). In eccrine sweat glands, K6 and K16 are found in the ductal portion, while K17 is found in the myoepithelial cells at the base of the glandular epithelium (Figure 3). In the nail, they are reportedly expressed in the matrix and/or the nail bed (Heid *et al.*, 1988; Moll *et al.*, 1988). With the exception of the thick epidermis of foot and sole skin, where their expression is patchy and heterogeneous (P. A. Coulombe, unpublished data), K6, K16, and K17 are absent from epidermis (Figure 3; Moll *et al.*, 1982a, b). This said, Moll *et al.*

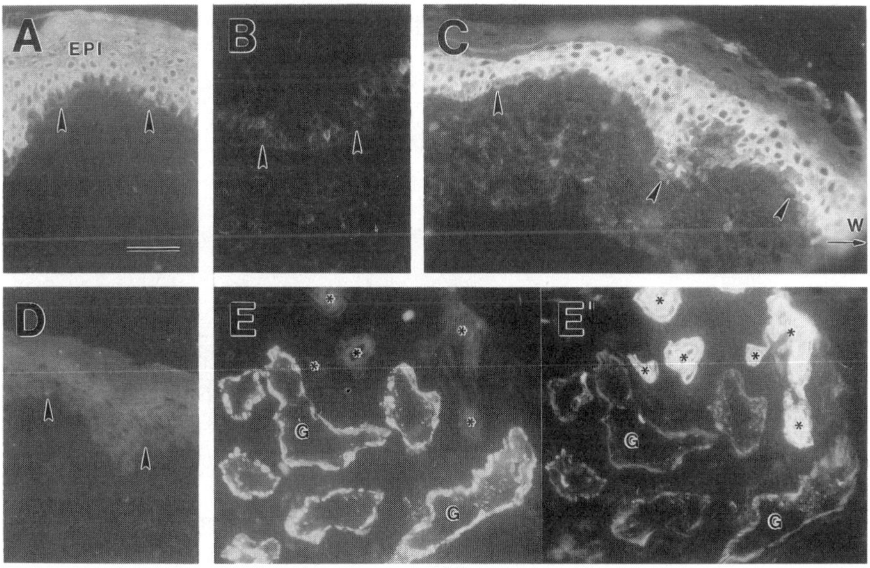

FIGURE 3. Expression of various keratin antigens in human skin, normal and wounded. Healthy volunteers were used as source of intact or wounded skin following the guidelines of the Joint Committee on Clinical Investigation at Johns Hopkins University as described (Paladini *et al.*, 1996). Four-mm punch biopsy samples were snap-frozen, and sections were obtained using a cryostat. The sections were incubated with rabbit polyclonal antisera directed against human K16 (Takahashi *et al.*, 1994) or mouse monoclonal antibodies against human K10 or K17 (Sigma Chemical Co., St. Louis, MO) followed by a fluorescence-conjugated goat anti-rabbit secondary. (A) The anti-K10 antibody intensely stains the suprabasal differentiating layers of normal human epidermis (EPI). In contrast, this tissue does not react appreciably with anti-K16 (B) or anti-K17 (D). At 12h following full-thickness injury to the skin, however, the suprabasal keratinocytes at and near the wound edge (w) are strongly positive for anti-K16 (C) and anti-K16 (data not shown; see Paladini *et al.*, 1996). (E, E′). Double-immunofluorescence staining of a human eccrine sweat gland with rabbit anti-K17 (E) and mouse anti-K16 (E′). Anti-K16 staining is restricted to the ducts (asterisks), while anti-K17 staining is restricted to the basal myoepithelial cells encapsulating the glandular epithelium (G). Arrowheads in panels A–D depict the dermoepidermal interface. Bar equals 50μm.

(1982b) detected transient expression of K6, K16, and K17 in the developing interfollicular epidermis of the 36-week human fetus, which ceases before birth. This finding may be of significant interest, since expression of K6, K16, and K17 "reemerges" in the mature (adult) epidermis subjected to injury or affected by chronic hyperproliferative diseases (as discussed below). Quite remarkably, the skin features a reservoir of K6, K16, and K17-expressing keratinocytes over the entire body surface under baseline conditions, another issue of potential significance to wound repair.

K6, K16, and K17 are also fairly ubiquitously expressed in various stratified epithelia of the oral mucosa and proximal gastrointestinal tract. Thus, they occur in many epithelia of the tongue, notably the filiform papillae (Takahashi *et al.*, 1994), in palate and in the gingiva (Collin *et al.*, 1992), in addition to the esophagus and the forestomach (Moll *et al.*, 1982a). Besides these "wet" stratified epithelia, expression of these proteins has been described in the thymus, trachea, anal canal epithelium, and exocervix (Moll *et al.*, 1983).

There are interesting differences in the expression of the two type I keratins, K16 and K17, in many of these epithelial tissues. In epithelia in which they both occur, they sometimes partition to distinct cell subpopulations, as illustrated in Figure 3 for human eccrine sweat glands. In other cases, such as in the outer root sheath of mouse hair follicles, they show a partially overlapping distribution, since K17 extends down to the bulb region, while K6 and K16 are restricted to the upper half of the follicle (K. McGowan and P. A. Coulombe, manuscript in preparation). Finally, K17 is expressed in the absence of K16 in myoepithelial cells at the base of glandular epithelia (salivary, sweat, sebaceous, esophagus, airway, breast, and prostate glands; see Figure 3), and some pseudostratified epithelia such as the bladder and trachea (Troyanovsky *et al.*, 1989). It is not clear whether K6 is coexpressed with K17 in these myoepithelial cells, raising the question of the identity of its type II pairing partner in those instances. K16, on the other hand, is coexpressed with K6 in most cases—in postmitotic cells featuring an established KIF network containing substantial amounts of K14. In contrast to most other established combinations of type I and type II keratins, therefore, there appears to be no obvious relationship between expression of K6, K16, and/or K17 and the status, fate, or function of a normal epithelial cell.

3.3. Involvement in Disease

Henry Sun, Werner Franke, Roland Moll and their colleagues made key contributions in documenting a phenomenon of "induction" of K6, K16, and K17 expression in a number of diseases associated with enhanced proliferation and/or abnormal differentiation of epithelial tissues (Moll *et al.*, 1983; Weiss *et al.*, 1984). Classic examples for skin include psoriasis and related diseases, basal and squamous cell carcinoma, human papilloma virus infection (Moll *et al.*, 1983; Weiss *et al.*, 1984; Stoler *et al.*, 1988), and even dermatofibroma, a disorder affecting primarily the dermis (Stoler *et al.*, 1989). Enhanced expression of K6, K16, and/or K17 also occurs in carcinomas of the esophagus (Grace *et al.*, 1985), lung airways (Blobel *et al.*, 1984), breast, bladder, and other epithelia (Moll *et al.*, 1983). In many of these cases, this "induction" occurs at the expenses of the normal differentiation-specific keratins being expressed, and correlates with alterations in the morphology and differentiation of epithelial cells.

The expression of K6, K16, and K17 in a plethora of chronic hyperproliferative disorders, along with the intriguing observation that they are induced as well when normal, freshly isolated epithelial cells from tissues such as epidermis, esophagus, cornea, and upper airways are seeded for culture *ex vivo* (Weiss *et al.*, 1984; Tyner and Fuchs, 1986; Kopan and Fuchs, 1989; Schermer *et al.*, 1989), has led to the widespread notion that these keratins are faithful biochemical "markers" of hyperproliferation. However, with the exception of selective cases involving K17, the expression of these keratins *in vivo* occurs primarily in postmitotic epithelial cells. Moreover, expression of K6 and K16 in cultured epidermal and corneal keratinocytes can be experimentally uncoupled from cell proliferation (Kopan and Fuchs, 1989; Schermer *et al.*, 1989; Choi and Fuchs, 1990). A similar phenomenon occurs during wound repair in human and mouse skin *in vivo*, given that expression of these keratins can be dissociated spatially as well as temporally from enhanced proliferation in the basal layer (Paladini *et al.*, 1996; Takahashi and Coulombe, 1997). Finally, "inappropriate" expression of K6 and K16 can be stimulated in the absence of enhanced proliferation in epidermis *in vivo*, for instance, by targeted expression of a dominant-negative retinoic acid receptor in transgenic mouse skin (Saitou *et al.*, 1995). As pointed out by Sun and his colleagues (Schermer *et al.*, 1989), therefore, "inappropriate" or "ectopic" expression of K6, K16, and K17 should be interpreted as indicative of the execution of an alternative program of epithelial differentiation.

3.4. Inheritable Disorders Involving Mutations in K6, K16, or K17

Missense mutations in the coding sequence of the K6a, K16, or K17 genes were recently discovered in patients suffering from pachyonychia congenita (PC; Munro *et al.*, 1994; *McLean et al.*, 1995; Bowden *et al.*, 1995). The position of these mutations along the sequence of these keratins is illustrated in Figure 4. PC consists of a heterogeneous group of (usually)

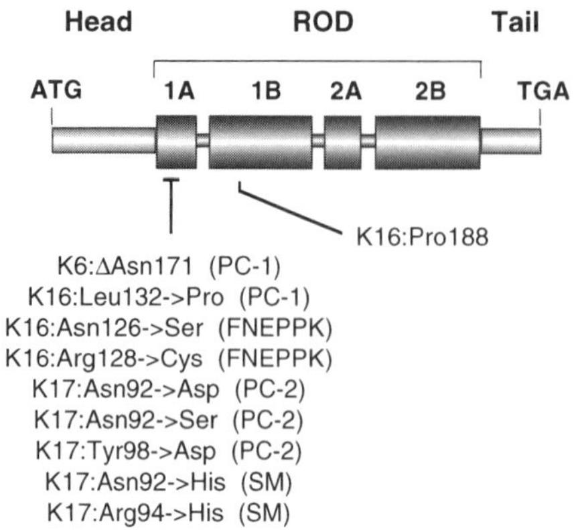

FIGURE 4. Distribution of disease-causing mutations in the coding sequence of K6, K16, and K17. Point mutations were discovered in the K6, K16, or K17 genes of patients suffering from clinically distinct subtypes of pachyonychia congenita (PC-1 and PC-2), from nonepidermolytic palmoplantar keratoderma (NEPPK), and from steatocystoma multiplex (SM). The positions of the various mutations discovered along the generic secondary structure shared by these three keratins (head, α-helical rod, tail) is depicted. Note that in reality the lengths of the head and tail domains vary significantly among these three keratins. The terminology is as follows: K16:Leu132 → Pro signifies a substitution of a leucine for a proline at codon 132 of the K16 coding sequence; Δ signifies a deletion. Codon numbering assumes the methionine translation start as 1 (see text for references). Note that all the mutations so far discovered in these keratins map to a narrow segment within subdomain 1A of the α-helical rod. The location of the amino acid residue responsible for the relatively unstable character of K16-containing tetramers, ^{188}Pro, is also indicated.

dominantly inherited genodermatoses that invariably involve hypertrophic and dyskeratotic changes to the nail and, depending on the clinical variant considered, may also feature lesions in a number of other stratified epithelia. The two major clinical variants of PC disease are Jadassohn–Lewandowsky or PC-1 and Jackson–Lawler or PC-2 (Munro *et al.*, 1994; McLean *et al.*, 1995). Consistent with the known distribution of K6, K16, and K17, all PC subtypes show some degree of involvement of hair follicles (ranging from follicular keratosis to alopecia), sweat glands (e.g., hyperhidrosis) and palmoplantar skin (hyperkeratosis). In addition to these, PC-1 patients typically show oral leukokeratosis, while PC-2 patients distinctly show neonatal teeth, multiple epidermal cysts, and recurrent infections in flexural body areas. Other manifestations, such as angular cheilosis, laryngeal lesions, and corneal dyskeratosis, are occasionally seen in PC patients (Feinstein *et al.*, 1988). Wound repair is apparently not significantly impaired in these patients, although this issue has not been formally examined in a controlled manner, for obvious reasons. Histologically, the main tissue affected in PC diseases is the nail bed, an epithelium lying beneath the nail proper that features abundant levels of K6, K16, and K17 (S. Mazzaluppo, K. McGowan and P. A. Coulombe, unpublished observations). The nail matrix, which produces the cells that terminally differentiate to form the nail plate, is reportedly not affected in PC diseases (Feinstein *et al.*, 1988).

The small number of mutation reports on the K6, K16, and K17 genes, along with a striking heterogeneity in the clinical presentation of the patients concerned (Figure 4), makes any attempt to draw genotype–phenotype correlations rather perilous. In their initial report McLean *et al.* (1995) suggested that PC-1 may be preferentially associated with mutations in K16, while the PC-2 clinical phenotype would generally be caused by mutations in K17. This latter association makes intuitive sense, as, for instance, K17 is characteristically expressed in glands (Troyanovsky *et al.*, 1989) and in the developing tooth (K. McGowan and P. A. Coulombe, manuscript in preparation). This said, mutations in K16 have been discovered in patients with a primary diagnosis of focal nonepidermolytic palmoplantar keratoderma (PPK) and showing very mild nail involvement (Shamsher *et al.*, 1995). Moreover, K17 mutations have been discovered in patients with a primary diagnosis of steatocystoma multiplex (Smith *et al.*, 1997). Finally, Bowden *et al.* (1995) found mutations in the K6a gene of patients suffering from PC-1. In all these instances the mutation involved localizes to a narrow segment of subdomain 1A in the rod domain (Figure 4). It is noteworthy that for K16 and K17 some of these mutations involve the arginine-encoding codon CGC (Figure 4), a mutational hot spot for type I keratins initially discovered in the K14 gene (Coulombe *et al.*, 1991a). Thus there is considerable heterogeneity in the overall clinical presentation in individuals with K16 or K17 mutations, even with respect to the presence and extent of nail lesions. Although clinical heterogeneity is common to all known keratin gene disorders (Table I), in these cases it is manifested mostly at the level of the severity of the disease phenotype.

A potentially interesting finding in patients is that K16 mutations are not conducive to the trauma-induced epithelial cytolysis typical of nearly all keratin gene disorders, while mutations in K17 do (McLean *et al.*, 1995b; Shamsher *et al.*, 1995). We reported in 1994 that overexpression of the human K16 gene in transgenic mouse produces striking skin lesions in the absence of cytolysis (Takahashi *et al.*, 1994). In retrospect, the phenotype seen in these mice is reminescent of PC disease, with the critical exception that even the animals expressing the phenotype most strongly do not show obvious nail changes (Coulombe *et al.*, 1995). The explanation for this discrepancy lies in the the absence of the transgene product in mouse nail tissue, a finding that remains unexplained (Coulombe *et al.*, 1995). As is the case in PC patients with K16 mutations, ultrastructural examination of lesional skin in phenotypic K16 transgenic mice revealed the presence of KIF aggregates near the nucleus and a reduction of the number of desmosomes at the cell surface, but no cytolysis (Takahashi *et al.*, 1994). While this line of evidence needs further investigation, it provides support to the emerging concept that among type I keratins, K16 stands out at biochemical and functional levels.

4. THE REGULATION OF K6, K16, AND K17 GENE EXPRESSION

Understanding the regulation of K6, K16, and K17 gene expression has important implications at two levels. First, these genes are often expressed whenever complex epithelia deviate from their normal program of differentiation. In the particular case of acute skin injury, the K16 and K17 genes are induced in wound-edge epidermis within hours after injury, preceding the onset of either migration or enhanced proliferation, and are silenced after the restoration of a functional epithelial lining (Mansbridge and Knapp, 1987; Takahashi and Coulombe, 1997). In the context of chronic hyperproliferative diseases such as psoriasis, these genes are chronically expressed in clinically involved skin, and successful pharmacological management of the lesions correlates with their reduced expression (e.g., Holland *et al.*, 1989; Berth-Jones *et al.*, 1992; Gerritsen *et al.*, 1993). Intriguingly, a number of nonkeratin genes appear consistently coregulated with K6, K16, and K17 during wound repair, psoriasis, and even skin carcinoma (Coulombe, 1997), raising the distinct possibility that related signaling pathways could be involved in these situations. Knowledge of the signals involved in activating the expression of K6, K16, and K17 in these contexts would likely apply to several other genes, and provide far-reaching insights into signaling mechanisms activated during wound repair and disease processes. Second, as outlined above, K6, K16, and K17 show a somewhat mysterious expression pattern in complex epithelia under resting conditions. K17, in particular, may be expressed in mitotic or postmitotic cells, in pseudostratified and in stratified epithelia, and is typically found in the myoepithelial cells of several glands. Paradoxically, therefore, these genes contain relevant information specifying expression in very defined cell subpopulations in a heterogeneous array of epithelial tissue types, and again this can be exploited to characterize the pathways involved.

4.1. *In Vitro* Promoter Analyses

A popular approach to identify critical regulatory elements within these genes is to analyze their promoter activities *in vitro*. This approach has the advantage of allowing for the measurement of reporter gene activation as a function of stimulation with specific compounds such as growth factors and cytokines. However, it is sometimes difficult to reconcile the results obtained with *in vitro* systems to the more complex *in vivo* context (Oshima, 1992). This is complicated by the redundancy and abundance of signaling molecules produced at the wound site (e.g., Clark, 1993; Martin, 1997) and further compounded by the fact that K6, K16, and K17 are often expressed at detectable levels in cultured epidermal cells (e.g., Weiss *et al.*, 1984; Tyner and Fuchs, 1986; Schermer *et al.*, 1989). This said, a number of laboratories have exploited this approach to examine the regulation of keratins 6, 16, and 17.

Surveying the proximal promoter region of human K6a (Takahashi *et al.*, 1995), human K6b (Tyner *et al.*, 1985), and bovine K6β (Navarro *et al.*, 1995) reveals consensus binding sites for several key transcription factors (Table II). Navarro and co-workers (Navarro *et al.*, 1995) performed extensive mapping of several of these sites using electrophoretic mobility shift analysis (EMSA) and DNAse I footprinting. Nuclear extracts prepared from bovine mammary gland epithelial cells bound to consensus elements for AP-1, AP-2, and an element designated E-box (Blessing *et al.*, 1987). This group also demonstrated the presence of a retinoid response element (RARE) within the bovine K6β gene. By testing the enhancer activity of these elements *in vitro* it was found that only the AP-1 site functioned as an enhancer when placed into a heterologous promoter. This has important implications for the regulation of K6 gene during wound healing, as it has been shown that elements of this transcription complex are rapidly induced in wound-edge keratinocytes (Martin and Nobes, 1992). Examining the response of human skin keratinocytes to stimulation with epidermal growth factor (EGF) or transforming growth factor-α (TGF-α), Jiang *et al.* (1993) indentified EGF-responsive elements in both the human K6 and K16 genes. These sequences were not homologous to other known EGF-REs and could not be detected in the 5′ upstream region of known EGF-responsive genes. Again, several EGF receptor ligands are elevated in the wound fluid early after injury (Martin, 1997).

The transcriptional control of keratin 17 has also been extensively studied, leading to the identification of several response elements within its promoter region. In particular, the response of this gene to interferon-gamma (IFN-γ), which upregulates K17 expression, has been intensely studied, but produced conflicting findings (Flohr *et al.*, 1992; Jiang *et al.*, 1994; Vogel *et al.*, 1995). Jiang and *et al.* (1994) examined the activity of a number of keratin gene promoters in cultured human skin keratinocytes and found that only K17 responds to IFN-γ. [Note: Barker *et al.* (1993) found that in human skin *in vivo*, subcutaneous administration of IFN-γ causes induction of K16. Moreover, the human K6a gene promoter features several gamma-interferon activating sequence (GAS) concensus binding sites; see Takahashi *et al.* (1995). It thus appears that the effects of IFN-γ

Table II
Transcriptional Regulatory Elements in the Promoter Regions of Keratins 6, 16, and 17

Gene	Element	Location[a]	Sequence	Reference
K6	AP-1	−331 to −309	5′-TGACTAA-3′	Navaro *et al.* (1995)
K6	AP-2	−594 to −582	5′-TTCCCAGGC-3′	Navaro *et al.* (1995)
K6	Epidermal box	−238 to −222	5′-AAACCAAA-3′	Navaro *et al.* (1995)
K6	?	−615 to −604	5′-GTGTCATGTCCC-3′	Navaro *et al.* (1995)
K6	EGF-RE	−138 to −115	5′-AAGCTCACCTTCC-AGGACTAG-3′	Jiang *et al.* (1993)
K16	EGF-RE	−212 to −192	5′-GGGCTAATAATCC-AGAGTGAG-3′	Jiang *et al.* (1993)
K17	E-box with AP-2	−126 to −103	5′-GGGAGCTGGCAG-GTGGCCAGTCC-3′	Troyanovsky and Leube (1994)
K17	SP-1	−155 to −144	5′-ATGGGCGGGGAG-3′	Troyanovsky and Leube (1994)
		−412 to −402	5′-TTTGGCGGGGGG-3′	
K17	C/EBP	−103 to −93	5′-TGATGAAAGCC-3′	Troyanovsky and Leube (1994)
K17	CRE	−310 to −305	5′-TGACGT-3′	Troyanovsky and Leube (1994)
K17	GAS	−1528 to −1515	5′-TTTCTAAGAATCAG-3′	Vogel *et al.* (1995)
K17	GAS	−183 to −171[b]	5′-TTTGGAGAAA-TGAG-3′	Vogel *et al.* (1995)
K17	GAS	−250 to −220[c]	5′-AGGTTTGGAGAAA-TGAGGCAAATT-3′	Jiang *et al.* (1994)

[a]These sequence motifs are located in the 5′ upstream region of the gene concerned, and the position of the nucleotides is generally defined with respect to the start of the coding sequence. See individual references for further details.
[b]This sequence did not shift in EMSA assay, but was shown by the authors to be critical for IFN-γ stimulation of reporter gene constructs.
[c]This element is identical to that of Vogel *et al.* (above), and was initially identified in the K17 pseudogene promoter. This oligomer was shown by the authors to interact with the STAT91 protein.

treatment on keratin gene expression may be context dependent.] Starting from a 600-base fragment derived from the 5′ upstream sequence of a human K17 pseudogene, Jiang *et al.* obtained evidence for three GAS elements in in the context of a CAT reporter gene assay involving human keratinocytes. DNA mobility shift assays with nuclear extracts prepared from IFN-γ-treated keratinocytes showed that the most proximal of these GAS sites (−210 bp) in the K17 pseudogene promoter binds STAT91, a component of a IFN-γ-responsive transcription factor (Jiang *et al.*, 1994). On the other hand, working with the entire 5′ upstream sequence (1.7 kb) of the functional human K17 gene, Vogel *et al.* (1995) showed that only the distalmost one (−1515 bp) among three consensus GAS elements binds GAF/STAT91 protein following stimulation of HeLa cells with IFN-γ, and, surprisingly, that mutation of this element does not effect the responsiveness of K17–5′-CAT to this cytokine. In contrast, mutation of the most proximal GAS element (−171 bp) significantly decreased the CAT response to IFN-γ. The discrepancy between the results obtained by these two groups may be due, in part, to the use of the K17 pseudogene promoter as a template for producing the expression constructs by Jiang *et al.* (1994), though many transcriptional elements are conserved between the K17 pseudogene and functional gene (Troyanovsky *et al.*, 1992; Troyanovsky and Leube, 1994).

Troyanovsky and Leube (1994) conducted an extensive analysis of the human K17 pseudogene and functional gene promoters. Their study showed that 1.7 kb of the upstream promoter region conferred high trancriptional activation of reporter gene constructs when transfected into K17-expressing HeLa cells. Similar transfections using a fibroblast cell line that does not express K17 (SV40-transformed SV80 cells) showed no activity, supporting the specificity of the promoter. They identified two major regulatory activities within this promoter. The proximal region from −456 to +13 contains several consensus binding elements for known transcription factors, some of which were protected from DNAse I digestion by proteins from HeLa cell nuclear extracts (Table II). The distal promoter region (−1762 to −457 bp) has enhancer activity. This was demonstrated convincingly by showing that this distal region could activate transcription when fused to the intrinsically inactive proximal promoter region of the pseudogene.

4.2. *In Vivo* Promoter Analysis of K6 Transcription

In vivo promoter analysis via transgenic mice represents a powerful tool for identifying critical regulatory elements within the promoter/enhancer regions of genes. Transgenic mice containing the upstream regulatory sequences of keratin genes fused to a reporter sequence allow for easy visualization of transgene activation (e.g., Abe and Oshima, 1990; Leask *et al.*, 1990). Although multiple pathways probably exist for activation of these genes, the identification of the *cis*-acting sequences responsible for the induction of K6, K16, and K17 following injury could provide insight into the signaling events during the early stages of wound repair. This approach may also serve to identify common mechanisms for activating keratin and nonkeratin genes alike early after injury to stratified epithelia (Coulombe, 1997, and references therein).

These types of experiments have recently been performed with the human keratin 6a gene (Takahashi and Coulombe, 1996, 1997) as well as the bovine K6β gene (Ramirez *et al.*, 1995). For the human K6a gene, constructs with either 5.20, 2.55, or 0.96 kb of 5′ upstream regulatory sequence were able to confer inducible expression to the LacZ reporter whenever trunk skin, footpads, oral mucosa, and cornea were subjected to experimental injury. Transgene expression also occured following the application of other stimuli known to cause K6 induction, such as topical application of phorbol esters or retinoic acid, and in a delayed-type hypersensitivity reaction. However, further deletion to a 550-bp construct resulted in a general loss of inducible transgene expression. Of note, no consistent expression of any of these K6a promoter-based LacZ constructs occurred in epithelia known to express K6 in a constitutive fashion, such as hair follicles, oral mucosa, tongue, forestomach, etc. (Takahashi and Coulombe, 1997). These findings suggest that critical elements necessary for conferring inducible expression to a heterologous reporter are located between bases −960 and 550 of the human K6a promoter sequence. Additionally, these elements are at least partially distinct from those involved in specifying constitutive expression in stratified epithelia under baseline conditions. It should now be possible, via gel-mobility shift analysis and DNAse I footprinting, to fully characterize the inducible element(s) within the proximal 5′ upstream segment of the K6a gene, as done by Ramirez *et al.* (1995) on the bovine K6β promoter.

The bovine K6β gene shows a different behavior when analyzed using a similar strategy in transgenic mice. Two groups observed specific expression of heterologous coding sequences in transgenic mouse skin when using either 5.2 kb or 8.8 kb of 5′ upstream sequence from the BK6β gene (Blessing *et al.*, 1993; Ramirez *et al.*, 1995). These findings clearly differ from those obtained with the human K6a gene. This is a surprising occurrence, given the extensive homology in the proximal 5′ upstream sequences of the human K6a, K6b, and bovine K6β genes (Navarro *et al.*, 1995; Takahashi and Coulombe, 1997). We do not know whether the bovine K6β gene is the orthologue of human K6a, and this issue could potentially explain the differences observed in the activity of the 5′ upstream sequence of these genes in transgenic mice. As is the case with the human K6a gene promoter, however, the bovine K6β-based LacZ construct is inducible by injury, phorbol esters, and retinoic acid (Ramirez *et al.*, 1995). The inducible nature of the K6 promoter offers an opportunity to develop expression vectors showing conditional activity in skin *in vivo* (e.g., Takahashi and Coulombe, 1996).

5. PROPERTIES AND FUNCTION OF K6 AND K16 IN NORMAL AND HEALING EPITHELIA

5.1. K16 Possesses Distinct Properties when Tested in a Variety of Experimental Contexts

Intrigued by the characteristic expression of K6 and K16 following injury to stratified epithelia, we initiated a study of their biochemical properties with the hope of identifying features unique to these keratins that would be suggestive of their function in wound repair. To do so, we compared the ability of the K6b–K16 and K5–K14 pairs to form stable tetramer assembly subunits and 10-nm filaments *in vitro* and assessed their behavior when transiently expressed in fibroblasts and epithelial cells in culture. The K5–K14 pair was selected as a reference because it is constitutively expressed in stratified epithelia and had already been characterized *in vitro*. Furthermore, the primary sequences of K5 and K14 are highly related to those of K6b and K16, respectively (see Section above). The properties of other human K6 isoforms, K6a in particular, and those of K17 have yet to be investigated in a detailed fashion.

We found that compared to K5–K14, K6b–K16 form less stable tetramer subunits that polymerize into shorter 10-nm filaments (Figure 5) with a lower efficiency *in vitro*. Expression of K6b–K16 in BHK fibroblasts produces IFs that do not extend radially from the nucleus as shown for K5–K14 filaments (Paladini *et al.*, 1996). Immunoelectron microscopy of these transfected fibroblasts show that as is the case *in vitro*, K6b–K16 filaments appear short compared to K5–K14 ones (Figure 5). Permutation experiments showed that K16 is mainly responsible for this behavior, as the K6b isoform behaves very similarly to K5 with respect to these parameters (Paladini *et al.*, 1996). Further studies in cultured PtK2 epithelial cells revealed that whereas transfected K14 integrates without perturbation into the preexisting KIF network in a majority of cases, K16 expresssion causes a dramatic perinuclear reorganization of the endogenous K8–K18 filaments in approximately 50% of transfected PtK2 cells (Paladini *et al.*, 1996).

The finding that K16 expression causes a KIF reorganization in PtK2 cells is of interest for two reasons: (a) we had already reported that overexpression of human K16 in transgenic mice caused skin alterations, the early manifestation of which is a reorganization of the KIF network in postmitotic keratinocytes (Takahashi *et al.*, 1994); and (b) the changes in the keratin pattern taking place after acute skin injury correlate with a retraction of KIFs from the cell periphery in suprabasal keratinocytes (see

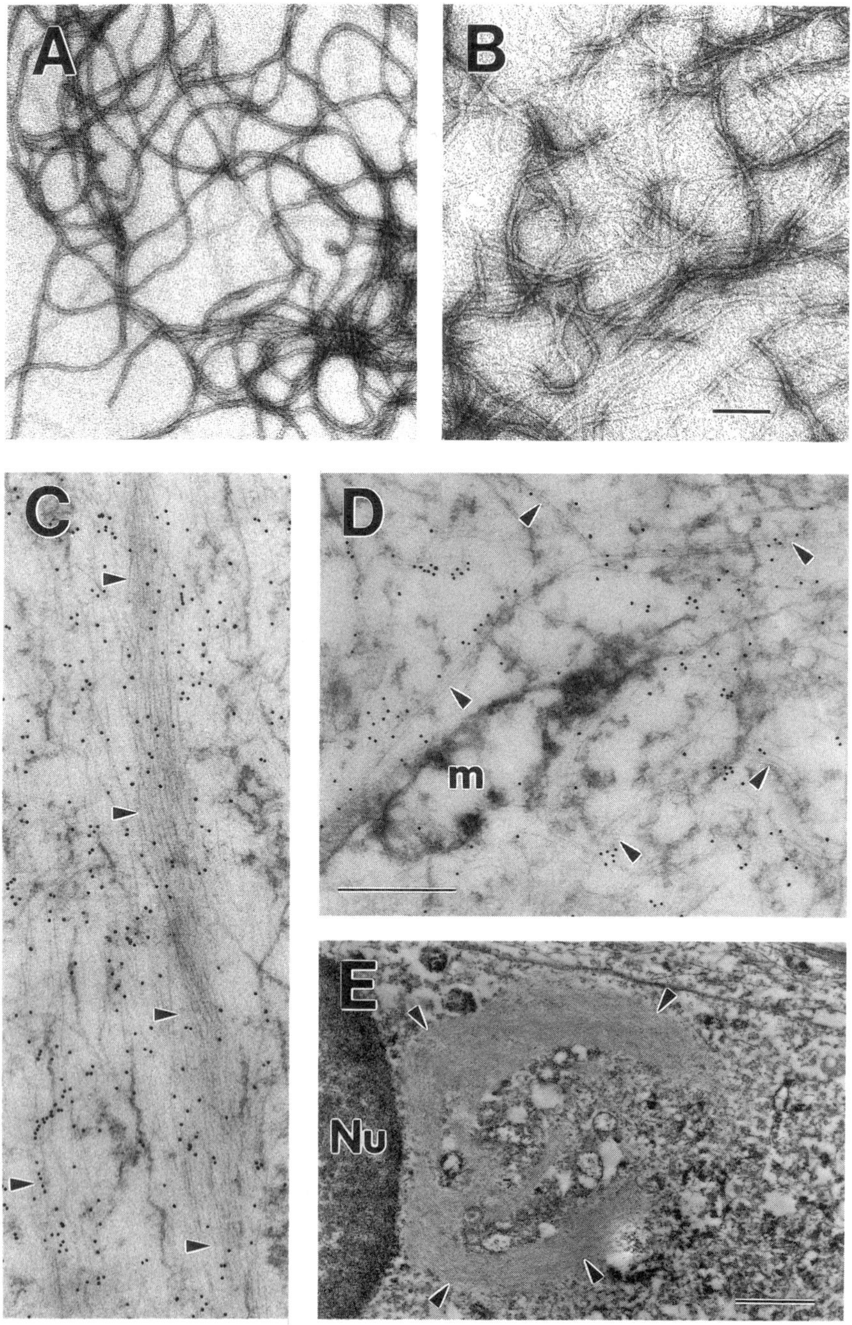
A
B
C
D
m
E
Nu

Section 2 above). From these findings we postulated that (a) the distinct properties of K14 and K16 are likely to translate into at least partially distinct function(s) for these keratins in epithelial cells, and (b) K16 is directly involved in promoting some of the cytoarchitectural changes in keratinocytes at the wound edge as they are being recruited for migration (Paladini *et al.*, 1996). The recent creation of a transgenic mouse with a null mutation in the K14 gene (Lloyd *et al.*, 1995), a manipulation that results in a severe skin blistering phenotype and rapid death, along with the availability of a K14 promoter-based cassette enabling tissue-specific expression in transgenic mouse (Vassar *et al.*, 1989; Saitou *et al.*, 1995), opens up the possibility of testing *in vivo* whether ectopically expressed K16 can compensate for the absence of K14 at a functional level. This experiment is in progress in our laboratory.

5.2. Mapping Biochemical Determinants in K16 Protein

Given their very high similarity at the primary sequence level, including the central α-helical rod domain (Figure 4), it is quite remarkable that K14 and K16 show marked differences in their assembly properties. We are exploiting this sequence relatedness to map the biochemical determinants that distinguish these two keratins. We constructed a series of K14–K16 chimeric cDNAs and are testing the properties of the resulting chimeric proteins at the level of tetramer stability, 10-nm filament formation *in vitro*, as well as their impact on KIF organization when transfected in cultured epithelial cells.

FIGURE 5. Comparison of K14 and K16 assembly properties *in vitro* and *in vivo*. (A, B). Purified heterotypic K6b–K14 and K6b–K16 complexes were induced to polymerize by dialysis against standard epidermal keratin assembly buffer and visualized by negative staining and electron microscopy. Under these conditions K6b–K14 form long 10-nm filaments (A) with high efficiency, whereas K6b–K16 form shorter 10-nm filaments (B) with a slightly lower efficiency [see Paladini *et al.*, (1996) for details]. (C–E). BHK-21 fibroblasts were cotransfected with either CMV–K5/CMV–K14 or CMV–K6b/CMV–K16 plasmid combinations. At 24 h posttransfection, the cells were fixed in methanol and processed for preembedding immunocytochemistry using 15-nm colloidal gold-conjugated secondary antibodies. (C) K5 and K14 coexpressing cell, with anti-K14 staining. Note the long 10-nm filament organized into bundles (arrowheads) extending throughout the cytoplasm. (D, E) K6b and K16 coexpressing cells, with anti-K16 staining. Panel D shows the peripheral cytoplasm of a transfected BHK-21 cell, which contains fewer and shorter filaments (arrowheads) compared to K5 and K14 coexpressing cells. Panel E shows a juxtanuclear whorl of densely packed filaments in a transfected BHK-21 cells at low magnification. Mock-transfected cells show only background levels of gold staining with either antisera (not shown). m, Mitochondria; Nu, Nucleus. Bar in panel B = 100 nm and applies for panel A as well; bar in panel C = 0.5 μm and applies for panel D as well; bar in panel E = 2 μm.

This strategy led us to the discovery that a single amino acid residue, ^{188}Pro, is virtually entirely responsible for the unstable character of K16-containing tetramers in the presence of urea (Wawersik *et al.*, 1997). Exceptionally, this proline residue occurs at a "d" position of the third heptad repeat of subdomain 1B of the α-helical rod domain of K16 (see Figure 4)—no other known IF sequence features a helix-disrupting proline within subdomains 1A, 1B, 2A, or 2B (Quinlan *et al.*, 1994). The corresponding position is occupied by a hydrophobic residue (Val, Ile) in all other type I keratins but one (Quinlan *et al.*, 1994). Introducing a proline substitution at the corresponding position of human K14 (^{186}Val → Pro) and K17 (^{155}Ile → Pro), which both form urea-stable heterotetramers with either K5, K6b, or K8 as type II pairing partner, renders them K16-like in their behavior. Conversely, replacing ^{188}Pro with a valine renders K16 capable of forming urea-stable tetramers with either K6b, K5, or K8, demonstrating that this specific residue accounts for this peculiar property of K16. We also showed that ^{188}Pro is partly responsible for the lower efficiency with which K16-containing subunits polymerize *in vitro* compared to K14 ones (Wawersik *et al.*, 1997). These findings have direct implications for the structure of the keratin tetramer [see Wawersik *et al.*, 1997 for discussion] and may have implications as well for the regulation of K16 *in vivo*.

The notion that K16 forms relatively unstable tetramers may represent a key element in attempting to solve the "K16 paradox." This paradox can be stated as follows: How could K16 be playing a role in keratinocyte activation after acute injury when its presence in a variety of epithelial tissues under resting conditions is reportedly not associated with a peculiar KIF organization and keratinocyte cytoarchitecture? We argued (Paladini *et al.*, 1996; Wawersik *et al.*, 1997) that this is perhaps made possible by the following facts. First, K16 is expressed in a cell context featuring many other type I keratins, which is indeed the case (cf. Section 3 above). Second, under resting conditions, K16 occurs at low levels relative to total type I keratins in the relevant epithelial cells—this has been shown to be the case in a few tissues, such as the outer root sheath of the hair (Yoshikawa *et al.*, 1995), sweat gland ducts (Moll *et al.*, 1983), and foot sole epidermis (Knapp *et al.*, 1986). After acute injury to stratified epithelia, however, the relative amounts of K16 among type I keratins would rise dramatically, a possibility that has yet to be examined at a biochemical level. Third, assuming that the subunit adding to the growing keratin IF polymer is larger than a dimer (Steinert, 1991; Stewart, 1993), unstable assembly subunits would be at a disadvantage for incorporation into the polymer pool (Wawersik *et al.*, 1997). Such incorporation into the polymer pool would be necessary for K16 to manifest its impact on KIF structure and organization. Obviously, there are alternative explanations for the K16 paradox. For instance, as is the case for all keratins investigated (Ku *et al.*, 1996), K16 is a substrate for phosphorylation *in vivo* (Gilmartin *et al.*, 1980), but the site(s) and role of this covalent modification for K16 are not known. Finally, the properties of K16 when incorporated into filaments could be modulated and even altered by the presence of other keratins such as K14 and K17, or by an interaction with "associated" proteins.

A distinct segment of the K16 protein appears to be involved in its ability to promote a reorganization of the KIF network in transfected PtK2 cells. Indeed, the K16 ^{188}Pro → Val mutant is nearly indistinguishable from K16 itself in this assay (Wawersik *et al.*, 1997). Studies in progress utilizing K14–K16 chimeras suggest that the determinant(s) responsible for this behavior reside(s) in the carboxy-terminal 100 amino acid residues of K16 (R. Paladini and P. A. Coulombe, unpublished observations). This is hardly surprising, as the primary sequences of K14 and K16 diverge mostly at the level of their nonhelical tail domains, particularly in the last 25 amino acid residues (out of 47–48 in the tail domain; data not shown). We hope to define the portion(s) of the carboxy-terminus of K16 and the mechanism(s) involved in this phenomenon in the future.

6. CONCLUSION

A general role for keratin intermediate filaments in providing epithelial cells with the structural resilience it needs to sustain the mechanical forces to which they are routinely subjected has been established. The multiplicity of keratin sequences, their highly regulated expression at a temporal and spatial level, and the degree to which these are conserved among mammalian species suggest that these proteins have specialized to meet the demands of different types of epithelia. How this family of proteins may contribute to the spectrum of epithelial differentiation and function remains a mystery. Because of their properties and highly intriguing pattern of expression, we think that continued studies on K6, K16, and K17 are likely to provide significant insights into the mechanisms via which keratin filaments and associated elements impact on the cytoarchitecture and properties of epithelial cells. In particular, the work in progress on keratin 16 suggests that it may represent an example of such specialization of a keratin sequence for a distinct functional purpose. Elucidating the logical principles that may link the regulation and function of K6, K16, and K17 in contrasting contexts such as normal (intact), healing, and diseased epithelia represents a formidable challenge. Fortunately, we and others are approaching the task at hand armed with the knowledge and tools produced by the pioneering efforts of a number of research laboratories over the past 15 years.

ACKNOWLEDGMENTS. Our research program is supported by National Institute of Health grants AR42047 and AR44232. P.A.C. is the recipient of a Junior Faculty Research Award from the American Cancer Society. K.M. is the recipient of National Research Service Award CA67513 from the National Cancer Institute.

7. REFERENCES

Abe, M., and Oshima, R. G., 1990, A single human keratin 18 gene is expressed in diverse epithelial cells of transgenic mice, *J. Cell Biol.* **111:**1197–1206.

Barker, J. N., Goodlad, J. R., Ross, E. L., Yu, C. C., Groves, R. W., and MacDonald, D. M., 1993, Increased epidermal cell proliferation in normal human skin *in vivo* following local administration of interferon-gamma, *Am. J. Pathol.* **142:**1091–1097.

Bereiter-Hahn, J., 1986, Epidermal cell migration and wound repair, in *Biology of the Integument, Vol. 2: Vertebrates* (J. Bereiter-Hahn, A. G. Matoltsy, and K. S. Richards, eds.), pp. 443–471, Springer-Verlag, Berlin.

Berth-Jones, J., Fletcher, A., and Hutchison, P. E., 1992, Epidermal cytokeratin and immunocyte responses during treatment of psoriasis with calcipotriol and betamethasone valerate, *Br. J. Dermatol.* **126:**356–361.

Blessing, M., Zentgraf, H., and Jorcano, J. L., 1987, Differentially expressed bovine cytokeratin genes. Analysis of gene linkage and evolutionary conservation of 5′-upstream sequences, *EMBO J.* **6:**567–575.

Blessing, M., Nanney, L. B., King, L. E., Jones, C. M., and Hogan, B. L., 1993, Transgenic mice as a model to study the role of TGF-beta-related molecules in hair follicles, *Genes Dev.* **7:**204–215.

Blobel, G. A., Moll, R., Franke, W. W., and Vogt-Moykopf, I., 1984, Cytokeratins in normal lung and lung carcinomas. 1. Adenocarcinomas, squamous cell carcinomas, cultured cell lines, *Virchows Arch. (Cell. Pathol.)* **45:**407–429.

Blumenberg, M., 1988, Concerted gene duplications in the two keratin gene families, *J. Mol. Evol.* **27:**203–211.

Bonifas, J. M., Rothman, A. L., and Epstein, E. J., 1991, Epidermolysis bullosa simplex: Evidence in two families for keratin gene abnormalities, *Science* **254:**1202–1205.

Bonifas, J. M., Matsumura, K., Chen, M. A., Berth-Jones, J., Hutchison, P. E., Zloczower, M., Fritsch, P. O., and Epstein, E. H., Jr., 1994, Mutations of keratin 9 in two families with palmoplantar epidermolytic hyperkeratosis, *J. Invest. Dermatol.* **103:**474–477.

Bowden, P. E., Haley, J. L., Kansky, A., Rothnagel, J. A., Jones, D. O., and Turner, R. J., 1995, Mutation of a type II keratin gene (K6a) in pachyonychia congenita, *Nat. Genet.* **10:**363–365.

Cheng, J., Syder, A. J., Yu, Q. C., Letai, A., Paller, A. S., and Fuchs, E., 1992, The genetic basis of epidermolytic hyperkeratosis: A disorder of differentiation-specific epidermal keratin genes, *Cell* **70:**811–819.

Chipev, C. C., Korge, B. P., Markova, N., Bale, S. J., DiGiovanna, J. J., Compton, J. G., and Steinert, P. M., 1992, A leucine → proline mutation in the H1 subdomain of keratin 1 causes epidermolytic hyperkeratosis, *Cell* **70:**821–828.

Choi, Y., and Fuchs, E., 1990, TGF-beta and retinoic acid: Regulators of growth and modifiers of differentiation in human epidermal cells, *Cell Regul.* **1:**791–809.

Clark, R. A. F., 1993, Mechanisms of cutaneous wound repair, in *Dermatology in General Medicine* (T. B. Fitzpatrick, A. Z. Eisen, K. Wolff, I. M. Freedberg, and M. D. Austen, eds.), Vol. I, pp. 473–478 McGraw-Hill, New York.

Collin, C., Ouhayoun, J. P., Grund, C., and Franke, W. W., 1992, Suprabasal marker proteins distinguishing keratinizing squamous epithelia: Cytokeratin 2 polypeptides of oral masticatory epithelium and epidermis are different, *Differentiation* **51:**137–148.

Coulombe, P. A., 1993, The cellular and molecular biology of keratins: Beginning a new era. *Curr. Opin. Cell Biol.* **5:**17–29.

Coulombe, P. A., 1997, Towards a molecular definition of keratinocyte activation after acute injury to stratified epithelia, *Biochem. Biophys. Res. Commun.* **236:**231–238.

Coulombe, P. A., and Fuchs, E., 1994, Molecular mechanisms of keratin gene disorders and other bullous diseases of the skin, in *Molecular Mechanisms in Epithelial Cell Junctions: From Development to Disease* (S. Citi, ed.), pp. 259–285, Landes, Austin, Texas.

Coulombe, P. A., and Takahashi, K., 1996, The cellular mechanism of keratinocyte migration into a skin wound site: An open question with important implication, *Cell Vision: J. Anal. Morphol.* **3:**217–223.

Coulombe, P. A., Hutton, M. E., Letai, A., Hebert, A., Paller, A. S., and Fuchs, E., 1991a, Point mutations in human keratin 14 genes of epidermolysis bullosa simplex patients: Genetic and functional analyses, *Cell* **66:**1301–1311.

Coulombe, P. A., Hutton, M. E., Vassar, R., and Fuchs, E., 1991b, A function for keratins and a common thread among different types of epidermolysis bullosa simplex diseases, *J. Cell Biol.* **115:**1661–1674.

Coulombe, P. A., Bravo, N. S., Paladini, R. D., Nguyen, D., and Takahashi, K., 1995, Overexpression of human keratin 16 produces a distinct phenotype in transgenic mouse skin, *Biochem. Cell Biol.* **73:**611–618.

Feinstein, A., Friedman, J., and Schewach, M. M., 1988, Pachyonychia congenita, *J. Am. Acad. Dermatol.* **19:**705–711.

Flohr, T., Buwitt, U., Bonnekoh, B., Decker, T., and Bottger, E. C., 1992, Interferon-gamma regulates expression of a novel keratin class I gene, *Eur. J. Immunol.* **22:**975–979.

Fuchs, E., and Coulombe, P. A., 1992, Of mice and men: Genetic skin diseases of keratin, *Cell* **69:**899–902.

Fuchs, E., and Weber, K., 1994, Intermediate filaments: Structure, dynamics, function, and disease, *Annu. Rev. Biochem.* **63:**345–382.

Fuchs, E., Esteves, R. A., and Coulombe, P. A., 1992, Transgenic mice expressing a mutant keratin 10 gene reveal the likely genetic basis for epidermolytic hyperkeratosis, *Proc. Natl. Acad. Sci. USA* **89:**6906–6910.

Fuchs, E., Coulombe, P. A., Cheng, J., Chan, Y.-M., Hutton, E., Syder, A., Degenstein, L., Yu, Q. C., Letai, A., and Vassar, R., 1994, Genetic bases of epidermolysis bullosa simplex and epidermolytic hyperkeratosis, *J. Invest. Dermatol.* **103:**25S-30S.

Garlick, J. A., and Taichman, L. B., 1994, Fate of human keratinocytes during reepithelialization in an organotypic culture model, *Lab. Invest.* **70:**916–924.

Gerritsen, M. J., Rulo, H. F., Van Vlijmen-Willems, I., Van Erp, P. E., and van de Kerkhof, P. C., 1993, Topical treatment of psoriatic plaques with 1,25-dihydroxyvitamin D3: A cell biological study, *Br. J. Dermatol.* **128:**666–673.

Gilmartin, M. E., Culbertson, V. B., and Freedberg, I. M., 1980, Phosphorylation of epidermal keratins, *J. Invest. Dermatol.* **75:**211–216.

Grace, M. P., Kim, K. H., True, L. D., and Fuchs, E., 1985, Keratin expression in normal esophageal epithelium and squamous cell carcinoma of the esophagus, *Cancer Res.* **45:**841–846.

Grinnell, F., 1992, Wound repair, keratinocyte activation and integrin modulation, *J. Cell Sci.* **101:**1–5.

Hanukoglu, I., and Fuchs, E., 1983, The cDNA sequence of a type II cytoskeletal keratin reveals constant and variable structural domains among keratins, *Cell* **33:**915–924.

Heid, H. W., Moll, I., and Franke, W. W., 1988, Patterns of expression of trichocytic and epithelial cytokeratins in mammalian tissues. II. Concomitant and mutually exclusive synthesis of trichocytic and epithelial cytokeratins in diverse human and bovine tissues (hair follicle, nail bed and matrix, lingual papilla, thymic reticulum), *Differentiation* **37**(3)**:**215–230.

Holland, D. B., Wood, E. J., Roberts, S. G., West, M. R., and Cunliffe, W. J., 1989, Epidermal keratin levels during oral 1-alpha-hydroxyvitamin D3 treatment for psoriasis, *Skin Pharmacol.* **2:**68–76.

Huang, T. H., Ann, D. K., Zhang, Y. J., Chang, A. T., Crabb, J. W., and Wu, R., 1994, Control of keratin gene expression by vitamin A in tracheobronchial epithelial cells, *Am. J. Respir. Cell Mol. Biol.* **10:**192–201.

Irvine, A. D., Corden, L., Swennsson, O., Swennsson, B., Moore, J. E., Frazer, D. G., Smith, F., Knowlton, R., Christophers, E., Rochels, R., Uitto, J., and McLean, W., 1997, Mutations in cornea-specific keratin K3 or K12 genes cause Meesmann's corneal dystrophy, *Nat. Genet.* **12:**184–187.

Jiang, C. K., Magnaldo, T., Ohtsuki, M., Freedberg, I. M., Bernerd, F., and Blumenberg, M., 1993, Epidermal growth factor and transforming growth factor alpha specifically induce the activation- and hyperproliferation-associated keratins 6 and 16, *Proc. Natl. Acad. Sci. USA* **90:**6786–6790.

Jiang, C. K., Flanagan, S., Ohtsuki, M., Shuai, K., Freedberg, I. M., and Blumenberg, M., 1994, Disease-activated transcription factor: Allergic reactions in human skin cause nuclear translocation of STAT-91 and induce synthesis of keratin K17, *Mol. Cell. Biol.* **14:**4759–4769.

Jones, J. C., and Green, K. J., 1991, Intermediate filament-plasma membrane interactions, *Curr. Opin. Cell Biol.* **3:**127–132.

Kimonis, V., DiGiovanna, J. J., Yang, J. M., Doyle, S. Z., Bale, S. J., and Compton, J. G., 1994, A mutation in the V1 subdomain of keratin 1 in non-epidermolytic palmoplantar keratoderma, *J. Invest. Dermatol.* **103:**764–769.

Knapp, A. C., Franke, W. W., Heid, H., Hatzfeld, M., Jorcano, J. L., and Moll, R., 1986, Cytokeratin No. 9, an epidermal type I keratin characteristic of a special program of keratinocyte differentiation displaying body site specificity, *J. Cell Biol.* **103:**657–667.

Knapp, B., Rentrop, M., Schweizer, J., and Winter, H., 1987, Three cDNA sequences of mouse type I keratins. Cellular localization of the mRNAs in normal and hyperproliferative tissues, *J. Biol. Chem.* **262:**938–945.

Kopan, R., and Fuchs, E., 1989, The use of retinoic acid to probe the relation between hyperproliferation-associated keratins and cell proliferation in normal and malignant epidermal cells, *J. Cell Biol.* **109:**295–307.

Ku, N.-O., Liao, J., Chou, C.-F., and Omary, M. B., 1996, Implications of intermediate filament phosphorylation, *Cancer Metast. Rev.* **15:**429–444.

Ku, N.-O., Wright, T. L., Terrault, N. A., Gish, R., and Omary, M. B., 1997, Mutation of human keratin 18 in association with cryptogenic cirrhosis, *J. Clin. Invest.* **99:**19–23.

Lane, E. B., Rugg, E. L., Navsaria, H., Leigh, I. M., Heagerty, A. H., Ishida, Y. A., and Eady, R. A., 1992, A mutation in the conserved helix termination peptide of keratin 5 in hereditary skin blistering, *Nature* **356:**244–246.

Leask, A., Rosenberg, M., Vassar, R., and Fuchs, E., 1990, Regulation of a human epidermal keratin gene: Sequences and nuclear factors involved in keratinocyte-specific transcription, *Genes Dev.* **4:**1985–1998.

Lloyd, C., Yu, Q. C., Cheng, J., Turksen, K., Degenstein, L., Hutton, E., and Fuchs, E., 1995, The basal keratin network of stratified squamous epithelia: Defining K15 function in the absence of K14, *J. Cell Biol.* **129:**1329–1344.

Mansbridge, J. N., and Knapp, A. M., 1987, Changes in keratinocyte maturation during wound healing, *J. Invest. Dermatol* **89:**253–263.

Marchuk, D., McCrohon, S., and Fuchs, E., 1984, Remarkable conservation of structure among intermediate filament genes, *Cell* **39:**491–498.

Martin, P., 1997, Wound healing—Aiming for the perfect skin regeneration, *Science* **276:**75–81.

Martin, P., and Nobes, C. D., 1992, An early molecular component of the wound healing response in rat embryos—Induction of c-fos protein in cells at the epidermal wound margin, *Mech. Dev.* **38:**209–215.

McLean, W. H. I., and Lane, E. B., 1995, Intermediate filaments in diseases, *Curr. Opin. Cell Biol.* **7:**118–125.

McLean, W. H., Morley, S. M., Lane, E. B., Eady, R. A., Griffiths, W. A., Paige, D. G., Harper, J. I., Higgins, C., and Leigh, I. M., 1994, Ichthyosis bullosa of Siemens—A disease involving keratin 2e, *J. Invest. Dermatol.* **103:**277–281.

McLean, W. H. I., Rugg, E. L., Lunny, D. P., Morley, S. M., Lane, E. B., Swensson, O., Dopping-Hepenstal, P. J. C., Griffiths, W. A. D., Eady, R. A. J., Higgins, C., Navsaria, H. A., Leigh, I. M., Stachan, T., Kunkeler, L., and Munro, C. S., 1995, Keratin 16 and keratin 17 mutations cause pachyonychia congenita, *Nat. Genet.* **9:**273–278.

Moll, R., Franke, W. W., Schiller, D. L., Geiger, B., and Krepler, R., 1982a, The catalog of human cytokeratins: Patterns of expression in normal epithelia, tumors and cultured cells, *Cell* **31:**11–24.

Moll, R., Moll, I., and Wiest, W., 1982b, Changes in the pattern of cytokeratin polypeptides in epidermis and hair follicles during skin development in human fetuses, *Differentiation* **23:**170–178.

Moll, R., Krepler, R., and Franke, W. W., 1983, Complex cytokeratin polypeptide patterns observed in certain human carcinomas, *Differentiation* **23:**256–269.

Moll, I., Heid, H. W., Franke, W. W., and Moll, R., 1988, Patterns of expression of trichocytic and epithelial cytokeratins in mammalian tissues. III. Hair and nail formation during human fetal development, *Differentiation* **39:**167–184.

Munro, C. S., Carter, S., Bryce, S., Hall, M., Rees, J. L., Kunkeler, L., Stephenson., A., and Strachan, T., 1994, A gene for pachyonychia congenita is closely linked to the keratin gene cluster on 17q12–q21, *J. Med. Genet.* **31:**675–678.

Navarro, J. M., Casatorres, J., and Jorcano, J. L., 1995, Elements controlling the expression and induction of the skin hyperproliferation-associated keratin K6, *J. Biol. Chem.* **270:**21362–21367.

O'Guin, W. M., Schermer, A., Lynch, M., and Sun, T.-T., 1990, Differentiation-specific expression of keratin pairs, in *Cellular and Molecular Biology of Intermediate Filaments* (R. D. Goldman and P. M. Steinert, eds.), pp. 301–334. Plenum Press, New York.

Oshima, R. G., 1992, Intermediate filament molecular biology, *Curr. Opin. Cell Biol.* **4:**110–166.

Paladini, R. D., Takahashi, K., Gant, T. M., and Coulombe, P. A., 1995, cDNA cloning and bacterial expression of the human type I keratin 16, *Biochem. Biophys. Res. Commun.* **215:**517–523.

Paladini, R. D., Takahashi, K., Bravo, N. S., and Coulombe, P. A., 1996, Onset of reepithelialization after skin injury correlates with a reorganization of keratin filaments in wound edge keratinocytes: Defining a potential role for K16, *J. Cell Biol.* **132:**381–397.

Porter, R. M., Leitgeb, S., Melton, D. M., Swensson, O., Eady, R. A. J., and Magin, T. M., 1996, Gene targeting at the mouse keratin 10 locus: Severe skin fragility and changes in cytokeratin expression in the epidermis, *J. Cell Biol.* **132:**925–936.

Quinlan, R., Hutchinson, C., and Lane, B., 1994, Intermediate filament proteins, *Protein Profile* **1:**779–911.

Ramirez, A., Vidal, M., Bravo, A., Larcher, F., and Jorcano, J. L., 1995, A 5′-upstream region of a bovine keratin 6 gene confers tissue-specific expression and hyperproliferation-related induction in transgenic mice, *Proc. Natl. Acad. Sci. USA* **92:**4783–4787.

RayChaudhury, A., Marchuk, D., Lindhurst, M., and Fuchs, E., 1986, Three tightly linked genes encoding human type I keratins: Conservation of sequence in the 5′-untranslated leader and 5′-upstream regions of coexpressed keratin genes, *Mol. Cell. Biol.* **6:**539–548.

Reis, A., Hennies, H.-C., Langbein, L., Digweed, M., Mischke, D., Drechsler, M., Schröck, E., Royer-Pokora, B., Franke, W. W., Sperling, K., and Küster, W., 1994, Keratin 9 gene mutations in epidermolytic palmoplantar keratoderma (EPPK), *Nat. Genet.* **6:**174–179.

Richard, G., Delaurenzi, V., Didona, B., Bale, S. J., and Compton, J. G., 1995, Keratin 13 point mutations underlies the hereditary mucosal epithelia disorder white sponge nevus, *Nat. Genet.* **11:**453–455.

Roop, D. R., Cheng, C. K., Titterington, L., Meyers, C. A., Stanley, J. R., Steinert, P. M., and Yuspa, S. H., 1984, Synthetic peptides corresponding to keratin subunits elicit highly specific antibodies, *J. Biol. Chem.* **259:**8037–8040.

Rosenberg, M., RayChaudhury, A., Shows, T. B., Le, B. M., and Fuchs, E., 1988, A group of type I keratin genes on human chromosome 17: Characterization and expression, *Mol. Cell. Biol.* **8:**722–736.

Rosenberg, M., Fuchs, E., Le, B. M., Eddy, R. L., and Shows, T. B., 1991, Three epidermal and one simple epithelial type II keratin genes map to human chromosome 12, *Cytogenet. Cell Genet.* **57:**33–38.

Rothnagel, J. A., Dominey, A. M., Dempsey, L. D., Longley, M. A., Greenhalgh, D. A., Gagne, T. A., Huber, M., Frenk, E., Hohl, D., and Roop, D. R., 1992, Mutations in the rod domains of keratins 1 and 10 in epidermolytic hyperkeratosis, *Science* **257:**1128–1130.

Rothnagel, J. A., Seki, T., Longley, M. A., Holder, R., Bundman, D. S., Adachi, K., and Roop, D. R., 1993a, Characterization of keratin 6 gene expression using transgenic mice, *J. Invest. Dermatol.* **101:**427.

Rothnagel, J. A., Greenhalgh, D. A., Wang, X. J., Sellheyer, K., Bickenbach, J. R., Dominey, A. M., and Roop, D. R., 1993b, Transgenic models of skin diseases, *Arch. Dermatol.* **129:**1430–1436.

Rothnagel, J. A., Traupe, H., Wojcik, S., Huber, M., Hohl, D., Pittelkow, M., Saeki, H., Ishibashi, Y., and Roop, D. R., 1994, Mutations in the rod domain of K2e in patients with ichtyosis bullosa of Siemens, *Nat. Genet.* **7:**485–489.

Rugg, E. L., Mclean, W. H. I., Allison, W. E., Lunny, D. P., Macleod, R. I., Felix, D. H., Lane, E. B., and Munro, C. S., 1995, A mutation in the mucosal keratin K4 is associated with oral white sponge nevus, *Nat. Genet.* **11:**450–452.

Saitou, M., Sugai, S., Tanaka, T., Shimouchi, K., Fuchs, E., Narumiya, S., and Kakizuka, A., 1995, Inhibition of skin development by targeted expression of a dominant-negative retinoic acid receptor, *Nature* **374:**159–162.

Schermer, A., Jester, J. V., Hardy, C., Milano, D., and Sun, T. T., 1989, Transient synthesis of K6 and K16 keratins in regenerating rabbit corneal epithelium: Keratin markers for an alternative pathway of keratinocyte differentiation, *Differentiation* **42:**103–110.

Schwarz, M. A., Owaribe, K., Kartenbeck, J., and Franke, W. W., 1990, Desmosomes and hemidesmosomes: Constitutive molecular components, *Annu. Rev. Cell Biol.* **6:**461–491.

Schweizer, J., 1993, Murine epidermal keratins, in *Molecular Biology of the Skin: The Keratinocyte* (M. Darmon and M. Blumemberg, eds.), pp. 33–78, Academic Press, San Diego, California.

Shamsher, M. K., Navsaria, H. A., Stevens, H. P., Ratnavel, R. C., Purkis, P. E., Kelsell, D. P., McLean, W. H. I., Cook, L. J., Griffiths, W. A. D., Gschmeissner, S., Spurr, N., and Leigh, I. M., 1995, Novel mutations in keratin 16 gene underly focal non-epidermolytic palmoplantar keratoderma in two families, *Hum. Mol. Genet.* **4:**1875–1881.

Smith, F. J., Corden, L. D., Rugg, E. L., Ratnavel, R., Leigh, I. M., Moss, C., Tidman, M. J., Hohl, D., Huber, M., Kunkeler, L., Munro, C. S., Lane, E. B., and McLean, W. H., 1997, Missense mutations in keratin 17 cause either pachyonychia congenita type 2 or a phenotype resembling steatocystoma multiplex, *J. Invest. Dermatol.* **108:**220–223.

Steinert, P. M., 1991, Analysis of the mechanism of assembly of mouse keratin 1/keratin 10 intermediate filaments *in vitro* suggests that intermediate filaments are built from multiple oligomeric units rather than a unique tetrameric building block, *J. Struct. Biol.* **107:**175–188.

Steinert, P. M., and Bale, S. J., 1993, Genetic skin diseases caused by mutations in keratin intermediate filaments, *Trends Genet.* **9:**280–284.

Stewart, M., 1993, Intermediate filament structure and assembly, *Curr. Opin. Cell Biol.* **5:**3–11.

Stoler, A., Kopan, R., Duvic, M., and Fuchs, E., 1988, Use of monospecific antisera and cRNA probes to localize the major changes in keratin expression during normal and abnormal epidermal differentiation, *J. Cell Biol.* **107:**427–446.

Stoler, A., Duvic, M., and Fuchs, E., 1989, Unusual patterns of keratin expression in the overlying epidermis of patients with dermatofibromas: Biochemical alterations in the epidermis as a consequence of dermal tumors, *J. Invest. Dermatol.* **93:**728–738.

Takahashi, K., and Coulombe, P. A., 1996, A transgenic mouse model with an inducible skin blistering disease phenotype, *Proc. Natl. Acad. Sci. USA* **93:**14776–14781.

Takahashi, K., and Coulombe, P. A., 1997, Defining a region of the human keratin 6a gene that confers inducible expression in stratified epithelia of transgenic mice, *J. Biol. Chem.* **272:**11979–11985.

Takahashi, K., Folmer, J., and Coulombe, P. A., 1994, Increased expression of keratin 16 causes anomalies in cytoarchitecture and keratinization in transgenic mouse skin, *J. Cell Biol.* **127:**505–520.

Takahashi, K., Paladini, R., and Coulombe, P. A., 1995, Cloning and characterization of multiple human genes and cDNAs encoding highly related type II keratin 6 isoforms, *J. Biol. Chem.* **270:**18581–18592.

Troyanovsky, S. M., and Leube, R. E., 1994, Activation of the silent human cytokeratin 17 pseudogene-promoter region by cryptic enhancer elements of the cytokeratin 17 gene, *Eur. J. Biochem.* **225:**61–69.

Troyanovsky, S. M., Guelstein, V. I., Tchipysheva, T. A., Krutovskikh, V. A., and Bannikov, G. A., 1989, Patterns of expression of keratin 17 in human epithelia: Dependency on cell position, *J. Cell Sci.* **93:**419–426.

Troyanovsky, S. M., Leube, R. E., and Franke, W. W., 1992, Characterization of the human gene encoding cytokeratin 17 and its expression pattern, *Eur. J. Cell Biol.* **59:**127–137.

Tyner, A. L., and Fuchs, E., 1986, Evidence for posttranscriptional regulation of the keratins expressed during hyperproliferation and malignant transformation in human epidermis, *J. Cell Biol.* **103:**1945–1955.

Tyner, A. L., Eichman, M. J., and Fuchs, E., 1985, The sequence of a type II keratin gene expressed in human skin: Conservation of structure among all intermediate filament genes, *Proc. Natl. Acad. Sci. USA* **82:**4683–4687.

Vassar, R., Rosenberg, M., Ross, S., Tyner, A., and Fuchs, E., 1989, Tissue-specific and differentiation-specific expression of a human K14 keratin gene in transgenic mice, *Proc. Natl. Acad. Sci. USA* **86:**1563–1567.

Vassar, R., Coulombe, P. A., Degenstein, L., Albers, K., and Fuchs, E., 1991, Mutant keratin expression in transgenic mice causes marked abnormalities resembling a human genetic skin disease, *Cell* **64:**365–380.

Vogel, U., Denecke, B., Troyanovsky, S. M., Leube, R. E., and Bottger, E. C., 1995, Transcriptional activation of psoriasis-associated cytokeratin K17 by interferon-gamma. Analysis of gamma-interferon activation sites, *Eur. J. Biochem.* **227:**143–149.

Wawersik, M., Paladini, R. D., Noensie, E., and Coulombe, P. A., 1997, A proline residue in the α-helical rod domain of type I keratin 16 destabilizes keratin heterotetramers, *J. Biol. Chem.* **272:**32557–32565.

Weiss, R. A., Eichner, R., and Sun, T. T., 1984, Monoclonal antibody analysis of keratin expression in epidermal diseases: A 48- and 56-kdalton keratin as molecular markers for hyperproliferative keratinocytes, *J. Cell Biol.* **98:**1397–1406.

Yoshikawa, K., Katagata, Y., and Kondo, S., 1995, Relative amounts of keratin 17 are higher than those of keratin 16 in hair-follicle-derived tumors in comparison with nonfollicular epithelial skin tumors, *J. Invest. Dermatol.* **104:**396–400.

Chapter 7

Cytokeratins as Markers of Differentiation in the Diagnosis of Epithelial Tumors

Roland Moll

1. INTRODUCTION

At the electron microscopic level, intermediate filaments (IFs) appear to be very simple structures. While these cytoskeletal elements attracted little serious scientific attention until the late 1970s, they have since been the subject of a considerable body of cell and molecular biological literature, whose scope is amply reflected by the contributions in the present volume. In addition to casting light on their importance with respect to basic biology, our growing understanding of IFs has also found practical applications in various fields of human medicine. The discipline in which such research has probably effected the most significant advances is tumor pathology, with the successful application of IFs leading to their routine employment as cell-lineage and differentiation markers in the histological diagnosis of malignant tumors. The cell-type specificity of IFs was first utilized for identifying glial fibrillary acidic protein (GFAP), which has long been rec-

Roland Moll Department of Pathology, Philipps University of Marburg Medical School, D-35033 Marburg, Germany.

Subcellular Biochemistry, Volume 31: Intermediate Filaments, edited by Herrmann and Harris. Plenum Press, New York, 1998.

ognized as a marker of astrocytic tumors (Deck *et al.*, 1978). Subsequently, other IFs, such as vimentin, desmin, and neurofilaments, have been shown to provide valuable data that may facilitate the diagnosis of mesenchymal, myogenic, and neuronal tumors, respectively (for a review, see Nagle, 1994).

Of the various IF classes, the (cyto)keratin (CK) filaments, which are typically expressed in epithelial cells, are of outstanding interest owing to the unusually high degree of diversity exhibited by their protein subunits, the expression of which is, within the various types of epithelia, again cell-type and differentiation specific (Franke *et al.*, 1981; Moll *et al.*, 1982a; Cooper *et al.*, 1985).

There are two ways in which CKs may be utilized when making histological diagnoses of epithelial tumors: first, using immunohistochemical staining procedures, broadly cross-reactive pan-CK antibodies are able to recognize CK filaments regardless of their CK subtypes, and can thus help to confirm or rule out the epithelial origin and nature of a tumor. This approach may be appropriate in such morphologically problematic cases as undifferentiated carcinomas occurring at various sites (e.g., nasopharyngeal carcinomas and undifferentiated lung carcinomas), in tumors growing in a dispersed, dissociated manner (e.g., diffuse-type gastric carcinomas and invasive lobular breast cancers), or when only small fragments or isolated cells are present, such as in fine-needle and cytological preparations (for reviews, see Nagle, 1989, 1994). Other areas of applicability for pan-CK antibodies include the differential diagnosis of small-cell lung carcinoma (versus lymphoma), thymoma (versus lymphoma), embryonal carcinoma of the testis (versus seminoma), and chordoma (versus chondroma). Among mesenchymal tumors (or tumors exhibiting a mesenchymal morphology), the strong and consistent expression of CKs in synovial and epithelioid sarcomas as well as in sarcomatoid mesotheliomas may be of diagnostic relevance. However, "ectopic" CK expression (almost entirely confined to the "primitive" CK 8 and CK 18) has been recorded in several other mesenchymal neoplasms, such as rhabdomyosarcomas, leiomyosarcomas, malignant fibrous histiocytomas, malignant Schwannomas, Ewing sarcomas, large-cell anaplastic lymphomas, and malignant melanomas (e.g., Miettinen and Franssila, 1989; Miettinen, 1991; Gustmann *et al.*, 1991; Litzky and Brooks, 1992). In such instances, though, the quantitative predominance of the "correct" IF protein, e.g., vimentin or desmin, is usually so clear and in such contrast to the inconsistent and focal expression of CKs that diagnostic misinterpretations are not to be expected (see also Swanson *et al.*, 1994).

The second, more sophisticated mode of applying CK analyses in tumor pathology involves the subtyping of carcinomas on the basis of their differential CK expression patterns. The diversity of such expression patterns in the various types of normal epithelia is conserved to some extent in their corresponding carcinomas, including the metastases of such tumors. Thus, in the event of one being confronted with a metastatic tumor whose primary tumor is unknown—a diagnostic problem frequently encountered by pathologists—CK subtyping may provide useful pointers with respect to the origin of the carcinoma. Moreover, CK-polypeptide patterns may be correlated with the specific subtype of a carcinoma of a given organ or with its degree of differentiation or malignancy. Several review articles have dealt with this field of research in considerable depth and detail (e.g., Corson, 1986; Moll, 1992, 1993; Schaafsma and Ramaekers, 1994; Nagle, 1994). Within the confines of the space available, the aim of the present contribution is to focus on particular aspects of CK subtyping in several frequently occurring and clinically important carcinoma types. It will be shown how analyses of individual CKs are capable of detecting tumor-cell differentiation regardless of and independent from the morphological appearance of primary and metastatic carcinomas. Further, the question will be considered of how conclusions may be drawn concerning the type and origin of a carcinoma as well as its particular differentiation features and even, perhaps, its biological behavior. Our own data based on biochemical and immunohistochemical analyses of frozen tumor-tissue specimens will be considered in the light of recently published investigations. From the data presented, two major concepts clearly emerge: first, although CK patterns may be modulated to varying degrees during malignant transformation and tumor progression, the cell-type-specific features of these patterns often remain sufficiently stable to allow a metastatic tumor to be traced back to its site of origin. Second, such modulations may give rise, at very different sites, to the seemingly paradoxical phenomenon of CK patterns becoming increasingly complex during dedifferentiation.

2. BASIC CYTOKERATIN EXPRESSION PATTERNS IN NORMAL HUMAN EPITHELIA

While their molecular diversity is striking, it is the cell-type- and differentiation-specific expression of individual CK polypeptides in particular kinds of normal epithelia that is a particularly distinctive feature of these IFs. The basic rules governing these expression patterns have been well established (Moll *et al.*, 1982a, 1990; Tseng *et al.*, 1982; Cooper *et al.*, 1985; Lane and Alexander, 1990; Schaafsma and Ramaekers, 1994). Table I provides a general survey of the major expression specificities of the various

Table I
General Principles of CK Expression in Simple and Stratified Squamous Epithelia

	Cytokeratin		
	Type II	Type I	
Simple epithelia	8	18	Primary pair; sole CK pair in most secretory and parenchymatous cells
	7	19	Ductal epithelia (bile and pancreatic ducts, renal collecting ducts); gastrointestinal epithelia[a]
		20	Gastrointestinal epithelia
Stratified squamous epithelia			
Basal cells	5	14	Primary pair
		15	
	(8)	(18) 19	Noncornifying stratified squamous epithelia
Suprabasal cells	1	10/11	Epidermis (entire suprabasal compartment)
		9	Epidermis (palmoplantar)
	2e		Epidermis (higher layers)
	2p		Gingiva, hard palate
	3	12	Corneal epithelium
	4	13	Noncornifying stratified squamous epithelia
	6	16, 17[b]	Hyperproliferative epidermis
	6	16	Noncornifying stratified squamous epithelia

[a]CK 19 (CK 7 essentially absent).
[b]In addition, CK 17 is most typically expressed in basal cells of complex epithelia.

CKs found in the two principal classes of human epithelia, i.e., simple and stratified squamous epithelia. This table further illustrates the principle of the pairwise expression of CKs, i.e., the typical coexpression of a type II and type I CK in pairs which are related to the particular differentiation pathways (Franke *et al.*, 1983; Sun *et al.*, 1984). These CK pairs constitute, as heterodimers, basic IF subunits (Hatzfeld and Weber, 1990). However, in many epithelia, the pairs of CKs that are actually coexpressed have yet to be determined directly.

Owing to the highly diverse nature of epithelial differentiation, many known cases do not accord fully with, or even represent exceptions to, the general principles outlined in Table I. Such exceptions include the heterogeneous expression of CK5 (as well as of vimentin and, focally, of desmin) in a particular simple epithelium, i.e., the mesothelium of serosal surfaces (van Muijen *et al.*, 1987a; Moll *et al.*, 1989), as well as the focal expression of the epidermal-maturation components CKs 1 and 10/11 in certain noncornifying stratified squamous epithelia, particularly in the vaginal and exocervical epithelium.

Both certain stratified epithelia, e.g., the urothelium, as well as various complex and pseudostratified epithelia exhibit strikingly complex CK patterns that are probably attributable to the different cell types that occur in such composite epithelia. For example, basal and myoepithelial cells present in several glands (salivary glands, prostate gland, mammary-gland ducts), basal cells of the respiratory epithelium, and reserve cells of the uterine endocervix express not only CK 5, CK 14, and CK 15, i.e., those CKs encountered in basal cells of stratified squamous epithelia, but also typically contain CK 17. The luminal cells of such epithelia usually correspond to simple ductal epithelia in expressing CK 7, CK 8, CK 18, and CK 19, but further components such as CK 4 and CK 13 may also be observable. The urothelium is characterized by a unique CK pattern, in that it exhibits certain squamous-epithelial features in its basal and intermediate cells, while also displaying simple-epithelial features in luminal umbrella cells that characteristically express CK 20.

3. CYTOKERATIN PATTERNS IN MAJOR HUMAN CARCINOMA TYPES

Analyses of CK expression patterns in carcinomas are important not only in terms of broadening our understanding of the pathogenesis, histogenesis, and differentiation processes of malignant tumors, but also, and more significantly, because CK typing may facilitate histopathological tumor diagnosis in such clinically and pathologically difficult cases as when metastases of an unknown primary carcinoma are encountered (Mackay and Ordoñez, 1993).

Owing to limitations of space, the scope of the present chapter will be confined to the consideration of several frequently occurring and clinically important types of carcinoma. For other organ and tumor systems, reference will be made to publications of key importance (see Section 5), while the interested reader is also referred to recent reviews (e.g., Schaafsma and Ramaekers, 1994; Nagle, 1994).

For each of the carcinoma types discussed below, the following approach has been adopted: after presenting the results of CK analysis of frozen tumor specimens obtained by myself and co-workers employing both two-dimensional gel electrophoresis and immunohistochemistry with a battery of monospecific CK antibodies (Moll, 1993), the results of these studies are then considered within the broader context of recent investigations, many of which have dealt with larger cohorts of tumors using routinely prepared paraffin-embedded tissues. Discussion is focused on the practical histodiagnostic usefulness of CKs for the carcinomas in question.

3.1. Colorectal Adenocarcinomas

Normal mucosa of the small and large intestine is characterized by the homogeneous expression of CK 8, CK 18, and CK 19. In addition, the typical gastrointestinal marker CK 20 is expressed, in a manner related to cellular differentiation, in both the enterocytes and goblet cells that line the small-intestinal villi as well as the upper crypts and surface epithelium of colorectal mucosa (Moll *et al.*, 1990, 1992b, 1993). In deep crypts, CK 20 is only detectable in scattered cells that appear to be undifferentiated. It is of interest that CK 7 is essentially absent in intestinal mucosa (Moll *et al.*, 1982a; Osborn *et al.*, 1986; Ramaekers *et al.*, 1987; van Niekerk *et al.*, 1991).

This unique CK phenotype of the intestinal epithelium, which may be referred to as the "gastrointestinal pattern" (for the stomach, see below), is quite reliably retained in primary colorectal adenocarcinomas and their metastases (Figure 1). Two-dimensional gel electrophoresis performed using cytoskeletal material from such tumors revealed the typical combination of CK 8, CK 18, CK 19, and CK 20 (Figure 1a). The first three of these were found to be uniformly expressed throughout the tumors, as demonstrated by immunohistochemical staining (Figures 1b, 1e). In contrast, although CK 20 was present in the vast majority of cases examined (96% of a series of 93 tumors; Moll *et al.*, 1992b) and sometimes exhibited a uniform distribution, it more commonly elicited a heterogeneous staining pattern involving variable numbers of tumor cells (Figures 1c, 1f, 1l). Interestingly, CK 20 expression was retained in most of the poorly differentiated tumors

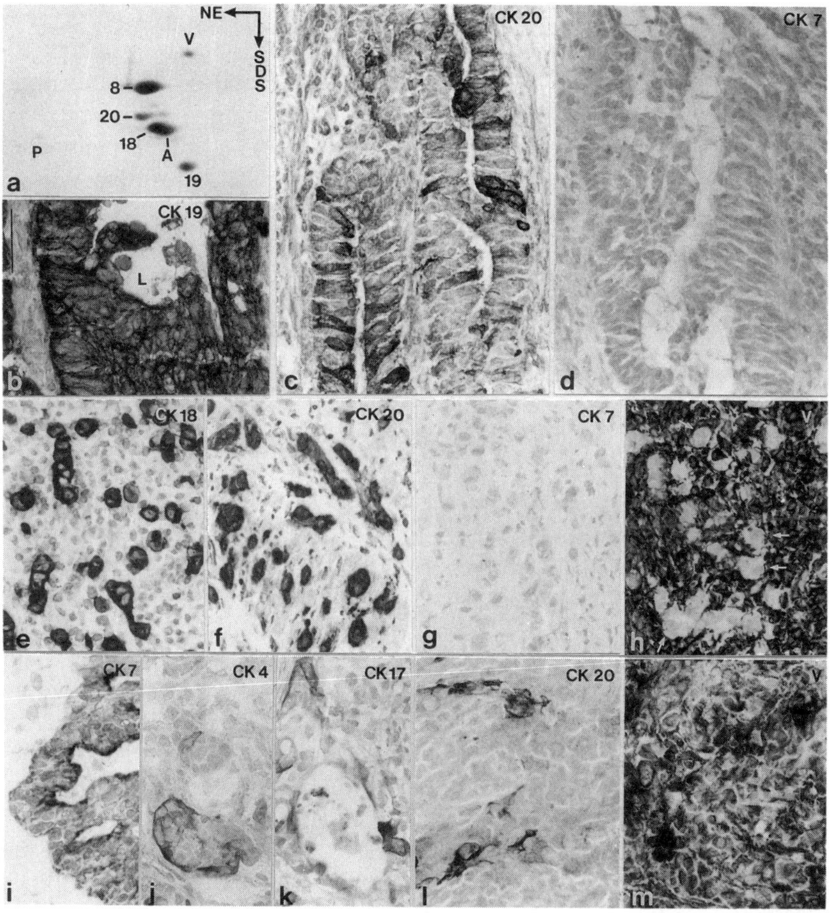

FIGURE 1. CK (and vimentin) expression in colon carcinomas. (a) Typical gel-electrophoretic pattern, showing CKs 8, 18, 19, and 20 (transplantation tumor CXF-1). Symbols: NE (-PHG) and SDS (-PAGE) indicate the electrophoretic dimensions of the two-dimensional gel electrophoresis. The numbers indicate the CK polypeptide (Moll *et al.*, 1982a, 1990). A, actin; V, vimentin. Added marker polypeptide: P, 3-phosphoglycerokinase from yeast. (b–m) Immunoperoxidase staining of frozen sections (bar, 50 μm): (b–d) Moderately differentiated adenocarcinoma with a typical CK profile, i.e., a homogeneously positive reaction for CK 19 (b; monoclonal antibody Ks 19.2), a heterogeneously positive reaction for CK 20 (c; GP-IT), and a negative reaction for CK 7 (d; CK-7). (e–h) This characteristic pattern is retained in a largely dedifferentiated portion of a colon carcinoma showing almost dissociated growth, i. e., positive staining for CK 18 (e; Ks 18.174) and CK 20 (f; GP-IT), and a negative reaction for both CK 7 (g; CK-7) and vimentin (h; V9; arrows, negative tumor cells among a positive stroma). (i–k) Unusually focal, primarily luminal staining for CK 7 (i; CK-7) in an ovarian metastasis; this case was also focally positive for CK 4 (j; poorly differentiated area; 6B10) and CK 17 (k; E3). (l, m) Poorly differentiated colon carcinoma exhibiting focal CK 20 expression (l; GP-IT), as well as, exceptionally, a positive reaction for vimentin in tumor cells (m; V9). Reprinted with permission from Moll (1993).

studied (Figures 1f, 1l), a finding which was unexpected in view of its differentiation-dependent expression in normal intestinal mucosa. As in normal mucosa, CK 7 was not detected in the vast majority of cases under investigation (Figures 1d, 1g), but 12% of the tumors did reveal focal expression of this CK (Figure 1i; Moll *et al.*, 1993). A more frequent observation was the presence of focal staining for certain stratification- or basal-cell-related CKs, including CK 14 and CK 17 (in 70–80% of cases) and, more sparsely distributed, CK 4 (24%; Figures 1j, 1k; Moll *et al.*, 1993). Interestingly, expression of the "ectopic" CKs, i.e., CK 4, CK 7, CK 14, and CK 17, was correlated with morphological signs of dedifferentiation, such as the flattening of cells, the occurrence of solid tumor-cell clusters, and the presence of dedifferentiated cells at the invasion front (Moll *et al.*, 1993). In almost all of the tumors examined, vimentin was entirely absent in the tumor cells (Figure 1h), so that the coexpression of this IF protein could be considered an extremely rare event (2%; Figure 1m). Table II presents a comparison of the typical CK and IF profiles of colorectal adenocarcinomas and those of other carcinomas.

Our finding of the essential retention and maintenance of the gastrointestinal CK pattern in primary and metastatic colorectal adenocarcinomas has been confirmed by a number of studies. Earlier investigations emphasized the absence of CK 7 as a factor of major relevance in the immunohistological differential diagnosis of adenocarcinomas (Moll and Franke, 1986; Osborn *et al.*, 1986; Voijtesek *et al.*, 1990; Ramaekers *et al.*, 1990; Bartek *et al.*, 1991), but exceptions may be encountered (Figure 1i). In fact, the proportion of CK 7-positive cases reported in recent studies varies from 5–10% (Wang *et al.*, 1995; Berezowski *et al.*, 1996; Loy *et al.*, 1996) through 15–21% (Loy and Calaluce, 1994; Savera *et al.*, 1996) to not less than about 30% (Wauters *et al.*, 1995; Maeda *et al.*, 1996), although the expression pattern is usually heterogeneous, often being confined to scattered cells. The differences in the published data may depend on the particular CK 7 antibody used, with monoclonal antibody OV-TL12/30 evidently eliciting the broadest span of immunoreactivity in both frozen and paraffin-embedded sections (Ramaekers *et al.*, 1990; van Niekerk *et al.*, 1991). In spite of such findings, though, the relatively low level of CK 7 expression in colorectal adenocarcinomas remains diagnostically relevant. The absence of CK 7 has also been shown to be a useful diagnostic criterion in cytological specimens (Sack and Roberts, 1997). However, the past few years have seen the emergence of CK 20 as a marker of even greater diagnostic significance, whose high degree of sensitivity has been convincingly demonstrated in a number of recent studies, several of which have also dealt with liver and ovarian metastases: the reported incidence of colorectal adenocaracinomas

Table II
Expression Patterns of Cytokeratin Polypeptides (and Vimentin) in Various Carcinomas[a]

		Simple-epithelial cytokeratins				Stratified-epithelial cytokeratins					
	Vimentin	CK 8, 18	CK 19	CK 7	CK 20	CK 5	CK 14, 17	CK 6, 16	CK 13	CK 4	CK 1, 10
Hepatocellular C	−	+++	−/+	−/+	−/(+)	n.d.	n.d.	n.d.	n.d.	n.d.	n.d.
Colorectal AC	−	+++	+++	−/+	++/+++	−/(+)	−//++	−[b]	−	−/+	−[b]
Gastric AC	−/+	+++	+++	−//++	−//++	−/(+)	−//++	n.d.	−//+	−//++	−
Pancreatic AC (ductal)	−/+	+++	+++	++/+++	−//++	−/+	−//++[c]	−//++[c]	−/+	(+)//++	−/+[c]
Lung AC	−//++	+++	+++	++/+++	−/(+)	−/(+)	n.d.	−[b]	−/(+)	−//++	−[b]
Breast, invasive ductal C: G1, 2	−/+	+++	+++	+/++	−	−/+	−/+	−//++	−/+	−/+	−[d]
G3	−//++	+++	+++	++/+++	−/+	−//++	−//++	−//++	−	−//++	−[d]
Endometrial AC	++/+++	+++	+++	+/++	−/(+)	(+)/+	n.d.	n.d.	−/(+)	(+)/+	−
Ovarian AC (nonmucinous)[e]	+//+++	+++	+++	++/+++	−	−/+	n.d.	n.d.	−/+	−//++	−
Renal cell C (clear-cell, chromophilic)	++/+++	+++	−//+++	−//++	−	−	n.d.	n.d.	−	−/+	n.d.
Renal cell C (chromophobe)	−	+++	+	++	−	−	n.d.	n.d.	−	−	n.d.
Malignant diffuse mesothelioma (epithelial, biphasic)	(+)/++	+++	+++	+//+++	−	++/+++	n.d.	−/+	−	−//++	−
Small cell lung C	−/+	+//+++	+//+++	−//++	−/(+)	−	n.d.	n.d.	−	−	−
Merkel cell C[f]	−	+++	−//+++	−	++/+++	−	n.d.	n.d.	−	−	−
Transitional cell C: G1, 2	−	+++	+++	+++	(+)//+++	−/+[g]	−/+[h]	−/(+)[g]	+//+++	(+)	−/(+)
G3	−	+++	+++	++/+++	(+)//+++	−[g]	−//+++[h]	n.d.	−/(+)	(+)	−/(+)
Squamous cell C (oropharyngeal): G1, 2	−/+	+/++	−//++	−	−	++/+++	+++	++/+++	−/+	−/+	(+)/+
G3	−//++	+//+++	++/+++	−/+	−/(+)	+//+++	+/+++	++/+++	−//++	−/+	−/±

[a]Data compiled from several studies (Moll *et al.*, 1988, 1989, 1991b, 1992b, 1993; Moll, 1993; Pitz *et al.*, 1987; Gillenkirch, 1994; Malzahn *et al.*, 1998; see also subsequent footnotes), and most of them have been obtained by immunohistochemistry. −, Negative; (+), +, ++, heterogeneous expression patterns with increasing proportions of positive tumor cells; +++, uniform strong expression. Bold symbols indicate the predominant expression pattern or range, as displayed by the majority of cases studied. Underlined symbols indicate the IF protein data with the highest diagnostic significance in each tumor category. AC, Adenocarcinoma; C, carcinoma; n.d., not determined.
[b]Huszar *et al.* (1986).
[c]Real *et al.* (1993).
[d]Santini *et al.* (1996).
[e]Mucinous ovarian tumors are distinctive by the expression of CK 20 (see Section 5).
[f]The presence of neurofilaments is another valuable diagnostic criterion.
[g]As determined by gel electrophoresis (Moll *et al.*, 1988).
[h]CK 14: Schaafsma *et al.* (1990).

positive for CK 20 has ranged from 84–88% (Wang *et al.*, 1995; Wauters *et al.*, 1995; Kaufmann *et al.*, 1996) to 96–100% (in confirmation of our own data; Loy and Calaluce, 1994; Miettinen, 1995; Maeda *et al.*, 1996; Berezowki *et al.*, 1996; Loy *et al.*, 1996; Savera *et al.*, 1996). With respect to minor CK components, the focal expression of stratified-epithelial CKs was found in 1 of the 20 colonic carcinomas investigated by Levy *et al.* (1992).

A drastic reduction in the level of CK 20 expression in colon cancers at a more advanced pT stage, as recorded by Cross *et al.* (1996) using Western blotting and immunofluorescence microscopy, is in accordance neither with our own data nor with most other published findings, for example, those recording the frequent occurrence of high CK 20 levels in pT4-stage tumors (Moll *et al.*, 1993) and in liver metastases (Maeda *et al.*, 1996). CK 20 can also be detected, with a high degree of sensitivity, in cytological specimens derived from effusion fluids and fine-needle aspirations (Ascoli *et al.*, 1995; Sack and Roberts, 1997).

The presence—as detected immunocytochemically—of "micrometastases" or disseminated tumor cells in the bone marrow has been shown to be of prognostic significance in patients suffering from colorectal cancer (Lindemann *et al.*, 1992). A highly sensitive and specific new approach for detecting such cells (although follow-up studies are required to confirm its prognostic relevance) is the analysis of CK 20 mRNA in bone-marrow aspirates using a nested reverse transcriptase–polymerase chain reaction (RT-PCR) method. Application of this procedure yielded positive results for 35% of the colorectal carcinoma patients examined, with a clear predominance (71%) emerging for tumors at stage IV (Soeth *et al.*, 1996).

3.2. Gastric Adenocarcinomas

CK analyses of normal stomach mucosa of both the corpus and pyloric regions demonstrate a clear-cut distinction between foveolar and glandular epithelium. The former exhibits the typical gastrointestinal pattern, comprising the expression of CK 8, CK 18, and CK 19, along with CK 20. In the glands, though, only the first three of these are expressed, with mosaiclike, heterogeneous staining patterns for CK 18 and CK 19 being observed in the specific corpus glands (Moll, 1993). Endocrine cells present in the upper zone of pyloric glands do strongly express CK 20 (Moll *et al.*, 1993). With respect to CK 7, we found that the monoclonal antibody CK-7 essentially yielded negative reactions, except for very focal staining in sporadically occurring, apparently morphologically altered glandular structures (Moll, 1993; cf. also van Niekerk *et al.*, 1991). The application of another CK 7 antibody, OV-TL12/30, has been found to produce positive staining of surface foveolar epithelium in frozen sections, but not in paraffin-embedded sections (van Niekerk *et al.*, 1991). Interestingly, the expression of CK 7 has been detected in gastric epithelia during fetal development (Stosiek *et al.*, 1991), as well as—as a neoexpression—in the event of reactive, benign epithelial alterations including such occurring in chronic gastritis (Stosiek and Kasper, 1990, 1993; Schwerer and Baczako, 1996a). In cases of intestinal metaplasia, intestinal metaplastic glands usually exhibit little CK 7 staining but particularly intense staining for CK 20 (being heterogeneous in deeper but homogeneous in more superficial portions) next to staining for CK 8, CK 18, and CK 19 (Moll, 1993) [for variations in CK 20 expression depending on the type of intestinal metaplasia, see Schwerer and Baczako (1996b)].

Although stomach carcinomas exhibit considerable variety with respect to morphological differentiation, their overall CK patterns appeared to be relatively uniform; however, the expression of some CKs in these tumors was found to be markedly heterogeneous, and interindividual variations were detected (for the general profiles, see Table II). Figure 2a shows a typical gel-electrophoresis pattern obtained for an intestinal-type tubular stomach carcinoma, revealing the complete spectrum of simple-epithelial CKs, including small amounts of both CK 7 and CK 20. Immunohistochemical staining yielded homogeneously positive results for CK 18 and CK 19 in all of the tumors studied (Figures 2b, 2h). In addition, 77% of the cases investigated were positive for CK 20 (somewhat more regularly in intestinal-type as compared to diffuse-type tumors), but the proportion of immunostained tumor cells was highly variable (Figures 2d, 2i). A similar percentage of cases (again including diffuse-type and signet-ring-cell carcinomas) was positive for CK 7, again with the staining patterns being heterogeneous (Figure 2c). In gastric carcinomas, the stratified-epithelial CKs proved to be only minor and irregularly expressed components: the focal expression of CK 17 (Figures 2e, 2j) was seen in the majority of cases, while focal staining for CK 4 (Figure 2l) and/or CK 13 (Figures 2f, 2k) was also frequently encountered, being most prominent in poorly differentiated regions of intestinal-type tumors. In 15% of our cases, low levels of vimentin coexpression were detectable (Moll *et al.*, 1993).

The variable expression of both CK 20 and CK 7 in gastric carcinomas, albeit usually in the majority of cases, has also been described in a number of investigations. The proportion of CK 20-positive stomach carcinomas has been reported to range from 33% (cardia and subcardial carcinomas; Sarbia *et al.*, 1993) through 50–60% (Wauters *et al.*, 1995; Kaufmann *et al.*, 1996) to 70% (Ascoli *et al.*, 1995 [cytological material]; Wang *et al.*, 1995; Miettinen 1995). In the case of CK 7, in some studies the reported proportion of tumors positive for this CK has been only ~30% (Osborn *et al.*, 1988; Ramaekers *et al.*, 1990) or 38% (Bartek *et al.*, 1991), with more recent

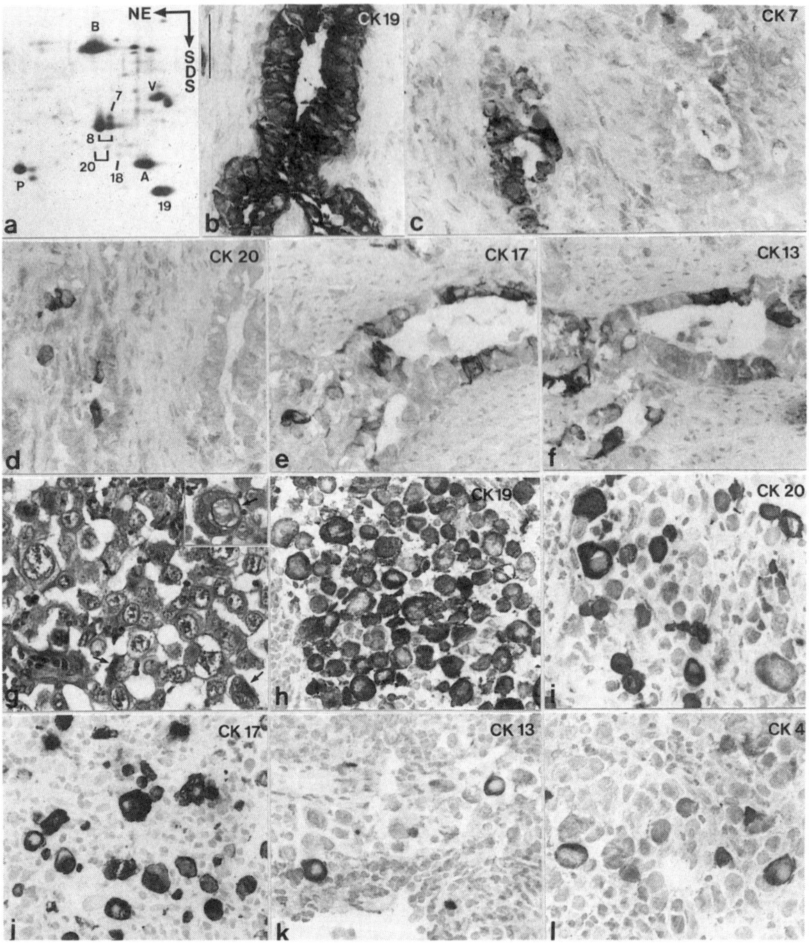

FIGURE 2. CK expression in stomach carcinomas. (a) Electrophoresis pattern of a moderately differentiated tubular adenocarcinoma; CK 20 is also present in significant (albeit small) amounts (for symbols, see the legend to Figure 1; B, bovine serum albumin added as another marker protein). (b–f) Moderately differentiated tubular adenocarcinoma of the gastric mucosa showing a typical CK staining constellation, with intense, homogeneous staining for CK 19 (b; monoclonal antibody Ks 19.2) as well as focal reactions for CK 7 (c; CK-7), CK 20 (d; GP-IT), CK 17 (e; E3), and CK 13 (f; 2D7). (g–l) Lymph-node metastasis of an undifferentiated gastric carcinoma (partially composed of signet-ring cells) exhibiting polymorphic tumor cells and evidence of focally arranged intracellular mucus as revealed by PAS staining (g, arrows; arrow in inset, large mucus vacuole). The tumor cells diffusely filling the lymph-node sinus are almost entirely positive for CK 19 (h; monoclonal antibody Ks 19.2); it is worthy of note that they are also frequently positive for CK 20 (i; GP-IT). In addition, a heterogeneous reaction for CK 17 is visible (j; E3), along with slight reactions for CK 13 (k; 2D7) and CK 4 (l; 6B10). Immunoperoxidase staining of frozen sections. Bar (in panel b), 50 μm. Reprinted with permission from Moll (1993).

studies recording positive findings in 55% (Wang *et al.*, 1995) and ~80% (van Niekerk *et al.*, 1991; Moll, 1993; Wauters *et al.*, 1995) of cases. No systematic differences between "intestinal" (tubular) and "diffuse" (undifferentiated/signet-ring-cell type) carcinomas have been detected (our own data; Osborn *et al.*, 1988). The presence of intestinal differentiation markers such as villin and brush-border enzymes has been demonstrated not only in intestinal, but also in diffuse-type stomach carcinomas (Osborn *et al.*, 1988). All these data suggest that the two major morphological types of stomach carcinoma do not represent sharply delineated, distinct qualities, but rather form continuous spectra. With respect to minor CK components, the focal expression of certain stratified-epithelial CKs (CK 13, CK 16) was observed by Levy *et al.* (1992) in a small proportion of stomach carcinomas (cf. Figures 2f, 2k). Regarding the coexpression of vimentin, our finding of just a few vimentin-positive cases is not compatible with that of Takemura *et al.* (1994), 60% of whose cases were positive for this IF protein. Rather, our data are comparable to those of Utsunomiya *et al.* (1996), whose study of solid-type, poorly differentiated adenocarcinomas revealed 6.3% to be vimentin-positive (with the coexpression of CK 20 in 27% of these), with such patients having poorer survival rates than those with vimentin-negative tumors. The gel-electrophoresis data recently presented by Gupta *et al.* (1997), allegedly demonstrating that only two high-molecular-weight type II CKs are present in gastric carcinomas, are impossible to reconcile with any other data currently available; indeed, if taken at their face value, these findings would lead one to the frankly incredible conclusion that such tumors contain no CK filaments at all. In summary, gastric adenocarcinomas exhibit a rather more complex CK pattern than colorectal carcinomas, with the ratio of CK 20 to CK 7 being more balanced in the former, whereas it is much more weighted in favor of CK 20 in the latter. With respect to the role of intestinal metaplasia in the formation of the various types of stomach carcinomas (Iida and Kusama, 1982), the CK data have yet to provide any significant contribution to our understanding of this question.

3.3. Adenocarcinomas of the Pancreas

In normal exocrine pancreas tissue, the distribution patterns of CK polypeptides in the acini and ducts exhibit marked differences. While such expression in acinar cells is essentially restricted to that of CK 8 and CK 18, along with the occasional, weak coexpression of CK 7, ductal epithelia manifest the "ductal pattern," i.e., CK 7 (somewhat heterogeneously), CK 8, CK 18, and CK 19 (Moll *et al.*, 1983a). In particular, CK 19 antibodies clearly delineate the entire ductal system, starting in the centroacinar parts.

In addition, ductal epithelia may heterogeneously express certain stratified-epithelial CKs, especially CK 4 (Kasper *et al.*, 1991; Schüssler *et al.*, 1992) and, to a lesser extent, CK 5 (Moll *et al.*, 1989), CK 14, and CK 17 (Kasper *et al.*, 1991; Schüssler *et al.*, 1992; Real *et al.*, 1993; Moll, 1993). Also, CK 13, CK 20, and vimentin have been observed in low amounts, albeit only in the epithelia of chronically damaged pancreatic ducts (Moll, 1993). Endocrine islet cells weakly express CK 8 and CK 18.

Our group of ductal adenocarcinomas of the pancreas were mainly characterized by the prominent expression of the "ductal-type" CKs comprising CK 8, CK 18, CK 19, and (somewhat heterogeneously) CK 7 (Figures 3a, 3b). In contrast, the expression of CK 20, although it was detectable in 58% of cases (Moll *et al.*, 1992b), was highly variable, the levels being usually low and the distribution focal (Figure 3c). In addition, the observed focal expression of stratified-epithelial CKs, including CK 17, CK 4, CK 13, and CK 5 (Figures 3a, 3d–3g), was significantly more pronounced than that seen in colorectal and gastric carcinomas; such expression appeared, in part, to be associated with squamous differentiation. Although the coexpression of vimentin was occasionally detected, most of the tumors investigated were vimentin-negative (Figure 3h).

The general retention and maintenance of the ductal CK pattern in ductal adenocarcinomas of the pancreas has also been emphasized by several other investigators (Osborn *et al.*, 1986; Schüssler *et al.*, 1992; Santini *et al.*, 1994; Herzig *et al.*, 1994). Further, the detection of the focal expression of certain stratified-epithelial CKs, particularly CK 4 (Schüssler *et al.*, 1992; Santini *et al.*, 1994) and CK 17 (Real *et al.*, 1993) as well as, to a lesser extent, CK 14 (Real *et al.*, 1993), CK 5 (Moll *et al.*, 1989; Real *et al.*, 1993), and more rarely, CK 10 (Real *et al.*, 1993), may be regarded as reflecting the maintenance of a feature of the normal epithelium. Obviously, though, the expression of these components becomes greatly enhanced in the event of squamous differentiation, thereby indicating a partial but not absolute correlation with this type of differentiation, which is not infrequently encountered in pancreatic carcinomas. The expression of such CKs in nonsquamous (simple) tumor cells may indicate their general developmental potential with respect to squamous metaplasia, this being clearly more pronounced in pancreatic tumors as compared to colonic and gastric carcinomas. The low-level, focal expression of CK 20, which has also been mentioned by other investigators (35% of cases in Kaufmann *et al.*, 1996; 75% of cases in Miettinen, 1995; 74% of cases in Wang *et al.*, 1995) should be regarded as neoexpression—since this CK is absent in normal pancreatic ducts except for its occasional expression in the event of chronic alterations (Moll, 1993)—and may be explicable in terms of the embryological relationship of pancreatic ducts with CK 20-expressing intestinal epithelium.

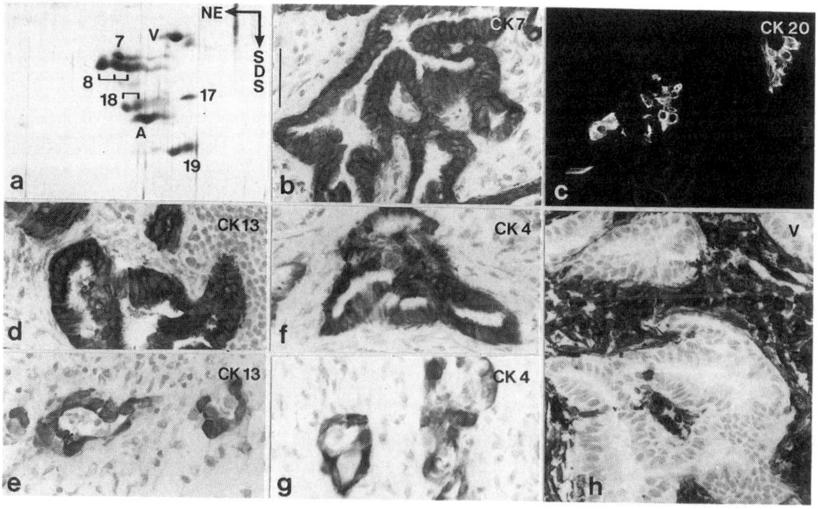

FIGURE 3. CK expression in adenocarcinomas of the pancreas. (a) Two-dimensional gel electrophoresis (for symbols, see legend to Figure 1) of an adenocarcinoma of the pancreas (G2–G3), showing the presence of CKs 7, 8, 18, and 19 as in normal pancreatic ducts, but the additional appearance of CK 17. (b–h) Immunoperoxidase (in panel c, immunofluorescence) staining of a G2–G3 tumor (lymph-node metastasis), showing homogeneous positive staining for CK 7 (b; monoclonal antibody CK-7), a slight, focal immunoreaction for CK 20 (c; GP-IT), and the expression to varying degrees of CK 13 (d, e; 2D7) and CK 4 (f, g; 6B10) in more (d, f) and less (e, g) differentiated portions of the tumor, which was negative for vimentin (h; V9; stroma-cell-positive). The close correspondence between the pattern for normal ductal epithelia (see text) and that for the tumor is evident. Frozen sections; bar, 50 μm. Reprinted with permission from Moll (1993).

All available studies agree that CK 7 is abundantly expressed in pancreatic carcinomas (Moll and Franke, 1986; Osborn *et al.*, 1986, Ramaekers *et al.*, 1990; Schüssler *et al.*, 1992; Herzig *et al.*, 1994; Santini *et al.*, 1994; Thorson *et al.*, 1996), so that the ratio of CK 20 to CK 7 always clearly favors the latter in this type of tumor.

3.4. Breast Carcinomas

In normal resting mammary glands, two-dimensional gel electrophoresis of microdissected glandular tissue reveals a complex CK pattern comprising the simple-epithelial components CK 7, CK 8, CK 18, and CK 19 and the stratified-epithelial components CK 5, CK 14, CK 15, and CK 17 (Moll *et al.*, 1983a). The monospecific CK antibodies that are now available have extended these older findings to the cellular level and have revealed what is essentially a dual expression pattern comprising luminal versus basal/myoepithelial cells, although the delineation between the two patterns is not strict in the distal lobular parts of the gland. Basically, luminal cells mainly express the simple-epithelial components CK 7, CK 8, CK 18, and CK 19, but not CK 20. In addition, they produce heterogeneous staining for the antibody Ks 8.12 directed against CK 13, CK 15, and CK 16, this positivity being less widespread when an antibody directed exclusively against CK 13 is applied (Moll, 1993). A small subgroup of luminal cells, particularly in the ducts, exhibits positive staining for CK 5, CK 14, and/or CK 17. In contrast, whereas basal/myoepithelial cells are devoid of simple-epithelial CKs, those of the ducts strongly express CK 6, CK 14, and CK 17, along with, less consistently, CK 5. It is worthy of note that myoepithelial cells coexpress vimentin and, heterogeneously, GFAP (V. E. Gould *et al.*, 1990). In the lobular acini, stratified-epithelial CKs are much less abundant and are more or less confined to a few luminal cells.

We analyzed a series of infiltrating ductal breast carcinomas (n = 101) by applying a battery of monospecific antibodies to cryostat sections of tumor tissues (Moll, 1993; Malzahn *et al.*, 1998). Figure 4 represents a good example of the data thus obtained (for overall profiles, see Table II). The main CK proteins detected using two-dimensional gel electrophoresis were CK 8, CK 18, and CK 19, while fairly minor amounts of CK 7 were also demonstrable (Figure 4a; Moll *et al.*, 1982a, 1983a); a small subpopulation of these tumors was also found to contain certain stratified-epithelial CKs (Moll *et al.*, 1982a, 1983a). Immunohistochemistry demonstrated the abundant expression and virtually uniform distribution of the simple-epithelial components CK 8, CK 18, and CK 19 in the majority of tumors (Figures 4b, 4f), while CK 7 expression was generally more heterogeneous (89% of tumors being positive; Figure 4c) as well as being somewhat more pronounced in G3 cases. The findings for CK 20 were mostly negative, although it was sometimes detectable in very rarely encountered single cells (Figure 4d); significant, more extended expression of this CK was only observed in 3 of a total of 101 cases, all of these 3 being G3 carcinomas. Among the cases studied, 27% (with some predominance of G3 tumors) exhibited a positive reaction with antibody Ks 8.12 directed against CKs 13/15/16 (Figure 4e). A *simple-epithelium* CK phenotype—defined as the exclusive expression of CK 7, CK 8, CK 18, and CK 19 (and occasionally CK 20; the reactivity with Ks 8.12 being ignored for the present purposes)—was detected in the vast majority of G1 and G2 carcinomas, as well as in a subgroup (18 of 45 cases, 40%) of G3 carcinomas.

The remainder (60%) of the G3 carcinomas (as well as a very few G1 and G2 carcinomas) exhibited a *bimodal* CK phenotype, as defined by the additional expression of one or more of the stratified-epithelial components CK 4, CK 14 (Figure 4g), and CK 17, almost always with focal and hetero-

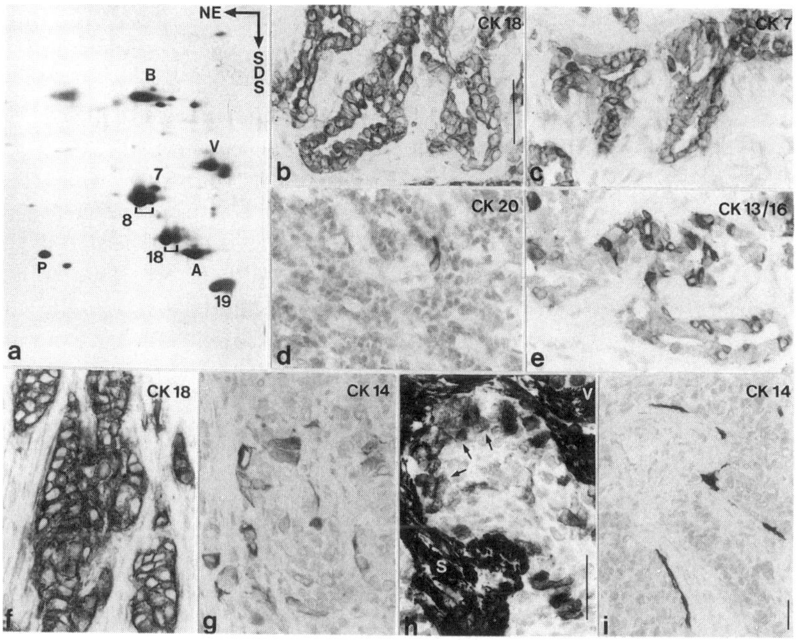

FIGURE 4. CK (and vimentin) expression in breast carcinomas. (a) Electrophoretic pattern of an invasive ductal carcinoma revealing the presence of CKs 7, 8, 18, and 19 (for explanation, see legend to Figure 1). (b–e) CK profile of an invasive ductal carcinoma (G1), showing homogeneous staining for CK 18 (b; monoclonal antibody Ks 18.174), heterogeneous staining for CK 7 (c; CK-7), a largely negative reaction for CK 20 (with the exception of one positive tumor cell; d; GP-IT), and a focal positive reaction with monoclonal antibody Ks 8.12 against CKs 13/15/16 (e). (f, g) Invasive ductal carcinoma (G3) exhibiting some CK 14-positive tumor cells (KA1; g) along with the usual homogeneous reaction for CK 18 (f; Ks 18.174). (h) Example of an invasive ductal carcinoma (G3) showing the focal expression of vimentin in tumor cells (arrows; S, strongly positive stroma cells). In this tumor, monoclonal antibody KA1 only decorates peripherally located remnants of myoepithelial cells (former *in situ* growth). Frozen sections, immunoperoxidase staining (bar, 50 μm). Reprinted with permission from Moll (1993).

geneous distribution patterns. Among these bimodal cases, it proved possible to distinguish a CK 4 subtype (predominantly expressing CK 4; vimentin-negative), a CK 4/17 subtype (mostly vimentin-positive), and a CK 14/17 subtype (vimentin-positive; Malzahn *et al.*, 1998). The expression of CK 4, CK 14, and CK 17 in the tumor cells of bimodal-type breast carcinomas was found to be statistically correlated with poor histological differentiation and the absence of steroid hormone receptors. The distribution of vimentin exhibited only partial overlapping with that of these three stratified-epithelial CKs. Figure 4h shows the focal expression of vimentin in a G3 carcinoma in which only residual myoepithelial cells, but not tumor cells were stained by the CK 14 antibody (Figure 4i). The myoepithelial remnants may indicate original *in situ* growth of the tumor. Prognostic analyses suggested that the presence of CK 4, CK 14, and/or CK 17 was associated with unfavorable overall and disease-free survival rates in certain subgroups, such as G3 tumors and estrogen-receptor-negative carcinomas. In node-positive tumors, the correlation between the presence of these three CKs and a shorter disease-free interval attained statistical significance (log rank, 0.0096; Malzahn *et al.*, 1998). Invasive lobular breast carcinomas were found mainly to express CK 8, CK 18, CK 19, and CK 7, with the same CKs also being detected in mucinous breast carcinomas, which, notably, were negative for CK 20.

For almost two decades, CK expression in breast carcinomas has been the subject of intense research efforts. Early on, it became clear that the simple-epithelial components CK 8, CK 18, CK 19, and, more variably, CK 7 were prevalent in such tumors (Moll *et al.*, 1982a, 1983a; Altmannsberger *et al.*, 1986; Nagle *et al.*, 1986; Jarasch *et al.*, 1988), thus suggesting the predominance of a luminal-cell phenotype. A correlation between the level of CK 7 expression and a higher degree of malignancy was also reported by Nathrath *et al.* (1994), who found a higher rate of recurrence among strongly CK 7-positive breast cancers as compared to tumors exhibiting little or no staining for CK 7. However, if one ignores the level of expression, the presence of CK 7 can be regarded as a feature of the vast majority (90–100%) of ductal breast cancers (Altmannsberger *et al.*, 1986; Ramaekers *et al.*, 1990; Bartek *et al.*, 1991; Moll, 1993; Wang *et al.*, 1995; Wauters *et al.*, 1995; Savera *et al.*, 1996; Santini *et al.*, 1996; Sack and Roberts, 1997 [cytological material]; Malzahn *et al.*, 1998). Whereas Takei *et al.* (1995) suggested that the presence of CK 8 is a favorable prognostic parameter, we are not alone (Altmannsberger *et al.*, 1986; Guelstein *et al.*, 1988) in having found that the mostly homogeneous diffuse staining obtained for the primary simple-epithelial CK 8 and CK 18 is inadequate for the unambiguous assignment of samples to particular subgroups. This apparent discrepancy may be a consequence of the different techniques used and/or of epitope masking. Recently, Schaller *et al.* (1996) found that intense CK 18 expression (detected using a particular antibody and immunohistochemical method) observed in a small subgroup of cases (16%) was correlated with a favorable prognosis. The significance of the very sporadic expression of CK 20 in infiltrating ductal breast carcinomas (in scattered tumor cells or, in rare G3 cases, more widespread) is not yet clear. Such very infrequent expression of CK 20, with CK 20 negativity being the far predominant feature of breast carcinomas, has also been reported by other investigators (Wauters *et al.*, 1995; Ascoli *et al.*, 1995 [cytological material];

Miettinen, 1995; Kaufmann *et al.*, 1996; Savera *et al.*, 1996). What seems at first glance to be an exceptionally high proportion of 19% of a group of breast carcinomas being reported as CK 20-positive by Wang *et al.* (1995) can be attributed to their positive assessment of scattered cells exhibiting often granular, equivocal CK 20 staining.

In early gel-electrophoresis studies, we reported that a minor subgroup of ductal breast carcinomas contains small amounts of stratified-epithelial CKs, including the basal-cell-type CK 14 and CK 17 (Moll *et al.*, 1982a, 1983a). In subsequent immunohistochemical studies, these findings have been confirmed and extended to the cellular level. The most widespread stratified-epithelial-CK reactivity is elicited by the monoclonal antibody Ks 8.12, indicating the presence of CK 13, CK 15, and/or CK 16 (see also Pellegrino *et al.*, 1988; Wetzels *et al.*, 1991). The significance of this reactivity and of the particular CK antigen(s) recognized remains a matter for conjecture. In stratified squamous epithelia, CK 16 is known to be a marker of hyperproliferation (Weiss *et al.*, 1984). In breast carcinomas, however, the low degree of concurrence between the distribution of CK 16 or CK 13/15/16 and that of the proliferation marker Ki 67 is noteworthy (Wetzels *et al.*, 1991). Our finding of focal CK 5 expression in high-grade breast carcinomas stands at variance with the proposal of Trask *et al.* (1990), based on cell-culture studies, that CK 5 is a marker of normal (as opposed to malignant) breast epithelium. However, the stratified-epithelial CK most frequently reported to be expressed in a subgroup of breast carcinomas is CK 14 (e.g., Altmannsberger *et al.*, 1986; Nagle *et al.*, 1986; Dairkee *et al.*, 1987; Jarasch *et al.*, 1988; Wetzels *et al.*, 1989; V. E. Gould *et al.*, 1990; Nathrath and Lane, 1990; Nathrath *et al.*, 1994; Santini *et al.*, 1996). Su *et al.* (1996a) detected CK 14 mRNA in 3 of their 12 cases of breast carcinoma, whereas CK 14 protein was found in only 2 of these tumors, suggesting certain disturbances in expression regulation. Dairkee *et al.* (1987) proposed that CK 14-positive breast carcinomas have an unfavorable prognosis. In comparison with CK 14, the related CK 17 is detectable in larger subgroups of infiltrating ductal carcinomas (Guelstein *et al.*, 1988; Wetzels *et al.*, 1991; Malzahn *et al.*, 1998). Another stratified-epithelial CK that is mainly but not exclusively expressed in (noncornifying) stratified squamous epithelia, CK 4, has not yet been detected in normal breast epithelium. This CK was detectable in one-fifth of our cases, a finding comparable to that reported by Heatley *et al.* (1995); in contrast, Santini *et al.* (1996) failed to detect CK 4 in any of their series of breast tumors.

The biological significance of the expression of the stratified-epithelial components CK 4, CK 14, and CK 17 in breast carcinomas remains an enigma. One hypothesis is that CK 14 and CK 17 expression might indicate derivation from, or differentiation toward, myoepithelial cells. Although in some studies other myoepithelial markers, such as GFAP and smooth-muscle-type actin, were essentially not detectable in the tumor cells of breast carcinomas (V. E. Gould *et al.*, 1990; Wada *et al.*, 1991; Böcker *et al.*, 1992b; Martin de las Mulas *et al.*, 1994), in another study positive staining for α-smooth-muscle actin has been reported in scarce tumor cells (Soini and Miettinen, 1992) and, in a recent study, in as many as 18% of the cases studied (mostly nuclear grade 3; Santini *et al.*, 1996). Another suggestion is that these particular tumors acquire a more general basal-cell phenotype (Wetzels *et al.*, 1991). It should be mentioned at this point, though, that CK 14 and CK 17 are also expressed in a subpopulation of luminal epithelial cells (see above), which raises the possibility that bimodal-type carcinomas develop some differentiation features of a particular "bimodal" luminal cell type (which would not necessarily indicate their derivation from such cells; cf. V. E. Gould, 1986). The expression of these stratified-epithelial CKs may also point to rudimentary squamous metaplastic processes, particularly when these components are intensely expressed; however, fully developed squamous metaplasia is rare in breast cancer. The finding of a positive correlation between CK 14 and CK 17 expression and a higher proliferation index (Ki-67 score; Santini *et al.*, 1996) is immediately relevant for the biological behavior of the tumors. From a general point of view, the correlation of the more complex bimodal CK phenotype with a poor degree of differentiation may reflect the principles underlying the increased CK complexity observed in other types of carcinomas upon reduction of the level of differentiation, e.g., in squamous-cell carcinomas (see Section 3.6). Such complex neoexpression may stem from the inactivation of repressive regulatory factors (Knapp and Franke, 1989).

The prognostic value of IF-protein expression in breast carcinomas has yet to be definitively established. Most relevant studies make reference to the coexpression of vimentin being related not only to a higher rate of proliferation, but also to an unfavorable overall survival rate (Raymond and Leong, 1989a, b; Domagala *et al.*, 1990). However, a recent broad-based (albeit retrospective) study has concluded that, despite being associated with other unfavorable prognosis parameters, vimentin expression is itself not genuinely correlated with a poor prognosis (Seshadri *et al.*, 1996). Nevertheless, in breast cancer cells *in vitro*, the coexpression of vimentin along with CKs has been directly linked with increasing invasiveness (Hendrix *et al.*, 1996).

In accordance with the earlier proposal of Dairkee *et al.* (1987), we have obtained preliminary data indicating the unfavorable prognostic implication of the expression of CK 4, CK 14, and/or CK 17 (Malzahn *et al.*, 1998). Studies employing larger series of tumor patients and multivariate statistical analyses are now required to test the validity of our findings.

Fortunately, thanks to the availability of appropriate antibodies and enhancement techniques, such investigations are now feasible using readily available, formalin-fixed, paraffin-embedded routine tissue (see below, Section 5). Recently, Santini *et al.* (1996) assigned 30% of their examined cases to group 3 (positive for CK 14, CK 17, vimentin, and α-smooth-muscle actin), a finding comparable to our delineation of bimodal-type carcinomas (except that they did not detect CK 4 in their group of tumors). Although correlations with markers of tumor aggressiveness were noted, prognostic data were not included within the scope of their investigation, in which the possible relationship of group 3 tumors with a myoepithelial phenotype was discussed.

From the diagnostic point of view, myoepithelial markers such as CK 14, CK 17, and GFAP may be useful for distinguishing sclerosing adenosis from neoplastic processes and for recognizing early invasive growth (Nagle *et al.*, 1986; Altmannsberger *et al.*, 1986; Jarasch *et al.*, 1988; V. E. Gould *et al.*, 1990; Böcker *et al.*, 1992a, b).

3.5. Transitional-Cell Carcinomas of the Urinary Tract

The transitional epithelium of the urinary tract is unique among epithelial tissues owing to the presence of specialized luminal "umbrella" cells, with their asymmetrical plasma membrane containing uroplakins, their fusiform vesicles, and their zonula occludens (X.R. Su *et al.*, 1994; Wu *et al.*, 1994; Moll *et al.*, 1995). Furthermore, the differentiation of both normal and malignant urothelium appears to be highly flexible, and thus various types of metaplasia may occur.

The urothelium also exhibits a unique pattern of CK expression. While the simple-epithelial components CK 8, CK 18, CK 19, and CK 7 are present in all urothelial layers, CK 20 is essentially confined to the umbrella cells (with some heterogeneity), so that this CK, in the context of other "urothelial" CKs, is a highly conspicuous marker of urothelial differentiation. In addition, CK 13 is found in the basal/intermediate cell compartment, but not in umbrella cells. Basal cells are the only cells of this epithelium that also express CK 5 and CK 17 (Moll *et al.*, 1988, 1990; Moll, 1993; Schaafsma *et al.*, 1990; Guelstein *et al.*, 1993).

CK analyses of transitional-cell carcinomas have pointed to the relative stability of several features typical of normal urothelium (Figure 5), for example, the predominance of CK 8, CK 18, CK 19, and CK 7 in such tumors (Figures 5d, 5e). It is particularly striking that the expression of the CK typically found in umbrella cells, i.e., CK 20, was found to be preserved in virtually all (96%) of our group of genuine transitional-cell carcinomas and their metastases (Figures 5c, 5e; Moll *et al.*, 1992b). However, only well-

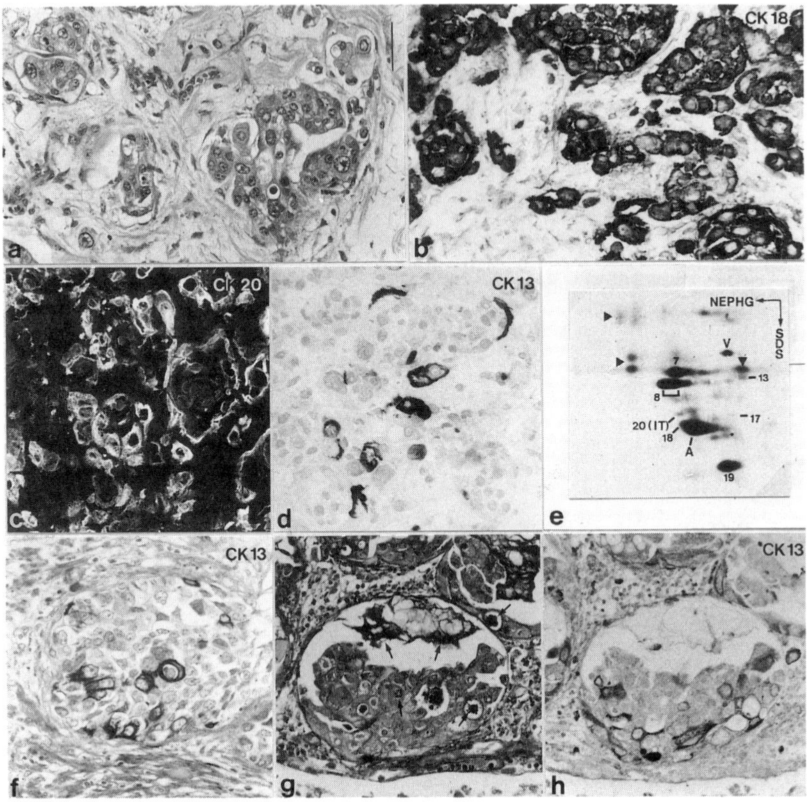

FIGURE 5. CK expression in transitional-cell carcinomas of grade G3 (a–e) as well as the demonstration of CK 13 in paraffin-embedded tissues taken from transitional-cell carcinomas (f–h; all from the urinary bladder). (a–e) Nonpapillary transitional cell carcinoma (G3) exhibiting invasive and, in places, dissociated growth (a, H & E staining; b–d, frozen sections). (b) Overall staining (immunoperoxidase) of the tumor cells by monoclonal antibody Ks 18.174 directed against CK 18. (c) Localization of CK 20 in many tumor cells (immunofluorescence microscopy; GP-IT). (d) A few tumor cells are positive for CK 13 (immunoperoxidase; 2D7). (e) Electrophoresis pattern, with CKs 7, 8, 18, and 19 predominant, and in addition a small, but significant amount of the CKs typical of transitional cell carcinomas, i.e., CK 13 (see panel d) and CK 20 (see panel c). For symbols, see the legend to Figure 1. (f–h) Demonstration of the focal expression of CK 13 in paraffin-embedded tissue (monoclonal antibody Ks 13.1; immunoperoxidase) of a papillary transitional cell carcinoma (G2; invasive; f) and a nonpapillary transitional cell carcinoma (G3) with glandular metaplasia (g, PAS staining with evidence of mucus, arrows; h, immunoperoxidase staining, serial section). In spite of partial glandular differentiation, the typical heterogenous CK 13 staining has been retained. Bar (in panel a), 50 μm. Reprinted with permission from Moll (1993).

differentiated papillary tumors still exhibited a staining pattern related to that observed in normal superficial cells; in less-differentiated and manifestly invasive tumors, the CK 20 expression pattern was either diffuse or irregular-heterogeneous (Figure 5c). Applying a cutoff level of 5% positive tumor cells, it emerged that 79% of the cases (comprising both noninvasive and invasive tumors, the latter predominating) could be assigned a CK 20-positive status, with the staining in these tumors being variable in extent, but often widespread (Moll *et al.*, 1992b). Of the stratified-epithelial CKs, CK 13 was the one most importantly and frequently encountered in transitional-cell carcinomas: in G1 and G2 tumors, the staining for this CK was either very extensive or conspicuous, but heterogeneous, the actual distribution patterns exhibiting considerable variety (Moll *et al.*, 1988; Moll, 1993). In G3 carcinomas, though, CK 13 levels were usually significantly lower, but this CK was still detectable in some cells of most of the tumors examined (Figures 5d–5f), even in cases of glandular metaplasia (Figures 5g, 5h). The other stratified-epithelial CKs detected in this group of tumors were CK 5, CK 14, and CK 17 (Moll *et al.*, 1988). Thus, in our series, poorly differentiated, invasive transitional-cell carcinomas were characterized by the overwhelming prevalence of simple-epithelial CKs, usually including CK 20, along with some low-level expression of CK 13 and CK 17. No differences between primary tumors and metastases were observed. As expected, transitional-cell carcinomas with squamous-epithelial metaplasia exhibited increased or novel expression of several stratified-epithelial CKs (see also Moll *et al.*, 1988; Schaafsma *et al.*, 1990, 1991).

Both our data and those of other investigators concerning CK expression in transitional-cell carcinomas (Ramaekers *et al.*, 1985; Cintorino *et al.*, 1988; Schaafsma *et al.*, 1990, 1991; Guelstein *et al.*, 1993) have revealed that the CK pattern of normal urothelium is relatively well conserved in these tumors, although it is also influenced by the degree of malignancy and by the occurrence of squamous-epithelial metaplasia. While dedifferentiation is mainly associated with a reduction of CK 13 levels, some residual expression of this CK is usually maintained, even in metastases (Moll *et al.*, 1988; Schaafsma *et al.*, 1991). This decrease is accompanied by an elevated level of immunostaining elicited by certain monoclonal antibodies directed against CK 18 (Ramaekers *et al.*, 1985). Increased expression of CK 8 and CK 18 was observed in tumor cells at the interface with peri- and intratumoral stroma (Schaafsma *et al.*, 1991), reminiscent of a similar "interface phenomenon" in squamous-cell carcinomas (Schaafsma *et al.*, 1993; see Section 3.6).

A particularly interesting feature of such carcinomas is their unexpected maintenance or even elevation of levels of CK 20 expression during the course of tumor progression. Thus, differentiation loss is not associated with the loss of this terminal-differentiation-related CK marker, but rather with its premature expression in lower cell layers. In this context, it is worth noting that recent studies have demonstrated that alterations in the normal umbrella-cell-related distribution pattern of CK 20 offer a reliable marker of urothelial dysplasia, while in noninvasive papillary transitional-cell carcinomas, preservation of the CK 20 pattern, as opposed to disturbed patterns (either homogeneous-diffuse or negative), may well point to a prognosis of nonrecurrence of disease (Harnden *et al.*, 1995, 1996). These authors have reported CK 20 expression in 65.5% of noninvasive papillary tumors of grades G1 or G2 (Harnden *et al.*, 1995), whereas the proportion of positive cases reported by Miettinen (1995) was 73%, and the corresponding figure obtained by Wang *et al.* (1995) was 89% (statements about the state of invasiveness were not provided in the last two studies). Thus, particularly when present together in a tumor, CK 13 and CK 20 are good markers of urothelial origin. Although CK 17 would appear to be somewhat more abundant than CK 13 in poorly differentiated transitional cell carcinomas (Guelstein *et al.*, 1993), its specificity for urothelial differentiation is lower. In this context, it should be noted that CK analysis can rarely claim to attain absolute specificity; in the case of transitional-cell carcinomas, such a role can be better fulfilled by those highly specific urothelial marker proteins, the uroplakins (Moll *et al.*, 1995). Although in invasive transitional-cell carcinomas, advanced tumor progression may be accompanied by certain alterations in CK expression, such as lower levels of CK 13 (Moll *et al.*, 1988) and the elevated expression of CK 14 (Schaafsma *et al.*, 1990; although being inconsistent, Schaafsma *et al.*, 1991), direct correlations between such changes and biological behavior or the disease prognosis have yet to be established. In contrast, the CK 20 pattern is a promising prognostic marker for the early stages of such tumors (see above; Harnden *et al.*, 1995).

3.6. Squamous-Cell Carcinomas of the Oropharynx and of Other Sites

Most regions of the oropharynx are lined by a nonkeratinizing stratified squamous epithelium that exhibits a typical stratum-related CK expression pattern (Moll *et al.*, 1982a, 1989; Bartek *et al.*, 1986; Wild and Mischke, 1986; Morgan *et al.*, 1987a; Bosch *et al.*, 1988; Purkis *et al.*, 1990; for esophagus, see van Muijen *et al.*, 1986). Basal cells express the primary stratified-epithelial components CK 5 and CK 14 as well as, less regularly, certain simple-epithelial CKs (Bosch *et al.*, 1988), with CK 19 being the most widespread of these (Bartek *et al.*, 1986; Morgan *et al.*, 1987a), while CK 8 and CK 18 are more focally expressed (Moll, 1993). In inflamed mucosa, the expression of CK 19 is more prominent and is found to spread into the suprabasal compartment (Bosch *et al.*, 1989; Moll, 1993). This terminal-differentiating compartment is characterized by the prominent expression of the mucosal maturation-associated components CK 13 and CK 4 (van Muijen *et al.*, 1986; Wild and Mischke, 1986), while the markers of epidermal maturation CK 1 and CK 10/11 may be focally and heterogeneously expressed in some suprabasal areas (Morgan *et al.*, 1987a; Moll, 1993). Further important CKs localized in this epithelium are CK 6 and CK 16, which, particularly in the epidermis, are regarded as markers of hyperproliferation, but are also constitutively expressed in mucosal stratified squamous epithelia.

The pattern of CK expression in squamous-cell carcinomas of the oropharyngeal region, and indeed of all other sites, is very complex. In general, the CK patterns of these tumors reflect, in part, the corresponding pattern of their epithelium of origin, but in practice a considerable degree of heterogeneity may also be encountered. This variability would appear to be related to several parameters, the most notable of which are the degree of differentiation and the tumor grade. This is illustrated in Figure 6, in which a well-differentiated, keratinizing squamous-cell carcinoma of the tongue (Figures 6a–6g) and a poorly differentiated, nonkeratinizing squamous-cell carcinoma of the hypopharynx (Figures 6h–6o) are shown together. In the well-differentiated tumor, two-dimensional gel electrophoresis revealed the basal-cell-type and hyperproliferative CKs to be predominant (Figure 6a), these being reciprocally expressed in the different layers of the tumor cell nests. While CK 5, a typical feature of basal cells, was immunohistochemically detectable in the peripheral cell layers of well-differentiated areas (Figure 6c; and, more homogeneously, in less-differentiated, smaller nests and trabeculae), the hyperproliferative CK 16 elicited intense staining in the central maturating regions of tumor cell nests (Figure 6d). In most cases, CK 17 was detectable, although its distribution pattern proved to be irregular (not shown; cf. Gillenkirch, 1994). Remarkably, the CKs typically indicative of orthologous mucosa maturation, CK 13 and CK 4, often occurred only sporadically (Figure 6e), while in well- to moderately differentiated tumors, epidermal-maturation CKs such as CK 10/11 were observable at a higher level, their presence displaying a good, but not absolute, correlation to horn-pearl formation (Figure 6f). All in all, the most prevalent components of maturating central tumor cells were found to be CK 6 and CK 16, a result also reflected by gel-electrophoresis data (Figure 6a). In this group of tumors, simple-epithelial CKs (CK 8, CK 18, CK 19) were detected at low to moderate levels (Figure 6g).

In poorly differentiated squamous-cell carcinomas (Figure 6h), CK 5 (Figure 6i) and CK 17 (Figure 6j) were found to be more widespread than in well- and moderately differentiated tumors, while heterogeneous expres-

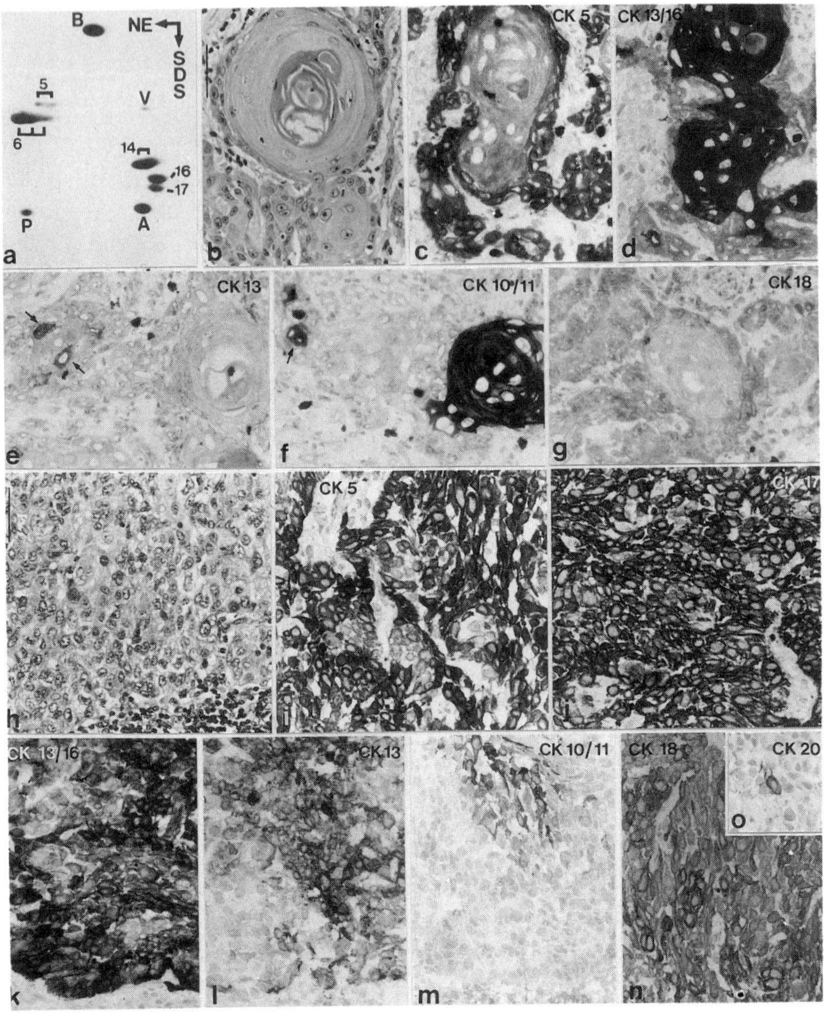

FIGURE 6. CK expression in squamous-cell carcinomas of the oral/pharyngeal mucosa. (a–g) Well-matured, cornifying squamous cell carcinoma of the tongue. (a) Typical electrophoretic pattern composed of stratified-epithelial CKs. For symbols, see the legend to Figure 1. (b) H & E staining showing horn pearls. (c–e) Immunoperoxidase staining (frozen sections): peripheral localization of CK 5 (c; monoclonal antibody AE14). Maturing tumor cells exhibit pronounced staining by monoclonal antibody Ks 8.12 (d); in this case, this should be taken to indicate primarily the presence of CK 16, as CK 13 (e; 2D7) is only expressed in a few of these cells (arrows). Maturing cells are only partially stained by monoclonal antibody Kk 8.60 against CK 10/11 (f), with a horn-pearl and single cells (arrow) being labeled. The reaction for CK 18 is very weak or negative (g; Ks 18.174). (h–o) Poorly differentiated, noncornifying squamous-cell carcinoma of the hypopharynx (h; H & E staining). Immunoperoxidase staining (i–o; frozen sections) demonstrated conspicuously positive reactions for CK 5 (i; monoclonal antibody AE14), CK 17 (j; E3), and CK 13/15/16 (k; Ks 8.12), the focal expression of CK 13 (l; 2D7), yet only slight CK 10/11 expression (m; Kk 8.60). In addition, CKs typical of simple epithelia are also demonstrable, including CK 18 to a pronounced degree (n; Ks 18.174), with CK 20 only being detectable in very few single cells (o; GP-IT). Bar (in panels b and h), 50 μm. Reprinted with permission from Moll (1993).

sion patterns were observed for CK 13 and CK 16 (Figures 6k, 6l). The epidermal CKs 10/11 exhibited a very focal distribution at markedly lower levels than those seen in the group of well- to moderately differentiated tumors (Figure 6m). In contrast, the overall levels of simple-epithelial CKs such as CK 18 were higher in poorly differentiated squamous-cell carcinomas (Figure 6n); elevated levels of expression were also observed for CK 8 and CK 19, the latter being the most prominent of all. On the other hand, CK 7 and CK 20 were detected extremely rarely in poorly differentiated tumors, and when at all, in a very sparse distribution pattern (Figure 6o).

The focal coexpression of vimentin was observed in about two-thirds of the cases regardless of differentiation status. No correlations emerged linking the expression patterns of CK polypeptides and vimentin with the tumor status, i.e., primary versus metastasis, or prior radiation therapy.

There is a considerable body of literature dealing with CK expression patterns in the various types of squamous-cell carcinomas of the oropharyngeal region as well as of other sites (for reviews, see Moll, 1993; Schaafsma and Ramaekers, 1994). In agreement with our own data, these studies have, in general, shown that squamous-cell carcinomas exhibit both (1) relatively constant features and (2) variable features that may be related to morphological parameters such as tumor grade. In addition, the site of the tumor would also appear to be reflected by certain features of the CK pattern. Among the constant features, the expression of the primary stratified-epithelial CKs, notably CK 5 and CK 14, is more or less generally maintained in squamous-cell carcinomas regardless of their origin. This was already demonstrable in early gel-electrophoresis studies (Moll *et al.*, 1982a; Nelson *et al.*, 1984) [the data presented by Vaidya *et al.* (1989, 1996), suggesting the loss of CK 5 and CK 14 in such tumors, are difficult to reconcile with either our data or the findings of other laboratories] and has been further confirmed in more recent immunohistochemical investigations (Moll *et al.*, 1989; Ivanyi *et al.*, 1990; Heyden *et al.*, 1992; Takahashi *et al.*, 1995; L. Su *et al.*, 1996b; Balm *et al.*, 1996). It is particularly worthy of note that the expression of these CKs is usually maintained even in poorly differentiated squamous-cell carcinomas (Figure 6i), although the reduction of expression of these CKs in poorly differentiated areas of cutaneous squamous-cell carcinomas has been described (Markey *et al.*, 1991). Thus, relevant antibodies (such as monoclonal antibodies AE 14 directed against CK 5 and D5/16B4 against CK 5/6; see Section 5) may be suitable for recognizing the squamous-epithelial character of metastases of dedifferentiated squamous-cell carcinomas when the primary tumor is unknown. Basal keratinocyte-type CKs are also constitutive components of squamous-cell carcinomas arising from pseudostratified epithelia (bronchus, endocervix with reserve cells; Moll *et al.*, 1982a, 1983b; Blobel *et al.*,

1984); the expression of such CKs in the normal basal (or reserve) cells of these epithelia (respiratory epithelium: Blobel *et al.*, 1984; Broers *et al.*, 1988; Moll *et al.*, 1989; Leube and Rustad, 1991; endocervix: Gigi-Leitner *et al.*, 1986; Weikel *et al.*, 1987; Smedts *et al.*, 1992b) may reflect their potential to give rise to squamous metaplasia and squamous-cell carcinomas. That such basal or reserve cells are in some way related to the basal cells of stratified squamous epithelia is further reflected by the fact that both cell types contain well-developed hemidesmosomes (P. R. Gould *et al.*, 1979; McDowell *et al.*, 1983).

Another related CK that is regularly expressed in squamous-cell carcinomas is CK 17, whose distribution may be confined to the peripheral tumor cell layer (cervical carcinomas; Senzaki *et al.*, 1997). In one study of oropharyngeal carcinomas, all 34 cases were found to display a positive reaction with antibody E3 directed against CK 17, with the distribution patterns being irregular in most cases (Gillenkirch, 1994).

Among the maturation-associated and therefore suprabasally expressed subtypes of the stratified-epithelial CKs, the expression of CK 6 and CK 16 in the epidermis has been found not only to be correlated with a higher proliferative status (Weiss *et al.*, 1984; Stoler *et al.*, 1988), but also to indicate an abnormal pathway of terminal differentiation (Sun, 1989). Both metaplastic squamous epithelium (Leube and Rustad, 1991) and most squamous-cell carcinomas (Figure 6a; Moll *et al.*, 1982a, 1983a) are rich in CK 6 and CK 16. Immunohistochemistry has confirmed the link between these CKs and squamous-epithelial maturation that proceeds in the central regions of tumor cell nests (Figure 6d). The specific monoclonal antibody directed against CK 6, KA 12 (R. B. Nagle *et al.*, unpublished; see Demirkesen *et al.*, 1995), performs well on formalin-fixed, paraffin-embedded material when the sections have been pretreated by microwave heating. Owing to its absence or low expression levels in many adeno-/nonsquamous carcinomas (Moll *et al.*, 1982a; Moll, 1993; cervical carcinomas: Smedts *et al.*, 1992a, 1993a), CK 6 may be used as another marker of squamous differentiation in poorly differentiated carcinomas, although it should be borne in mind that this CK is (like CK 5) expressed in certain adenocarcinoma-related tumors, such as high-grade breast carcinomas (see Section 3.4 and Malzahn *et al.*, 1998).

Early gel-electrophoresis data (Moll *et al.*, 1982a, 1983a) suggested that orthologous maturation-associated CKs (CK 4, CK 13) are not present in mucosal squamous-cell carcinomas, a finding that we then interpreted as evidence of the maintenance of the undifferentiated basal-cell phenotype. However, sensitive immunohistochemical techniques have since shown that squamous-cell carcinomas may indeed express some of these differentiation-associated CKs, albeit only focally. The reported incidence of CK 13 expression in squamous-cell carcinomas has ranged from 42% (Klijanienko *et al.*, 1993) to 67% (Malecha and Miettinen, 1991) of the cases examined. In a limited series (34 cases) of oropharyngeal squamous-cell carcinomas studied in our laboratory, CK 4 was found to be expressed to a variable and usually minor degree in 35% of the G1/G2 carcinomas, but in 70% of the G3 carcinomas. In both of these tumor groups, CK 13 was detectable in about two-thirds of the cases. For both CK 4 and CK 13, the expression level was found to be, on average, more elevated in the poorly differentiated G3 cases (Gillenkirch, 1994). These findings suggest a rather unexpected, albeit relatively weak, tendency toward the elevated expression of CK 13 and, in particular, CK 4 in the event of dedifferentiation. Results of other studies have pointed rather to a reduction in the levels of these CKs in association with diminished tumor differentiation (Morgan *et al.*, 1987b; Malecha and Miettinen, 1991). In skin carcinomas, though, the levels of CK 4 and CK 13 have been reported to rise when dedifferentiation occurs (Watanabe *et al.*, 1995), whereas other studies have emphasized the essential absence of these CKs in cutaneous squamous-cell carcinomas (Kuruc *et al.*, 1989; Lavrijsen *et al.*, 1989; Markey *et al.*, 1991). Two investigations using monoclonal antibody KS-1A3 have recorded the highly sporadic expression or essential absence of CK 13 in oral and esophageal squamous-cell carcinomas (Heyden *et al.*, 1992; Takahashi *et al.*, 1995). At any rate, there is evidence that a limited degree of orthologous terminal differentiation may take place in squamous-cell carcinomas, although this is of a limited and confused nature.

The disordered maturation processes occurring in squamous-cell carcinomas are further reflected by the additional manifestation of some "ectopic" epidermis-type differentiation, as indicated by the focal expression of CK 1 and CK 10/11. Although it is well established that these CKs are associated with terminal epidermal differentiation (Fuchs and Green, 1980; Moll *et al.*, 1982a; for gingiva and hard palate, see Bosch *et al.*, 1989), they are also focally expressed in internal noncornifying stratified squamous epithelia (Moll *et al.*, 1983b; Morgan *et al.*, 1987a). In a study performed in our laboratory (Gillenkirch, 1994), these CKs were found to occur focally in 85% of the well- and moderately differentiated oropharyngeal squamous-cell carcinomas and in at least 50% of the poorly differentiated tumors examined (see also Figures 6f, 6m). Although their presence is usually manifestly associated with maturation and horn-pearl formation, CK 1 and CK 10/11 are detectable in some morphologically immature tumor cells, thereby indicating a very early stage of keratotic maturation. A positive correlation between these CKs and the degree of differentiation has also been reported in other studies (Huszar *et al.*, 1986; Morgan *et al.*, 1987b; Cintorino *et al.*, 1990; Williams *et al.*, 1997). As long ago as 1980, an immunohistochemical study using polyclonal antibodies reported similar results (Löning *et al.*, 1980). Thus, a single squamous-cell carcinoma may undergo all three possible pathways of keratinocyte maturation (hyperproliferative = CK 6 and CK 16; epidermal = CK 1 and CK 10/11; mucosal = CK 4 and CK 13), with the hyperproliferative pathway usually being clearly predominant. It is remarkable that, in well- and moderately differentiated oropharyngeal carcinomas, the epidermal pathway is relatively more pronounced than the orthologous mucosal pathway. It should also be noted that expression "pairs" such as CK 4 and CK 13 may be uncoupled in carcinomas. Apparently, there are also certain site-specific differences such as the elevated levels of CK 13 in squamous-cell carcinomas of the cervix (see below). The degree of keratinization is principally reflected by the presence and extent of the epidermal-maturation CKs. Particularly prominent differences between nonkeratinizing and keratinizing squamous-cell carcinomas have been reported for cervical tumors (Smedts *et al.*, 1994), in which higher levels of CK 10 and CK 13 are seen in the keratinizing variant.

A large amount of published data has accumulated concerning the, initially unexpected, expression of simple-epithelial CKs in squamous-cell carcinomas. Highly sensitive immunocytochemical techniques as well as *in situ* hybridization have convincingly demonstrated that basal cells in normal noncornifying stratified squamous epithelia may express CK 8, CK 18, and CK 19 (Bosch *et al.*, 1988; Moll, 1993; L. Su *et al.*, 1994). In our series of oropharyngeal squamous-cell carcinomas, the focal expression of CK 8, CK 18, and/or CK 19 (usually all three) was detected in all 34 cases studied, with a higher percentage of positive cells usually being observed in the poorly differentiated tumors (Gillenkirch, 1994). In other studies, too, the expression of simple-epithelial CKs has been found to be related to the malignancy grade of oropharyngeal and laryngeal squamous-cell carcinomas (Terry, 1986; Morgan *et al.*, 1987b; Wild *et al.*, 1987; Henzen-Logmans *et al.*, 1988; Lindberg and Rheinwald, 1989; Mallofré *et al.*, 1993; Schaafsma *et al.*, 1993 [laryngeal tumors]; Kannan *et al.*, 1994; L. Su *et al.*, 1994, 1996b). Similar correlations have been reported for squamous-cell carcinomas arising at certain transition sites, such as the anal region (Williams *et al.*, 1997) and the vulva (Weikel *et al.*, 1996), as well as for squamous-cell carcinomas of the skin (Markey *et al.*, 1991). However, Ogden *et al.* (1993) and Suo *et al.* (1993) have reported a weak correlation only or no correlation at all, respectively, between the expression of simple-epithelial CKs and the grade of oropharyngeal squamous-cell carcinomas. It should be borne in mind that, according to all published studies, the correlation linking the levels of CK 8, CK 18, and CK 19 with the tumor grade is by no means strict and, moreover, that reproducible grading of squamous-cell carcinomas is problematic per se (Bryne *et al.*, 1989). Considered as a whole, these data demonstrate that the CK-polypeptide patterns are more complex in poorly differentiated than in well-differentiated squamous-cell carcinomas, a finding that also applies to a number of other types of carcinoma (see Section 4).

In individual squamous-cell carcinomas, CK 8 and CK 18 may be preferentially expressed at the tumor invasion front. This "interface phenomenon" was observed in about one-half of the squamous-cell carcinomas studied by Schaafsma *et al.* (1993), and its occurrence may be related to increased proliferation and/or environmental influences. In addition, it might reflect the fact that, in many squamous-cell carcinomas, the tumor front exhibits a lower degree of differentiation than the tumor center; indeed, tumor grading based on the appearance of the tumor-front regions may yield data and correlations of greater prognostic value than those obtained by overall grading (Bryne *et al.*, 1989). A similar interface phenomenon has also been described in squamous-cell carcinomas of the skin (Markey *et al.*, 1991).

Published data directly concerned with the prognostic significance of simple-epithelial-CK expression in squamous-cell carcinomas are in short supply. In one study of vulvar squamous-cell carcinomas, CK 8 expression appeared to be related to shorter survival periods, albeit only in the case of recurrent tumors (Weikel *et al.*, 1996). A similar correlation between the expression of a simple-epithelial CK, CK 18, and a poorer prognosis has recently been reported for squamous-cell carcinomas of the lung (Cohen *et al.*, 1997). Clearly, larger scale prognostic studies, including multivariate analyses, are necessary before any definite claims can be made relating to the clinical value of the CK (and IF) typing of squamous-cell carcinomas.

While the expression of CK 8, CK 18, and CK 19 is quite common in squamous-cell carcinomas, the other two simple-epithelial CKs, CK 7 and CK 20, are absent in normal stratified squamous epithelia and are thus much less frequently detected in such tumors. Although it has in fact been suggested that the detection of CK 7 provides a criterion for distinguishing adenocarcinomas (CK 7-positive) from squamous-cell carcinomas (CK 7-negative) of the lung (van de Molengraft *et al.*, 1993; paraffin-embedded sections), one study of frozen tissue specimens found some pulmonary squamous-cell carcinomas (predominantly poorly differentiated ones) to be positive for CK 7 (Broers *et al.*, 1988). In (nonkeratinizing) squamous-cell carcinomas of the uterine cervix, the presence of CK 7 is a more common finding (Moll *et al.*, 1983b; Moll, 1993; Smedts *et al.*, 1992a, 1993a, 1994; Senzaki *et al.*, 1997). In contrast, the expression of CK 20 in squamous-cell carcinomas is a very rare occurrence, observable only, if at all, in very sporadically occurring single cells in poorly differentiated tumors (Figure 6o; Moll *et al.*, 1992b). In an investigation of squamous-cell carcinomas obtained from various locations, Suo *et al.* (1993) were unable to detect CK 20 staining in any of their 42 cases, while Miettinen (1995) similarly found only 1 case exhibiting scattered CK 20-positive cells among a group of 93 squamous-cell tumors. Somewhat exceptionally, rather more prominent CK 20 expression has been reported in 1 of 13 large-cell nonkeratinizing squamous-cell carcinomas of the cervix (Smedts *et al.*, 1993a) as well as in 3 of 12 recurrent squamous-cell carcinomas of the vulva, the disease course of these last CK 20-positive cases being unfavorable (Weikel *et al.*, 1996).

In squamous-cell carcinomas, the reported incidence of the (focal) coexpression of the mesenchymal IF protein vimentin with CKs ranges from 20% (Fischer *et al.*, 1989) to about one-half of the cases examined (Schaafsma *et al.*, 1993; de Araujo *et al.*, 1993; Weikel *et al.*, 1996). It has been suggested that in such tumors vimentin expression is related to a higher grade of malignancy (de Araujo *et al.*, 1993; Weikel *et al.*, 1996), tumor–stroma interactions ("interface phenomenon"; see above; Schaafsma *et al.*, 1993), and prior radiation and/or chemotherapy (Fischer *et al.*, 1989). In a study performed in our laboratory, similar percentages of about 60% of both G1/2 and G3 oropharyngeal squamous-cell carcinomas were found to express vimentin, but the overall level of expression was higher in the poorly differentiated group (Gillenkirch, 1994). In a series of vulvar squamous-cell carcinomas, a small percentage of cases exhibiting widespread vimentin immunoreactivity were found to have a significantly worse prognosis (Weikel *et al.*, 1996).

Another unexpected form of neoexpression is the occasional occurrence of neurofilaments in some squamous-cell carcinomas of the upper aerodigestive tract (Fischer *et al.*, 1989) and of the lung (van Muijen *et al.*, 1984; Gatter *et al.*, 1986; Moll, 1993); in the latter, this serves to underline the multidirectional differentiation potential observable in lung carcinomas in general.

Owing to the broad diversity of CK expression patterns that make it possible to distinguish a variety of different epithelia, the question naturally arises of whether CKs might be able to indicate the site of origin of a metastatic squamous-cell carcinoma. All data currently available suggest that there are indeed differences between the expression of certain CKs in squamous-cell carcinomas of different organs; however, such differences are usually no more than indicative of tendencies, with overlaps that preclude definite correlations between particular sites and CK expression patterns. Such site-related differences would appear to be capable of delineating—although the distinctions are somewhat blurred—squamous-cell carcinomas of the skin, oral mucosa, lung, and cervix uteri. In skin tumors, several reports have indicated that, of the maturation-associated CKs, only the epidermal ones, i.e., CK 1 and CK 10/11, are expressed, whereas those typical of the mucosa, i.e., CK 4 and CK 13, are not (Moll *et al.*, 1984; Lavrijsen *et al.*, 1989; Kuruc *et al.*, 1989; Malecha and Miettinen, 1991; Markey *et al.*, 1991). However, there is one report claiming that especially poorly differentiated cutaneous squamous-cell carcinomas may in actual fact express CK 4 and CK 13 (Watanabe *et al.*, 1995), i.e., components of fetal epidermis (Moll *et al.*, 1982b; van Muijen *et al.*, 1987b). CK 13 may also be found in skin carcinomas that have been experimentally induced in mice (Nischt *et al.*, 1988). Nevertheless, these mucosal CKs do seem to occur much more frequently in squamous-cell carcinomas of inner organs. Particularly high levels of CK 13 seem to be typical of squamous-cell carcinomas of the cervix uteri (Moll *et al.*, 1982a, 1983b; Smedts *et al.*, 1992a; Suo *et al.*, 1993; Senzaki *et al.*, 1997). Another CK that is conspicuous in cervical squamous-cell carcinomas is that minor component typical of basal cells, CK 15, which is even detectable using relatively insensitive gel-electrophoresis procedures (Moll *et al.*, 1983b). More recently, immunohistochemical investigations have also revealed the consistent presence of this CK in keratinizing squamous-cell carcinomas of the cervix, with a more variable expression being reported for nonkeratinizing tumors (Smedts *et al.*, 1993a). Another feature that is fairly typical of cervical squamous-cell carcinomas is their relatively high levels of expression of simple-epithelial CKs (Moll *et al.*, 1983b; Debus *et al.*, 1984; Smedts *et al.*, 1992a), which, in addition to CK 8, CK 18, and CK 19, may include CK 7 (Moll *et al.*, 1983b; Moll, 1993; Smedts *et al.*, 1992a, 1994; Senzaki *et al.*, 1997). The levels of CK 8 and CK 18 in squamous-cell carcinomas of the larynx and the lung have been reported to be higher than those observed in tumors originating from the floor of the mouth, the tonsil, and the esophagus (Schaafsma *et al.*, 1993). The relatively high levels of simple-epithelial CKs recorded in squamous-cell carcinomas of the lung (Blobel *et al.*, 1984; Broers *et al.*, 1988) may also be interpreted as reflecting the presence of a glandular differentiation component, this being in accordance with well-documented findings of composite differentiation in such tumors (Müller and Fisseler-Eckhoff, 1989). Altogether, these comparative differences suggest that squamous-cell carcinomas arising from pseudostratified epithelia (of which simple-epithelial CKs are major components) via processes of squamous epithelial metaplasia are intrinsically prone to higher expression levels of these CKs. A surprisingly clear-cut delimitation of CK 18-positive squamous-cell carcinomas of the larynx and hypopharynx as opposed to essentially CK 18-negative squamous-cell carcinomas of the oral cavity has been proposed by Balm *et al.* (1996); however, their study dealt with a relatively small number of cases and failed to take morphological features and tumor grading into account. In conclusion, although differences between the CK expression patterns of squamous-cell carcinomas from different sites of origin appear to indicate tendencies, any attempt to utilize such differences as *specific* site markers is ruled out not only by their overlapping, but also by the superimposition of modulations related to tumor differentiation [for CK differences between nonkeratinizing and keratinizing cervical squamous cell carcinomas, see Smedts *et al.* (1992a, 1993a)] and tumor grade.

Another field of interest concerns the premalignant stages of squamous-cell carcinomas, i.e., the question of whether diagnostically relevant alterations in CK expression occur in squamous epithelial dysplasia and carcinomas *in situ*. During hyperplasia, metaplasia, and dysplasia of the bronchial epithelium and during the early development of squamous-cell carcinomas, shifts in the expression patterns of certain stratified-epithelial CKs, particularly CK 17 and CK 10, have been reported (Fisseler-Eckhoff *et al.*, 1996). Another observation that is of relevance in this context is the reduced expression of maturation-typical CKs in the event of dysplastic alterations of mucosal stratified epithelia (Klijanienko *et al.*, 1989; Kuruc *et al.*, 1989; Malecha and Miettinen, 1991). In oral stratified squamous epithelia, this change is paralleled by the increased expression of CK 19 and its spread into suprabasal regions (Lindberg and Rheinwald, 1989; Cintorino *et al.*, 1990; Copper *et al.*, 1993). However, the diagnostic value of this phenomenon seems to be limited, since a similar increase in CK 19 levels may occur upon benign conditions such as inflammation (Bosch *et al.*, 1989; Moll, 1993) and hyperplasia (van der Velden *et al.*, 1996). Although in the event of oral epithelial dysplasia, suprabasal CK 19 expression has been consistently observed at the mRNA level, the protein level has been found to be much more variable, indicating disturbances in expression regulation and thereby suggesting that CK 19 protein is not a reliable marker of premalignancy (L. Su *et al.*, 1996b). Also for CK 18, an increase in the level of this CK has been reported in the case of oral dysplasia (Kannan *et al.*, 1994).

Particularly extensive studies have dealt with the expression of CK 8 in preneoplastic and early neoplastic lesions of the uterine cervix [for reviews concerned with CKs and tumor progression in cervical neoplasia, see Smedts *et al.* (1993b) and Nair *et al.* (1995)]. A specific correlation of the CK 8 level with malignant transformation or invasive potential has been suggested by some investigators (Bobrow *et al.*, 1986; Angus *et al.*, 1988), but as immature squamous metaplasia is also (at least focally) CK 8-positive (Malecha and Miettinen, 1992; Moll *et al.*, 1992a; Smedts *et al.*, 1993a), this CK, too, cannot be regarded as a specific marker of malignancy. Nonetheless, immunohistochemical studies using frozen tissue have demonstrated that an increasing grade of cervical intraepithelial neoplasia (CIN) is accompanied by the increased expression of CK 8, CK 18, and also CK 17 (Smedts *et al.*, 1990). The expression of these CKs probably reflects some relationship with endocervical reserve cells [for CK expression in reserve cells, see Gigi-Leitner *et al.* (1986), Weikel *et al.* (1987) and Smedts *et al.* (1992b, 1993a)], but it remains to be clarified whether they could be utilized as specific markers of the progression potential of CIN. The dual CK expression patterns (stratified- *and* simple-epithelial components) observable in endocervical reserve cells (as well as in respiratory basal cells) reflect their bipotent character and histogenetic potential. The question of the possible existence of a relationship linking human papilloma viruses with CK expression in cervical carcinomas and their premalignant stages has been considered by Syrjänen *et al.* (1988).

In conclusion, CK analyses have shown that certain CKs (e.g., CK 5 and CK 14, CK 6 and CK 16) may serve as general markers of squamous-cell carcinomas, as their expression is broadly maintained even in poorly differentiated and metastatic tumors. Other CKs, such as those associated with squamous maturation and those typical of simple epithelia, are more variably expressed, so that their expression may be related to tumor differentiation, keratinization, the grade of malignancy, and, in part, to the site of origin. Practical diagnostic aspects will be discussed in Section 5.

4. THE SIGNIFICANCE OF CYTOKERATIN DIVERSITY IN CARCINOMAS

In the first years after the diversity of IFs had become apparent, the view that the expression of IF proteins is strictly conserved and maintained in the event of malignant transformation, and that with these proteins invariable histogenetic tumor markers had become available, enjoyed widespread acceptance (Gabbiani *et al.*, 1981; Osborn and Weber, 1983). It is indeed true that, in several instances, the specific CK pattern does remain essentially stable even when major cellular alterations such as malignant transformation have occurred. Examples of such conservation include the intestinal epithelium (see Section 3.1) and Merkel cells (for a review, see Moll *et al.*, 1997) and their respective carcinomas, with CK 20 being a prominent, largely stable lineage marker in both instances. Such cases are thus characterized by the high degree of genetic stability of cell-type-specific programs of IF-protein expression. In other epithelia, the expression of at least some key CKs—such as CK 20 along with CK 13 in transitional-cell carcinomas (see Section 3.5), and certain primary stratified-epithelial CKs in squamous-cell carcinomas (see Section 3.6)—remains relatively constant. In other cases, though, the CK (and IF) expression patterns may undergo various modulations and alterations during the course of various exogenous and endogenous processes, such as inflammation, toxic damage, malignant transformation, and tumor progression (see, for example, Gröne *et al.*, 1987; Bosch *et al.*, 1989; Smedts *et al.*, 1990). According to current concepts, the differentiation of a tumor is not equal to its histogenesis (V. E. Gould, 1986), and it appears that IF-protein expression is more closely coupled to the former than to the latter.

Two general principles can be said to have emerged. The first and more expected implies that CK expression diminishes in parallel with a decrease in the degree of differentiation. Examples include the comparative decline—in poorly differentiated as opposed to well-differentiated tumors—in the levels of CK 13 in transitional-cell carcinomas as well as of the keratinization markers CK 1 and CK 10/11 in squamous-cell carcinomas (see Sections 3.5 and 3.6). In highly dedifferentiated or anaplastic carcinomas, CK expression may even be entirely lost. This has been observed in some dedifferentiated renal-cell carcinomas (Beham *et al.*, 1992) and anaplastic thyroid carcinomas (Schröder *et al.*, 1986); in such cases, vimentin may be the only detectable residual IF protein. The second general principle of CK expression in carcinomas is rather more contrary to expectation: upon reduction in the degree of differentiation resulting from nonneoplastic or neoplastic epithelial alteration, the complexity of the expression patterns of CK (and IF) proteins may increase conspicuously. Examples already discussed in the present chapter include the neoexpression of CK 7 in stomach mucosa in the event of chronic inflammatory processes (see Section 3.2), the apparent neoexpression of certain nongastrointestinal CKs in dedifferentiated areas of colorectal carcinomas (Figures 1j, 1k), the increased expression of stratified-epithelial CKs and vimentin in certain poorly differentiated ductal breast carcinomas (Figure 4g; see Section 3.4), and the increase in the levels of simple-epithelial CKs as well as in the overall complexity of the CK expression patterns in poorly differentiated squamous-cell carcinomas (Figure 6; see Section 3.6). A further well-characterized example concerns proximal-tubular epithelial cells of the kidney: while normal adult cells have a very simple CK cytoskeleton comprising just CK 8 and CK 18, chronic damage induces the expression of vimentin, various CKs, and GFAP (Gröne *et al.*, 1987; Moll *et al.*, 1991a), with similar changes occurring after malignant transformation and the development of clear-cell and chromophilic renal-cell carcinomas (Pitz *et al.*, 1987; Moll, 1993). The examples mentioned here illustrate the intriguing principle that, in many instances, nonneoplastic and neoplastic cellular alterations are accompanied by similar changes in CK and IF profiles. It follows that such alterations of CK expression are, in fact, not tumor-specific, but may also be the outcome of nonneoplastic alterations (such as suprabasal CK 19 expression in stratified squamous epithelia; see Section 3.6). In some cases, the changes in CK and IF expression may be explicable in terms of oncofetal atavisms, during which embryonal differentiation is recapitulated (such as the expression of CK 7 in stomach epithelium, and that of vimentin and CK 19 in kidney tubular epithelium). Some forms of novel CK (or IF-protein) expression occurring in the event of decreased differentiation could also be regarded as reflecting (rudimentary) metaplasia, e.g., ductular metaplasia in hepatocytes (van Eyken *et al.*, 1989), and putative rudimentary squamous metaplasia in poorly differentiated breast carcinomas (see Section 3.4; Malzahn *et al.*, 1998).

Similar phenomena of increase of CK proteins (and filaments) upon dedifferentiation may occur in cell-culture models, examples of which include the increased expression of CKs and other cytoskeletal proteins along with the reduced synthesis of liver-specific proteins such as albumin in rat hepatocyte cultures (Ben-Ze´ev *et al.*, 1988), as well as the increase in CK expression observable during the dedifferentiation of cultured type II pneumonocytes (Woodcock-Mitchell *et al.*, 1989). A possible explanation for such occurrences would be a loss of control governing the expression of certain IF proteins, thereby permitting their abnormal neoexpression. It would indeed appear that, in some epithelia, certain IF-protein genes are subject to only rather loose regulatory control, so that they can easily be induced, possibly via the loss of inhibitory regulatory factors causing derepression of the activity of IF-protein genes (Knapp and Franke, 1989). Finally, it is important to remember that, in spite of the variety of possible changes of CK expression and neoexpressions, such alterations are not random, but rather, like basic CK expression patterns in normal tissues, appear to be essentially cell-type-specific. Thus, it is possible to assign to each epithelium an "expression framework" outlining the general limits within which alterations and neoexpressions are more or less confined (Moll, 1993). For example, in the case of the intestinal epithelium, this framework is relatively restricted, including CK 8, CK 18, CK 19, CK 20, and, secondarily, CK 7 and CK 17 (see Section 3.1; Moll *et al.*, 1993); in contrast, noncornifying stratified squamous epithelia have a much broader and very different framework that includes many stratified-epithelial CKs as well as several simple-epithelial CKs (see Section 3.6). Within the confines of such frameworks, modulations of CK and IF expression may in part be determined by epigenetic factors or the influence of the microenvironment. Furthermore, such modifications of IF expression often occur irregularly, resulting in phenotypical heterogeneity or, in the case of neoplasms, tumor-cell heterogeneity (Nicolson, 1987).

It is still unclear which molecular biological mechanisms are responsible for the observed alterations in cellular CK-protein composition. Control of the expression of CK genes is mainly transcriptional (Oshima, 1992; van de Klundert *et al.*, 1993), but in some instances, posttranscriptional mechanisms may be in operation (Tyner and Fuchs, 1986; Knapp and Franke, 1989; L. Su *et al.*, 1994, 1996b). Recently, direct induction of transcription of the CK 18 gene in keratinocytes by an oncogene (activated Ha-Ras) has been demonstrated (Pankov *et al.*, 1994), a finding which may be related to the frequent occurrence of simple-epithelial CKs in squamous-cell carcinomas (see Section 3.6). It remains an open question whether such cytoskeletal changes occurring during tumor progression may influence the biological behavior of tumor cells. One might speculate that in poorly differentiated squamous-cell carcinomas the partial shift to a simple-epithelial CK pattern might result in a cytoskeleton that is easier to remodel (Franke *et al.*, 1983) and would confer on the cells greater cytoplasmic plasticity and higher mobility. However, the reverse shift toward elevated stratified-epithelial CKs observed in some poorly differentiated breast carcinomas (see Section 3.4), which also behave aggressively, suggests that the cellular CK equipment is not of prime importance for tumor cell behavior. From a clinical point of view, it would be particularly interesting—but still remains to be determined—whether or not the CK- and IF-protein changes observed in malignant tumors are prognostically relevant; the preliminary data concerning ductal breast carcinomas presented above (see Section 3.4) do indeed suggest that at least in some instances such alterations may be of some prognostic value.

5. CYTOKERATINS AS DIAGNOSTIC MARKERS IN TUMOR PATHOLOGY

For the majority of pathological tumor diagnoses, conventional morphological analyses based on the application of standard staining procedures to paraffin-embedded sections are perfectly adequate. Not infrequently, though, problematic cases require the employment of further techniques—such as those involving histological tumor markers like CKs—before a correct diagnosis is obtainable. Important areas in which the diagnostic application of CK typing may be of great value include accurate determination of the tumor type in the case of morphologically undifferentiated tumors, along with the identification of the origin of tumors whose spread renders their starting point unclear, or most importantly, in cases of metastases of an unknown primary tumor (Mackay and Ordoñez, 1993; Muir, 1994). The correct identification of such tumors is particularly important in view of the differential tumor-specific therapeutic schemes currently in use; for example, the chemotherapy schedules appropriate for colorectal and ovarian adenocarcinomas are vastly different.

When considered as a group, CKs—as localized using pan-CK antibodies—represent a well-established carcinoma marker capable of demonstrating the epithelial nature of morphologically undifferentiated tumors, such as diffuse stomach carcinomas and lobular breast carcinomas with dissociated growth, nasopharyngeal carcinomas, lymphoepithelial carcinomas, small-cell carcinomas, and embryonal carcinomas (Corson, 1986; Nagle, 1989). Up to now, though, the routine diagnostic application of CK subtyping remains less widespread than it might be, partly perhaps because the procedure is perceived as being rather complex, and also because specific antibodies suitable for routine employment have only recently become available (for further reviews, see Moll, 1993; Nagle, 1994; Schaafsma and Ramaekers, 1994).

The basis for the histodiagnostic application of CK subtyping is the finding that in many cases the expression profiles of CK (and IF) proteins are, despite some overlapping, sufficiently characteristic to allow the identification of certain carcinoma types. In some cases, such as transitional-cell carcinomas, colorectal adenocarcinomas, and Merkel-cell carcinomas, the CK pattern is close to being specific to a particular tumor type. More frequently, though, particular CK expression patterns occur in more than one type of carcinoma, enabling one to narrow down the range of possibilities by ruling out certain carcinoma types or at least rendering them unlikely. When interpreting CK data, the influences of tumor differentiation and grade as well as inter- and intratumoral heterogeneity must always be taken into account. In general, the CK data should be regarded as signposts that, in conjunction with clinical, morphological, and other immunohistochemical tumor-marker data, can guide one toward a correct diagnosis; by their very nature, such data only rarely allow absolute certainty with respect to the nature and origin of a carcinoma, but rather they enable one to make statements regarding the probability of the tumor under examination being of a particular type. In the following, diagnostic guidelines for the subclassification of carcinomas based on typically expressed and diagnostically relevant CKs will be presented; such CKs whose presence (or absence) is of particular diagnostic importance are underlined in Table II.

The identification of *squamous-cell carcinomas*, including even nonkeratinizing tumors exhibiting poor differentiation, is usually possible on the basis of the prevalent expression of stratified-epithelial CKs along with the absence (or low levels) of CK 7 and CK 20 (see Section 3.6). These criteria are in most cases sufficient to distinguish such tumors from poorly differentiated adenocarcinomas and transitional-cell carcinomas as well as from neuroendocrine carcinomas. The most sensitive markers for squamous-cell carcinomas are the primary stratified-epithelial CKs such as CK 5. Monoclonal antibody D5/16B4 directed against CK 5/6 (Lobeck *et al.*, 1989) can be effectively applied to paraffin-embedded sections, preferably after these have undergone microwave-oven pretreatment (Demirkesen *et al.*, 1995; Weikel *et al.*, 1996), as simple protease treatment appears to result in a lower degree of sensitivity as is suggested by the unexpectedly inconsistent CK 5/6 staining results in squamous-cell carcinomas reported by Suo *et al.* (1993). Another antibody that can be used with paraffin-embedded material is KA 12 directed against CK 6, although the sensitivity achieved is inferior to that attainable with D5/16B4 (Weikel *et al.*, 1996). Questions arising concerning the significance of CKs with respect to the site of origin of squamous-cell carcinomas as well as the possible prognostic value of the coexpression of simple-epithelial CKs in such tumors have been discussed in Section 3.6.

The coexpression of CK 7 and CK 20 allows *transitional-cell carcinomas* of the urinary tract to be distinguished from poorly differentiated squamous-cell carcinomas (Figure 2 of Schaafsma and Ramaekers, 1994), while the presence of CK 13 and CK 14 serves to separate such tumors from poorly differentiated adenocarcinomas (see Section 3.5).

The diagnosis of the extensive group of carcinomas exhibiting a simple-epithelial phenotype, notably *adenocarcinomas* of various origins, is a matter of major clinical importance, not least because such tumors are very common. Indeed, about one-half of the metastases encountered whose primary tumor is unknown belong to this group of carcinomas (Muir, 1994). When attempting the subclassification of such tumors, CK typing may be very helpful in conjunction with precise morphological analysis and, if necessary, with other immunohistochemical tumor markers (Krüger *et al.*, 1992; Kaufmann *et al.*, 1996). Obviously, this category of tumors is, as a whole, characterized by the predominance of simple-epithelial CKs. With respect to the more important question of their subtyping, CK 7 and CK 20 have turned out to be especially helpful markers (van Niekerk *et al.*, 1991; Moll *et al.*, 1992b; Loy and Calaluce, 1994; Miettinen, 1995; Wauters *et al.*, 1995; Wang *et al.*, 1995; Kaufmann *et al.*, 1996; Maeda *et al.*, 1996; Berezowski *et al.*, 1996; Loy *et al.*, 1996; Savera *et al.*, 1996), while CK 19, CK 5, and vimentin may also be useful. Features relevant to the diagnosis of *colorectal*, *gastric*, and *pancreatic adenocarcinomas*, including their directly contrary expression ratios of CK 20 and CK 7, have already been discussed (see Sections 3.1–3.3). Among liver tumors, *hepatocellular carcinomas*, like hepatocytes, typically exhibit the predominant expression of CK 8 and CK 18 alone, with CK 7, CK 19, and CK 20 sometimes also being present, but usually only focally (Fischer *et al.*, 1987; van Eyken *et al.*, 1988; Moll *et al.*, 1992b; Maeda *et al.*, 1995, 1996; Schirmacher *et al.*, 1998). Carcinomas of the biliary system, including *cholangiocarcinomas* of the liver as well as *bile-duct* and *gall-bladder carcinomas* (Maeda *et al.*, 1996), are distinguished by much higher levels and usually diffuse expression of CK 7, as well as the (more variable, but often quite extensive) expression of CK 20 (Moll *et al.*, 1992b; Maeda *et al.*, 1996). While these tumors cannot be distinguished from pancreatic and gastric adenocarcinomas with any certainty, the more distinctive features of colorectal adenocarcinomas (CK 20-positive, CK 7-negative) and hepatocellular carcinomas (low levels or absence of CK 7, CK 19, and CK 20) are useful in the differential diagnosis of primary and metastatic liver tumors (Maeda *et al.*, 1996).

Adenocarcinomas of the lung mainly express CK 7, CK 8, CK 18, and CK 19 (Blobel *et al.*, 1984), while CK 20 is usually not detectable in such tumors, although it may be expressed very sparsely or in a tiny minority of cases (Moll *et al.*, 1992b; Miettinen, 1995; Kaufmann *et al.*, 1996; Savera *et al.*, 1996). In addition, stratified-epithelial CKs, such as CK 4, CK 5, and CK 13, are often focally expressed in these adenocarcinomas (Broers *et al.*, 1988; Moll, 1993). The incidence of focal coexpression of vimentin is comparably frequent, being encountered in roughly one-half of cases (Mullink *et al.*, 1986; Moll, 1993). Such vimentin coexpression is a less common finding in gastrointestinal adenocarcinomas. CK 7 may be a particularly useful marker (when applied with paraffin-embedded sections) for distinguishing adenocarcinomas of the lung from pulmonary squamous-cell carcinomas when only small lung-tumor biopsies are available (van de Molengraft *et al.*, 1993). It is important, however, to note that a clear-cut distinction between pulmonary adenocarcinomas and, for example, adenocarcinomas of the epigastrium is rarely possible. The CK patterns of *breast carcinomas* have been described above (see Section 3.4). In the case of metastatic tumors of an unknown primary tumor, the absence of CK 20, in conjunction with the presence of other markers such as steroid hormone receptors and GCDFP-15, may be diagnostically helpful in establishing the breast as the site of origin (Kaufmann *et al.*, 1996). The possible prognostic value of the coexpression of stratified-epithelial CKs and vimentin has been considered above (see Section 3.4).

Particularly pronounced coexpression of vimentin with CKs is a feature quite typical of both columnar epithelia of mesodermal origin and their corresponding tumors, so that this can serve as a useful diagnostic indicator. Thus, while *adenocarcinomas of the endometrium* typically coexpress several simple-epithelial CKs (CK 20 is essentially absent), they also exhibit focal stratified-epithelial-CK expression, as well as the coexpression of vimentin (Moll *et al.*, 1991b, 1992b; Miettinen, 1995; Wang *et al.*, 1995; Wauters *et al.*, 1995; Savera *et al.*, 1996). The focal presence of stratified-epithelial CKs probably reflects the potential of these tumors to develop squamous metaplasia. *Nonmucinous ovarian adenocarcinomas* (including *serous* and *endometrioid types*) display similar IF features, i.e., frequent vimentin expression along with the expression of simple-epithelial CKs, with CK 7 being prominent among the latter, while CK 20 is very rarely encountered (Moll *et al.*, 1991b, 1992b; Ascoli *et al.*, 1995; Miettinen, 1995;

Wang *et al.*, 1995; Wauters *et al.*, 1995; Kaufmann *et al.*, 1996). The finding of CK 20 in about one-half of a group of endometrioid ovarian carcinomas by Berezowski *et al.* (1996) and in 60% of a series of serous ovarian carcinomas by Loy *et al.* (1996) is therefore at variance with most other reported data. These ovarian tumors, like endometrial adenocarcinomas, may occasionally even express GFAP (Moll *et al.*, 1991b). Thus, adenocarcinomas of the female genital tract exhibit a quite typical phenotype of IF proteins, which are of diagnostic value as supplementary markers, in addition to other tumor markers such as the presence of CA125 and steroid hormone receptors. Interestingly, the *mucinous variant of ovarian carcinomas* is very distinctive in that these tumors very frequently express CK 20 (Moll *et al.*, 1991b, 1992b; Ascoli *et al.*, 1995; Miettinen, 1995; Wang *et al.*, 1995; Wauters *et al.*, 1995; Berezowki *et al.*, 1996; Loy *et al.*, 1996). These tumors also usually express CK 7, thereby allowing many, but not all, such cases to be distinguished from metastatic mucinous colorectal adenocarcinomas. Generally, mucinous adenocarcinomas of the ovary express both CK 7 and CK 20, with those of the colorectum only containing CK 20, while such tumors of the breast only exhibit CK 7 (Moll *et al.*, 1992b; Wang *et al.*, 1995).

Renal-cell carcinomas, another example of tumors of mesodermal origin, also coexpress vimentin with simple-epithelial CKs (Pitz *et al.*, 1987; Beham *et al.*, 1992; for further references, see Schaafsma and Ramaekers, 1994): CK 20 is again absent (Moll *et al.*, 1992b; Miettinen, 1995; Wang *et al.*, 1995; Kaufmann *et al.*, 1996; Savera *et al.*, 1996), and the CK pattern is somewhat simpler than in the tumors discussed above (Pitz *et al.*, 1987; Moll, 1993). If vimentin is undetectable in a renal-cell tumor, this argues in favor of it being a tumor of the chromophobe-cell type or a benign oncocytoma (Pitz *et al.*, 1987; Thoenes *et al.*, 1988). In the case of *malignant diffuse mesotheliomas* showing epithelial differentiation, the expression of simple-epithelial CKs in conjunction with both the frequent expression of CK 5 and the coexpression of vimentin may be diagnostically useful for separating them from several types of adenocarcinoma (Blobel *et al.*, 1985; Moll *et al.*, 1989; for further references on mesotheliomas see Schaafsma and Ramaekers, 1994). The typical CK patterns seen in further types of *adenocarcinomas*, such as those of the *prostate gland* (e.g., Okada *et al.*, 1992; Verhagen *et al.*, 1992; Schaafsma and Ramaekers, 1994; Nagle, 1996; Yang *et al.*, 1997) and *thyroid gland* (Schaafsma and Ramaekers, 1994; Schröder *et al.*, 1996; Fonseca *et al.*, 1997; see there for further references), as well as in *salivary-gland tumors* (Schaafsma and Ramaekers, 1994) have been described and discussed in detail by other investigators.

A relatively rare differential diagnostic problem in which, though—in the case of skin tumors—CK typing is very reliable and helpful is the distinction of *Merkel-cell carcinomas*, expressing CK 20, from metastatic *small-cell anaplastic carcinomas of the lung*, which lack this CK, although both tumors are neuroendocrine in nature and may be morphologically similar (Moll *et al.*, 1997).

It should again be borne in mind that, on the basis of CK analysis alone, it is usually not possible to make a definite distinction between benign and malignant lesions (see Section 3). However, in certain cases, this question may be answered indirectly as a result of the specific staining, and thus identification, of nonneoplastic basal cells of certain glandular epithelia by antibodies directed against primary stratified-epithelial (basal-cell-typical) CKs. This may, for example, aid in the diagnosis of prostatic adenocarcinomas, in which invasive neoplastic glands lack such basal cells (Shah *et al.*, 1991; Nagle, 1996). In our experience, antibody D5/16B4 against CK 5/6 is well suited here in that it clearly delineates these basal cells in routine biopsies (see Section 3.4 for the specific staining by such antibodies of nonneoplastic basal/myoepithelial cells of the mammary gland).

Although CKs are both numerous and diverse, just a small panel of antibodies directed against a limited range of "key CKs" and against vimentin are sufficient to perform most of the above-mentioned differential diagnoses (see the underlined symbols in Table II). Monospecific, monoclonal antibodies that perform satisfactorily when applied to routine paraffin-embedded sections, preferably after microwave-oven enhancement (Shi *et al.*, 1991), have been already mentioned (cf. also Cattoretti *et al.*, 1993; Hazelbag *et al.*, 1995); for suitable antibodies directed against CK 5/6, CK 7, CK 8, CK 10/11, CK 13, CK 17, CK 19, and CK 20, see Moll *et al.* (1993), Demirkesen *et al.* (1995), and Weikel *et al.* (1996).

A further diagnostic application of CK antibodies involves their use in serological assays to detect soluble CK fragments released from carcinoma cells; in fact, tissue polypeptide antigen (TPA), which has long served as a serological tumor marker, corresponds to such fragments (Moll, 1994). It has been demonstrated that the detectable presence of fragments of CK 19 is of clinical value for the assessment of the clinical response and the prognosis in squamous-cell carcinomas of the lung (e.g., Pujol *et al.*, 1993; Takada *et al.*, 1995).

6. CONCLUSIONS AND PERSPECTIVES

In the fields of cell biology and pathology, CKs as a group have gained widespread acceptance as very sensitive epithelial-cell and tumor markers, even though their possible occurrence in certain mesenchymal cells and tumors needs to be borne in mind. Indeed, pan-CK antibodies are now to be found among the basic reagents of antibody panels routinely used in most pathology laboratories.

As already pointed out, the subtyping of individual CK proteins—as opposed to pan-CK analyses—is a procedure that greatly increases the potential discriminatory capability of these IFs with respect to the identification of carcinoma types; however, until recently, the practical application of such procedures was greatly hampered by the lack of antibodies suitable for routine use. Antigen-retrieval techniques developed in the past few years, most notably microwave-oven pretreatment (Shi *et al.*, 1991) and pressure cooking, are now capable of achieving sensitivity levels in formalin-fixed, paraffin-embedded tissue that are superior even to those attainable in frozen sections. Several newly established CK antibodies yield highly sensitive as well as specific staining results of excellent quality when applied in conjunction with such techniques. Thus, in recent years, effective reagents have been evolved that make it possible to perform much more subtle CK analyses on routine paraffin-embedded blocks than was previously feasible using native tissue and elaborate two-dimensional gel electrophoresis. Since these reagents are widely available and can be applied without difficulty in virtually any standard pathology laboratory, it will very soon be evident for which indications and differential diagnostic questions specific CK antibodies will become as diagnostically indispensable as pan-CK antibodies. It is thus foreseeable that the increased application of such antibodies will result in a significant and clinically relevant improvement in the diagnosis of carcinomas, as compared to the present situation using morphological criteria and other immunohistochemical markers. However, if monospecific CK antibodies are to achieve widespread acceptance and be applied for diagnostic purposes in laboratories throughout the world, it is crucial that a general consensus of agreement be reached with respect to which particular antibodies are most effective, while the standardization of immunohistochemical methods, particularly enhancement techniques, is important.

In addition to their value in characterizing the type and origin of carcinomas as well as features of differentiation and tumor-cell heterogeneity, there are grounds for thinking that CK patterns may well exhibit correlations with both the biological behavior of tumors and their prognosis. However, the use of specific CKs as first-choice prognostic markers would require proof of their clinical relevance as revealed by multivariate statistical analyses. With the exception of their regulatory links with certain oncogenes revealed in recent studies (Pankov *et al.*, 1994), CK proteins appear to be only indirectly connected with the molecular events that occur during tumorigenesis and tumor progression.

Ongoing cell-biological studies aimed at elucidating the incompletely understood functions of the diverse CKs can be expected to increase our knowledge of epithelial differentiation and may also contribute to a broader comprehension of the pathological processes that take place during the development and growth of malignant tumors.

ACKNOWLEDGMENTS. The help of Andrew Cowin (Heidelberg) in correcting the English grammar and the assistance of Iris Haag and Karin Greinwald with the typing of this manuscript are gratefully acknowledged.

7. REFERENCES

Altmannsberger, H., Dirk, T., Droese, M., Weber, K., and Osborn, M., 1986, Keratin polypeptide distribution in benign and malignant breast tumors: Subdivision of ductal carcinomas using monoclonal antibodies, *Virchows Arch. B Cell Pathol.* **51:**265–275.

Angus, B., Kiberu, R., Purvis, J., Wilkinson, K., and Horne, C. H. W., 1988, Cytokeratins in cervical dysplasia and neoplasia: A comparative study of immunohistochemical staining using monoclonal antibodies NCL-5D3, CAM 5.2, and PKK1, *J. Pathol.* **155:**71–75.

Ascoli, V., Taccogna, S., Scalzo, C. C., and Nardi, F., 1995, Utility of cytokeratin 20 in identifying the origin of metastatic carcinomas in effusions, *Diagn. Cytopathol.* **12:**303–308.

Balm, A. J. M., Hageman, P. C., van Doornewaard, M. H., Groeneveld, E. M., and Ivanyi, D., 1996, Cytokeratin 18 expression in squamous cell carcinoma of the head and neck, *Eur. Arch. Otorhinolaryngol.* **253:**227–233.

Bartek, J., Bartkova, J., Taylor-Papadimitriou, J., Rejthar, A., Kovarik, J., Lukas, Z., and Vojtesek, B., 1986, Differential expression of keratin 19 in normal human epithelial tissues revealed by monospecific monoclonal antibodies, *Histochem. J.* **18:**565–575.

Bartek, J., Vojtesek, B., Staskova, Z., Batkova, J., Kerekes, Z., Rejthar, A., and Kovarik, J., 1991, A series of 14 new monoclonal antibodies to keratins: Characterization and value in diagnostic histopathology, *J. Pathol.* **164:**215–224.

Beham, A., Ratschek, M., Zatloukal, K., Schmid, C., and Denk, H., 1992, Distribution of cytokeratins, vimentin and desmoplakins in normal renal tissue, renal cell carcinomas and oncocytoma as revealed by immunofluorescence microscopy, *Virchows Arch. A Pathol. Anat.* **421:**209–215.

Ben-Ze'ev, A., Robinson, G. S., Bucher, N. L. R., and Farmer, S. R., 1988, Cell–cell and cell–matrix interactions differentially regulate the expression of hepatic and cytoskeletal genes in primary cultures of rat hepatocytes, *Proc. Natl. Acad. Sci. USA* **85:**2161–2165.

Berezowski, K., Stastny, J. F., and Kornstein M. J., 1996, Cytokeratins 7 and 20 and carcinoëmbryonic antigen in ovarian and colonic carcinoma, *Mod. Pathol.* **9**:426–429.

Blobel, G. A., Moll, R., Franke, W. W., and Vogt-Moykopf, I., 1984, Cytokeratins in normal lung and lung carcinomas: I. Adenocarcinomas, squamous cell carcinomas and cultured cell lines, *Virchows Arch. B Cell Pathol.* **45:**407–429.

Blobel, G. A., Moll, R., Franke, W. W., Kayser, K. W., and Gould, V. E., 1985, The intermediate filament cytoskeleton of malignant mesotheliomas and its diagnostic significance, *Am. J. Pathol.* **121:**235–247.

Bobrow, L. G., Makin, C. A., Law, S., and Bodmer, W. F., 1986, Expression of low molecular weight cytokeratin proteins in cervical neoplasia, *J. Pathol.* **148:**135–140.

Böcker, W., Bier, B., Freytag, G., Brömmelkamp, B., Jarasch, E.-D., Edel, G., Dockhorn-Dworniczak, B., and Schmid, K. W., 1992a, An immunohistochemical study of the breast using antibodies to basal and luminal keratins, alpha-smooth muscle actin, vimentin, collagen IV and laminin. Part I: Normal breast and benign proliferative lesions, *Virchows Arch. A Pathol. Anat.* **421:**315–322.

Böcker, W., Bier, B., Freytag, G., Brömmelkamp, B., Jarasch, E.-D., Edel, G., Dockhorn-Dworniczak, B., and Schmid, K. W., 1992b, An immunohistochemical study of the breast using antibodies to basal and luminal keratins, alpha-smooth muscle actin, vimentin, collagen IV and laminin. Part II: Epitheliosis and ductal carcinoma *in situ*, *Virchows Arch. A Pathol. Anat.* **421:**323–330.

Bosch, F. X., Leube, R. E., Achtstätter, T., Moll, R., and Franke, W. W., 1988, Expression of simple epithelial type cytokeratins in stratified epithelia as detected by immunolocalization *in situ*, *J. Cell Biol.* **106:**1635–1648.

Bosch, F. X., Ouhayoun, J.-P., Bader, B. L., Collin, C., Grund, C., Lee, I., and Franke, W. W., 1989, Extensive changes in cytokeratin expression patterns in pathologically affected human gingiva, *Virchows Arch. B Cell Pathol.* **58:**59–77.

Broers, J. L. V., Ramaekers, F. C. S., Klein Rot, M., Oostendorp, T., Huysmans, A., Van Muijen, G. N. P., Wagenaar, S. S., and Vooijs, G. P., 1988, Cytokeratins in different types of human lung cancer as monitored by chain-specific monoclonal antibodies, *Cancer Res.* **48:**3221–3229.

Bryne, M., Koppang, H. S., Lilleng, R., Stene, T., Bang, G., and Dabelsteen, E., 1989, New malignancy grading is a better prognostic indicator than Broders' grading in oral squamous cell carcinomas, *J. Oral Pathol. Med.* **18:**432–437.

Cattoretti, G., Pileri, S., Parravicini, C., Becker, M. H. G., Poggi, S., Bifulco, C., Key, G., D'Amato, L., Sabattini, E., Feudale, E., Reynolds, F., Gerdes, J., and Rilke, F., 1993, Antigen unmasking on formalin-fixed, paraffin-embedded tissue sections, *J. Pathol.* **171:**83–98.

Cintorino, M., Del Vecchio, M. T., Bugnoli, M., Petracca, R., and Leoncini, P., 1988, Cytokeratin pattern in normal and pathological urothelium: Immunohistochemical investigation using monoclonal antibodies, *J. Urol.* **139:**428–432.

Cintorino, M., Petracca, R., Vindigni, C., Tripodi, S. A., and Leoncini, P., 1990, Topography-related expression of individual cytokeratins in normal and pathological (non-neoplastic and neoplastic) human oral mucosa, *Virchows Arch. A Pathol. Anat.* **417:**419–426.

Cohen, R., Guber, A., Siegal, A., Bruderman, I., Huszar, M., Yellin, A., Marom, Z., and Geiger, B., 1997, The prognostic application of cytokeratin typing of nonsmall cell lung carcinoma. A retrospective study, *Cancer* **79:**468–473.

Cooper, D., Schermer, A., and Sun, T.-T., 1985, Classification of human epithelia and their neoplasms using monoclonal antibodies to keratins: Strategies, applications and limitations, *Lab. Invest.* **52:**243–256.

Copper, M. P., Braakhuis, B. J., de Vries, N., van Dongen, G. A., Nauta, J. J., and Snow, G. B., 1993, A panel of biomarkers of carcinogenesis of the upper aerodigestive tract as potential intermediate endpoints in chemoprevention trials, *Cancer* **71:**825–830.

Corson, M., 1986, Keratin protein immunohistochemistry in surgical pathology practice, in: *Pathology Annual*, Vol. 21, Part 2 (S. C. Sommers, P. P. Rosen, and R. E. Fechner, eds.), pp. 49–81, Appleton-Century-Crofts, Norwalk, Connecticut.

Cross, H. S., Bajna, E., Bises, G., Genser, D., Kállay, E., Pötzi, R., Wenzl, E., Wrba, F., Roka, R., and Peterlik, M., 1996, Vitamin D receptor and cytokeratin expression may be progression indicators in human colon cancer, *Anticancer Res.* **16:**2333–2338.

Dairkee, S. H., Ljung, B. M., Smith, H., and Hackett, A., 1987, Immunolocalization of a human basal epithelium specific keratin in benign and malignant breast disease, *Breast Cancer Res. Treat.* **10:**11–20.

De Araujo, V. C., Pinto, D. S., Jr., de Sousa, S. O. M., Nunes, F. D., and de Araujo, N. S., 1993, Vimentin in oral squamous cell carcinoma, *Eur. Arch. Otorhinolaryngol.* **250:**105–109.

Debus, E., Moll, R., Franke, W. W., Weber, K., and Osborn, M., 1984, Immunohistochemical distinction of human carcinomas by cytokeratin typing with monoclonal antibodies, *Am. J. Pathol.* **114:**121–130.

Deck, J. H. N., Eng, L. F., Bigbee, J., and Woodcock, S. M., 1978, The role of glial fibrillary acidic protein in the diagnosis of central nervous system tumors, *Acta Neuropathol.* **42:**183–190.

Demirkesen, C., Hoede, N., and Moll, R., 1995, Epithelial markers and differentiation in adnexal neoplasms of the skin: An immunohistochemical study including individual cytokeratins, *J. Cutan. Pathol.* **22:**518–535.

Domagala, W., Lasota, J., Bartkowiak, J., Weber, K., and Osborn, M., 1990, Vimentin is preferentially expressed in human breast carcinomas with low estrogen receptor and high Ki-67 growth fraction, *Am. J. Pathol.* **136:**219–227.

Fischer, H.-P., Altmannsberger, M., Weber, K., and Osborn, M., 1987, Keratin polypeptides in malignant epithelial liver tumors. Differential diagnostic and histogenetic aspects, *Am. J. Pathol.* **127:**530–537.

Fischer, H.-P., Wallner, F., Maier, H., Weber, K., Osborn, M., and Altmannsberger, M., 1989, Coexpression of intermediate filaments in squamous cell carcinomas of upper aerodigestive tract before and after radiation and chemotherapy, *Lab. Invest.* **61:**433–439.

Fisseler-Eckhoff, A., Erfkamp, S., and Müller, K.-M., 1996, Cytokeratin expression in preneoplastic lesions and early squamous cell carcinoma of the bronchi, *Pathol. Res. Pract.* **192:**552–559.

Fonseca, E., Nesland, J. M., Höie, J., and Sobrinho-Simões, M., 1997, Pattern of expression of intermediate cytokeratin filaments in the thyroid gland: An immunohistochemical study of simple and stratified epithelial-type cytokeratins, *Virchows Arch.* **430:**239–245.

Franke, W. W., Schiller, D. L., Moll, R., Winter, S., Schmid, E., Engelbrecht, I., Denk, H., Krepler, R., and Platzer, B., 1981, Diversity of cytokeratins. Differentiation specific expression of cytokeratin polypeptides in epithelial cells and tissues, *J. Mol. Biol.* **153:**933–959.

Franke, W. W., Schiller, D. L., Hatzfeld, M., and Winter, S., 1983, Protein complexes of intermediate-sized filaments: Melting of cytokeratin complexes in urea reveals different polypeptide separation characteristic, *Proc. Natl. Acad. Sci. USA* **80:**7113–7117.

Fuchs, E., and Green, H., 1980, Changes in keratin gene expression during terminal differentiation of the keratinocytes, *Cell* **19:**1033–1042.

Gabbiani, G., Kapanci, Y., Barazzone, O., and Franke, W. W., 1981, Immunochemical identification of intermediate-sized filaments in human neoplastic cells, *Am. J. Pathol.* **104:**206–216.

Gatter, K. C., Dunnill, M. S., van Muijen, G. N. P., and Mason, D. Y., 1986, Human lung tumours may coexpress different classes of intermediate filaments, *J. Clin. Pathol.* **39:**950–954.

Gigi-Leitner, O., Geiger, B., Levy, R., and Czernobilsky, B., 1986, Cytokeratin expression in squamous metaplasia of the human uterine cervix, *Differentiation* **31:**191–205.

Gillenkirch, R., 1994, Biochemische und immuncytochemische Analyse der Cytokeratin-Polypeptid-Muster von Plattenepithelcarcinomen verschiedener Herkunft, M.D. thesis, Johannes Gutenberg University of Mainz, Germany.

Gould, P. R., Barter, R. A., and Papadimitriou, J. M., 1979, An ultrastructural, cytochemical, and autoradiographic study of the mucous membrane of the human cervical canal with reference to subcolumnar reserve cells, *Am. J. Pathol.* **95:**1–9.

Gould, V. E., 1986, Histogenesis and differentiation. A re-evaluation of these concepts as criteria for the classification of tumors, *Hum. Pathol.* **17:**212–215.

Gould, V. E., Koukoulis, G. K., Jansson, D. S., Nagle, R. B., Franke, W. W., and Moll, R., 1990, Coexpression patterns of vimentin and glial filament protein with cytokeratins in the normal, hyperplastic and neoplastic breast, *Am. J. Pathol.* **137:**1143–1155.

Gröne, H. J., Weber, K., Gröne, E., Helmchen, U., and Osborn, M., 1987, Coexpression of keratin and vimentin in damaged and regenerating tubular epithelia of the kidney, *Am. J. Pathol.* **129:**1–8.

Guelstein, V. I., Tchypysheva, T. A., Ermilova, V. D., Litvinova, L. V., Troyanovsky, S. M., and Bannikov, G. A., 1988, Monoclonal antibody mapping of keratins 8 and 17 and of vimentin in normal human mammary gland, benign tumors, dysplasias and breast cancer, *Int. J. Cancer* **42:**147–153.

Guelstein, V. I., Tchypysheva, T. A., Ermilova, V. D., and Troyanowsky, S. M., 1993, Immunohistochemical localization of cytokeratin 17 in transitional cell carcinomas of the human urinary tract, *Virchows Arch. B Cell Pathol.* **64:**1–5.

Gupta, A., Malati, T., and Gupta, P.D., 1997, Differential reduction in the expression of keratin polypeptides in human gastric carcinomas, *Cancer Detect. Prevent.* **21:**129–134.

Gustmann, C., Altmannsberger, M., Osborn, M., Griesser, H., and Feller, A. C., 1991, Cytokeratin expression and vimentin content in large cell anaplastic lymphomas and other non-Hodgkin's lymphomas, *Am. J. Pathol.* **138:**1413–1422.

Harnden, P., Allam, A., Joyce, A. D., Patel, A., Selby, P., and Southgate, J., 1995, Cytokeratin 20 expression by non-invasive transitional cell carcinomas: Potential for distinguishing recurrent from non-recurrent disease, *Histopathology* **27:**169–174.

Harnden, P., Eardley, I., Joyce, A. D., and Southgate, J., 1996, Cytokeratin 20 as an objective marker of urothelial dysplasia, *Br. J. Urol.* **78:**870–875.

Hatzfeld, M., and Weber, K., 1990, The coiled coil of *in vitro* assembled keratin filaments is a heterodimer of type I and type II keratins; use of site-specific mutagenesis and recombinant protein expression, *J. Cell Biol.* **110:**1199–1210.

Hazelbag, H. M., v. d. Broek, L. J. C. M., van Dorst, E. B. L., Offerhaus, G. J. A., Fleuren, G. J., and Hogendoorn, P. C. W., 1995, Immunostaining of chain-specific keratins on formalin-fixed, paraffin-embedded tissues: A comparison of various antigen retrieval systems using microwave heating and proteolytic pre-treatments, *J. Histochem. Cytochem.* **43:**429–437.

Heatley, M., Maxwell, P., Whiteside, C., and Toner, P., 1995, Cytokeratin intermediate filament expresssion in benign and malignant breast disease, *J. Clin. Pathol.* **48:**26–32.

Hendrix, M. J. C., Seftor, E. A., Chu, Y.-W., Trevor, K. T., and Seftor, R. E. B, 1996, Role of intermediate filaments in migration, invasion and metastasis, *Cancer Metast. Rev.* **15:**507–525.

Henzen-Logmans, S. C., Balm, A. J. M., van der Waal, I., Mullink, H., Snow, G. B., and Meyer, C. J. L. M., 1988, The expression of intermediate filaments and MAM-6 antigen in relation to the degree of morphologic differentiation of carcinoma of the head and neck: Diagnostic implications, *Otolaryngol. Head Neck Surg.* **99:**539–547.

Herzig, K. H., Altmannsberger, M., and Folsch, U. R., 1994, Intermediate filaments in rat pancreatic acinar tumors, human ductal carcinomas, and other gastrointestinal malignancies, *Gastroenterology* **106:**1326–1332.

Heyden, A., Huitfeldt, H. S., Koppang, H. S., Thrane, P. S., Bryne, M., and Brandtzaeg, P., 1992, Cytokeratins as epithelial differentiation markers in premalignant and malignant oral lesions, *J. Oral Pathol. Med.* **21:**7–11.

Huszar, M., Gigi-Leitner, O., Moll, R., Franke, W. W., and Geiger, B., 1986, Monoclonal antibodies to various acidic (type I) cytokeratins of stratified epithelia: Selective markers for stratification and squamous cell carcinomas, *Differentiation* **31:**141–153.

Iida, F., and Kusama, J., 1982, Gastric carcinoma and intestinal metaplasia. Significance of types of intestinal metaplasia upon development of gastric carcinoma, *Cancer* **50:**2854–2858.

Ivanyi, D., Groeneveld, E., van Doornewaard, G., Mooi, W. J., and Hageman, P. C., 1990, Keratin subtypes in carcinomas of the uterine cervix: Implications for histogenesis and differential diagnosis, *Cancer Res.* **50:**5143–5152.

Jarasch, E.-D., Nagle, R. B., Kaufmann, M., Maurer, C., and Böcker, W. J., 1988, Differential diagnosis of benign epithelial proliferations and carcinomas of the breast using antibodies to cytokeratins, *Hum. Pathol.* **19:**276–289.

Kannan, S., Balaram, P., Chandran, G. J., Pillai, M. R., Mathew, B., Nalinakumari, K. R., and Nair, M. K., 1994, Differential expression of cytokeratin proteins during tumour progression in oral mucosa, *Epithelial Cell Biol.* **3:**61–69.

Kasper, M., Hahn von Dorsche, H., and Stosiek, P., 1991, Changes in the distribution of intermediate filament proteins and collagen IV in fetal and adult human pancreas, *Histochemistry* **96:**271–277.

Kaufmann, O., Deidesheimer, T., Muehlenberg, M., Deicke, P., and Dietel, M., 1996, Immunohistochemical differentiation of metastatic breast carcinomas from metastatic adenocarcinomas of other common primary sites, *Histopathology* **29:**233–240.

Klijanienko, J., Micheau, C., Carlu, C., and Caillaud, J. M., 1989, Significance of keratin 13 and 6 expression in normal, dysplastic and malignant squamous epithelium of pyriform fossa, *Virchows Arch. A Pathol. Anat.* **416:**121–124.

Klijanienko, J., El-Naggar, A., de Braud, F., Micheau, Ch., Janot, F., Luboinski, B., Gentile, A., Russo, A., and Cvitkovic, E., 1993, Keratins 6, 13 and 19: Differential expression in squamous cell carcinoma of the head and neck, *Anal. Quant. Cytol. Histol.* **15:**335–340.

Knapp, A. C., and Franke, W. W., 1989, Spontaneous losses of control of cytokeratin gene expression in transformed, non-epithelial human cells occurring at different levels of regulation, *Cell* **59:**67–79.

Krüger, R., de Leon, F., and Maihoff, J., 1992, Der Wert der Immunhistochemie in der histologischen Routinediagnostik von Metastasen unbekannter Primärtumoren, *Pathologe* **13:**65–72.

Kuruc, N., Leube, R. E., Moll, I., Bader, B. L., and Franke, W. W., 1989, Synthesis of cytokeratin 13, a component characteristic of internal stratified epithelia, is not induced in human epidermal tumors, *Differentiation* **42:**111–123.

Lane, E. B., and Alexander, C. M., 1990, Use of keratin antibodies in tumor diagnosis, *Semin. Cancer Biol.* **1:**165–179.

Lavrijsen, A. P. M., Tieben, L. M., Ponec, M., van der Schroeff, J. G., and van Muijen, G. N. P., 1989, Expression of EGF receptor, involucrin, and cytokeratins in basal cell carcinomas and squamous cell carcinomas of the skin, *Arch. Dermatol. Res.* **281:**83–88.

Leube, R. E., and Rustad, T. J., 1991, Squamous cell metaplasia in the human lung: Molecular characteristics of epithelial stratification, *Virchows Arch. B Cell Pathol.* **61:**227–253.

Levy, R., Czernobilsky, B., and Geiger, B., 1992, Cytokeratin polypeptide in gastrointestinal adenocarcinomas displaying squamous differentiation, *Hum. Pathol.* **23:**695–702.

Lindberg, K., and Rheinwald, J. G., 1989, Suprabasal 40 kd keratin (K 19) expression as an immunohistologic marker of premalignancy in oral epithelium, *Am. J. Pathol.* **134:**89–98.

Lindemann, F., Schlimok, G., Dirschedl, P., Witte, J., and Riethmüller, G., 1992, Prognostic significance of micrometastatic tumour cells in bone marrow of colorectal cancer patients, *Lancet* **340**(8821)**:**685–689.

Litzky, L. A., and Brooks, J. J., 1992, Cytokeratin immunoreactivity in malignant fibrous histiocytoma and spindle cell tumors: Comparison between frozen and paraffin-embedded tissues, *Mod. Pathol.* **5:**30–34.

Lobeck, H., Bartke, I., Naujoks, K., Müller, D., Bornhöft, G., Mischke, D., and Wild, G., 1989, Verteilungsmuster der Zytokeratinpolypeptide 4 und 5 im normalen und neoplastischen Epithel unter Verwendung neuer paraffingängiger monoklonaler Antikörper (eine immunhistochemische Untersuchung), *Verh. Dtsch. Ges. Pathol.* **73:**645.

Löning, T., Staquet, M.-J., Thivolet, J., and Seifert, G., 1980, Keratin polypeptide distribution in normal and diseased human epidermis and oral mucosa, *Virchows Arch. A Pathol. Anat.* **388:**273–288.

Loy, T. S., and Calaluce, R. D., 1994, Utility of cytokeratin immunostaining in separating pulmonary adenocarcinomas from colonic adenocarcinomas, *Am. J. Clin. Pathol.* **102:** 764–767.

Loy, T. S., Calaluce, R. D., and Keeney, G. L., 1996, Cytokeratin immunostaining in differentiating primary ovarian carcinoma from metastatic colonic adenocarcinoma, *Mod. Pathol.* **9:**1040–1044.

Mackay, B., and Ordoñez, N. G., 1993, Pathological evaluation of neoplasms with unknown primary origin, *Semin. Oncol.* **20:**206–228.

Maeda, T., Adachi, E., Kajiyama, K., Sugimachi, K., and Tsuneyoshi, M., 1995, Combined hepatocellular and cholangiocarcinoma: Proposed criteria according to cytokeratin expression and analysis of clinicopathologic features, *Hum. Pathol.* **26:**956–964.

Maeda, T., Kajiyama, K., Adachi, E., Takenaka, K., Sugimachi, K, and Tsuneyoshi M, 1996, The expression of cytokeratins 7, 19, and 20 in primary and metastatic carcinomas of the liver, *Mod. Pathol.* **9:**901–909.

Malecha, M. J., and Miettinen, M., 1991, Expression of keratin 13 in human epithelial neoplasms, *Virchows Arch. A Pathol. Anat.* **418:**249–254.

Malecha, M. J., and Miettinen, M., 1992, Patterns of keratin subsets in normal and abnormal uterine cervical tissues. An immunohistochemical study, *Int. J. Gynecol. Pathol.* **11:**24–29.

Mallofré, C., Cardesa, A., Campo, E., Condom, E., Palacin, A., Garin-Chesa, P., and Traserra, J., 1993, Expression of cytokeratins in squamous cell carcinomas of the larynx: Immunohistochemical analysis and correlation with prognostic factors, *Pathol. Res. Pract.* **189:**275–282.

Malzahn, K., Mitze, M., Thoenes, M., and Moll, R., 1998, Biological and prognostic significance of stratified epithelial cytokeratins in infiltrating ductal breast carcinomas, *Virchows Arch.*, in press.

Markey, A. C., Lane, E. B., Churchill, L. J., MacDonald, D. M., and Leigh, I. M., 1991, Expression of simple epithelial keratins 8 and 18 in epidermal neoplasia, *J. Invest. Dermatol.* **97:**763–770.

Martin de las Mulas, J., Espinosa de los Monteros, A., Bautista, M. J., Gomez-Villamandos, J. C., and Morales, C., 1994, Immunohistochemical distribution pattern of intermediate filament proteins and muscle actin in feline and human mammary carcinomas, *J. Comp. Pathol.* **111:**365–381.

McDowell, E. M., Combs, J. W., and Newkirk, C., 1983, A quantitative light and electron microscopic study of hamster tracheal epithelium with special attention to so-called "intermediate cells," *Exp. Lung Res.* **4:**205–226.

Miettinen, M., 1991, Keratin subsets in spindle cell sarcomas. Keratins are widespread but synovial sarcoma contains a distinctive keratin polypeptide pattern and desmoplakins, *Am. J. Pathol.* **138:**505–513.

Miettinen, M., 1995, Keratin 20: Immunohistochemical marker for gastrointestinal, urothelial, and Merkel cell carcinomas, *Mod. Pathol.* **8:**384–388.

Miettinen, M., and Franssila, K., 1989, Immunohistochemical spectrum of malignant melanoma. The common presence of keratins, *Lab. Invest.* **61:**623–628.

Moll, R., 1992, Molecular diversity of cytokeratins: Significance for cell and tumor differentiation, *Acta Histochem.* **41:**117–127.

Moll, R., 1993, *Cytokeratins as Markers of Differentiation: Expression Profiles in Epithelia and Epithelial Tumors*, Gustav Fischer, Stuttgart.

Moll, R., 1994, Cytokeratins in the histological diagnosis of malignant tumors, *Int. J. Biol. Markers* **9:**63–69.

Moll, R., and Franke, W. W., 1986, Cytochemical cell typing of metastatic tumors according to their cytoskeletal proteins, in *Biochemistry and Molecular Genetics of Cancer Metastasis* (K. Lapis, L. A. Liotta, and A. S. Rabson, eds.), pp. 101–114, Martinus Nijhoff, Boston.

Moll, R., Franke, W. W., Schiller, D. L., Geiger, B., and Krepler, R., 1982a, The catalog of human cytokeratins: Patterns of expression in normal epithelia, tumors and cultured cells, *Cell* **31:**11–24.

Moll, R., Moll, I., and Wiest, W., 1982b, Changes in the pattern of cytokeratin polypeptides in epidermis and hair follicles during skin development in human fetuses, *Differentiation* **23:**170–178.

Moll, R., Krepler, R., and Franke, W. W., 1983a, Complex cytokeratin polypeptide patterns observed in certain human carcinomas, *Differentiation* **23:**256–269.

Moll, R., Levy, R., Czernobilsky, B., Hohlweg-Majert, P., Dallenbach-Hellweg, G., and Franke, W. W., 1983b, Cytokeratins of normal epithelia and some neoplasms of the female genital tract, *Lab. Invest.* **49:**599–610.

Moll, R., Moll, I., and Franke, W. W., 1984, Differences of expression of cytokeratin polypeptides in various epithelial skin tumors, *Arch. Dermatol. Res.* **276:**349–363.

Moll, R., Achtstätter, T., Becht, E., Balcarova-Ständer, J., Ittensohn, M., and Franke, W. W., 1988, Cytokeratins in normal and malignant transitional epithelium: Maintenance of expression of urothelial differentiation features in transitional cell carcinomas and bladder carcinoma cell culture lines, *Am. J. Pathol.* **132:**123–144.

Moll, R., Dhouailly, D., and Sun, T.-T., 1989, Expression of keratin 5 as a distinctive feature of epithelial and biphasic mesotheliomas: An immunohistochemical study using monoclonal antibody AE 14, *Virchows Arch. B Cell Pathol.* **58:**129–145.

Moll, R., Schiller, D. L., and Franke, W. W., 1990, Identification of protein IT of the intestinal cytoskeleton as a novel type I cytokeratin with unusual properties and expression patterns, *J. Cell Biol.* **111:**567–580.

Moll, R., Hage, C., and Thoenes, W., 1991a, Expression of intermediate filament proteins in fetal and adult human kidney: Modulations of intermediate filament patterns during development and in damaged tissue, *Lab. Invest.* **65:**74–86.

Moll, R., Pitz, S., Levy, R., Weikel, W., Franke, W. W., and Czernobilsky, B., 1991b, Complexity of expression of intermediate filament proteins, including glial filament protein, in endometrial and ovarian adenocarcinomas, *Hum. Pathol.* **22:**989–1001.

Moll, R., Baumann, R., and Hage, C., 1992a, Differentiation markers in the early detection of cancer, in *Cancer Diagnosis: Early Detection* (P. Bannasch, ed.), pp. 163–171, Springer, Berlin.

Moll, R., Löwe, A., Laufer, J., and Franke, W. W., 1992b, Cytokeratin 20 in human carcinomas: A new histodiagnostic marker detected by monoclonal antibodies, *Am. J. Pathol.* **140:**427–447.

Moll, R., Zimbelmann, R., Goldschmidt, M. D., Keith, M., Laufer, J., Kasper, M., Koch, P. J., and Franke, W. W., 1993, The human gene encoding cytokeratin 20 and its expression during fetal development and in gastrointestinal carcinomas, *Differentiation* **53:**75–93.

Moll, R., Wu, X.-R., Lin, J.-H., and Sun, T.-T., 1995, Uroplakins, specific membrane proteins of urothelial umbrella cells, as histological markers of metastatic transitional cell carcinomas, *Am. J. Pathol.* **147:**1383–1397.

Moll, R., Moll, I., and Gould, V. E., 1997, Neuroendocrine–Merkel cells of the skin and their neoplasms, in *Bloodworth's Endocrine Pathology*, 3rd ed. (J. Lechago and V. E. Gould, eds.), pp. 641–661, Williams & Wilkins, Baltimore.

Morgan, P. R., Leigh, I. M., Purkis, P. E., Gardner, I. D., van Muijen, G. N. P., and Lane, E. B., 1987a, Site variation in keratin expression in human oral epithelia—An immunocytochemical study of individual keratins, *Epithelia* **1:**31–43.

Morgan, P. R., Shirlaw, P. J., Johnson, N. W., Leigh, I. M., and Lane, E. B., 1987b, Potential applications of anti-keratin antibodies in oral diagnosis, *J. Oral Pathol.* **16:**212–222.

Muir, C., 1994, Cancer of unknown primary site, *Cancer* **75:**353–356.

Müller, K. M., and Fisseler-Eckhoff, A., 1989, What's new in lung tumor heterogeneity? *Pathol. Res. Pract.* **184:**108–115.

Mullink, H., Henzen-Logmans, S. C., Alons-van Kordelaar, J. J. M., Tadema, T. M., and Meijer, C. J. L. M., 1986, Simultaneous immunoenzyme staining of vimentin and cytokeratins with monoclonal antibodies as an aid in the differential diagnosis of malignant mesothelioma from pulmonary adenocarcinoma, *Virchows Arch. B Cell Pathol.* **52:**55–65.

Nagle, R. B., 1989, Intermediate filaments: Efficacy in surgical pathologic diagnosis, *Am. J. Clin. Pathol.* **91:**S14–S18.

Nagle, R. B., 1994, A review of intermediate filament biology and their use in pathologic diagnosis, *Mol. Biol. Rep.* **19:**3–21.

Nagle, R. B., 1996, Intermediate filament expression in prostate cancer, *Cancer Metast. Rev.* **15:**473–482.

Nagle, R. B., Böcker, W., Davis, J. R., Heid, H. W., Kaufmann, K., Lucas, D. O., and Jarasch, E.-D., 1986, Characterization of breast carcinomas by two monoclonal antibodies distinguishing myoepithelial from luminal epithelial cells, *J. Histochem. Cytochem.* **34:**869–881.

Nair, S. A., Lakshmi, S., Kannan, S., Nair, M. K., and Pillai, M. R., 1995, Cytokeratins and tumor progression in the uterine cervix. *J. Exp. Clin. Cancer Res.* **14:**131–137.

Nathrath, W. B. J., and Lane, E. B., 1990, Immunhistologische Untersuchung der Verteilung von Cytokeratinen in 100 Mammacarcinomen, *Verh. Dtsch. Ges. Pathol.* **74:**596.

Nathrath, W. B. J., Reiss, B., and Lane, E. B., 1994, Bedeutung verschiedener, insbesondere plattenepithelialer Keratinmuster in Mammakarzinomen, *Verh. Dtsch. Ges. Pathol.* **78:**291.

Nelson, W. G., Battifora, H., Santana, H., and Sun, T.-T., 1984, Specific keratins as molecular markers for neoplasms with stratified epithelial origin, *Cancer Res.* **44:**1600–1603.

Nicolson, G. L., 1987, Tumor cell instability, diversification, and progression to the metastatic phenotype: From oncogene to oncofetal expression, *Cancer Res.* **47:**1473–1487.

Nischt, R., Roop, D. R., Mehrel, T., Yuspa, S. H., Rentrop, M., Winter, H., and Schweizer, J., 1988, Aberrant expression during two-stage mouse skin carcinogenesis of a type 1 47-kDa keratin, K13, normally associated with terminal differentiation of internal stratified epithelia, *Mol. Carcinogenesis* **1:**96–108.

Ogden, G. R., Chisholm, D. M., Adi, M., and Lane, E. B., 1993, Cytokeratin expression in oral cancer and its relationship to tumor differentiation, *J. Oral Pathol. Med.* **22:**82–86.

Okada, H., Tsubura, A., Okamura, A., Senzaki, H., Naka, Y., Komatz, Y., and Mori, S., 1992, Keratin profiles in normal/hyperplastic prostates and prostate carcinoma, *Virchows Arch. A Pathol. Anat.* **421:**157–161.

Osborn, M., and Weber, K., 1983, Tumor diagnosis by intermediate filament typing: A novel tool for surgical pathology, *Lab. Invest.* **48:**372–394.

Osborn, M., van Lessen, G., Weber, K., Klöppel, G., and Altmannsberger, M., 1986, Differential diagnosis of gastrointestinal carcinomas by using monoclonal antibodies specific for individual keratin polypeptides, *Lab. Invest.* **55:**497–504.

Osborn, N., Mazzolein, G., Santini, D., Marrano, D., Martinelli, G., and Weber, K., 1988, Villin, intestinal brush border hydrolases and keratin polypeptides in intestinal metaplasia and gastric cancer: An immunohistologic study emphasizing the different degrees of intestinal and gastric differentiation in signet ring cell carcinomas, *Virchows Arch. A Pathol. Anat.* **413:**303–312.

Oshima, R. G., 1992, Intermediate filament molecular biology, *Curr. Opin. Cell Biol.* **4:**110–116.

Pankov, R., Umezawa, A., Maki, R., Der, C. J., Hauser, C. A., and Oshima, R. G, 1994, Oncogene activation of human keratin 18 transcription via the Ras signal transduction pathway, *Proc. Natl. Acad. Sci. USA* **91:**873–877.

Pellegrino, D. B., Asch, B. B., Connolly, J. L., and Asch, H. L., 1988, Differential expression of keratins 13 and 16 in normal epithelium, benign lesions, and ductal carcinomas of the human breast determined by the monoclonal antibody Ks8.12, *Cancer Res.* **48:**5831–5836.

Pitz, S., Moll, R., Störkel, S., and Thoenes, W., 1987, Expression of intermediate filament proteins in subtypes of renal cell carcinomas and in renal onocytomas, *Lab. Invest.* **56:**642–653.

Pujol, J.-L., Grenier, J., Daurès, J.-P., Daver, A., Pujol, H., and Michel, F.-B., 1993, Serum fragment of cytokeratin subunit 19 measured by CYFRA 21-1 immunoradiometric assay as a marker of lung cancer, *Cancer Res.* **53:**61–66.

Purkis, P. E., Steel, J. B., Mackenzie, I. C., Nathrath, W. B. J., Leigh, I. M., and Lane, E. B., 1990, Antibody markers of basal cells in complex epithelia, *J. Cell Sci.* **97:**39–50.

Ramaekers, F. C. S., Huysmans, A., Moesker, O., Schaart, G., Herman, C., and Vooijs, P., 1985, Cytokeratin expression during neoplastic progression of human transitional cell carcinomas as detected by a monoclonal and a polyclonal antibody, *Lab. Invest.* **52:**31–38.

Ramaekers, F. C. S., Huysmans, A., Schaart, G., Moesker, O., and Vooijs, P., 1987, Tissue distribution of keratin 7 as monitored by a monoclonal antibody, *Exp. Cell Res.* **170:**235–249.

Ramaekers, F. C. S., van Niekerk, C., Poels, L., Schaafsma, E., Huijsmans, A., Robben, H., Schaart, G., and Vooijs, P., 1990, Use of monoclonal antibodies to keratin 7 in the differential diagnosis of adenocarcinomas, *Am. J. Pathol.* **136:**641–655.

Raymond, W. A., and Leong, A. S.-Y., 1989a, Co-expression of cytokeratin and vimentin intermediate filament proteins in benign and neoplastic breast epithelium, *J. Pathol.* **157:**299–306.

Raymond, W. A., and Leong, A. S.-Y., 1989b, Vimentin—A new prognostic parameter in breast carcinoma? *J. Pathol.* **158:**107–114.

Real, F. X., Vilá, M. R., Skoudy, A., Ramaekers, F. C. S., and Corominas, J. M., 1993, Intermediate filaments as differentiation markers of exocrine pancreas. II. Expression of cytokeratins of complex and stratified epithelia in normal pancreas and in pancreas cancer, *Int. J. Cancer* **54:**720–727.

Sack, M. J., and Roberts, S. A., 1997, Cytokeratins 20 and 7 in the differential diagnosis of metastatic carcinoma in cytologic specimens, *Diagn. Cytopathol.* **16:**132–136.

Santini, D., Ceccarelli, C., Martinelli, G. N., Pasquinelli, G., Leone, O., Marrano, D., and Mancini, A. M., 1994, Expression of intermediate filaments in normal and neoplastic exocrine pancreas, *Zentralbl. Pathol.* **140:**247–258.

Santini, D., Ceccarelli, C., Taffurelli, M., Pileri, S., and Marrano, D., 1996, Differentiation pathways in primary invasive breast carcinoma as suggested by intermediate filament and biopathological marker expression, *J. Pathol.* **179:**386–391.

Sarbia, M., Borchard, F., and Hengels, K. J., 1993, Histogenetical investigations on adenocarcinomas of the esophagogastric junction: An immunohistochemical study, *Pathol. Res. Pract.* **189:**530–535.

Savera, A. T., Torres, F. X., Linden, M. D., Bacchi, C. E., Gown, A. M., and Zarbo, R. J., 1996, Primary versus metastatic pulmonary adenocarcinoma: An immunohistochemical study using villin and cytokeratins 7 and 20, *Appl. Immunohistochem.* **4:**86–94.

Schaafsma, H. E., and Ramaekers, F. C. S., 1994, Cytokeratin subtyping in normal and neoplastic epithelium: Basic principles and diagnostic applications, *Pathol. Annu. Part 1* **29:**21–62.

Schaafsma, H. E., Ramaekers, F. C. S., van Muijen, G. N. P., Lane, E. B., Leigh, I. M., Robben, H., Huijsmans, A., Ooms, E. C. M., and Ruiter, D. J., 1990, Distribution of cytokeratin polypeptides in human transitional cell carcinomas, with special emphasis on changing expression patterns during tumor progression, *Am. J. Pathol.* **136:**329–343.

Schaafsma, H. E., Ramaekers, F. C. S., van Muijen, G. N. P., Robben, H., Lane, E. B., Leigh, I. M., Ooms, E. C. M., Schalken, J. A., Van Moorselaar, R. J. A., and Ruiter, D. J., 1991, Cytokeratin expression patterns in metastatic transitional cell carcinoma of the urinary tract. An immunohistochemical study comparing local tumor and autologous metastases, *Am. J. Pathol.* **139:**1389–1400.

Schaafsma, H. E., van der Velden, L.-A., Manni, J. J., Peters, H., Link, M., Ruiter, D. J., and Ramaekers, F. C. S., 1993, Increased expression of cytokeratins 8, 18 and vimentin in the invasion front of mucosal squamous cell carcinoma, *J. Pathol.* **170:**77–86.

Schaller, G., Fuchs, I., Pritze, W., Ebert, A., Herbst, H., Pantel, K., Weitzel, H., and Lengyel, E., 1996, Elevated keratin 18 protein expression indicates a favorable prognosis in patients with breast cancer, *Clin. Cancer Res.* **2:**1879–1885.

Schirmacher, P., Dienes, H.-P., and Moll, R., 1998, *De novo* expression of non-hepatocellular cytokeratins in Mallory-body formation, *Virchows Arch.* **432:**143–152.

Schröder, S., Dockhorn-Dworniczak, B., Kastendieck, H., Böcker, W., and Franke, W. W., 1986, Intermediate-filament expression in thyroid gland carcinomas, *Virchows Arch. A Pathol. Anat.* **409:**751–766.

Schröder, S., Wodzynski, A., and Padberg, B., 1996, Cytokeratin expression of benign and malignant epithelial thyroid tumours. An immunohistochemical study of 154 neoplasias using 8 different monoclonal anti-cytokeratin antibodies, *Pathologe* **17:**425–432.

Schüssler, M. H., Skoudy, A., Ramaekers, F., and Real, F. X., 1992, Intermediate filaments as differentiation markers of normal pancreas and pancreas cancer, *Am. J. Pathol.* **140:**559–568.

Schwerer, M. J., and Baczako, K., 1996a, Expression of cytokeratins typical for ductal and squamous differentiation in the human stomach: An immunohistochemical study of normal foveolar epithelium, *Helicobacter pylori* gastritis and intestinal metaplasia, *Histopathology* **29:**131–137.

Schwerer, M. J., and Baczako, K., 1996b, Immunohistochemical evaluation of keratin 20 expression in intestinal metaplasia types I to III, *J. Clin. Pathol.* **49:**791–794.

Senzaki, H., Ogura, E., Iwamoto, S., Nambu, H., Uemura, Y., Shikata, N., and Tsubura, A., 1997, Keratin expression in normal uterine cervical epithelium and carcinomas of cervical origin, *Oncol. Rep.* **4:**79–83.

Seshadri, R., Raymond, W. A., Leong, A. S.-Y., Horsfall, D. J., and McCaul, K., 1996, Vimentin expression is not associated with poor prognosis in breast cancer, *Int. J. Cancer* **67:**353–356.

Shah, I. A., Schlageter, M. O., Stinnett, P., and Lechago, J., 1991, Cytokeratin immunohistochemistry as a diagnostic tool for distinguishing malignant from benign epithelial lesions of the prostate, *Mod. Pathol.* **4:**220–224.

Shi, S. R., Key, M. E., and Kalra, K. L., 1991, Antigen retrieval in formalin-fixed, paraffin embedded tissues: An enhancement method for immunohistochemical staining based on microwave oven heating of tissue section, *J. Histochem. Cytochem.* **39:**741–748.

Smedts, F., Ramaekers, F., Robben, H., Pruszczynski, M., van Muijen, G., Lane, B., Leigh, I., and Vooijs, P., 1990, Changing patterns of keratin expression during progression of cervical intraepithelial neoplasia, *Am. J. Pathol.* **136:**657–668.

Smedts, F., Ramaekers, F., Troyanovsky, S., Pruszczynski, M., Link, M., Lane, B., Leigh, I., Schijf, C., and Vooijs, P., 1992a, Keratin expression in cervical cancer, *Am. J. Pathol.* **141:**497–511.

Smedts, F., Ramaekers, F., Troyanovsky, S., Pruszcynski, M., Robben, H., Lane, B., Leigh, I., Plantema, F., and Vooijs, G. P., 1992b, Basal-cell keratins in reserve cells and a comparison with their expression in cervical intraepithelial neoplasia, *Am. J. Pathol.* **140:**601–612.

Smedts, F., Ramaekers, F., Leube, R. E., Keijser, K., Link, M., and Vooijs, P., 1993a, Expression of keratins 1, 6, 15, 16, and 20 in normal cervical epithelium, squamous metaplasia, cervical intraepithelial neoplasia, and cervical carcinoma, *Am. J. Pathol.* **142:**403–412.

Smedts, F., Ramaekers, F. C. S., and Vooijs, P. G., 1993b, The dynamics of keratin expression in malignant transformation of cervical epithelium: A review, *Obstet. Gynecol.* **82:**465–474.

Smedts, F., Ramaekers, F., Link, M., Lauerova, L., Troyanovsky, S., Schijf, C., and Vooijs, G. P., 1994, Detection of keratin subtypes in routinely processed cervical tissue: Implications for tumour classification and the study of cervix cancer aetiology, *Virchows Arch.* **425:**145–155.

Soeth, E., Röder, C. Juhl, H., Krüger, U., Kremer, B., and Kalthoff, B., 1996, The detection of disseminated tumor cells in bone marrow from colorectal-cancer patients by a cytokeratin-20-specific nested reverse-transcriptase-polymerase-chain reaction is related to the stage of disease, *Int. J. Cancer (Pred. Oncol.)* **69:**278–282.

Soini, Y., and Miettinen, M., 1992, Immunohistochemical evaluation of the cytoarchitecture of benign and malignant breast lesions, *APMIS* **100:**901–907.

Stoler, A., Kopan, R., Duvi, M., and Fuchs, E., 1988, Use of monospecific antisera and cRNA probes to localize the major changes in keratin expression during normal and abnormal epidermal differentiation, *J. Cell Biol.* **107:**427–466.

Stosiek, P., and Kasper, M., 1990, Neoexpression von Cytokeratin 7 bei der chronisch-atrophischen Gastritis mit perniziöser Anämie, *Pathologe* **11:**14–17.

Stosiek, P., and Kasper, M., 1993, Transitorisches Auftreten von Zytokeratin 7 bei der Entstehung des Magenkarzinoms, *Pathologe* **14:**71–73.

Stosiek, P., Bräutigam, E., and Kasper, M., 1991, Expression of cytokeratin 7 in human glandular epithelium of fetal stomach, *Acta Histochem.* **91:**21–23.

Su, L., Morgan, P. R., and Lane, E. B., 1994, Protein and mRNA expression of simple epithelial keratins in normal, dysplastic, and malignant oral epithelia, *Am. J. Pathol.* **145:**1349–1357.

Su, L., Morgan, P. R., and Lane, E. B., 1996a, Expression of cytokeratin messenger RNA versus protein in the normal mammary gland and in breast cancer, *Hum. Pathol.* **27:**800–806.

Su, L., Morgan, P. R., and Lane, E. B., 1996b, Keratin 14 and 19 expression in normal, dysplastic and malignant oral epithelia. A study using *in situ* hybridization and immunohistochemistry, *J. Oral Pathol. Med.* **25:**293–301.

Su, X. R., Lin, J. H., Walz, T., Häner, M., Yu, J., Aebi, U., and Sun, T.-T., 1994, Mammalian uroplakins, *J. Biol. Chem.* **18:**13716–13724.

Sun, T.-T. 1989, Pathways of keratinocyte differentiation, in *Cytoskeletal Proteins in Tumor Diagnosis* (M. Osborn, and K. Weber, eds.), pp. 121–126, Cold Spring Harbor Laboratory, Cold Spring Harbor, New York.

Sun, T.-T., Eichner, R., Schermer, A., Cooper, D., Nelson, W. G., and Weiss, R. A., 1984, Classification, expression and possible mechanisms of evolution of mammalian epithelial keratins: An unifying model, in *Cancer Cells, 1, The Transformed Phenotype* (A. J. Levine, G. F. Vande Woude, W. C. Topp, and J. D. Watson, eds.), pp. 169–176, Cold Spring Harbor Laboratory Press, Cold Spring Harbor, New York.

Suo, Z., Holm, R., and Nesland, J. M., 1993, Squamous cell carcinomas. An immunohistochemical study of cytokeratins and involucrin in primary and metastatic tumours, *Histopathology* **23:**45–54.

Swanson, P. E., Dehner, L. P., Sirgi, K. E., and Wick, M. R., 1994, Cytokeratin immunoreactivity in malignant tumors of bone and soft tissue: A reappraisal of cytokeratin as a reliable marker in diagnostic immunohistochemistry, *Appl. Immunohistochem.* **2:**103–112.

Syrjänen, S., Cintorino, M., Armellini, D., Del Vecchio, M. T., Leoncini, P., Bugnoli, M., Pallini, V., Silvestri, S., Tosi, P., Mäntyjärvi, R., and Syrjänen, K., 1988, Expression of cytokeratin polypeptides in human papillomavirus (HPV) lesions of the uterine cervix: 1. Relationship to grade of CIN and HPV-type, *Int. J. Gynecol. Pathol.* **7:**23–38.

Takada, M., Masuda, N., Matsuura, E., Kusunoki, Y., Matui, K., Nakagawa, K., Yana, T., Tuyuguchi, I., Oohata, I., and Fukuoka, M., 1995, Measurement of cytokeratin 19 fragments as a marker of lung cancer by CYFRA 21-1 enzyme immunoassay, *Br. J. Cancer* **71:**160–165.

Takahashi, H., Shikata, N., Senzaki, H., Shintaku, M., and Tsubura, A., 1995, Imunohistochemical staining patterns of keratins in normal oesophageal epithelium and carcinoma of the oesophagus, *Histopathology* **26:**45–50.

Takei, H., Iino, Y., Horiguchi, J., Kanoh, T., Takao, Y., Oyama, T., and Morishita, Y., 1995, Immunohistochemical analysis of cytokeratin #8 as a prognostic factor in invasive breast carcinoma, *Anticancer Res.* **15:**1101–1106.

Takemura, K., Hirayama, R., Hirokawa, K., Inagaki, M., Tsujimura, K., Esaki, Y., and Mishima, Y., 1994, Expression of vimentin in gastric cancer: A possible indicator for prognosis, *Pathobiology* **62:**149–154.

Terry, R. M., Gray, C., and Bird, C. C., 1986, Aberrant expression of low molecular weight cytokeratins in primary and secondary squamous carcinoma of the head and neck, *J. Laryngol. Otol.* **100:**1283–1287.

Thoenes, W., Störkel, S., Rumpelt, H. J., Moll, R., Baum, H. P., and Werner, S., 1988, Chromophobe cell renal carcinoma and its variants—A report on 32 cases, *J. Pathol.* **155:**277–287.

Thorson, P., Shevlin, D., and Swanson, P. E., 1996, Cytokeratin 7 and 20 immunoreactivity in carcinomas of the pancreaticobiliary tree, *Mod. Pathol.* **9:**66A.

Trask, D. K., Band, V., Zajchowski, D. A., Yaswen, P., Suh, T., and Sager, R., 1990, Keratins as markers that distinguish normal and tumor-derived mammary epithelial cells, *Proc. Natl. Acad. Sci. USA* **87:**2319–2323.

Tseng, S. C. G., Jarvinen, M. J., Nelson, W. G., Huang, J.-W., Woodcock-Mitchell, J., and Sun, T.-T., 1982, Correlation of specific keratins with different types of epithelial differentiation: Monoclonal antibody studies, *Cell* **30:**361–372.

Tyner, A. L., and Fuchs, E., 1986, Evidence for posttranscriptional regulation of the keratins expressed during hyperproliferation and malignant transformation in human epidermis, *J. Cell Biol.* **103:**1945–1955.

Utsunomiya, T., Yao, T., Masuda, K., and Tsuneyoshi, M., 1996, Vimentin-positive adenocarcinomas of the stomach: Co-expression of vimentin and cytokeratin, *Histopathology* **29:**507–516.

Vaidya, M. M., Borges, A. M., Pradhan, S. A., Rajpal, R. M., and Bhisey, A. N., 1989, Altered keratin expression in buccal mucosal squamous cell carcinoma, *J. Oral Pathol. Med.* **18:**282–286.

Vaidya, M. M., Borges, A. M., Pradhan, S. A., and Bhisey, A. N., 1996, Cytokeratin expression in squamous cell carcinomas of the tongue and alveolar mucosa, *Eur. J. Cancer B Oral. Oncol.* **32:**333–336.

Van de Klundert, F. A. J. M., Raats, J. M. H., and Bloemendal, H., 1993, Intermediate filaments: Regulation of gene expression and assembly, *Eur. J. Biochem.* **214:**351–366.

Van de Molengraft, F. J. J. M., van Niekerk, C. C., Jap, P. H. K., and Poels, L. G., 1993, OV-TL 12/30 (keratin 7 antibody) is a marker of glandular differentiation in lung cancer, *Histopathology* **22:**35–38.

Van der Velden, L.-A., Schaafsma, H. E., Manni, J. J., Link, M., Ruiter, D. J., Ramaekers, F. C. S., and Kuijpers, W., 1996, Cytokeratin and vimentin expression in normal epithelium and benign lesions of the vocal cords, *Acta Otolaryngol.* (Stockh.) **116:**325–331.

Van Eyken, P., Sciot, R., Paterson, A., Callea, F., Kew, M. C., and Desmet, V. J., 1988, Cytokeratin expression in hepatocellular carcinoma: An immunohistochemical study, *Hum. Pathol.* **19:**562–568.

Van Eyken, P., Sciot, R., and Desmet, V. J., 1989, A cytokeratin immunohistochemical study of cholestatic liver disease: Evidence that hepatocytes can express "bile duct-type" cytokeratins, *Histopathology* **15:**125–135.

Van Muijen, G. N. P., Ruiter, D. J., and van Leeuwen, C., 1984, Cytokeratin and neurofilament in lung carcinomas, *Am. J Pathol.* **116:**363–369.

Van Muijen, G. N. P., Ruiter, D. J., Franke, W. W., Achtstätter, T., Haasnoot, W. H. B., Ponec, M., and Warnaar, S. O., 1986, Cell type heterogeneity of cytokeratin expression in complex epithelia and carcinomas as demonstrated by monoclonal antibodies specific for cytokeratins nos. 4 and 13, *Exp. Cell Res.* **162:**97–113.

Van Muijen, G. N. P., Ruiter, D. J., and Warnaar, S. O., 1987a, Coexpression of intermediate filament polypeptides in human fetal and adult tissues, *Lab. Invest.* **57:**359–369.

Van Muijen, G. N. P., Warnaar, S. O., and Ponec, M., 1987b, Differentiation related changes of cytokeratin expression in cultured keratinocytes and in fetal, newborn and adult epidermis, *Exp. Cell Res.* **171:**331–345.

Van Niekerk, C. C., Jap, P. H. K., Ramaekers, F. C. S., van de Molengraft, F., and Poels, L. G., 1991, Immunohistochemical demonstration of keratin 7 in routinely fixed paraffin-embedded human tissues, *J. Pathol.* **165:**145–152.

Verhagen, A. P. M., Ramaekers, F. C. S., Aalders, T. W., Schaafsma, H. E., Debruyne, F. M. J., and Schalken, J. A., 1992, Colocalization of basal and luminal cell-type cytokeratins in human prostate cancer, *Cancer Res.* **52:**6182–6187.

Voijtesek, B., Staskova, Z., Nenutil, R., Bartkova, J., Kovarik, J., Rejthar, A., and Bartek, J., 1990, A panel of monoclonal antibodies to keratin No. 7: Characterization and value in tumor diagnosis, *Neoplasma* **3:**333–342.

Wada, T., Yasutomi, M., Yamada, K., Hashimura, K., Kunikata, M., Tanaka, T., Jianwen, H., and Mori, M., 1991, Heterogeneity of keratin expression and actin distribution in benign and malignant mammmary diseases, *Anticancer Res.* **11:**1983–1994.

Wang, N. P., Zee, S., Zarbo, R. J., Bacchi, C. E., and Gown, A. M., 1995, Coordinate expression of cytokeratins 7 and 20 defines unique subsets of carcinoma, *Appl. Immunohistochem.* **3:**99–107.

Watanabe, S., Ichikawa, E., Takahashi, H., and Otsuka, F., 1995, Changes of cytokeratin and involucrin expression in squamous cell carcinomas of the skin during progression to malignancy, *Br. J. Dermatol.* **132:**730–739.

Wauters, C. C. A. P., Smedts, F., Gerrits, L. G. M., Bosman, F. T., and Ramaekers, F. C. S., 1995, Keratins 7 and 20 as diagnostic markers of carcinomas metastatic to the ovary, *Hum. Pathol.* **26:**852–855.

Weikel, W., Wagner, R., and Moll, R., 1987, Characterization of subcolumnar reserve cells and other epithelia of human uterine cervix: Demonstration of diverse cytokeratin polypeptides in reserve cells, *Virchows Arch. B Cell Pathol.* **54:**98–110.

Weikel, W., Moll, R., Brumm, C., Wilkens, C., and Knapstein, P. G., 1996, Cytokeratin and vimentin expression in primary and recurrent carcinoma of the vulva: Correlations with prognostic factors and the course of disease, *Int. J. Gynecol. Pathol.* **15:**326–337.

Weiss, P. A., Eichner, R., and Sun, T.-T., 1984, Monoclonal antibody analysis of keratin expression in epidermal disease: A 48-kD and a 56-kD keratin as molecular markers for hyperproliferative keratinocytes, *J. Cell Biol.* **98:**1397–1406.

Wetzels, R. H. W., Holland, R., van Haelst, U. J. G. M., Lane, E. B., Leigh, I. M., and Ramaekers, F. C. S., 1989, Detection of basement membrane components and basal cell keratin 14 in noninvasive and invasive carcinomas of the breast, *Am. J. Pathol.* **134:**571–579.

Wetzels, R. H. W., Kuijpers, H. J. H., Lane, E. B., Leigh, I. M., Troyanowsky, S. M., Holland, R., van Haelst, U. J. G. M., and Ramaekers, F. C. S., 1991, Basal cell-specific and hyperproliferation-related keratins in human breast cancer, *Am. J. Pathol.* **138:**751–763.

Wild, G. A., and Mischke, D., 1986, Variation and frequency of cytokeratin polypeptide patterns in human squamous non-keratinizing epithelium, *Exp. Cell Res.* **162:**114–126.

Wild, G. A., Lobeck, H., and Mischke, D., 1987, Keratinexpression im normalen und maligne transformierten Plattenepithel der Kopfdarmschleimhaut, *Laryngol. Rhinol. Otol.* **66:**615–621.

Williams, G. R., Talbot, I. C., and Leigh, I. M., 1997, Keratin expression in anal carcinoma: An immunohistochemical study, *Histopathology* **30:**443–450.

Woodcock-Mitchell, J., Rannels, S. R., Mitchell, J., Rannels, D. E., and Low, R. B., 1989, Modulation of keratin expression in type II pneumocytes by the extracellular matrix, *Am. Rev. Respir. Dis.* **139:**343–351.

Wu, X.-R., Lin, J.-H., Walz, T., Häner, M., Yu, J., Aebi, U., and Sun, T.-T., 1994, Mammalian uroplakins: A group of highly conserved urothelial differentiation-related membrane proteins, *J. Biol. Chem.* **269:**13716–13724.

Yang, Y., Hao, J., Liu, X., Dalkin, B., and Nagle, R. B., 1997, Differential expression of cytokeratin mRNA and protein in normal prostate, prostatic intraepithelial neoplasia, and invasive carcinoma, *Am. J. Pathol.* **150:**693–704.

Chapter 8

Molecular Dissection of Desmosomal Assembly and Intermediate Filament Anchorage

Sergey M. Troyanovsky and Rudolf E. Leube

1. INTRODUCTION

Desmosomes are prominent cell–cell adhesion structures that are present in almost all mammalian epithelial cells. The extracellular desmosomal portion, which is characterized by a defined 20- to 30-nm gap filled with the electron-dense desmoglea and is partitioned by a central midline, mediates the contact between two neighboring plasma membranes, whereas the intracellular desmosomal region, the electron-dense plaque, serves as an anchorage site for intermediate filament (IF) bundles. Desmosomes are therefore placed at a critical position from which the IF organization can be regulated not only within an individual cell, but also across cell boundaries, resulting in a transcellular filament network that is involved in maintaining the structural integrity of epithelial tissues. This stabilizing function is underscored by the observed blister formation in autoimmune diseases of the

Sergey M. Troyanovsky Department of Dermatology, Washington University Medical School, St. Louis, Missouri 63110. **Rudolf E. Leube** Department of Anatomy, Johannes Gutenberg University, D-55099 Mainz, Germany.

Subcellular Biochemistry, Volume 31: Intermediate Filaments, edited by Herrmann and Harris. Plenum Press, New York, 1998.

pemphigus foliaceus (PF) and pemphigus vulgaris (PV) type in which tissue disintegration is induced by autoantibodies against desmosomal transmembrane proteins. On the other hand, desmosomes must be highly dynamic structures to facilitate and coordinate normal epithelial morphogenesis. The regulation of this delicate equilibrium is therefore of direct relevance to tumor formation and metastasis. The contribution of desmosomes and their constituting molecules to intracellular signaling, which has become apparent only recently, adds another degree of complexity to the multiple functions of these structures for tissue homeostasis in developing and mature epithelia.

Desmosomes are composed of a tightly clustered assembly of only a few protein types. In the context of this review we will restrict the presentation to four obligatory polypeptide components: the desmosomal cadherins of the desmoglein (Dsg) and desmocollin (Dsc) type that are transmembrane glycoproteins presumably mediating the calcium-sensitive cell–cell adhesion, and the desmosomal plaque constituents desmoplakin (Dp) I and II and plakoglobin (Pg) as mediators between the transmembrane proteins and IFs. Other recently described desmosomal plaque proteins are only facultative and often minor components (see Table I). Since there are no experimental data that indicate a contribution of these molecules to desmosome formation, we will not discuss them in any detail.

Recent reviews on desmosomes have focused on their structure and molecular composition (Schwarz *et al.*, 1990; Magee and Buxton, 1991; Buxton and Magee, 1992; Garrod, 1993; Schäfer *et al.*, 1993, Koch and Franke, 1994), their interaction with IFs (Green and Jones, 1990, 1996; Schmidt *et al.*, 1994; Cowin and Burke, 1996), and their role in blistering diseases and signaling (Amagai, 1995, Stanley, 1995; Garrod *et al.*, 1996). In this chapter we will place emphasis on basic molecular mechanisms and principles of desmosome assembly.

2. TOOLS TO EXAMINE DESMOSOME ASSEMBLY IN CULTURED CELLS

Since a desmosome is a complex and dynamic structure consisting of several tightly associated molecules that interact in multiple ways at multiple positions, any *in vitro* approach using purified polypeptides and mutants thereof will only help to elucidate very limited aspects of the overall assembly mechanism. Researchers have therefore turned to the examination of desmosome formation in living cells. Common to most of these approaches is the desire to construct and build up functional

Table I
Molecular Composition of Desmosomes

Transmembrane region		Intracellular plaque	
Proteins	Isoforms	Proteins	Isoforms
Desmogleins[a]	Dsg1	*Plakoglobin*[b]	
	Dsg2	*Plakophilin 1*	
	Dsg3	*Plakophilin 2*	
		p0071	
Desmocollins[a]	Dsc1a	Desmoplakin[c]	DpI
	Dsc1b	IFAP300	DpII
	Dsc2a	Plectin	
	Dsc2b	Envoplakin	
	Dsc3a	Pinin[d]	
	Dsc3b	Desmocalmin[e]	
		Lamin B-related (140-kD) protein[f]	

[a]Nine different cell type-specific isoforms of Dsg and Dsc have been identified.Dsg2,Dsc2a, and Dsc2b are present in most epithelia, whereas expression of the other known desmosomal cadherins is restricted to certain stratified epithelia (Schafer *et al.*, 1996; Garrod *et al.*, 1996; Nuber *et al.*, 1996; North *et al.*, 1996).
[b]Members of the "arm" family are underlined (for references see section 4.3).
[c]Italics denote proteins that are related to desmoplakin (for references see section 4.2).
[d]See Quyang and Sugrue, 1996.
[e]See Tsukita and Tsukita, 1985.
[f]See Cartaud *et al.*, 1990.

desmosomelike structures by combining the correct essentials in the correct location. We will briefly summarize three major experimental designs that have contributed significantly to the understanding of desmosome assembly processes during the recent years, employing manipulated cell culture systems.

The most straightforward idea was the rebuilding of a desmosome from scratch, i.e., by successive addition of all necessary ingredients to cells that lack any desmosomal components. Suitable cells appeared to be fibroblastic murine L-cells, a classic model system used to examine adhesive properties of cadherin-type transmembrane molecules. However, expression of desmosomal cadherins, either alone or in combination, resulted in surprisingly little cell clustering (Amagai *et al.*, 1994; Kowalczyk *et al.*, 1994a, b, 1996; Chidgey *et al.*, 1996). Even the addition of further desmosomal plaque proteins such as Pg did not significantly enhance the rather weak adhesive properties of these cells nor did it result in clustering of the ectopically expressed molecules in distinct membrane regions (Kowalczyk *et al.*, 1994a, b, 1996). Somewhat stronger adhesive properties, however, could be induced by the expression of chimeras consisting of the cytoplasmic domain of the classic E-type cadherin fused to the extracellular domain of desmosomal cadherins (Kowalczyk *et al.*, 1994a; Chidgey *et al.*, 1996). Taken together these observations suggested that additional factors are needed to confer competence for desmosome formation in L-cells.

Another experimental design was chosen by us, employing epithelial A-431 cells that are particularly rich in mature desmosomes, thereby providing all the components for desmosome assembly and IF anchorage (Leube *et al.*, 1988; Troyanovsky *et al.*, 1993; Bornslaeger *et al.*, 1996, and references therein). The key question asked here was whether it would be possible to redirect desmosomal components to other, ultrastructurally well-defined membrane domains. To accomplish this we expressed fusion proteins that contained molecular segments known to enrich in defined membrane regions together with selected parts of desmosomal polypeptides. We had previously shown that a group of integral membrane proteins with four membrane-spanning segments was able to target foreign molecular segments to such circumscribed membrane regions where these fusion proteins accumulated in tightly spaced homomultimeric clusters (Leube, 1995). In particular, synaptophysin, a major component of the small, electron-translucent, transmitter-containing secretory vesicles in neurons, is sorted to cytoplasmic microvesicles in nonneuroendocrine cells (Leube *et al.*, 1989, 1994), and the gap junction component connexin32 is targeted to and induces the formation of typical gap junctions at distinct cell contact sites, generating dye-permeable channels to neighboring cells (e.g., Troyanovsky *et al.*, 1994a; Leube, 1995; and references therein). We have shown that these properties rely solely on the presence of the four transmembrane regions and that alteration of the cytoplasmic domains, most notably of the carboxy terminus, does not influence the characteristic intracellular topogenesis of either molecule (Troyanovsky *et al.*, 1993, 1994a, b, 1996; Leube, 1995; Chitaev *et al.*, 1996). These tools can therefore be used to direct any given molecular segment in an assembly of closely spaced homomultimers to specific intracellular regions either in the cytoplasm as small vesicles or in the cell cortex as specialized cell–cell contact sites of the gap junction type. As these sites are morphologically well defined and distinct from desmosomes, they are therefore ideal means to use as harpoons for the initiation of complex assembly processes. Using this approach, the interaction between the cytoplasmic domains of desmosomal cadherins and the plaque protein Pg was characterized in the context of IF recruitment and Dp association (Troyanovsky *et al.*, 1993, 1994a, b, 1996; Chitaev *et al.*, 1996, see also below). Even more importantly, with this strategy we were able to generate chimeric structures exploiting the fact that the cell provided the perfect *in vivo* environment for desmosomal plaque formation. These phenomena and their relationship to endogenous desmosome formation as well as dominant-negative effects of certain mutants could be monitored easily by immunohistological methods and by electron microscopy. Despite the fascinating possibilities of this experimental approach, a major drawback was the fact that it was not possible to identify and define all the factors precisely that are needed for *de novo* desmosome formation, as they are provided by the genetically altered cell itself.

An intermediate approach between the two aforementioned systems was made possible by using human fibrosarcoma-derived HT-1080 cells, which maintain specific cell contacts of the adherens type, but do not make desmosomes; although they constitutively synthesize the desmosomal protein Dsg2, they lack other desmosomal proteins including Dp, Dscs, and Pg (Chitaev and Troyanovsky, 1997). Furthermore, Dsg is distributed evenly throughout the cell surface in these cells (Chitaev and Troyanovsky, 1997). It was speculated that these cells, in contrast to the murine L-fibroblasts, would allow desmosome formation, as stable cell contacts already existed and an interrelationship between desmosome formation and the presence of classic cadherin-mediated cell adhesion had been observed in other cell systems (e.g., Lewis *et al.*, 1994, 1997; Amagai *et al.*, 1995; van Hengel *et al.*, 1997). Technically, the preexisting expression of at least one major desmosomal component would make a complementation attempt somewhat easier. Using these cells, it was shown that the additional expression of the second desmosomal cadherin type, Dsc, resulted in the recruitment of Dsg into calcium-sensitive clusters that probably correspond to specialized cell contact sites and also incorporate added Pg (Chitaev and Troyanovsky, 1997). It is hoped that with the help of these cells one will learn how to build mature desmosomes in a junction-competent cell type.

3. STRUCTURE OF THE EXTRACELLULAR DESMOSOMAL REGION

3.1. Extracellular Cadherin Domains as Adhesive Units

Dsgs and Dscs are members of the supergene family of calcium-sensitive cell adhesion molecules, the cadherins (Koch *et al.*, 1990, 1991). Based on the specific localization in distinct adhesion structures, two major groups can be defined, the "classic" cadherins in adherens junctions mediating actin filament anchorage (Takeichi, 1990; Geiger and Ayalon, 1992; Grunwald, 1993) and "desmosomal" cadherins in desmosomes as the attachment sites of IF to the plasma membrane (e.g., Green and Jones, 1990; Buxton and Magee, 1992; Garrod, 1993; Koch and Franke, 1994; Schmidt *et al.*, 1994; Cowin and Burke, 1996). Molecular cloning demonstrated that the mammalian genome contains at least three different genes coding for Dsg 1–3 and three separate genes for Dsc 1–3 that are differentially expressed in a cell-type-specific but partially overlapping pattern (reviewed by Koch and Franke, 1994; Garrod *et al.*, 1996; for nomenclature see Buxton *et al.*, 1993). Moreover, each Dsc gene gives rise to two different splice forms, a and b, containing alternative cytoplasmic carboxy termini (Figure 1A).

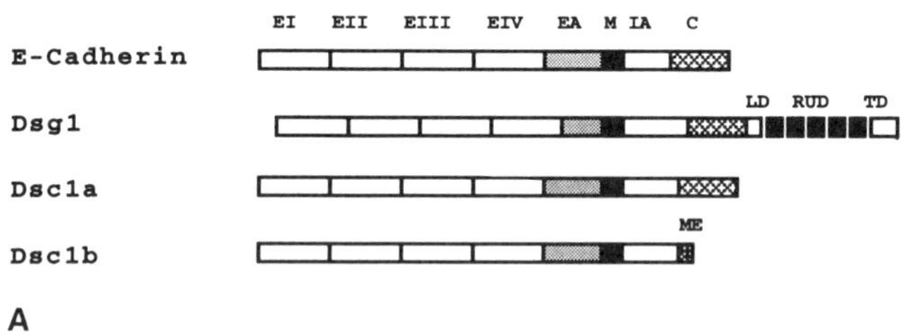

FIGURE 1. Cadherin domain organization (A) and amino Acid comparison (B, C). (A) Structural domains of E-cadherin, Dsg1, and the two splice forms of Dsc1 (Dsc1a and Dsc1b) are shown: EI–EIV, four extracellular cadherin-typical repeats; EA, extracellular anchor domain; M, transmembrane domain; IA, intracellular anchor domain; C, C-domain. Dsg1 contains additional intracellular domains: proline-rich linker domain (LD), RUD domain consisting of five repeating units, and terminal domain (TD). In Dsc1b variant a short sequence (ME) substitutes for the C-domain. (B) Comparison of the amino acid sequences (one-letter code) of the EI and EII domains of the mouse E-cadherin (Ec; Ringwald *et al.*, 1987), bovine Dsg1 (Dg; Koch *et al.*, 1990), and bovine Dsc1 (Dc; Koch *et al.*, 1991) in an optimized alignment showing amino acid identities between all cadherins in boldface. Note that all but one residue that are involved in Ca^{2+}-binding in E-cadherin (indicated by asterisks, Nagar *et al.*, 1996) are conservative in all cadherins. The single amino acid residue needed for adhesive properties of E-cadherin (Ozawa *et al.*, 1990a) is denoted by a dot. (C) Comparison of the amino acid sequences of the IA-domains within classic cadherins: mouse E-cadherin (EC, Ringwald *et al.*, 1987); human N-cadherin (NC, Reid and Hemperly, 1990), human P-cadherin (PC, Shimoyama *et al.*, 1989), Dsgs: human Dsg1 (Dg1, Wheeler *et al.*, 1991), human Dsg2 (Dg2, Schäfer *et al.*, 1994), and human Dsg3 (Dg3, Amagai *et al.*, 1991); Dscs: human Dsc1 (Dc1, Theis *et al.*, 1993); human Dsc2 (Dc2, Parker *et al.*, 1991); and human Dsc3 (Dc3, Kawamura *et al.*, 1994). Note that IA-domains are cadherin-type specific and relatively divergent, but that the juxtamembraneous regions are comparatively conserved within each group (indicated by asterisks). These latter regions were shown to be important for the dominant negative effect of N-cadherin (Kintner, 1992), for the activation of the adhesion by E-cadherin extracellular region in a Dsg–E-cadherin chimera (Roh and Stanley, 1995), and for recruitment of Dp to chimeric connexin32–Dsc gap junctions (Troyanovsky *et al.*, 1994b).

Domain EI

```
Ec  DWVIPPISCPENEKGEFPKNLVQIKSNRDKETKVFYSITGQGADKPPVGVFIIERETGWLKVTQPLDREAIAKYILYSHAVSSNGEAVEDPMEIVITVTDQNDNR
Dg  EWIKFAAACREGEDNSKRNPIAKIHSDCAANQQVTYRISGVGIDQPPYGIFVINQKTGEINITSIVDREVTPFFIIYCRALNSLGQDLEKPLELRVRVLDINDNP
Dc  RWAPIPCSLMENSLGPFPQHVQQVQSDAAQNYTIFYSISGPGVDKEPFNLFFIEKDTGDIFCTRSIDREQYQEFPIYAYATTADGYAPEYPLPLVFKVEDDNDNA
              *                                                       * *                              *****
```

Domain EII

```
                                ●
Ec  PEFTQEVFEGSVAEGAVPGTSVMKVSATDADDDVNTYNAAIAYTIVSQDPELPHKNMFTVNRDTGVISVLTSGLDRESYPTYTLVVQAADLQGE---GLSTTAKAVITVKDDINDNA
Dg  PVFSMSTFVGEIEENSNANTLVMVLNATDADE-PNNLNSKIAFKIIRQEPSDSP--MFIINRYTGEIRTMNNFLDREQYGQYSLAVRGSDRDGGAD-GMSAECECNIKILD-VNDNI
Dc  PYFENKLTVFTVPENCRTGTSVGKVTAIDLDE-PDTLHTRLKYKILQQIPNNPR--HFTVHPDTGVITTTTPLLDRTKCDTYKLIMEVRDMGGQPFGLFNT-GTITISLED-ENDNA
                                * **     *
```

B

```
EC.-RRRTVVKEP--LLP-PDDDTRDNVYYYDEEGGGEEDQDFDLSQLHRGLDAR-PEVT------RNDVAPTLMSVPQYRPRPA--NPDEIGNFIDENL
NC.-KRRDKERQAKQLLIDPEDDVRDNILKYDEEGGGEEDQDYDLSQLQ-QPDTVEPDAIKPVGIRRLDERPIHAE-PQYPVRSAAPHPGDIGDFINEGL
PC.-RKKRKVKEP--LLL-PEDDTRDNVFYYGEEGGGEEDQDYDITQLHRGLEAR-PEVVL-----RNDVVPTFIPTPMYRPRPA--NPDEIGNFIIENL
           ***********************

Dg1-CDCG-GAPRS-AAGFEPVPKCSDGAIESWAVEGPQPEPRDITTV-IPQIPPDNA---NIIECIDNSGVYTNEYGG----------REMQDLGG---------GERMTGFELTEGVKTSGMPEICQEYSGTLRRNSMRECREGGLNMNFMESYFCQ
Dg2-CHCGKGAKA-----FTPIPGTIE-MLEPWNNEGAPPEDKVVPSF-LPVDQGGSLVGRNGVG----GMAKEATMKGSSSASIVKGQHEMSEMDGRWEEHRSLLSGRATQFTGATGAIMTTETTKTARATGASRDMAGAQAAAVALNEEFLRNYFTD
Dg3-CDCGAGSTGGVTGGFIPVPDGSEGTIEQWGIEGAHPEDKEITNICVPPVTANGADFMESSEVCTNTYARGTAVEGTSGMEMTTKLGAATESGGAAGFATGTVSGAASGFGAATGVGICSSGQSGTMRTRHSTGGTNKDYADGAISMNFLDSYFSQ
    *******************************************

Dc1-K--KCFPEDIAQQNLIVSNTEGPGEE-VTEANIRLPMQTSN-----ICDTSMSVGTVGGQGIKTQQSFEMVKGGY-TLDSNKGGGEQT-LESVKGVGQGDTGRYAYT--DWQSFTQPRLGE
Dc2-KQPKVIPDDLAQQNLIVSNTEAPGDDKVYSANG-FTTQTVGASAQGVCGTVGSGIKNGGQ-----ETIEMVKGGHQTSESCRGAGEHHTLDSCRG-GHTEVDNCRYTYSEWHSFTQPRLGE
Dc3-G--KRFPEDLAQQNLIISNTEAPGDDRVCSANG-FMTQTTNNSSQGFCGTMGSGMKNGGQ-----ETIEMMKGGNQTLESCRGAGEHHTLDSCRG-GHTEVDNCRYTYSEWHSFTQPRLGE
              **************
```

C

FIGURE 1. (*Continued*).

Recent experiments demonstrated that different Dsc isoforms are localized in the same desmosome (North *et al.*, 1996; Nuber *et al.*, 1996).

The extracellular regions of Dsc and Dsg each encompasses four repeats of approximately 110 amino acids characteristic for all cadherins (denoted EI, EII, EIII, and EIV in Figure 1). X-ray crystallography of the EI-domain of N-cadherin, a classic cadherin abundantly expressed in mesenchymal cells and neurons, and of the EI–EII-domains of E-cadherin, a classic cadherin produced by epithelial cells, provided valuable information about the spatial structure of these conserved domains (Shapiro *et al.*, 1995; Nagar *et al.*, 1996). It was shown that three calcium ions can be inserted into the cleft between two adjacent E-domains (Figure 1B) resulting in rigidification and a rodlike arrangement of the extracellular domains. Conversely, removal of the calcium ions by chelating agents led to a collapse of these regions (Pokutta *et al.*, 1994). In addition, a point mutation within one of the presumptive Ca^{2+}-binding sites (denoted by dot in Figure 1) in the EII domain of E-cadherin abolished the adhesion properties of this protein (Ozawa *et al.*, 1990a).

The structural analyses of cadherin domains derived from N- or E-cadherins also suggested that the EI-domain is able to form side-to-side homodimers (Shapiro *et al.*, 1995; Nagar *et al.*, 1996). That such a lateral interaction occurs and may be of relevance to the stabilization of cell contacts is further supported by the recent experiments of Brieher and colleagues (1996), who demonstrated dimerization of the extracellular portion of *Xenopus* C-cadherin in solution and a positive effect of this dimerization on homophilic adhesion. It was proposed that each of these dimers may interact in a head-to-head fashion with a partner dimer emanating from the surface of an adjacent cell, thus forming a complex zipperlike structure (Shapiro *et al.*, 1995). Consequently, this model predicts that the EI-cadherin domain may be involved in the formation of side-to-side as well as head-to-head complexes, being in agreement also with experimental observations that demonstrated the critical role of the EI-domain in adhesion and led to the identification of the tripeptide HAV as a cell adhesion recognition (CAR) sequence (Blaschuk *et al.*, 1990; Nose *et al.*, 1990). The proposed model, however, is far from being complete. There is no experimental evidence demonstrating the existence of the cadherin dimers at the cell surface and it is not clear whether other domains contribute to intercadherin interactions. In fact, monoclonal antibody DECMA-1, which is able to block cell–cell adhesion, binds to the juxtamembraneous portion of E-cadherin, suggesting that this region may also be involved in cell–cell adhesion (Ozawa *et al.*, 1991). Another important question is whether intercadherin interactions and calcium binding to cadherins are interdependent processes. Finally, while it is well established that intracellular interactions of the classic cadherins with the plaque proteins Pg and β- and α-catenin are required for adherens junction assembly (Nagafuchi and Takeichi, 1988; Ozawa *et al.*, 1990b), the exact molecular mechanism leading to cadherin clustering is not understood.

Mechanisms proposed for the classic cadherins may also apply to desmosomal cadherins, based on the high degree of sequence similarity in the extracellular domains (approximately 50% similarity in the EI-domain), including identical residues at the proposed calcium-binding sites (see Figure 1B). Furthermore, desmosomal cadherins also need calcium for formation and stabilization of desmosomes (Demlehner *et al.*, 1995). Yet, while the basic principles of oligomerization envisaged for the classic cadherins may be valid for the desmosomal cadherins as well, there are clearly also significant differences. Most notably, all desmosomes appear to contain both Dscs and Dsgs, whereas it is generally assumed that the classic cadherins occur individually in a given adhesion structure. Furthermore, different motifs must direct the interaction between the desmosomal cadherins of neighboring cells, as the CAR tripeptide HAV is lacking in these molecules.

3.2. Heterodimer Formation by Extracellular Desmocollin and Desmoglein Regions

The direct involvement of desmosomal cadherins in facilitating cell–cell adhesion between desmosomal halves was originally proposed from cell culture experiments in which antibodies against the extracellular regions of Dsc interfered with the formation of a contiguous epithelial sheet (Cowin *et al.*, 1984). This idea was recently further substantiated by knockout experiments in which the loss of Dsg3 led to the appearance of half-desmosomes at the cell surface that no longer adhered to the other desmosomal half in the neighboring cell (Koch *et al.*, 1997). Furthermore, the resulting perturbation of epidermal cell–cell interactions in transgenic mice producing a dominant-negative form of Dsg (Allen *et al.*, 1996) and the occurrence of blistering diseases that are caused by autoantibodies against the extracellular portion of either Dsg or Dsc both point to the importance of desmosomal cadherins for cell adhesion (for review see Stanley, 1995; Amagai 1995). Interestingly, most of the pathogenic autoantibodies found in PF or PV patients react with the EI–EII-domains of Dsg (Dmochowski *et al.*, 1994; Olague-Alcala *et al.*, 1994; Kowalczyk *et al.*, 1995), i.e., those regions that correspond to the domains of the classic cadherins directly involved in cell–cell adhesion. Yet, desmosomal cadherins, in contrast to the classic cadherins, are not able, either alone or in combination, to induce cell–cell adhesion upon expression in nonadhesive fibroblastlike cells (Amagai *et al.*, 1994; Chidgey *et al.*, 1996; Kowalczyk *et al.*, 1996), suggesting fundamental functional differences despite overall structural homology between these polypeptide families.

No experimental data are available on the oligomeric state of either Dsg or Dsc and the contribution of any of their extracellular regions to desmosome formation. Efforts may have been hampered, however, by the insolubility of Dsg and Dsc multimers once they are formed.

Recently we have been able to demonstrate interactions between Dsg and Dsc (Chitaev and Troyanovsky, 1997). In these experiments we used human fibrosarcoma HT-1080 cells that lack mature desmosomes, but express Dsg2 in the absence of any other major desmosomal protein. When these cells were forced to express bovine Dsc1a, a dramatic redistribution of the endogenous Dsg2 was noted, resulting in a shift from a diffuse distribution in the plasma membrane to distinct clusters close to preexisting adherens junctions where both desmosomal cadherins colocalized. This relocation was shown to be dependent on the presence of the extracellular cadherinlike repeats EI and EII of Dsc1a (Arg^{1}–Thr^{170}), as a mutant in which this region was replaced by the homologous Dsg region did not show this effect in transfected cells. On the other hand, the complete deletion of the intracellular region of Dsc1a did not abolish the relocation of Dsg2 in distinct clusters. Furthermore, coimmunoprecipitation experiments on lysates from cells stably expressing Dsc1a using anti-Dsc or anti-Dsg antibodies demonstrated a physical association between both molecules that was weakened by chelation of Ca^{2+} ions. Direct Ca^{2+}-dependent binding of the EI–EII-region of Dsc1a to the extracellular portion of Dsg1 was also demonstrated by solid-phase binding experiments. Future experiments will be needed to define further the topology of these Dsg/Dsc complexes to show their existence in "normal" epithelial cells and tissues, and to analyze the mechanisms of their assembly in mature desmosomes. Clearly, other intracellular desmosomal plaque proteins, such as Pg and Dp, may be necessary for these processes (see below).

Nothing is known about interactions within the plane of the membrane, although it is conceivable that the densely clustered transmembrane molecules may interact through their rigid and hydrophobic membrane-spanning domains.

4. STRUCTURE OF THE DESMOSOMAL PLAQUE

4.1. Contribution of Intracellular Desmosomal Cadherin Domains to Desmosome Assembly

The intracellular portions of both desmosomal cadherins are integrated into the desmosomal plaque, the electron-dense structure that is directly traversed by associating IFs. Based on sequence analyses, the intracellular regions of the desmosomal cadherins can be subdivided into several major domains (Koch and Franke, 1994; see also Figure 1A). A short, cadherin-type-specific intracellular anchoring domain (IA-domain; see Fig. 1C) is located next to the membrane-spanning portion. In Dsgs and the larger Dsc splice forms, i.e., variants a, the approximately 72-amino acid-long intracellular cadherin-typical domain (C-domain) is next. This region is conserved among all cadherins (approximately 30% identity) with the exception of the Dsc splice variant b, in which this domain is replaced by a short, 11-amino acid carboxy-terminal sequence that is almost identical in all known Dsc isoforms (Troyanovsky *et al.*, 1993). In contrast to the Dscs and the similarly structured classic cadherins that end with the C-domain, Dsgs contain long and unique carboxy-terminal extensions that are specific for each Dsg isoform [for amino acid comparison see Schäfer *et al.* (1994)] consisting of the intracellular linker domain (LD), the repeated unit domain (RUD) with variable numbers of the very divergent Dsg-specific repeats (five in Dsgs 1, and two and three in Dsg 3), and a short terminal domain (TD) which is also highly specific for each Dsg isoform (for review see, e.g., Koch and Franke, 1994; Schmidt *et al.*, 1994). The function and significance of the unique carboxy-terminal Dsg extension are completely unknown, but specific activities have been assigned to the other two intracellular domains of the desmosomal cadherins (see below).

Although direct binding of proteins to the IA domain of either the desmosomal cadherins or the classic cadherins has not been demonstrated, several observations suggest a crucial contribution of this domain to desmosome formation. We have observed that the IA-domain of Dsc is important for desmosome formation by recruiting Dp to chimeric gap junctions (Troyanovsky *et al.*, 1994b). Roh and Stanley (1995) showed that the 40 amino acids of Dsg 3 that are located directly next to the hydrophobic transmembrane domain (see Figure 1C) are sufficient to confer adhesive properties to an E-cadherin mutant lacking the entire intracellular region. Interestingly, a truncated E-cadherin lacking only the C-domain but still retaining the corresponding juxtamembraneous amino acids could not activate cell–cell adhesion (Nagafuchi and Takeichi, 1988; Ozawa *et al.*, 1990b).

Two alternative explanations could be given for this effect. Either the IA domain of Dsg induced oligomerization of the intracellular regions, thereby stabilizing the extracellular cadherinlike zipper (see above) or this domain is essential for the establishment of a specific "active" conformation of the extracellular E-domains.

There is compelling evidence for the involvement of the C domain of desmosomal cadherins in the assembly of mature desmosomes. By expressing chimeric molecules consisting of the topogenic four transmembrane domains of rat liver connexin32 fused to the cytoplasmic domain of bovine Dsc1a at the carboxy terminus, we were able to induce the formation of novel chimeric structures in epithelial A-431 cells (Troyanovsky *et al.*, 1993, 1994b). These structures represented a hybrid composed of a functional gap junction that was associated with desmosomal plaque material incorporating Dp and Pg and was competent for IF attachment. The chimeric structure formation was critically dependent on the presence of the C-domain, and hence the truncated cytoplasmic region of Dsc splice variant b or mutants with an incomplete Dsc C-domain were not able to generate such chimeric structures (Troyanovsky *et al.*, 1993, 1994b). Interestingly, these latter mutants were still associated with Dp (see above), while Pg was no longer attached (Troyanovsky *et al.*, 1993, 1994b). *In vitro* binding assays further pinpointed the C-domain as the module mediating the direct binding of Dsc to Pg (Chitaev *et al.*, 1996; Troyanovsky *et al.*, 1996).

Furthermore, we demonstrated that overexpression of the chimeric protein CoDsg, comprising the amino-terminal transmembrane domain of rat liver connexin32 and the intracellular region of bovine Dsg1, inhibits the formation of endogenous desmosomes in epithelial A-431 cells (Troyanovsky *et al.*, 1993, 1994a). In the resulting desmosome-deficient cells production of endogenous desmosomal cadherins was not altered, but the distribution changed to a random localization along the cell surface. A similar strong dominant-negative effect on desmosome assembly was obtained upon overexpression of a mutant form of Dsg lacking an extensive part of the extracellular region (Allen *et al.*, 1996). These observations suggest that interactions of the Dsg tail with other proteins are required for lateral clustering of desmosomal cadherins. Deletion of the C-domain, but not of other segments of the intracellular Dsg region, completely abolished the dominant negative effect of CoDsg chimeras (Troyanovsky *et al.*, 1994a). Since coimmunoprecipitation experiments and *in vitro* binding studies indicated that the C-domain of Dsg1 interacts similarly to the corresponding region in Dsc1a directly with Pg (Troyanovsky *et al.*, 1994a; Chitaev *et al.*, 1996; Troyanovsky *et al.*, 1996), we proposed that the inhibition of desmosome assembly by an excess of the Dsg tail domains was due to their competition with the binding of endogenous Dsg for Pg. To further characterize the Dsg–Pg interactions, we fine–mapped the Pg-binding site by alanine scanning mutagenesis of the Dsg C-domain and testing the mutants in solid-phase binding assays. By this strategy we located the Pg-binding site to the row of large hydrophobic residues at the end of the Dsg C-domain. If we then expressed a CoDsg chimera which had only point mutations at these critical residues, no dominant-negative effect was observed (manuscript in preparation). We take these data as compelling evidence for the important role of desmosomal cadherin C-domains in desmosome assembly by their specific and direct interaction with Pg.

4.2. Desmoplakin as a Potential Linker between Desmosomal Plaque and Intermediate Filaments

The two major desmosomal plaque proteins, Dp and Pg, are thought to be directly involved in both lateral clustering of desmosomal cadherins and in anchorage of cadherins to IFs. Above we summarized our experiments with CoDsg and CoDsc chimeras demonstrating a functional role of Pg–Dsg interactions in these processes. A crucial contribution of Dp to desmosome formation and maintenance is suggested from the observations of Bornslaeger *et al.* (1996), who expressed the aminoterminal Dp-domain in epithelial A-431 cells, which led not only to disruption of desmosomes, but also to missorting of desmosomal cadherins and endogenous Dp to adherens junctions.

Dp as well as its close structural relatives plectin, envoplakin, BPAG1, and IFAP300 have been proposed to act as bridging proteins between different elements of the cytoskeleton and/or of the cytoskeleton to the plasma membrane (see Green and Jones, 1996, and Chou *et al.*, 1997, for recent reviews). The two Dps I and II are encoded by a single gene from which they are generated by alternative splicing (Green *et al.*, 1990; Virata *et al.*, 1992). Dps probably form large, dumb bell-like molecules (O'Keefe *et al.*, 1989). Based on amino acid sequence analysis, electron microscopy, and functional studies (Stappenbeck *et al.*, 1993), Dps can be divided into three major domains. The amino-terminal globular domain is inserted in the desmosomal plaque, the central rod domain is involved in homodimerization by formation of stable coiled-coil helices, and the globular carboxyl-terminal domain interacts with IFs. DpII differs from DpI only by its central rod domain. The binding partner(s) of Dp in the desmosomal plaque has not been identified. The major difficulty for the identification of such interactions in mature desmosomes is due to the insolubility of the complexes that contain cadherins together with Dp and other plaque components [for characterization of different pools of desmosomal components see, e.g., Pasdar and Nelson (1989) and Pasdar *et al.* (1991)]. However, a short, juxtamembraneous amino acid sequence of Dsc is needed for recruitment of Dp to chimeric junctions formed in transfected cells (Troyanovsky *et al.*, 1994b). The observation of desmosome disruption by overexpression of the amino-terminal domain of Dp in transfected cells further indicates the binding to a limiting factor in the desmosomal plaque (Bornslaeger *et al.*, 1996). The binding of Dp to cytokeratins has been mapped to the carboxy terminus of Dp by direct *in vitro* binding assays identifying a conserved 18-amino acid (aa)-long stretch binding motif in the amino terminus of cytokeratins (Kouklis *et al.*, 1994). It is not understood, however, why only some of the cytokeratins exhibit this binding, while others that may be expressed alone with desmosomes do not. It is possible that other Dp-related molecules such as plectin or envoplakin are also engaged in the anchorage of IFs by binding these cytokeratins to desmosomes. The *in vitro* binding experiments also showed a strong affinity of Dp to vimentin IFs (Kouklis *et al.*, 1994), suggesting that Dp may participate in IF anchorage to desmosomes in vimentin-containing cells such as arachnoidal and meningioma cells (Kartenbeck *et al.*, 1984).

The emerging picture suggests that Dp is the primary candidate to link desmosomal cadherins with IFs, possibly by direct interaction with both. These binding activities, however, must be regulated in a very specific fashion to avoid Dp binding to IFs in the cytoplasm and to limit this association to the desmosomal plaque region. These mechanisms do not operate when carboxyl-terminal Dp mutants are expressed in transfected cells or when Dp is overexpressed resulting in the decoration of IF bundles with the exogenously added molecules (Stappenbeck and Green, 1992). Such a "spillover" may even occur under normal circumstances, as shown by filamentous Dp staining in some cultured cell lines (e.g., Trevor and Steben, 1992). It was proposed that phosphorylation may be one of the regulating mechanisms, since activation of protein kinase A or inactivation of the corresponding phosphorylation site within the carboxyl-terminal domain of Dp abolished Dp–IF association (Stappenbeck *et al.*, 1994). Involvement of protein kinases and phosphatases in desmosome assembly was also suggested by experiments with specific inhibitors of these enzymes (Sheu *et al.*, 1989; Pasdar *et al.*, 1995). It is possible that the binding of Dp to IFs may also be activated by Dp–Dsc and/or Dp–Pg interactions in desmosomes. Accordingly, we found that gap junctions made from a CoDsc chimera lacking the Dsc C-domain efficiently recruited Dp, but was not able to anchor Pg and did not recruit IFs (Troyanovsky *et al.*, 1994b).

4.3. Multiple Cadherin-Binding Sites of Plakoglobin

Pg is, in contrast to Dp and the aforementioned desmosomal cadherins, not a specific desmosomal component, as it is found in a wide array of other plaque structures localized at cell–cell contacts, also including those that are involved in the anchorage of actin filaments (Cowin *et al.*, 1986; for review see Schmidt *et al.*, 1994). This universal plaque protein entered the limelight when it was shown that it fulfills, in addition to its functions in practically all known cell–cell adhesion plaques, also crucial functions in signal transduction from the cell surface to the nucleus (for review see Klymkovsky and Parr, 1995; Miller and Moon, 1996; Nusse, 1997). Furthermore, recent gene ablations of Pg have demonstrated the essential role of this molecule for the assembly of myocardial desmosomes, although desmosomes are still formed in epithelial cells of these Pg-deficient mouse embryos (Ruiz *et al.*, 1996).

The globular Pg molecule consists of a central domain of approximately 560 residues with twelve 40- to 42-aa-long repeats that are interrupted by a short spacer between repeats 10 and 11, and are flanked by a 120-aa-long amino terminus and a 60-aa-long carboxy terminus [for primary amino acid determination see Franke *et al.* (1989)]. The central repeat motif, termed "arm repeat," was first recognized in the *Drosophila armadillo* gene product that is involved in segment polarity formation (Peifer *et al.*, 1994). This arm repeat is a common motif found in a diverse group of proteins including other desmosomal plaque proteins of the plakophilin multigene family such as plakophilin 1 and 2 and p0071 (Heid *et al.*, 1994; Hatzfeld *et al.*, 1994; Hatzfeld and Nachtsheim, 1996; Mertens *et al.*, 1996; see also Table I), proteins present in the plaques of the adherens junctions such as p120, a substrate of the $pp60^{v\text{-}src}$ kinase, and β-catenin (McCrea *et al.*, 1991; Butz *et al.*, 1992; Reynolds *et al.*, 1992), as well as several other polypeptides with diverse functions and intracellular localizations (for review see, e.g., Schmidt *et al.*, 1994; Peifer *et al.*, 1994, 1995). The closest Pg relative is β-catenin, with which it shares 80% sequence identity in the central domain and 20% in the flanking end domains. This central repeat domain of Pg is also highly conserved in evolution, the Pg/β-catenin homologues in flies, sea urchin, and euchuroid exhibiting more than 60% identity to human Pg in this region (Rosenthal, 1993).

Pg/β-catenin are both able to interact with a number of different proteins. Both bind to classic cadherins (Korman *et al.*, 1989; Ozawa *et al.*, 1989; Knudsen and Wheelock, 1992; Troyanovsky *et al.*, 1994a, b; Kowalczyk *et al.*, 1994a; Mathur *et al.*, 1994; Jou *et al.*, 1995); to the arm-containing, cytoplasmic tumor suppressor gene product APC (Rubinfeld *et al.*, 1993, 1995; Su *et al.*, 1993); to the actin-binding proteins α-catenin and fascin (Nagafuchi and Takeichi, 1988; Ozawa *et al.*, 1989; Aberle *et al.*, 1994; Hinck *et al.*, 1994; Tao *et al.*, 1996); to the tight junction protein ZO-1 (Rajasekaran *et al.*, 1996); to the epidermal growth factor receptor (Hoschuetzky *et al.*, 1994); and to the nuclear architectural transcription factor Lef-1 (Molenaar *et al.*, 1996; Huber *et al.*, 1996; Behrens *et al.*, 1996). The major difference in binding is the inability of β-catenin to bind desmosomal cadherins, though it is difficult to exclude that these interactions could exist when Dsg is sorted into an insoluble pool. The central arm-repeat region is apparently involved in most if not all of these interactions (Figure 2A), which, however, appear to be regulated by distinct and often separate parts of the arm-repeat domain (Aberle *et al.*, 1994, 1996; Sacco *et al.*, 1995; Chitaev *et al.*, 1996; Fagotto *et al.*, 1996; Orsulic and Peifer, 1996; Troyanovsky *et al.*, 1996; Wahl *et al.*, 1996; Witcher *et al.*, 1996; van de Wetering *et al.*, 1997).

The linkage of Pg to the intracellular regions of Dsg and the larger Dsc isoforms has been investigated in detail. Within the cytoplasmic regions of Dsc and Dsg the C-domains were identified as the regions responsible for direct interaction with Pg. A corresponding 72-aa-long domain (catenin-binding domain) in the classic cadherins was shown to

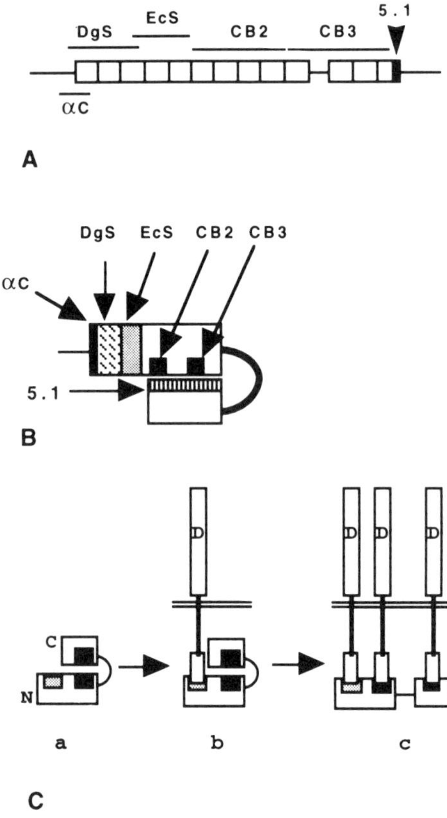

FIGURE 2. Plakoglobin structure and its hypothetical role in cadherin clustering. (A) Localization of Pg-binding sites for α-catenin (αC), desmoglein (DgS), E-cadherin (EcS), and monoclonal antibody 5.1 (5.1). Two cryptic cadherin-binding sites are denoted as CB2 and CB3. The 13 central arm repeats are represented by open boxes, the short 22-amino acid insertion and the flanking end domains by solid lines. (B) Model of Pg showing how the cryptic CB2 and CB3 sites could be masked by intramolecular interactions involving the 5.1 recognition sequence (see text for details). (C) Possible interaction between Pg and Dsg. Newly synthesized Pg exposes a single cadherin-binding site (CB1, stippled box) on its surface, while other cadherin-binding sites (black) are masked (form a). After binding to Dsg (D) the complex (form b) is transported to the cell–cell interface, where cryptic cadherin-binding sites are activated and association with other cadherins can proceed (form c).

also interact with either Pg or β-catenin (Ozawa *et al.*, 1990b). The short Dsc isoform Dsc b lacking this domain is not able to bind Pg (Troyanovsky *et al.*, 1994b; Wahl *et al.*, 1996). Pg and β-catenin both interact with the vinculin-related protein α-catenin when it is present in a complex with classic cadherins. Whether a similar bridging function is fulfilled by Pg in desmosomes and whether Pg-binding proteins with similarities to α-catenin exist in desmosomes remains to be shown and is at this point rather questionable.

Results of our *in vitro* binding experiments using recombinant intracellular domains of Dsg, Dsc, and E-cadherin with Pg or Pg fragments revealed that Pg has three cadherin-binding (CB) regions (Chitaev *et al.*, 1996; Troyanovsky *et al.*, 1996; see Figure 2A). The N-terminal proximal CB1 region (Gln^{114}–Leu^{292}) has the same binding affinity to cadherins as the entire Pg molecule. It binds primarily to Dsg and very weakly to Dsc and E-cadherin. The isolated CB2 and CB3 regions (Ile^{305}–Ile^{505}, Ile^{505}–Phe^{672}, respectively) interact equally well with all tested cadherins. However, in intact Pg, their binding activities are masked, so that Pg with a deleted CB1 sequence has only a residual affinity to cadherins. At least one of the cryptic sites (located in the CB2 or CB3 region) is activated by deletion of the C-terminal sequence of Pg (Troyanovsky *et al.*, 1996). The fine mapping of this region by deletion mutagenesis showed that it corresponds exactly to the epitope for monoclonal anti-plakoglobin antibody 5.1 (Lys^{673}–Gln^{683} "5.1 sequence"). While it is not known whether cryptic cadherin-binding sites of Pg found *in vitro* interact with cadherins *in vivo*, our data present evidence that the 5.1 sequence is implicated in intramolecular interactions of Pg (Figure 2B).

4.4. Involvement of Plakoglobin in Cadherin Clustering and in Intermediate Filament Anchorage to Desmosomes

We recently described the chimeric protein SyPg in which the vesicular protein synaptophysin is fused to the entire Pg molecule (Chitaev *et al.*, 1996). This chimera integrated efficiently into small vesicles of transfected epithelial cells and bound to E-cadherin, Dsg, and Dsc similar to wild-type Pg. These results allowed us to develop a new approach that we refer to as synaptophysin insertion mutagenesis to study the modular structure of Pg (Troyanovsky *et al.*, 1996). In this type of mutagenesis, synaptophysin is inserted at different positions of Pg to investigate the effects of this insertion on binding of the resulting mutant to partner molecules such as desmosomal cadherins, E-cadherin, α-catenin, and p120 and on their localization in transfected epithelial A-431 cells. This approach in conjunction with more conventional strategies led to the identification of two functionally distinct domains. One domain, the "binding" domain, contains distinct E-cadherin, Dsg, and α-catenin binding sites and mediates the direct linkage of Pg to cadherins (Figure 2B). This domain corresponds exactly to the CB1 region mapped by the solid-phase assay *in vitro* (Troyanovsky *et al.*, 1996; see also above). Detailed point mutagenesis of this domain is now in progress to determine whether binding sites for α-catenin, desmosomal cadherins, and classic cadherins overlap. The second "targeting" domain, containing the CB2 and CB3 regions, is responsible for the efficient incorporation of Pg/cadherin complexes into cell–cell junctions (Troyanovsky *et al.*, 1996; Figure 2). Two groups of mutants were obtained. Mutants of the first group were localized exclusively in the cytoplasm despite their capacity to bind to both cadherins, whereas mutants of the second group that also bound to cadherins accumulated in cell–cell contact regions, but not in desmosomes. Mutants of the second group affected the formation of endogenous desmosomes in the transfected A-431 cells, reducing their number and inducing a more diffuse distribution of Dsg at the cell surface (Troyanovsky *et al.*, 1996).

Interestingly, the epitope for monoclonal Pg antibody 5.1 is masked when Pg is bound to Dsg, but not when it is associated with E-cadherin (Wahl *et al.*, 1996, and our unpublished observations). Since the Pg region including this epitope does not exhibit direct binding activity to Dsg (Chitaev *et al.*, 1996; Troyanovsky *et al.*, 1996), it must have an indirect effect on Pg conformation, and one may speculate that such a conformational change induced by Dsg binding is important for the insertion of Pg–Dsg complexes into desmosomes.

Two fundamentally different models of Pg function for desmosome formation can be deduced from the experimental observations. In the first model, Pg works as a motor in the lateral clustering of desmosomal cadherins. The binding of the C-domain of Dsg to Pg via the CB1 site may be a first step occurring at an early stage of desmosome assembly. The Pg–Dsg complex may then be delivered to cell–cell contacts, where latent CB2 and CB3 sites are activated to mediate association with other Dsg and/or Dsc molecules. This activation could be induced by conformational changes in the C-domain of Dsg, by dimerization of Dsg, or by Dsg–Dsc interactions (Figure 2C). Head-to-head intercadherin interactions could induce further structural changes that are required for the proceeding of lateral clustering of desmosomal cadherins and the stable attachment to IFs. Alternatively, the second model takes into account the involvement of Pg in signaling cascades. In such a scenario Pg bound to the C-domain of Dsg may regulate the efficiency of desmosome assembly by signaling mechanisms. Thus, Pg could be involved in the transduction of the signals induced by adherens junctions that are needed for desmosome assembly (Lewis *et al.*, 1997). By this model, an excess of Pg–Dsg complexes may downregulate the desmosome assembling machinery, which, however, does not require Pg as a structural component. This second model could explain both the Pg-dependent dominant-negative effect of Dsg C-domains conferred by expression of CoDsg chimeras (see above) and the persisting desmosome formation in murine epithelial cells lacking Pg (Ruiz *et al.*, 1996).

5. INTERPLAY AMONG DISTINCT DESMOSOMAL ASSEMBLY PROCESSES

At least three distinct processes can be distinguished that are needed for desmosome formation: (i) the interactions between cadherins of neighboring cells, (ii) the lateral clustering of cadherins, and (iii) the recruitment of desmosomal plaque proteins and IFs to the clusters. These processes may, at least in part, be interdependent. For example, process (i) resulting in formation of cadherin hetero- and/or homodimers may be required for activation of processes (ii) and (iii). Since the latter processes implicate intracellular proteins, this assumption postulates in turn the existence of a transmembrane cross-talk between extra- and intracellular cadherin domains. The existence of such a type of signaling was demonstrated for another class of adhesion molecules, integrins (for recent review see Humphries, 1996). It was shown that overall integrin conformation is changed upon binding to extracellular ligands and that interactions of the extracellular integrin portion with divalent cations are essential for these changes. Conformational changes of integrins are thought to activate the assembly of integrins in cell–extracellular matrix junctional structures. Additional experiments are needed to show whether transmembrane signaling leading to activation of intracellular cadherin domains exists and, if so, whether Ca^{2+} binding to cadherins is involved in such a process. In this respect, the observation reported by the laboratory of W. Franke (Demlehner *et al.*, 1995) is particularly intriguing, showing that desmosome assembly is possible without adhesive interactions between cadherins of neighboring cells. It was demonstrated that epithelial cells grown in low-Ca^{2+} medium efficiently form half-desmosomes. One of the possible interpretations of this finding is that changes in extracellular cation concentrations might lead to altered occupancy of cation-binding sites within desmosomal cadherins, which, in turn, mimics conformational changes involved in their activation.

In conclusion, we discussed evidence that desmosome assembly is a complex process implicating protein–protein interactions in both intracellular and extracellular compartments. Undoubtedly, elucidation of the precise mechanisms of desmosome assembly remains a major challenge. This knowledge will possibly give us an important tool to regulate cell–cell adhesion and intracellular distribution of IFs using pharmacological drugs that are applied extracellularly.

6. REFERENCES

Aberle, H., Butz, S., Stappert, J., Weissing, H., Kemler, R., and Hoschuetzky, H., 1994, Assembly of the cadherin-catenin complex *in vitro* with recombinant proteins, *J. Cell Sci.* **107:**3655–3663.

Aberle, H., Schwartz, H., Hoschuetzky, H., and Kemler, R., 1996, Single amino acid substitutions in proteins of the armadillo gene family abolish their binding to alpha-catenin, *J. Biol. Chem.* **271:**1520–1526.

Allen, E., Yu, Q.-C., and Fuchs, E., 1996, Mice expressing a mutant desmosomal cadherin exhibit abnormalities in desmosomes, proliferation, and epithelial differentiation, *J. Cell Biol.* **133:**1367–1382.

Amagai, M., 1995, Adhesion molecules. I: Keratinocyte–keratinocyte interactions; cadherins and pemphigus, *J. Invest. Dermatol.* **104:**146–152.

Amagai, M., Klaus-Kovtun, V., and Stanley, J. R., 1991, Autoantibodies against a novel epithelial cadherin in pemphigus vulgaris, a disease of cell adhesion, *Cell* **67:**869–877.

Amagai, M., Karpati, S., Klaus-Kovtun, V., Udey, M. C., and Stanley, J. R., 1994, Extracellular domain of pemphigus vulgaris antigen (desmoglein 3) mediates weak homophilic adhesion, *J. Invest. Dermatol.* **103:**609–615.

Amagai, M., Fujimori, T., Masunaga, T., Shimizu, H., Nishikawa, T., Shimuzu, N., Takeichi, M., and Hashimoto, T., 1995, Delayed assembly of desmosomes in keratinocytes with disrupted classic-cadherin-mediated cell adhesion by a dominant negative mutant, *J. Invest. Dermatol.* **104:**27–32.

Behrens, J., von Kries, J. P., Kuhl, M., Bruhn, L., Wedlich, D., Grosschedl, R., and Birchmeier, W., 1996, Functional interaction of beta-catenin with the transcription factor LEF-1, *Nature* **382:**638–642.

Blaschuk, O. W., Sullivan, R., David, S., and Pouliot, Y., 1990, Identification of a cadherin cell adhesion recognition sequence, *Dev. Biol.* **139:**227–229.

Bornslaeger, E. B., Corcoran, C. M., Stappenbeck, T. S., and Green. K. J., 1996, Breaking the connections: Displacement of the desmosomal plaque protein desmoplakin from cell–cell interfaces disrupts anchorage of intermediate filament bundles and alters intercellular junction assembly, *J. Cell Biol.* **134:**985–1001.

Brieher, W. M., Yap, A. S., and Gumbiner, B., 1996, Lateral dimerization is required for the homophilic binding activity of C-cadherin, *J. Cell Biol.* **135:**487–496.

Butz, S., Stappert, J., Weissig, H., and Kemler. R., 1992, Plakoglobin and β-catenin: Distinct but closely related, *Science* **257:**1142–1144.

Buxton, R. S., and Magee, A. I., 1992, Structure and interactions of desmosomal and other cadherins, *Semin. Cell Biol.* **3:**157–167.

Buxton, R. S., Cowin, P., Franke, W. W., Garrod, D. R., Green, K. J., King, I. A., Koch, P. J., Magee, I. A., Rees, D. A., Stanley, J. R., and Steinberg, M. S., 1993, Nomenclature of the desmosomal cadherins, *J. Cell Biol.* **121:**481–483.

Cartaud, A., Ludosky, M. A., Courvalin, J. C., and Cartaud, J., 1990, A protein antigenically related to nuclear lamin B mediates the association of intermediate filaments with desmosomes, *J. Cell Biol.* **111:**581–588.

Chidgey, M. A., Clarke, J. P., and Garrod, D. R., 1996, Expression of full-length desmosomal glycoproteins (desmocollins) is not sufficient to confer strong adhesion on transfected L929 cells, *J. Invest. Dermatol.* **106:**689–695.

Chitaev, N. A., and Troyanovsky, S. M., 1997, Direct Ca^{2+}-dependent heterophilic interaction between desmosomal cadherins, desmoglein and desmocollin, contributes to cell–cell adhesion, *J. Cell Biol.* **138:**193–201.

Chitaev, N. A., Leube, R. E., Troyanovsky, R. B., Eshkind, L. G., Franke, W. W., and Troyanovsky, S. M., 1996, The binding of plakoglobin to desmosomal cadherins: Patterns of binding sites and topogenic potential, *J. Cell Biol.* **133:**359–369.

Chou, Y.-H., Skalli, O., and Goldman, R. D., 1997, Intermediate filaments and cytoplasmic networking: New connections and more functions, *Curr. Opin. Cell Biol.* **9:**49–53.

Cowin, P., and Burke, B., 1996, Cytoskeleton–membrane interactions, *Curr. Opin. Cell Biol.* **8:**56–65.

Cowin, P., Mattey, D., and Garrod, D. R., 1984, Identification of desmosomal surface components (desmocollins) and inhibition of desmosome formation by specific FAB′, *J. Cell Sci.* **70:**41–60.

Cowin, P., Kapprell, H. -P., Franke, W. W., Tamkun, J., and Hynes. R. O., 1986, Plakoglobin: A protein common to different kinds of intracellular adhering junctions, *Cell* **46:**1063–1073.

Demlehner, M. P., Schäfer, S., Grund, C., and Franke, W. W., 1995, Continual assembly of half-desmosomal structures in the absence of cell contacts and their frustrated endocytosis: A coordinated sisyphus cycle, *J. Cell Biol.* **131:**745–760.

Dmochowski, M., Hashimoto, T., Amagai, M., Kudoh, J., Shimizu, N., Koch, P., Franke, W. W., and Nishikawa, T., 1994, The extracellular aminoterminal domain of bovine desmoglein 1 (Dsg1) is recognized only by certain pemphigus sera, whereas its intracellular domain is recognized by both pemphigus vulgaris and pemphigus foliaceus sera, *J. Invest. Dermatol.* **103:**173–177.

Fagotto, F., Funayama, N, Gluck, U., and Gumbiner, B., 1996, Binding to cadherins antagonizes the signalling activity of β-catenin during axis formation in *Xenopus*, *J. Cell Biol.* **132:**1105–1114.

Franke, W. W., Goldschmidt, M. D., Zimbelmann, R., Mueller, H. M., Schiller, D. L., and Cowin, P., 1989, Molecular cloning and amino acid sequence of human plakoglobin, the common junctional plaque protein, *Proc. Natl. Acad. Sci. USA* **86:**4027–4031.

Garrod, D. R., 1993, Desmosomes and hemidesmosomes, *Curr. Opin. Cell Biol.* **5**:30–40.

Garrod, D., Chidgey, M., and North, A., 1996, Desmosomes: Differentiation, development, dynamics and disease, *Curr. Opin. Cell Biol.* **8:**670–678.

Geiger, B., and Ayalon. O., 1992, Cadherins, *Annu. Rev. Cell Biol.* **8:**307–332.

Green, K. J., and Jones, J. C. R., 1990, Interaction of intermediate filaments with the cell surface, in *Cellular and Molecular Biology of Intermediate Filaments* (R. D. Goldman and P. M. Steinert, eds.), pp. 147–174, Plenum Press. New York.

Green, K. J., and Jones, J. C. R., 1996, Desmosomes and hemidesmosomes: Structure and function of molecular components, *FASEB J.* **10:**871–881.

Green, K.J., Parry, D. A. D., Steinert, P. M., Virata, M. L. A., Wagner, R. M., Angst, B. D., and Nilles, L. A., 1990, Structure of the human desmoplakin: Implications for function in the desmosomal plaques, *J. Biol. Chem.* **265:**2603–2612.

Grunwald, G. B., 1993, The structural and functional analysis of cadherin calcium-dependent cell adhesion molecules, *Curr. Opin. Cell Biol.* **5:**797–805.

Hatzfeld, M., and Nachtsheim, C., 1996, Cloning and characterization of a new armadillo family member, p0071, associated with the junctional plaque: Evidence for a subfamily of closely related proteins, *J. Cell Sci.* **109:**2767–2778.

Hatzfeld, M., Kristjansson, G. I., Plessmann, U., and Weber, K., 1994, Band 6 protein, a major constituent of desmosomes from stratified epithelia, is a novel member of the *armadillo* multigene family, *J. Cell Sci.* **107:**2259–2270.

Heid, H. W., Schmidt, A., Zimbelmann, R., Schäfer, S., Winter-Simanowski, S., Stumpp, S., Keith, M., Figge, U., Schnölzer, M., and Franke, W. W., 1994, Cell type-specific desmosomal plaque proteins of the Plakoglobin family: Plakophilin 1 (band 6 protein), *Differentiation* **58:**113–131.

Hinck, L., Näthke, I. S., Papkoff, J., and Nelson, W. J., 1994, Dynamics of cadherin/catenin complex formation: Novel protein interactions and pathways of complex assembly, *J. Cell Biol.* **125:**1327–1340.

Hoschuetzky, H., Aberle, H., and Kemler, R., 1994, β-Catenin mediates the interaction of the cadherin–catenin complex with epidermal growth factor receptor, *J. Cell Biol.* **127:**1375–1380.

Huber, O., Korn, R., McLaughlin, J., Ohsugi, M., Herrmann, B. G., and Kemler, R., 1996, Nuclear localization of β-catenin by interaction with transcription factor LEF-1, *Mech. Dev.* **59:**3–10.

Humphries, M. J., 1996, Integrin activation: The link between ligand binding and signal transduction, *Curr. Opin. Cell Biol.* **8:**632–640.

Jou, T.-S., Stewart, D. B., Stappert, J., Nelson, W. J., and Marrs, J. A., 1995, Genetic and biochemical dissection of protein linkages in the cadherin–catenin complex, *Proc. Natl. Acad. Sci. USA* **92:**5067–5071.

Kartenbeck, J., Schwechheimer, K., Moll, R., and Franke, W. W., 1984, Attachment of vimentin filaments to desmosomal plaques in human meningiomal cells and arachnoidal tissue, *J. Cell Biol.* **98:**1072–1081.

Kawamura, K., Watanabe, K., Suzuki, T., Yamakawa, T., Kaniyama, T., Nakagawa, H., and Tsurufuji, S., 1994, cDNA cloning and expression of a novel human desmocollin, *J. Biol. Chem.* **269:**26295–26302.

Kintner, C., 1992, Regulation of embryonic cell adhesion by the cadherin cytoplasmic domain, *Cell* **69:**225–236.

Klymkowsky, M. W., and Parr. B., 1995, A glimpse into the body language of cells: The intimate connection between cell adhesion and gene expression, *Cell.* **83:**5–8.

Knudsen, K. A., and Wheelock, M. J., 1992, Plakoglobin, or an 83-kD homologue distinct from β-catenin, interacts with E-cadherin and N-cadherin, *J. Cell Biol.* **118:**671–679.

Koch, P. J., and Franke, W. W., 1994, Desmosomal cadherins: Another growing multigene family of adhesion molecules, *Curr. Opin. Cell Biol.* **6:**682–287.

Koch, P. J., Walsh, M. J., Schmelz, M., and Franke, W. W., 1990, Identification of desmoglein, a constitutive desmosomal glycoprotein, as a member of the cadherin family of cell adhesion molecules, *Eur J. Cell Biol.* **53:**1–12.

Koch, P. J., Goldschmidt, M. D., Walsh, M. J., Zimbelmann, R., Schmelz, M., and Franke, W. W., 1991, Amino acid sequence of bovine muzzle epithelial desmocollin derived from cloned cDNA: A novel subtype of desmosomal cadherins, *Differentiation* **47:**29–36.

Koch, P. J., Mahoney, M. G., Ishikawa, H., Pulkkinen, L., Uitto, J., Shultz, L., Murphy, G. F., Whitaker-Menezes, D., and Stanley, J. R., 1997, Targeted disruption of the pemphigus vulgaris antigen (desmoglein 3) gene in mice causes loss of keratinocyte cell adhesion with a phenotype similar to pemphigus vulgaris, *J. Cell Biol.* **137:**1091–1102.

Korman, N. J., Eyre, R. W., Klaus-Kovtun, V., and Stanley, J. R., 1989, Demonstration of an adhering-junction molecule (plakoglobin) in the autoantigens of pemphigus foliaceus and pemphigus vulgaris, *N. Engl. J. Med.* **321:**631–635.

Kouklis, P. D., Hutton, E., and Fuchs, E., 1994, Making a connection: Direct binding between keratin intermediate filaments and desmosomal proteins, *J. Cell Biol.* **127:**1049–1060.

Kowalczyk, A. P., Palka, E., Luu, H. H., Nilles, L. A., Anderson, J. E., Wheelock, M. J., and Green, K. J., 1994a, Posttranslational regulation of plakoglobin expression. Influence of the desmosomal cadherins on plakoglobin metabolic stability, *J. Biol. Chem.* **269:**31214–31223.

Kowalczyk, A. P., Stappenbeck, T. S., Parry, D. A. D., Palka, H. L., Virata, M. L. A., Bornslaeger, E. A., Nilles, L. A., and Green, K. J., 1994b, Structure and function of desmosomal transmembrane core and plaque molecules, *Biophys. Chem.* **50:**97–112.

Kowalczyk, A. P., Anderson, J. E., Borgwardt, J. E., Hashimoto, T., Stanley, J. R., and Green, K. J., 1995. Pemphigus sera recognize conformationally sensitive epitopes in the amino-terminal region of desmoglein-1, *J. Invest. Dermatol.* **105:**147–152.

Kowalczyk, A. P., Borgwardt, J. E., and Green, K. J., 1996, Analysis of desmosomal cadherin-adhesive function and stoichiometry of desmosomal cadherin–plakoglobin complexes, *J. Invest. Dermatol.* **107:**293–300.

Leube, R. E., 1995, The topogenic fate of the polytopic transmembrane proteins, synaptophysin and connexin, is determined by their membrane-spanning domains, *J. Cell Sci.* **108:**883–894.

Leube, R. E., Bader, B. L., Bosch, F. X., Zimbelmann, R., Achtstaetter, T., and Franke, W. W., 1988, Molecular characterization and expression of the stratification-related cytokeratins 4 and 15, *J. Cell Biol.* **106:**1249–1261.

Leube, R. E., Wiedenmann, B., and Franke. W. W., 1989, Topogenesis and sorting of synaptophysin: Synthesis of a synaptic vesicle protein from a gene transfected into nonneuroendocrine cells, *Cell* **59:**433–446.

Leube, R. E., Leimer, U., Grund, C., Franke, W. W., Harth, N., and Wiedenmann. B., 1994, Sorting of synaptophysin into special vesicles in nonneuroendocrine epithelial cells, *J. Cell Biol.* **127:**1589–1601.

Lewis, J. E., Jensen, P. J., and Wheelock. M. J., 1994, Cadherin function is required for human keratinocytes to assemble desmosomes and stratify in response to calcium, *J. Invest Dermatol.* **102:**870–877.

Lewis, J. E., Wahl III, J. K., Sass, K. M., Jensen, P. J., Johnson, K. R., and Wheelock, M. J., 1997, Cross-talk between adherens junctions and desmosomes depends on plakoglobin, *J. Cell Biol.* **136:**919–934.

Magee, A. I., and Buxton, R. S., 1991, Transmembrane molecular assemblies regulated by the greater cadherin family, *Curr. Opin. Cell Biol.* **3:**854–861.

Mathur M., Goodwin, L., and Cowin, P., 1994, Interactions of the cytoplasmic domain of the desmosomal cadherin Dsgl with plakoglobin, *J. Biol. Chem.* **269:**14075–14080.

McCrea, P. D., Turck, C. W., and Gumbiner, B., 1991, A homolog of the *armadillo* protein in *Drosophila* (plakoglobin) associated with E-cadherin, *Science* **254:**1359–1361.

Mertens, C., Kuhn, C., and Franke, W. W., 1996, Plakophilins 2a and 2b: Constitutive proteins of dual location in the karyoplasm and the desmosomal plaque, *J. Cell Biol.* **135:**1009–1025.

Miller, J. R., and Moon, R. T., 1996, Signal transduction through β-catenin and specification of cell fate during embryogenesis, *Genes Dev.* **10:**2527–2539.

Molenaar, M., van de Wetering, M., Oosterwegel, M., Peterson-Maduro, J., Godsave, S., Korinek., V., Roose, J., Destree, O., and Clevers, H., 1996, XTcf-3 transcription factor mediates β-catenin-induced axis formation in *Xenopus* embryos, *Cell* **86:**391–400.

Nagafuchi, A., and Takeichi, M., 1988, Cell binding function of E-cadherin is regulated by the cytoplasmic domain, *EMBO J.* **7:**3679–3684.

Nagar, B., Overduin, M., Ikura, M., and Rini, J. M., 1996, Structural basis of calcium-induced E-cadherin rigidification and dimerization, *Nature* **380:**360–364.

North, A. J., Chidgey, M. A. J., Clarke, J. P., Bardsley, W. G., and Garrod, D. R., 1996, Distinct desmocollin isoforms occur in the same desmosomes and show reciprocally graded distribution in bovine nasal epidermis, *Proc. Natl. Acad. Sci. USA* **93:** 7701–7705.

Nose, A., Tsuji, K., and Takeichi. M., 1990, Localization of specificity determining sites in cadherin cell adhesion molecules, *Cell* **61:**147–155.

Nuber, U. A., Schäfer, S., Stehr, S., Rackwitz, H.-R., and Franke, W. W., 1996, Patterns of desmocollin synthesis in human epithelia: Immunolocalization of desmocollins 1 and 3 in special epithelia and in cultured cells, *Eur. J. Cell Biol.* **73:**1–13.

Nusse, R, 1997, A versatile transcriptional effector of wingless signalling, *Cell* **89:**321–323.

O'Keefe, E. J., Erickson, H. P., and Bennett, V., 1989, Desmoplakin I and desmoplakin II. Purification and characterization, *J. Biol. Chem.* **264:**8310–8318.

Olague-Alcala, M., Giudice, G. J., and Diaz, L. A., 1994, *Pemphigus foliaceus* sera recognize an N-terminal fragment of bovine desmoglein 1, *J. Invest. Dermatol.* **102:**882–885.

Orsulic, S., and Peifer, M., 1996, An *in vivo* structure–function study of Armadillo, the β-catenin homologue, reveals both separate and overlapping regions of the protein required for cell adhesion and for wingless signalling, *J. Cell Biol.* **134:**1283–1300.

Ouyang, P., and Sugrue, S. P., 1996, Characterization of pinin, a novel protein associated with the desmosome-intermediate filament complex, *J. Cell Biol.* **135:**1027–1042.

Ozawa, M., Baribault, H., and Kemler, R., 1989, The cytoplasmic domain of the cell adhesion molecule uvomorulin associates with three independent proteins structurally related in different species, *EMBO J.* **8:**1711–1717.

Ozawa, M., Engel, J., and Kemler, R., 1990a, Single amino acid substitution in one Ca^{2+} binding site of uvomorulin abolishes the adhesive function, *Cell* **63:**1033–1038.

Ozawa, M., Ringwald, M., and Kemler, R., 1990b, Uvomorulin-catenin complex formation is regulated by a specific domain in the cytoplasmic region of the cell adhesion molecule, *Proc. Natl. Acad. Sci. USA* **87:**4246–4250.

Ozawa, M., Hoschutzky, H., Herrenknecht, K., and Kemler, R., 1991, A possible new adhesion site in the cell-adhesion molecule uvomorulin, *Mech. Dev.* **33:**49–56.

Parker, A. E., Weeler, G. N., Amemann, J., Pidsley, S. C., Ataliotis, P., Thomas, C. L., Rees, D. A., Magee, A. I., and Buxton, R. S., 1991, Desmosomal glycoproteins II and III. Cadherin-like junctional molecules generated by alternative splicing, *J. Biol. Chem.* **266:**10438–10445.

Pasdar, M., and Nelson, W. J., 1989, Regulation of desmosome assembly in epithelial cells: Kinetics of synthesis, transport, and stabilization of desmoglein 1, a major protein of the membrane core domain, *J. Cell Biol.* **109:**163–177.

Pasdar, M., Krzeminski, K. A., and Nelson, W. J., 1991, Regulation of desmosome assembly in MDCK epithelial cells: Coordination of membrane core and cytoplasmic plaque domain assembly at the plasma membrane, *J. Cell Biol.* **113:**645–655.

Pasdar, M., Li, Z., and Chan, H., 1995, Desmosome assembly and disassembly are regulated by reversible protein phosphorylation in cultured epithelial cells, *Cell. Motil. Cytoskel.* **30:**108–121.

Peifer, M., 1995, Cell adhesion and signal transduction: The Armadillo connection, *Trends Cell Biol.* **5:**224–229.

Peifer, M., Berg, S., and Reynolds, A. B., 1994, A repeating amino acid motif shared by proteins with diverse cellular roles, *Cell* **76:**789–791.

Pokutta, S., Herrenknecht, K., Kemler, R., and Engel, J., 1994, Conformational changes of the recombinant extracellular domain of E-cadherin upon calcium binding, *Eur. J. Biochem.* **223:**1019–1026.

Rajasekaran, A. K., Hojo, M., Huima, T., and Rodriguez-Boulan, E., 1996, Catenins and zonula occludens-1 form a complex during early stages in the assembly of tight junctions, *J. Cell Biol.* **132:**451–463.

Reid, R. A., and Hemperly, J. J., 1990, Human N-cadherin: Nucleotide and deduced amino acid sequence, *Nucleic Acids Res.* **18:**5896–5899.

Reynolds, A. B., Herbert, L., Cleveland, J. L., Berg, S. T., and Gaut, J. R., 1992, p120, a novel substrate of protein tyrosine kinase receptors and of $p60^{v\text{-}src}$, is related to cadherin-binding factors b-catenin, plakoglobin and armadillo, *Oncogene* **7:**2439–2445.

Ringwald, M., Schuh, R., Vestweber, D., Eistetter, H., Lottspeich, F., Engel, J., Dolz, R., Jahnig, F., Epplen, J., Mayer, S., Muller, C., and Kemler, R., 1987, The structure of cell adhesion molecule uvomorulin. Insights into the molecular mechanism of Ca^{2+}-dependent cell adhesion, *EMBO J.* **6:**3647–3653.

Roh, J.-Y., and Stanley, J. R., 1995, Intracellular domain of desmoglein 3 (pemphigus vulgaris antigen) confers adhesive function on the extracellular domain of E-cadherin without binding catenins, *J. Cell Biol.* **128:**939–947.

Rosenthal, E., 1993, Identification of homologues to β-catenin/plakoglobin/armadillo in two invertebrates, *Urechis caupo* and *Tripneustes gratilla*, *Biochim. Biophys. Acta* **1173:**337–341.

Rubinfeld, B., Souza, B., and Polakis, I., 1993, Association of the APC gene product with beta-catenin, *Science* **262:**1731–1734.

Rubinfeld, B., Souza, B., Albert, I., Munemitsu, S., and Polakis, P., 1995, The APC protein and E-cadherin form similar but independent complexes with α-catenin, β-catenin, and plakoglobin, *J. Biol. Chem.* **270:**5549–5555.

Ruiz, P., Brinkmann, V., Ledermann, B., Behrend, M., Grund, C., Thalhammer, C., Vogel, F., Birchmeier, C., Gunthert, U., Franke, W. W., and Birchmeier, W., 1996, Targeted mutation of plakoglobin in mice reveals essential functions of desmosomes in the embryonic heart, *J. Cell. Biol.* **135:**215–225.

Sacco, P. A., McGranahan, T. M., Wheelock, M. J., and Johnson, K. R., 1995, Identification of plakoglobin domains required for association with N-cadherin and α-catenin, *J. Biol. Chem.* **270:**20201–20206.

Schäfer, S., Troyanovsky, S., Eshkind, L., Koch, P., and Franke, W. W., 1993, Cytoskeletal architecture and epithelial differentiation: Molecular determinants of cell interaction and cytoskeletal filament anchorage, *C. R. Acad. Sci. Paris Sci. Vie/Life Sci* **316:**1316–1323.

Schäfer S., Koch, P., and Franke, W. W., 1994, Identification of the ubiquitous human desmoglein, Dsg2, and the expression catalogue of the desmoglein subfamily of desmosomal cadherins, *Exp. Cell Res.* **211:**391–399.

Schäfer, S., Stumpp, S., and Franke. W. W., 1996, Immunological identification and characterization of the desmosomal cadherin Dsg2 in coupled and uncoupled epithelial cells and in human tissues, *Differentiation* **60:**99–108.

Schmidt, A., Heid, H. W., Schäfer, S., Nuber, U. A., Zimbelmann, R., and Franke, W. W., 1994, Desmosomes and cytoskeletal architecture in epithelial differentiation. Cell type-specific plaque components and intermediate filament anchorage, *Eur. J. Cell Biol.* **65:**229–245.

Schwarz, M. A., Owaribe, K., Kartenbeck, J., and Framke, W. W., 1990, Desmosomes and hemidesmosomes: Constitutive molecular components, *Annu. Rev. Cell Biol.* **6:**461–491.

Shapiro, L., Fannon, A. M., Kwong, P. D., Thompson, A., Lehman, M. S., Grubel, G., Legran, J.-F., Als-Neilsen, J., Colman, D. R., and Hendrickson, W. A., 1995, Structural basis of cell–cell adhesion by cadherins, *Nature* **374:**327–337.

Sheu, H.-M., Kitajima, Y., and Yaoita, H., 1989, Involvement of protein kinase C in translocation of desmoplakins from cytosol to plasma membrane during desmosome formation in human squamous cell carcinoma cells grown in low to normal calcium concentration, *Exp. Cell Res.* **185:**176–190.

Shimoyama, Y., Yoshida, T., Terada, M., Shimosata, Y., Abe, O., and Hirohashi, S., 1989, Molecular cloning of a human Ca^{2+}-dependent cell–cell adhesion molecule homologous to mouse placenta cadherin: Its low expression in human placental tissues, *J. Cell Biol.* **109:**1787–1794.

Stanley, J. R., 1995, Autoantibodies against adhesion molecules and structures in blistering skin diseases, *J. Exp. Med.* **181:**1–4.

Stappenbeck, T., and Green, K. J., 1992, The desmoplakin carboxyl terminus coaligns with and specifically disrupts intermediate filament networks when expressed in cultured cells, *J. Cell Biol.* **116:**1197–1209.

Stappenbeck, T., Bornslaeger, E. A., Corcoran, C. M., Luu, H. H., Virata, M. L. A., and Green, K. J., 1993, Functional analysis of desmoplakin domains: Specification of the interaction with keratin versus vimentin intermediate filament networks, *J. Cell Biol.* **123:**691–705.

Stappenbeck, T., Lamb, J. A., Corcoran, C. M., Luu, H. H., Virata, M. L. A., and Green, K. J., 1994, Phosphorylation of the desmoplakin COOH terminus negatively regulates its interaction with keratin intermediate filament networks, *J. Biol. Chem.* **269:**29351–29354.

Su, L. K., Vogelstein, B., and Kinzler, K. W., 1993, Association of the APC tumor suppressor protein with catenins, *Science* **262:**1734–1737.

Takeichi, M., 1990, Cadherins: A molecular family important in selective cell–cell adhesion, *Annu. Rev. Biochem.* **59:**237–252.

Tao, Y. S., Edwards, R. A., Tubb, B., Wang, S., Bryan, J., and McCrea, P. D., 1996 β-Catenin associates with actin-bundling protein fascin in a noncadherin complex, *J. Cell Biol.* **134:**1271–1281.

Theis, D. G., Koch, P. J., and Franke, W. W., 1993, Differential synthesis of type 1 and type 2 desmocollin mRNAs in human stratified epithelia, *Int. J. Dev. Biol.* **37:**101–110.

Trevor, K. T., and Steben, L. S., 1992, Distribution of desmosomal proteins in F9 embryonal carcinoma cells and epithelial cell derivatives, *J. Cell Sci.* **103:**69–80.

Troyanovsky, S. M., Eshkind, L. G., Troyanovsky, R. B., Leube, R. E., and Franke, W. W., 1993, Contributions of cytoplasmic domains of desmosomal cadherins to desmosome assembly and intermediate filament anchorage, *Cell.* **72:**561–574.

Troyanovsky, S. M., Troyanovsky, R. B., Eshkind, L. G., Krutovskikh, V. A., Leube, R. E., and Franke, W. W., 1994a, Identification of the plakoglobin-binding domain in desmoglein and its role in plaque assembly and intermediate filament anchorage, *J. Cell Biol.* **127:**151–160.

Troyanovsky, S. M., Troyanovsky, R. B., Eshkind, L. G., Leube, R. E., and Franke, W. W., 1994b, Identification of amino acid sequence motifs in the desmosomal glycoprotein, desmocollin, that are required for plakoglobin binding and plaque formation, *Proc. Natl. Acad. Sci. USA.* **91:**10790–10794.

Troyanovsky, R. B., Chitaev, N. A., and Troyanovsky, S. M., 1996, Cadherin binding sites of plakoglobin: Localization, specificity and role in targeting to adhering junctions, *J. Cell Sci.* **109:**3069–3078.

Tsukita, S., and Tsukita, S., 1985, Desmocalmin: A calmodulin-binding high molecular weight protein isolated from desmosomes, *J. Cell Biol.* **98:**1565–1571.

van de Wetering, M., Cavallo, R., Dooijes, D., van Beest, M., van Es, J., Loureiro, J., Ypma, A., Hursh, D., Jones, T., Bejsovec, A., Peifer, M., Mortin, M., and Clevers, H., 1997, Armadillo coactivates transcription driven by the product of the *Drosophila* segment polarity gene dTCF, *Cell* **88:**789–799.

van Hengel, L., Gohon, L., Bruyneel, E., Vermeulen, S., Mareel, M., and van Roy, F., 1997, Protein kinase C activation upregulates intercellular adhesion of α-catenin-negative human colon cancer cell variants via induction of desmosomes, *J. Cell Biol.* **137:**1103–1116.

Virata, M. L. A., Wagner, R. M., Parry, D. A. D., and Green, K. J., 1992, Molecular structure of the human desmoplakin I and II amino terminus, *Proc. Natl. Acad. Sci. USA* **89:**544–548.

Wahl, J. K., Sacco, P. A., McGranahan-Sadler, T. M., Sauppe, L. M., Weelock, M. J., and Johnson, K. R., 1996, Plakoglobin domains that define its association with the desmosomal cadherins and the classical cadherins: Identification of unique and shared domains, *J. Cell. Sci.* **109:**1143–1154.

Wheeler, G. N., Parker, A. E., Thomas, C. L., Ataliotis, P., Poynter, D., Arnemann, J., Rutman, A. J., Pidsley, S. C., Watt, F. M., Rees, D. A., Buxton, R. S., and Magee, A. I., 1991, Desmosomal glycoprotein DG1, a component of intercellular desmosome junctions, is related to the cadherin family of cell adhesion molecules, *Proc. Natl. Acad. Sci. USA* **88:**4796–4800.

Witcher, L. L., Collins, R., Puttagunta, S., Mechanic, S. E., Munson, M., Gumbiner, B., and Cowin, P., 1996, Desmosomal cadherin binding domains of plakoglobin, *J. Biol. Chem.* **271:**10904–10909.

Chapter 9

Lens Intermediate Filament Proteins

Aileen Sandilands, Shigeo Masaki, and
Roy A. Quinlan

1. THE VERTEBRATE LENS—A BRIEF OVERVIEW

The vertebrate lens is a transparent tissue whose sole purpose is to transmit and focus light onto the retina. The vast bulk of the lens is comprised of highly elongated lens fiber cells which are derived from the differentiation of a single layer of epithelial cells which overlie the anterior face of the lens. Crystallins (α-, β-, and γ-crystallin) comprise the vast majority of the protein within the fiber cells and contribute to the refractive index of the lens. The fiber cells are formed by differentiation from a single layer of epithelial cells which underlie the anterior lens capsule. This differentiation process continues throughout the life of the individual with new lens fiber cells laid down over preexisting fiber cells. The lens fiber cells are thus arranged as concentric layers, with the oldest fiber cells found within the center of the lens. The lens is subdivided into a cortex which contains the most recently differentiated fiber cells, and a nucleus which contains the oldest lens fiber cells, those fiber cells laid down during embryonic develop-

---

Aileen Sandilands and Roy A. Quinlan Department of Biochemistry, University of Dundee, Dundee DD1 4HN, Scotland. **Shigeo Masaki** Department of Biochemistry, Institute for Developmental Research, Aichi Human Service Center, Kasugai, Aichi 480-03, Japan.

Subcellular Biochemistry, Volume 31: Intermediate Filaments, edited by Herrmann and Harris. Plenum Press, New York, 1998.

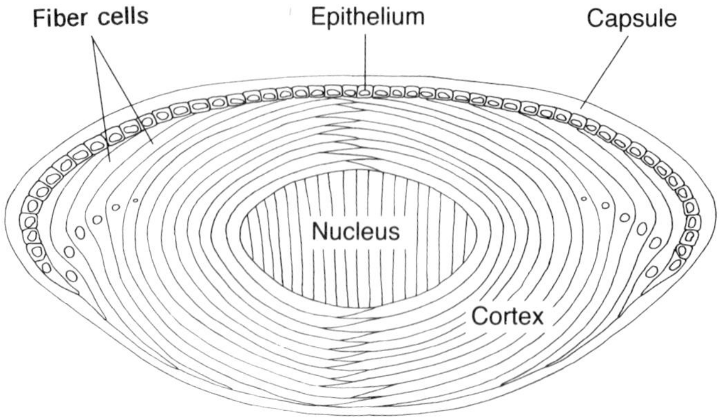

FIGURE 1. Diagrammatic representation of the mammalian lens. A cross section of the mammalian lens reveals the elongated lens fiber cells arranged as concentric layers with the oldest lens fiber cells at the center of the lens (nucleus). Differentiation of lens epithelial cells into fiber cells continues throughout the life of the individual and occurs within the bow region.

ment (primary lens fiber cells). Figure 1 shows a diagrammatic representation of the mammalian lens.

One of the most striking features of the lens differentiation pathway is the removal of all intracellular organelles including the nucleus from fiber cells as they differentiate, a process essential to reduce light scattering. Differentiation occurs within the equatorial (or bow region) of the lens and this process is characterised by rapid cellular elongation, the formation of numerous cell–cell interactions (ball-and-socket joints), and the removal of cytoplasmic organelles. Accompanying these morphological changes are biochemical changes such as the expression of lens fiber cell-specific crystallins (e.g., the β- and γ-crystallins) and plasma membrane-associated proteins (e.g., membrane protein 26), reviewed by Piatigorsky (1981). Not surprisingly, the cytoskeleton of the lens, and the intermediate filament network in particular, is also significantly remodeled during this event.

2. INTERMEDIATE FILAMENT PROTEINS OF THE LENS

During the early stages of lens embryonic development keratins 8 and 18 are transiently expressed (Kasper and Viebahn, 1992). However, despite the ectodermal origins of the lens, the type III intermediate filament protein vimentin is abundantly expressed throughout the lens (Ellis *et al.*, 1984;

Ramaekers *et al.*, 1980, 1982) and the lens has often been used as a source for purifying this protein (Geisler and Weber, 1981). More recently, however, two other intermediate filament proteins have been identified within the lens, and on the basis of Northern blot (Hess *et al.*, 1993; Merdes *et al.*, 1993) and immunological analyses (Remington, 1993; Ireland and Maisel, 1984) are considered specific to the lens. These intermediate filament proteins were initially isolated from urea-soluble extracts of chick lenses and were subsequently named cytoskeletal protein 49 (CP49) and cytoskeletal protein 95 (CP95) (Ireland and Maisel, 1983), where 49 and 95 denote their apparent sizes in kDa by sodium dodecyl sulfate–polyacrylamide gel electrophoresis (SDS–PAGE). Later investigations revealed the presence of immunologically related proteins within the lenses of other vertebrates ranging from squid to human (FitzGerald and Casselman, 1991).

CP49 is also referred to as phakinin (Merdes *et al.*, 1993), and CP95 has been identified as CP115 in the bovine lens (FitzGerald and Gottlieb, 1989) and CP94 within the rat lens (Masaki and Watanabe, 1992). CP94, CP95, and CP115 are also collectively referred to as filensin (Gounari *et al.*, 1993; Merdes *et al.*, 1991).

3. SEQUENCE CHARACTERISTICS OF LENS-SPECIFIC INTERMEDIATE FILAMENT PROTEINS

cDNA sequences have been obtained for bovine (Gounari *et al.*, 1993), rat (Masaki and Watanabe 1992), and chicken filensin (Remington, 1993) and for human (Hess *et al.*, 1996), bovine (Merdes *et al.*, 1993), mouse (partial cDNA; Hess *et al.*, 1993), and chicken CP49 (Sawada *et al.*, 1995). Primary and secondary sequence analyses confirm that both proteins are members of the intermediate filament protein family. However, both filensin and CP49 reveal unusual sequence characteristics, illustrated in Figure 2, which makes assigning either protein to a particular intermediate filament protein type problematic.

3.1. Filensin

Over the rod region, the currently known filensins share over 70% amino acid homology (Masaki and Quinlan, 1997). Comparison of the primary amino acid sequence of filensin with that of other intermediate filament proteins, however, reveals a variable degree of homology. Weak homology with the type II (basic) keratins has been reported (Remington, 1993; Masaki and Watanabe, 1992). Not surprisingly the greatest degree of homology resides within the central rod domain; for example, coil 1A

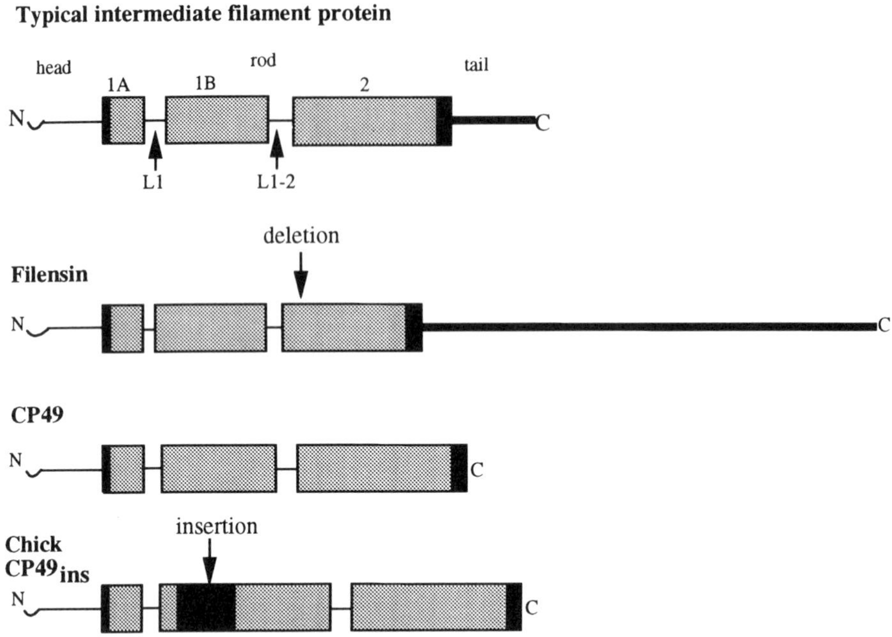

FIGURE 2. Comparison of the secondary structure of lens intermediate filament proteins with a typical intermediate filament protein. Intermediate filament proteins can be subdivided into an α-helical central rod domain flanked on either side by non-α-helical head and tail domains. Consensus sequences important in the assembly of intermediate filament proteins (LNDR and TYRKLLEGEE) are denoted by black boxes at either end of the rod domain. Filensin has a truncated rod domain as a result of a deletion within helix 2 which is predicted to be 2–3 nm shorter than other intermediate filament proteins. Notice also the extensive C-terminal tail domain of filensin. CP49 is a completely tail less intermediate filament protein. Also shown is chick CP49*ins*, which has an insertion of 49 amino acids within coil 1B.

of bovine filensin shows 43% identity with coil 1A of neurofilament L, whereas coil 2 shows 25% identity to coil 2 of nestin (Gounari *et al.*, 1993).

Analysis of the primary sequence and secondary structure of filensin (Gounari *et al.*, 1993; Remington, 1993) reveals a tripartite organization which is typical of all intermediate filament proteins. Filensin contains a central rod domain of between 279 (bovine; Gounari *et al.*, 1993) and 280 amino acids (chicken; Remington, 1993). The rod domain can be further subdivided into coils 1A, 1B, and 2 (which are predicted to be α-helical in conformation) separated by short non-α-helical linker regions. Coil 2 also contains an interruption in the heptad repeat pattern, the so-called "stutter." The striking feature of filensin therefore is its shortened rod domain (~2–3 nm shorter than the rod domain of other cytoplasmic intermediate filament proteins) which results from a truncation within coil 2 immediately after the fourth heptad repeat (Gounari *et al.,* 1993; Remington, 1993). This truncation means that filensin has the shortest rod domain of all intermediate filament proteins. The size of the truncation can vary, as in the mouse, the truncation is only 43 residues due to an insertion of 18 bp (Masaki and Quinlan, 1997). Flanking the rod domain is an amino-terminal head domain of between 35 (chicken; Remington, 1993) to 38 amino acids (bovine; Gounari *et al.*, 1993) in length. Contained within the head domain of all known filensin sequences is a conserved diarginine/aromatic residue motif typical of type III intermediate filament proteins and which contains a consensus protein kinase A phosphorylation site (Gounari *et al.*, 1993). Filensin also possesses an extensive C-terminal tail domain which ranges from 339 (chicken) (Remington, 1993) to 428 (bovine) amino acids (Gounari *et al.*, 1993). Within the tail domain of bovine filensin there is a sequence motif of 14 amino acids which is repeated 6.5 times (Figure 3) and which is predicted to form β-turns (Gounari *et al.*, 1993). This motif has only been found within bovine filensin and corresponds to sequences found in the tail domains of the nuerofilament triplet protein of medium (NF-M) and high (NF-H) molecular weight (Gounari *et al.*, 1993). The most divergent region of filensin is the C-terminal tail domain, yet comparison of the amino acid sequences from the C-terminal tail domain of mouse and rat (Masaki and Quinlan, 1997), bovine (Gounari *et al.,* 1993), and chicken (Remington, 1993) filensin reveals two motifs which are conserved between all four species (Masaki and Quinlan, 1997). These are illustrated in Figure 3. The motifs appear to be important for filensin and CP49 filament assembly (Masaki and Quinlan, 1997; Goulielmos *et al.*, 1996b). Figure 4 compares the consensus motifs located at the start of coil 1A (LNDR) and at the end of coil 2 (TYRKLLEGEE) of various intermediate filament proteins from all five sequence homology classes with those of filensin (Gounari *et al.*, 1993; Remington, 1993).

Given the weak degree of sequence homology and distinct assembly properties, it is indeed difficult to classify filensin into one of the known classes of intermediate filament proteins.

3.2. CP49

Analysis of the primary and secondary sequence of CP49 reveals two interesting features. First, CP49 is a completely tailless intermediate filament protein. A stop codon is located immediately after the final amino acid in coil 2 (Hess *et al.*, 1996; Merdes *et al.*, 1993). Second, the highly conserved consensus sequence located within the start of coil 1A of the rod

```
       320                                                    369
Mouse  gsRLsSvfie TPIsLiTpsh gaslslgssv kDlarAvQDI taaKPRqKaL
Rat    gnRLsSvfie TPIsLiTpsh gaslslgsnv kDltrAvQDI taaKPRqKaL
Cow    gnRLsSafie TPItLyTash gaslsprhgg kDltrAvQDI taaKPRlKgL
Chick  dsRLnSaiag TPVtLfTqiy rpvqpqasrg rDitqAmQEI asvKPRqKaL
       --RL-S---- TP--L-T--- ---------- -D---A-Q-I ---KPR-K-L

       370                                                    419
Mouse  pKsLpkrKEi iaqDkveetl eDapl.kppq trkalqverk aEGgsqpgaG
Rat    pKsLpkrKEi iaqDkvdetl eDapl.ktlq epkavqgelt gDGdsqlgaG
Cow    pKnLprkKEm vakDradeil eEtll.rgpe dmkpgrvvik eEGesklepG
Chick  tKkLsrkKEi mskDitdgls pEklyertve vfdqdqlefr hEGsvtcepG
       -K-L---KE- ---D------ ---------- ---------- --G------G

       420                                                    469
Mouse  gghgvsptqE ggPEDVPDGg qISKAFgKLc kvvKErVsgh KepePEpptn
Rat    gghevsptqE ggPEDVPDGs qISKAFgKLc kvvKErVsgh KepvPEppad
Cow    deeaspptqE gaPEDVPDGg kISKAFeKLg kmiKEkVkgp Ke..PEppad
Chick  qeelelvekE avPEDVPDGa qISKAFdKLc nlvKEkIrvy Kr..PEakvd
       ---------E --PEDVPDG- -ISKAF-KL- ---KE-V--- K---PE----

       470                                                    519
Mouse  lftKGRhvlV tGessfvdPe FyssSIPArG GvviSieeds mhhdghVEPs
Rat    lftKGRhilV tGessfvdPe FyssSIPArG GvvvSieeds mhhdghVEPs
Cow    lytKGRyvmV sGdgsfvdPg FcvfSVPAkG GvvvSkgdds vppdsgVEPs
Chick  shpKGRyvlV tGeegyeePc F..sSIPAgG GitvStsngk vtiggdVEPi
       ---KGR---V -G------P- F---SIPA-G G---S----- ------VEP-

       520                                                    569
Mouse  PgqPmPpven gqgvp..... .......... .......... ..........
Rat    PgqPmPpven gqgvp..... .......... .......... ..........
Cow    PqqPePplee gqgppqeked glkeeggppe gkgeppegkg dsvkeeggpp
Chick  PelPePseps ekekr..... .......... .......... ..........
       P--P-P---- ---------- ---------- ---------- ----------

       570                                                    619
Mouse  .......... .......... .......... .......... ..........
Rat    .......... .......... .......... .......... ..........
Cow    egkgdgvkee ggppegkgdg vkeeggppeg kgdgvkkege ppegkgeglk
Chick  .......... .......... .......... .......... ..........
       ---------- ---------- ---------- ---------- ----------

       620                                                    669
Mouse  ......qgre gdhsnhqqgt dknglr.akE pkdLeekddd gkkeaegsrr
Rat    ......qgre gahsnhqqvt dkngir.akE pkdLeekddd srkddeagrr
Cow    eeegplqeke dgqsptphpa dkgdeknakE lkgLqgkqdd qk..eegarg
Chick  .......... .dicerrdef etqdklkeeE kedLfewgki rgkieqvtky
       ---------- ---------- ---------E ---L------ ----------

       670                                                    719
Mouse  Pcpviipgpd ePstshsqts gsnqggpvgp askssslL.a kgpskalsik
Rat    Pcpviipgpd gPstthsqts gsnqggpegp gskssslL.a kspskalsfk
Cow    Pcpmvapgpe gPstprsqgp qvilggsegh garsgsrL.a rspprklaye
Chick  Pdvsepeavp sPglispaep gvlqetdhdr edkqgllFre aglpgsvsye
       P--------- -P-------- ---------- ---------- ----------

       720                                                    769
Mouse  KVEVvESIEK iStesIQTYE ETsvIVETlI gKskgn.KKl GeKslpdtra
Rat    KVEVvESIEK iStesIQTYE ETsvIVETvI gKtkgn.KKp GeKsssnaka
Cow    KVEVmESIEK fStesIQTYE ETavIVETmI eKtkankKKl GeKgsssa..
Chick  KVEVvESIEK fSddrIQTYE ETamIVETmI eKt..skKKp GdKgs.....
       KVEV-ESIEK -S---IQTYE ET--IVET-I -K-----KK- G-K-------
```

domain (LNDR) is highly divergent, resulting in the sequence LGGC rather than LNDR (Hess *et al.*, 1996, 1993; Merdes *et al.*, 1993). This sequence divergence (Figure 4) is seen in human, bovine, and mouse CP49 (Hess *et al.*, 1996, 1993; Merdes *et al.*, 1993). In other intermediate filament proteins this LNDR motif has been shown to be critical in filament assembly (for review, see Fuchs and Weber, 1994). Even more surprising is that in the inherited skin-blistering disease Dowling Meara epidermolysis bullosa simplex (EBS), the same arginine-to-cysteine change is seen within the LNDR motif of keratin 14 (Coulombe *et al.*, 1991).

Human CP49 shares 89% amino acid identity with bovine CP49 and 85.8% amino acid identity with murine CP49 (Hess *et al.*, 1996). Sequence comparisons of the known CP49 sequences are presented in Figure 5. Analysis of the primary amino acid sequence of CP49 with other intermediate filament proteins reveals greatest homology to the type I (acidic) keratins and in particular to human keratin 18 (Hess *et al.*, 1996; Merdes *et al.*, 1993; Orii *et al.*, 1993). However, the degree of homology is low; for example, the rod domain of human CP49 shares 34.1% identity with human keratin 18 (Orii *et al.*, 1993; Sawada *et al.*, 1995; Hess *et al.*, 1996), and is more consistent with the homologies seen between intermediate filament classes than within a particular class. Amino acid analysis of CP49 reveals a central rod domain of 311 amino acids in bovine (Merdes *et al.*, 1993) and human (Hess *et al.*, 1996). The rod domain can be subdivided into coils 1A, 1B, and 2, which is predicted to be α-helical in conformation, separated by non-α-helical linker regions (Hess *et al.*, 1996; Merdes *et al.*, 1993). Consistent with other intermediate filament proteins is the presence of the "stutter" in coil 2 which interupts the heptad repeats. The rod domain is flanked by an amino-terminal head domain which varies between 95 (Merdes *et al.*, 1993) and 103 amino acids (Hess *et al.*, 1996).

FIGURE 3. Alignment of the C-terminal non-α-helical tail sequences of filensin. Sequences for mouse (Masaki and Quinlan, 1997; accession number AB003147), rat (Masaki and Quinlan, 1997; accession number AB003104), cow (Gounari *et al.*, 1993; accession number X72388), and chick (Remington, 1993; accession number X72873) were aligned using the GCG program PILEUP with default settings. The alignment shows all identical residues in capitals and a consensus sequence is included beneath each block of the alignment. Two motifs are highlighted in bold and underlined. These are PEDVPDG––ISKAF and KVEV–ESIEK–S–––IQTYEET––IVET. As the filensin tail sequence is one of the most divergent yet found across species of the published vertebrate intermediate filament proteins, this suggests these conserved motifs are of functional importance. The motif KVEV–ESIEK–S–––IQTYEET––IVET is not required for assembly *in vitro* or apparently in the lens. The motif PEDVPDG––ISKAF is close to a physiological cleavage site which generates a 53-kDa fragment of filensin, which is a prominent component of the intermediate filament and beaded filament fractions of the lens.

	Helix IA LNDR motif	Helix IIB YRKLLEGE motif
chick CP49	LNgc	YhlLLDGE
cow CP49	Lggc	YhaLLDrE
human CP49	Lggc	YhaLLDrE
mouse CP49	----	YhaLLDrE
cow filensin	LgER	YhRiiEnE
chick filensin	LNER	YKRiiEnE
rat filensin	LgER	YhRiiEiE
Mouse filensin	LgER	YhRiiEiE
Type I		
keratin 18	LNDR	YRRLLEdg
keratin 19	LNDR	YRsLLEGq
Type II		
keratin 8	LNnk	YRKLLEGE
keratin 1	LNnq	YRKLLEGE
Type III		
Vimentin	LNDR	YRKLLEGE
GFAP	LNDR	YRKLLEGE
Type IV		
NF-L	LNDR	YRKLLEGE
NF-M	LNDR	YRKLLEGE
NF-H	LNDR	YRKLLEGE
Nestin	LNrR	YRtLLEaE
rat internexin	LNDR	YRKLLEGE
Type V		
lamin A-type	LNDR	YRKLLEGE
lamin B-type	LNDR	YRKLLEGE

FIGURE 4. Comparison of filensin and CP49 LNDR/TYRKLLEGE motifs. Notice the replacement of the arginine residue for cysteine in the LNDR motif of CP49 and the substitution of the acidic residue for glycine in the same motif. In filensin, these changes are not seen in the LNDR motif. Both filensin and CP49 have changes in the YRKLLEGE motif. For instance, in CP49 the dibasic "RK" residues are substituted by "H(A/L)," which would reduce the positive charge in this region. Also, in filensin, histidine usually replaces one of these basic residues and the "LL" is replaced by "II" residues. The significance of these changes in the motifs has yet to determined for both filensin and CP49. Accession numbers and references to the intermediate filament sequences used in this figure can be found in Quinlan *et al.* (1995).

More recently, a unique splice variant of chicken CP49 has been identified (Figures 2 and 5). This variant, which has been called CP49*ins*, contains a 49-amino acid-residue insertion into coil 1b of the rod domain (Sawada *et al.*, 1995), a feature characteristic of the nuclear lamins (Quinlan *et al.*, 1995). It is also a feature present in the lens-specific proteins of squid and octopus (Tomarev *et al.*, 1993), providing a direct evolutionary link with the avian CP49*ins* protein, as has also been found for the neurofilament proteins (Way *et al.*, 1992). Whether the CP49*ins* gene product exists in the chicken lens and whether such CP49 splice variants exist in species other than the chicken are not yet known. Nevertheless, the identification of this variant further illustrates the diversity and complexity of intermediate filament proteins as a whole and in particular the lens-specific proteins.

4. GENE STRUCTURE OF LENS INTERMEDIATE FILAMENT PROTEINS

In addition to primary sequence and secondary structural characteristics, the organization of the gene for a particular intermediate filament protein is also important for classification (Quinlan *et al.*, 1995). Intron position and number are typically diagnostic of a particular intermediate filament type. Recently, Hess and co-workers have reported the gene structure for human CP49 (Hess *et al.*, 1996). Analysis of the intron positions within the CP49 gene (Figure 6) reveals both an identical number and position of introns with the type I keratins (Hess *et al.*, 1996). For example, intron B, which is unique to type I keratins, is also found in human CP49 (Hess *et al.*, 1996). It is therefore suggested that CP49 is a type I intermediate filament protein, albeit a distantly related member (Hess *et al.*, 1996).

The gene structure for rat filensin has recently been determined (Masaki and Quinlan, 1997). The rat filensin gene spans 31 kb and contains seven introns. Analysis of the intron positions reveals that the first six introns are conserved with respect to those of type II, type III, and type V intermediate filament proteins (Figure 5). The seventh intron is unique to filensin. From these results it appears that the filensin gene is most similar to type III intermediate filament genes, since none of the intron positions of filensin coincide with those introns unique to type I, type II, or type IV genes (Masaki and Quinlan, 1997).

```
        1                                                   50
Cow     mstrrvvvda pagasssMPL qRhkaSFraa qspSSldgl. .pasRtvavs
Human   mserrvvvdl ptsasssMPL qRrraSFrgp rssSSlesp. .pasRtnams
Mouse   .......... .......... .......... .......... ..........
Chick   .......... .......MPL pRrrsSFlgq qapSSatesv ggpgRrvsvg
        ---------- -------MPL -R---SF--- ---SS----- ----R-----

        51                                                 100
Cow     gLvRtPrVYV GmaPsGptgg LGaRVtRRAL GISSVFLQGL RSSglAtapA
Human   gLvRaPgVYV GtaPsGcigg LGaRVtRRAL GISSVFLQGL RSSglAtvpA
Mouse   .......... .......... .......... .......... ..........
Chick   sLsRpPgVYV GavPtGgvss LGtRVsRRAL GISSVFLQGL RSScsAvplA
        -L-R-P-VYV G--P-G---- LG-RV-RRAL GISSVFLQGL RSS--A---A

        101                                                150
Cow     psLErdlG.a vEdLgGCLVE YmaKVhALEk VsQELEaqlR mhLesKatrs
Human   pgLErdhG.a vEdLgGCLVE YmaKVhALEq VsQELEtqlR mhLesKatrs
Mouse   .......... .......... .......... .......... ..........
Chick   agLEkgrGls yEsLnGCLVE YieKVrALEq VnQELEehiR vyLdkKsssv
        --LE---G-- -E-L-GCLVE Y--KV-ALE- V-QELE---R --L--K----
                           HELIX IA

        151                                                200
Cow     enWGALRaSW assCqQ.... .......... .......... ..........
Human   gnWGALRaSW assCqQ.... .......... .......... ..........
Mouse   .......... .......... .......... .......... ..........
Chick   gsWGALReSW eavYhQvsyl qqepmqshpr rkmalwelrt thsrhehqif
        --WGALR-SW ---C-Q---- ---------- ---------- ----------
                              HELIX IB

        201                                                250
Cow     .......... .....VGEAV LENARLMLqT EnIQAgADDF KERYENEQPF
Human   .......... .....VGEAV LENARLMLqT EtIQAgADDF KERYENEQPF
Mouse   .......... .......... .......... .......... ........PF
Chick   lgsrkkmtlr rtadmVGEAV LENARLMLhT EnIQAsAEDF KDRYENEQPF
        ---------- -----VGEAV LENARLML-T E-IQA-ADDF KERYENEQPF
                              HELIX IB
```

FIGURE 5. Comparison of CP49 sequences. Cow (Merdes *et al.*, 1993; accession numbers U12016 and X75160), human (Hess *et al.*, 1996; accession number U48224), mouse (Hess *et al.*, 1993; accession number S55549), and chicken (Sawada *et al.*, 1995; accession number X84807) CP49 sequences have been aligned using the GCG program PILEUP. The helices 1A, 1B, and 2 have been boxed. Note that the insert sequences present in helix IB found in a variant, CP49*ins*, of chicken CP49 are in bold. The relationship between CP49*ins* and CP49 at the genomic level has yet to be established as well as the presence of the CP49*ins* gene product in the chicken lens. All identical residues are in capitals and these are also printed below the alignment so that regions of greatest sequence identity can be easily recognized.

```
         251                                                    300
Cow      RKAaEEEiNS LYKVIDeANs sKmDLEsQIE SlKEELgfLS rsyEEDVKmL
Human    RKAaEEEiNS LYKVIDeANl tKmDLEsQIE SlKEELgsLS rnyEEDVKlL
Mouse    RKAaEEEvsS LYKVIDeANl tKtDLEhQIE SlKEELgfLS rsyEEDVKvL
Chick    RKAvEDEiNS LYKVIDdANl tKmDLEsQIE SmKEELtlLS kthEEDVKvL
         RKA-EEE-NS LYKVID-AN- -KmDLE-QIE S-KEEL--LS ---EEDVK-L
                             HELIX IB
```

```
         301                                                    350
Cow      yKQLAGseLE QlnvPiGtGL DdiLEtIRih WERDvEKNRl qaGALLqaKQ
Human    hKQLAGceLE QmDaPiGtGL DdiLEtIRiq WERDvEKNRv eaGALLqaKQ
Mouse    yKQLAGseLE QaDvPmGtGL DdvLEtIRvq WERDvEKNRa eaGALLqaKQ
Chick    yKQLAGsqLE elDvPlGsGL DniLEkIRih WERDiEKNRa eaGALLrtKQ
         -KQLAG--LE Q-DvP-G-GL D--LE-IR-- WERD-EKNR- --GALL--KQ
                                          HELIX II
```

```
         351                                                    400
Cow      QaElarraqt QEEklaaaLr vElHnTacqi QSLQAETESL RaLKRGLENt
Human    QaEvahmsqt QEEklaaaLr vElHnTacqv QSLQAETESL RaLKRGLENt
Mouse    QtEvvhvsqt QEEklaaaLs vElHdTarqv QSLQAETESL RaLKRGLENs
Chick    QaEataavrs QEEeLvegLr tEFHeTacki QSLQAETESL RtLKRGLENs
         Q-E------- QEE-----L- -E-H-T-c-- QSLQAETESL R-LKRGLEN-
                             HELIX II
```

```
         401                                                    450
Cow      LhDakHWHdi ELQNLGaVvs RLEAElrEmR aEaEQQlqaR ehLLshKcQL
Human    LhDakHWHdm ELQNLGaVvg RLEAElrEiR aEaEQQQqeR ahLLarKcQL
Mouse    LhDtqHWHdm ELQNLGaVvg RLEAElaEiR sEtEQQQqeR ahLLacKsQL
Chick    LyDakHWHei ELQNLGsVis kLEAEmaEik aEtEQQQrdR enLLtnKqQL
         L-DAKHWH-- ELQNLG-V-- RLEAE--E-R -E-EQQQ--R --LL--K-QL
                             HELIX II
```

```
         451           467
Cow      qrDvAsYHaL LDrEess
Human    qkDvAsYHaL LDrEesg
Mouse    akDvAsYHaL LDrEenn
Chick    ekDiAaYHcL LDgEqss
         --D-A-YH-L LD-E---
              HELIX II
```

FIGURE 5. (*Continued*).

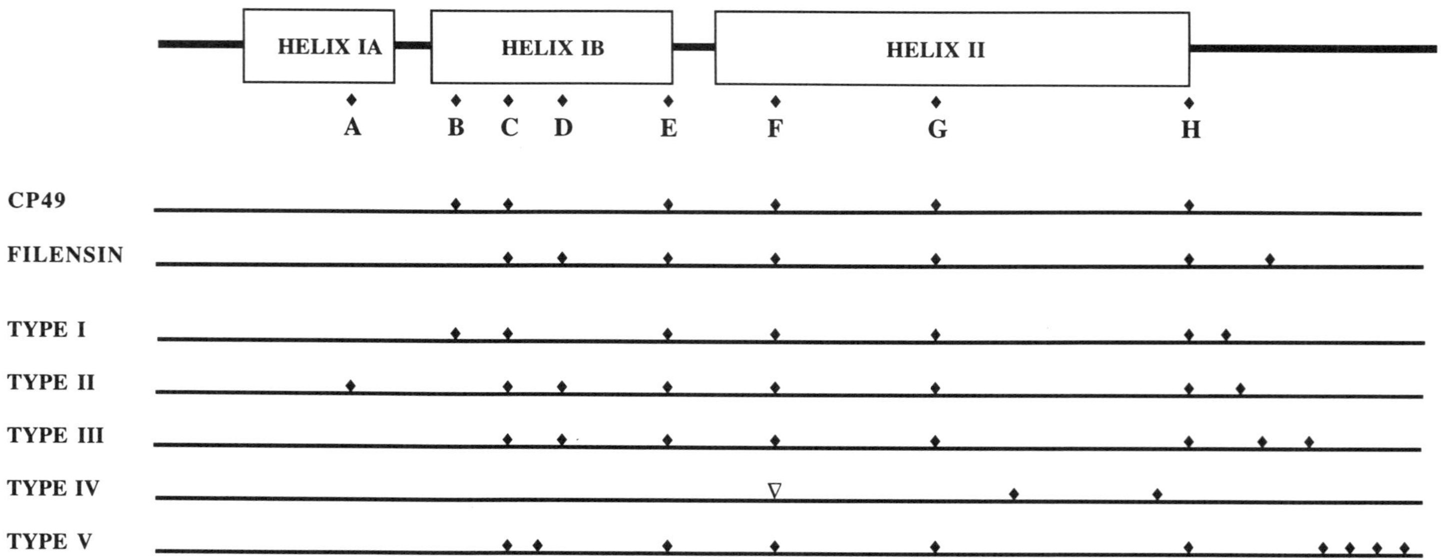

FIGURE 6. Gene structure of CP49 and filensin. The intron (filled diamonds) pattern of CP49 (Hess *et al.*, 1996) and filensin (Masaki and Quinlan, 1997) genes are compared to other intermediate filament genes. To construct this figure, gene data for human keratin 14, human keratin 8, hamster vimentin, human NF-H, and *Xenopus* lamin B3 were used (see Quinlan *et al.*, 1995, for references and accession numbers). Coincident intron positions also indicate preservation of the phase of the intron. Where the position is conserved, but the phase is not as for NF-H, then this is denoted by an inverted triangle. To aid comparison of intron positions in the helical domains, the lettering scheme of Hess *et al.* (1996) has been adopted. Notice that the CP49 gene possesses intron B, which is typical of type I genes. Intron D is also missing from both the human CP49 and type I genes. Filensin, on the other hand, shows most similarity in exon–intron pattern to type III genes. Filensin possesses none of the unique introns typical of either type I, II, IV, or V genes. In terms of gene structure, these data indicate CP49 and filensin are most similar to type I and type III genes, respectively.

5. *IN VITRO* ASSEMBLY OF LENS INTERMEDIATE FILAMENT PROTEINS

Several *in vitro* assembly studies with lens-purified filensin and CP49 have shown that each protein is individually capable of limited filament assembly (Goulielmos *et al.*, 1996a; Carter *et al.*, 1995, 1993; Quinlan *et al.*, 1992; Lieska *et al.*, 1991b; Merdes *et al.*, 1991). CP49 is capable of assembling into thin filaments 4–6 nm in diameter (Lieska *et al.*, 1991b), but does show a tendancy to associate laterally to form aggregates (Goulielmos *et al.*, 1996a; Carter *et al.*, 1995; Merdes *et al.*, 1993). Filament aggregation *in vitro* is often a feature of intermediate filament proteins which have deletions or truncations within the tail domain (Kouklis *et al.*, 1992; Coulombe *et al.*, 1990; Kaufmann *et al.*, 1985), suggesting that the tail domain is important in controlling lateral associations between subunits (Heins and Aebi, 1994). Since CP49 is a tailless intermediate filament protein, this most likely explains its tendency to aggregate (e.g., Lu and Lane, 1990). Filensin can only form short fibrils of 10 nm diameter *in vitro* (Goulielmos *et al.*, 1996a, b; Carter *et al.*, 1995, 1993; Merdes *et al.*, 1993). However, when both filensin and CP49 are coassembled together *in vitro*, long and smooth 10 nm filaments with a diameter of 10-nm are formed (Carter *et al.*, 1995, 1993; Merdes *et al.*, 1993). Filensin is unable to coassemble with vimentin (Carter *et al.*, 1995), whereas CP49 does not coassemble with type I or type II keratins (Merdes *et al.*, 1993), despite its relatedness to type I keratins (Hess *et al.*, 1996). Together these results illustrate the specificity of the interaction between filensin and CP49. Unlike the keratins, the optimal ratio of filensin to CP49 required for filament assembly is not equimolar, but is reported to vary from 1 mole of filensin to 2 moles of CP49 (Carter *et al.*, 1995), to 1 mole of filensin to 3 moles of CP49 (Merdes *et al.*, 1993).

The roles of the head, rod, and tail domains of filensin in filament assembly have been recently investigated through the construction of filensin/vimentin chimeric proteins (Goulielmos *et al.*, 1996b). Under normal assembly conditions, filensin can only form short, 10-nm fibrils (Carter *et al.*, 1995; Merdes *et al.*, 1993). If, however, the rod domain of filensin is substituted by the rod domain of vimentin, then extended filaments are formed (Goulielmos *et al.*, 1996b). This result suggests that the rod domain of filensin exerts a negative influence on filament elongation. The rod domain of filensin also appears to be important in the recognition of CP49 as an assembly partner, since under normal conditions CP49 does not coassemble with vimentin (Carter *et al.*, 1995). However, if the head and tail domains of filensin are substituted with those of vimentin and then coassembled with CP49, filaments are formed (Goulielmos *et al.*,

1996b). The C-terminal tail domain of filensin is important in counteracting the tendency of CP49 to associate laterally, since the complete removal of this domain results in aggregated structures (Goulielmos *et al.*, 1996b).

CP49 can also coassemble with a naturally occurring proteolytic fragment derived from filensin, called the 53 kDa fragment (Quinlan *et al.*, 1992), to form intermediate-sized filaments *in vitro* (Carter *et al.*, 1993). This proteolytic fragment encompasses the N-terminal head and rod domain of filensin (Quinlan *et al.*, 1992). However, the 53 kDa fragment is not completely tailless, since part of the C-terminal tail domain is still associated with this fragment (Sandilands *et al.*, 1995b). These results therefore identify a region within the C-terminal tail domain of filensin which contains sequences that are important in the regulation of filament assembly (Masaki and Quinlan, 1997), since if the entire filensin tail domain is deleted, filensin and CP49 form aggregates rather than filaments (Goulielmos *et al.*, 1996b).

6. *IN VIVO* ASSEMBLY OF LENS INTERMEDIATE FILAMENT PROTEINS

Both filensin and CP49 are only expressed in lens fiber cells and not in lens epithelial cells. Since lens fiber cells cannot be maintained in culture, determining the assembly characteristics of filensin and CP49 outside the confines of the lens is possible by transfection into other cell types. Recent transfection experiments have shown that filensin and CP49, when introduced independently into various cultured epithelial cells (both lens- and non-lens-derived), do not form filament networks (Goulielmos *et al.*, 1996a). Typically, aggregates of protein are seen which do not appear to interfere with the endogenous intermediate filament networks (Goulielmos *et al.*, 1996a). However, when both proteins are cotransfected into cells, then heteropolymeric networks are formed which extend throughout the cytoplasm (Goulielmos *et al.*, 1996a). The type of filamentous arrays formed upon cotransfection, however, appears to be dependent upon the host cell background, and the endogenous intermediate filament network influences the type of networks formed. For example, transfection into cells which contained an endogenous vimentin filament network (CHO cells) resulted in colocalization of the filensin/CP49 network with the vimentin network (Goulielmos *et al.*, 1996a), whereas transfection into MCF-7 cells (which contain keratins 8, 18 and 19, but not vimentin) resulted in a sub-plasma membrane filament network (Goulielmos *et al.*, 1996a). These cotransfection experiments also demonstrated that filensin and CP49 were capable of *de novo* filament assembly when introduced into cell lines which lack any cytoplasmic intermediate filament proteins (Goulielmos *et al.*, 1996a).

7. ASSEMBLY OF FILENSIN AND CP49 INTO BEADED FILAMENTS

Although *in vitro* filensin and CP49 coassemble to form intermediate filaments, *in vivo* filensin and CP49 localize immunologically to beaded filaments (FitzGerald and Gottlieb, 1989; Ireland and Maisel, 1989, 1984). Beaded filaments were originally described in 1972 (Maisel and Perry, 1972) and have been identified in all vertebrate lenses as well as in squid (FitzGerald and Casselman, 1991). Beaded filaments are apparently unique to the lens and are restricted to the lens fiber cells. The beaded filament (Figure 7) consists of a filamentous backbone 4–5 nm in diameter which is decorated by 15- to 20-nm diameter globular "beads" (Maisel and Perry, 1972). Filensin and CP49 *in vitro* readily coassemble to form intermediate filaments. It has been argued, however, that other proteins are likely to be associated with these unique structures *in vivo*. The "beads" of the beaded filament were originally proposed to consist of α-crystallin particles, since they share similar morphology and size at the electron microscope level (Benedetti *et al.*, 1976). Furthermore, α-crystallins are always associated with filensin and CP49-enriched extracts from the lens, and antibodies to α-crystallin immunolocalize to the lens fiber cell cytoskeleton (FitzGerald and Graham, 1991). Previous studies also reveal the association with and modulation of the assembly of intermediate filaments *in vitro* by α-crystallin (Nicholl and Quinlan, 1994). Not surprisingly, therefore, beaded filaments strikingly similar to those seen *in vivo* can be formed *in vitro* if filensin and CP49 are coassembled in the presence of α-crystallin (Carter *et al.*, 1995).

More recently, Goulielmous and co-workers reported that beaded filaments can be reconstituted from filensin and CP49 alone *in vitro* (Goulielmos *et al.*, 1996a). Rotary shadowing of *in vitro* assembled filensin and CP49 revealed a 19- to 21-nm axial beading pattern which had not been previously detected using conventional staining (Goulielmos *et al.*, 1996a). Further analysis using scanning transmission electron microscopy of *in vitro* assembled filensin and CP49 filaments indicated a bead diameter of 12–15 nm arranged along the filament backbone. However, given that this beading pattern could be removed by dilution of the sample preparation and that the beads were not regularly arranged along the filament at defined intervals, they concluded that the beads were peripherally associated with

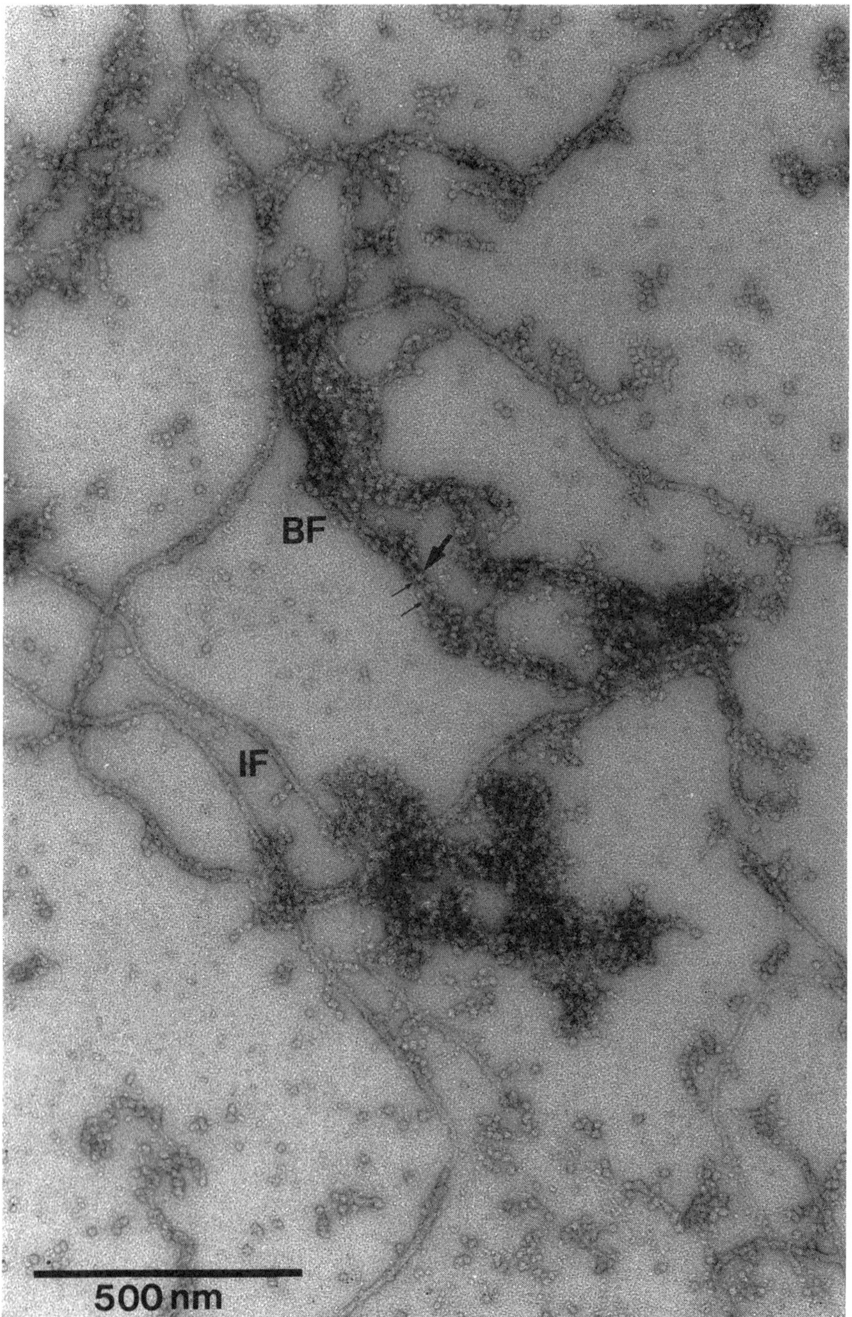
BF
IF
500 nm

the filament backbone (Goulielmos *et al.*, 1996a). On the basis of these observations and mass-per-length measurements, a model for the assembly of beaded filaments from filensin and CP49 has been proposed (Goulielmos *et al.*, 1996a) which is very much reminiscent of the assembly of neurofilament proteins. In this model, CP49 homodimers form a central core filament, which is then surrounded by peripherally associated filensin and CP49 heterodimers. They propose that the beads of the beaded filament are projections of the long carboxy-terminal tail domain of filensin (Goulielmos *et al.*, 1996a).

In vivo, however, it is likely that the α-crystallins enhance the beading pattern of the filensin/CP49 filament. For example, in the oldest fiber cells of the lens nucleus, beaded filaments are still observed (Bradley *et al.*, 1979; Ireland *et al.*, 1978), but by this stage the C-terminal tail domain of filensin has been cleaved from the parent molecule and no full-length filensin can be detected (Sandilands *et al.*, 1995b).

The assembly of filensin and CP49 into filaments, whether beaded or not, presents many difficult questions concerning filament formation. For example, how does the truncated rod domain of filensin interact with the longer rod domain of CP49 to form heteropolymers? How does the LGGC motif of CP49 influence the assembly, given that this motif is so important in the assembly of other intermediate filament proteins? What effect does the CP49 splice variant CP49*ins*, with its longer rod domain, have on filament assembly? The answers to these questions should provide interesting and novel insights into the diversity of the mechansims of intermediate filament assembly and assist in devising strategies to treat genetic diseases caused by altered intermediate filament assemblies.

8. BEADED FILAMENTS AND VIMENTIN INTERMEDIATE FILAMENT NETWORKS IN THE LENS

In vivo, the vimentin intermediate filament network is cleared from the lens fiber cells as they mature. The removal of vimentin intermediate

◄

FIGURE 7. Native lens intermediate filaments and beaded filaments. Transmission electron micrograph of negatively stained intermediate filaments (IF) and beaded filaments (BF) isolated from the bovine lens. The small arrows denote the thin filament backbone of the beaded filament, whereas the globular "beads" are indicated by a large arrow. Scale bar = 500 nm.

filaments can be detected both biochemically and immunologically (Sandilands *et al.*, 1995a; Ellis *et al.*, 1984; Maisel and Ellis, 1984), and in the bovine lens occurs after the removal of the nuclei from the differentiating lens fiber cells (Sandilands *et al.*, 1995a). In contrast, filensin/CP49 beaded filaments are retained in all fiber cells, including the fiber cells of the lens nucleus (Sandilands *et al.*, 1995a; Bradley *et al.*, 1979; Ireland *et al.*, 1978).

The filensin/CP49 beaded filament network is not static, but is actively remodeled during lens fiber cell differentiation and aging. The lens is the perfect system to explore how this network is remodeled, since lens fiber cells at all the different stages of differentiation and maturation can be visualized in a single section cut through the lens. During the very early stages of lens fiber cell differentiation, filensin and CP49 are associated with the fiber cell plasma membranes, in particular the short faces of the membranes (Sandilands *et al.*, 1995b). However, antibodies raised against epitopes within the C-terminal tail domain of filensin (but not antibodies raised against the N-terminal head and rod domain) also label the fiber cell cytoplasm (Sandilands *et al.*, 1995b). At a specific point in the differentiation pathway, which coincides with the removal of all cytoplasmic organelles from the lens fiber cells, the filensin/CP49 beaded filament network changes (Sandilands *et al.*, 1995a). At this stage in the differentiation pathway all filensin domains (and CP49) are distributed throughout the fiber cell cytoplasm as well as associated with the plasma membrane (Sandilands *et al.*, 1995b). CP49 is also seen to enter the nuclear compartment of denucleating fiber cells at this time (Sandilands *et al.*, 1995a). As the lens fiber cells mature, the C-terminal tail domain-derived fragments of filensin associate exclusively with the plasma membrane of the fiber cell, whereas the N-terminal head-and-rod domain-derived fragments (and CP49) adopt a cytoplasmic distribution as well (Sandilands *et al.*, 1995b).

These observations at the light microscope level would appear to reflect the distribution of beaded filaments observed at the ultrastructural level. Beaded filaments are dispersed throughout the cytoplasm of the fiber cells, but as these cells age they become progressively more coiled and aggregated (Bradley *et al.*, 1979; Ireland *et al.*, 1978) and are often seen closely opposed to the fiber cell plasma membrane in nuclear fiber cells (Ireland *et al.*, 1978).

The mechanisms that account for the selective removal of the vimentin intermediate filament network and the remodeling of the filensin/CP49 beaded filament network are unknown, but are very likely to be influenced by the factors discussed in the following section.

9. FACTORS INFLUENCING VIMENTIN AND FILENSIN/CP49 FILAMENT NETWORKS WITHIN THE LENS

9.1. Proteolysis

Cytoskeletal proteins of the lens such as vimentin and spectrin are substrates for calcium-dependent proteolytic enzymes (Truscott *et al.*, 1990) and hence proteolysis is likely to be a major influence on filament networks in the lens. Filensin is also a target of the calcium-activated protease calpain (Marcantonio, 1992) and indeed filensin is extensively proteolyzed within the lens (Sandilands *et al.*, 1995b). The proteolysis of bovine filensin generates two major species which encompass the N-terminal head/rod domain (termed the 53-kDa fragment; Quinlan *et al.*, 1992) and the C-terminal tail domain (~62 kDa; FitzGerald, 1988), indicating that a major site for filensin proteolysis is situated at the junction of the head/rod domain with the tail domain. In addition to cleavage of filensin between these domains, there is also proteolysis, or "nibbling," from either end of the parent molecule, indicating that several pathways of posttranslational processing exist *in vivo* (Sandilands *et al.*, 1995b).

Analysis by two-dimensional electrophoresis of proteolytic fragments derived from the N-terminal head/rod domain and the C-terminal tail domain of filensin reveal distinctive isoelectric points, with fragments derived from the N-terminal head/rod domain more basic and fragments from the the C-terminal tail domain more acidic than the parent molecule (Sandilands *et al.*, 1995b). Accompanying the proteolytic processing of filensin is an increase in its degree of insolubility during fiber cell maturation with compartmentalization to the water-insoluble (or plasma membrane) fraction of the fiber cells (Sandilands *et al.*, 1995b; FitzGerald, 1988).

Similarly, CP49 is also proteolytically processed during lens fiber differentiation, yielding a major breakdown species of 40 kDa (Quinlan, 1991). CP49 lacks a C-terminal tail domain and consequently the extent of processing is less as compared to filensin. Analysis of the solubility properties of CP49 during lens fiber cell maturation also reveals a progressive compartmentalization to the insoluble fraction (or membrane fraction) of the lens (Sandilands *et al.*, 1995b).

During the processing of filensin, removal of the C-terminal tail domain would not be expected to have a detrimental effect on filament stability, since the naturally occuring 53-kDa fragment (which encompasses the N-terminal head/rod domain of filensin) is itself sufficient for coassemby with CP49 (Masaki and Quinlan, 1997; Carter *et al.*, 1993). Thus the N-terminal head/rod domain of filensin and CP49 would be expected to remain in a filamentous form during lens fiber cell aging. The C-terminal tail domain of filensin, however, becomes exclusively associated with the plasma membrane of the fiber cell (Sandilands *et al.*, 1995b). Filensin has an extensive tail domain which is proposed to protrude from the beaded filament backbone (Goulielmos *et al.*, 1996a). Since filensin and CP49 first appear during lens fiber cell differentiation closely associated with the plasma membrane of the fiber cell, it is plausible that the plasma membrane (or plasma membrane-associated proteins) acts as nucleating site for filament assembly and/or filament anchorage. Previous investigations have revealed that the C-terminal tail domain of filensin can interact tightly with the fiber cell plasma membrane (Brunkener and Georgatos, 1992); however, the acceptor molecule has yet to be identified.

9.2. α-Crystallins as Modulators of Intermediate Filament Networks

The α-crystallins share significant sequence homologies with the ubiquitous small heat shock protein (sHSP) family (Merck *et al.*, 1993a, b) and since αA- and αB-crystallin are expressed in a diverse range of tissues outside the lens (Srinivasan *et al.*, 1992; Bhat *et al.*, 1991; Bhat and Nagineni, 1989; Dubin *et al.*, 1989), a chaperone activity has been proposed. αB crystallin is also associated with a variety of non-lens pathological conditions (Noort *et al.*, 1995; Pineda *et al.*, 1993; Iwaki *et al.*, 1989). Within the lens the α-crystallins are intimately associated with the lens cytoskeleton (Nicholl and Quinlan, 1994; FitzGerald and Graham, 1991) and the "beads" of beaded filaments have been proposed to be associated α-crystallin particles (Carter *et al.*, 1995). α-Crystallins can bind to and directly inhibit the assembly of the type III intermediate filament proteins vimentin and glial fibrillary acidic protein (GFAP) *in vitro* (Nicholl and Quinlan, 1994). Furthermore, the addition of α-Crystallins to preformed vimentin filaments elevates the soluble pool of vimentin subunits, implying that they can induce filament disassemby. In the lens, it is conceivable that such a mechanism may also contribute to the selective removal of the vimentin intermediate filament network. Although both αA-, αB-crystallin, and double-knockout mice have recently been established (Brady *et al.*, 1997; Brady and Wawrousek, 1997), the effect of the removal of αA-crystallin on the lens cytoskeleton, in particular on the vimentin network and the filensin/CP49 network, has yet to be assessed.

9.3. Phosphorylation

Phosphorylation of intermediate filament proteins regulates filament assembly and disassembly (reviewed by Inagaki *et al.*, 1996). Given the importance of phosphorylation in regulating intermediate filament networks, it is conceivable that phosphorylation may influence the vimentin and filensin/CP49 networks in the lens. Indeed phosphorylation has been identified as the mechanism by which chick CP49 undergoes changes in its subcellular distribution during fiber cell maturation, with a shift from the soluble to the insoluble fraction of the lens (Ireland *et al.*, 1993). Similarily, analysis of the isoelectric properties of bovine CP49 reveals that the more-basic isoforms are identified within the soluble fraction of the lens, whereas more-acidic isoforms remain in the insoluble compartment (Sandilands *et al.*, 1995b). The N-terminal head domain of filensin contains a potential phosphorylation site (Gounari *et al.*, 1993) which is also present in type III intermediate filament proteins. However, there is no evidence to suggest that filensin is extensively phosphorylated *in vivo*. Proteolytic cleavage of filensin gives rise to fragments with widely different charge properties (Sandilands *et al.*, 1995b) which are derived from protein domains with quite different functions. Proteolysis is therefore a key regulator of filensin function in the lens, whereas the contribution of phosphorylation to these functions has yet to be determined.

9.4. Intermediate Filament-Associated Proteins

In addition to the effects of proteolysis and phosphorylation, intermediate filament networks are also believed to be regulated through the actions of a diverse group of proteins which associate with intermediate filament proteins. These are known as intermediate filament–associated proteins (IFAPs; reviewed by Foisner and Wiche, 1991). For exampe, IFAP-300 kD, identified within the lens, associates with vimentin intermediate filaments (Lieska *et al.*, 1991a). IFAP-300 kD localizes to the fiber cell plasma membrane and during lens fiber cell differentiation there is loss of IFAP-300 kD expression (Lieska *et al.*, 1991a). Consequently, alterations in the distribution of IFAPs may also contribute to the remodeling of the vimentin and filensin/CP49 filament networks during lens fiber cell differentiation. In addition to IFAPs, the lens fiber cell also possesses an extensive submembranous network of other cytoskeletal proteins, namely spectrin (Aster *et al.*, 1984), band 4.1 protein (Aster *et al.*, 1986), band 3 protein (Allen *et al.*, 1987), and ankyrin (Allen *et al.*, 1987). Interactions between this membrane network and the cytoplasmic cytoskeletal network are believed to be mediated via interlinking proteins such as plectin (Weitzer and

Wiche, 1987). It has also been suggested that filensin serves such a role since its tail domain interacts with both the plasma membrane and with vimentin (Georgatos *et al.*, 1994; Merdes *et al.*, 1991), despite the inability of filensin to coassemble with vimentin (Carter *et al.*, 1995; Merdes *et al.*, 1993).

10. FUNCTIONS OF LENTICULAR INTERMEDIATE FILAMENT PROTEINS

10.1. Transgenic Studies

In order to determine the role of intermediate filament proteins within the lens, several lines of transgenic mice have been established. Together these transgenic studies would appear to implicate directly intermediate filament proteins in the loss of lens transparency, i.e., cataract formation (Bloemendal, 1991). For example, the overexpression of vimentin results in the perturbation of normal lens cell differentiation and the formation of lens cataract (Capetanaki *et al.*, 1989a, b). The level of expressed vimentin is indeed proportional to the severity of the lens opacity and the age of onset. The lenses of such animals show extensive fiber cell degeneration with areas of vacuolated tissue, abnormal fiber cell elongation, and the retention of nuclei within the fiber cells (Capetanaki *et al.*, 1989a). Furthermore, the expression of a vimentin–desmin hybrid gene construct is also associated with the disruption of normal plasma membrane–cytoskeleton interactions and the subsequent formation of lens opacities, i.e., cataract (Bloemendal, 1991; Dunia *et al.*, 1990; Pieper *et al.*, 1989). Cataract formation is also observed in the lenses of transgenic mice overexpressing the small neurofilament triplet protein (NF-L) (Monteiro *et al.*, 1990). Taken together these transgenic studies would suggest an important role for intermediate filament proteins in normal lens development and differentiation.

More recently, however, transgenic mice lacking vimentin were described (Colucci-Guyon *et al.*, 1994). The mice showed normal lens cell development and differentiation with no apparent abnormalities. These results imply that the absence of vimentin from the lens is indeed tolerable. It is possible that there is a degree of redundancy between intermediate filament proteins within the lens due to the presence of the filensin and CP49 filament network. During normal lens fiber cell differentiation vimentin is cleared from the fiber cells, whereas filensin and CP49 are retained (Sandilands *et al.*, 1995a). Consequently, it may not be so surprising that removing vimentin has such little effect. If this is the case, then removing filensin or CP49 or both from the lens would be expected to have a detrimental effect on the lens (see, however, Chapter 1 of this book for the general expression of vimentin in the vertebrate lens).

10.2. Mouse Mutants

The position of mouse filensin has been mapped to chromosome 2 (Masaki and Watanabe, 1994). Chromosome 2 of the mouse also contains the loci for several congenital eye disorders and the filensin gene has been mapped close to the *blind-sterile* locus (Masaki and Watanabe, 1994). The *blind-sterile* mutant mouse displays both cataract and microphthalmia (Varnum, 1983). *Blind-sterile* and the filensin loci are in fact separate, as revealed by a backcross experiment where 6 recombinants between CP94 and *bs* were found out of a total of 103 mice (S. Maski and N. Jenkins, unpublished data). No direct link between cataract and filensin has been discovered.

The deficient expression of filensin has been described in the inherited microphthalmic mutant mouse *Elo* (eye lens obsolescence) (Masaki *et al.*, 1991). The lens of the *Elo* mouse, although transparent , fails to develop to the normal size, as a result of abnormal fiber cell elongation. The *Elo* mouse mutant, however, has defective γE-crystallin synthesis (Cartier *et al.*, 1992) and consequently it is not clear how this defective filensin expression contributes to the *Elo* phenotype. Despite this, it may prove to be a useful model for determining the regulation of filensin gene expression within the lens.

11. FUTURE DIRECTIONS

The lens-specific proteins filensin and CP49 represent highly divergent intermediate filament proteins with unique sequence characteristics and assembly properties. In the future, the use of transgenic strategies will be important in determining the functions of these proteins within the lens. Furthermore, the gene structures of filensin and CP49 will be important in understanding the evolutionary relationship of these lens-specific proteins to other intermediate filament proteins.

NOTE ADDED IN PROOFS. Since the writing of this review article, articles have been published on the structure of the mouse filensin gene (Gounari *et al.*, 1997) and the chicken CP49 gene (Wallace *et al.*, 1998).

12. REFERENCES

Allen, D. P., Low, P. S., Dola, A., and Maisel, H., 1987, Band 3 and ankyrin homologues are present in eye lens: Evidence for all major erythrocyte membrane components in same non-erythroid cell, *Biochem. Biophys. Res. Commun.* **149:**266–275.

Aster, J. C., Welch, M. J., Brewer, G. J., and Maisel, H., 1984, Identification of spectrin and band 4.1-like proteins in the mammalian lens, *Biochim. Biophys. Res. Commun.* **119:**726–734.

Aster, J. C., Brewer, G. J., and Maisel, H., 1986, The 4.1-like proteins of the bovine lens: Spectrin-binding proteins closely related in structure to red blood cell protein 4.1, *J. Cell Biol.* **103:**115–122.

Benedetti, E. L., Dunia, I., Bentzel, C. J., Vermorken, A. J. M., Kibbelaar, M., and Bloemendal, H., 1976, A portrait of plasma membrane specialisations in eye lens epithelium and fibers, *Biochem. Biophys. Acta* **457:**353–384.

Bhat, S. P., and Nagineni, C. N., 1989, αB subunit of lens-specific protein α-crystallin is present in other ocular and non-ocular tissues, *Biochem. Biophys. Res. Commun.* **158:**319–325.

Bhat, S. P., Horwitz, J., Srinivasan, A., and Ding, L., 1991, Alpha B-crystallin exists as an independent protein in the heart and in the lens, *Eur. J. Biochem.* **202:**775–781.

Bloemendal, H., 1991, Proctor lecture. Disorganization of membranes and abnormal intermediate filament assembly lead to cataract, *Invest. Ophthalmol. Vis. Sci.* **32:**445–455.

Bradley, R. H., Ireland, M. E., and Maisel, H., 1979, Age changes in the skeleton of the human lens, *Acta Ophthal.* **57:**461–469.

Brady, J. P., and Wawrousek, E. F., 1997, Targeted disruption of the mouse αB-crystallin gene, *Invest. Ophthal. Vis. Sci.* **38:**4352.

Brady, J. P., Garland, D., Duglas-Tabor, Y., Robison, W. G., Groome, A., and Wawrousek E. F., 1997, Targeted disruption of the mouse αA crystallin gene induces cataract and cytoplasmic inclusion bodies containing the small heat shock protein αB-crystallin, *Proc. Natl. Acad. Sci. USA* **94:**884–889.

Brunkener, M., and Georgatos, S. D., 1992, Membrane-binding properties of filensin, a cytoskeletal protein of the lens fiber cells, *J. Cell Sci.* **103:**709–718.

Capetanaki, Y., Smith, S., and Heath, J. P., 1989a, Overexpression of the vimentin gene in transgenic mice inhibits normal lens cell differentiation, *J. Cell Biol.* **109:**1653–1664.

Capetanaki, Y., Starnes, S., and Smith, S., 1989b, Expression of the chicken vimentin gene in transgenic mice: Efficient assembly of the avian protein into the cytoskeleton, *Proc. Natl. Acad. Sci. USA* **86:**4882–4886.

Carter, J. M., Hutcheson, A. M., and Quinlan, R. A., 1993, *In vitro* assembly studies of purified bovine lens specific 49 and 115 kDa proteins, *Invest. Ophthalmol. Vis. Sci. Suppl. (Spring Meet. Assoc. Res. Vis. Ophthalmol.)*, **1993:**1339.

Carter, J. M., Hutcheson, A. M., and Quinlan, R. A., 1995, *In vitro* assembly properties of lens specific intermediate filament proteins CP49 and Filensin: Coassembly with α-crystallins but not vimentin, *Exp. Eye Res.* **60:**181–192.

Cartier, M., Breitman, M. L., and Tsui, L. C., 1992, A frameshift mutation in the γE-crystallin gene of the *Elo* mouse, *Nat. Genet.* **2:**42–45.

Colucci-Guyon, E., Portier, M. M., Dunia, I., Paulin, D., Pourin, S., and Babinet, C., 1994, Mice lacking vimentin develop and reproduce without an obvious phenotype, *Cell* **79:**679–694.

Coulombe, P. A., Chan, Y. M., Albers, K., and Fuchs, E., 1990, Deletions in epidermal keratins leading to alterations in filament organization *in vivo* and in intermediate filament assembly *in vitro*, *J. Cell Biol.* **111:**3049–3064.

Coulombe, P. A., Hutton, M. E., Letai, A., Hebert, A., Paller, A. S., and Fuchs, E., 1991, Point mutations in human keratin 14 genes of epidermolysis bullosa simplex patients: Genetic and functional analyses, *Cell* **66:**1301–1311.

Dubin, R. A., Wawrousek, E. F., and Piatigorsky, J., 1989, Expression of the murine αB crystallin gene is not restricted to the lens, *Mol. Cell. Biol.* **9:**1083–1091.

Dunia, I., Pieper, F., Manenti, S., van de Kemp, A., Devilliers, G., Benedetti, E. L., and Bloemendal, H., 1990, Plasma membrane-cytoskeleton damage in eye lenses of transgenic mice expressing desmin, *Eur. J. Cell Biol.* **53:**59–74.

Ellis, M., Alousi, S., Lawniczak, J., Maisel, H., and Welsh, M., 1984, Studies on lens vimentin, *Exp. Eye Res.* **38:**195–202.

FitzGerald, P. G., 1988, Age-related changes in a fiber cell specific extrinsic membrane protein, *Curr. Eye Res.* **7:**1255–1262.

FitzGerald, P. G., and Casselman, J., 1991, Immunologic conservation of the fiber cell beaded filament, *Curr. Eye Res.* **10:**471–478.

FitzGerald, P. G., and Gottlieb, W., 1989, The Mr 115kd fiber cell-specific protein is a component of the lens cytoskeleton, *Curr. Eye Res.* **8:**801–811.

FitzGerald, P. G., and Graham, D., 1991, Ultrastructural localization of alpha A-crystallin to the bovine lens fiber cell cytoskeleton, *Curr. Eye Res.* **10:**417–436.

Foisner, R., and Wiche, G., 1991, Intermediate filament-associated proteins, *Curr. Opin. Cell Biol.* **3:**75–81.

Fuchs, E., and Weber, K., 1994, Intermediate filaments: Structure, dynamics, function and disease, *Annu. Rev. Biochem.* **63:**345–382.

Geisler, N., and Weber, K., 1981, Isolation of polymerisation-competent vimentin from porcine eye lens tissue, *FEBS Lett.* **125:**253–256.

Georgatos, S. D., Gounari, F., and Remington, S., 1994, The beaded intermediate filaments and their potential functions in the eye lens, *Bioessays* **16:**413–418.

Goulielmos, G., Gounari, F. Remington, S., Müller, S., Häner, M., Aebi, U., and Georgatos, S. D., 1996a, Filensin and phakinin form a novel type of beaded intermediate filaments and coassemble *de novo* in cultured cells, *J. Cell Biol.* **132:**643–655.

Goulielmos, G., Remington, S., Schwesinger, F., Georgatos, S. D., and Gounari, F., 1996b, Contributions of the structural domains of filensin in polymer formation and filament distribution, *J. Cell Sci.* **109:**447–456.

Gounari, F., Merdes, A., Quinlan, R., Hess, J., FitzGerald, P. G., Ouzounis, C. A., and Georgatos, S. D., 1993, Bovine filensin possesses primary and secondary structure similarity to intermediate filament proteins, *J. Cell Biol.* **121:**847–853.

Gounari, F., Karagianni, N., Mincheva, A., Lichter, P., Georgatos, S. D., and Schirrmacher, V., 1997, The mouse filensin gene: Structure and evolutionary relation to other intermediate filament genes, *FEBS Letters* **413:**371–378.

Harding, J., 1991, *Cataract: Biochemistry, Epidemiology and Pharmacology,* Chapman and Hall, London.

Heins, S., and Aebi, U., 1994, Making heads and tails of intermediate filament assembly, dynamics and networks, *Curr. Opin. Cell Biol.* **6:**25–33.

Hess, J. F., Casselman, J. T., and FitzGerald, P. G., 1993, cDNA analysis of the 49kDa lens fiber cell cytoskeletal protein: A new, lens-specific member of the intermediate filament family? *Curr. Eye Res.* **12:**77–88.

Hess, J. F., Casselman, J. T., and FitzGerald, P. G., 1996, Gene structure and cDNA sequence identify the beaded filament protein CP49 as a highly divergent type I intermediate filament protein, *J. Biol. Chem.* **271:**6729–6735.

Inagaki, M., Matsuoka, Y., Tsujimura, K., Ando, S., Tokui, T., Takahashi, T., and Inagaki, N., 1996, Dynamic property of intermediate filaments: Regulation by phosphorylation. *BioEssays* **18:**481–487.

Ireland, M., and Maisel, H., 1983, Identification of native actin filaments in chick lens fiber cells, *Exp. Eye Res.* **36:**531–536.

Ireland, M., and Maisel, H., 1984, A cytoskeletal protein unique to lens fiber cell differentiation, *Exp. Eye Res.* **38:**637–645.

Ireland, M., and Maisel, H., 1989, A family of lens fiber cell specific proteins, *Lens Eye Toxicity Res.* **6:**623–638.

Ireland, M., Maisel, H., and Bradley, R. H., 1978, The rabbit lens cytoskeleton: An ultrastructural study, *Ophthalmic Res.* **10:**231–236.

Ireland, M. E., Klettner, C., and Nunlee, W., 1993, Cyclic AMP-mediated phosphorylation and insolubilization of a 49-kDa cytoskeletal marker protein of lens fiber terminal differentiation, *Exp. Eye Res.* **56:**453–461.

Iwaki, T., Kume-Iwaki, A., Liem, R. K. H., and Goldman, J. E., 1989, αB crystallin is expressed in non-lenticular tissues and accumulates in Alexander's disease brain, *Cell* **57:**71–78.

Kasper, M., and Viebahn, C., 1992, Cytokeratin expression and early lens development, *Anat. Embryol.* (Berlin) **186:**285–290.

Kaufmann, E., Weber, K., and Geisler, N., 1985, Intermediate filament forming ability of desmin derivatives lacking either the amino-terminal 67 or the carboxy-terminal 27 residues, *J. Mol. Biol.* **185:**733–742.

Kouklis, P. D., Traub, P., and Georgatos, S. D., 1992, Involvement of the consensus sequence motif at coil-2b in the assembly and stability of vimentin filaments, *J. Cell Sci.* **102:**31–41.

Lieska, N., Shao, D., Kriho, V., and Yang, H. Y., 1991a, Expression and distribution of cytoskeletal IFAP-300kD as an index of lens cell differentiation, *Curr. Eye Res.* **10:**1165–1174.

Lieska, N., Yang, H. Y., and Maisel, H., 1991b, Reconstitution of the filamentous backbone of lens beaded-chain filaments from a purified 49kD polypeptide, *Curr. Eye Res.* **10:**1037–1048.

Lu, X., and Lane, E. B. 1990, Retrovirus mediated transgenic keratin expression in cultured fibroblasts: Specific domain functions in keratin stabilization and filament formation, *Cell* **62:**681–696.

Maisel, H., and Ellis, M., 1984, Cytoskeletal proteins of the aging human lens, *Curr. Eye Res.* **3:**369–381.

Maisel, H., and Perry, M. M., 1972, Electron microscope observations on some structural proteins of the chick lens, *Exp. Eye Res.* **14:**7–12.

Marcantonio, J. M., 1992, Susceptibility of the bovine lens 115kDa beaded filament protein to degradation by calcium and calpain, *Curr. Eye Res.* **11:**103–108.

Masaki, S., and Quinlan, R. A., 1997, Gene structure and sequence comparisons of the eye lens specific protein, filensin, from rat and mouse: Implications for protein classification and assembly, *Gene* **201:**11–20.

Masaki, S., and Watanabe, T., 1992, cDNA sequence analysis of CP94: Rat lens fiber cell beaded-filament structural protein shows homology to cytokeratins, *Biochem. Biophys. Res. Commun.* **186:**190–198.

Masaki, S., and Watanabe, T., 1994, The gene *Bfsp1* for the lens fiber cell beaded-filament structural protein CP94 maps to mouse chromosome 2, *Genomics* **22:**449–450.

Masaki, S., Katsuyuki, T., Shoji, R., and Watanabe, T. , 1991, Defect of a fiber cell-specific 94KDa protein in the lens of inherited microthalmic mutant mouse *Elo*, *Biochem. Biophys. Res. Commun.* **179:**1175–1180.

Merck, K. B., Groenen, P. J. T. A., Voorter, C. E. M., Haard-Hoekman, W. A., Horwitz, J.,

Bloemendal, H., and de Jong, W. W., 1993a, Structural and functional similarities of bovine α-crystallin and mouse small heat-shock protein, *J. Biol. Chem.* **268:**1046–1052.

Merck, K. B., Horwitz, J., Kersten, M., Overkamp, P., Gaestel, M., Bloemendal, H., and de Jong, W. W., 1993b, Comparison of the homologous carboxy-terminal domain and tail of α-crystallin and small heat shock protein, *Mol. Biol. Rep.* **18:**209–215.

Merdes, A., Brunkener, M., Horstmann, H., and Georgatos, S. D., 1991, Filensin: A new vimentin-binding, polymerization-competent, and membrane-associated protein of the lens fiber cell, *J. Cell Biol.* **115:**397–410.

Merdes, A., Gounari, F., and Georgatos, S. D., 1993, The 47-kD lens-specific protein phakinin is a tailless intermediate filament protein and an assembly partner of filensin, *J. Cell Biol.* **123:**1507–1516.

Monteiro, M. J., Hoffman, P. N., Gearhart, J. D., and Cleveland, D. W., 1990, Expression of NF-L in both neuronal and nonneuronal cells of transgenic mice: Increased neurofilament density in axons without affecting caliber, *J. Cell Biol.* **111:**1543–1557.

Nicholl, I. D., and Quinlan, R. A., 1994, Chaperone activity of α-crystallins modulates intermediate filament assembly, *EMBO J.* **13:**945–953.

Noort, J. M., van Sechel, A. C., Bajramovic, J. J., Ouagmiri, M. E., Polman, C. H., Lassmann, H., and Ravid, R., 1995, The small heat shock protein αB-crystallin as candidate autoantigen in multiple sclerosis, *Nature* **375:**798–801.

Orii, H., Agata, K., Sawada, K., Eguchi, G., and Maisel, H., 1993, Evidence that the chick lens cytoskeletal protein CP 49 belongs to the family of intermediate filament proteins, *Curr. Eye Res.* **12:**583–588.

Piatigorsky, J., 1981, Lens differentiation in vertebrates, *Differentiation* **19:**134–153.

Pieper, F. R., Schaart, G., Krimpenfort, P. J., Henderik, J. B., Mostiage, H. J., van de Kemp, A., Ramaekers, F. C., Berns, A., and Bloemendal, H., 1989, Transgenic expression of the muscle-specific intermediate filament protein desmin in nonmuscle cells, *J. Cell Biol.* **108:** 1009–1024.

Pineda, R., Chan, C. C., Ni, M., Hayden, B. J., Johnson, M. A., Nickerson, J., and Chader, G. J., 1993, Human retinoblastoma cells express αB-crystallin *in vivo* and *in vitro*, *Curr. Eye Res.* **12:**239–245.

Quinlan, R. A., 1991, The soluble plasma membrane–cytoskeleton complexes and aging in the lens, in *Eye Lens Membranes and Aging* (G. F. J. M. Vrensen and J. Clauwaert, eds.), pp. 171–184, Eurage, Leiden, The Netherlands.

Quinlan, R. A., Carter, J. M., Hutcheson, A. M., and Campbell, D. G., 1992, The 53kDa polypeptide component of the bovine fiber cell cytoskeleton is derived from the 115kDa beaded filament protein: Evidence for a fiber cell specific intermediate filament protein, *Curr. Eye Res.* **11:**909–921.

Quinlan, R., Hutchison, C., and Lane, B., 1995, Intermediate filament proteins, *Protein Profile* **2:**801–952.

Ramaekers, F. C. S., Osborn, M., Schmid, E., Weber, K., Bloemendal, H., and Franke, W. W., 1980, Identification of the cytoskeletal proteins in lens-forming cells, a special epithelioid cell type, *Exp. Cell Res.* **127:**309–327.

Ramaekers, F. C. S., Dunia, I., Dodemont, H. J., Benedetti, E. L., and Bloemendal, H., 1982, Lenticular intermediate-sized filaments: Biosynthesis and interaction with plasma membrane, *Proc. Natl. Acad. Sci. USA* **79:**3208–3212.

Remington, S. G., 1993, Chicken filensin: A lens fiber cell protein that exhibits sequence similarity to intermediate filament proteins, *J. Cell Sci.* **105:**1057–1068.

Sandilands, A., Prescott, A. R., Hutcheson, A. M., Carter, J. M., Quinlan, R. A., Richards, J., and FitzGerald, P. G., 1995a, Vimentin and CP49/filensin form distinct networks in the lens which are independently modulated during lens fiber cell differentiation *J. Cell Sci.* **108:**1397–1406.

Sandilands, A., Prescott, A. R., Hutcheson, A. M., Quinlan, R. A., Casselman, J. T., and FitzGerald, P. G., 1995b, Filensin is proteolytically processed during lens fiber cell differentiation by multiple independant pathways, *Eur. J. Cell Biol.* **65:**238–253.

Sawada, K., Agata, J., Eguchi, G., Quinlan, R., and Maisel, H., 1995, The predicted structure of chick lens CP49 and a variant thereof, CP49*ins*, the first vertebrate cytoplasmic intermediate filament protein with a lamin-like insertion in helix 1B, *Curr. Eye Res.* **14:**545–553.

Srinivasan, A. N., Nagineni, C. N., and Bhat, S. P., 1992, Alpha A-crystallin is expressed in non-ocular tissues, *J. Biol. Chem.* **267:**23337–23341.

Tomarev, S. I., Zinovieva, R. D., and Piatigorsky, J., 1993, Primary structure and lens-specific expression of genes for an intermediate filament protein and a beta-tubulin in cephalopods, *Biochim. Biophys. Acta* **1216:**245–254.

Truscott, R. J. W., Marcantonio, J. M., Tomlinson, J., and Duncan, G., 1990, Calcium induced opacification and proteolysis in the intact rat lens, *Invest. Ophthal. Vis. Sci.* **31:**2405–2411.

Varnum, D. S., 1983, *Blind-sterile*: A new mutation on chromosome 2 of the house mouse, *J. Hered.* **74:**206–207.

Wallace, P., Signer, E., Paton, I. R., Burt, D., and Quinlan, R., 1998, The chicken CP49 gene contains an extra exon compared to the human CP49 gene which identifies an important step in the evolution of the eye lens intermediate filament proteins, *Gene* **211:**19–27.

Way, J., Hellmich, M. R., Jaffe, H., Szaro, B., Pant, H. C., Gainer, H., and Battey, J., 1992, A high-molecular-weight squid neurofilament protein contains a lamin-like rod domain and a tail domain with Lys-Ser-Pro repeats, *Proc. Natl. Acad. Sci. USA* **89:**6963–6967.

Weitzer, G., and Wiche G., 1987, Plectin from bovine lenses, *Eur. J. Biochem.* **169:**41–52.

Chapter 10

Structure, Assembly, and Dynamics of Intermediate Filaments

Harald Herrmann and Ueli Aebi

1. INTRODUCTION

The recognition that indirect immunofluorescence microscopy is, in combination with appropriate fixation protocols, a powerful tool to identify individual filaments within a cultured cell or in cut tissue sections came along at the right time for the study of intermediate filaments (IFs) in relation to the other cytoplasmic structures (Dahl and Bignami, 1973; Lazarides and Weber, 1974; Osborn *et al.*, 1978). IFs measuring between 7 and 11 nm in diameter were, as revealed by electron microscopy, clearly different from the two other known filaments of the cytoskeleton, microfilaments and microtubules, since they could not be decorated with heavy meromyosin like the former, and did not disintegrate by the action of colchicine like the latter (Ishikawa *et al.*, 1968). Due to their size between microfilaments (5–7 nm) and thick filaments of muscle (15 nm), consisting of actin and myosin, respectively, and microtubules (25 nm), they were distin-

Harald Herrmann Division of Cell Biology, German Cancer Research Center, D-69120 Heidelberg, Germany. **Ueli Aebi** Maurice E. Müller Institute, Biozentrum, University of Basel, CH-4056 Basel, Switzerland.

Subcellular Biochemistry, Volume 31: Intermediate Filaments, edited by Herrmann and Harris. Plenum Press, New York, 1998.

guished from the other fiber types as intermediate filaments. IFs were found to be present in nearly every tissue or cell of higher animals, being highly enriched in the insoluble high-salt-resistant residue after homogenization (Franke *et al.*, 1978a,b; Starger *et al.*, 1978), and it was shown that relatively drastic conditions, such as 8 M urea, had to be applied to solubilize these filaments. From the fully denatured state, filaments are obtained simply by dialyzing the protein back into buffers of physiological ionic strength (Figure 1), or, as in the case of cytokeratins,* into low-ionic-strength buffers (e.g., 5 mM Tris-HCl, pH 7.5) without the need for any cofactor (Steinert *et al.*, 1976; Renner *et al.*, 1981). The filaments produced from the various IF proteins looked, by electron microscopy, very much the same, although their biochemical properties, including their behavior in one- and two-dimensional gel electrophoresis (apparent molecular mass of 40–210 kDa), differed considerably. This perplexing situation was solved step by step with the help of specific antibodies which allowed the development of a histologically meaningful expression concept that was paralleled by known principles of embryonic differentiation. Thus, the five "presequencing" subclasses were defined as neurofilaments in neural cells, glial fibrillarly acidic protein (GFAP) filaments of glia, vimentin filaments predominantly found in mesenchymal cells, desmin filaments in muscle cells, and cytokeratin (CK) filaments in epithelial cells, and therefore specific antibodies were used extensively for tissue typing (Osbern and Weber, 1983). Since, in particular, the latter gave rise to "transcellular networks" via their connection to cell–cell junctions, i.e., desmosomes, or to the extracellular matrix via hemidesmosomes, it was only logical to postulate them to be the "mechanical integrators of cellular space" (Lazarides, 1980). Finally, with the generation of a monoclonal antibody called IFA that reacted with all five classes of IF proteins, the IF proteins were tentatively shown to be molecularly related (Pruss *et al.*, 1981).

Early on, IFs were also found in invertebrates. For instance, snails contained three neurofilament (NF) proteins, as do mammals, although their amino acid sequences differ considerably (Liem, 1993). Worms also may have complex patterns of IFs as demonstrated recently for *Caenorhabditis elegans*, which expresses at least 11 different IF proteins in its 959 cells (Dodemont *et al.*, 1994). By electron microscopic criteria, filaments resembling IFs have even been found, sometimes in conjunction with desmosomelike structures, in the "living fossil" *Priapulus* (Bartnik and

*Often also called keratins. Historically, the term keratin was used, however, for a very heterogeneous fraction of proteins isolated from hoof or hair, for instance. For further details see Franke and Kartenbeck (1993).

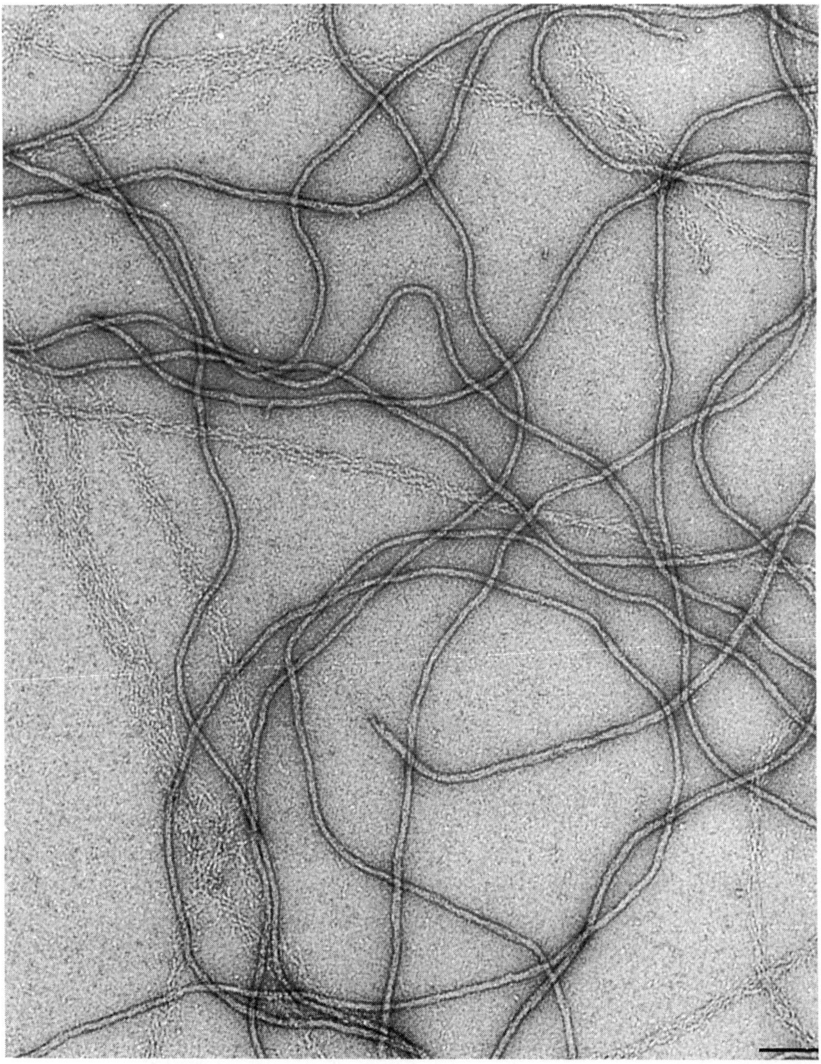

FIGURE 1. Electron micrograph showing negatively stained filaments assembled *in vitro* at 28°C for 1 hr (Herrmann *et al.*, 1993) from recombinant *Xenopus laevis* vimentin. Assembly was carried out in 25 mM Tris-HCl, pH 7.5, 50 mM NaCl. Note that long, unraveled filaments are found side by side with normal IFs. Bar, 100 nm.

Weber, 1989). Members of an apparently very related genus were widely distributed in the prevertebrate era of the Cambrian, hiding in the sand under the sea. It may be interesting to note that both *Caenorhabditis* and *Priapulus* are grouped together as Nemathelminthes.

2. DOMAIN ORGANIZATION

2.1. The Tripartite Structure

Biochemical characterization of chicken desmin, the muscle-specific IF protein, revealed that it consisted of a central protease-resistant domain with high α-helical content (80%) and two nonhelical flanking domains (Geisler *et al.*, 1982). The amino-terminal "head" is highly sensitive to proteases, which is the reason for the observed "acidic staircase" pattern of desmin (and its "sister" molecule vimentin) upon two-dimensional gel electrophoresis indicating that with progessive shortening of the molecule, basic peptides are lost (Quax-Jeuken *et al.*, 1983). When low amounts of chymotrypsin are employed, the carboxy-terminal "tail" domain can be isolated intact, which led to the determination of its amino acid sequence by protein chemical methods (Geisler and Weber, 1982). After limited proteolysis of the whole molecule, the central domain was purified under denaturing conditions and reconstituted into rodlike particles of about 50 nm length and 2–3 nm diameter (Geisler *et al.*, 1982). The complete sequence of the muscle-specific desmin allowed, in combination with various partial sequences, the proposal of a general structural model for IF proteins emphasizing a central α-helical rod, interrupted by three nonhelical linker regions (Geisler and Weber, 1982; and Figure 2A). The helical domains revealed a repeating sequence motif such that hydrophobic residues were present in a 3-4-3-4 spacing. Thus, if the primary sequence is ordered into consecutive heptads $(abcdefg)_n$, then positions *a* and *d* are occupied by hydrophobic residues leading to the formation from two α-helices of a left-handed superhelix. This structure was deduced by Crick (1953) from X-ray diffraction patterns obtained with keratin material by William Astbury in the 1930s.

The general validity of the above concept was very soon confirmed by various laboratories reporting amino acid sequences of many IF proteins from different species, mostly mammals. A comparison of the primary sequences, underscored the principally identical tripartite structure of these proteins and suggested together with the gene structure and the expression principles, that they constituted a multigene family (Weber and Geisler, 1984; Conway and Parry, 1988; Parry, 1990). However, deviations from the rule also became evident very soon. Within the cytokeratins a nearly "tail-less" member was identified (CK 19; Bader *et al.*, 1986), soon after followed by two really tailless homologues in the pseudotetraploid clawed toad, *Xenopus laevis* (Franz, 1987). This extensive "sequencing phase" also made clear that cytokeratins indeed existed in two distinct branches, as originally proposed from two-dimensional gel electrophoresis experiments (Moll *et*

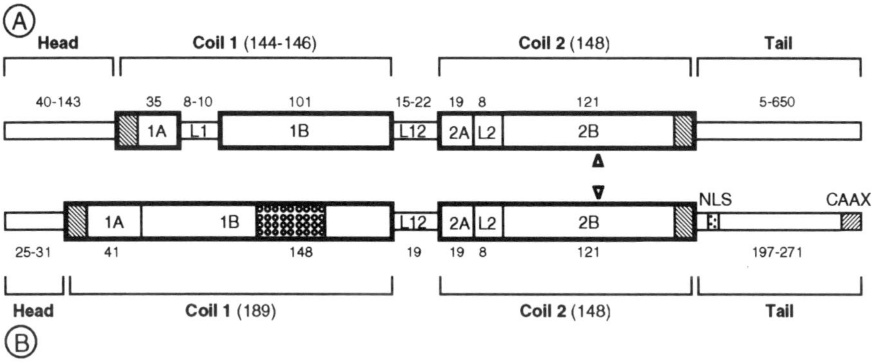

FIGURE 2. Schematic representation of the domain structure common to cytoplasmic (A) and nuclear (B) IF proteins. A central rod-forming domain, consisting of α-helical subdomains coil 1 and coil 2, is flanked by non-α-helical amino-terminal "head" and carboxy-terminal "tail" domains. Helical regions are boxed and are connected by nonhelical "linker" (L) regions. The position of linker L2 between a presumptive coil 2A and coil 2B is not absolutely clear (see text). A discontinuity in the helical organization of coil 2B (*"stutter"*) is indicated by an arrowhead. The numbers of amino acids found with the individual domains is indicated. The dotted area in coil 1 of (B) represents the 42-amino acid insertion found in general in nuclear IF proteins, also called nuclear lamins, but also in cytoplasmic IFs of lower vertebrates (Fuchs and Weber, 1994). Note that coil 1 of lamins does not contain a linker. NLS, Nuclear localization signal. The darkened regions at the ends of coil 1A and coil 2B indicate the evolutionarily highly conserved "consensus" regions. The light grey marking at the end of the tail domain of nuclear IF proteins (B) represents the so-called "CAAX box," which is a target for isoprenylation (for more details see Parry and Steinert, 1995).

al., 1982): the acidic (type I) and the basic (type II) representatives. It was shown then that each of the type I cytokeratins could coassemble with any of the type II ones (Hatzfeld amd Franke, 1985), the basis of which was later demonstrated to lie in the need for cytokeratins to form heterodimers in order to assemble into IFs (Hatzfeld and Weber, 1990; Steinert, 1990). These dimers, representing coiled coils, are stable even under rather high urea concentrations (>4 M), and associate into tetramers when dialyzed from such solutions into low-salt/high-pH buffers.

A major surprise to the IF world was the notion that nuclear lamins, demonstrated to form regular networks underlying the nuclear membrane inside the huge nuclei of *Xenopus* oocytes (Aebi *et al.*, 1986), were also IF proteins (Fisher *et al.*, 1986; McKeon *et al.*, 1986). Even more surprisingly, one helix of the central rod (coil 1B) contained an additional 42 amino acids (6 heptads) compared to the cytoplasmic IF proteins (Figure 2B). This new type of IF protein completed the family to five classes according to sequence homology (Franke, 1987b), the cytoplasmic members being differentially expressed with regard to developmental programs, whereas the lamins are present much more uniformly throughout the generation of various cell types (Table I).

Since then, new IF proteins have been identified, including nestin, which originally was supposed to constitute a sixth class, but is probably better grouped into class IV together with the neurofilament triplet proteins, and the lens proteins phakinin and filensin, which form a unique type

Table I
The Intermediate Filament Protein Multigene Family

Name	Sequence homology class	M_r ($\times 10^{-3}$)	Typical occurrence in mammals
Cytoplasmic IF proteins			
Acidic cytokeratins: CK 9–20	I	40–64	All epithelia
Basic cytokeratins: CK 1–8	II	52–68	All epithelia
Vimentin[a]	III	55	Mesenchymal cells
Desmin	III	53	Muscle cells
Glial filament protein	III	50–52	Glia cells, astrocytes
Peripherin	III	54	Diverse neuronal cells
Nestin	[III/IV][b]	240[c]	Neuroepithelial stem cells, muscle cells
α-Internexin	IV	56[c]	Neurons
Neurofilament triplet proteins			
NF-L	IV	68[c]	Neurons
NF-M	IV	110[c]	Neurons; copolymer with NF-L
NF-H	IV	130[c]	Neurons; copolymer with NF-L
Phakinin	[I?][d]	46	Lens; copolymer with filensin
Filensin	[IV?][d]	83[c]	Lens; copolymer with phakinin
Nuclear IF Proteins			
Lamins			Nuclear lamina
Type A	V	62–72	Most, but not all cells
Type B	V	65–68	All cell types

[a]Vimentin is expressed in nearly every cultured cell due to the presence of a "serum-response element" in its promoter.
[b]Nestin exhibits sequence homology to both type III and type IV IF proteins. According to its gene structure, it should be placed into class IV (Dahlstrand *et al.*, 1992).
[c]Molecular weights are given as deduced from the amino acid sequence. "Apparent" molecular weight sobtained by SDS–polyacrylamide electrophoresis may be considerably higher.
[d]Phakinin shows highest sequence identity with CK18; however, it does not coassemble with cytokeratins. Filensin exhibits some sequence identity with NF-L (in coil 1) as well as to nestin and vimentin (in coil 2) (for a recent review, see Georgatos *et al.*, 1997).

of "beaded chain filament" and do not fit well into any of the four cytoplasmic sequence homology classes and therefore would probably better deserve an extra class (Goulielmos *et al.*, 1996; and Table I). Here, as in IF proteins of nematodes, the standard tripartite order is modified in one of the proteins. Moreover, the cytoplasmic IF proteins of invertebrates contain the extra 42 amino acids of coil 1B as found with vertebrates only in lamins, making it very likely that lamins were the first IF protein to occur in evolution. Insects apparently do not possess cytoplasmic IF proteins; their exoskeleton seems to allow a different conception of cellular stabilization. However, typical lamins have indeed been isolated from *Drosophila melanogaster* (Gruenbaum *et al.*, 1988; Bossie and Sanders, 1993). Interestingly, yeast does not contain lamins, as revealed from the sequencing of the complete genome of *Saccharomyces cerevisiae* (The Yeast Genome Directory, 1997). The MDM1 protein, called an IF protein in this report, does not show the tripartite structural organization as specified in Figure 2 and is therefore probably not related to IF proteins (for discussion see Herrmann *et al.*, 1993).

2.2. The Nonhelical Amino-Terminal "Head" Domain

This part of the IF proteins is highly variable in length as well as in primary sequence among the five sequence classes and without any obvious significant common organization. When orthologous forms of certain IF proteins are compared, however, it becomes clear that these heads are far from being fortuitous, random sequence stretches. In vimentin from various species, trout to human, for instance, three structural principles have been retained (Figure 3A): (1) a nonapeptide motif at the very amino terminus (box 1) shown to be essential for regular IF assembly to occur (Herrmann *et al.*, 1992); (2) six aromatic amino acids similarly spaced interrupting the hydrophilic amino acids dominating this domain (arrows); and (3) 12 arginine residues (boxed and numbered), also similarly spaced within the various sequences, which are part of known recognition motifs of various protein kinases (see Chapter 14 by Matsuzawa *et al.* in this book) and may loop out from a complex of the stacking benzene rings of the aromatic residues (Figure 3B). Aromatic amino acids are also abundant in the head domain of various cytokeratins. In particular, a motif of the type GGG(G/S)(F/Y) is repeated extensively in human CK10 (Rieger and Franke, 1988). A three-dimensional computer-simulated reconstruction suggests that the phenylalanines form a kind of hydrophobic channel from which the glycine/serine-containing chains loop out (Rieger, 1988). The occurrence of pairs or networks of aromatic side chains has been shown to occur in a variety of

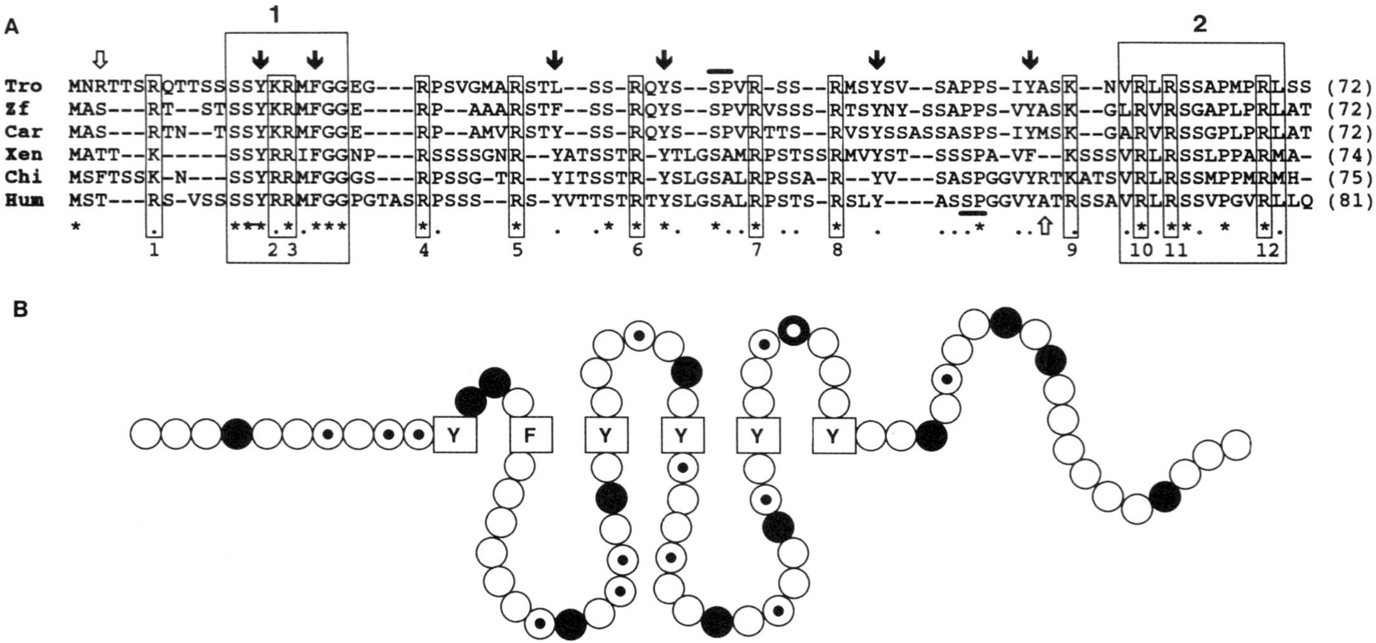

FIGURE 3. (A) Amino acid sequence comparison of the nonhelical head domain of vimentin from trout (Tro), zebrafish (Zf), carp (Car), the toad *Xenopus laevis* (Xen), chicken (Chi), and human (Hum). (B) A hypothetical structural model for the human vimentin head domain. In (A) black arrows mark aromatic amino acids, numbered boxes indicate arginines (or lysines), and the boxed regions 1 and 2 indicate highly conserved sequence motifs. The two open arrows indicate one additional arginine each found in trout and chicken, respectively. The bar above and below the sequences indicate potential cdc2/p34 kinase recognition sites. Asterisks indicate amino acid positions identical in all species; dots indicate identical amino acids in at least three species. The proline in the potential cdc2/p34 site of fish is replaced by alanine in all three higher vertebrate species. In (B) amino acids are represented by circles or boxes, aromatic amino acids are boxed, basic ones are filled in, and potential phosphorylation sites are dotted (for a similar representation of the aromatic residues see also Parry and Steinert, 1995, and Figure 5 in Chapter 3 of this volume). The circle with the clear center indicates the proline of the potential cdc/p34 Kinase recognition site. For sequence information see Herrmann *et al.* (1989a, 1996b); Zf: J. Cerdà *et al.* (in preparation).

different proteins and has been postulated to stabilize protein structure (Huber, 1979; Burley and Petsko, 1985).

To test functionally if different heads could be substituted for one another, we swapped the head domains of human and *Xenopus* vimentin. After cDNA transfection, both the human and the frog chimera behaved similar to the wild-type molecules, indicating that they could to some extent functionally replace one another (Figures 4A–4D). This experiment also showed that the demonstrated thermosensitivity of *Xenopus* vimentin (Herrmann *et al.*, 1993) is not contributed by the head domain.

The recombinant human vimentin with the *Xenopus* head behaved also during *in vitro* assembly similar to the human wild-type (wt) vimentin (Figure 4E). The reciprocal chimera, however, assembled only under low-salt conditions (50 mM, Figure 4F); at physiological salt concentrations, however, open filaments side by side with aggregated masses predominated (data not shown).

Another highly conserved motif (box 2) could be replaced by other, rather different amino acid sequences as long as the basic character was retained. For instance, the presumed nuclear localization signal of lamin B did not interfere with the protein's ability to form IFs. In addition, it completely carried this recombinant vimentin into the nucleus after cDNA transfection into vimentin-free cells (Herrmann *et al.*, 1993; see below). The replacement of the two arginines by serines, however, created a recombinant vimentin with very poor assembly properties. The deletion of box 2 (Δ60–70) generated a vimentin that assembled into apparently normal IFs, which, however, formed densely aggregated masses of filaments as revealed by ultrathin sections through pelleted and fixed assembled material. Thus, the elimination of this motif impedes the ability of vimentin IFs to retain a certain distance between the individual filaments, as best seen in sections from pelleted filaments (Hofmann *et al.*, 1991; Herrmann *et al.*, 1992). Interestingly, the formation of such dense IF masses has also been observed *in vivo* in mitotic cells (Franke *et al.*, 1982b, 1984), when vimentin is known to be phosphorylated in the head domain (Evans, 1988; Chou *et al.*, 1990; Inagaki *et al.*, 1997). A more pronounced deletion (Δ18–54) produced severely distorted fibrils that preferentially aggregated into large masses (Figure 5A). A larger deletion between box 1 and box 2, Δ18–61 in *Xenopus* vimentin, however, abolished filament-forming ability only slightly, as the mutated protein assembled into compact filaments side by side with open structures, indicating it is not the exact number of arginines, but certain structural principles that are important for assembly to occur (Figure 5B).

Finally, a completely headless vimentin did not assemble with various buffers tested. Under conditions where both the mutated and wt vimentin

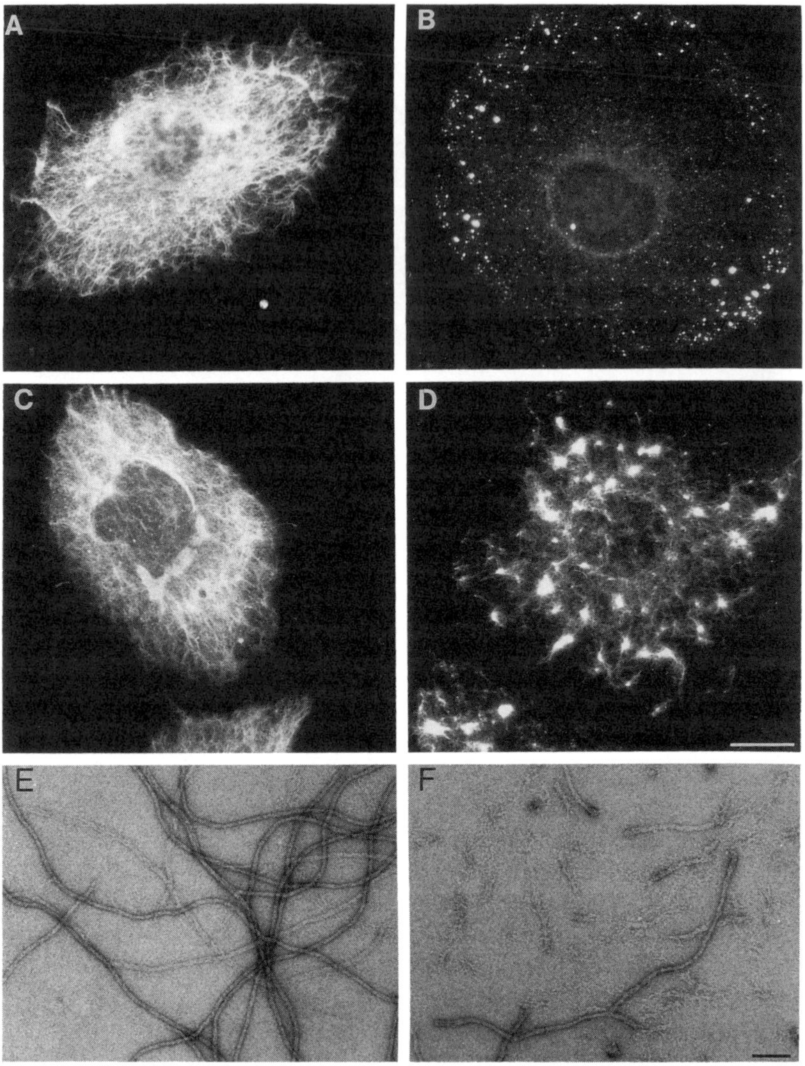

FIGURE 4. Analysis of chimeric vimentins consisting of human vimentin containing a frog head domain (A, C) or frog vimentin with the corresponding human head domain (B, D) after cDNA transfection into vimentin-free bovine mammary gland epithelial cells grown at 37°C (A, B) or 28°C (C, D), respectively. Note that the chimeric frog vimentin is temperature sensitive, forming no IF system at 37°C (B), whereas fibers are present at 28°C (D). The corresponding recombinant proteins also behave quite differently during *in vitro* assembly: chimeric human vimentin (E) forms extensive filament arrays even at high salt concentrations (160 mM NaCl); chimeric frog vimentin (F) forms some compacted IFs at low salt concentration (50 mM) in addition to unraveled filaments. At the higher salt concentration, this protein forms aggregated masses only. Bar, 100 nm.

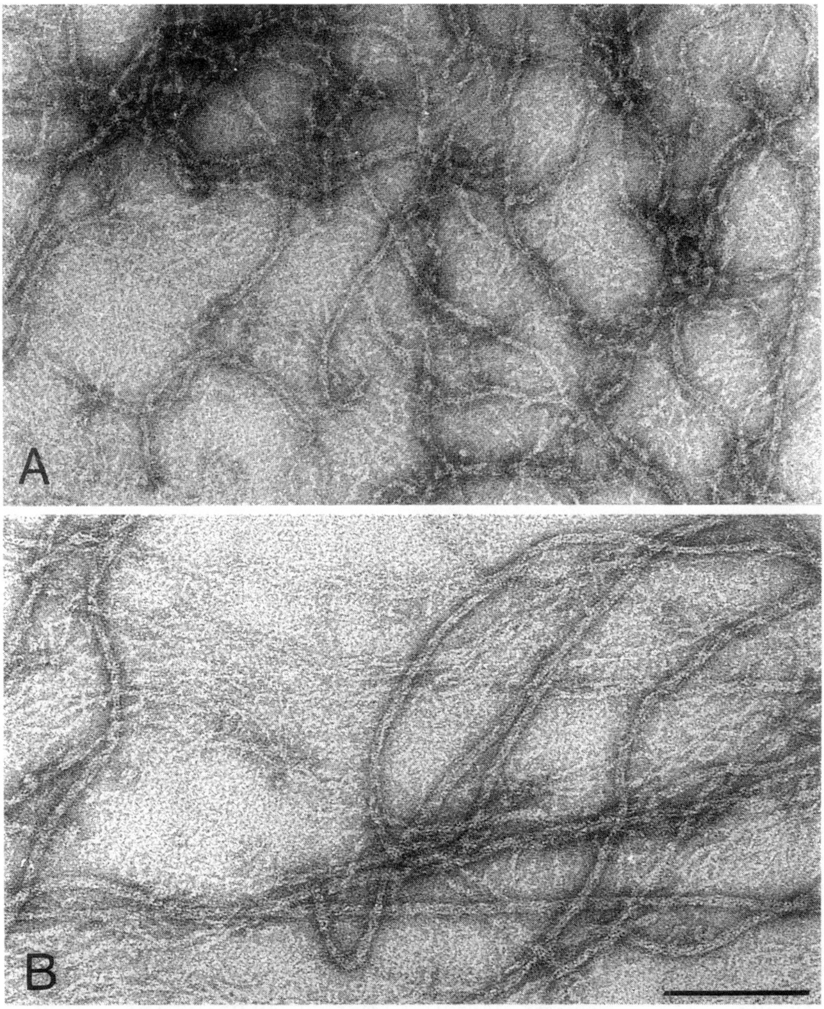

FIGURE 5. Reconstituted filaments from the deletion mutants (A) *Xenopus* vimentin (Δ18–54) and (B) *Xenopus* vimentin (Δ18–61), as visualized by negative staining. Note the pronounced presence of unraveled fibers. Bar, 100 nm.

are soluble, it sedimented with a significantly different S value in the ultracentrifuge than wt vimentin (5.2 vs. 4.0), thus indicating that it was present in smaller aggregates. This was confirmed by sedimentation equilibrium centrifugation runs at different speeds. The headless vimentin was present in monomeric to tetrameric complexes, whereas, the wt protein formed tetrameric to octameric complexes (Herrmann *et al.*, 1996a). Therefore, the head domain is apparently already engaged in the reconstitution of the soluble complexes. Similar results have been obtained with partial head-deleted derivatives of desmin and cytokeratins (Kaufmann *et al.*, 1985; Hatzfeld and Burba, 1994). These soluble complexes are obviously "prepared" to assemble, but the conformation is not compatible with the appropriate interaction to occur. Therefore, amino acid substitutions, deletions, and phosphorylation alike (see Chapter 14 by Matsuzawa *et al.* in this book) are able to change the conformation of the head such that orderly assembly and an independent display of individual filaments in solution is abrogated.

2.3. The α-Helical "Rod" Domain

At both the amino- and carboxy-terminal end a high degree of sequence conservation has been observed over stretches of 30–40 amino acids between sequences from various IF proteins, especially if conservative amino acids changes are allowed (see Figure 2). The absolute identity, even when five sequences from one species are compared (i.e., human CK18, CK8, vimentin, NF-L, lamin B) is very low: 35 of 314. Even in the conserved coil 1A stretch, significant differences between the class members are evident. Type I (acidic) cytokeratins characteristically show a LASYL motif that is FASFI in type II (basic) cytokeratins, FAAFI in type II trichocytic (hair) keratins (Rogers *et al.*, 1997), FANYI in vimentin, FASFI in NF-L, and LAVYI in lamin B. Immediately upstream in the sequence, a highly conserved tripeptide (NDR) is observed. Point mutations in this motif, i.e., NDC or NDH, in CK5 have been demonstrated to be the cause of inherited epidermolytic diseases (for review see Fuchs, 1997). The basic cytokeratins, however, exhibit two changes in this motif, NNK, and the type II trichocytic keratins contain NSR instead.

Despite these differences, the general composition in coil 1A is indeed very similar. One principal difference with regard to the overall structure is observed in the "pre-rod" domain of vimentin and desmin. This part, designated the H1 domain (Parry and Steinert, 1995), may in type III proteins already adopt α-helical structure, since in 26 residues that precede the general coil 1A part, no proline or glycine is found, and putative heptad positions do indeed possess hydrophobic residues (Quax-Yeuken *et al.*, 1983). Also, using a limited proteolysis approach, Geisler and Weber (1982) assumed that these 26 amino acids do not form β-turn structures, as they were not readily proteolyzed, but are engaged in forming the rod particle, named due to its appearance similar to the myosin rod in electron microscopy. The first linker sequence, L1, between coil 1A and coil 1B is rather short, 11 amino acids in lamin B, 10 in CK8 and CK18, and 8 in vimentin. The next linker, L12, between coil 1B and 2A carries a few more residues: 19 in lamin B, 17 in CK8 and CK18, and 16 in vimentin. Linker L2, between coil 2 A and B, is positioned differently by various authors and it may be α-helical, as no prolines and glycines are found in it (Fuchs and Weber, 1994; Parry and Steinert, 1995). In total, vimentin misses three amino acids in the linkers compared to cytokeratins 8 and 18, and it has been argued that this may explain the different rod length of vimentin compared to CK8/18: 43.9 versus 46.2 nm (Steinert *et al.*, 1993). In order to test how amino acid insertions into linkers may be tolerated, in particular, when amino acids with different physical qualities are used, we chose two approaches to modify linker L1: (1) insertion of two new amino acids and change of one amino acid compatible with α-helix and possibly coiled-coil formation; (2) corresponding insertion of two prolines and replacement of one amino acid by proline resulting in three consecutive prolines. The former mutant vimentin assembled only into short, loosely associated fibers, whereas the latter formed bona fide IFs with assembly kinetics indistinguishable from wt vimentin. This indicates that the linker indeed contributes a certain flexibility, and that a few more amino acids which may increase this flexibility do not disturb it. In contrast, interference with the helical ends by elongating coil 1A, for instance, has a drastic impact on the elongation of the short fibers formed (H. Herrmann *et al.*, manuscript in preparation). In accordance with these observations, the addition of "nonsense" sequences to linker L12 in NF-L was also tolerated to some extent (Gill *et al.*, 1990).

This led our interest to the next remarkable sequence feature in the rod, the so-called *stutter*. The exact position of this discontinuity in the phasing of the heptad repeat in coil 2B is not completely clear (Quax *et al.*, 1983; Fuchs and Weber, 1994; Steinert *et al.*, 1994), however, the introduction of three amino acids compatible with helix formation generated a primary amino acid sequence with theoretically ideal continuous heptads (Figure 6A) able to form an α-helix with optimal hydrophobic coiled-coil interaction (Herrmann, 1995). The assembly behavior of this mutated vimentin, Vim(FASAT), was followed by viscometry (see Chapter 11 by Hofmann in this book). Vim(FASAT) generated only little viscosity when compared with wt vimentin, independent of the temperature (22°C vs. 37°C) at which assembly was carried out (Figure 6B). Apparently, short fibrils of normal thickness formed, which elongated only slowly from 5 to 60 min of assembly at both temperatures tested (Figure 7). Therefore, a first important step in assembly, the formation of short "unit-length" fibrils, occurs, but elongation of these short IFs into longer filaments is slowed down considerably.

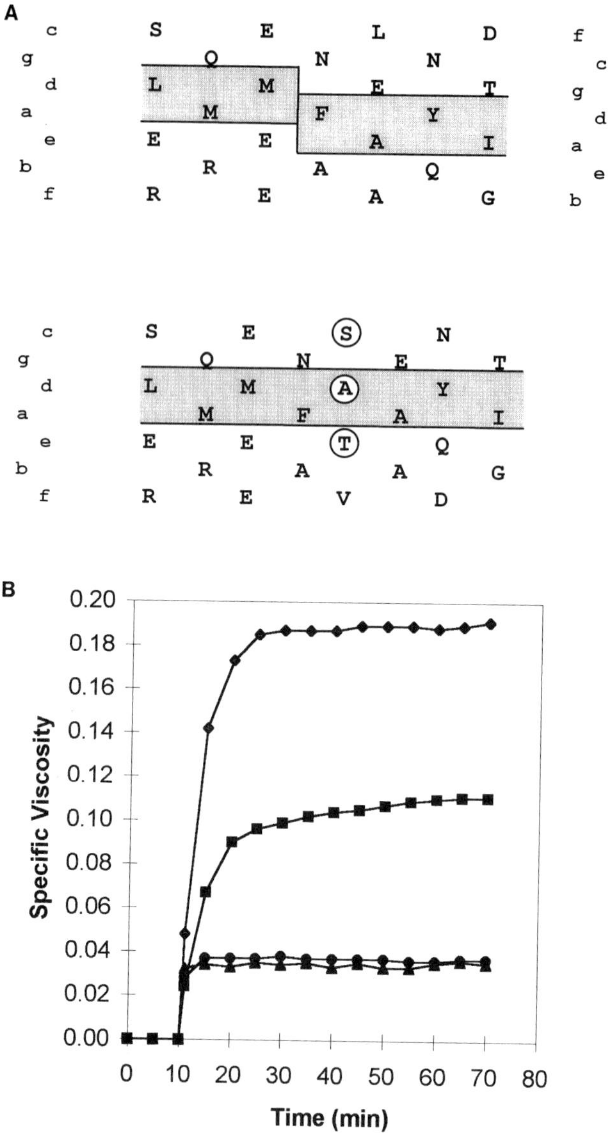

FIGURE 6. (A) Insertion mutagenesis into the "*stutter*" region of coil 2B of human vimentin apparently removing the discontinuity in the phasing of the heptad pattern. Inserted amino acids are circled. The sequence is written from top to bottom and left to right, and the respective position in the heptad repeat (*abcdefg*) is indicated at the left and right sides, respectively. The resulting hydrophobic stretch is shaded. Note that the insertion of Ser Ala Thr leads to a continuous heptad phasing. (B) The assembly kinetics of the resulting Vim(FASAT) is drastically altered: At both 22°C (triangles) and 37°C (circles) viscosity reaches a plateau of relatively low values 5 min after initiation of assembly. In contrast, human wt vimentin rapidly gains viscosity during the first 15 min of assembly, both at 22°C (squares) and at 37°C (diamonds). Apparently, the principal viscosity increase caused by the formation of unit-length filaments is actually represented by Vim(FASAT). Since its viscosity no longer increases after 5 min, one may speculate that unit-length filament formation is finished by and large at that time point.

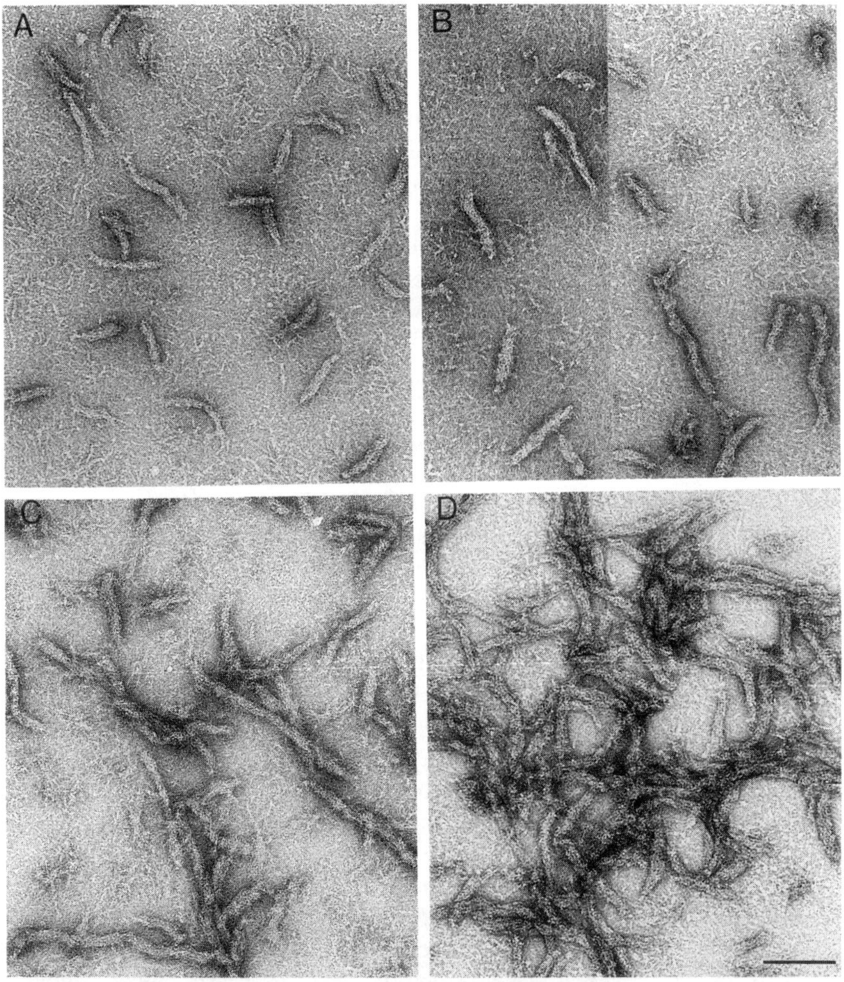

FIGURE 7. Assembly of Vim(FASAT) at 22°C (A, C) and 37°C (B, D) as observed at 5 min (A, B) and 60 min (C, D). Ionic conditions were as for viscometry (see Figure 5), i.e., 25 mM Tris-HCl, pH 7.5, 50 mM NaCl. Bar, 100 nm.

The consensus motif at the carboxy-terminal end of coil 2B is a sensitive part of the molecule, point mutations within this region of CKs also causing the above-mentioned epidermolytic diseases in human. Early on, deletions into this region were demonstrated to provoke serious defects in IF assembly, in particular *in vivo* after transfection into appropriate cells (Albers and Fuchs, 1989). Correspondingly, such mutated proteins formed aberrant filaments *in vitro*, presumably elongating more slowly than the wt proteins. To verify this assumption in a homopolymeric system, we engineered a *Xenopus* vimentin missing the last 12 amino acids of the conserved motif (YRKLLEGEESRI) in addition to the nonhelical tail domain, therefore ending with the amino acids IAT (K. R. Rogers *et al.*, unpublished). The recombinant protein, XenVim(IAT), assembled quickly, very similar to wt vimentin, but viscosity dropped to lower values by 15–30 min, which is indicative for filament aggregation. Indeed, electron microscopy performed in parallel revealed that long filaments, preferentially aggregating laterally, had formed (Figure 8A). This rotary shadowing image nicely shows the pronounced 21-nm repeat along the filaments. The repeat was also clearly seen in negatively stained preparations, which is mostly not evident in wt-IFs with this mode of visualization. The laterally associated filaments resembled the paracrystalline arrays formed from lamins or GFAP (Gieffers and Krohne, 1991; Stewart *et al.*, 1989), and therefore we may conclude that the elongation of filaments is not influenced by the truncation, but that these filaments differ from normal IFs in their pronounced lateral interaction.

The helical rod domain is truly the building block of IFs. It is puzzling, therefore, that rods like headless IF proteins do not readily interact with one another. The data from the analytical ultracentrifugation indicated that the rod even under optimal filament-forming conditions (25 mM Tris-HCl, pH 7.5, 160 mM NaCl) was present in its minimal, i.e., tetrameric, form (Herrmann *et al.*, 1996a). Also other conditions, such as those employed by Geisler *et al.* (1982) yielding filamentous structures from proteolytic desmin fragments, did not support the formation of higher aggregates.

Despite the lack of apparent interaction of rods, the data obtained by ultracentrifugation indicated that some change occurred when rods kept in 5 mM Tris-HCl, pH 8.4, were transferred into filament buffer (25 mM Tris-HCl, pH 7.5, 160 mM NaCl). Therefore, we tested the assembly of wt vimentin in the presence and absence of the rod by viscometry. Figure 9 clearly shows that the presence of the rod has a drastic inhibitory influence on assembly, indicating the rod indeed is able to interact with wt vimentin. Electron microscopy revealed that in addition to short filaments of normal appearance, many "open" structures were obtained. Formation of aggregates, another possible cause for obtaining reduced viscosity, was not observed. Thus, the rod domain, present as dimer or tetramer, obviously reacts with wt vimentin tetramers to form an assembly–inhibitory complex during the transition from low-salt/high-pH to filament buffer conditions. Centrifugation experiments furthermore showed that coil 1 alone, which forms dimers under low-salt conditions, is sufficient to trap wt vimentin in an assembly-incompetent state. This is not the case, however, for coil 2

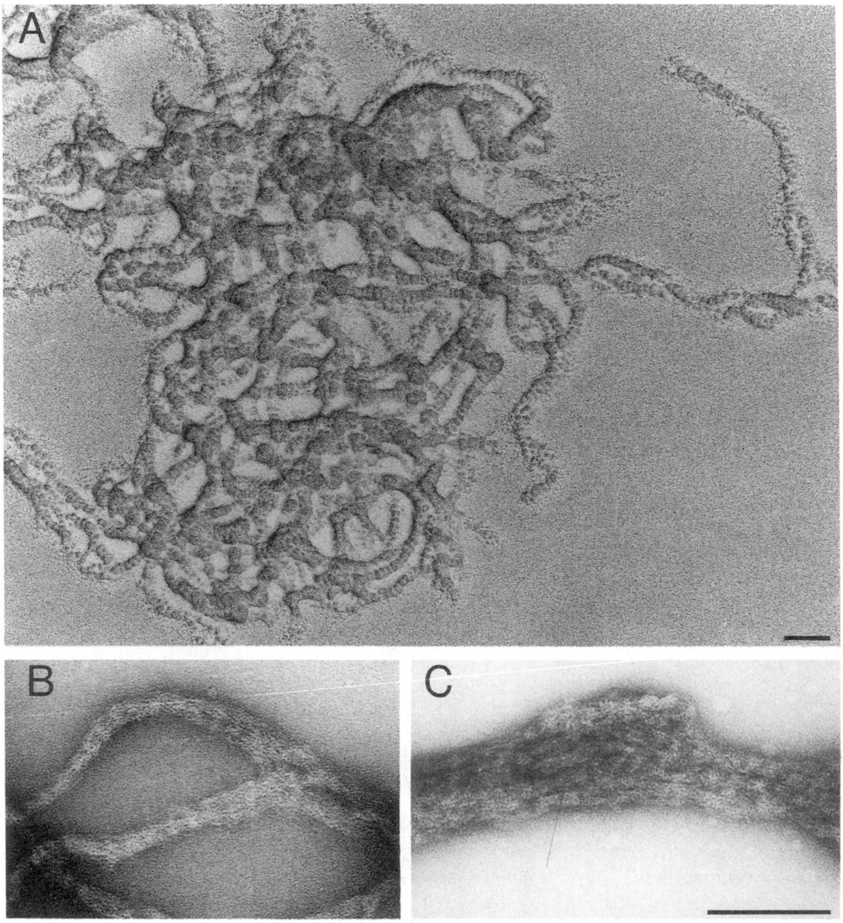

FIGURE 8. Electron microscopic observation of Vim(IAT), a deletion mutant lacking the carboxy-terminal domain including a highly conserved sequence motif of coil 2B. (A) Rotary-shadowed preparation; (B, C) negatively stained preparation. Note the prominent beading clearly visible in both types of samples. The filaments have diameters up to 30 nm. Bars, 100 nm.

(H. Herrmann *et al.*, manuscript in preparation). For this reason we were hoping that methods successfully employed for other protein complexes to yield two-dimensional crystals would also be applicable for the vimentin rods in order to get some insight into the association of the four protein chains in the tetrameric complex. Unfortunately, we have not obtained ordered crystal arrays. Instead, under the conditions applied, they only

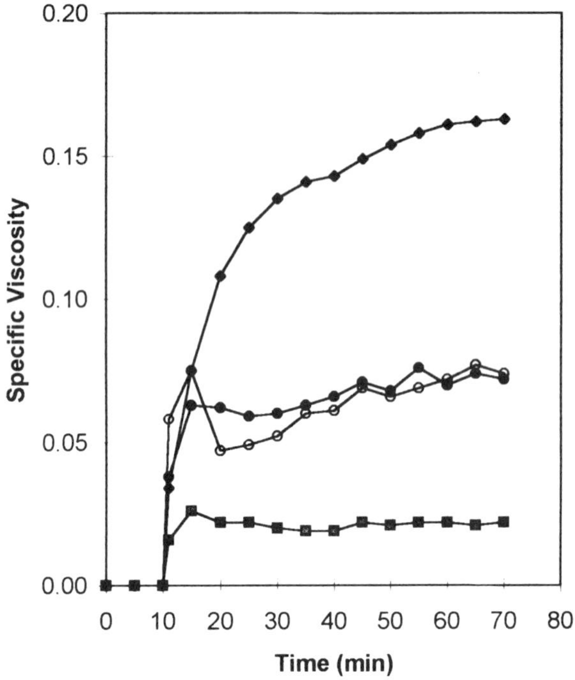

FIGURE 9. Interference of the human rod domain with the assembly of wt vimentin as determined by viscometry. Vimentin was assembled at 0.5 mg/ml either alone (diamonds) or in the presence of 0.5 mg vimentin rod (circles). Note that the rod has a drastic effect on the filament formation of wt vimentin, i.e., 5 min after the start of assembly, viscosity drops considerably and rises only slightly during the following 50 min. The vimentin rod generates only little viscosity (squares), but the switch to assembly conditions (carried out at 10 min) apparently increases the viscosity slightly, indicating the generation of higher order complexes. This change is indeed observed in ultracentrifugation experiments (Herrmann *et al.*, 1996a).

formed irregular associations (Figure 10; J. R. Harris *et al.*, unpublished observations; see also Harris, 1991).

2.4. The Nonhelical Tail Domain

The tail domain of vimentin of various species is highly conserved (Figure 11) and is also very similar to that of desmin from various species (Herrmann *et al.*, 1989a, b). To highlight this, we boxed three regions with distinct character. Box I starts with four identical amino acids LRE(S/T), then revealing distinct differences between fishes and the other vertebrates (indicated by arrows). Box II is nearly identical over nine amino acids and ends in a motif DG that is continued as DGH(I/V)I in fish and DGQV(V/

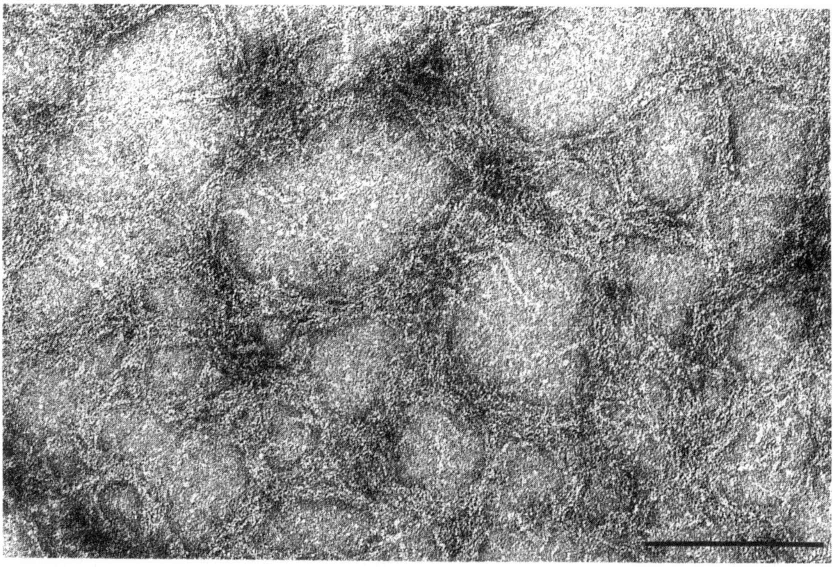

FIGURE 10. Electron microscopic observation of the human vimentin rod assembled at 0.5 mg/ml in the presence of 0.1% PEG (M_r 6000) and 1% ammonium molybdate, pH 6, in order to generate two-dimensional crystals on freshly cleaved mica. For further details see Harris (1991). Note that no uniform associations occurred. Bar, 100 nm.

```
Tro   TTPMPNFSSFN|LRESMLEARPM|IDNLSKK-VV|IKTIETRDG|HVINESTQNH|DDLE|
Zf    TSPFPNFSSLS|LRESMKETRPL|IDNLSKK-VV|IKTIETRDG|HIINESTQN-|EDLE|
Car   TTPFPNLSSLT|LRETMKETRPA|MDSLSKK-VV|IKTIETRDG|HIINESSQN-|DDLE|
Xen   SLPVHSFSTMS|LRETNLDSHPA|-ETHSKRTVL|IKTVETRDG|QVVNESSQHH|DDFE|
Chi   NMPIPTFASLN|LRETNIESQPI|VDTHSKRTLL|IKTVETRDG|QVINETSQHH|DDLE|
Hum   SLPLPNFSSLN|LRETNLDSLPL|VDTHSKRTFL|IKTVETRDG|QVINETSQHH|DDLE|
        *  •••••  ***•  •   *   •  **  ••  *** *****  ••**••*  • •*•*
        ▲ ▲             I   ▲               II                III
```

FIGURE 11. Amino acid sequence comparison of the vimentin tail from different species (see Figure 3 for abbreviations). Three conserved sequence motifs are boxed. Conserved prolines are indicated by arrowheads. Open arrows designate amino acid positions decisively differing between fish and higher vertebrates. The bracket indicates the motif FSSL (in human and zebrafish) modified to LSSL and FSSF in carp and trout, respectively.

I) in tetrapodes. The same motif is found in CK8 and 18 at a corresponding position in the tail (Franke, 1987a). The following sequence is also highly conserved ending in box III, which is very characteristic for vimentins: (D/E)D(L/F)E. In the second vimentin gene of the pseudotetraploid *Xenopus* (Vim4 of Herrmann *et al.*, 1989a) it carries an insertion (in brackets) reading DD(LDLD)LE. It may be interesting to note that this protein is a "bad assembler." As it probably is coexpressed with Vim1 in every cell, cellular filaments very likely are obligate copolymers of Vim1 and Vim4 (which are only 92% identical at the amino acid level). Here we are apparently watching the coevolution of IF proteins diverging from one another, but still forming the same structure.

Tails of IF proteins of the various sequence classes vary highly in length (13 to more than 600 amino acids) and composition. Cytokeratins may contain repeated serine/glycine clusters, neurofilament proteins may contain highly acidic stretches of unknown function, and lamins contain nuclear localization signals and membrane attachment motifs. It was speculated that the tails may therefore be engaged in specific cellular interactions, but not in IF assembly as such. This seemed to be indeed the case (Bader *et al.*, 1991; Eckelt *et al.*, 1992; Heins and Aebi, 1994; Rogers *et al.*, 1995). We finally learned, however, that, although the tail is not needed for assembly as such, the tail truncation has an influence on the mass composition of the IFs (Herrmann *et al.*, 1996a; for details on the method see Engel *et al.*, 1985). This is nicely exemplified when scanning transmission electron microscopy (STEM) data of wt vimentin are compared side by side with those for tailless vimentin, XenVim(ΔT), and two carboxy-terminal truncation mutants, ΔC35 and ΔC47 (Figure 12). Truncation of the tail leads to the formation of IFs with different filament diameter and therefore different mass per length (MPL) values (Figures 12B and 12D). However, the filament type preferentially formed by VimΔT has 54 kDa/nm, a species also formed by Vim(wt), though much less pronounced. The same MPL value was obtained for VimΔC47, having a seven-amino acid tail stump. In stark contrast, VimΔC35, which retains 19 of the 54 amino acids of the tail domain, exhibited two MPL peaks one of 41 ± 4 kDa/nm, nearly identical to the main peak of wt vimentin, and a second of 50 ± 4 kDa/nm, reasonably similar to the main value for VimΔT, 54 ± 5 kDa/nm (compare Figure 12E with Figures 12C and 12D). From this assembly behavior we conclude that the tail domain is critically engaged in the control of subunits per diameter, the ΔC35 mutant being on the edge between vimentin wt and ΔT, generating filaments either with 32 or 40 molecules per diameter, respectively. Assuming the subunits are packed in a analogous way, the increase by mass of 25% would result in a filament diameter increase from 10 to 11 nm, which is indeed observed (Table II in Herrmann *et al.*, 1996a).

To our surprise, the systematic tail truncation carried out with *Xenopus* vimentin (Rogers *et al.*, 1995) revealed relatively few differences in the biophysical behavior, such as viscoelasticity and kinetics of assembly, of the recombinant proteins. In particular, in light of the many reports on the aberrant properties of tail-truncated or modified IF proteins *in vitro* and

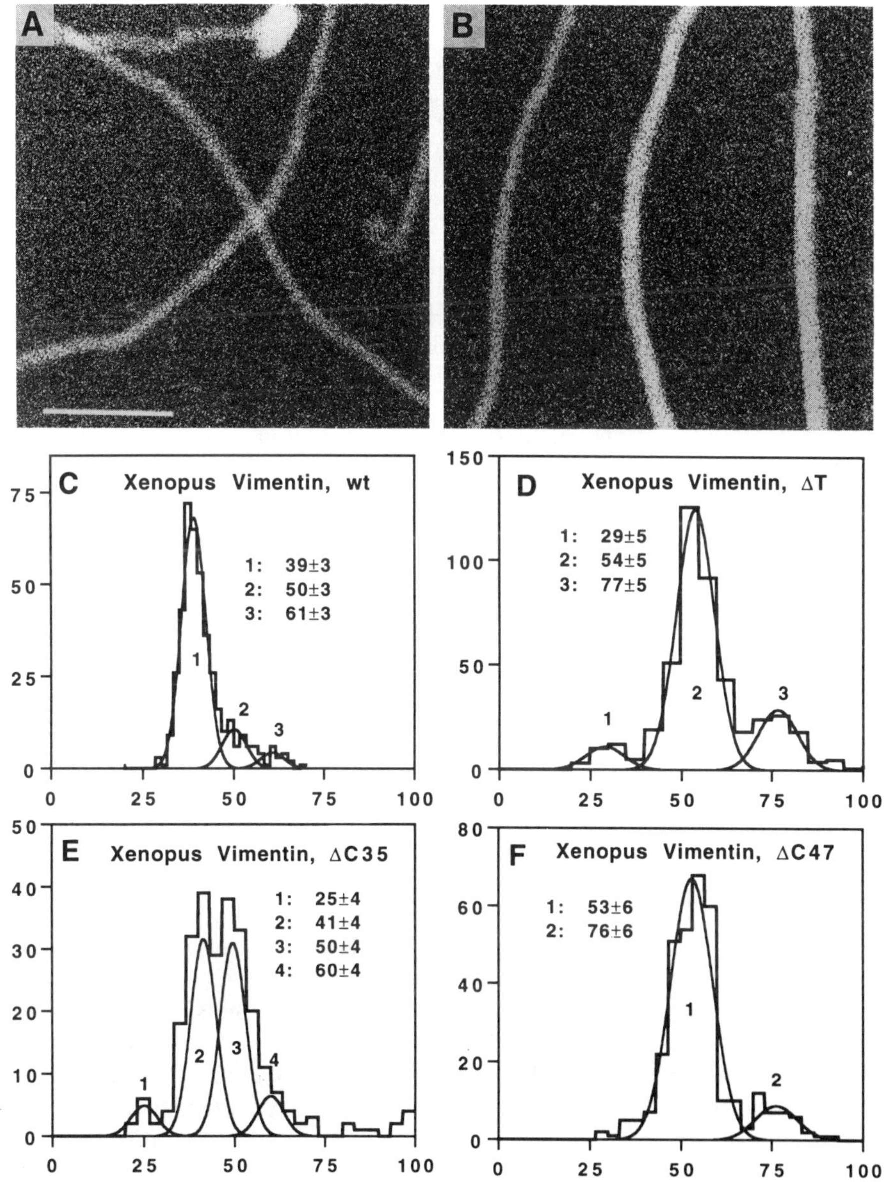

FIGURE 12. Heterogeneity of apparent filament diameter (A, B) and mass per length (MPL) values (C–F) of *Xenopus* wt vimentin (A, C), tailless vimentin (B, D), and tail-truncated mutants XenVim(Δ35) (E) and XenVim(Δ47) (F). In (C–F) the main peaks are numbered and the corresponding MPL values are given in kDa/nm. Ordinate, number of segments; abscissa, mass per length (kDa/nm). Bar, 100 nm. (Reprinted with permission from the *Journal of Molecular Biology*.)

in vivo, i.e., in transfected cells (e.g., Lu and Lane, 1990; Kouklis *et al.*, 1991, 1993; McCormick *et al.*, 1993), the positive results of Bader *et al.* (1991), Hatzfeld and Weber (1990), and Eckelt *et al.* (1992) were not easily understood (see also Heins and Aebi, 1994). During the polymerase chain reaction (PCR) cloning of our deletion mutants, however, we identified one clone that contained an additional mutation changing the fifth amino acid of the tail from His to Arg. The intended mutation, ΔC11, already generated a protein that easily aggregated when challenged by rapid ionic change to high-salt conditions (Rogers *et al.*, 1995). With the additional mutation, however, this VimΔC11 (His 405 → Arg) formed filaments that notoriously lost "contact inhibition," thus forming masses of aggregated filaments (Figure 13A) that in ultrathin sections of pelleted filaments appeared as lateral assemblies of hundreds of individual filaments (Figure 13B). Even under relaxed ionic conditions, filaments interacted laterally to form bundles of several twisted IFs never observed with wt vimentin (Figure 13C). Thus, point mutations as well as the addition of "sequence tags" may easily introduce conformational changes that are detrimental to the orderly formation of filaments (see also Makarova *et al.*, 1994). This argues against the use of sequence tags added "somehow" to the carboxy-terminal domain. We are not even sure if a tag fused to the very last amino acid via a proline/glycine linker would be without influence on filament structure.

3. FILAMENT ASSEMBLY

IFs are among the most insoluble cellular structures, and they are easily purified from cells or tissues by extraction with buffers of neutral pH containing 1% Triton X-100 and 1.5 M potassium chloride (Starger *et al.*, 1978; Franke *et al.*, 1978b). Under conditions minimizing proteolysis, presumptive associated proteins may be coisolated as shown for the high-molecular-weight protein plectin (Pytela and Wiche, 1980; Wiche *et al.*, 1983). In skin or in certain pathological conditions, the situation is very complex, since numerous components are associated with and cross-linked to IFs (for reviews see Fraser, 1969; Osborn and Weber, 1989). Some of these associated factors may indeed influence the way particular IF proteins assemble *in vivo* leading to the formation of "macrofibrils" (reviewed by Yang *et al.*, 1990).

3.1. Soluble IF Subunits

Since newly synthesized subunits assemble very rapidly into filaments—in cultured cells a half-life of 7 min for vimentin staying in the

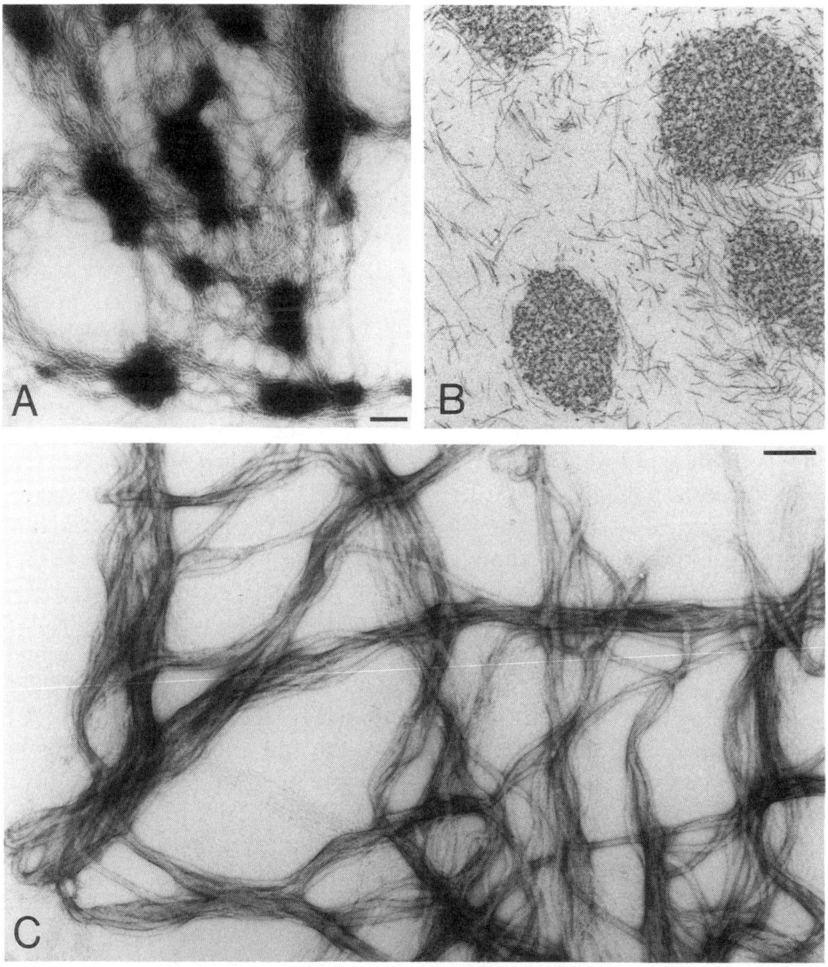

FIGURE 13. Electron microscopy showing aberrant assembly of a tail-truncated vimentin (ΔC11) with a point mutation in the fifth position of the tail domain (His → Arg). This amino acid is conserved in all other species as a proline (see Figure 11), indicating an important structural role. (A) At physiological salt concentration, seemingly normal IFs are generated that, however, form centers of tight association. (B) A corresponding ultrathin section of pelleted material as shown in (A). (C) At reduced salt concentration (50 mM), IFs align laterally and proceed as twisted bundles. Bar, 100 nm.

soluble pool has been determined (Blikstad and Lazarides, 1983)—it has not been easy to obtain enough material in order to characterize the soluble form of IF proteins. It was even suggested that IF proteins may be synthesized off the ribosome directly into the cellular IF system (Fulton and Wan,

1983). However, with the help of sucrose density centrifugation experiments, Soellner *et al.* (1985) managed to demonstrate the existence in cultured cells of discrete complexes of vimentin sedimenting with an S value similar to immunoglobulin G (6.5S). This was very close to the data obtained by ultracentrifugation for various IF proteins reconstituted from the denatured state (i.e., 8–9.5 M urea) into low-salt/high-pH buffers (5.2S). From sedimentation velocity analysis it was concluded that these complexes were indeed tetramers (for references see Soellner *et al.*, 1985). We recently confirmed and extended these results with recombinant vimentin showing that also higher aggregates such as hexamers and octamers are present (Herrmann *et al.*, 1996a). At higher pH, i.e., 5 mM Tris-HCl, pH 9.5, only tetramers are detected (data not shown).

These subunits, although not interacting with one another, are perfectly competent for assembly, since small changes in the ionic milieu as facilitated by rapid addition of buffered salt solutions may immediately lead to the formation of filamentous structures (Huiatt *et al.*, 1980; Steinert *et al.*, 1982). Coiled coils are formed already at rather high urea concentration. For recombinant vimentin, dimeric species are clearly seen at 6 M urea, and at 5 M urea tetramers are the predominant species (Herrmann *et al.*, 1996a; see also Fuchs and Weber, 1994). The formation of the tetrameric subunits proceeds probably through the interaction of two dimeric coiled-coil molecules via their coil 1 domains (A_{11}; Parry and Steinert, 1995). Two other orientations may occur, interaction-mediated by the coil 2 domains (A_{22}) and the antiparallel, in-register binding of the coil 1 to the coil 2 domains (A_{12}). Our cross-linking data of the single cysteine in vimentin indeed favor the notion that most of the vimentin is, at 5 mM Tris-HCl, pH 8.4, in the A_{11} orientation (Rogers *et al.*, 1996). However, the type of orientation is probably not fixed, and a dynamic equilibrium generated by "tetramer switching" may prevail instead (Aebi *et al.*, 1988).

3.2. Mechanics of IF Assembly

How then are these complexes arrested so that they do not interact with one another to form filaments? Since it had been noted early on that the very N-terminal part of type III IF proteins is essential for filament assembly (Traub and Vorgias, 1983; Kaufmann *et al.*, 1985), we tested the idea that the head domain may be "locked" and that peptides corresponding to the critical sequences could induce a polymerization-competent configuration of the protein chains within the tetramer (Hofmann and Herrmann, 1992). Already when a decameric peptide covering the evolutionarily conserved nonapeptide motif (Herrmann *et al.*, 1992) was added at a fourfold molar excess over vimentin under the low-salt/high-pH conditions, short fibrils of 50–200 nm length and 12–24 nm diameter formed, but the formation of longer filaments was not observed. Only after shifting the ionic conditions to 160 mM NaCl, 25 mM Tris-HCl, pH 7.5, were these peptide-induced short fibrils apparently fused to long filaments that were obviously thicker than normal IFs. At higher peptide concentration, long, smooth filaments formed in addition to aggregated masses of filaments, indicating that regular assembly was somehow disturbed.

Short, filamentous material has been observed frequently during various filament assembly experiments, especially when conditions were employed that do not favor filament formation or even dissolve them (Renner *et al.*, 1981; Steinert, 1981; Franke *et al.*, 1982a; Milam and Erickson, 1984; Eichner *et al.*, 1985; Ip *et al.*, 1985; Stromer *et al.*, 1987; Coulombe *et al.*, 1990; Coulombe and Fuchs, 1990; Heins *et al.*, 1993).

The absence of extensive filament masses on grids processed for negative staining may, however, only mean that they have been drawn off the carbon film together with the buffer, which has to be removed for drying the specimen. We observed such a case with one of the head-deletion mutants of vimentin, which was missing exactly the conserved nonapeptide motif, Vim(Δ6–14) (Herrmann *et al.*, 1992). Negatively stained preparations mostly showed short, irregular, nonfilamentous material. Cut sections through pelleted assembled protein, however, revealed that most of the material resided in huge aggregates from which seemingly normal filaments emanated. These aggregates were notoriously lost in negative staining since they probably destroyed the carbon film support, thereby being lost for microscopic observations. In order to visualize early assembly stages, hoping that they would not be as big, we prepared, 1 min after initiation of assembly, a specimen for rotary shadowing, stopping the assembly process by adding glycerol and spraying the protein solution onto mica. We observed large masses of tightly packed sheets with a substructure reminiscent of the 21-nm repeat pattern observed with sprayed standard filaments.

To be able to look at very early assembly intermediates, we developed a procedure to "freeze" assembling IF proteins instantaneously. Assembly was started by addition of an equal volume of twice-concentrated filament buffer (40 mM Tris-HCl, pH 7.0, 100 mM NaCl) to a solution of soluble vimentin followed directly by the addition of two volumes of filament buffer containing 0.2% glutaraldehyde. This stops any further interaction, as shown in control experiments by the inclusion of glutaraldehyde into twice-concentrated filament buffer. In this case, no assembly of tetramers/octamers into any discernible structure occurred. For this reason we were confident that what we observed, when assembly was allowed to proceed for 1 sec, indeed represented the structures present at around that time point. Surprisingly, very homogeneous populations of particles with a length of 50–65 nm and an average diameter of 16 nm resulted (Figure 14; Herrmann *et al.*, 1996a). Similar structures were also observed with CK8/18, desmin, and NF-L; however, cytokeratins appear to assemble much faster than vimentin (data not shown), and NF-L forms filaments which seem to be of smaller diameter (Figure 14C). At later time points, e.g., at 10 sec, longer structures were observed that appeared to be segmented as if the above "unit-length" fibrils had fused longitudinally. This segmentation became particularly clear with tailless vimentin as if the C-terminal tail domains "obscured" the ~50 nm-spaced junction points. With time the filament diameter appeared to be reduced, suggesting a distinct compaction event. Such a compaction phenomenon has also been observed with desmin

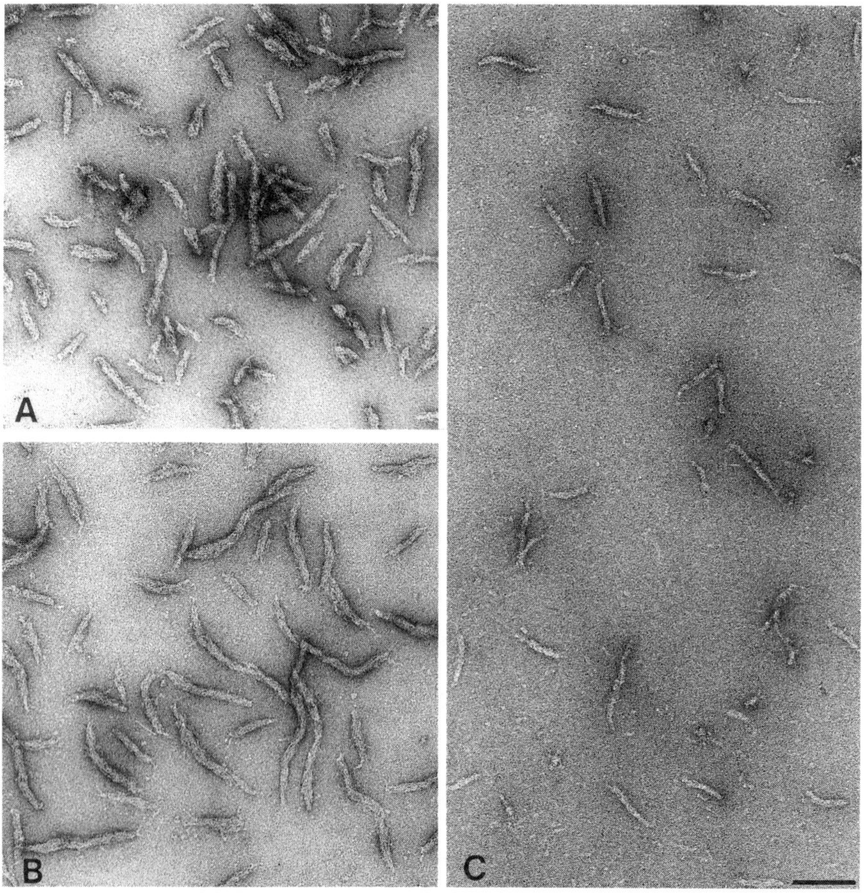

FIGURE 14. Early intermediates of assembly (10 sec) of (A) human vimentin, (B) human desmin, and (C) mouse low-molecular-weight neurofilament protein (NF-L). Bar, 100 nm.

under assembly conditions slightly different from ours, and it has been measured at 5 and 60 min after initiation of assembly (Stromer *et al.*, 1987). As documented, this compaction seems to occur in a stochastic manner early after filaments have reached a certain length (Herrmann *et al.*, 1996a), and it becomes most pronounced during the first 5 min of assembly (H. Herrmann *et al.*, manuscript in preparation). Hence we may describe IF assembly by a novel three-step process:

1. Rapid formation of ~60-nm-long "unit-length" filaments by lateral association of tetramers/octamers, which, in turn, are formed by antiparallel, approximately half-staggered association of two/four dimers.
2. Elongation by longitudinal annealing of "unit-length" filaments into several-micrometer-long "immature" (i.e., ~16 nm in diameter) filaments.
3. Compaction of "immature" filaments into "mature" (i.e., ~11 nm in diameter) IFs. Immature filaments may persist for extended times, as they are often found side by side with long, mature IFs (see Figure 1).

The compaction event may go hand in hand with a rearrangement of dimers or tetramers/octamers (i.e., in terms of their relative stagger), so that, under certain conditions, IFs partially unravel into octameric protofibrils (Aebi *et al.*, 1983; Herrmann *et al.*, 1996a). However, based on our kinetic-type assembly experiments, IFs do not form by lateral association of extended protofibrils (Herrmann *et al.*, 1996a).

By this type of assembly regime it was also possible to follow the kinds of structures into which mutated forms of vimentin assemble in a first instance, and how these structures further develop. For example, very much like intact vimentin, the deletion mutant Vim(Δ6–14) (see above) first assembles into ~60-nm-long unit-length filaments that then anneal longitudinally into longer filaments (Figure 15A, arrowheads). Similarly, the truncation mutant Vim(IAT), i.e., missing the complete tail in addition to the 12 carboxy-terminal amino acids of coil 2, first forms ~60-nm-long fibers, albeit of larger diameter (big arrowhead in Figure 15B), which then anneal very quickly into long filaments of markedly increased diameter. Moreover, insertion mutations such as the one with a PPGG motif interrupting the amino-terminal end of the rod, Vim(PPGG) (Hofmann and Herrmann, 1992), and another one with three amino acids, SAT, introduced into the *stutter* region, Vim(FASAT), both yielded unit-length filaments by 10 sec; however, those produced by Vim(PPGG) revealed a significantly increased diameter (Figures 15C and 15D). After further incubation, during which Vim(PPGG) aggregated into masses of thick filaments (i.e., after 1 min),

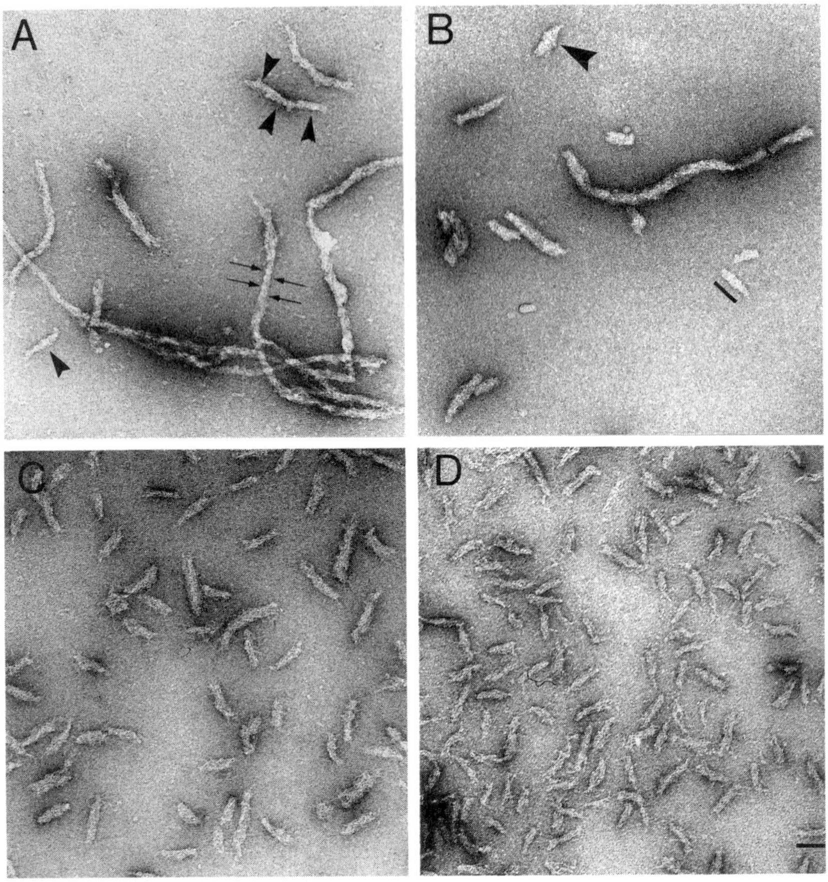

FIGURE 15. Early intermediates of assembly (10 sec) of (A) XenVim(Δ6–14), (B) XenVim(IAT), (C) XenVim(PPGG), and (D) HumVim(FASAT). Arrowheads in (A) point to normal-looking unit-length filaments. Arrows indicate beaded appearance of a filament. The big arrowhead in (B) points to a thick unit-length filament. The bar is positioned next to a seemingly normal unit-length fiber. Bars, 50 nm.

Vim(FASAT) failed to elongate even after 1 hr of assembly (see Figure 7). Therefore, the two mutated proteins behaved completely differently with regard to their lateral and longitudinal assembly: Vim(PPGG) obviously assembled into thicker unit-length filaments that appeared to elongate normally, and thus mainly phase 1 of the three-step assembly process was affected. Vim(FASAT) formed normal-sized unit-length filaments that did, however, fail to anneal into elongated structures, so in this case phase 2 of the assembly process was slowed down nearly completely.

4. SUPRAMOLECULAR ORGANIZATION—FUTURE PROSPECTS

As discussed in Section 3.2., short peptides may be used to interfere with IF structure and/or IF assembly. Peptides representing the consensus sequence motif at the carboxy-terminal end of coil 2B cause complete disassembly of authentic IFs when added at sufficiently high concentrations to preformed filaments (Hatzfeld and Weber, 1992). Similar studies were carried out by Steinert and colleagues with various rod peptides, thus demonstrating that synthetic coil 1A peptides, for example, may disassemble vimentin IFs down to the monomer level (see Chapter 5 in Parry and Steinert, 1995). The specificity of such interactions is demonstrated by the fact that single-amino acid replacements in such peptides can completely abolish their ability to interact with IFs (Chipev *et al.*, 1992). Therefore, such peptides offer a powerful tool to investigate the importance of distinct subdomains or single amino acids for the individual phases of IF assembly as described above.

Heterologous peptides also may exert specific effects on the assembly of IF proteins, such as shown for a peripherin carboxy-terminal 35-mer on vimentin assembly (Kouklis *et al.*, 1991). In contrast, we were unable to demonstrate a pronounced influence of the corresponding carboxy-terminal vimentin peptide on vimentin assembly, by neither viscometry nor electron microscopy (data not shown). This observation may indicate that the peripherin tail domain indeed interacts with the vimentin system. Conversely, the vimentin tail domain has been reported to interact with actin-containing structures (Cary *et al.*, 1994).

As mentioned above, we employed amino-terminal vimentin peptides to investigate the role of the evolutionarily conserved nonapeptide motif in filament assembly (Herrmann *et al.*, 1989b; Hofmann and Herrmann, 1992). This approach was followed up using longer peptides assuming they may have a more pronounced effect on assembly. When *Xenopus* vimentin was preincubated with a 23-mer peptide, representing vimentin's amino terminus, in the soluble state and then induced to assemble, thick and irregular filaments were formed (Figure 16A), indicating that its presence led to additional subunit interactions. A control peptide with the two arginines in the nonapeptide motif replaced by threonines (Figure 3A, box 1) had only little influence on filament assembly as judged by viscometry (data not shown), although the presence of partially unraveled filaments appeared to be enhanced (Figure 16B). Since we assumed that the original amino-terminal head peptide mediated some important interaction during the assembly process, we wondered if it would still interact with headless vimentin. Accordingly, at a 20-fold molar excess over the protein, the

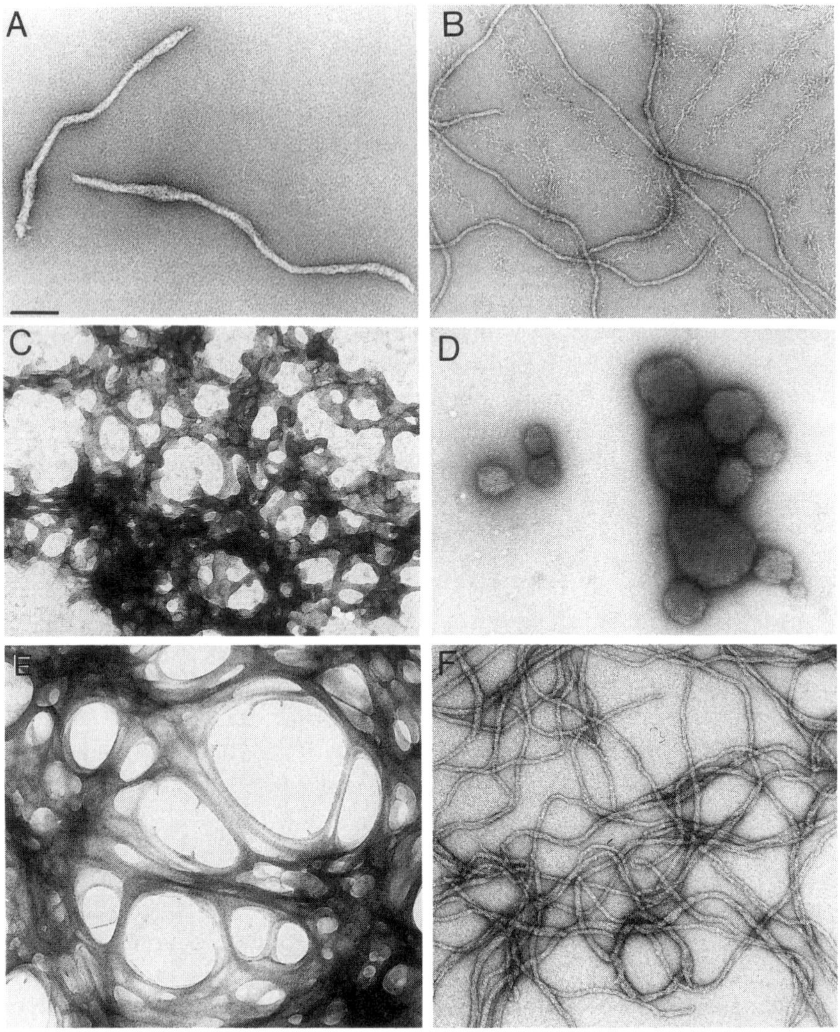

FIGURE 16. Interference of a 23-mer peptide representing the *Xenopus laevis* vimentin amino-terminal region (plus a carboxy-terminal cysteine) with the assembly of (A) *Xenopus* vimentin under standard assembly conditions (tenfold molar excess of peptide). The same peptide was incubated in 5 mM Tris-HCl, pH 8.4, 1 mM EDTA, with (C) headless *Xenopus* vimentin and (E) human cytokeratin 8/18 at 20-fold molar excess. In (B) a control peptide with two threonine residues instead of the relevant diarginine motif (see box I in Figure 3) was employed. In (D) headless *Xenopus* vimentin and in (F) cytokeratin 8/18 were assembled in the presence of a tenfold molar excess of the 23-mer peptide oxidatively cross-linked via the carboxy-terminal cysteine. Note that headless vimentin, not assembling above the tetrameric state under standard filament assembly conditions, forms massive aggregates in its presence even at low-salt/high-pH conditions. Accordingly, the cytokeratins form filamentous arrays under the same conditions as a result of the addition of peptide.

peptide induced the formation of thick interconnected fibrils, even under low-salt/high-pH conditions (Figure 16C). A corresponding dipeptide formed by oxidative cross-linking via the carboxy-terminal cysteines induced the formation of even higher, "ball-like" aggregates (Figure 16D). To our great surprise, these amino-terminal vimentin peptides also heavily interacted with cytokeratins CK8/18 at 5 mM Tris-HCl, pH 8.4, 1 mM EDTA, for example, producing extensive filament bundles at a 20-fold molar excess of the dipeptide over the keratins (Figure 16F). As expected, the control peptide (i.e., that with the two arginines in the nonapeptide motif replaced by threonines; see above) did not cause any significant aggregation of the cytokeratins. Taken together, these date indicate that the vimentin head domain is instrumental during phase 1 of the assembly process. Moreover, since the vimentin head peptide also induced assembly of cytokeratin filaments at alkaline pH, a similar mechanistic principle to rearrange the soluble subunits from their "*arrested*" into an *assembly-competent* state may control both vimentin and cytokeratin filament formation.

The amino-terminal nonapeptide does not interact with preformed vimentin IFs (Hofmann and Herrmann, 1992). In particular, it does not disassemble or cross-link vimentin IFs even at very high concentrations, suggesting that its binding site is masked in the mature filament. Therefore, it is conceivable that the diarginine motif of the peptide does not simply cause nonspecific ionic cross-linking of the filaments. Hence, unlike with the carboxy-terminal peptides of coil 2 discussed above, this amino-terminal nonapeptide does not cause dissociation of preformed IFs; instead it stimulates association of IF oligomers into higher order structures. Therefore, it may be important to include the head peptides in any structural consideration of IF chain oligomer formation, for example, already at the level of tetramer/octamer formation, since the differences in the sedimentation behavior of headless versus wild-type IF oligomers should be taken into account.

Further evidence regarding the specificity of the action of the head peptides stems from observations with unrelated basic proteins on IF integrity. In this context, we incubated the basic non-arm-repeat head domain of plakophilin 1, a member of the plakoglobin family (Schmidt *et al.*, 1994), with preformed IFs. Accordingly, the 331-amino acid-long recombinant subdomain bundled vimentin IFs into huge fibers (Figure 17A), resembling the IF "macrofibrils" observed in the presence of short synthetic peptides derived from fillagrin (Mack *et al.*, 1993). Similarly, full-length recombinant plakophilin 1 also caused marked bundling of vimentin IFs. As depicted in less densely packed regions, small complexes seemed to "glue" individual IFs together (Figure 17B). Since the recombinant plakophilin 1 is mono-

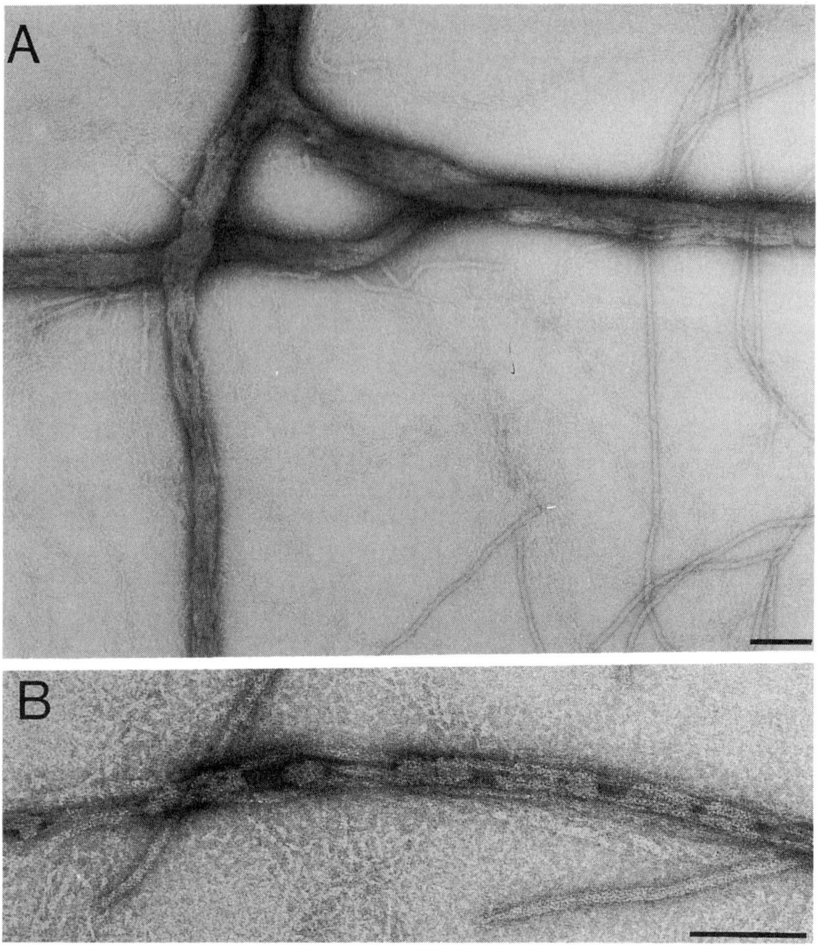

FIGURE 17. Bundling of preformed human vimentin filaments by the addition of recombinant human plakophilin 1 (PP1) fragment (1–331) (A) and complete PP1 (B) as observed after addition of the proteins (in a 1 : 1 ratio) by electron microscopy of negatively stained preparations. Bar, 100 nm.

meric under the buffer conditions employed, the observed cross-linking event may be mediated by individual polypeptides. Indeed, centrifugation experiments suggested that recombinant plakophilin 1 bound to IFs in a stoichiometric, apparently 1:1, manner, since even at eightfold molar excess of plakophilin 1, only 1/8 cosedimented with IFs (H. Herrmann *et al.*, unpublished observations). Whether this *in vitro* cross-linking of IFs represents a physiological function of plakophilin 1 remains to be determined. Although plakophilin 1 (i.e., band 6 protein) was originally isolated from desmosomal fractions of bovine snout, transfection into epithelial as well as fibroblast cells revealed that most of the expressed protein was translocated into the nucleus (Schmidt *et al.*, 1997). The functional significance of a nuclear protein being recruited to desmosomes in certain tissues is of high interest. Moreover, at this stage we do not know whether the recombinant protein is indeed folded in the same way as the authentic protein. Last but not least, the IF-binding domain of plakophilin 1 may in fact be "cryptic" or even not be used in the native molecule.

A similarly complex situation was encountered with several presumptive intermediate filament-associated proteins (IFAPs). For example, it was demonstrated that plectin, or high-molecular-weight protein (270,000–340,000), copurifies with vimentin IFs (Pytela and Wiche, 1980), and that this protein coexists in the cell in a cytosolic and a cytoskeleton-associated form (Herrmann and Wiche, 1983). In solid-phase binding assays plectin bound strongly to α-fodrin and to microtubule-associated proteins (MAPs), as well as to vimentin (Herrmann and Wiche, 1987). In cycled brain microtubule preparations, plectin efficiently cocycled with tubulin and MAPs, which had previously not been observed, since plectin comigrates with MAPs in conventional gel systems (Wiche *et al.*, 1982; Koszka *et al.*, 1985). For these reasons we argued that plectin may be a universal and ubiquitous cytoskeleton cross-linking molecule (Wiche *et al.*, 1982; Herrmann and Wiche, 1987). Probably because most of the cellular plectin was removed during specimen preparation, immunoelectron microscopy revealed plectin mainly associated with IFs (Foisner *et al.*, 1988). This problem has now been overcome in an impressive study from the laboratory of G. Borisy (Svitkina *et al.*, 1996). These authors showed that plectin indeed cross-links IFs with microtubules and microfilaments. Differentially spliced plectin isoforms, in combination with site-specific phosphorylation reactions, may explain the molecule's unique, versatile binding properties (Elliot *et al.*, 1997; Herrmann and Wiche, 1987). Moreover, plectin appears to react to physiological stimuli, as recently shown for epithelial cells incubated under standard compared to Ca^{2+}-free conditions (Eger *et al.*, 1997): in the latter situation, when desmosomes are internalized and uncoupled from IFs, plectin redistributed from a subdesmosonal to a more uniform cytoplasmic localization.

While plectin is able to span a cytoplasmic network also in the absence of vimentin, it readily connects to the vimentin IF system when the latter is introduced by cDNA transfection as in the case of vimentin-free bovine mammary gland epithelial (BMGE + H) cells (Herrmann, 1995). Incubation of such bovine epithelial cells containing a transgenic (human) vimentin IF system with colcemid induced the reorganization of both the IFs and plectin (Figure 18). Very likely this represents not simply a physical entrapment of plectin within the IF system, as microfilaments are not abundantly found in such perinuclear vimentin tangles. It will be of great interest to learn how the vimentin and the microtubule system are coupled, at what stage of IF assembly plectin comes onto the scene, and how its interaction with microfilaments proceeds (Hollenbeck *et al.*, 1989; Gyoeva and Gelfand, 1991). In this context, one would also like to know how IFs are connected to desmosomes and motor proteins, respectively.

We have recently shown by transfection experiments that a lamin-type nuclear localization signal introduced into the *Xenopus* vimentin head domain at position 60–70 is able to guide vimentin quantitatively into the nucleus of vimentin-free BMGE + H cells (Herrmann *et al.*, 1993). After 1- to 4-hr incubation at 28°C, the permissive temperature for assembly of the temperature-sensitive amphibian vimentin, the aggregated protein reorganized into filament bundles seemingly interconnecting the remnants of such dotlike nuclear inclusions (Figures 19D and 19E). This was particularly evident in stable cell lines, whereas in transiently transfected cells, nuclei could be observed that had accumulated so much of the transgene product that it was hard to believe that normal cellular physiology remained unaffected (Figures 19A–19C). Meanwhile, we have also generated stably transfected lines of the human vimentin-free line SW-13 and "painted"

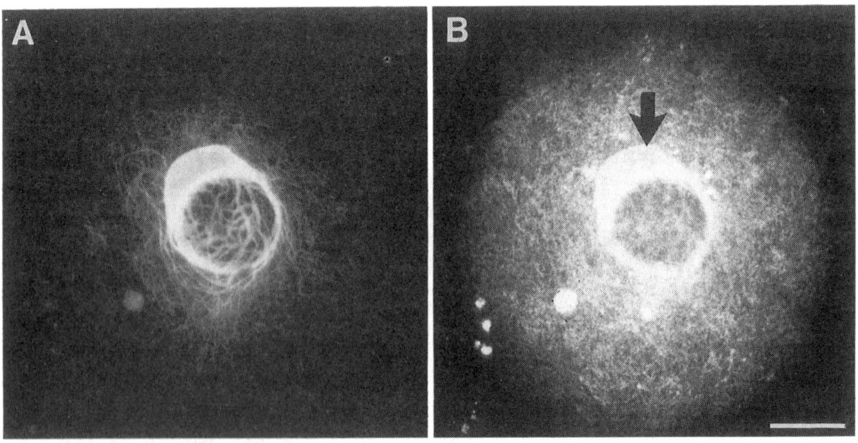

FIGURE 18. Double immunofluorescence localization of (A) vimentin and (B) plectin in vimentin-free bovine mammary gland epithelial cells, stably transfected with human vimentin cDNA, after prolonged colcemid treatment (10^{-6} M for 16 hr). Note that the plectin partially redistributes (arrow in panel B) with the reorganized vimentin system. Bar, 10 μm.

individual chromosomes side by side with the nuclear *Xenopus* vimentin IF system (Bridger *et al.*, 1998). These experiments indicate that this artificial nuclear IF system may be a useful topological marker for the presumed "interchromosomal domain" compartment (Kurz *et al.*, 1996), since it colocalizes with RNA tracks, coiled bodies, and promyelocytic leukemia (PML) bodies, but is absent from chromosomal territories. This experimental scenario, in turn, may provide the opportunity to use the *Xenopus* NLS-vimentin as a nuclear "space shuttle" to target protein domains of interest for interaction studies into this nuclear compartment. The known assembly characteristics of IFs provide hints concerning the site(s) within the vimentin cDNA where foreign cDNA segments might be integrated to produce functional chimeric molecules.

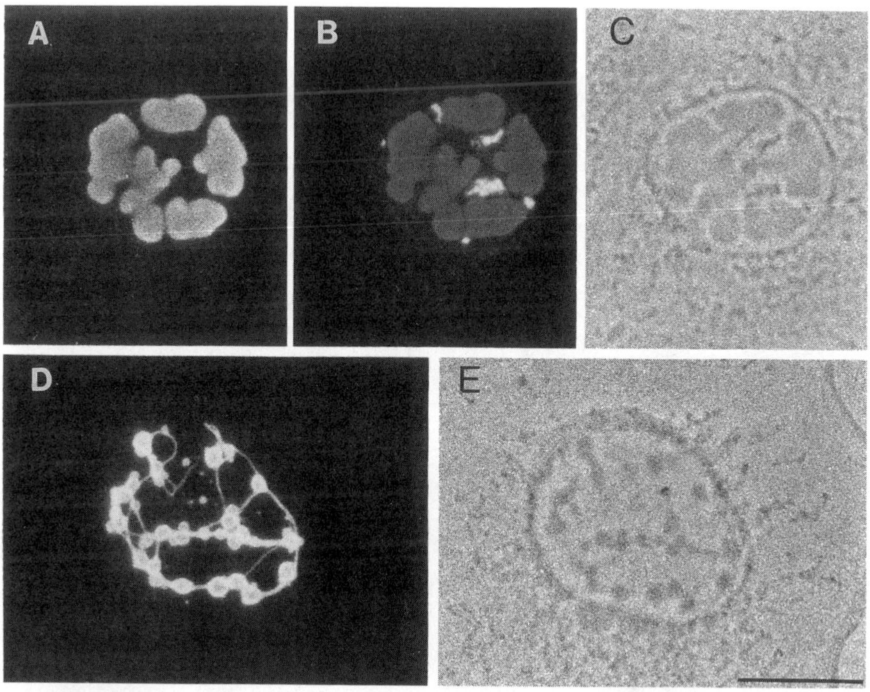

FIGURE 19. Localization of recombinant *Xenopus* vimentin, carrying a nuclear localization signal (Herrmann *et al.*, 1993), after (A–C) transient and (D, E) stable cDNA transfection into bovine mammary gland epithelial cells. (A, D) vimentin; (B) nucleolus-specific immunostaining; (C, E) corresponding phase contrast microscopy. In (D) cells were briefly (4 hr) incubated at 28°C, which allows the temperature-sensitive vimentin to reorganize into normal IFs traversing the nucleus in defined orientations, thereby connecting the dotlike masses, seen even in phase contrast, like pearls on a string. Bar, 10 μm.

5. CONCLUSIONS

We hope we have convinced the reader that for a meaningful biochemical and molecular understanding of IF assembly, a thorough structural analysis such as is only possible with the electron microscope is crucial. In particular, the STEM mass measurements were absolutely essential for a more quantitative understanding of the effects that particular mutations or truncations introduced into the IF building blocks by recombinant DNA techniques had on IF structure and/or assembly. Also, after transfection of cDNAs into living cells, the sometimes subtle effects of "protein engineering" on IF structure, organization, and network formation can only be depicted at the EM level. It has been shown several times that "dots" seen at the immunofluorescent level may well consist of IFs. Last but not least,

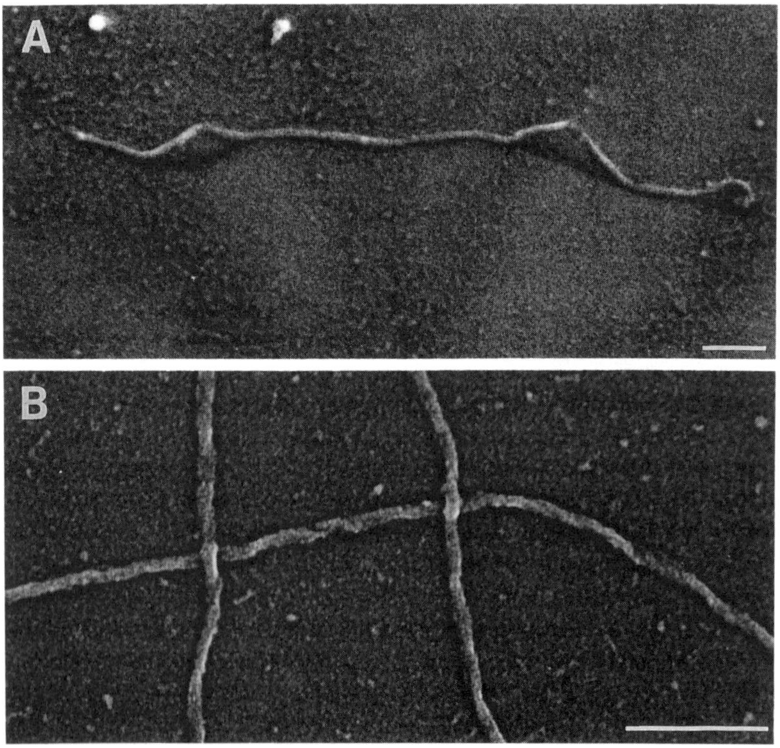

FIGURE 20. Field emission scanning electron microscopy of freeze-dried human vimentin showing that the filaments are able to stand out from the support (A). (B) The partially twisted appearance of the IF is apparent. Bars, 100 nm.

the modification of standard procedures may allow new insights even into very dynamic processes as documented by us for IF assembly. Therefore, the potential of new or less frequently used EM techniques (Figure 20) should not be underrated; on the contrary, they should be carefully explored in the context of the particular question one is trying to answer.

ACKNOWLEDGMENTS. We thank Monika Brettel and Markus Häner for continuous excellent technical assistance, Ralf Zimbelmann for performing sequence comparisons, and Jutta Müller-Osterholt for photography. Robert Wyss helped with the design and production of several schematic figures and computer drawings. Birgit Stamm patiently and carefully typed various versions of the manuscript. Dr. Kevin R. Rogers provided electron micrographs of tailless vimentin, Dr. Roger Wepf those recorded by field-emission scanning electron microscopy, and Dr. J. Robin Harris worked on the two-dimensional crystals of the human vimentin rod. The financial support of the German Research Foundation (to H.H.) and of the Swiss National Research Foundation, the M.E. Mueller Foundation of Switzerland, and the Kanton Basel-Stadt (to U.A.) is gratefully acknowledged.

6. REFERENCES

Aebi, U., Fowler, W. E., Rew, P., and Sun, T.-T., 1983, The fibrillar substructure of keratin filaments unravelled, *J. Cell Biol.* **97:**1131–1143.

Aebi, U., Cohn, J., Buhle, L., and Gerace, L., 1986, The nuclear lamina is a meshwork of intermediate-type filaments, *Nature* **323:**560–564.

Aebi, U., Häner, M., Troncoso, J., Eichner, R., and Engel, A., 1988, Unifying principles in intermediate filament (IF) structure and assembly, *Protoplasma* **145:**73–81.

Albers, K., and Fuchs, E., 1989, Expression of mutant keratin cDNAs in epithelial cells reveals possible mechanisms for initiation and assembly of intermediate filaments, *J. Cell Biol.* **108:**1477–1493.

Bader, B. L., Magin, T. M., Hatzfeld, M., and Franke, W. W., 1986, Amino acid sequence and gene organization of cytokeratin no. 19, an exceptional tailless intermediate filament protein: Change in protein conformation corresponds to presence of a landmark intron, *EMBO J.* **5:**1865–1875.

Bader, B. L., Magin, T. M., Freudenmann, M., Stumpp, S., and Franke, W. W., 1991, Intermediate filaments formed *de novo* from tailless cytokeratin in the cytoplasm and in the nucleus, *J. Cell Biol.* **115:**1293–1307.

Bartnik, E., and Weber, K., 1989, Widespread occurrence of intermediate filaments in invertebrates; common principles and aspects of diversion, *Eur. J. Cell Biol.* **50:**17–33.

Blikstad, I., and Lazarides, E., 1983, Vimentin filaments are assembled from a soluble precursor in avian erythroid cells, *J. Cell Biol.* **96:**1803–1808.

Bossie, C. A., and Sanders, M. M., 1993, A cDNA from *Drosophila melanogaster* encodes a lamin C-like intermediate filament protein, *J. Cell Sci.* **104:**1263–1272.

Bridger, J., Herrmann, H., Münkel, C., and Lichter, P., 1998, Identification of an interchromosomal compartment by polymerization of nuclear-targeted vimentin, *J. Cell Sci.* **111:**1241–1253.

Burley, S. K., and Petsko, G. A., 1985, Aromatic–aromatic interaction: A mechanism of protein structure stabilization, *Science* **229:**23–28.

Cary, R. B., Klymkowsky, M. W., Evans, R. M., Domingo, A., Dent, J. A., and Backhus, I. E., 1994, Vimentin's tail interacts with actin-containing structures *in vivo*, *J. Cell Sci.* **107:**1609–1622.

Chipev, C. C., Korge, B. P., Markova, N., Bale, S. J., DiGiovanna, J. J., Compton, J. G., and Steinert, P. M., 1992, A leucine → proline mutation in the H1 subdomain of keratin 1 causes epidermolytic hyperkeratosis, *Cell* **70:**821–828.

Chou, Y.-H., Bischoff, R., Beach, D., and Goldman, R. D., 1990, Intermediate filament reorganization during mitosis is mediated by p34cdc2 phosphorylation of vimentin, *Cell* **62:**1063–1071.

Conway, E. Y., and Parry, D. A. D., 1988, Intermediate filament structure: 3. Analysis of sequence homologies, *Int. J. Biol. Macromol.* **10:**79–98.

Coulombe, P. A., and Fuchs, E., 1990, Elucidating the early stages of keratin filament assembly, *J. Cell Biol.* **111:**153–169.

Coulombe, P. A., Chan, Y., Albers, K., and Fuchs, E., 1990, Deletions in epidermal keratins leading to alterations in filament organization *in vivo* and in intermediate filament assembly *in vitro*, *J. Cell Biol.* **111:**3049–3064.

Crick, F. H. C., 1953, The packing of α-helices: Simple coiled-coils, *Acta Crystallogr.* **6:**689–697.

Dahl, D., and Bignami, A., 1973, Immunochemical and immunofluorescence studies of the glial fibrillary acidic protein in vertebrates, *Brain Res.* **61:**279–293.

Dahlstrand, J., Eimmerman, L. B., McKay, R. D. G., and Lendahl, U., 1992, Characterization of the human nestin gene reveals a close evolutionary relationship to neurofilaments, *J. Cell Sci.* **103:**589–597.

Dodemont, H., Riemer, D., Ledger, N., and Weber, K., 1994, Eight genes and alternative RNA processing pathways generate an unexpectedly large diversity of cytoplasmic intermediate filament proteins in the nematode, *Caenorhabditis elegans*, *EMBO J.* **13:**2625–2638.

Eckelt, A., Herrmann, H., and Franke, W. W., 1992, Assembly of a tailless mutant of the intermediate filament protein, vimentin, *in vitro* and *in vivo*, *Eur. J. Cell Biol.* **58:**319–330.

Eger, A., Stockinger, A., Wiche, G., and Foisner, R., 1997, Polarisation-dependent association of plectin with desmoplakin and the lateral submembrane skeleton in MDCK cells, *J. Cell Sci.* **110:**1307–1316.

Eichner, R., Rew, P., Engel, A., and Aebi, U., 1985, Human epidermal keratin filaments: Studies on their structure and assembly, *Ann. N. Y. Acad. Sci.* **455:**381–402.

Elliot , C. E., Becker, B., Oehler, S., Castañon, M. J., Hauptman, R., and Wiche, G., 1997, Plectin transcript diversity: Identification and tissue distribution of variants with distinct first coding exons and rodless isoforms, *Genomics* **42:**115–125.

Engel, A., Eichner, R., and Aebi, U., 1985, Polymorphism of reconstituted human epidermal keratin filaments: Determination of their mass-per-unit-length and width by scanning transmission electron microscopy (STEM), *J. Ultrastruct. Res.* **90:**323–335.

Evans, R. M., 1988, The intermediate-filament proteins vimentin and desmin are phosphorylated in specific domains, *Eur. J. Cell Biol.* **46:**152–160.

Fisher, D. Z., Chaudhary, N., and Blobel, G., 1986, cDNA sequencing of nuclear lamin A and C reveals primary and secondary structural homology to intermediate filament proteins, *Proc. Natl. Acad. Sci. USA* **83:**6450–6454.

Foisner, R., Leichtfried, F. E., Herrmann, H., Small, J. V., Lawson, D., and Wiche, G., 1988, Cytoskeleton-associated plectin: *in situ* localization, *in vitro* reconstitution, and binding to

immobilized intermediate filament proteins, *J. Cell. Biol.* **106:**723–733.
Franke, W. W., 1987a, Homology of a conserved sequence in the tail domain of intermediate filament proteins with the loop region of calcium binding proteins, *Cell Biol. Int. Rep.* **11:**831.
Franke, W. W., 1987b, Nuclear lamins and cytoplasmic intermediate filament proteins: A growing multigene family, *Cell* **48:**3–4.
Franke, W. W., and Kartenbeck, J., 1993, Cytokeratins, in *Guidebook to the Cytoskeletal and Motor Proteins* (T. Kreis and R. Vale, eds.), pp. 145–148, Oxford University Press, Oxford.
Franke, W. W., Schmid, E., Osborn, M., and Weber, K., 1978a, Different intermediate-sized filaments distinguished by immunofluorenscence microscopy, *Proc. Natl. Acad. Sci. USA* **75:**5034–5038.
Franke, W. W., Schmid, E., Osborn, M., and Weber, K., 1978b, The intermediate-sized filaments in rat kangaroo PtK_2 cells. II. Structure and composition of isolated filaments, *Cytobiologie* **17:**392–411.
Franke, W. W., Schiller, D. L., and Grund, C., 1982a, Protofilamentous and annular structures as intermediates during reconstitution of cytokeratin filaments *in vitro*, *Biol. Cell* **46:**257–268.
Franke, W. W., Schmid, E., and Grund, C., 1982b, Intermediate filament proteins in nonfilamentous structures: Transient disintegration and inclusion of subunit proteins in granular aggregates, *Cell* **30:**103–113.
Franke, W. W., Schmid, E., Schiller, D. L., Winter, S., Jarasch, E. D., Moll, R., Denk, H., Jackson, B. W., and Illmensee, K., 1982c, Differentiation-related patterns of expression of proteins of intermediate-size filaments in tissues and cultured cells, *Cold Spring Harbor Symp. Quant. Biol.* **46:**431–453.
Franke, W. W., Schmid, E., Mittnacht, S., Grund, C., and Jorcano, J. L., 1984, Integration of different keratins into the same filament system after microinjection of mRNA for epidermal keratins into kidney epithelial cells, *Cell* **36:**813–825.
Franz, J. K.-H., 1987, Charakterisierung und Klonierung von Cytokeratinen aus Oocyten von *Xenopus laevis*, Dissertation, University of Heidelberg, Germany.
Fraser, R. D. B., 1969, Keratins, *Sci. Am.* **221:**86–96.
Fuchs, E., 1997, Of mice and men: Genetic disorders of the cytoskeleton, *Mol. Biol. Cell* **8:** 189–203.
Fuchs, E., and Weber, K., 1994, Intermediate filaments structure, dynamics, function, and disease, *Annu. Rev. Biochem.* **63:**345–382.
Fulton, A. B., and Wan, K. M., 1983, Many cytoskeletal proteins associate with the HeLa cytoskeleton during translation *in vitro*, *Cell* **32:**619–625.
Geisler, N., and Weber, K., 1982, The amino acid sequence of chicken muscle desmin provides a common structural model for intermediate filament proteins, *EMBO J.* **12:**1649–11656.
Geisler, N., Kaufmann, E., and Weber, K., 1982, Protein chemical characterization of three structurally distinct domains along the protofilament unit of desmin 10 nm filaments, *Cell* **30:**277–286.
Georgatos, S. D., Gounari, F., Goulielmos, G., and Aebi, U., 1997, To bead or not to bead? Lens-specific intermediate filaments revisited, *J. Cell Sci.* **110:**2629–2634.
Gieffers, C., and Krohne, G., 1991, *In vitro* reconstitution of recombinant lamin A and a lamin A mutant lacking the carboxy-terminal tail, *Eur. J. Cell Biol.* **55:**191–199.
Gill, S. R., Wong, P. C., Monteiro, M. J., and Cleveland, D. W., 1990, Assembly properties of dominant and recessive mutations in the small mouse neurofilament (NF-L) subunit, *J. Cell Biol.* **111:**2005–2019.

Goulielmos, G., Gounari, F., Remington, S., Müller, S., Häner, M., Aebi, U., and Georgatos, S. D., 1996, Filensin and phakinin form a novel type of beaded intermediate filaments and coassemble *de novo* in cultured cells, *J. Cell Biol.* **132:**643–655.

Gruenbaum, Y., Landesman, Y., Drees, B., Bare, J. W., Saumweber, H., Paddy, M. R., Sedat, J. W., Smith, D. E., Benton, B. M., and Fisher, P. A., 1988, *Drosophila* nuclear lamin precursor Dm_0 is translated from either of two developmentally regulated mRNA species apparently encoded by a single gene, *J. Cell Biol.* **106:**585–596.

Gyoeva, F. K., and Gelfand, V. I., 1991, Coalignment of vimentin intermediate filaments with microtubules depends on kinesin, *Nature* **353:**445–448

Harris, J. R., 1991, The negative staining–carbon film procedure: Technical considerations and a survey of macromolecular applications, *Micron Microscop. Acta* **22:**341–359.

Hatzfeld, M., and Burba, M., 1994, Function of type I and type II keratin head domains: Their role in dimer, tetramer and filament formation, *J. Cell Sci.* **107:**1959–1972.

Hatzfeld, M., and Franke, W. W., 1985, Pair formation and promiscuity of cytokeratins: Formation *in vitro* of heterotypic complexes and intermediate-sized filaments by homologous and heterologous recombinations of purified polypeptides, *J. Cell Biol.* **101:**1826–1841.

Hatzfeld, M., and Weber, K., 1990, Tailless keratins assemble into regular intermediate filaments *in vitro*, *J. Cell Sci.* **97:**317–324.

Hatzfeld, M., and Weber, K., 1992, A synthetic peptide representing the consensus sequence motif at the carboxy-terminal end of the rod domain inhibits intermediate filament assembly and disassembles preformed filaments, *J. Cell Biol.* **116:**157–166.

Heins, S., and Aebi, U., 1994, Making heads and tails of intermediate filament assembly, dynamics and networks, *Curr. Opin. Cell Biol.* **6:**25–33.

Heins, S., Wong, P. C., Muller, S., Goldie, K., Cleveland, D. W., and Aebi, U., 1993, The rod domain of NF-L determines neurofilament architecture, whereas the end domains specify filament assembly and network formation, *J. Cell Biol.* **123:**1517–1533.

Herrmann, H., 1995, Strukturbildende und topogene Domänen von Intermediärfilamentproteinen, Habilitationsschrift, University of Heidelberg, Germany.

Herrmann, H., and Wiche, G., 1983, Specific *in situ* phosphorylation of plectin in detergent resistant cytoskeletons from cultured Chinese hamster ovary cell, *J. Biol. Chem.* **258:**4610–4618.

Herrmann, H., and Wiche, G., 1987, Plectin and IFAP-300K are homologous proteins binding to microtubule-associated proteins 1 and 2 and to the 240 kilodalton subunit of spectrin, *J. Biol. Chem.* **262:**1320–1325.

Herrmann, H., Fouquet, B., and Franke, W. W., 1989a, Expression of intermediate filament proteins during development of *Xenopus laevis*. I. cDNA clones encoding different forms of vimentin, *Development* **105:**279–298.

Herrmann, H., Fouquet, B., and Franke, W. W., 1989b, Expression of intermediate filament proteins during development of *Xenopus laevis*. II. Identification and molecular characterization of desmin, *Development* **105:**299–307.

Herrmann, H., Hofmann, I., and Franke, W. W., 1992, Identification of a nonapeptide motif in the vimentin head domain involved in intermediate filament assembly, *J. Mol. Biol.* **223:**637–650.

Herrmann, H., Eckelt, A., Brettel, M., Grund, C., and Franke, W. W., 1993, Temperature-sensitive intermediate filament assembly alternative structures of *Xenopus laevis* vimentin *in vitro* and *in vivo*, *J. Mol. Biol.* **234:**99–113.

Herrmann, H., Häner, M., Brettel, M., Müller, S. A., Goldie, K. N., Fedtke, B., Lustig, A., Franke, W. W., and Aebi, U., 1996a, Structure and assembly properties of the intermediate filament protein vimentin: The role of its head, rod and tail domains, *J. Mol. Biol.* **264:**933–953.

Herrmann, H., Münick, D., Brettel, M., Fouquet, B., and Markl, J., 1996b, Vimentin in a cold-water fish, the rainbow trout: Highly conserved primary structure but unique assembly properties, *J. Cell Sci.* **109:**569–578.

Hofmann, I., and Herrmann, H., 1992, Interference in vimentin assembly *in vitro* by synthetic peptides derived from the vimentin head domain, *J. Cell Sci.* **101:**687–700.

Hofmann, I., Herrmann, H., and Franke, W. W., 1991, Assembly and structure of calcium-induced thick vimentin filaments, *Eur. J. Cell Biol.* **56:**328–341.

Hollenbeck, P. J., Bershadsky, A. D., Pletjushkina, O. Y., Tint, I. S., and Vasilev, J. M., 1989, Intermediate filament collapse is an ATP-dependent and actin-dependent process, *J. Cell Sci.* **92:**621–631.

Huber, R., 1979, Conformational flexibility and its functional significance in some protein molecules, *Trends Biochem. Sci.* **4:**271–276.

Huiatt, T. W., Robson, R. M., Arakawa, N., and Stromer, M. H., 1980, Desmin from avian smooth muscle: Purification and partial characterization, *J. Biol. Chem.* **255:**6981–6989.

Inagaki, M., Inagaki, N., Takahashi, T., and Takai, Y., 1997, Phosphorylation-dependent control of structures of intermediate filaments: A novel approach using site- and phosphorylation state-specific antibodies, *J. Biochem* **121:**407–414.

Ip, W., Hartzer, M. K., Pang, Y.-Y. S., and Robson, R. M., 1985, Assembly of vimentin *in vitro* and its implications concerning the structure of intermediate filaments, *J. Mol. Biol.* **183:**365–375.

Ishikawa, H., Bischoff, R., and Holtzer, H., 1968, Mitosis and intermediate-sized filaments in developing skeletal muscle, *J. Cell Biol.* **38:**538–555.

Kaufmann, E., Weber, K., and Geisler, N., 1985, Intermediate filament forming ability of desmin derivatives lacking either the amino-terminal 67 or the carboxy-terminal 27 residues, *J. Mol. Biol.* **185:**733–742.

Koszka, C., Leichtfried, G. E., and Wiche, G., 1985, Identification and spatial arrangement of high molecular weight proteins (Mr 300,000–330,000) co-assembling with microtubules from a cultured cell line (rat glioma C6), *Eur. J. Cell Biol.* **38:**149–156.

Kouklis, P. D., Papamarcaki, T., Merdes, A., and Georgatos, S. D., 1991, A potential role for the COOH-terminal domain in the lateral packing of type III intermediate filaments, *J. Cell Biol.* **114:**773–786.

Kouklis, P. D., Hatzfeld, M., Brunkener, M., Weber, K., and Georgatos, S. D., 1993, *In vitro* assembly properties of vimentin mutagenized at the B-site tail motif, *J. Cell Sci.* **106:**919–928.

Kurz, A., Lampel, S., Nickolenko, J. E., Bradl, J., Benner, A., Zirbel, R. M., Cremer, T., and Lichter, P., 1996, Active and inactive genes localize preferentially in the periphery of chromosome territories, *J. Cell Biol.* **135:**1195–1205.

Lazarides, E., 1980, Intermediate filaments as mechanical integrators of cellular space, *Nature* **283:**249–256.

Lazarides, E., and Weber, K., 1974, Actin antibody: The specific visualization of actin filaments in nonmuscle cells, *Proc. Natl. Acad. Sci. USA* **71:**2268–2272.

Levine, A. J., Vande Woude, G. F., Topp, W. C., and Watson, J. D., eds., 1984, *Cancer Cells, 1, The transformed phenotype*, Cold Spring Harbor Laboratory, Cold Spring Harbor, New York.

Liem, R. K. H., 1993, Molecular biology of neuronal intermediate filaments, *Curr. Opin. Cell Biol.* **5:**12–16.

Lu, X., and Lane, E. B., 1990, Retrovirus-mediated transgenic keratin expression in cultured fibroblasts: Specific domain functions in keratin stabilization and filament formation, *Cell* **62:**681–696.

Mack, J. W., Steven, A. C., and Steinert, P. M., 1993, The mechanism of interaction of filaggrin with intermediate filaments: The ionic zipper hypothesis, *J. Mol. Biol.* **232:**50–66.

Makarova, I., Carpenter, D., Khan, S., and Ip, W., 1994, A conserved region in the tail domain of vimentin is involved in its assembly into intermediate filaments, *Cell Motil. Cytoskel.* **28:**265–277.

McCormick, M. B., Kouklis, P. D., Syder, A., and Fuchs, E., 1993, The roles of the rod end and tail in vimentin IF assembly and IF network formation, *J. Cell Biol.* **122:**395–407.

McKeon, F., Kirschner, M. W., and Caput, D., 1986, Homologies in both primary and secondary structure between nuclear envelope and intermediate filament proteins, *Nature* **319:**463–468.

Milam, L., and Erickson, H. P., 1984, Structural characteristics of the desmin protofilament, *J. Ultrastruct. Res.* **89:**179–186.

Moll, R., Franke, W. W., and Schiller, D. L., 1982, The catalog of human cytokeratins: Patterns of expression in normal epithelia, tumors and cultured cells, *Cell* **31:**11–24.

Osborn, M., and Weber, K., 1983, Tumor diagnosis by intermediate filament typing: A novel tool for surgical pathology, *Lab. Invest.* **48:**372–394.

Osborn, M., and Weber, K., 1989, Keratins, transglutaminase, and Mallory bodies—The really insoluble stuff, *Lab. Invest.* **61:**585–587.

Osborn, M., Webster, R. E., and Weber, K., 1978, Individual microtubules viewed by immunofluorescence and electron-microscopy in the same PtK_2 cell, *J. Cell Biol.* **77:**R27–R34.

Parry, D. A. D., 1990, Primary and secondary structure of IF protein chains and modes of molecular aggregation, in *Cellular and Molecular Biology of Intermediate Filaments* (R. D. Goldman and P. M. Steinert, eds.), pp. 175–204, Plenum Press, New York.

Parry, D. A. D., and Steinert, P. M., 1995, *Intermediate Filament Structure*, Springer-Verlag, Berlin.

Pruss, R. M., Mirsky, R., Raff, M. C., Thorpe, R., Dowding, A. J., and Anderton, B. H., 1981, All classes of intermediate filaments share a common antigenic determinant defined by a monoclonal antibody, *Cell* **27:**419–427.

Pytela, R., and Wiche, G., 1980, High molecular weight polypeptides (270,000–340,000) from cultured cells are related to hog brain microtubule-associated proteins, but copurify with intermediate filaments, *Proc. Natl. Acad. Sci. USA* **77:**4808–4812.

Quax, W., Vree Egberts, W., Hendriks, W., Quax-Jeuken, Y., and Bloemendal, H., 1983, The structure of the vimentin gene, *Cell* **35:**215–223.

Quax-Jeuken, Y. E. F. M., Quax, W. J., and Bloemendal, H., 1983, Primary and secondary structure of hamster vimentin predicted from the nucleotide sequence, *Proc. Natl. Acad. Sci. USA* **80:**3548–3552.

Renner, W., Franke, W. W., Schmid, E., Geisler, N., Weber, K., and Mandelkow, E., 1981, Reconstitution of intermediate-sized filaments from denatured monomeric vimentin, *J. Mol. Biol.* **149:**285–306.

Rieger, M., 1988, Struktur eines Cytokeratin-Gens: Das Cytokeratin 10-Gen des Menschen und seine Orthologen in anderen Säugerarten, Dissertation, University of Heidelberg, Germany.

Rieger, M., and Franke, W. W., 1988, Identification of an orthologous mammalian cytokeratin gene. High degree of intron sequence conservation during evolution of human cytokeratin 10, *J. Mol. Biol.* **204:**841–856.

Rogers, K. R., Eckelt, A., Nimmrich, V., Janssen, K.-P., Schliwa, M., Herrmann, H., and

Franke, W. W., 1995, Truncation mutagenesis of the non-α-helical carboxy-terminal tail domain of vimentin reveals contributions to cellular localization but not to filament assembly, *Eur. J. Cell Biol.* **66:**136–150.

Rogers, K. R., Herrmann, H., and Franke, W. W., 1996, Characterization of disulfide crosslink formation of human vimentin at the dimer, tetramer, and intermediate filament levels, *J. Struct. Biol.* **117:**55–69.

Rogers, M. A., Langbein, L., Prätzel, S., Moll, I., Krieg, T., Winter, H., and Schweizer, J., 1997, Sequences and differential expression of three novel human type-II hair keratins, *Differentiation* **61:**187–194.

Schmidt, A., Heid, H. W., Schäfer, S., Nuber, U. A., Zimbelmann, R., and Franke, W. W., 1994, Desmosomes and cytoskeletal architecture in epithelial differentiation: Cell type-specific plaque components and intermediate filament anchorage, *Eur. J. Cell Biol.* **65:**229–245.

Schmidt, A., Langbein, L., Rode, M., Prätzel, S., Zimbelmann, R., and Franke, W. W., 1997, Plakophilins 1a and 1b: Widespread nuclear proteins recruited in specific epithelial cells as desmosomal plaque components, *Cell Tis.* **290:**481–499.

Soellner, P., Quinlan, R. A., and Franke, W. W., 1985, Identification of a distinct soluble subunit of an intermediate filament protein: Tetrameric vimentin from living cells, *Proc. Natl. Acad. Sci. USA* **82:**7929–7933.

Starger, J., Brown, W. E., Goldmann, A. E., and Goldman, R. D., 1978, Biochemical and immunological analysis of rapidly purified 10 nm filaments from baby hamster kidney (BHK-21) cells, *J. Cell Biol.* **78:**93–109.

Steinert, P.M., 1981, Intermediate filaments, in *Electron microscopy of proteins*, Vol. 1 (J. R. Harris, ed.), pp. 126–166, Academic Press, New York.

Steinert, P. M., 1990, The two-chain coiled-coil molecule of native epidermal keratin intermediate filaments is a type I–type II heterodimer, *J. Biol. Chem.* **265:**8766–8774.

Steinert, P. M., Idler, W. W., and Zimmerman, S. B., 1976, Self-assembly of bovine epidermal keratin filaments *in vitro*, *J. Mol. Biol.* **108:**547–567.

Steinert, P., Idler, W., Aynardi-Whitman, M., Zackroff, R., and Goldman, R. D., 1982, Heterogeneity of intermediate filaments assembled *in vitro*, *Cold Spring Harbor Symp. Quant. Biol.* **46:**465–474.

Steinert, P. M., Marekov, L. N., and Parry, D. A. D., 1993, Diversity of intermediate filament structure: Evidence that the alignment of coiled-coil molecules in vimentin is different from that in keratin intermediate filaments, *J. Biol. Chem.* **3268:**24916–24925.

Steinert, P. M., North, A. C. T., and Parry, D. A. D., 1994, Structural features of keratin intermediate filaments, *J. Invest. Dermatol.* **103:**19S–24S.

Stewart, M., Quinlan, R. A., and Moir, R. D., 1989, Molecular interactions in paracrystals of a fragment corresponding to the α-helical coiled-coil rod portion of glial fibrillary acidic protein: Evidence for an antiparallel packing of molecules and polymorphism related to intermediate filament structure, *J. Cell Biol.* **109:**225–234.

Stromer, M. H., Ritter, M. A., Pang, Y.-Y. S., and Robson, R. M., 1987, Effect of cations and temperature on kinetics of desmin assembly, *Biochem. J.* **246:**75–81.

Svitkina, T. M., Verkhovsky, A. B., and Borisy, G. G., 1996, Plectin sidearms mediate interaction of intermediate filaments with microtubules and other components of the cytoskeleton, *J. Cell Biol.* **135:**991–1007.

The Yeast Genome Directory, 1997, *Nature* (*Suppl.*) **387:**1–105.

Traub, P., Vorgias, C. E. 1983, Involvement of the N-terminal polypeptide of vimentin in the formation of intermediate filaments, *J. Cell Sci.* **63:**43–67.

Weber, K., and Geisler, N., 1984, Intermediate filaments—From wool α-keratin to neurofilaments: A structural overview, in *Cancer Cells, I, The Transformed Phenotype*

(A. J. Levine, G. F. Vande Woode, W. C. Topp, and J. D. Watson, eds.), pp. 153–159, Cold Spring Harbor Laboratory, Cold Spring Harbor, New York.

Wiche, G., Herrmann, H., Leichtfried, F., and Pytela, R., 1982, Plectin: A high-molecular-weight cytoskeletal polypeptide component that copurifies with intermediate filaments of the vimentin type, *Cold Spring Harbor Symp. Quant. Biol.* **46:**475–482.

Yang, H.-Y., Lieska, N., and Goldman, R. D., 1990, Intermediate filament-associated proteins, in *Cellular and Molecular Biology of Intermediate Filaments* (R. D. Goldman and P. M. Steinert, eds.), pp. 371–391, Plenum Press, New York.

Chapter 11

Measuring the Assembly Kinetics and Binding Properties of Intermediate Filament Proteins

Ilse Hofmann

1. INTRODUCTION

Intermediate filament proteins (IFs) form one of the three major cytoplasmic fibrillar systems. They constitute a large multigene family which can be classified into five different types on the basis of sequence homology, cell-type expression, and gene exon–intron patterns (for review see Fuchs and Weber, 1994). Although IF amino acid sequence homology is relatively low and restricted to certain segments of their central "rod" domain, they all share a common principle of molecular organization. Furthermore, nearly all cytoplasmic members of the family can form IFs *in vivo* and *in vitro* (Aebi *et al.*, 1983; Renner *et al.*, 1981). According to the currently favored view of IF assembly, first two molecules align in axial register as two coiled-coils (e.g., Steinert, 1990; Quinlan *et al.*, 1986; Parry *et al.*, 1985) which then form tetramers. Subsequent steps in IF assembly are, however, still poorly understood (for reviews see Fuchs and Weber, 1994; Stewart, 1993).

Ilse Hofmann Division of Cell Biology, German Cancer Research Center, D-69120 Heidelberg, Germany.

Subcellular Biochemistry, Volume 31: Intermediate Filaments, edited by Herrmann and Harris. Plenum Press, New York, 1998.

IFs are highly insoluble, and denaturing agents such as urea or guanidinium hydrochloride are required to solubilize their subunits and the constituent proteins. Filaments can be reconstituted *in vitro* by dialysis into buffers of physiological pH and ionic conditions depending on the IF protein under investigation (Eichner *et al.*, 1986, Hatzfeld and Franke, 1985; Renner *et al.*, 1981; Rueger *et al.*, 1979).

This review will mainly focus on how IF assembly kinetics can be followed by Ostwald viscometry, how this simple assay can be used to address various aspects of IF assembly, and how mutual affinities between IF proteins can be estimated by surface plasmon resonance (SPR).

2. ASSEMBLY KINETICS OF INTERMEDIATE FILAMENT PROTEINS

2.1. Principle of Measurements

Protein polymerization can be monitored by various techniques. With regard to actin assembly, a comprehensive overview of these techniques, their advantages, and drawbacks has been published by Cooper and Pollard (1982).

A rapid and simple method for measuring polymer assembly is by Ostwald viscometry. The viscometer is a U-shaped glass tube that includes a section with a narrow diameter through which the sample flows driven by hydrostatic pressure (Figure 1). The length and diameter of the capillary can be varied yielding different shear rates and flow times. Routinely we use hand-made viscometers with a flow time of 30 sec and a sample volume of 800 μl. They can also be obtained commercially from the Cannon Instrument Company (State College, PA). The viscometer is filled with soluble precursors of a polymer (e.g., G-actin) and the solution is sucked, by a rubber tube attached to the small opening of the viscometer, above the upper mark. Suction is released, and the time required for the solution to pass between the upper and lower marks on the glass capillary is taken every 5 min. Specific viscosity is calculated by the equation

$$\eta_{sp} = (t_s - t_b)/t_b$$

where t_s is the flow time of the sample and t_b is the flow time of the buffer.

Assembly is started by the addition and mixing of 1/10 volume of a buffer such that the desired pH and ionic strength is obtained in the viscometer. Growing filaments induce an increase in viscosity, which in-

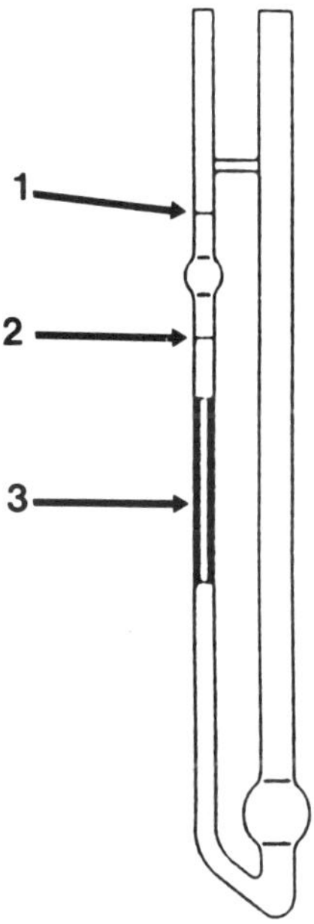

FIGURE 1. An Ostwald viscometer. The sample is added to the U-shaped glass tube through the wide opening. The fluid is sucked above the upper mark (1) and the time required for the solution to pass through the upper mark (1) and the lower mark (2) on the glass capillary (3) is taken every 5 min. Specific viscosity is calculated by the equation $\eta_{sp} = (t_s - t_b)/t_b$, where t_s is the flow time of the sample and t_b is the flow time of the buffer.

creases the flow time needed to pass the two marks. Air bubbles should not be included into the capillary, as this strongly influences the flow time of the solution and thereby changes the calculated values for the specific viscosity. After each measurement the viscometer is cleaned with 10% nitric acid followed by extensive washing with distilled water.

2.2. Vimentin Assembly Kinetics

Vimentin together with desmin and glial filament protein (GFP) belong to the type III class of IF proteins. All three proteins form homopolymeric IFs and are the most soluble of the IF proteins. Even at higher protein concentrations they remain soluble in the absence of denaturing agents, in low ionic strength buffers, and at a slightly alkaline pH (Ip *et al.*, 1985; Huiatt *et al.*, 1980; Rueger *et al.*, 1979).

For these reasons, vimentin, which in addition can be easily purified in large amounts from bovine lens, was chosen for our initial kinetic measurements (Hofmann *et al.*, 1991). Urea-solubilized bovine lens vimentin was dialyzed against and kept soluble in 5 mM Tris-HCl (pH 8.4). As a first parameter the intrinsic viscosity was determined over a period of 10 min. The values were reproducible and proportional to the protein concentration (Figure 2a). At time point 10 min, IF assembly was started by the

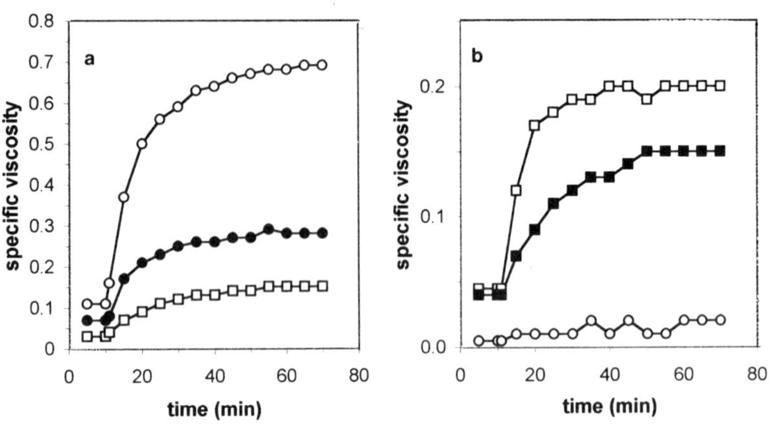

FIGURE 2. Time course of viscosity increase during assembly of bovine lens vimentin in assembly buffer, as measured in an Ostwald viscometer. The assembly was started at 10 min by the addition of a tenfold buffer (final concentration 160 mM NaCl, 25 mM Tris-HCl, pH 7.5). (a) Open circles: 2.9 mg/ml; filled circles: 1.5 mg/ml; open squares: 0.8 mg/ml at 22°C. Note that intrinsic viscosity is strongly dependent on protein content. (b) Open squares: 37°C; filled squares: 22°C; open circles: 10°C at 0.75 mg/ml.

addition of a tenfold buffer (final concentration 160 mM NaCl, 25 mM Tris-HCl, pH 7.5). Viscosity increased for the first 30 min and leveled off by 70 min. The half-maximal value of specific viscosity was reached within 5 min. Increase in protein concentration correlated with an increase in specific viscosity. Samples reconstituted in the viscometer can be recovered, and negatively stained specimens prepared for analysis in the transmission electron microscope. Smooth-surfaced filaments of diameter 8–14 nm and several micrometers in length were obtained 70 min after the start of assembly and they were indistinguishable from filaments reconstituted by dialysis against assembly buffer (Hofmann *et al.*, 1991).

Moreover, the assembly kinetics was also affected by temperature (Figure 2b), a parameter that can be easily changed by using a water bath. Comparing the assembly kinetics of bovine lens vimentin at temperatures of 10°C, 22°C, and 37°C revealed that the highest values of specific viscosity were measured at 37°C. At 10°C no increase of specific viscosity within the monitored period of time was observed, and electron microscopic analysis of negatively stained specimens revealed only protofilamentous material and a few short filaments. At 22°C and 37°C smooth-surfaced, several-micrometer-long filaments were visible. This clearly showed that IF assembly of bovine lens vimentin was accelerated at elevated temperatures (Hofmann, 1991).

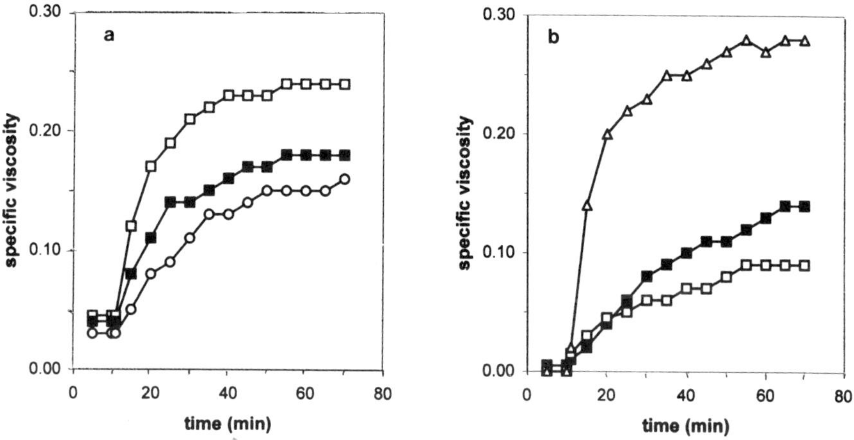

FIGURE 3. Viscometric analyses of the effects of monovalent and divalent ions on the assembly of (a) bovine lens and (b) recombinant amphibian vimentin. Assembly was started at time point 10 min by the addition of a tenfold buffer such that the indicated salt concentrations were added. (a) Open squares: 160 mM NaCl; filled squares: 100 mM NaCl; open circles: 80 mM NaCl. (b) Open triangles: 50 mM NaCl; filled squares: 1 mM $CaCl_2$; open squares: 5 mM $CaCl_2$. Protein concentrations were (a) 1.5 mg/ml or (b) 0.75 mg/ml. Note in panel b that the specific viscosity does not correlate with increasing $CaCl_2$ concentrations.

The viscometric assay was also used to compare the effects of mono- and divalent cations on the kinetics of IF assembly (Figure 3). For bovine lens vimentin higher concentrations of monovalent ions (160 mM compared to 80 mM NaCl) yielded higher values of specific viscosity. Recombinant *Xenopus* vimentin, however, could be studied only in lower ion concentrations (50 mM NaCl), because higher salt concentrations resulted in the formation of extremely viscous gels which unavoidably plugged the capillary of the viscometer. Moreover, the assembly of bovine lens vimentin and *Xenopus* vimentin to IFs could also be induced by the addition of divalent cations (e.g., $MgCl_2$), and here again increasing concentrations of $MgCl_2$ (5 mM compared to 1 mM) yielded higher values of specific viscosity (Hofmann *et al.*, 1991). In contrast, the increase in specific viscosity in the presence of $CaCl_2$ showed a lack of correlation to increasing ion concentrations (Figure 3b). Electron microscopic analysis of vimentin filaments assembled in 5 mM $CaCl_2$ revealed that they were considerably shorter and thicker than those produced by NaCl (Figure 4).

These data clearly show that Ostwald viscometry is a valuable tool for analysis of vimentin assembly. It allows the rapid and efficient comparison of factors influencing assembly such as protein content, temperature, and ionic conditions.

The method of Ostwald viscometry has several advantages compared to other techniques applied for monitoring polymer assembly (Cooper and Pollard, 1982). Early assembly stages cannot be analyzed by centrifugation studies (Inagaki *et al.*, 1988), while turbidometric measurements (Yang and Babitch, 1988; Stromer *et al.*, 1987; Zackroff and Goldman, 1979; Steinert *et al.*, 1976) are insensitive to polymer length and do not discriminate between aggregates and filaments. The use of fluorescently labeled proteins (Ip and Fellows, 1990; Angelides *et al.*, 1989) requires protein modification, which may influence assembly properties. The major drawback of viscometry, i.e., the disturbance of the reaction by shearing forces, appears to be of only minor importance for IFs, which seem to be rather stable structures. Thus viscometry is the preferred method to measure IF assembly.

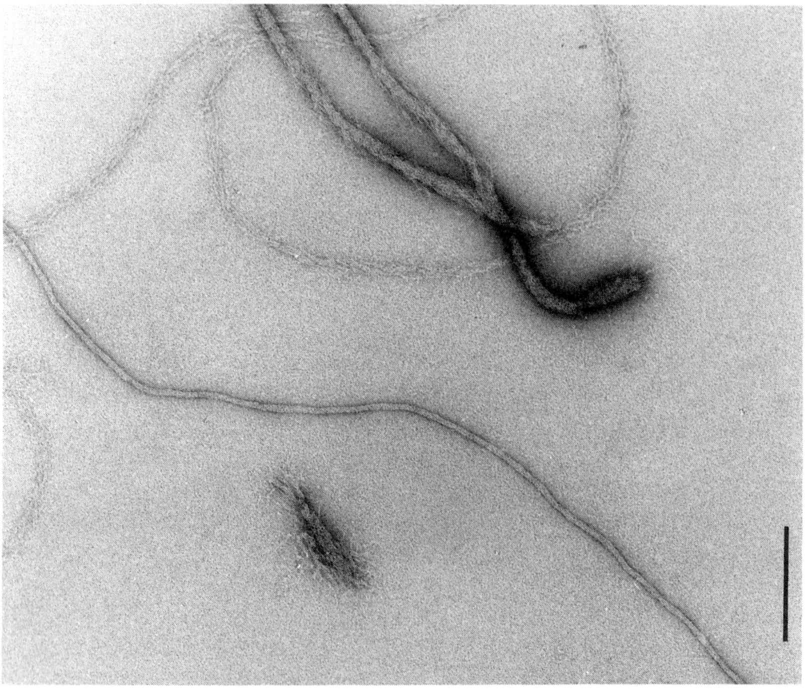

FIGURE 4. Electron micrograph showing direct comparison of NaCl- and $CaCl_2$-induced filaments reconstituted *in vitro* from recombinant amphibian vimentin. Filaments were reconstituted *in vitro* separately and mixed approximately 1:1 prior to negative staining. Note that $CaCl_2$-induced filaments are much thicker than NaCl-induced ones. Bar represents 200 nm.

2.3. Cytokeratin Assembly Kinetics

Cytokeratins (CKs) represent the most complex subgroup of IF proteins. In contrast to other IF proteins, which can assemble into homopolymeric IFs *in vitro* and *in vivo*, CK IFs occur in two distinct subfamilies, the type I and type II CKs. For CK IF assembly at least one polypeptide from each CK subfamily is required to form IFs (Steinert, 1990; Hatzfeld and Franke, 1985). Attempts to measure CK assembly kinetics have been hampered by the observation that intermediates of CK IF assembly have only been described under partial denaturing conditions (Quinlan *et al.*, 1986, 1984) or at very low protein concentrations (Steinert, 1991). In contrast, simple epithelia CKs, e.g., CK8 and CK18, can be kept soluble in 2 mM Tris-HCl, pH 9.0, up to a concentration of 0.5 mg/ml, which allowed viscometric measurements to be made (Hofmann and Franke, 1997). The CK pairs 8 and 18 were mixed in near equimolar amounts in 9.5 M urea, and dialyzed to 2 mM Tris-HCl (pH 9.0), and assembly kinetics was monitored by Ostwald viscometry (Figure 5a). Upon addition of assembly buffer (final concentration 10 mM Tris-HCl, pH 7.5), the specific viscosity value increased rapidly, with a further gradual increase over a period of another 20 min, followed by an extended plateau. No increase in specific viscosity

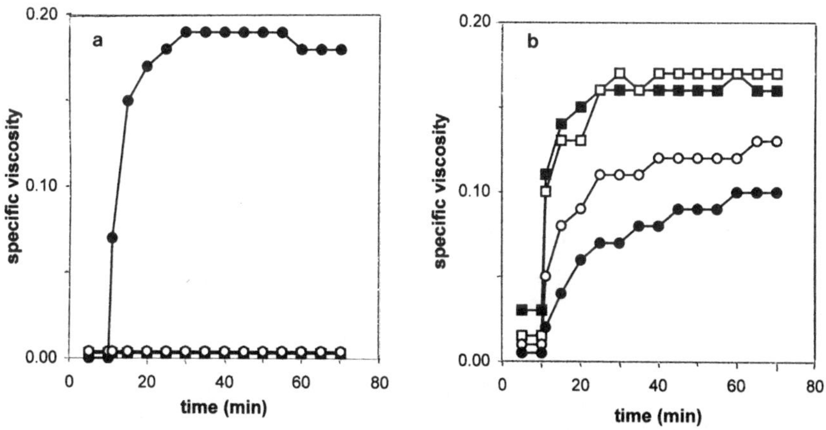

FIGURE 5. Time course of viscosity increase during assembly of recombinant type II CK8 and various type I CKs as measured in an Ostwald viscometer. The assembly reaction was started at 10 min by the addition of a tenfold buffer (final concentration 10 mM Tris-HCl, pH 7.5). Protein concentration was 0.4 mg/ml. (a) Closed circles: CK8 and CK18 were mixed in near-equimolar amounts; open circles: CK8 alone; filled squares: CK18 alone. (b) Open squares: CK8 and CK20; filled squares: CK8 and CK18; open circles: CK8 and CK13; filled circles: CK8 and CK19. In panel a, neither CK8 and CK18 alone develops any considerable viscosity and therefore the two symbols merge with one another.

was observed when either CK8 or CK18 alone was analyzed separately, indicating that no elongated structures were formed.

Cytokeratins are expressed as specific type I–type II pairs in a differentiation-specific fashion (Sun *et al.*, 1984; Moll *et al.*, 1982). In recent years it has become clear that the specific CK complement present is important for cell and tissue functions (e.g., Paladini *et al.*, 1996; Blessing *et al.*, 1993). The specific assembly properties of various CKs might regulate the CK composition in a certain IF array. To compare the assembly kinetics of various CK pairs, the human nonepidermal CK8, CK13, CK18, CK19, and CK20 were expressed in *Escherichia coli*, extracted, and purified. The various type I CKs were mixed in near equimolar amounts with type II CK8, dialyzed against 2 mM Tris-HCl (pH 9.0), and assembly kinetics measured utilizing Ostwald viscometry (Figure 5b). The CK pairs 8:18 and 8:20 attained the highest viscosity values, whereas the CK pair 8:19 displayed a relatively low value. The viscometric assay allowed the *in vitro* analysis of CK filament formation and growth in quantitative terms and thereby the comparison of the reaction of CKs in the assembly process. Differences in IF assembly kinetics may be of great biological importance in all situations in which various CKs occur in the same cell.

2.4. Applications

In this section several useful applications of the viscometric assay will be discussed. Of course, one possibility is to analyze mutated IF proteins and compare their assembly kinetics with the corresponding wild-type proteins (Rogers *et al.*, 1995; Eckelt *et al.*, 1992; Herrmann *et al.*, 1992). Moreover, by mixing experiments with wild-type protein, the severity of IF-forming defects of mutated proteins can be monitored. Herrmann and coworkers (1992) restored and quantified the influence of deletions and point mutations in *Xenopus* vimentin on IF assembly by such experiments. For a mutant missing an amino-terminal nonapeptide motif (SSYRRIFGG) in the head domain, Del, as little as 25% of wild-type vimentin was sufficient to approach viscosity values obtained with wild-type vimentin alone (Figure 6a). Alternatively, for another mutant, $(YIF)^-$, with point mutations within this motif, 50% wild-type vimentin had to be added to yield values comparable to wild-type vimentin (Figure 6b). Electron microscopic analysis of mixtures of both mutants with 10% wild-type vimentin revealed that both preparations contained filaments, although the $(YIF)^-$ mixtures often showed partly irregularly formed filaments, which then took on the typical smooth-surfaced IF form (Figure 7).

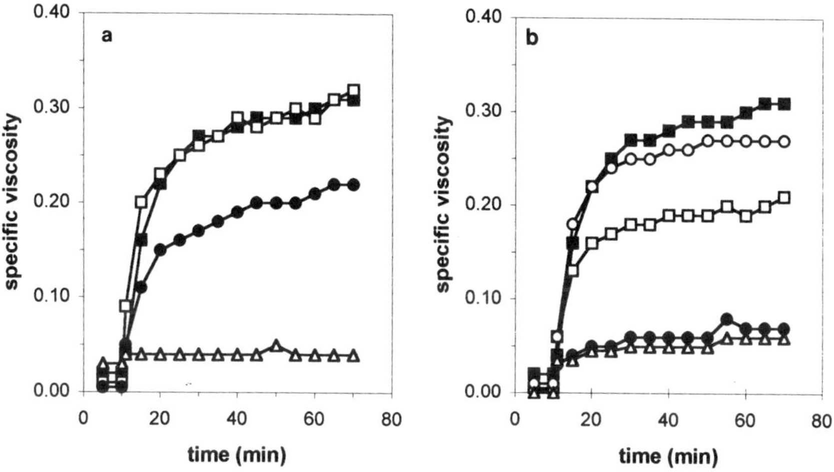

FIGURE 6. Viscosity measurements of various mixtures of wild-type and mutated vimentins. (a) The mutant Del, missing a nonapeptide motif in the head, and (b) the mutant $(YIF)^-$, replacing hydrophobic amino acids by serines within this motif, were mixed with increasing proportions of wild-type vimentin in 5 mM Tris-HCl (pH 8.5). At 10 min the assembly was started by the addition of a tenfold buffer (final concentration 50 mM NaCl, 25 mM Tris-HCl, pH 7.5). Open triangles: 0%; filled circles: 10%; open squares: 25%; open circles: 50%; closed squares: 100% wild-type vimentin.

Another application is to study the influence of assembly-inhibiting compounds. For example, synthetic peptides derived from IF proteins can be used to gain insight into the significance and possible function of these protein motifs for the self-assembly of IFs (Hofmann and Herrmann, 1992). A synthetic peptide corresponding to the above-mentioned conserved amino-terminal motif was used to study the influence on vimentin assembly (Figure 8). Preincubation with an eightfold molar excess of peptide did not significantly influence the kinetics of viscosity increase. At a 25-fold molar excess of peptide, specific viscosity reached only one third of the control value after 70 min, and at a 50-fold molar excess the peptide efficiently inhibited assembly.

As this simple viscometric assay also allows a rapid change of the assembly temperature, Herrmann and co-workers were able to study comparatively the temperature optimum of human, *Xenopus laevis*, and rainbow trout vimentin (Herrmann *et al.*, 1996, 1993). The mammalian and amphibian vimentin assembled most efficiently at temperatures of 37°C and

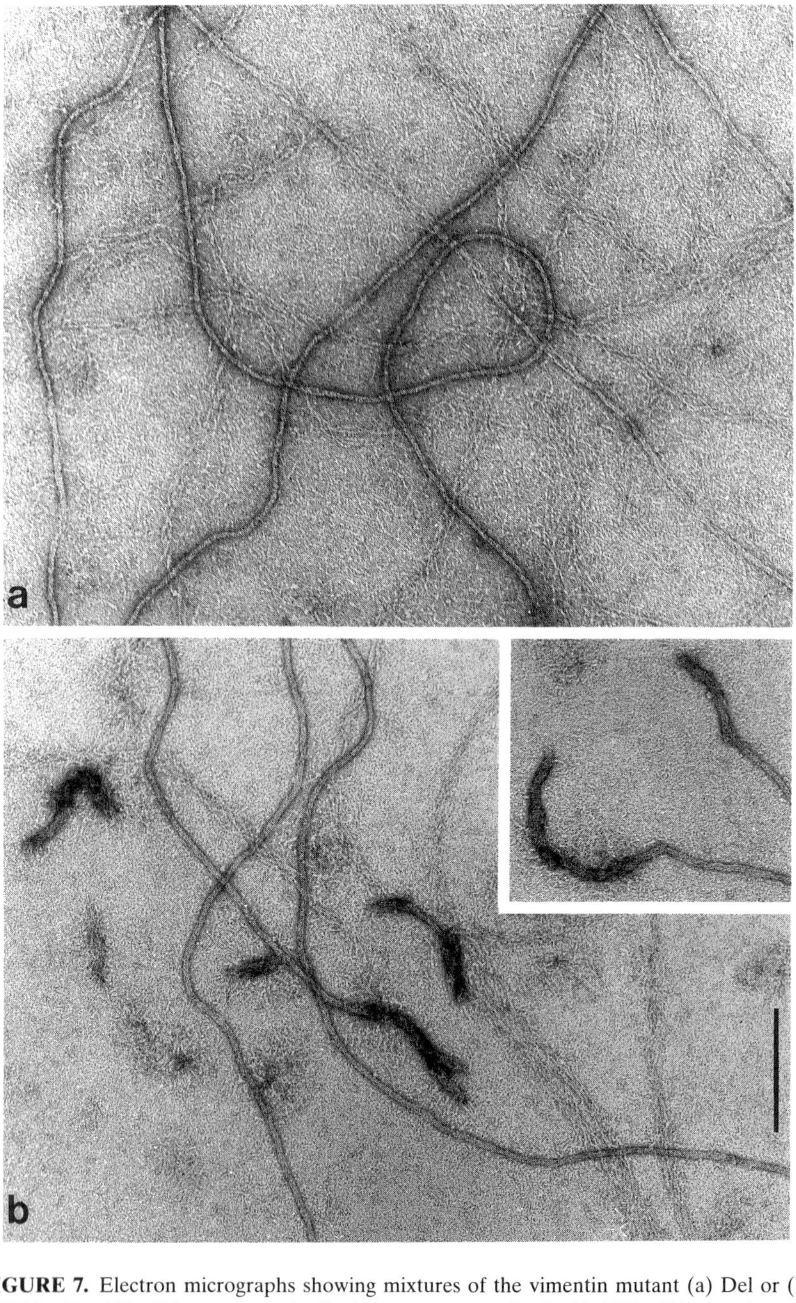

FIGURE 7. Electron micrographs showing mixtures of the vimentin mutant (a) Del or (b) $(YIF)^-$ with 10% wild-type vimentin. Both mutants were mixed with 10% wild-type vimentin in 5 mM Tris-HCl (pH 8.5). Assembly was started by the addition of a tenfold buffer (final concentration 50 mM NaCl, 25 mM Tris-HCl, pH 7.5). Bar represents 200 nm.

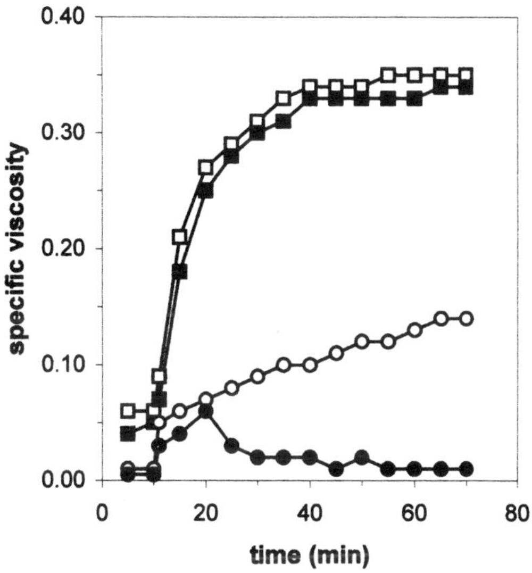

FIGURE 8. Viscometric analysis of the effect of various amounts of the peptide SSSYRRMFGG on bovine vimentin assembly. Various amounts of the peptide were added to soluble vimentin (30 μM) and assembly was started at 10 min by the addition of a tenfold buffer (final concentration 160 mM NaCl, 25 mM Tris-HCl, pH 7.5); open squares: control; closed squares: 8-fold excess; open circles: 25-fold molar excess; filled circles: 50-fold molar excess of peptide.

28°C, respectively, whereas the fish vimentin had a temperature optimum between 18°C and 24°C. All temperature optima correspond to the temperatures at which each of the species survives in nature or keeps the body temperature constant. These data again demonstrate that such an *in vitro* assay is a valuable test to characterize assembly behavior of IF proteins, and that these observations can be correlated with conditions in the living organism.

Posttranslational modifications of IF proteins might interfere with IF assembly. It is therefore interesting to compare such influences on IF assembly, either to control the synthetically modified protein or to analyze *in vivo* occurring modifications. Rogers *et al.* (1996) introduced cysteine cross-links into wild-type vimentin and various cysteine mutants and compared the assembly behavior of cross-linked versus untreated proteins. They found that the conditions used for cross-linking as well as the position of the cysteine residues influenced the assembly kinetics of vimentin.

Finally, the influence of intermediate filament-associated proteins (IFAPs) on IF assembly can be easily quantified by viscometry. Bianchi and co-workers presented evidence utilizing viscometry and a sedimentation assay that S-100 protein and calpactin I act in concert to regulate the assembly state of GFP IFs (Bianchi *et al.*, 1995, 1994). Recently, the study was extended to desmin and it was shown that a muscle-specific protein (S-100a_0) inhibited desmin assembly (Garbuglia *et al.*, 1996).

3. BINDING PROPERTIES OF INTERMEDIATE FILAMENT PROTEINS

3.1. Principle of Surface Plasmon Resonance Detection

With the optical technique of surface plasmon resonance (SPR) detection, one can monitor changes in refractive index in the vicinity of a surface [for detailed description of the method see Szabo *et al.* (1995), Malmqvist (1993), and Karlsson *et al.* (1991)]. The chemical modification of a carboxylated dextran surface coupled to a thin gold surface allows the covalent coupling of a ligand (Johnsson *et al.*, 1991; see Figure 9). An integrated liquid handling system was developed for the exact transport of sample to the adsorption and detection spot (BIACORE AB, Uppsala, Sweden). When a soluble analyte binds to the immobilized ligand this is typically accompanied by an increase in protein concentration at the sensor surface, with a corresponding increase in refractive index. Binding is measured in arbitrary units (response units or RU) and is presented as a graph termed the "sensorgram" of response versus time (Figure 9). Monitored response units are directly related to the concentration of analyte bound to the surface layer.

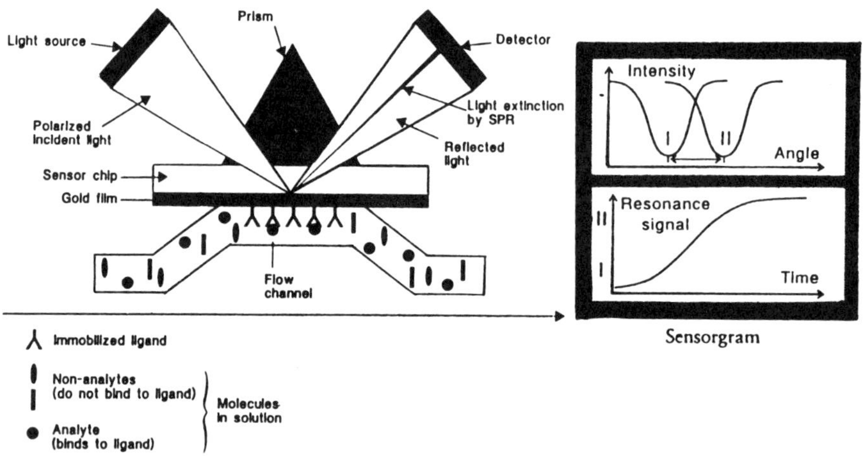

FIGURE 9. Configuration of a surface plasmon resonance (SPR) detector, sensor chip, and integrated fluidic cartridge in the BIACORE system (BIACORE AB, Uppsala, Sweden). The detector features polarized light from a light-emitting diode which is reflected by the gold film; extinction of light can be detected in the reflected beam for a special angle of incidence. This SPR angle varies with refractive index changes on and in the vicinity of the surface as a result of adsorption of molecules. Binding is monitored in arbitrary units (response units) and presented as a "sensorgram" of response versus time. (According to Szabo *et al.*, 1995.)

The major advantage of this system is the possibility to analyze association and dissociation of an analyte of interest to the ligand without the use of labels in real time. As SPR detection is independent of the chemical nature of the analyzed sample, in principle all types of molecules such as proteins, nucleic acids, lipids, carbohydrates, or low-molecular-weight compounds can be used to monitor interactions.

3.2. Estimation of Relative Affinities of Intermediate Filament Proteins

SPR detection was used to study the interaction between type I and type II CKs (Hofmann and Franke, 1997). The type II CK8 was immobilized on the sensor surface, various amounts of IF proteins were allowed to react, and the resonance increase, expressed in response units, was followed. When the association curves of various IF proteins were compared (Figure 10) the combination of vimentin and CK8 did not result in any

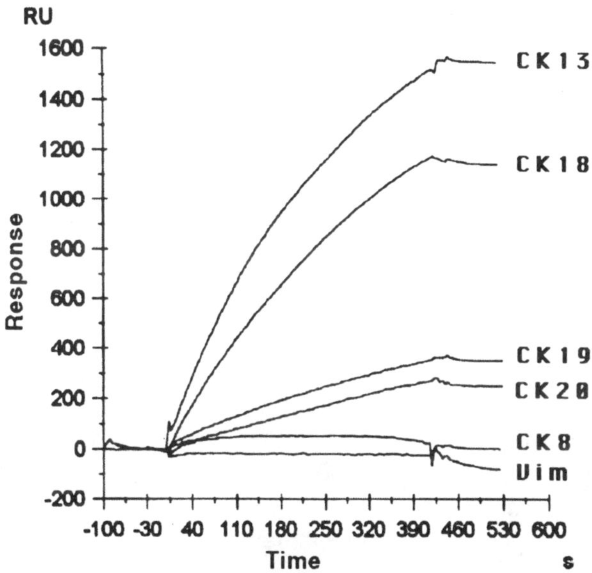

FIGURE 10. Binding curves of various IF proteins to CK 8 as measured by surface plasmon resonance. CK8 was immobilized on the carboxylated layer of a sensor chip. Binding of CK8, CK13, CK18, CK19, CK20, and vimentin to CK8 was measured by using identical protein concentrations (10 μg/ml) using a BIACORE Biosensor (BIACORE AB, Uppsala, Sweden). Resonance was plotted on the ordinate in arbitrary response units. The binding curves obtained were corrected for nonspecific binding to the dextrane matrix. Note marked differences of binding to CK8, with especially high rates for CK13 and CK18. (According to Hofmann and Franke, 1997.)

increase. Surprisingly, the various type I CKs examined (i.e., CKs 13, 18, 19, and 20) differed markedly in their relative binding rates: For example, CK18 and CK13 displayed much higher resonance signals than CK19 and CK20. Using the new method of SPR, the principle of heterotypic binding specificity as determined by solid-phase assays (Hatzfeld *et al.*, 1987) was confirmed. In addition, this method allowed the estimation of quantitative differences of binding rate and strength between various type I CKs and CK8.

4. FUTURE PROSPECTS

The two methods described here for measuring the interaction between IFs in quantitative terms allow for the first time the *in vitro* analysis of the formation of IF oligomers as well as of filament formation and growth, and thereby the comparison of the reaction of individual IF proteins in the assembly process.

In order to understand how a certain IF composition contributes to tissue stability it would be important to analyze not only the assembly behavior of type III IF proteins and nonepidermal CKs, but also the assembly of epidermal CKs, as in epidermis complex sets of CK pairs are expressed. During vertical differentiation of epidermal keratinocytes the synthesis of CK5 and CK14, which are expressed in basal, mitotically active cells, is downregulated, and expression of the genes encoding CK1 and CK10 is induced followed by the expression of CK2e in upper cell layers (for a recent review see Fuchs, 1995). In addition, the CK pair 6/16 that is expressed during wound healing in epidermis (Paladini *et al.*, 1996) seems to be a likely candidate for finding differences in oligomer and filament formation. Hair-specific cytokeratins, which are typical of trichocytes (Rogers *et al.*, 1997; Heid *et al.*, 1988, 1986), appear to have special features regarding their assembly conditions (Winter *et al.*, 1997).

Moreover, the relatively simple and efficient viscometric protocol could also give important information on the assembly kinetics of type IV IFs and allow a quantitative comparison of the influence of NF-H and/or NF-M on NF-L assembly (Lee *et al.*, 1993; Hisanaga and Hirokawa, 1990).

Mutations affecting specific CK genes have been described that cause, or underlie, several inheritable disorders such as epidermolysis bullosa simplex, epidermolytic hyperkeratosis, and palmoplantar keratoderma (for reviews see Fuchs, 1997; McLean and Lane, 1995). The assays described here would also allow the comparison of assembly kinetics of mutated CKs and thereby probably contribute important information on the influence of certain mutations on affinity and assembly behavior.

In recent years evidence has accumulated that IFs exist in a state of dynamic equilibrium *in vivo*. Posttranslational modifications such as phosphorylation (for review see Takai *et al.*, 1996) have emerged as one regulator of such dynamic events. Elevated levels of IF phosphorylation have been associated with increased pools of IF subunits. However, the exact stochiometry of subtle changes in posttranslational modifications and their influence on IF assembly kinetics have not yet been estimated.

In recent years, an increasing number of proteins has been shown to bind reversibly to IFs and thereby modulate assembly characteristics. Examples are α-crystallins in case of vimentin (Nicholl and Quinlan, 1994), and S-100 protein and calpactin I for GFP (Bianchi *et al.*, 1995, 1994), and heat shock protein 70 (Liao *et al.*, 1995) and 14-3-3 proteins (Liao and Omary, 1996) for CK8/18. The viscometric protocol could help to screen for and characterize new IFAPs that either promote or inhibit polymer formation.

The two assays described here are valuable tools which can help to understand how the IF network is built and regulated either by the affinity and assembly kinetics of IF subunits, by posttranslational modifications of IF subunits, or by the influence of IFAPs on IF assembly. These three parameters probably modulate and determine the composition and properties of the IF cytoskeleton and thus may play a functional role in regulating the cytoskeleton and cell architecture in general.

ACKNOWLEDGMENTS. I thank Dr. W. W. Franke for encouragement and generous support and Drs. H. Herrmann and M. Schnölzer for comments on the manuscript.

5. REFERENCES

Aebi, U., Fowler, W. E., Rew, P., and Sun, T.-T., 1983, The fibrillar substructure of keratin filaments unraveled, *J. Cell Biol.* **97:**1131–1143.

Angelides, K. J., Smith, K. E., and Takeda, M., 1989, Assembly and exchange of intermediate filament proteins of neurons: Neurofilaments are dynamic structures, *J. Cell Biol.* **108:**1495–1506.

Bianchi, R., Verzini, M., Garbuglia, M., Giambanco, I., and Donato, R., 1994, Mechanism of S100 protein-dependent inhibition of glial fibrillary acidic protein (GFAP) polymerization, *Biochim. Biophys. Acta* **1223:**354–360.

Bianchi, R., Garbuglia, M., Verzini, M., Giambanco, I., Spreca, A., and Donato, R., 1995, S-100 protein and annexin II_2-$p11_2$ (calpactin I) act in concert to regulate the state of assembly of GFAP intermediate filaments *in vitro*, *Biochem. Biophys. Res. Commun.* **208:**910–918.

Blessing, M., Rüther, U., and Franke, W. W., 1993, Ectopic synthesis of epidermal cytokeratins in pancreatic islet cells of transgenic mice interferes with cytoskeletal order and insulin production, *J. Cell Biol.* **120:**743–755.

Cooper, J. A., and Pollard, T. D., 1982, Methods to measure actin polymerization, *Meth. Enzymol.* **85:**182–210.

Eckelt, A., Herrmann, H., and Franke, W. W., 1992, Assembly of a tail-less mutant of the intermediate filament protein, vimentin, *in vitro* and *in vivo*, *Eur. J. Cell Biol.* **58:**319–330.

Eichner, R., Sun, T.-T., and Aebi, U., 1986, The role of keratin subfamilies and keratin pairs in the formation of human epidermal intermediate filaments, *J. Cell Biol.* **102:**1767–1777.

Fuchs, E., 1995, Keratins and the skin, *Annu. Rev. Cell Dev.* **11:**123–153.

Fuchs, E., 1997, Of mice and men: Genetic disorders of the cytoskeleton, *Mol. Biol. Cell* **8:**189–203.

Fuchs, E., and Weber, K., 1994, Intermediate filaments: Structure, dynamics, function, and disease, *Annu. Rev. Biochem.* **63:**345–382.

Garbuglia, M., Verzini, M., Giambanco, I., Spreca, A., and Donato, R., 1996, Effects of calcium-binding proteins (S-100a_0, S-100a, S-100b) on desmin assembly *in vitro*, *FASEB J.* **10:**317–324.

Hatzfeld, M., and Franke, W. W., 1985, Pair formation and promiscuity of cytokeratins: Formation *in vitro* of heterotypic complexes and intermediate-sized filaments by homologous and heterologous recombinations of purified polypeptides, *J. Cell Biol.* **101:**1826–1841.

Hatzfeld, M., Maier, G., and Franke, W. W., 1987, Cytokeratin domains involved in heterotypic complex formation determined by *in-vitro* binding assays, *J. Mol. Biol.* **197:**237–255.

Heid, H. W., Werner, E., and Franke, W. W., 1986, The complement of native α-keratin polypeptides of hair-forming cells: A subset of eight polypeptides that differ from epithelial cytokeratins, *Differentiation* **32:**101–119.

Heid, H. W., Moll, I., and Franke, W. W., 1988, Patterns of expression of trichocytic and epithelial cytokeratins in mammalian tissues. I. Human and bovine hair follicles, *Differentiation* **37:**137–157.

Herrmann, H., Hofmann, I., and Franke, W. W., 1992, Identification of a nonapeptide motif in the vimentin head domain involved in intermediate filament assembly, *J. Mol. Biol.* **223:**637–650.

Herrmann, H., Eckelt, A., Brettel, M., Grund, C., and Franke, W. W., 1993, Temperature-sensitive intermediate filament assembly alternative structures of *Xenopus laevis* vimentin *in vitro* and *in vivo*, *J. Mol. Biol.* **234:**99–113.

Herrmann, H., Münick, M. D., Brettel, M., Fouquet, B., and Markl, J., 1996, Vimentin in a cold-water fish, the rainbow trout: Highly conserved primary structure but unique assembly properties, *J. Cell Sci.* **109:**569–578.

Hisanaga, S.-I., and Hirokawa, N., 1990, Molecular architecture of the neurofilament. II. Reassembly process of neurofilament L protein *in vitro*, *J. Mol. Biol.* **211:**871–882.

Hofmann, I., 1991, Aufbau aund Anwendung eines Bestimmungsverfahrens zur *in vitro* Polymerisation von Intermediärfilamentproteinen, Ph.d. thesis. Faculty of Biology, University of Heidelberg, Germany.

Hofmann, I., and Franke, W. W., 1997, Heterotypic interactions and filament assembly of type I and type II cytokeratins *in vitro*: Viscometry and determinations of relative affinities, *Eur. J. Cell Biol.* **72:**122–132.

Hofmann, I., and Herrmann, H., 1992, Interference in vimentin assembly *in vitro* by synthetic peptides derived from the vimentin head domain, *J. Cell Sci.* **101:**687–700.

Hofmann, I., Herrmann, H., and Franke, W. W., 1991, Assembly and structure of calcium-induced thick vimentin filaments, *Eur. J. Cell Biol.* **56:**328–341.

Huiatt, T. W., Robson, R. M., Arakawa, N., and Stromer, M. H., 1980, Desmin from avian smooth muscle, *J. Biol. Chem.* **255:**6981–6989.

Inagaki, M., Gonda, Y., Matsuyama, M., Nishizawa, K., Nishi, Y., and Sato, C., 1988, Intermediate filament reconstitution *in vitro*, *J. Biol. Chem.* **263:**5970–5978.

Ip, W., and Fellows, M. E., 1990, Fluorescent measurement of desmin intermediate filament assembly, *Anal. Biochem.* **185:**10–16.

Ip, W., Hartzer, M. K., Pang, Y.-Y., and Robson, R.M., 1985, Assembly of vimentin *in vitro* and its implications concerning the structure of intermediate filaments, *J. Mol. Biol.* **183:**365–375.

Johnsson, B., Löfas, S., and Lindquist, G., 1991, Immobilization of proteins to a carboxymethyldextran-modified gold surface for biospecific interaction analysis in surface plasmon resonance sensors, *Anal. Biochem.* **198:**268–277.

Karlsson, R., Michaelsson, A., and Mattsson, L., 1991, Kinetic analysis of monoclonal antibody–antigen interactions with a new biosensor based analytical system, *J. Immunol. Meth.* **145:**229–240.

Lee, M. K., Xu, Z., Wong, P. C., and Cleveland, D. W., 1993, Neurofilaments are obligate heteropolymers *in vivo*, *J. Cell Biol.* **122:**1337–1350.

Liao, J., and Omary, M. B., 1996, 14-3-3 proteins associate with phosphorylated simple epithelial keratins during cell cycle progression and act as a solubility cofactor, *J. Cell Biol.* **133:**345–357.

Liao, J., Lowthert, L. A., Ghori, N., and Omary, M. B., 1995, The 70-kDa heat shock proteins associate with glandular intermediate filaments in an ATP-dependent manner, *J. Biol. Chem.* **270:**915–922.

Malmqvist, M., 1993, Biospecific interaction analysis using biosensor technology, *Nature* **361:**186–187.

McLean, W., and Lane, E. B., 1995, Intermediate filaments in diseases, *Curr. Opin. Cell Biol.* **7:**118–125.

Moll, R., Franke, W. W., Schiller, D. L., Geiger, B., and Krepler, R., 1982, The catalog of human cytokeratins: Patterns of expression of specific cytokeratins in normal epithelia, tumors and cultured cells, *Cell* **31:**11–24.

Nicholl, I. D., and Quinlan, R. A., 1994, Chaperone activity of α-crystallins modulates intermediate filament assembly, *EMBO J.* **13:**945–953.

Paladini, R. D., Takahashi, K., Bravo, N. S., and Coulombe, P. A., 1996, Onset of re-epithelialization after skin injury correlates with a reorganization of keratin filaments in wound edge keratinocytes: Defining a potential role for keratin 16, *J. Cell Biol.* **132:**381–397.

Parry, D. A. D., Steven, A. C., and Steinert, P. M., 1985, The coiled-coil molecules of intermediate filaments consist of two parallel chains in exact axial register, *Biochem. Biophys. Res. Commun.* **127:**1012–1018.

Quinlan, R. A., Cohlberg, J. A., Schiller, D. L., Hatzfeld, M., and Franke, W. W., 1984, Heterotypic tetramer (A_2D_2) complexes of non-epidermal keratins isolated from cytoskeletons of rat hepatocytes and hepatoma cells, *J. Mol. Biol.* **178:**365–388.

Quinlan, R. A., Hatzfeld, M., Franke, W. W., Lustig, W., Schultheiss, T., and Engel, J., 1986, Characterization of dimer subunits of intermediate filament proteins, *J. Mol. Biol.* **192:**337–349.

Renner, W., Franke, W. W., Schmid, E., Geisler, N., Weber, K., and Mandelkow, E., 1981, Reconstitution of intermediate-sized filaments from denatured monomeric vimentin, *J. Mol. Biol.* **149:**285–306.

Rogers, K. R., Eckelt, A., Nimmrich, V., Janssen, K.-P., Schliwa, M., Herrmann, H., and Franke, W. W., 1995, Truncation mutagenesis of the non-α-helical carboxyterminal tail domain of vimentin reveals contributions to cellular localization but not to filament assembly, *Eur. J. Cell Biol.* **66:**136–150.

Rogers, K. R., Herrmann, H., and Franke, W. W., 1996, Characterization of disulfide crosslink formation of human vimentin at the dimer, tetramer, and intermediate filament levels, *J. Struct. Biol.* **117:**55–69.

Rogers, M. A., Langbein, L., Praetzel, S., Moll, I., Krieg, T., Winter, H., and Schweizer, J., 1997, Sequences and differential expression of three novel human type-II hair keratins, *Differentiation* **61:**187–194.

Rueger, D. C., Huston, J. S., Dahl, D., and Bignami, A., 1979, Formation of 100 Å filaments from purified glial fibrillary acidic protein *in vitro*, *J. Mol. Biol.* **135:**53–68.

Sun, T.-T., Eichner, R., Schermer, A., Cooper, D., Nelson, W. G., and Weiss, R. A., 1984, Classification, expression and possible mechanism of evolution of mammalian epithelial keratins: A unifying model, in *Cancer Cells, 1, The Transformed Phenotype* (A. J. Levine, G. F. Vande Woude, W. C. Topp, and J. D. Watson, eds.), pp. 169–176, Cold Spring Harbor Laboratory Press, Cold Spring Harbor, New York.

Steinert, P. M., 1990, The two-chain coiled-coil molecule of native epidermal keratin intermediate filaments is a type I–type II heterodimer, *J. Biol. Chem.* **265:**8766–8774.

Steinert, P. M., 1991, Organization of coiled-coil molecules in native mouse keratin 1/keratin 10 intermediate filaments: Evidence for alternating rows of antiparallel in-register and antiparallel staggered molecules, *J. Struct. Biol.* **107:**157–174.

Steinert, P. M., Idler, W. W., and Zimmerman, S. B., 1976, Self-assembly of bovine epidermal keratin filaments *in vitro*, *J. Mol. Biol.* **108:**547–567.

Stewart, M., 1993, Intermediate filament structure and assembly, *Curr. Opin. Cell Biol.* **5:**3–11.

Stromer, M. H., Ritter, M. A., Pang, Y.-Y. S., and Robson, R. M., 1987, Effect of cations and temperature on kinetics of desmin assembly, *Biochem. J.* **246:**75–81.

Szabo, A., Stolz, L., and Granzow, R., 1995, Surface plasmon resonance and its use in biomolecular interaction analysis (BIA), *Curr. Opin. Struct. Biol.* **5:**699–705.

Takai, Y., Inagaki, N., Tsutsumi, O., and Inagaki, M., 1996, Visualization and regulation of intermediate filament kinase activities, *Cell Dev. Biol.* **7:**741–749.

Winter, H., Hofmann, I., Langbein, L., Rogers, M. A., and Schweizer, J., 1997, A splice site mutation in the gene of the human type I hair keratin hHa1 results in the expression of a tailless keratin isoform, *J. Biol. Chem.* **272:**32345–32352.

Yang, Z. W., and Babitch, J. A., 1988, Factors modulating filament formation by bovine glial fibrillary acidic protein, the intermediate filament component of astroglial cells, *Biochemistry* **27:**7038–7045.

Zackroff, R. V., and Goldman, R. D., 1979, *In vitro* assembly of intermediate filaments from baby hamster kidney (BHK-21) cells, *Biochemistry* **76:**6226–6230.

Chapter 12

Viscoelasticity of Intermediate Filament Networks

Paul A. Janmey, Jagesh V. Shah, Klaus-Peter Janssen, and Manfred Schliwa

1. INTRODUCTION

1.1. Mechanical Properties of Cells

The mechanical properties of cells allow them to maintain their shapė or respond to external forces. In addition, mechanical forces generated within the cell are necessary for basic cellular processes such as locomotion and cytokinesis. Loss of the mechanical integrity of a cell results in cell death or dysfunction and underlies a variety of disease states (Fuchs, 1996; Janmey, 1991a). The structure responsible for the mechanical properties of the cell is the cytoskeleton, a network of protein filaments that permeate throughout the volume of the cell. The three major components of the cytoskeleton are actin filaments, microtubules, and intermediate filaments. Each of these

Paul A. Janmey Division of Experimental Medicine, Brigham and Women's Hospital, and Program in Biological and Biomedical Sciences, Harvard Medical School, Boston, Massachusetts 02115. **Jagesh V. Shah** Division of Experimental Medicine, Brigham Women's Hospital, Boston, Massachusetts 02115. **Klaus-Peter Janssen and Manfred Schliwa** Adolf-Butenandt-Institut für Zellbiologie, Ludwig-Maximilians-Universität, D-80336 Munich, Germany.

Subcellular Biochemistry, Volume 31: Intermediate Filaments, edited by Herrmann and Harris. Plenum Press, New York, 1998.

filamentous systems plays a role in providing macroscopic mechanical properties to the cell. In particular, intermediate filaments have the unique potential to withstand large deformations that would otherwise rupture F-actin and microtubules, implying that they are important in maintaining tissue integrity. Furthermore, disruption of intermediate filaments in disease often results in phenotypic deficits that are mechanical in nature.

Mechanical forces alter cell shape and function in several distinct ways. One example is the deformation of tissues by external forces such as those to which the epidermal surface of organisms is exposed. In this case specialized covalently reinforced networks of the intermediate filament keratin (Baden *et al.*, 1973; Breuer, 1968; Christensen *et al.*, 1977) form the basis for many protective surface layers such as hooves, nails, hair, and skin. Cells also generate internal stresses, the most apparent of which is the contractile force produced by muscles. Intermediate filaments, often of the desmin type, contribute along with titin to the passive elastic recovery of muscle, and link sarcomeres to each other and to structures in the cell surface. Loss of desmin results in a number of defects in muscle function and structure, in particular with heart muscle (Li *et al.*, 1996; Milner *et al.*, 1996; see also Chapter 16 in this book), although somewhat surprisingly, desmin is not required for muscle formation *in vivo* even though it is one of the earliest markers for myocyte differentiation (Li *et al.*, 1994). A third way in which mechanical force impacts cell function is through activation of intracellular signaling pathways.

Mechanotransduction is a feature of both acute responses, such as changes in intracellular Ca^{2+} (Glogauer *et al.*, 1995), and of gene regulation (Bloom *et al.*, 1996; Ingber and Folkman, 1989; Traub, 1995; Traub and Shoeman, 1994). Recent work has shown that internal structures, such as the nucleus, are linked elastically to the cell membrane even when actin and microtubules have been destabilized (Maniotis *et al.*, 1997), suggesting that intermediate filaments may play a dominant role in cellular response to mechanical forces.

1.2. Modeling the Cytoskeleton

The fundamental structural unit of the eukaryotic cytoskeleton is a filamentous protein polymer composed of one of three chemically distinct subunits, actin, tubulin, or one of several classes of intermediate filament protein (Schliwa, 1986). The molecular structures of these filaments have been extensively characterized by light and electron microscopy, and in the case of actin filaments, a molecular model based on the crystal structure of the actin monomer is available (Lorenz *et al.*, 1995). The viscoelastic properties of macroscopic samples of each filament type have also been characterized by rheologic methods (Janmey, 1991a; Janmey *et al.*, 1991), and the microscopic elasticity of individual filaments, determined by micromanipulation or analysis of their thermal motions, is available in some cases (Gittes *et al.*, 1993; Isambert *et al.*, 1995; Kas *et al.*, 1993; Kojima *et al.*, 1994; Ott *et al.*, 1993). The three kinds of cytoskeletal polymers represent the three distinct classes of viscoelastic polymers: rodlike (microtubules), semiflexible (F-actin), and flexible (intermediate filaments) (Kas *et al.*, 1996). One implication of this difference is that macroscopic samples of each type of cytoskeletal filament will have radically different viscoelastic properties, and composite networks composed of more than one filament type may have unique material properties not achievable with a single type of polymer.

Intermediate filaments, of all types studied to date, are qualitatively more flexible than F-actin even though they have a larger diameter (10 nm). They are therefore better modeled as flexible polymers, although they are far stiffer than the random-coil polymers common in many synthetic polymer gels.

2. VISCOELASTICITY

2.1. Theory

Rheology is broadly defined as the study of deformation and flow. Viscoelasticity is a quantitative measure of the rate and extent of deformation (usually called strain) of a material in response to an imposed force (usually expressed as stress or force per area). In the case of a simple liquid, deformation is resisted by dissipating the energy applied to the material during viscous flow, where the resistance depends on the rate of deformation, and the strain can increase without limit. The ratio of the applied stress to the rate of strain of a liquid defines the viscosity η, and the material remains deformed after the stress is released. One has the relation $\sigma = \eta\, d\gamma/dt$, where σ is the stress, η is the viscosity, and γ is the strain.

In the other extreme of a perfectly elastic material, the force applied is resisted to a degree that depends only on the extent of deformation (strain), but not the rate, and the work done to deform the material is recovered as the sample returns to its original shape when the stress is removed. The ratio of stress to the strain produced in an elastic material is termed the elastic modulus, and when the stress is applied parallel to one surface of the material (a shear stress) the relevant quantity is denoted as G, the shear modulus: $\sigma = G\gamma$.

Nearly all biological materials are characterized by both viscous and elastic properties. Neither η nor G is constant, and both viscous and elastic contributions depend on the duration and magnitude of the deformation. A conventional method to measure such properties is by applying oscillatory shear deformations and measuring the energy stored or dissipated during such deformation to obtain the storage and loss shear moduli $G'(\gamma,\omega)$ and $G''(\gamma,\omega)$, respectively, where each quantity is a function of strain γ and time (or equivalently frequency, ω) (Janmey *et al.*, 1988). The viscoelasticity of an aqueous solution is dominated by the largest polymeric solutes, and in the case of the cytoplasm, these polymers are generally the components of the cytoskeleton.

2.2. Measurement Techniques

Most of the mechanical measurements of intermediate filaments (IFs) have been made on the relatively strong, partially dehydrated, and covalently bound materials composed largely of intact tissues or of processed keratins. There is a wealth of rheologic information related both to their biological functions and to their practical use (Baden *et al.*, 1973; Breuer, 1968; Christensen *et al.*, 1977). Most of the measurement methods and instrumentation used for such studies are not applicable to the much more delicate and dynamic networks formed by IFs in aqueous solutions at the concentrations typical of most cytoskeletons. This chapter will focus on recent measurements of reconstituted aqueous solutions and gels of IF proteins using methods designed to minimize sample perturbation.

Several different types of instrument have been applied to IF gels. The simplest method is that of low-shear viscometry done by monitoring the movement of a steel ball under the force of gravity (approximately 14 Pa shear stress) down a capillary tube filled with the IF sample. Such measurements have proven useful for following the kinetics of IF assembly and of neurofilament (NF) gelation (Eyer and Leterrier, 1988; Leterrier and Eyer, 1987). This method is perhaps the most convenient for initial studies and qualitative comparison, but the nonuniform stress field, the large strains imposed on the samples, and the strong shear-rate dependence of IF viscosity limit quantitative analyses. Two other methods that measure both the elastic and dissipative (viscous) resistance of gel deformation have used either free or forced oscillations of disc-shaped samples held between parallel plates (Janmey *et al.*, 1991; Leterrier *et al.*, 1996; Rogers *et al.*, 1995).

Oscillatory measurements of viscoelastic materials are made by placing samples between two parallel plates separated by a known distance. One of the plates is deformed by a small-amplitude torsional deformation, and the resistance to the deformation is measured by free or forced oscillations. For free oscillatory measurements, one plate of the sample is held fixed while the other is momentarily displaced through a small angular deformation. Subsequent measurements of the free rotational trajectory of the deformed plate can be used to calculate the rheologic parameters of the sample. For viscoelastic materials, such as cytoskeletal gels, the angular position after the deformation takes the form of a damped sinusoidal oscillation as seen in Figure 1. Intuitively, the oscillatory motion is due to the elasticity of the material and the damping of the oscillations is due to viscous losses in the sample. The combination of the frequency of oscillation and rate of damping can be related to the elastic (G') and loss (G'') moduli (Ferry, 1980) described above. Free oscillatory measurements are conducted on a torsion pendulum (Janmey, 1991b; Plazek *et al.*, 1958), a simple instrument that has been used for a variety of fragile biological materials, such as fibrin clots (Janmey *et al.*, 1983) and cell suspensions (Eichinger *et al.*, 1996). Forced oscillatory measurements apply a known sinusoidal stress or strain to one plate of the sample and measure the induced strain or stress, respectively, in the other plate. The stress and strain waveforms can be used directly to calculate the elastic (G') and loss (G'') moduli of the sample.

Forced oscillatory instruments have the advantage over free oscillatory instruments of probing a large range of frequencies of deformation. For many biomaterials, including IF networks, frequency-dependent measurements can be useful in probing the molecular mechanisms of viscoelasticity (Isambert and Maggs, 1996; MacKintosh *et al.*, 1995). Both forced and free oscillatory instruments allow measurements of elastic and loss moduli at increasing deformations and therefore the dependence of viscoelastic parameters on the extent of strain. These measurements are of particular interest for the study of IF networks, since it is at large strains that IF networks exhibit drastically different behavior when compared to other cytoskeletal elements (Janmey *et al.*, 1991; Leterrier *et al.*, 1996).

3. INTERMEDIATE FILAMENT VISCOELASTICITY: GENERAL PROPERTIES AND COMPARISON WITH OTHER BIOPOLYMERS

The rheologic feature that distinguishes IF gels from those formed by other cytoskeletal polymers is the relatively high compliance at low strains combined with their increasing stiffness at larger strain and the ability to withstand deformations that rupture networks formed by most other types of polymers. This feature is illustrated in Figure 1, which shows the free oscillations of a vimentin IF gel when measured by applying a small oscilla-

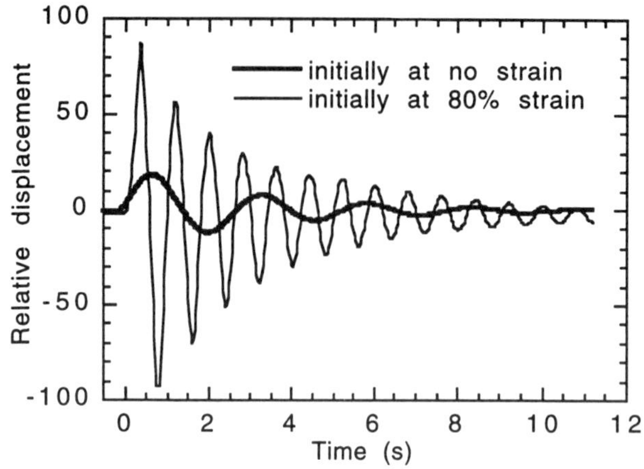

FIGURE 1. Deformation increases the stiffness of vimentin gels. Free oscillations of vimentin gels with and without imposition of a constant strain. The angular displacement of a torsion pendulum, proportional to the shear strain of the sample, is shown in arbitrary units after imposition of a momentary displacement of a few percent to a 1.6 mg/ml sample of polymerized vimentin initially at rest (heavy line) or held at a strain of 80% (thin line). The storage shear modulus G' is proportional to the square of the oscillation frequency.

tory strain (<5%) to a sample held initially at rest and when held at a large (80%) strain.

Figure 2 shows in more detail the dynamic shear modulus G' of a vimentin network at increasing strains and compares it to those of actin, fibrin, and microtubule networks. At small strains, F-actin (from rabbit skeletal muscle) has a much higher elastic storage modulus (by one order of magnitude) than an equal weight concentration of vimentin. At low strains, IF networks respond to mechanical deformations in a more viscous, softer manner than the highly elastic and relatively rigid actin filament networks. Similar results have been obtained from rheometric measurements of vimentin from different origins (*Xenopus* and cultured cells of human origin) and essentially the same is true for desmin (from mouse), another type III IF protein from muscle cells (see also Figure 3). Microtubules (from porcine brain, stabilized with taxol) show a slightly lower elastic storage modulus than vimentin at equal weight concentrations; however, the difference is less pronounced (a factor of 1.5, which is within experimental variation).

The resistance of IF networks to deformation changes dramatically when the sample is deformed to a constant extent prior to the measurements made by a small additional oscillatory deformation, that is, by prestraining the sample. At high degrees of strain, the storage moduli of

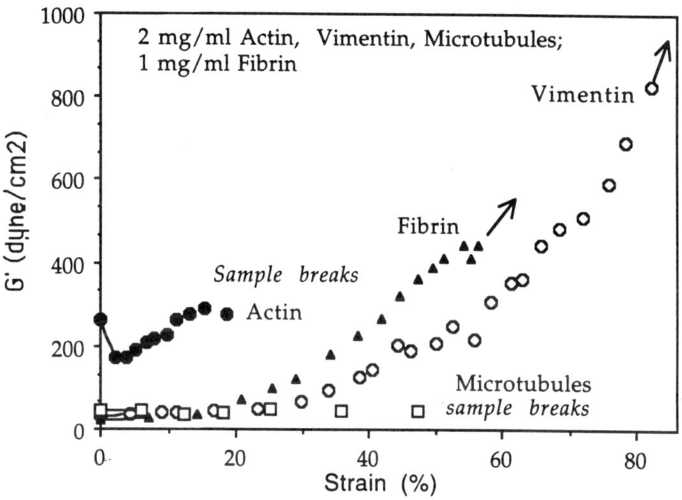

FIGURE 2. Differential responses of biopolymer networks to large deformations. The storage shear modulus G' determined from measurements similar to those shown in Figure 1 are compared for polymerized actin, tubulin, vimentin, and fibrin. Data are adapted from Janmey *et al.* (1991).

vimentin and desmin are even higher than those of comparable, unstrained samples of F-actin. Both F-actin and the very stiff and rigid microtubules would rupture when prestrained to such large extents. Cross-linking of actin filaments by the actin-binding protein ABP-280 to form highly elastic gels does not prevent rupture at large strains, suggesting that the fragility is inherent in the actin filament. IF networks are not only able to resist high strains, but they acquire intriguing, new mechanical properties when deformed, a phenomenon known as strain hardening. Figure 3 shows that both vimentin and desmin exhibit strain hardening, as do neurofilament gels (Leterrier *et al.*, 1996), demonstrating that strain hardening is a feature of all IFs tested. The molecular basis for strain hardening in not known, and could involve reorientation of individual filaments within the IF networks as well as intrafilament reorganization, possibly by movements of the subunit packing as detected by X-ray studies of stretched, dehydrated keratin (G. A. Wilson, 1972).

The biopolymer gels to which IFs bear the most general resemblance are those formed by the extracellular proteins fibrin (illustrated in Figure 2) and collagen. Both of these materials at approximately physiological concentrations form networks with elastic moduli somewhat larger than those of an IF network, but also exhibit high degrees of strain hardening. Since many types of epithelial and other cells interact with extracellular matrices

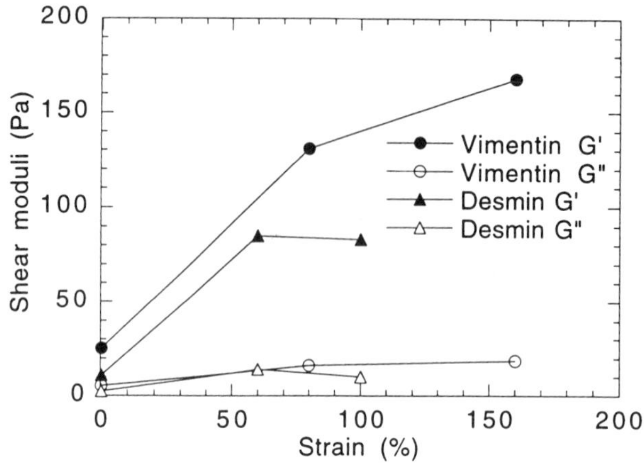

FIGURE 3. Strain dependence of vimentin and desmin networks. Storage (G', filled symbols) and loss (G'', open symbols) shear moduli of 1.6 mg/ml human vimentin (circles) and 0.75 mg/ml murine desmin (triangles) are shown at various extents of constant strain. Methods and conditions are given in Rogers *et al.* (1995).

by connections linking the cytoplasmic IF network to extracellular lattices via integrins or cadherins, the physical compatibility of both intracellular and extracellular materials may be intimately tied to their ability to accommodate mechanical stresses or to transmit mechanical signals.

4. ASSEMBLY KINETICS

Another intriguing difference between microfilaments and IFs is observed when the increase in the elastic storage modulus is monitored during filament assembly (Figure 4). Whereas F-actin reaches half-maximal elasticity only after 15–20 min upon the addition of salt (Mg^{2+},K^{+}), desmin and to a similar extent vimentin show a very rapid increase in G' upon addition of salt, which is almost completed after 5 min, followed by a very slow phase of continued increase, for 60 min and more. Gelation of neurofilaments qualitatively exhibits a similar slow phase of increasing storage modulus, but displays a larger net increase in G' over this phase (Leterrier *et al.*, 1996). In the case of the IFs the rise in elastic modulus is not clearly related to changes in filament length which dominate the elasticity of F-actin at these concentrations, but to increased numbers of contacts between filaments. This difference may point to major differences in the structural architecture of IF and F-actin networks, suggesting a greater degree of entanglement in the more flexible IFs.

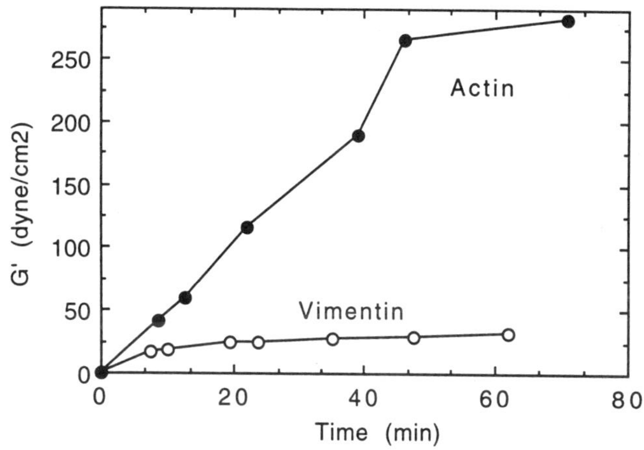

FIGURE 4. Vimentin networks develop elasticity earlier in the course of polymerization. The storage shear modulus *G'* of 2 mg/ml actin and vimentin, polymerized by addition of 2 mM $MgCl_2$ and 150 mM KCl (actin) or 150 mM KCl (vimentin), is measured by low-strain oscillatory measurements during the process of gelation.

5. STRUCTURAL BASIS OF IF VISCOELASTICITY: INSIGHTS FROM MUTATIONAL ANALYSIS

Vimentin and desmin belong to the type III IF proteins and show the classical three-domain structure: a non-α-helical domain at the N-terminus, a central α-helical rod, and a non-α-helical C-terminal domain. It has been shown that tail-truncated *Xenopus* vimentin is able to form IF *in vivo* and *in vitro* (Eckelt *et al.*, 1992). Rheologic measurements with a torsion pendulum on a series of vimentin mutants with varying degrees of tail truncation revealed no evidence for the involvement of the C-terminal non-α-helical domain in IF assembly. The viscoelastic properties of the mutants were essentially unchanged when compared to the wild type. These findings were confirmed by *in vivo* transfection studies (Rogers *et al.*, 1995).

In agreement with this finding, tail truncation of type I and II IF proteins does not interfere with assembly *in vitro* and *in vivo* (Bader *et al.*, 1991; Hatzfeld and Weber, 1990; A. K. Wilson *et al.*, 1992). The tail rather seems to be involved in width control of filament formation (Herrmann *et al.*, 1996). However, removal of the amino-terminal head domain yielded assembly-incompetent proteins (Herrmann *et al.*, 1996). In conclusion, the central, stiff rod domain seems to determine, together with the more flexible head domain, the overall viscoelastic properties of assembled filamentous networks.

6. INTEGRATED MECHANICAL PROPERTIES

6.1. IF–Actin Mixtures

Among the most interesting features of some IF types is their interaction with other cytoskeletal filaments both *in vivo* and *in vitro*. It has long been shown that disruption of the actin or microtubule system can in many cases also lead to IF network disruption, and vice versa. Structures linking these filaments are beginning to be identified by biochemical and genetic studies. Two examples of such interactions with direct effects on rheology are evident in mixtures of either vimentin or NFs with F-actin.

Mixed macroscopic solutions of two different types of cytoskeletal filaments, when assayed *in vitro* for their viscoelastic properties at low strains, showed no cooperative effects as compared to solutions of only one polymer type alone (Figure 5). The overall viscoelasticity in vimentin/F-actin mixtures seemed to be determined largely by the F-actin properties. Vimentin filaments, if anything, have a softening effect on the actin filament network.

However, when shorter, gelsolin-capped actin filaments are mixed with vimentin and subjected to a range of strains, so that the intrinsic elasticity of F-actin does not dominate overall rheology, interesting features characteristic of both filament types that cannot be achieved by either protein alone are observed. Figure 6 shows stress–strain curves of equal concentrations of F-actin and vimentin and compares them to a mixture of both filaments polymerized together. The composite network shows features of each type of filament. Part of the elastic resistance of the vimentin–actin composite is lost at strains around 100%, consistent with rupture or slippage of the short actin filaments, but significant elastic resistance persists at very large strains characteristic of vimentin networks.

When short (0.7 μm) actin filaments are present during NF gelation, the resulting gel is weaker than that formed by purified NF alone (Figure 7). This interaction illustrates that filament–filament interactions can have specific effects other than steric and linkage effects that tend to strengthen composite networks. A possible explanation of the weakening effect seen with NFs and short actin filaments is that an association between F-actin and NF sidearms competes with the self-association of these cross-bridging NF extensions and thereby reduces the number of elastically active cross-links between NFs that dominate some aspects of NF rheology. The possibility that such links between NFs and either F-actin or microtubules (Runge *et al.*, 1981) may be specifically regulated by phosphorylation and other signaling mechanisms may reveal part of the complex dynamics of these systems. Moreover, the recently discovered cellular factors with binding capacities for both actin and IFs (e.g., BPAG-1; Yang *et al.*, 1996)

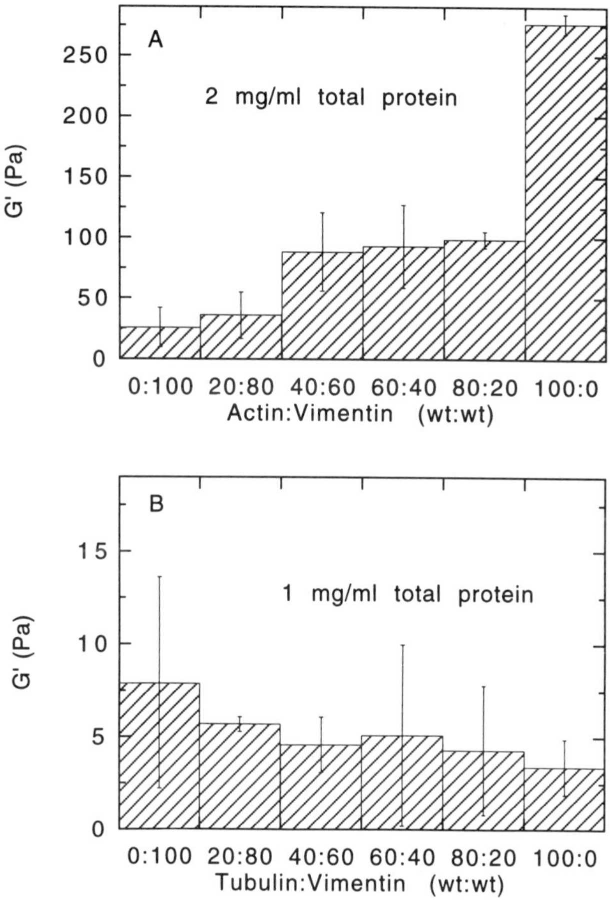

FIGURE 5. Rheology of mixtures containing two types of cytoskeletal filaments. Vimentin was copolymerized with either actin (A) or tubulin (B) at constant total protein concentration at the weight ratios shown. Elastic moduli were measured using a torsion pendulum as shown in Figure 1.

might drastically alter the mechanical properties of mixed systems by adding stable, regulated cross-links to the otherwise only loosely connected polymers. These interfilament cross-linking proteins might create compound materials with novel viscoelastic properties.

6.2. Vimentin–Microtubule Composite Networks

In mixtures of purified vimentin with microtubules (MTs), no cooperativity is observed at varying compositions of the mixed system (Figure

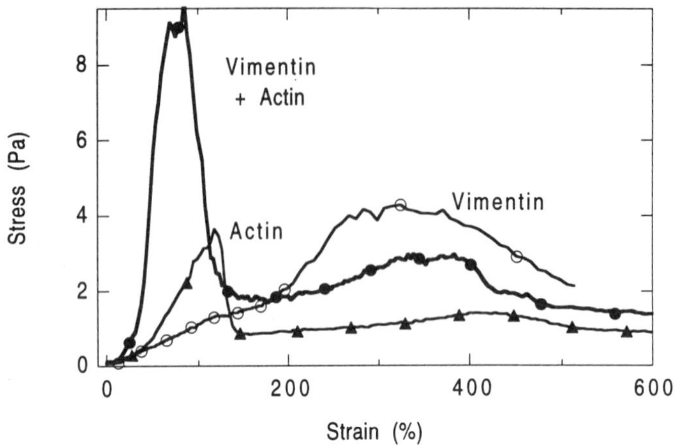

FIGURE 6. Rheology of an actin–vimentin composite network at large strains. Vimentin was polymerized alone or with actin in a solution that also contained gelsolin (ratio of gelsolin to actin, 1:2000) to limit actin filament length to an average below 6 μm. Increasing strain was applied using a Rheometrics RFSII instrument at a rate of 0.5% sec^{-1} and the resulting stress is shown for purified vimentin (open circles), actin:gelsolin (triangles), or the mixture of the two filament types (closed circles).

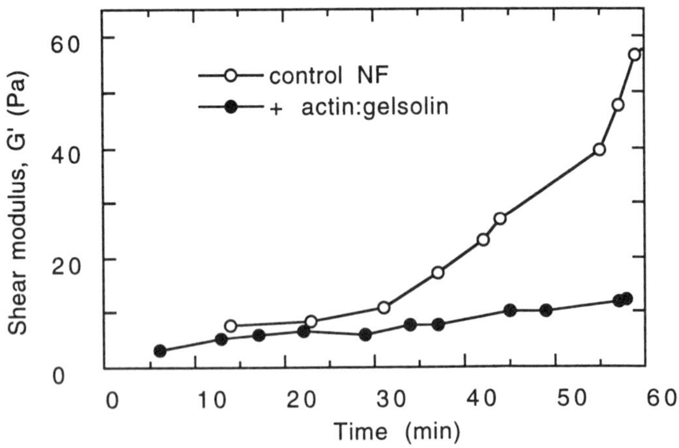

FIGURE 7. Effect of actin of neurofilament gelation. The increase of G' during gelation of 2.6 mg/ml neurofilaments is shown with and without the presence of 1.7 mg/ml actin: gelsolin 200:1 (0.7 μm average length). Methods and conditions are given in Leterrier *et al.* (1996).

5B). The apparently noninteracting mixtures of these purified filaments *in vitro* contrast with their frequent functional dependence *in vivo*, and suggest that, as with actin filaments, specific proteins (Gurland and Gundersen, 1995; Gyoeva and Gelfand, 1991; Svitkina *et al.*, 1996), perhaps microtubule-associated proteins (MAPs) (Hirokawa *et al.*, 1988; Leterrier *et al.*, 1982), linking these two filament types remain to be described.

7. STRUCTURAL ORIGINS OF IF VISCOELASTICITY

The molecular basis of IF viscoelasticity, notably its compliance at small strains and extreme ability to withstand large deformations, may be related to the rather loosely structured connections between IF subunits that have made elucidation of their packing by diffraction methods difficult. As demonstrated by a number of electron and fluorescence microscope studies, IF polymers are notably more flexible than either F-actin or MTs. Recent theoretical treatments (Kroy and Frey, 1996; MacKintosh *et al.*, 1995) of semiflexible polymers such as the cytoskeletal filaments predict that the low-strain shear modulus depends on approximately the square of the individual polymer stiffness, assuming an otherwise topologically identical cross-linked network. The extent of deformation that can be accommodated at increasing strains likewise depends on stiffness because the flexibility of the filaments determines the excess contour length of polymer present between adjacent cross-link or entanglement points. Once the slack has been pulled out of the polymer, other modes of reorganization must take place as the deformation increases. In the case of F-actin and MTs, this deformation leads to collapse of the elastic network, but in the case of IF gels, it does not. As the strain is increased, extensions and slippage of molecular contacts may be possible in IFs, but not in F-actin or microtubules, where the contacts between their globular subunits cannot accommodate spatial rearrangements. An alternative possibility is that accommodation to strain involves the unfolding of secondary structures within the IF subunit, as has recently been shown to occur for the even more dramatically strain-accommodating protein titin (Maruyama, 1997; Tskhovrebova *et al.*, 1997).

8. CONCLUSIONS AND FUTURE DIRECTIONS

The science of cells is justifiably dominated by cell and molecular biologists who decipher the extraordinarily complex networks of biochemical pathways that maintain cellular integrity and function. But cells are physical entities as well as biochemical factories whose mechanical properties are established and maintained by the elements of the cytoskeleton. However, the study of the physics of cells (rather than their chemistry) is still in its infancy. Nevertheless, as outlined in this chapter, many of the basic features of the rheologic properties of cytoskeletal fibers have already been determined, and we have learned that intermediate filaments possess intriguing and indeed unique mechanical properties. A major goal for the near future should be the analysis of the molecular basis of the fascinating property of intermediate filaments to withstand large deformations that rupture networks formed by other cytoskeletal fibers. It can only be speculated what the advantages might be that this property imparts to cells. A second promising area for future research is the study of composite networks formed by two or more cytoskeletal elements, and the effect associated proteins or cross-linkers have on these mixtures. Of course, random polymer networks are only a weak representation of the highly ordered cytoskeleton of a living cell, but we need to start with *in vitro* systems and models that are experimentally feasible and comprehensible. What we have learned so far seems to suggest that intermediate filaments play a key role in the orchestra of cytoskeletal elements that determine the mechanical properties of the cell.

9. REFERENCES

Baden, H. P., Goldsmith, L. A., and Fleming, B., 1973, A comparative study of the physicochemical properties of human keratinized tissues, *Biochim. Biophys. Acta* **322:**269–278.

Bader, B. L., Magin, T. M., Freudenmann, M., Stumpp, S., and Franke, W. W., 1991, Intermediate filaments formed *de novo* from tail-less cytokeratins in the cytoplasm and in the nucleus, *J. Cell Biol.* **115:**1293–1307.

Bloom, S., Lockard, V. G., and Bloom, M., 1996, Intermediate filament-mediated stretch-induced changes in chromatin: A hypothesis for growth initiation in cardiac myocytes, *J. Mol. Cell. Cardiol.* **28:**2123–2127.

Breuer, M. M., 1968, Strain–stress isotherm of keratin fibers, *Biopolymers* **6:**1503–1506.

Christensen, M. S., Hargens, C. W. d., Nacht, S., and Gans, E. H., 1977, Viscoelastic properties of intact human skin: Instrumentation, hydration effects, and the contribution of the stratum corneum, *J. Invest. Dermatol.* **69:**282–286.

Eckelt, A., Herrmann, H., and Franke, W. W., 1992, Assembly of a tail-less mutant of the intermediate filament protein, vimentin, *in vitro* and *in vivo*, *Eur. J. Cell Biol.* **58:**319–330.

Eichinger, L., Koppel, B., Noegel, A. A., Schleicher, M., Schliwa, M., Weijer, K., Witke, W., and Janmey, P. A., 1996, Mechanical perturbation elicits a phenotypic difference between *Dictyostelium* wild-type cells and cytoskeletal mutants, *Biophys. J.* **70:**1054–1060.

Eyer, J., and Leterrier, J. F., 1988, Influence of the phosphorylation state of neurofilament proteins on the interactions between purified filaments *in vitro*, *Biochem. J.* **252:**655–660.

Ferry, J., 1980, *Viscoelastic Properties of Polymers*, 3rd ed., Wiley, New York.

Fuchs, E., 1996, The cytoskeleton and disease: Genetic disorders of intermediate filaments, *Annu. Rev. Gen.* **30:**197–231.

Gittes, F., Mickey, B., Nettleton, J., and Howard, J., 1993, Flexural rigidity of microtubules and actin filaments measured from thermal fluctuations in shape, *J. Cell Biol.* **120:**923–934.

Glogauer, M., Ferrier, J., and McCulloch, C. A., 1995, Magnetic fields applied to collagen-coated ferric oxide beads induce stretch-activated Ca^{2+} flux in fibroblasts, *Am. J. Physiol.* **269:**C1093–C1114.

Gurland, G., and Gundersen, G. G., 1995, Stable, detyrosinated microtubules function to localize vimentin intermediate filaments in fibroblasts, *J. Cell Biol.* **131:**1275–1290.

Gyoeva, F. K., and Gelfand, V. I., 1991, Coalignment of vimentin intermediate filaments with microtubules depends on kinesin, *Nature* **353:**445–448.

Hatzfeld, M., and Weber, K., 1990, Tailless keratins assemble into regular intermediate filaments *in vitro*, *J. Cell Sci.* **99:**317–324.

Herrmann, H., Haner, M., Brettel, M., Muller, S. A., Goldie, K. N., Fedtke, B., Lustig, A., Franke, W. W., and Aebi, U., 1996, Structure and assembly properties of the intermediate filament protein vimentin: The role of its head, rod and tail domains, *J. Mol. Biol.* **264:**933–953.

Hirokawa, N., Hisanaga, S., and Shiomura, Y., 1988, MAP2 is a component of crossbridges between microtubules and neurofilaments in the neuronal cytoskeleton: Quick-freeze, deep-etch immunoelectron microscopy and reconstitution studies, *J. Neurosci.* **8:**2769–2779.

Ingber, D., and Folkman, J., 1989, Tension and compression as basic determinants of cell form and function: Utilization of a cellular tensegrity mechanism, in *Cell Shape: Determinants, Regulation, and Regulatory Role* (W. Stein and F. Bronner, eds.), pp. 3–31, Academic Press, San Diego.

Isambert, H., and Maggs, A., 1996, Dynamics and rheology of actin solutions, *Macromolecules* **29:**1036–1040.

Isambert, H., Venier, P., Maggs, A. C., Fattoum, A., Kassab, R., Pantaloni, D., and Carlier, M. F., 1995, Flexibility of actin filaments derived from thermal fluctuations. Effect of bound nucleotide, phalloidin, and muscle regulatory proteins, *J. Biol. Chem.* **270:**11437–11444.

Janmey, P. A., 1991a, Mechanical properties of cytoskeletal polymers, *Curr. Opin. Cell Biol.* **3:**4–11.

Janmey, P. A., 1991b, A torsion pendulum for measurement of the viscoelasticity of biopolymers and its application to actin networks, *J. Biochem. Biophys. Meth.* **22:**41–53.

Janmey, P., Amis, E., and Ferry, J., 1983, Rheology of fibrin clots. VI. Stress relaxation, creep, and differential dynamic modulus of fine clots in large shearing deformations, *J. Rheol.* **27:**135–153.

Janmey, P. A., Hvidt, S., Peetermans, J., Lamb, J., Ferry, J. D., and Stossel, T. P., 1988, Viscoelasticity of F-actin and F-actin/gelsolin complexes, *Biochemistry* **27:**8218–8227.

Janmey, P. A., Euteneuer, U., Traub, P., and Schliwa, M., 1991, Viscoelastic properties of vimentin compared with other filamentous biopolymer networks, *J. Cell Biol.* **113:**155–160.

Kas, J., Strey, H., Barmann, M., and Sackmann, E., 1993, Measurement of the wave-vector-dependent bending stiffness of freely flickering actin filaments, *Europhys. Lett.* **21:**865–870.

Kas, J., Strey, H., Tang, J. X., Finger, D., Ezzell, R., Sackmann, E., and Janmey, P. A., 1996, F-Actin, a model polymer for semiflexible chains in dilute, semidilute, and liquid crystalline solutions, *Biophys. J.* **70:**609–625.

Kojima, H., Ishijima, A., and Yanagida, T., 1994, Direct measurement of stiffness of single actin filaments with and without tropomyosin by *in vitro* nanomanipulation, *Proc. Natl. Acad. Sci. USA* **91:**12962–12966.

Kroy, K., and Frey, E., 1996, Force–extension relation and plateau modulus for wormlike chains, *Phys. Rev. Lett.* **77:**306–309.

Leterrier, J. F., and Eyer, J., 1987, Properties of highly viscous gels formed by neurofilaments *in vitro*. A possible consequence of a specific inter-filament cross-bridging, *Biochem. J.* **245:**93–101.

Leterrier, J. F., Liem, R. K., and Shelanski, M. L., 1982, Interactions between neurofilaments and microtubule-associated proteins: A possible mechanism for intraorganellar bridging, *J. Cell Biol.* **95:**982–986.

Leterrier, J. F., Kas, J., Hartwig, J., Vegners, R., and Janmey, P. A., 1996, Mechanical effects of neurofilament cross-bridges. Modulation by phosphorylation, lipids, and interactions with F-actin, *J. Biol. Chem.* **271:**15687–15697.

Li, H., Choudhary, S. K., Milner, D. J., Munir, M. I., Kuisk, I. R., and Capetanaki, Y., 1994, Inhibition of desmin expression blocks myoblast fusion and interferes with the myogenic regulators MyoD and myogenin, *J. Cell Biol.* **124:**827–841.

Li, Z., Colucci-Guyon, E., Pincon-Raymond, M., Mericskay, M., Pournin, S., Paulin, D., and Babinet, C., 1996, Cardiovascular lesions and skeletal myopathy in mice lacking desmin, *Dev. Biol.* **175:**362–366.

Lorenz, M., Poole, K. J., Popp, D., Rosenbaum, G., and Holmes, K. C., 1995, An atomic model of the unregulated thin filament obtained by X-ray fiber diffraction on oriented actin–tropomyosin gels, *J. Mol. Biol.* **246:**108–119.

MacKintosh, F., Kas, J., and Janmey, P., 1995, Elasticity of semiflexible biopolymer networks, *Phys. Rev. Lett.* **75:**4425–4428.

Maniotis, A. J., Chen, C. S., and Ingber, D. E., 1997, Demonstration of mechanical connections between integrins, cytoskeletal filaments, and nucleoplasm that stabilize nuclear structure, *Proc. Natl. Acad. Sci. USA* **94:**849–854.

Maruyama, K., 1997, Connectin/titin, giant elastic protein of muscle, *FASEB J.* **11:**341–345.

Milner, D. J., Weitzer, G., Tran, D., Bradley, A., and Capetanaki, Y., 1996, Disruption of muscle architecture and myocardial degeneration in mice lacking desmin, *J. Cell Biol.* **134:**1255–1270.

Ott, A., Magnasco, M., Simon, A., and Libchaber, A., 1993, Measurement of the persistence length of polymerized actin using fluorescence microscopy, *Phys. Rev. E* **48:**r1642–r1645.

Plazek, D., Vrancken, M., and Berge, J., 1958, A torsion pendulum for dynamic and creep measurements of soft viscoelastic materials, *Trans. Soc. Rheol.* **2:**39–51.

Rogers, K. R., Eckelt, A., Nimmrich, V., Janssen, K. P., Schliwa, M., Herrmann, H., and Franke, W. W., 1995, Truncation mutagenesis of the non-alpha-helical carboxyterminal tail domain of vimentin reveals contributions to cellular localization but not to filament assembly, *Eur. J. Cell Biol.* **66:**136–150.

Runge, M., Laue, T., Yphantis, D., Lifsics, M., Saito, A., Altin, M., Reinke, K., and Williams, R., 1981, ATP-induced formation of an associated complex between microtubules and neurofilaments, *Proc. Natl. Acad. Sci. USA* **78:**1431–1435.

Schliwa, M., 1986, *The Cytoskeleton: An Introductory Survey*, Springer-Verlag, Vienna.

Svitkina, T. M., Verkhovsky, A. B., and Borisy, G. G., 1996, Plectin sidearms mediate interaction of intermediate filaments with microtubules and other components of the cytoskeleton, *J. Cell Biol.* **135:**991–1007.

Traub, P., 1995, Intermediate filaments and gene regulation, *Physiol. Chem. Phys. Med. NMR* **27:**377–400.

Traub, P., and Shoeman, R. L., 1994, Intermediate filament proteins: Cytoskeletal elements with gene-regulatory function? *Int. Rev. Cytol.* **154:**1–103.

Tskhovrebova, L., Trinick, J., Sleep, J. A., and Simmons, R. M., 1997, Elasticity and unfolding of single molecules of the giant muscle protein titin, *Nature* **387:**308–312.

Wilson, A. K., Coulombe, P. A., and Fuchs, E., 1992, The roles of K5 and K14 head, tail, and R/K L L E G E domains in keratin filament assembly *in vitro*, *J. Cell Biol.* **119:**401–414.

Wilson, G. A., 1972, Low-angle x-ray diffraction studies of the effect of extension on macromolecular organisation in keratin, *Biochim. Biophys. Acta* **278:**440–449.

Yang, Y., Dowling, J., Yu, Q. C., Kouklis, P., Cleveland, D. W., and Fuchs, E., 1996, An essential cytoskeletal linker protein connecting actin microfilaments to intermediate filaments, *Cell* **86:**655–665.

Chapter 13

Transient Electric Birefringence in the Study of Intermediate Filament Assembly

Herbert van Amerongen, Martin Kooijman, and Michael Bloemendal

1. TRANSIENT ELECTRIC BIREFRINGENCE AND THE CYTOSKELETON—PERSPECTIVE

The cytoskeleton of most vertebrate cells contains fibrillar protein structures of various thickness. On the basis of their diameter *d* they are called microtubules (*d* about 5 nm), intermediate filaments (*d* about 10 nm), or microfilaments (*d* about 25 nm). The microtubules and microfilaments have been shown to be involved in a variety of cellular events, including cell division, locomotion, protoplasmic streaming, cellular polarity, and anchorage (Bloemendal and Pieper, 1989). For intermediate filaments (IFs) no general function, apart from a structural one leading to mechanical integrity, has unequivocally been proven. Some diseases have been corre-

Herbert van Amerongen, Martin Kooijman, and Michael Bloemendal Department of Physics and Astronomy, and Institute for Molecular Biological Sciences, Vrije Universiteit, 1081 HV Amsterdam, The Netherlands. *Present address for MK:* Centre for Protein Technology TNO-WAU, 6700 EV Wageningen, The Netherlands.

Subcellular Biochemistry, Volume 31: Intermediate Filaments, edited by Herrmann and Harris. Plenum Press, New York, 1998.

lated with defects in IF proteins, but these seem to be caused by disrupted filament structure (Fuchs and Weber, 1994). A role in differentiation (Heins and Aebi, 1994) or nucleic acid binding (Shoeman and Traub, 1990; Kooijman *et al.,* 1992) also has been suggested, but the evidence is far from conclusive.

The common feature of the three types of fibrillar structure in the cytoskeleton is the formation of chains and networks. Whatever their function, it is related to or affected by assembly and disassembly of the filaments. Thus, much effort has been put into the unraveling of filamentous structure and dynamics (for recent reviews see Heins and Aebi, 1994; Fuchs and Weber, 1994; Quinlan *et al.*, 1995), mostly by biochemical cross-linking, electron microscopy, and gel-permeation chromatography. An alternative technique that can be used to this end is transient electric birefringence (TEB), which can provide information on molecular dimensions (and hence aggregational size), intramolecular flexibility, electric properties, and optical anisotropy. The added value of TEB is mainly its high sensitivity to the size of the particles, in particular when these particles are rod-shaped like IF proteins. In the past TEB has been used to study the structure of filamentous proteins like bacterial flagellar filaments (Gerber *et al.*, 1975), actin (Kobayashi *et al.*, 1964), and myosin (Highsmith *et al.*, 1977; Highsmith and Eden, 1985, 1986). Recently we have applied TEB to determine electric and hydrodynamic properties of type III IF molecules at several stages of assembly. From these data we were able to distinguish between suggested models for IF structure and to characterize intermediate steps in the aggregational building.

In this chapter we present the potentials of TEB for IF research. After a general introduction about the technique of TEB, we discuss its application for the study of IF protein structure. Advantages and disadvantages of the method are given. The chapter ends with a model for IF aggregation based on TEB measurements, conclusions, and prospects.

2. TRANSIENT ELECTRIC BIREFRINGENCE—THE TECHNIQUE

2.1. Historical Overview

The technique of transient electric birefringence has a long history, nicely overviewed by Fredericq and Houssier (1973). The phenomenon of electric birefringence was first observed in 1875 by Kerr (Kerr, 1875), who noticed that certain solutions exhibit a difference in refractive index for light polarized parallel and perpendicular to the direction of an applied electric field. In 1880 he discovered that the electric birefringence is proportional to the square of the field strength (Kerr, 1880). Later it was established that this so-called Kerr law only holds at low field strength (Bergholm and Björnstahl, 1920). The theoretical explanation for the phenomenon was given by Langevin (1910) and Born (1918), who described the birefringence in terms of the orientation of optically anisotropic molecules under the action of the electric field. Already in the 1930s electric birefringence was used for biological research; for example, Bawden *et al.* (1936) and Lauffer (1939) studied the large tobacco mosaic virus, which is very suitable for electric birefringence measurements because of its large size and strong asymmetry. However, it was only after the development of pulsed fields for electric birefringence measurements (Kaye and Devaney, 1947; Tolstoi and Feofilov, 1949; Benoit, 1949; O'Konski and Zimm, 1950) that the phenomenon could be widely monitored and interpreted for biological macromolecules like DNA and proteins in aqueous solutions. Recent results are reviewed by Hagerman (1996).

2.2. Theoretical Background

Solutions can become birefringent by the application of an electric field E due to ordering of the molecules of the solution by the electrostatic interaction between the field and the dipole moments of the molecules. The dipole moment may be an intrinsic property of the molecules (the permanent dipole moment μ_p), induced by the electric field ($\mu_{ind} = \alpha E$, where α is the electric polarizability), or a combination of the two. For example, water molecules are oriented by their permanent dipole moment, double-stranded DNA by its induced dipole moment. If the molecules are optically anisotropic, which is the case for almost all macromolecules, their orientation due to the applied field will cause the solution to become optically anisotropic and therefore birefringent: the refractive index $n_{//}$ for light polarized parallel to the electric field is different from the refractive index $n_{\perp}$ for light polarized perpendicular to the field. The difference $\Delta n = n_{//} - n_{\perp}$ is called the (amount of) birefringence and can be measured accurately (see Section 2.3). Although even pure water shows measurable electric birefringence, the effect is often many orders of magnitude larger for solutions of macromolecules.

In order to facilitate the discussion in this chapter we assume that the complexes under study can be considered as rods with the long axis being an axis of rotational symmetry. The long axis will be called the *a* axis and the short one the *b* axis. For the discussion of the results of the elongated IF protein structures with their coiled-coil basic units, this assumption is justified and the following treatment suffices. For a detailed discussion of the theory of electric birefringence we refer to Fredericq and Houssier (1973) and references therein.

The electric birefringence for rods can be written as a product of four terms:

$$\Delta n = \kappa C(g_a - g_b)\Phi(E) \tag{2.1}$$

where $\varkappa$ is a proportionality constant, and C is the molar concentration of the molecules contributing to the birefringence. The term $(g_a - g_b)$ is proportional to the intrinsic anisotropy of the contributing molecules and we will simply call it the intrinsic anisotropy. Here Φ is the orientation factor, which indicates how well the molecules are oriented, i.e., how well they have lined up with the electric field. Its value is zero for an isotropic solution, i.e., in the absence of an electric field. Upon raising the field strength E, the orientation factor increases and at very high values of E, Φ saturates and approaches the limiting value of 1. If, for instance, two rodlike molecules associate to form a rodlike dimer, the concentration C is reduced by a factor of two, whereas the intrinsic anisotropy increases by a factor of two. Whether the electric birefringence will change will thus depend on what happens to Φ. The dependence of Φ on E was studied extensively by O'Konski *et al.* (1959) and it was shown that for low field strength

$$\Phi(E) \sim (P+Q)E^2 \tag{2.2}$$

with

$$P = \mu_p^2/k^2T^2 \tag{2.3}$$

and

$$Q = (\alpha_a - \alpha_b)/kT \tag{2.4}$$

where k and T are the Boltzmann constant and the absolute temperature, respectively, and α_a and α_b are the electric polarizabilities along the a and b axes, respectively. Note that for sufficiently low field strengths the orientation factor and thus the birefringence scale linearly with the square of the electric field (Kerr's law). This allows the definition of the Kerr constant $K = (\Delta n)/E^2$, which is proportional to $C\ (g_a - g_b)(P + Q)$. The range of E values where the law holds is called the Kerr regime. Also note the occurrence of the thermal energy kT in these expressions, reflecting the fact that Brownian motion tends to disorient the molecules. Q is proportional to the induced dipole moment and hence to the polarizability, which can have different origins: (1) All the charges in a molecule contribute to its polarizability. This causes the intrinsic polarizability, which scales more or less linearly with the mass of the molecules. (2) In the case of charged rods, like, e.g., DNA, there are usually counterions surrounding the molecule. These also contribute to the polarizability and this contribution scales more or less with the surface area of the molecules. Hence, in general aggregation leads to an increased value of Q (and thus of μ_{ind} and Δn). In contrast, the dependence of the contribution of the permanent dipole moment component (P) on aggregation is less straightforward, since the total dipole moment is the vectorial sum of the individual dipole moments. Hence, antiparallel aggregation leads to cancellation of P, whereas parallel aggregation leads to a doubling of P.

Knowledge about P and Q can provide relevant information about the aggregation of IF proteins. However, from equations (1) and (2) it appears that Δn is proportional to the sum of P and Q only. In order to separate the contributions of P and Q, use can be made of TEB. Transient electric birefringence is measured by applying short, high-field pulses across a sample (see Figure 1). As soon as the field is switched on, the molecules start to orient toward a quasiequilibrium state where the electrostatic orienting forces and the disorienting Brownian motions balance each other. The corresponding birefringence is termed Δn_s. After the field is switched off, only the disorienting Brownian motion remains and the birefringence disappears. It can be shown (Fredericq and Houssier, 1973) that, as long as E is within the Kerr regime, P/Q is related to the ratio of the areas enclosed by the rise and decay curves (S_1 and S_2 in Figure 1) as

$$\frac{S_1}{S_2} = \frac{4(P/Q) + 1}{P/Q + 1} \tag{2.5}$$

The decay of the TEB signal is also a very useful source of information about the dimensions of molecules and complexes, since it depends directly on the rotational diffusion coefficient D_r. For rodlike molecules the decay of the birefringence signal is single exponential in form and it is described by

$$\Delta_n(t) = \Delta n_s e^{-D_r t} \tag{2.6}$$

where the time t starts when the field is switched off.

Accurate hydrodynamic expressions are available which relate D_r to the dimensions of the molecules or complexes. A general expression for a rigid rod is given by (Broersma, 1960)

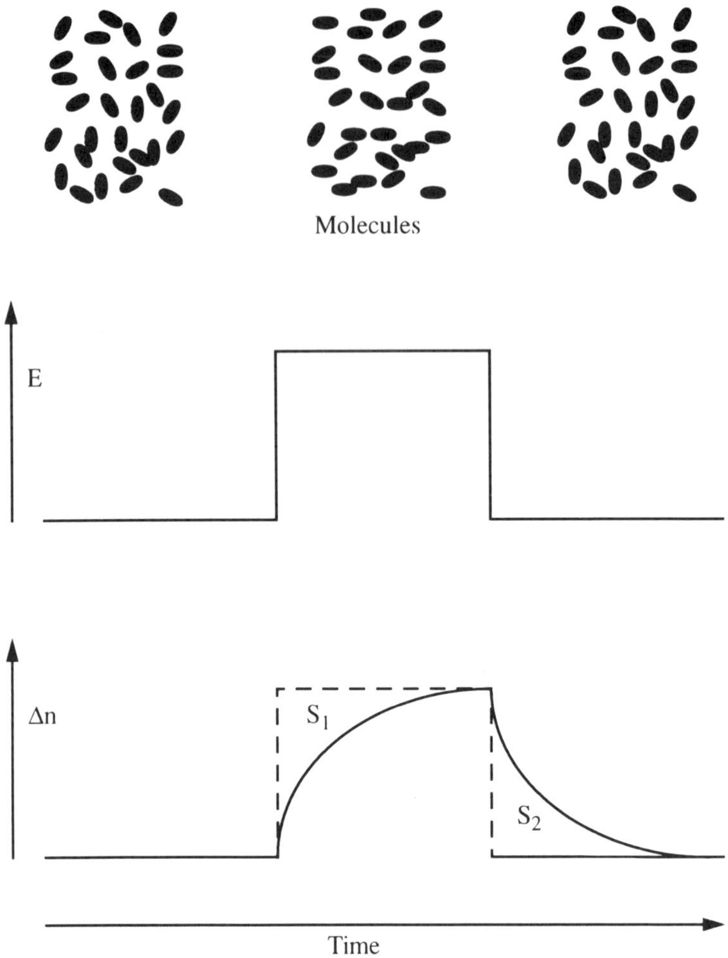

FIGURE 1. Electric field-induced transient birefringence: the molecules (elliptical blobs), the electric field (E), and the birefringence (Δn) as function of time. The areas S_1 and S_2 are the areas enclosed by the rise and decay curves, respectively, from which the ratio P/Q can be calculated.

$$D_r = \frac{3kT}{\eta_0 \pi L^3}\left[\ln\left(\frac{L}{d}\right) - \gamma_R\right] \tag{2.7}$$

where L and d are the length and the diameter of the rod, respectively, η_0 is the solvent viscosity, and γ_R is a frictional parameter that accounts for end effects. Two expressions based on comparison with experimental results, and that have appeared in literature most recently, are those by Broersma (1981)

$$\gamma_R = 0.757 - 7\left[\frac{1}{\ln(2L/d)} - 0.27\right]^2 \tag{2.8}$$

and by Mandelkern *et al.* (1981)

$$\gamma_R = 0.577 + 6.51\left[\frac{1}{\ln(2L/d)} - 0.49\right]^2 \tag{2.9}$$

Both expressions, which lead to similar results, will be used in the rest of this chapter.

From equations (2.7)–(2.9) it appears that the rotational diffusion depends strongly on the length and much less on the diameter of the molecules. Hence, when the diameter is known within certain limits, it is possible to obtain from TEB a rather accurate estimate of the length. However, this also implies that the diameter should be known from other techniques and thus that TEB is not sufficient on its own to determine unambiguously the sizes of molecules.

2.3. Experimental Background

2.3.1. Cell and Sample

A schematic view of the experimental setup is given in Figure 2. At the heart of the setup is the Kerr cell which contains the sample. As the birefringence of a sample is proportional to the length of the cell (see below), rather long cells are used (almost 5 cm in our case). The signal increases linearly with the concentrations of the samples. However, too high concentrations cannot be used, as this leads to hindered rotation of the molecules. Therefore, the concentration dependence of observed decay times always has to be checked.

2.3.2. The Laser

The cell has glass windows on opposite sides to allow light from a laser to shine through the cell onto a detector (photomultiplier), which is connected to a transient recorder. Wavelength and power of the laser are not very critical for the experiments, provided that no absorption of the light occurs. (In our experiments we have used an argon-ion laser at 514.5 nm operating at 100–200 mW). Laser stability between differ-

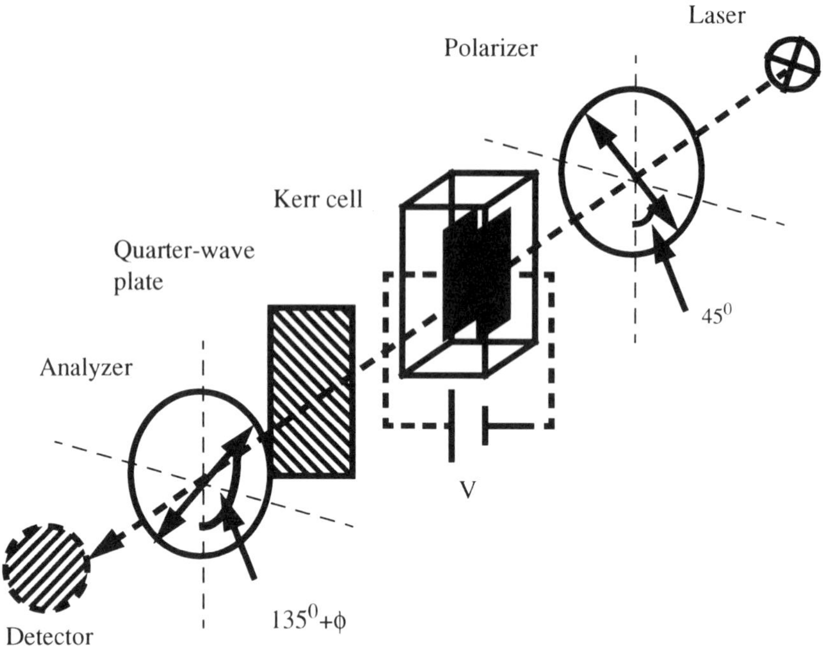

FIGURE 2. Schematic drawing of the TEB setup. Light from the laser is directed toward the detector and its path is indicated by the dashed line. The two polarizers (polarizer and analyzer) are almost crossed. Applying a high voltage *V* to the electrodes in the Kerr cell leads to partial orientation of the IF proteins in the cell. The IF-containing solution becomes birefringent and the amount of light on the detector changes. For the role of the quarter-wave plate see text and indicated reference.

ent pulses is not required, but stability of the intensity during the pulse is essential.

2.3.3. The Electric Field

The cell contains two electrodes that are positioned perpendicular to the windows and parallel to each other. In our case the distance between them is 1.63 mm. This short distance is used to obtain a high field. The electrodes are connected to a high-voltage pulse unit, which gives rectangular pulses of microseconds to milliseconds in length and up to 2500 V in strength. Every pulse causes a current through the sample, thereby heating the preparation, which may damage the biomolecules. Therefore, the repetition frequency of the pulses is kept low (several seconds between pulses). Moreover, the electric current depends on the salt concentrations in the sample, which consequently cannot be too high (in our setup the salt concentration has to be kept below 15 mM; we used low concentrations of Mg^{2+} to induce aggregation of IFs). In order to obtain accurate values for the decay time and S_1/S_2 [see equation (2.5)], it is important that the rise and decay time of the electric field pulses are sufficiently small. They are always kept below 0.1 μsec.

2.3.4. The Polarizers

Before the light hits the sample it passes through a polarizer P that polarizes the light at an angle of 45 deg with respect to the direction of the electric field. Behind the sample is another polarizer (called the analyzer, A), which is almost crossed with respect to polarizer P. In the absence of an electric field, no light reaches the detector when the polarizers are perfectly crossed. When an electric pulse is applied to the sample, the molecules orient and the refractive index parallel to the electric field usually becomes different from the one perpendicular to it (birefringence Δn). The incoming light vector can be decomposed into two equal components parallel and perpendicular to the electric field. Because the two components experience a different refractive index in the sample due to the birefringence, they will leave the sample with a phase difference δ between them which is equal to

$$\delta(t) = \frac{2\pi l}{\lambda} \Delta n(t) \tag{2.10}$$

where λ is the wavelength of the laser light, l is the length of the Kerr cell, and t denotes the time in the transient experiment. The phase difference causes the light to be elliptically polarized after passing the sample, and hence the analyzer no longer blocks all the laser light. Consequently, the changing birefringence causes a variation in the intensity of the light that reaches the detector. Since no intensity is detected in the absence of birefringence (zero background), intensity variations due to the birefringence can be measured very accurately.

2.3.5. The Quarter-Wave Plate

In order to improve the signal-to-noise ratio even more and to discriminate between positive and negative birefringence, the analyzer A is not perfectly perpendicular to the polarizer P, but is rotated over a small angle ϕ and in addition a quarter-wave plate with its slow axis at 135 deg with respect to the electric field direction is placed just in front of the analyzer. For a detailed discussion of the improvements made by these adaptations of the setup, we refer to Fredericq and Houssier (1973). The result is that light intensities in the absence, I_ϕ, and presence, $I_\delta(t)$, of birefringence can be accurately measured. These intensities are related to the phase difference δ by

$$\frac{I_\delta(t) - I_\phi}{I_\phi} = \frac{\sin^2\left(\phi + \frac{1}{2}\delta(t)\right) - \sin^2\phi}{\sin^2\phi} \tag{2.11}$$

Combining equations (2.10) and (2.11), it is now possible to calculate the (transient) birefringence from the observed intensity variations.

For small values of ϕ, and for $\delta(t)$ much smaller than ϕ (which is usually the case), equation (2.11) can be simplified to

$$\Delta I(t) = I_\delta(t) - I_\phi \sim \phi\delta(t) \tag{2.12}$$

where $\Delta I(t)$ is the measured change in intensity, which is more or less proportional to the birefringence. As equation (2.12) shows, an increase in the angle ϕ leads to an increase of the signal. However, too large values of ϕ cannot be used, since the linearity of the detector is lost in case of high intensities. Usually ϕ is set below 1 deg.

2.3.6. Temperature Control

The rotational diffusion time, which is deduced from the measurements, depends critically on the temperature, because of the marked increase of the viscosity η_0 with the decrease of temperature. Therefore, good temperature control is essential. Low temperatures can be used to slow down the fast decay times in order to better resolve them.

2.3.7. Further Experimental Details and Data Analysis

Further improvement in the signal-to-noise ratio is achieved by the use of an offset voltage to compensate to a large extent for the output voltage of the photomultiplier which is due to the background signal I_ϕ. In this way it is possible to increase the dynamic range of the setup. Furthermore, averaging is performed over many measurements.

The decay of the signals is analyzed in our case with the computer programs DISCRETE (Provencher, 1976a, b) and CONTIN (Provencher, 1982a, b). DISCRETE fits the decay to a discrete sum of exponentials, whereas CONTIN analyzes data in terms of a continuous distribution of exponential relaxation rates by determining the smoothest distribution that adequately fits the data.

2.4. Validation of the Method: Myosin

The reliability of TEB results on filamentous systems can be inferred from the work of Highsmith and co-workers on myosin (Highsmith *et al.*, 1977; Highsmith and Eden, 1985, 1986). TEB measurements were performed on light meromyosin (LMM), heavy meromyosin (HMM), myosin rod, myosin subfragment-2 (S2), and meromyosin subfragment 1 (MM-S1). From the data it was concluded that LMM behaves as a rigid cylinder, whereas myosin rod and S2 are flexible. The flexibility of the myosin rod appeared to be localized in the hinge region between the S1 and S2 parts. For MM-S1 a structure was indicated with two tertiary structural domains connected by a flexible linkage with substantial restoring force. These conclusions are in agreement with results from electron microscopy, light scattering, X-ray diffraction, nuclear magnetic resonance, proteolysis, fluorescence, electron spin resonance, heat denaturation, and cross-linking.

3. CURRENT KNOWLEDGE ABOUT INTERMEDIATE FILAMENT ASSEMBLY

A detailed discussion of the current knowledge of IF assembly is given by Herrmann and Aebi in Chapter 10 of this book. In this section and Figure 3 we summarize the information that is relevant for the discussion of the application of TEB to IF assembly.

All IF proteins contain a central α-helical region of approximately 310 amino acids flanked by nonhelical amino- and carboxy-end domains of variable size and charge (Steinert and Roop, 1988). The α-helical regions of two IF-protein chains assemble in parallel and exactly in register to form a coiled-coil dimer with four segments designated 1A, 1B, 2A, and 2B linked by short linker segments L1, L2, and L12 (Figure 3a). The total length of such a rod domain is approximately 45 nm, while the net lengths of the end domains are uncertain (e.g., Ip *et al.*, 1985; Kaufmann *et al.*, 1985; Quinlan *et al.*, 1986; Steinert, 1991a, b).

The next level of organization is a tetramer formed by association of two dimers. Essentially four alignments of the constituting dimers have been suggested (reviewed by Heins and Aebi, 1994, and Fuchs and Weber, 1994): antiparallel in-register forming a rod of approximately 45 nm, head-

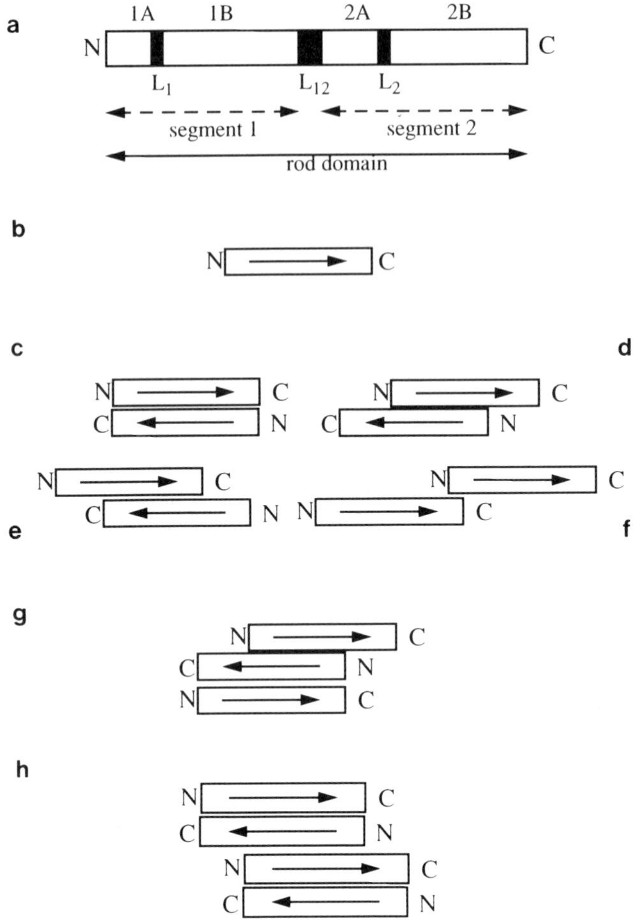

FIGURE 3. Previously proposed models for oligomers of intermediate filament chains (see text): (a) The α-helical domains 1A, 1B, 2A, and 2B are connected by linkers L_1, L_{12}, and L_2 and flanked by N- and C-terminal end domains. (b) Schematic representation of the coiled-coil dimer that is present in the higher oligomers in the rest of this figure (~45 nm). The arrow symbolizes polarity of the dimer. (c) In-register antiparallel tetramer (~45 nm). (d, e) Staggered antiparallel tetramers (~65 nm). (f) Staggered parallel tetramer (~125 nm). (g) Hexamer (~65 nm). (h) Octamer (~65 nm).

to-tail with slight overlap, and antiparallel staggered in two different ways yielding a rod length of approximately 65 nm (Figures 3c–3f).

For the next stages of IF assembly, various, sometimes conflicting models have been given (reviewed by Kooijman *et al.*, 1994). The occurrence of hexamers (Figure 3g) has been proposed, and evidence has been presented for the existence of octamers as shown in Figure 3h. Also, the number of monomers in the cross section of a full-width filament has been a matter of debate (Herrmann *et al.*, 1996), although generally it is assumed that the cross section contains approximately 32 monomers leading to a diameter of 10 nm (Aebi *et al.*, 1983, 1988; Fuchs and Weber, 1994; Parry, 1996). In the next section we will show how TEB results can discriminate among suggested models and what details it can add.

4. THE USE OF TRANSIENT ELECTRIC BIREFRINGENCE FOR THE STUDY OF INTERMEDIATE FILAMENT ASSEMBLY

In this section, the application of TEB for the study of the different levels of IF protein assembly is discussed. A summary of relevant results in given in Table I. As a model for the very low molecular assembly states of IF proteins, the N-terminal-truncated so-called T-vimentin was used and this is described in Section 4.1. TEB results on lower order aggregates of intact IF proteins vimentin, desmin, and glial fibrillary acidic protein (GFAP) in low-ionic-strengths solutions are presented in Section 4.2. The formation of higher order vimentin and GFAP IF assembly states as studied by TEB is discussed in Section 4.3.

4.1. T-Vimentin

As described in Section 3, essentially four arrangements of dimers in tetramers have been suggested in the literature (Figures 3c–3f). In order to discriminate among these models, we performed TEB measurements on T-vimentin (mouse vimentin missing the first 70 amino acids) (Kooijman *et al.*, 1995a). The absence of higher oligomers in populations of N-terminal-missing IF proteins makes this protein very suitable to study the properties of the dimeric and tetrameric forms. In low-salt solution they are generally assumed to form dimers only (Potschka *et al.*, 1990). Assuming a mean axial rise per residue in a coiled-coil of 0.1485 nm (Steinert *et al.*, 1993) and compact globular particles with a specific volume of 0.73 cm^3/g for the end domains yields the length of a stretched rigid T-vimentin rod of 48.3 nm (Kooijman *et al.*, 1995a).

Field-free decay analysis of the TEB signal of T-vimentin in low-salt solution with CONTIN and DISCRETE (see Section 2.3.7) yields two decay times of 1.5 and 0.3 μsec, respectively. Applying the hydrodynamic model calculations for rods (see Section 2.2) and using a dimer diameter of 2.3 nm (Potschka *et al.*, 1990), we find that the decay time of 1.5 μsec

Table I
Summary of Results of TEB on IF Proteins[a]

Protein	Condition	$\tau_{Average}$(μsec)	S_1/S_2	$K(m^2V^{-2})$	Interpretation
T-Vimentin	Low salt	0.3, 1.5	3.0	0.6×10^{-13}	Flexible or bent structure; 40- to 43-nm dimers
	High salt	5.1	1.1	1.0×10^{-13}	~65-nm staggered antiparallel tetramers
Vimentin	Low salt	1.3, 5.4	2.6	1.2×10^{-13}	~45-nm dimers; ~65-nm antiparallel tetramers, hexamers, and possibly octamers
	0.3 mM $MgCl_2$	1.3, 5.4	1.1	0.9×10^{-13}	~45-nm dimers; ~65-nm antiparallel tetramers and possibly octamers
	0.5 mM $MgCl_2$	1.3, 5.4, 28	1.1	4.0×10^{-13}	Flexibility of higher oligomer; ~65-nm antiparallel tetramers and octamers, ~100-nm full-width particles
	1.0 mM $MgCl_2$	5.4, 28, 63	1.1	13.2×10^{-13}	~65-nm antiparallel tetramers and octamers; ~100-nm full-width particles; ~135-nm full-width particles
	1.4 mM $MgCl_2$	28, 63	—	—	~100-nm full-width particles; ~35-nm full-width particles
	1.8 mM $MgCl_2$	115	—	—	~170-nm full-width particles
	2.5 mM $MgCl_2$	115	—	—	~170-nm full-width particles
GFAP	Low salt	5.7, 17	1.5	1.4×10^{-13}	~65-nm staggered antiparallel tetramers, hexamers, and possibly octamers; small-diameter 100-nm particles
	0.2 mM $MgCl_2$	5.7, 17	1.2	1.4×10^{-13}	~65-nm staggered antiparallel tetramers and/or octamers; small-diameter 100-nm particles
	0.3 mM $MgCl_2$	30, 69	1.1	12.4×10^{-13}	~100-nm full-width particles; ~135-nm full-width particles
	0.5 mM $MgCl_2$	30, 69	1.1	17.0×10^{-13}	~100-nm full-width particles; ~135-nm full-width particles
	1.0 mM $MgCl_2$	102	—	—	~170-nm full-width particles
	2.0 mM $MgCl_2$	102, 433	—	—	~170-nm full-width particles; larger structures
	2.5 mM $MgCl_2$	102, 433	—	—	~170-nm full-width particles; larger structures
Desmin	Low salt	1.6, 5.9, 24	1.6	2.5×10^{-13}	~45-nm dimers; ~65-nm antiparallel tetramers and octamers (~100-nm full-width particles?)

[a]Full results are given in Kooijman *et al.* (1994, 1995a, b, 1997).

corresponds to a 40- to 43-nm-long vimentin molecule. As the theoretical length for a rigid T-vimentin dimer is 48.3 nm, it can be concluded that the dimers are not straight and rigid, but bent and/or flexible. Additional evidence for this hypothesis is provided by the observation of the shortest decay time of 0.3 μsec. In the case of flexibility, the short time describes the relaxation of the flexible segment. In the case of a stable bent dimer, the short decay time could be related to longitudinal rotation of the rod. Also, the increase of the Kerr constant upon going from dimers to tetramers (Table I) cannot readily be explained if dimers are rigid and straight. This lends further support to the bent/flexible structure of T-vimentin dimers.

The N-truncated IF proteins in solution with a higher salt concentration are assumed to consist essentially of tetramers (Kaufmann *et al.*, 1985; Potschka *et al.*, 1990). When 0.5 mM $MgCl_2$ is added to the T-vimentin solution, the field-free decay of the TEB yields a large (>85%) contribution of a decay time of 5.1 ± 0.3 μsec. A rigid rod with this end-over-end rotation time has a length between 63 and 68 nm. Hence, the dimers appear to associate mainly as staggered tetramers.

TEB measurements also give information on the charge distribution of T-vimentin aggregates. The TEB signal measured in the absence of Mg^{2+} yields $S_1/S_2 = 3.0$, which corresponds to $P/Q = 2.2 \pm 0.7$. This shows that the driving force of the dimer orientation is due to both a permanent and an induced dipole moment. On the other hand, when 0.5 mM Mg^{2+} is present, $S_1/S_2 = 1.1 \pm 0.1$ ($P/Q = 0.04 \pm 0.04$), which shows that the permanent-dipole contribution to the orientation of the particles is negligible. This indicates an antiparallel alignment of two T-vimentin dimers within a tetramer, which leads to cancellation of the permanent dipole moments of the constituting dimers.

Thus, in the study of the assembly state of T-vimentin, TEB measurements can clearly discriminate among conflicting models concerning the length of tetramers which were presented in the literature. Moreover, it can be concluded that the tetramers have an antiparallel arrangement. The measurements also indicate that T-vimentin dimers are bent and/or flexible.

4.2. Type III Intermediate Filament Proteins in Low-Ionic Strength Solution

As a next step, TEB measurements have been performed on intact vimentin, desmin, and GFAP in low-ionic strength solutions (Kooijman *et al.*, 1994, 1995b). When vimentin was dissolved in 5 mM Tris. HCl, the decay was dominated by two relaxation times (see Table I). At pH 8.5, the TEB decay signal consists predominantly of a component of 4.0 ± 1.0 μsec corresponding to a rod length of 63 ± 9 nm (Kooijman *et al.*, 1994). The S_1/S_2

value of 2.6 ± 0.5 shows that at least part of the particles are oriented by a permanent dipole moment. The smallest vimentin oligomer with such a length and dipole moment is a hexamer (Figure 3g) as proposed by Steinert (1991a, b), since even fully stretched dimers are 48.3 nm only and tetramers (Figures 3d and 3e) are antiparallel (thus without permanent dipole moment; see Section 4.1) or significantly longer (Figure 3f). It was proposed by Steinert (1991a, b) that tetramers, hexamers, and octamers coexist. This is compatible with our results, although we cannot prove that tetramers and octamers are indeed present. When the pH was lowered to 6.8 the tendency for aggregation was less, but the same species were formed.

When the low-ionic strength structures of desmin and GFAP were studied by TEB (Kooijman *et al.*, 1995b), flexible or bent dimers with a length around 45 nm, particles with a length of 68 ± 6 nm (presumably tetramers and hexamers), and aggregates of 108 ± 19 nm were indicated. In contrast to vimentin, GFAP has an increased tendency for aggregation at lower pH, while the aggregation state of desmin does not change between pH 6.8 and 8.5.

Thus, by studying the size and electrical properties of the type III IF proteins vimentin, desmin, and GFAP in low-ionic strength solutions, a model is supported in which single dimers, antiparallel staggered tetramers, hexamers, and larger aggregates coexist. The pH dependence of the aggregation, which is easy to study by TEB, is different for the three proteins. In particular, the hexameric structures are difficult to recognize by electron microscopy, as their length is identical to that of antiparallel tetramers.

4.3. The Formation of Higher Order Structures of Type III Intermediate Filament Proteins

Also at the level of higher order IF organization, on which very little is known, TEB can provide useful information (Kooijman *et al.*, 1997). IF formation can be induced by relatively high concentrations of NaCl (50–150 mM). However, since these salt concentration are too high for TEB measurements, we used (sub)millimolar concentrations of $MgCl_2$, which also induce filament formation.

Vimentin at very low $MgCl_2$ concentrations (0.3 mM) and pH 7.5 gives similar decay components as in the absence of $MgCl_2$ (see Table I). However, the permanent dipole moment almost disappears ($S_1/S_2 = 1.1 \pm 0.1$), which can be explained by the disappearance of ~65-nm-long hexamers and the formation of antiparallel ~65-nm-long tetramers and/or octamers (Figure 3). In addition, the disappearance of the dipole moment may partly be due to the binding of Mg^{2+} to the protein, leading to a compensation of its negative charges.

Between 0.5 and 1.4 mM Mg^{2+} larger aggregates are formed. Three decay components of the TEB signal appear with average values of 5.4 ± 1.0, 28 ± 4, and 63 ± 9 μsec (see Table I). The longer decay times start to contribute more when the concentration of Mg^{2+} is increased. In addition to the increased TEB decay time, a drastic increase of the Kerr constant from 1.2×10^{-13} to $13.2 \times 10^{-13}\,m^2V^{-2}$ was observed upon going from 0.0 to 1.0 mM $MgCl_2$. The observed well-defined and more or less constant decay times show that vimentin IF assembly occurs via a specific pathway rather than a random lateral and/or longitudinal growth of aggregates. The 5.4 ± 1.0 μsec relaxation time again reflects the presence of 65-nm particles which will be considered as the building blocks for larger aggregates. It is impossible to determine whether they are tetramers or octamers. The 28-μsec decay time cannot be due to a 65-nm full-width (10 nm diameter) particle, which would have a decay time of 10 μsec. Neither can it be caused by a head-to-tail arrangement of 65-nm tetramers or octamers, since this would lead to a decay time of about 40 μsec. Thus the length of the 28-μsec particle is intermediate, indicating overlap of the building blocks. In combination with the increase of the Kerr constant at higher concentrations of Mg^{2+}, this shows that the complex contains at least 31 ± 8 monomers, is partly full-width, and has a length of 97 ± 6 nm [the complicated proof is given by Kooijman *et al.* (1997)]. The 63-μsec decay time belongs to a particle of 136 ± 8 nm length, when a diameter of 10 nm is assumed. At 1.8 and 2.5 mM $MgCl_2$ the dominating decay time is 115 ± 11 μsec, which corresponds to a length of 174 ± 6 nm. Apparently, longitudinal growth of vimentin IFs from 65 to 174 nm occurs in more or less discrete steps of ~35 nm. Further elongation of vimentin is impossible to follow quantitatively with TEB, since too many complexes with different lengths seem to coexist and the individual decay times cannot be resolved adequately.

For GFAP IF assembly at pH 6.8 (as in Inagaki *et al.*, 1990) we obtained comparable results (Kooijman *et al.*, 1997) (see also Table I). In general, the aggregation behavior of vimentin and GFAP is very similar, although the latter assembles more readily into larger aggregates under the conditions that were used. The observed decay times of 30 ± 3, 69 ± 4, and 102 ± 6 μsec lead to calculated lengths (assuming a diameter of 10 nm) of 100 ± 4, 142 ± 4, and 166 ± 4 nm, respectively, rather similar to what was calculated for vimentin. However, an additional 17-μsec component was found at low $MgCl_2$ concentrations, which could be due to two staggered 65-nm building units (tetramers or octamers) with a total length of ~100 nm and a diameter of ~5 nm. In Figure 4 the different structures are summarized schematically, assuming that the building blocks are either tetramers or octamers.

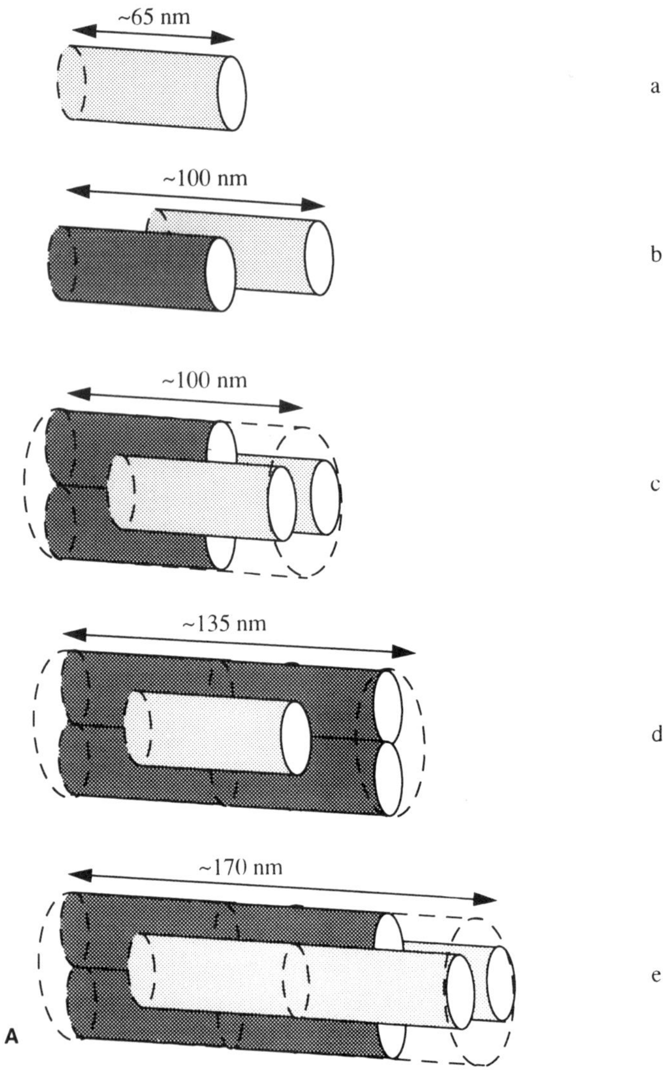

FIGURE 4. (A) A model for vimentin and GFAP IF formation from octameric building blocks. (a) Octamer, (b) ~100-nm aggregate, (c) smallest full-width particle, (d, e) elongation of the particle shown in panel c by the subsequent addition of two octamers. For more details see Kooijman *et al.* (1997). (B) A model for IF formation from tetrameric building blocks. (a) Tetramer, (b) ~100-nm aggregate, (c) smallest full-width particle, (d, e) elongation of the particle shown in panel c by the subsequent addition of four tetramers. For more details see Kooijman *et al.* (1997).

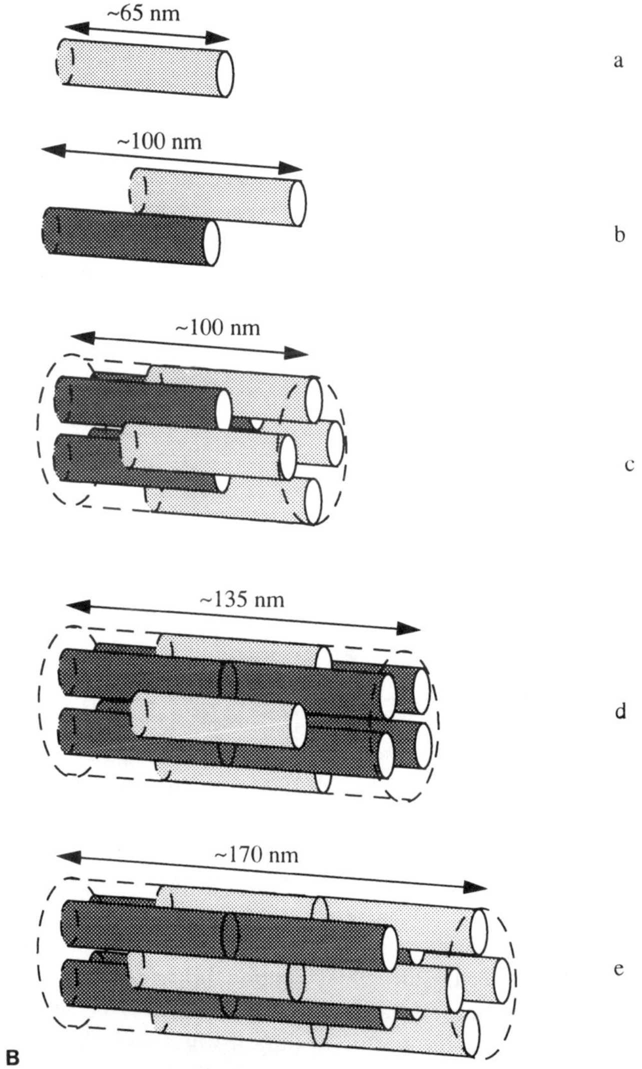

FIGURE 4. (*Continued*).

Thus, by TEB measurements the oligomerization of vimentin and GFAP IF proteins into larger structures can be monitored. The data are consistent with a model in which at the first stage of vimentin and GFAP aggregation, hexameric oligomers disappear to form tetramers and/or octamers. About 100-nm-long structures are then formed from 65-nm IF building blocks (tetramers or octamers) which have a strong tendency to laterally assemble into full-width IF structures. Subsequently, two more or less discrete elongation steps occur that each increases the length of the filaments with about 35 nm. The occurrence of full-width ~65-nm-long particles and ~130-nm-long head-to-tail associated building blocks which have been suggested in the literature could not be detected.

This section illustrates that also for more complicated IF aggregates, TEB is able to verify or falsify hypotheses on their structure, and that the state of aggregation of IF proteins in solution can easily be compared under various conditions (provided that the salt concentration is not too high).

5. CONCLUSIONS AND PROSPECTS

TEB is a very powerful technique for the study of elongated structures like intermediate filaments. Rather accurate estimates of the length of IF particles can be obtained provided that (a rough estimate of) the diameter is known. Additional information that may be obtained from TEB is on permanent and inducible electric properties of the molecules. So far, TEB has only be used in IF research to study the type III IF proteins vimentin, desmin, and GFAP. From the results it was possible to deduce a model for IF growth (Figure 4). IF dimers, appearing to occur in low-salt solutions, were proven to be flexible or bent. Tetramers appeared to be almost exclusively antiparallel and staggered. At low salt concentrations hexamers could be detected. At higher salt strength the hexamers are converted into tetramers and/or octamers. When the pH is varied between 8.5 and 6.8, the tendency to aggregate decreases for vimentin, increases for GFAP, and is constant for desmin. In general, GFAP aggregates more readily than vimentin.

Advantages of TEB are its noninvasive character (no chemical modification of the sample is required), its applicability to solutions (various conditions can easily be compared, no crystallization or precipitation is required), the obtained combination of dimensional and electrical information, the wide range of molecular fimensions that can be studied (a few nanometers up to several micrometers), and its ability to monitor both internal and external motions. Disadvantages of TEB are that it needs information from other techniques to obtain reliable conclusions, and it uses rather high electric fields to orient the molecules. The latter may influence the conformation or integrity of the molecules and restricts the range of ionic strengths that may be used in solutions. Moreover, TEB is nonspecific and the measured signal reflects the sum of all the mobilities of the system. This sometimes complicates interpretation of the results. The problem of nonspecificity can (partly) be overcome by the development of advanced data processing methods and by variation of the pulse length. Much higher salt strength can be used when an alternating electric field is used. An apparatus based on this principle with nanosecond resolution and applicability at physiological salt strength has been described by Pörschke and Obst (1991). It may be anticipated that with these developments TEB can further develop into a valuable tool in IF research.

6. REFERENCES

Aebi, U., Fowler, W. E., Rew, P., and Sum, T.-T., 1983, The fibrillar substructure of keratin filaments unravelled, *J. Cell Biol.* **97:**1131–1143.

Aebi, U., Häner, M., Troncoso, J., Eichner, R., and Engel, A., 1988, Unifying principles in intermediate filament (IF) structure and assembly, *Protoplasma* **145:**73–81.

Bawden, F. C., Pirie, N. U., Bernal, J. D., and Fankuchen, I., 1936, Liquid crystalline substances from virus-infected plants, *Nature* **138:**1051–1052.

Benoit, H., 1949, Apparatus for measuring the Kerr effect for single electric pulses, *Compt. Rend.* **228:**1716–1718.

Bergholm, C., and Björnstahl, Y., 1920, Electrical double refraction, *Phys. Z.* **21:**137–141.

Bloemendal, H., and Pieper, F. R., 1989, Intermediate filaments: Known structure, unknown function, *Biochim, Biophys. Acta* **1007:**245–253.

Born, M., 1918, Elektronentheorie des natürlichen optischen Drehungsvermögens isotroper und anisotroper Flüssigkeiten, *Ann. Phys.* (Leipzig) **55:**177–240.

Broersma, S., 1960, Rotational diffusion constant of a cylindrical particle, *J. Chem. Phys.* **32:**1626–1631.

Broersma, S., 1981, Viscous force and torque constants for a cylinder, *J. Chem. Phys.* **74:**6989–6990.

Fredericq, E., and Houssier, C., 1973, *Electric Dichroism and Electric Birefringence*, Clarendon Press, Oxford.

Fuchs, E., and Weber K., 1994, Intermediate filaments: Structure, dynamics, function and disease, *Annu. Rev. Biochem.* **63:**345–382.

Gerber, B. R., Minakata, A., and Kahn, L. D., 1975, Electric birefringence of bacterial flagellar protein filaments: Evidence for field-induced interactions, *J. Mol. Biol.* **92:**507–528.

Hagerman, P. J., 1996, Sometimes a great motion: The application of transient electric birefringence to the study of macromolecular structure, *Curr. Opin. struct. Biol.* **6:**643–649.

Heins, S., and Aebi, U., 1994, Making heads and tails of intermediate filament assembly, dynamics and networks, *Curr. Opin. Cell Biol.* **6:**25–33.

Herrmann, H., Häner, M., Brettel, M., Müller, S., Goldie, K. N., Fedtke, B., Lustig, A., Franke, W. W., and Aebi, U., 1996, Structure and assembly properties of the intermediate filament protein vimentin: The role of its head, rod and tail domains. *J. Mol. Biol.* **264:**933–953.

Highsmith, S., and Eden, D., 1985, Transient electric birefringence characterization of heavy meromyosin, *Biochemistry* **24:**4917–4924.

Highsmith, S., and Eden, D., 1986, Myosin subfragment 1 has tertiary structural domains, *Biochemistry* **25:**2237–2242.

Highsmith, S., Kretzschmar, K. M., O'Konski, C. T., and Morales, M. F., 1977, Flexibility of myosin rod, light meromyosin, and myosin subfragment-2 in solution, *Proc. Natl. Acad. Sci. USA* **74:**4986–4990.

Inagaki, M., Gonda, Y., Nishizawa, K., Kitamura, S., Sato, C., Ando, S., Tanabe, K., Kikuchi, K., Tsuiki, S., and Nishi, Y., 1990, Phosphorylation sites linked to glial filament disassembly *in vitro* locate in a non-α-helical head domain, *J. Biol. Chem.* **265:**4722–4729.

Ip, W., Hartzer, M. K., Susana Pang, Y.-Y., and Robson, R. M., 1985, Assembly of vimentin *in vitro* and its implications concerning the structure of intermediate filaments, *J. Mol. Biol* **183:**365–375.

Kaufmann, E., Weber, K., and Geisler, N., 1985, Assembly of vimentin *in vitro* and its implications concerning the structure of intermediate filaments, *J. Mol. Biol.* **185:**733–742.

Kaye, W., and Devaney, R., 1947, An apparatus for the observation of the Kerr effect with microsecond electrical pulses, *J. Appl. Phys.* **18:**912–918.

Kerr, J., 1875, A new relation between electricity and light: Dielectric media birefringent, *Phil. Mag.* [*4*] **50:**337–348.

Kerr, J., 1880, Measurements and law in electrooptics, *Phil. Mag.* [*5*] **9:**157–174.

Kobayashi, S., Hiroshi, A., and Oosawa, F., 1964, Electric birefringence of actin, *Biochim. Biophys. Acta* **88:**528–540.

Kooijman, M., Bloemendal, M., Traub, P., and van Grondelle, R., 1992, Fluorescence study of the nucleic acid binding of vimentin, *FEBS Lett.* **302:**177–180.

Kooijman, M., Bloemendal, M., van Amerongen, H., Traub, P., and van Grondelle, R., 1994, Characterization of multiple oligomeric vimentin intermediate filament units by transient electric birefringence measurements, *J. Mol. Biol.* **236:**1241–1249.

Kooijman, M., Bloemendal, M., Traub, P., van Grondelle, R., and van Amerongen, H., 1995a, Hydrodynamic and electrical characterization of T-vimentin dimers and tetramers by transient electric birefringence measurements, *J. Biol. Chem.* **270:**2931–2937.

Kooijman M., van Amerongen, H., Traub, P., van Grondelle, R., and Bloemendal, M., 1995b, The assembly state of the intermediate filament proteins desmin and glial fibrillary acidic protein at low ionic strength, *FEBS Lett.* **358:**185–188.

Kooijman, M., Bloemendal, M., Traub, P., van Grondelle, R., and van Amerongen, H., 1997, Transient electric birefringence study of intermediate filament formation from vimentin and glial fibrillary acidic protein, *J. Biol. Chem.* **272:**22548–22555.

Langevin, P., 1910, Electric and magnetic double refraction, *Compt. Rend.* **7:**475–478.

Lauffer, M. A., 1939, Electro-optical effects in certain viruses, *J. Am. Chem. Soc.* **61:**2412–2416.

Mandelkern, M., Elias, J. G., Eden, D., and Crothers, D. M., 1981, The dimensions of DNA in solution, *J. Mol. Biol.* **152:**153–161.

O'Konski, C. T., and Zimm, B. H., 1950, New method for studying electrical orientation and relaxation effects in aqueous colloids: Preliminary results with tobacco mosaic virus, *Science* **111:**113–116.

O'Konski, C. T., Yoshioka, K., and Orttung, W. H., 1959, Electric properties of macromolecules. IV. Determination of electric and optical paramenters from saturation of electric birefringence in solution, *J. Phys. Chem.* **63:**1558–1565.

Parry, D. A. D., 1996, Hard α-keratin intermediate filaments: An alternative interpretation of the low-angle equatorial X-ray diffraction pattern, and the axial disposition of putative disulphide bonds in the intra- and inter-protofilamentous networks, *Int. J. Biol. Macromol.* **19:**45–50.

Pörschke, D., and Obst, A., 1991, An electric field jump apparatus with ns time resolution for electro-optical measurements at physiological salt concentrations, *Rev. Sci. Instrum.* **62:**818–820.

Potschka, M., Nave, R., Weber, K., and Geisler, N., 1990, The two coiled coils in the isolated rod domain of the intermediate filament protein desmin are staggered, *Eur. J. Biochem.* **190:**503–508.

Provencher, S. W., 1976a, A Fourier method for the analysis of exponential decay curves, *Biophys. J.* **16:**27–41.

Provencher, S. W., 1976b, An eigenfunction expansion method for the analysis of exponential decay curves, *J. Chem. Phys.* **64:**2772–2777.

Provencher, S. W., 1982a, A constrained regularization method for inverting data represented by linear algebraic or integral equations, *Comp. Phys. Commun.* **27:**213–227.

Provencher, S. W., 1982b, A general purpose constrained regularization program for inverting noisy linear algebraic and integral equations, *Comp. Phys. Commun.* **27:**229–242.

Quinlan, R. A., Hatzfeld, M., Franke, W. W., Lustig, A., Schulthess, T., and Engel, J., 1986, Characterization of dimer subunits of intermediate filament proteins, *J. Mol. Biol.* **192:**337–349.

Quinlan, R. A., Hutchinson, C., and Lane, B., 1995, Intermediate filament proteins, *Protein Profile* **2:**801–952.

Shoemann, R. L., and Traub, P., 1990, The *in vitro* DNA-binding properties of purified nuclear lamin proteins and vimentin, *J. Biol. Chem.* **265:**9055–9061.

Steinert, P. M., 1991a, Organization of coiled-coil molecules in native mouse keratin 1/keratin 10 intermediate filaments: Evience for alternative rows of antiparallel in-register and antiparallel staggered molecules, *J. Struct. Biol.* **107:**157–174.

Steinert, P. M., 1991b, Analysis of the mechanism of assembly of mouse keratin 1/keratin 10 intermediate filaments *in vitro* suggests that intermediate filaments are built from multiple oligomeric units rather than a unique tetrameric building block, *J. Struct. Biol.* **107:**175–188.

Steinert, P. M., and Roop, D. R., 1988, Molecular and cellular biology of intermediate filaments, *Annu. Rev. Biochem.* **57:**593–625.

Steinert, P. M., Marekow, L. N., Fraser, R. D. B., and Parry, D. A. D., 1993, Keratin intermediate filament structure crosslinking studies yield quantitative information on molecular dimensions and mechanism of assembly, *J. Mol. Biol.* **230:**436–452.

Tolstoi, N. A., and Feofilov, P. P., 1949, A new method of studying relaxation processes, *Zh. Eksperim. Teor. Fiz.* **19:**421–430.

Chapter 14

Possible Regulation of Intermediate Filament Proteins by Rho-Binding Kinases

Kaori Matsuzawa, Hidetaka Kosako, Ichiro Azuma, Naoyuki Inagaki, and Masaki Inagaki

1. INTRODUCTION

Intermediate filaments (IFs) are major elements of the cytoskeleton and nuclear envelope in eukaryotic cells (Steinert and Roop, 1988; Eriksson *et al.*, 1992; Fuchs and Weber, 1994). Although IFs were thought to be relatively stable compared to other cytoskeletal elements such as microtubles and actin filaments, accumulating evidence indicates that IFs undergo dynamic reorganization of their filament structures, accompanied by site-specific phosphorylation of the constituent proteins. Site-specific phosphorylations of IF proteins are spatially and temporally regulated during

Kaori Matsuzawa Laboratory of Biochemistry, Aichi Cancer Center Research Institute, Chikusa-ku, Nagoya, Aichi 464, Japan; and Institute of Immunological Science, Hokkaido University, Sapporo 060, Japan. **Hidetaka Kosako, Naoyuki Inagaki, and Masaki Inagaki** Laboratory of Biochemistry, Aichi Cancer Center Research Institute, Chikusa-ku, Nagoya, Aichi 464, Japan. **Ichiro Azuma** Institute of Immunological Science, Hokkaido University, Sapporo 060 Japan.

Subcellular Biochemistry, Volume 31: Intermediate Filaments, edited by Herrmann and Harris. Plenum Press, New York, 1998.

cellular events such as cell signaling and cell cycle, and some protein kinases responsible for the phosphorylation *in vivo* have been identified (Inagaki *et al.*, 1996). Thus, the intracellular organization of the IF network appears to be under the control of protein kinases and phosphatases.

The small GTP-binding protein Rho is thought to play a critical role in the control of cytoskeletal structures, cell adhesion, cell morphology, and cytokinesis (Machesky and Hall, 1996). Upon stimulation with certain signals, the GDP-bound inactive form of Rho may be converted to the GTP-bound active form, which binds to specific downstream targets and exerts its biological functions. Several putative target molecules for Rho have been identified, including p128 PKN (Amano *et al.*, 1996a; Watanabe *et al.*, 1996), p164 Rho-kinase (Matsui *et al.*, 1996) [also called ROK (Leung *et al.*, 1995) or ROCK (Ishizaki *et al.*, 1996)], p138 myosin-binding subunit (MBS) of myosin phosphatase (Kimura *et al.*, 1996), rhophilin (Watanabe *et al.*, 1996), citron (Madaule *et al.*, 1995), rhotekin (Reid *et al.*, 1996), and p140mDia (Watanabe *et al.*, 1997) (Figure 1). Recently, we have found that p128 PKN and p164 Rho-kinase are responsible for the phosphorylation of IF proteins *in vitro* (Matsuzawa *et al.*, 1997; Kosako *et al.*, 1997). In this chapter, we

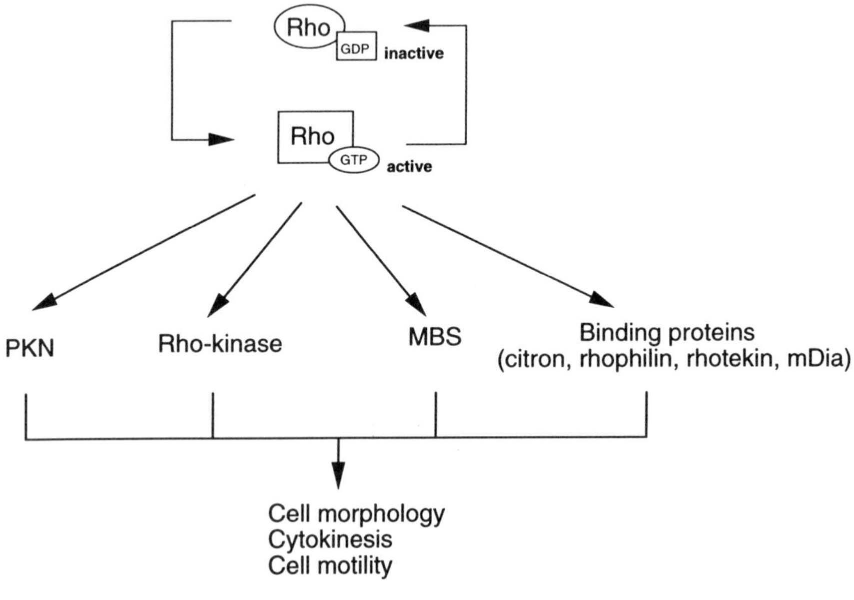

FIGURE 1. Rho and its putative downstream target proteins. Rho is a small GTPase, which has GDP-bound inactive and GTP-bound active forms. Recently, the putative targets for Rho have been identified, including PKN, Rho-kinase, myosin-binding subunit (MBS) of myosin phosphatase, citron, rhophilin, rhotekin, and mDia.

describe the possibility that structures of IFs are regulated by Rho-binding protein kinases.

2. IDENTIFICATION OF INTERMEDIATE FILAMENT KINASES

Before directing attention to the subject of Rho-binding kinases, we first describe briefly examples of how protein kinases responsible for IF phosphorylation and regulation have been identified. The most basic criterion for an IF kinase is that it phosphorylates IF proteins *in vitro*, at a reasonable rate and stoichiometry. IFs phosphorylated by a kinase often undergo structural alteration *in vitro* (Inagaki *et al.*, 1987). Such observations support the idea that the kinase may be involved in regulating IF structures in the cell. Because kinases that phosphorylate IF proteins *in vitro* do not always phosphorylate IF proteins *in vivo*, it is essential that *in vitro* phosphorylation sites of IFs by a putative IF kinase completely overlap those *in vivo*. Two-dimensional maps of vimentin phosphopeptides showed that the sites phosphorylated in mitotic cells were similar to those phosphorylated *in vitro* by Cdc2 kinase (Chou *et al.*, 1990). Vimentin sites phosphorylated *in vivo* by glutamate receptor activation were also similar to those phosphorylated *in vitro* by Ca^{2+}/calmodulin-dependent protein kinase II (CaMKII) (Yano *et al.*, 1994). These results suggested that Cdc2 kinase and CaMKII may directly phosphorylate vimentin *in vivo*, but this cannot be concluded unless the phosphorylation sites completely overlap.

An alternative strategy to identify IF kinases is to utilize site- and phosphorylation state-specific antibodies (Inagaki *et al.*, 1996, 1997). Among the *in vitro* phosphorylated sites, there are those specifically phosphorylated by a single kinase (Figure 2). For example, Ser-33, Ser-55, and Ser-82 of vimentin are sites apparently specific for protein kinase C (PKC), Cdc2 kinase, and CaMKII, respectively. Such a specific site serves as a pertinent substrate for detection of *in vivo* phosphorylation by each kinase. We have developed monoclonal antibodies that recognize the phosphorylation of these specific sites and showed that these sites were phosphorylated *in vivo* during mitosis and by cell signaling (Tsujimura *et al.*, 1994; Ogawara *et al.*, 1995; Takai *et al.*, 1996).

Another example is cleavage furrow (CF) kinase (Nishizawa *et al.*, 1991; Matsuoka *et al.*, 1992; Sekimata *et al.*, 1996). Site- and phosphorylation state-specific antibodies could be used to detect unidentified *in vivo* IF kinase activity. When an antibody recognizes the phosphorylation of a Ser/Thr residue that is a target of redundant kinases, it may detect IF phosphorylation not only by a known kinase, but also by an unidentified kinase. CF kinase activity was detected using such antibodies (see below).

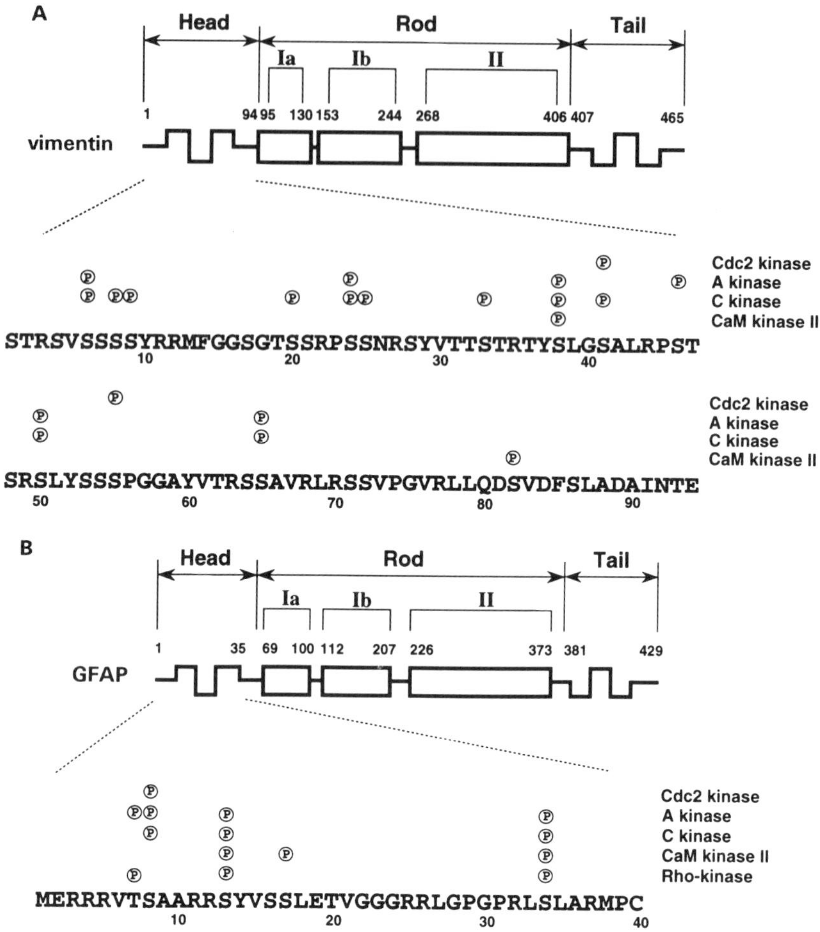

FIGURE 2. Maps of vimentin and glial fibrillary acidic protein (GFAP) molecules showing sites of phosphorylation by kinases identified in *in vitro* studies. The phosphorylation sites are indicated by P within a circle. (A) Vimentin; (B) GFAP.

3. RHO-BINDING PROTEIN KINASES AS POSSIBLE INTERMEDIATE FILAMENT KINASES

By developing site- and phosphorylation state-specific antibodies, we have previously reported that Thr-7, Ser-13, and Ser-34 of glial fibrillary acidic protein (GFAP) were phosphorylated at the cleavage furrow during cytokinesis (Nishizawa *et al.*, 1991; Matsuoka *et al.*, 1992; Sekimata *et al.*, 1996) (Figure 3). This type of GFAP phosphorylation was observed not only in astroglial cells, but also in other cultured cells transfected with

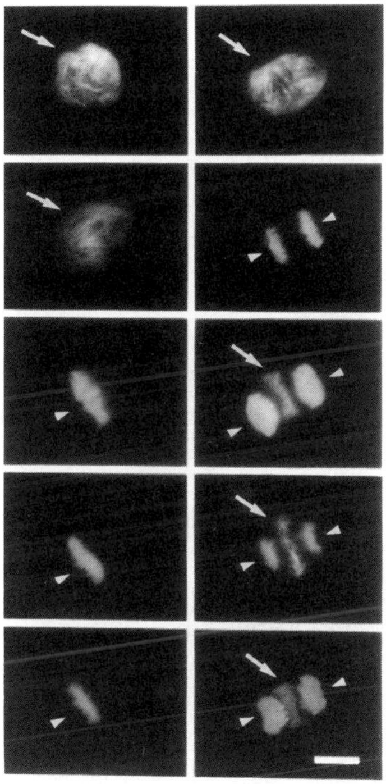

FIGURE 3. Fluorescent photomicrographs showing cell cycle-dependent phosphorylation of GFAP. Mitotic U251 human astrocytoma cells were stained with the antibody MO389 (anti-GFAP), YC10 (anti-phosphoSer^{8}), PG1-T (anti-phosphoThr^{7}), KT13 (anti-phosphoSer^{13}), or KT34 (anti-phosphoSer^{34}). The arrows indicate immunoreactions of each antibody and arrowheads the chromosomes stained with propidium iodide. The bar represents 10 μm. MO389 stained filamentous structures in both mitotic and interphase cells, while YC10 stained filamentous structures throughout the cytoplasm of metaphase, but not anaphase cells. In contrast, the immunoreactivities of PG1-T, KT13, and KT34 were specifically observed between the daughter nuclei and at the cleavage furrow of anaphase cells. Modified with permission from Kosako *et al.* (1997).

GFAP cDNA (Sekimata *et al.*, 1996). These findings suggest the existence of a protein kinase that may play an important role in cytokinesis, and we tentatively termed this CF kinase (Sekimata *et al.*, 1996). However, the molecular identity, regulation, and function of CF kinase remained to be examined. Rho was shown to play a critical role in cytokinesis by inducing and maintaining the contractile ring, an actin-based cytoskeletal structure (Kishi *et al.*, 1993; Mabuchi *et al.*, 1993; Drechsel *et al.*, 1997). In addition, Rho was reported to be translocated from the cytoplasm to the cleavage furrow during cytokinesis (Takaishi *et al.*, 1995). These observations raised the possibility that Rho may be located upstream of CF kinase, and we examined the possibility that IF proteins may be phosphorylated and regulated by Rho-binding kinases.

3.1. Protein Kinase N

Protein kinase N (PKN) is a fatty acid- and protease-activated Ser/Thr kinase with a catalytic domain homologous to the PKC family. The amino-terminal region of PKN contains repeats of the leucine zipper-like motif presumed to be critical for regulating biological activity (Mukai and Ono, 1994; Mukai *et al.*, 1994; Kitagawa *et al.*, 1995). This enzyme is reported to be one of the Rho-binding kinases (Amano *et al.*, 1996a; Watanabe *et al.*, 1996). It has been demonstrated that the amino-terminal region of PKN associates with the head-rod domain of neurofilament (NF) protein and that the phosphorylation of NF protein by this enzyme inhibited its polymerization (Mukai *et al.*, 1996).

IF proteins are grouped into the five major types (Steinert and Roop, 1988; Fuchs and Weber, 1994). NF protein is one of the type IV IF proteins, and is expressed specifically in neuronal cells. Vimentin, one of the type III IF proteins, is the most common IF, and is expressed during development in a wide range of cells, in mescenchymal cells, and in a variety of cultured cell lines and tumors. To better understand the relationships between PKN and IF proteins, we examined interactions between PKN and vimentin. Using analyses with the yeast two-hybrid system, we found that the head-rod domain of vimentin specifically interacted with the amino-terminal region of PKN. This notion was confirmed by an *in vitro* binding assay. The *in vitro* translated amino-terminal region of PKN bound to the GST-fused head-rod domain of vimentin (Matsuzawa *et al.*, 1997).

Furthermore, vimentin and GFAP, another type III IF protein, were found to be phosphorylated by PKN purified from rat testis *in vitro*, and the level of phosphorylation increased by arachidonic acid, an activator of PKN. Digestion of radiolabeled proteins by lysyl endopeptidase and subsequent sodium dodecyl sulfate–polyacrylamide gel electrophoresis (SDS–PAGE) analysis revealed that the phosphorylation occurred in head domains of these IF proteins (Matsuzawa *et al.*, 1997).

To examine the effect of phosphorylation of these proteins by PKN on filament formation, soluble vimentin or GFAP was phosphorylated by PKN and then incubated under conditions of polymerization. Although unphosphorylated vimentin did form 10-nm filaments, vimentin phosphorylated by PKN lost its filament-forming ability (Figure 4). Similar results were also obtained with GFAP. Thus, the phosphorylation of vimentin and GFAP by PKN apparently inhibits the polymerization of these proteins *in vitro*.

Phosphorylation sites of vimentin by PKN were analyzed using Western blotting and site- and phosphorylation state-specific antibodies. As shown in Figure 5, PKN phosphorylated Ser-6, Ser-33, and Ser-50, but not Ser-55 or Ser-82 on vimentin, the same sites phosphorylated by PKC. We also examined phosphorylation sites of GFAP by PKN using site- and phosphorylation state-specific antibodies. PKN phosphorylated mainly Thr-7, Ser-8, and Ser-34 on GFAP (data not shown). This result indicates that PKN is not CF kinase, since CF kinase phosphorylates Thr-7, Ser-13,

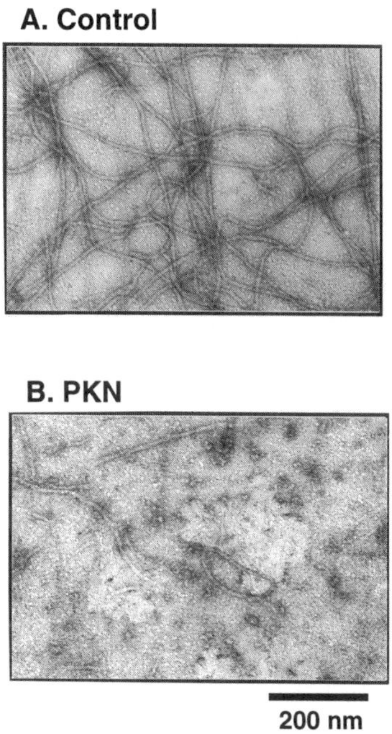

FIGURE 4. Electron microscopy of negatively stained vimentin samples. Recombinant murine vimentin was phosphorylated by PKN, and then polymerized with 100 mM NaCl. Samples were negatively stained with 2% uranyl acetate and examined under an electron microscope. Although unphosphorylated vimentin (control) did form 10-nm filaments (A), vimentin phosphorylated by PKN lost its filament-forming ability (B).

and Ser-34 on GFAP. However, PKN may possibly be involved in phosphorylation of IF proteins during some cellular events. All these observations suggested that PKN functions as a protein kinase responsible for phosphorylation and regulation of various types of IF proteins.

3.2. Rho-Kinase

Rho-kinase is a p164 Ser/Thr kinase, which has a catalytic domain in the amino-terminal region, a coiled-coil domain in the middle region, and a zinc finger-like motif in the carboxy-terminal region. The catalytic domain shares 72% sequence homology with that of myotonic dystrophy kinase and the coiled-coil domain contains a Rho-interacting interface (Matsui *et al.*, 1996). To search for the putative CF kinase responsible for the cleavage furrow-specific phosphorylation of GFAP, we asked whether Rho-kinase purified from bovine brain would phosphorylate GFAP *in vitro*. Rho-kinase was shown to phosphorylate GFAP in a GTP·Rho-dependent manner. GTPγS-bound Rho (the constitutively active form of Rho) enhanced the phosphorylation of GFAP by Rho-kinase about 300-fold (Kosako *et al.*, 1997). The effect of phosphorylation of GFAP by Rho-kinase on the

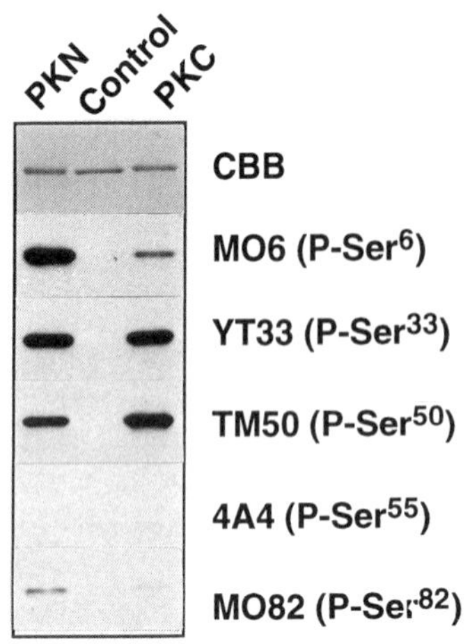

FIGURE 5. Analysis of phosphorylation sites of vimentin by PKN and PKC. Recombinant murine vimentin was phosphorylated by PKN or PKC, then resolved by SDS–PAGE and immunoblotted with the site- and phosphorylation state-specific antibodies MO6 (anti-phosphoSer^{6}), YT33 (anti-phosphoSer^{33}), TM50 (anti-phosphoSer^{50}), 4A4 (anti-phosphoSer^{55}), or MO82 (anti-phosphoSer^{82}), using the ECL Western blotting detection system. PKN phosphorylated vimentin at Ser-6, Ser-33, and Ser-50, but not at Ser-55 or Ser-82.

filament-forming ability was then examined. Soluble GFAP was preincubated with or without Rho-kinase and subjected to further incubation under conditions of polymerization. The phosphorylation of GFAP by Rho-kinase resulted in a near complete inhibition of its filament formation *in vitro* (Figure 6).

Western blotting with site- and phosphorylation state-specific antibodies showed that the phosphorylation sites of GFAP by this kinase were Thr-7, Ser-13, and Ser-34, which are the same sites that CF kinase phosphorylates at the cleavage furrow during cytokinesis (Figure 7). On the other hand, cAMP-dependent protein kinase (PKA) phosphorylated Thr-7, Ser-8, Ser-13, and Ser-34, and Cdc2 kinase weakly phosphorylated only Ser-8. These phosphorylation sites of Rho-kinase on GFAP were confirmed by amino acid sequence analysis of tryptic phosphopeptides and phosphoamino acid analysis (Kosako *et al.*, 1997).

These results indicate that Rho-kinase is one candidate for CF kinase. We think that Rho translocates from the cytoplasm to the cleavage furrow during cytokinesis (Takaishi *et al.*, 1995), and then the activated Rho-kinase phosphorylates GFAP in the cleavage furrow. If GFAP phosphorylation at Thr-7, Ser-13, and Ser-34 during cytokinesis induces the fragmentation of glial filaments at the cleavage furrow, then this event may function for efficient separation of the glial filament network into two daughter cells.

4. FUTURE PROSPECTS

There is accumulating evidence that intracellular IF organization is dynamically regulated by site-specific phosphorylation. Several protein

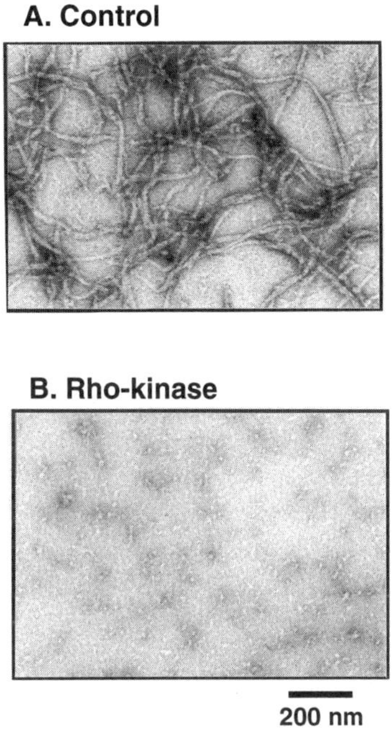

FIGURE 6. Electron microscopy of negatively stained GFAP samples. Recombinant human GFAP was phosphorylated by Rho-kinase and then polymerized with 100 mM NaCl. Samples were negatively stained with 2% uranyl acetate and examined under an electron microscope. Although unphosphorylated GFAP (control) did form 10-nm filaments (A), GFAP phosphorylated by Rho-kinase lost its filament-forming ability (B). Modified with permission from Kosako *et al.* (1997).

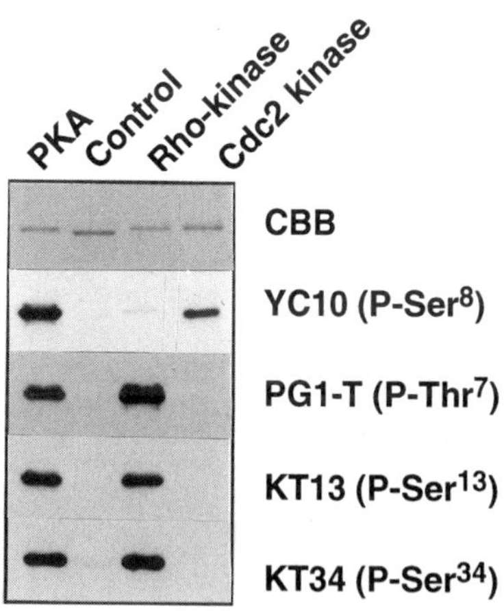

FIGURE 7. Analysis of phosphorylation sites of GFAP by several protein kinases. Recombinant human GFAP was phosphorylated by PKA, Rho-kinase, or Cdc2 kinase. Each sample was resolved by SDS–PAGE and then immunoblotted with the site- and phosphorylation state-specific antibodies YC10 (anti-phosphoSer^{8}), PG1-T (anti-phosphoThr^{7}), KT13 (anti-phosphoSer^{13}), or KT34 (anti-phosphoSer^{34}), using the ECL Western blotting detection system. Rho-kinase phosphorylated GFAP at Thr-7, Ser-13, and Ser-34, but not at Ser-8.

kinases act as IF kinases *in vivo*. Here we report that PKN and Rho-kinase are new candidates for IF kinases. The possibility was considered that the regulation of IF phosphorylation occurs downstream of the signaling cascade of the small GTPase Rho.

Although vimentin and GFAP are excellent substrates for PKN *in vitro*, whether this kinase phosphorylates these IF proteins in Rho dependent-manner *in vivo* is unknown. Since the catalytic characteristics of PKN toward vimentin and GFAP seem to be similar to those of PKC, one needs to distinguish the phosphorylation by PKN from that by PKC in order to elucidate relationships between IFs and PKN *in vivo*. It is necessary to identify PKN-specific phosphorylation sites and to visualize IF phosphorylation by this kinase *in vivo* using site- and phosphorylation state-specific antibodies.

Rho is implicated in the formation of stress fibers and focal adhesions in fibroblasts stimulated by extracellular signals such as lysophosphatidic acid (Ridley and Hall, 1992, 1994). Because Rho-kinase was reported to act as a mediator of signals from Rho in the formation of stress fibers and focal adhesions (Leung *et al.*, 1996; Amano *et al.*, 1997) and in the regulation of myosin phosphorylation (Kimura *et al.*, 1996; Amano *et al.*, 1996b), Rho-kinase may also mediate the regulation of cytokinesis by Rho. If Rho-kinase is CF kinase itself, GFAP phosphorylation at the cleavage furrow will not occur in cells transfected with cDNA encoding dominant-negative forms of Rho-kinase and these cells may undergo an abnormal cytokinesis. Whether Rho-kinase is activated during cytokinesis and whether the regulation of IF phosphorylation by Rho-kinase occurs only in cytokinesis or in various cellular events are subjects of ongoing studies.

ACKNOWLEDGMENTS. We are grateful to M. Ohara for critical comments on the manuscript and to K. Kuromiya for secretarial assistance. This work was supported in part by Grants-in-Aid for Scientific Research and Cancer Research from the Ministry of Education, Science, Sports and Culture of Japan, and special coordination funds from the Science and Technology Agency of the Government of Japan.

5. REFERENCES

Amano, M., Mukai, H., Ono, Y., Chihara, K., Matsui, T., Hamajima, Y., Okawa, K., Iwamatsu, A., and Kaibuchi, K., 1996a, Identification of a putative target for Rho as the serine-threonine kinase protein kinase N, *Science* **271:**648–650.

Amano, M., Ito, M., Kimura, K., Fukata, Y., Chihara, K., Nakano, T., Matsuura, Y., and Kaibuchi, K., 1996b, Phosphorylation and activation of myosin by Rho-associated kinase (Rho-kinase), *J. Biol. Chem.* **271:**20246–20249.

Amano, M., Chihara, K., Kimura, K., Fukata, Y., Nakamura, N., Matsuura, Y., and Kaibuchi, K., 1997, Formation of actin stress fibers and focal adhesions enhanced by Rho-kinase, *Science* **275:**1308–1311.

Chou, Y.-H., Bischoff, J. R., Beach, D., and Goldman, R. D., 1990, Intermediate filament reorganization during mitosis is mediated by $p34^{cdc2}$ kinase phosphorylation of vimentin, *Cell* **62:**1063–1071.

Drechsel, D. N., Hyman, A. A., Hall, A., and Glotzer, M., 1997, A requirement for Rho and Cdc42 during cytokinesis in *Xenopus* embryos, *Curr. Biol.* **7:**12–23.

Eriksson, J. E., Opal, P., and Goldman, R. D., 1992, Intermediate filament dynamics, *Curr. Opin. Cell Biol.* **4:**99–104.

Fuchs, E., and Weber, K., 1994, Intermediate filaments: Structure, dynamics, function, and disease, *Annu. Rev. Biochem.* **63:**345–382.

Inagaki, M., Nishi, Y., Nishizawa, K., Matsuyama, M., and Sato, C., 1987, Site-specific phosphorylation induces disassembly of vimentin filaments *in vitro*, *Nature* **328:**649–652.

Inagaki, M., Matsuoka, Y., Tsujimura, K., Ando, S., Tokui, T., Takahashi, T., and Inagaki, N., 1996, Dynamic property of intermediate filaments: Regulation by phosphorylation, *BioEssays* **18:**481–487.

Inagaki, M., Inagaki, N., Takahashi, T., and Takai, Y., 1997, Phosphorylation-dependent control of structures of intermediate filaments: A novel approach using site- and phosphorylation state-specific antibodies, *J. Biochem.* **121:**407–414.

Ishizaki, T., Maekawa, M., Fujisawa, K., Okawa, K., Iwamatsu, A., Fujita, A., Watanabe, N., Saito, Y., Kakizuka, A., Morii, N. and Narumiya, S., 1996, The small GTP-binding protein Rho binds to and activates a 160 kDa Ser/Thr protein kinase homologous to myotonic dystrophy kinase, *EMBO J.* **15:**1885–1893.

Kimura, K., Ito, M., Amano, M., Chihara, K., Fukata, Y., Nakafuku, M., Yamamori, B., Feng, J., Nakano, T., Okawa, K., Iwamatsu, A., and Kaibuchi, K., 1996, Regulation of myosin phosphatase by Rho and Rho-associated kinase (Rho-kinase), *Science* **273:**245–248.

Kishi, K., Sasaki, T., Kuroda, S., Itoh, T., and Takai, Y., 1993, Regulation of cytoplasmic division of *Xenopus* embryo by *rho* p21 and its inhibitory GDP/GTP exchange protein (*rho* GDI), *J. Cell Biol.* **120:**1187–1195.

Kitagawa, M., Mukai, H., Shibata, H., and Ono, Y., 1995, Purification and characterization of a fatty acid-activated protein kinase (PKN) from rat testis, *Biochem. J.* **310:**657–664.

Kosako, H., Amano, M., Yanagida, M., Tanabe, K., Nishi, Y., Kaibuchi, K., and Inagaki, M., 1997, Phosphorylation of glial fibrillary acidic protein at the same sites by cleavage furrow kinase and Rho-associated kinase, *J. Biol. Chem.* **272:**10333–10336.

Leung, T., Manser, E., Tan, L., and Lim, L., 1995, A novel serine/threonine kinase binding the Ras-related RhoA GTPase which translocates the kinase to peripheral membranes, *J. Biol. Chem.* **270:**29051–29054.

Leung, T., Chen, X. Q., Manser, E., and Lim, L., 1996, The p160 RhoA-binding kinase ROKα is a member of a kinase family and is involved in the reorganization of the cytoskeleton, *Mol. Cell. Biol.* **16:**5313–5327.

Mabuchi, I., Hamaguchi, Y., Fujimoto, H., Morii, N., Mishima, M., and Narumiya, S., 1993, A rho-like protein is involved in the organization of the contractile ring in dividing sand dollar eggs, *Zygote* **1:**325–331.

Machesky, L. M., and Hall, A., 1996, Rho: A connection between membrane receptor signalling and the cytoskeleton, *Trends Cell Biol.* **6:**304–310.

Madaule, P., Furuyashiki, T., Reid, T., Ishizaki, T., Watanabe, G., Morii, N., and Narumiya, S., 1995, A novel partner for the GTP-bound forms of *rho* and *rac*, *FEBS Lett.* **377:**243–248.

Matsui, T., Amano, M., Yamamoto, T., Chihara, K., Nakafuku, M., Ito, M., Nakano, T., Okawa, K., Iwamatsu, A., and Kaibuchi, K., 1996, Rho-associated kinase, a novel serine/threonine kinase, as a putative target for the small GTP binding protein Rho, *EMBO J.* **15:**2208–2216.

Matsuoka, Y., Nishizawa, K., Yano, T., Shibata, M., Ando, S., Takahashi, T., and Inagaki, M., 1992, Two different protein kinases act on a different time schedule as glial filament kinases during mitosis, *EMBO J.* **11:**2895–2902.

Matsuzawa, K., Kosako, H., Inagaki, N., Shibata, H., Mukai, H., Ono, Y., Amano, M., Kaibuchi, K., Matsuura, Y., Azuma, I., and Inagaki, M., 1997, Domain-specific phosphorylation of vimentin and glial fibrillary acidic protein by PKN, *Biochem. Biophys. Res. Commun.* **234:**621–625.

Mukai, H., and Ono, Y., 1994, A novel protein kinase with leucine zipper-like sequences: Its catalytic domain is highly homologous to that of protein kinase C, *Biochem. Biophys. Res. Commun.* **199:**897–904.

Mukai, H., Kitagawa, M., Shibata, H., Takanaga, H., Mori, K., Shimakawa, M., Miyahara, M., Hirao, K., and Ono, Y., 1994, Activation of PKN, a novel 120-kDa protein kinase with leucine zipper-like sequences, by unsaturated fatty acids and by limited proteolysis, *Biochem. Biophys. Res. Commun.* **204:**348–356.

Mukai, H., Toshimori, M., Shibata, H., Kitagawa, M., Shimakawa, M., Miyahara, M., Sunakawa, H., and Ono, Y., 1996, PKN associates and phosphorylates the head-rod domain of neurofilament protein, *J. Biol. Chem.* **271:**9816–9822.

Nishizawa, K., Yano, T., Shibata, M., Ando, S., Saga, S., Takahashi, T., and Inagaki, M., 1991, Specific localization of phosphointermediate filament protein in the constricted area of dividing cells, *J. Biol. Chem.* **266:**3074–3079.

Ogawara, M., Inagaki, N., Tsujimura, K., Takai, Y., Sekimata, M., Ha, M. H., Imajoh-Ohmi, S., Hirai, S., Ohno, S., Sugiura, H., Yamauchi, T., and Inagaki, M., 1995, Differential targeting of protein kinase C and CaM kinase II signalings to vimentin, *J. Cell Biol.* **131:**1055–1066.

Reid, T., Furuyashiki, T., Ishizaki, T., Watanabe, G., Watanabe, N., Fujisawa, K., Morii, N., Madaule, P., and Narumiya, S., 1996, Rhotekin, a new putative target for Rho bearing homology to a serine/threonine kinase, PKN, and rhophilin in the Rho-binding domain, *J. Biol. Chem.* **271:**13556–13560.

Ridley, A. J., and Hall, A., 1992, The small GTP-binding protein Rho regulates the assembly of focal adhesions and actin stress fibers in response to growth factors, *Cell* **70:**389–399.

Ridley, A. J., and Hall, A., 1994, Signal transduction pathways regulating Rho-mediated stress fibre formation: Requirement for a tyrosine kinase, *EMBO J.* **13:**2600–2610.

Sekimata, M., Tsujimura, K., Tanaka, J., Takeuchi, Y., Inagaki, N., and Inagaki, M., 1996, Detection of protein kinase activity specifically activated at metaphase–anaphase transition, *J. Cell Biol.* **132:**635–641.

Steinert, P. M., and Roop, D. R., 1988, Molecular and cellular biology of intermediate filaments, *Annu. Rev. Biochem.* **57:**593–625.

Takai, Y., Ogawara, M., Tomono, Y., Moritoh, C., Imajoh-Ohmi, S., Tsutsumi, O., Taketani, Y., and Inagaki, M., 1996, Mitosis-specific phosphorylation of vimentin by protein kinase C coupled with reorganization of intracellular membranes, *J. Cell Biol.* **133:**141–149.

Takaishi, K., Sasaki, T., Kameyama, T., Tsukita, S., Tsukita, S., and Takai, Y., 1995, Translocation of activated *Rho* from the cytoplasm to membrane ruffling area, cell–cell adhesion sites and cleavage furrows, *Oncogene* **11:**39–48.

Tsujimura, K., Ogawara, M., Takeuchi, Y., Imajoh-Ohmi, S., Ha, M. H., and Inagaki, M., 1994, Visualization and function of vimentin phosphorylation by cdc2 kinase during mitosis, *J. Biol. Chem.* **269:**31097–31106.

Watanabe, G., Saito, Y., Madaule, P., Ishizaki, T., Fujisawa, K., Morii, N., Mukai, H., Ono, Y., Kakizuka, A., and Narumiya, S., 1996, Protein kinase N (PKN) and PKN-related protein rhophilin as targets of small GTPase Rho, *Science* **271:**645–648.

Watanabe, N., Madaule, P., Reid, T., Ishizaki, T., Watanabe, G., Kakizuka, A., Saito, Y., Nakao, K., Jockusch, B. M., and Narumiya, S., 1997, p140mDia, a mammalian homolog of *Drosophila* diaphanous, is a target protein for Rho small GTPase and is a ligand for profilin, *EMBO J.* **16:**3044–3056.

Yano, S., Fukunaga, K., Ushio, Y., and Miyamoto, E., 1994, Activation of Ca^{2+}/calmodulin-dependent protein kinase II and phosphorylation of intermediate filament proteins by stimulation of glutamate receptors in cultured rat cortical astrocytes, *J. Biol. Chem.* **269:**5428–5439.

Chapter 15

Vimentin and Lipid Metabolism

Susan C. Schweitzer and Robert M. Evans

1. INTRODUCTION

Although all IF proteins possess certain canonical structural features, there are significant differences in the assembly and organizational properties of individual IF proteins (Klymkowsky *et al.*, 1989; Quinlan *et al.*, 1994). The importance of this diversity of IF proteins is not understood, and how this variation in the characteristics of IFs might influence cell structure or function is only beginning to be characterized. While it has been speculated that IF diversity may reflect differences in the functional requirements of specific cell types, there is a notable variation in the pattern of IF proteins that are expressed in given tissues among species (Herrmann *et al.*, 1996).

Over the past decade, there have been major advances in our understanding of the functional roles of the IFs that are characteristic of the epidermis and muscle. Numerous studies have shown that in these tissues IF proteins provide a mechanical strengthening function that protects cells against physical injury (Fuchs and Weber, 1994; Milner *et al.*, 1996; Li *et al.*, 1996). In addition, recent studies have provided strong evidence that in neurons, IFs are an important factor that influence axonal diameter (Eyer

Susan C. Schweitzer and Robert M. Evans Department of Pathology, University of Colorado Health Sciences Center, Denver, Colorado 80262.

Subcellular Biochemistry, Volume 31: Intermediate Filaments, edited by Herrmann and Harris. Plenum Press, New York, 1998.

and Peterson, 1994; Marszalek *et al.*, 1996). However, the functional significance of IF proteins that are characteristic of other tissues has been more difficult to demonstrate. This has been particularly true in the case of vimentin. In "higher" vertebrates, vimentin IFs are expressed in many mesenchymal cell types during development and are the only IF type found in a variety of cells including fibroblasts, endothelial cells, adipocytes, macrophages, neutrophils, and lymphocytes. Although cytoplasmic IFs are clearly not required for the survival of individual cells, there is no consensus on the functional role of this IF protein (Evans, 1998). Observations from a number of laboratories have raised the possibility that vimentin IFs are involved in some aspect, as yet undefined, of the intracellular metabolism of lipids.

2. *IN VITRO* VIMENTIN–LIPID INTERACTIONS

Direct interactions of vimentin with a variety of lipid preparations have been studied *in vitro*. Initially, it was observed that preparations of vimentin isolated from Ehrlich ascites tumor cells were contaminated with lipids (Traub *et al.*, 1985). Both phospholipids and neutral lipids were found in association with purified vimentin, although neutral lipids were a more prominent component (Traub *et al.*, 1985). Traub and co-workers have studied the interaction of vimentin and lipids using cosedimentation in sucrose density gradients. It was found that the interaction of vimentin IFs with lipid vesicles at low temperature (0°C) was stable during sucrose density gradient centrifugation (Perides *et al.*, 1986). This vimentin-lipid vesicle interaction appeared to be labile at 37°C, apparently resulting in fragmentation of the filaments during centrifugation, and the separation of the lipid vesicles from vimentin. However, the fragmented filaments were still associated with lipids in a nonvesicular form (Perides *et al.*, 1986). The presence of lipids strongly inhibits *in vitro* filament assembly, and the level of this inhibition of vimentin assembly is affected by the composition of the lipid used. Negatively charged phospholipids were more effective inhibitors than uncharged phospholipids (Perides *et al.*, 1986). Other experimental approaches have been used to demonstrate a direct vimentin–lipid interaction and to "map" the structural region of vimentin that interacts with lipid molecules. The interaction between vimentin and lipids appears to involve vimentin's NH_2-terminal "head" domain (Perides *et al.*, 1987). In studies that measured the release of carboxyfluorescein from loaded lipid vesicles, the presence of a purified NH_2-terminal domain fragment of vimentin resulted in a release of the impermeant fluorescent probe, suggesting that this region of the vimentin molecule created an irregularity in the lipid bilayer, resulting in a "leakiness" (Horkovics-Kovats and Traub, 1990). These results show that under *in vitro* conditions, the NH_2-terminal domain of vimentin can directly interact with phospholipid bilayers.

While not as extensively studied as vimentin–phospholipid interactions, an association of vimentin with neutral lipids has also been described. Neutral lipid incorporation into phospholipid bilayers is reported to increase the affinity of vimentin for phospholipid vesicles (Traub *et al.*, 1987). When purified vimentin was incubated with radiolabeled neutral lipids and the mixture separated by size-exclusion chromatography, a significant amount of the labeled lipid coeluted with vimentin (Traub *et al.*, 1987). The most stable interactions appeared to involve cholesteryl esters, although a significant association of vimentin was also observed with diacylglycerols, cholesterol, oxysterols, vitamin D_3, and the uncharged methyl ester derivatives of fatty acids (Traub *et al.*, 1987). This ambivalent nature of vimentin interactions with phospholipids and neutral lipids was also reflected in a difference in the domain of vimentin that appears to be involved in these interactions. Thus, while phospholipid interactions appear to require the NH_2-terminal "head" domain, the interaction of vimentin with neutral lipids was reported to involve the central α-helical "rod" domain (Traub *et al.*, 1987).

The significance of this interaction of vimentin with various lipid preparations *in vitro* is not entirely clear. Traub and colleagues (Perides *et al.*, 1987; Horkovics-Kovats and Traub, 1990) have speculated that the affinity of vimentin for lipid components could represent the molecular basis for the observed associations of vimentin IFs with cell membranes and membranous organelles. However, the nature of vimentin–membrane interactions in cells has not been well characterized, and there is some evidence that the association of vimentin with the plasma membrane involves an interaction of filament-associated proteins with membrane adhesion protein complexes (Seifert *et al.*, 1992; Svitkina *et al.*, 1996), rather than a direct interaction with the lipid bilayer. Whether vimentin IFs may directly interact with other lipid components *in vivo* remains an open question.

3. VIMENTIN ASSOCIATIONS WITH LIPID-RICH STRUCTURES IN CELLS

3.1. Adipocytes

A number of studies have indicated an association of vimentin IFs with lipid droplets in intact cells. The most dramatic reported example occurs in adipose cells, where vimentin IFs appear to have a unique organization that reflects a specialized association with lipid droplets. Franke *et al.* (1987) have shown that during the differentiation of 3T3-L1 preadipocyte cells in culture, the familiar fibroblastic vimentin IF organization is reorganized to form cagelike structures surrounding nascent lipid droplets. Electron microscopy of preadipose cells following adipose conversion showed regularly spaced parallel arrays of filaments (Franke *et al.*, 1987). It is not completely clear if these vimentin cagelike structures surround all the lipid droplets in adipose cells. Blanchette-Mackie *et al.* (1995) observed that while vimentin was present around small to medium-sized lipid droplets in preadipose cells, vimentin was more difficult to detect or absent from the surface of large lipid droplets. This formation of vimentin IFs surrounding lipid droplets does not appear to be a unique feature of the adipose conversion of cells in culture. The presence of a regular structure of filaments surrounding membrane-free lipid droplets has been described in adipose cells (Luckenbill and Cohen, 1966; Wood, 1967; Novikoff *et al.*, 1980; Franke *et al.*, 1987), cholesterol-loaded macrophages (McGookey and Anderson, 1983), and the foam cells of atherosclerotic lesions (Amanuma *et al.*, 1986).

The mechanism responsible for the reorganization of vimentin filaments observed during adipose conversion of preadipose cells is not known. Increased phosphorylation of IF proteins is known to alter their assembly properties *in vitro*, and is temporally associated with changes in filament organization in many cell types during cell division (reviewed in Inagaki *et al.*, 1996). However, it is not clear whether the change in vimentin IF organization during adipose conversion is associated with increased phosphorylation. Franke *et al.* (1987) reported that vimentin phosphorylation was not observably increased in 3T3-L1 cells during adipose conversion. We have analyzed detergent-insoluble proteins from 3T3-L1 cells by two-dimensional gel electrophoresis and have not observed an obvious increase in the acidic isoelectric focusing variants of vimentin (J. G. Lieber and R. M. Evans, unpublished observation) that are characteristic of increased phosphorylation (Evans and Fink, 1982). In contrast, Brandes *et al.* (1993) reported that an increase in the phosphorylation of a protein that appears to be vimentin is an early event during adipose conversion. The potential role of specific filament-associated proteins that could alter vimentin organization is another possibility that remains to be examined.

3.2. Steroidogenic Cells

Lipid droplets are also a prominent feature of the cytoplasm of steroidogenic cells (Hall, 1992). However, in contrast to adipose tissue, the lipid droplets of adrenal cells are enriched in cholesteryl esters rather than triglycerides. Cholesteryl ester-rich lipid droplets represent a stored form of cholesterol that is mobilized and transported to the mitochondria, where it is converted to pregnenolone that is then used for the synthesis of steroid hormones. Almahbobi and Hall (1990; Almahbobi *et al.*, 1992a) have shown that in cultured steroidogenic cells, there is a close physical association between lipid droplets and vimentin IFs. This IF organization seems to be distinct from the cagelike structures observed in adipose cells. In these studies, fluorescence microscopy of lipid droplets and vimentin IFs in detergent-extracted adrenal tumor cells indicated that the lipid droplets often colocalize with the vimentin IFs. Electron microscopy of intact and detergent-extracted adrenal and Leydig cells has shown what appear to be direct contacts between the IF network, lipid droplets, and mitochondria (Almahbobi *et al.*, 1992b, 1993; Almahbobi, 1995). Because of the proximity of vimentin, lipid droplets, and mitochondria, Almahbobi *et al.* (1992a) postulated that vimentin could have a role in the movement of cholesterol involved in steroidogenesis to the mitochondria.

4. PERTURBATION OF VIMENTIN ORGANIZATION IN PREADIPOSE CELLS AND TRIGLYCERIDE STABILITY

During the adipose conversion of 3T3-L1 preadipose cells, the formation of cagelike arrays of vimentin filaments at the surface of nascent lipid droplets, ensheathed by endoplasmic reticulum cisterna (Franke *et al.*, 1987), occurs simultaneously with a large increase in the cell's lipid synthetic capacity. In an effort to understand if this change in IF organization had a role in the accumulation of lipid droplets, Lieber and Evans (1996) studied the effects of perturbation of vimentin organization during adipose conversion of 3T3-L1 cells. Initial experiments showed that aggregation of IFs by treatment of the cells with nocodazole, during or shortly after induction of adipose conversion, dramatically reduced the lipid droplet content of 3T3-L1 cells (Lieber and Evans, 1996). A more specific, but transient disruption of vimentin IFs in cells injected with an antivimentin antibody also resulted in a decrease in lipid droplet formation during adipose conversion (Table I). To specifically and stably affect IF organization, a 3T3-L1 cell line was established that expressed a dominant-negative mutant vimentin cDNA under the regulation of a glucocorticoid-inducible hormone response element. Treatment of these cells with dexamethasone resulted in expression of the dominant-negative mutant vimentin, which led to the disruption of the endogenous IF network (Figures 1E and 1G). Induction of adipose conversion in these cells in the absence of the dominant-negative mutant vimentin led to the formation of lipid droplets that were comparable to those seen in untransfected 3T3-L1 cells. In con-

Table I
Effect of Anti-Intermediate Filament Antibody Microinjection on Lipid Droplet Accumulation in 3T3-L1 Cells[a]

Experiment	Time after microinjection (hr)	Lipid droplet formation: Uninjected, Percent $(n)^b$	B IgG-injected: Percent $(n)^b$	B IgG-injected: Percent of control[c]	IFA-injected: Percent $(n)^b$	IFA-injected: Percent of control[c]
1	48	26 (444)	22 (577)	85	7 (1006)	27
	72	43 (954)	42 (1112)	98	21 (1044)	49
2	48	52 (676)	47 (626)	90	30 (612)	58
	72	50 (1009)	57 (1029)	114	28 (947)	56
3	48	32 (718)	39 (1345)	121	18 (959)	56
	72	56 (1288)	63 (1900)	112	36 (1478)	64

[a]Cells were induced to differentiate with 2 μg/ml ADD 4743 and 10^{-7} M insulin for 24 hr prior to microinjection. Cells in adjacent defined fields were microinjected with either purified bovine IgG (B IgG) or anti-intermediate filament IFA (IFA) antibodies. At the indicated times after injection, the cells werestained with oil red O and the number of cells in each injected field with lipid droplet accumulations was determined. Reproduced from Lieber and Evans (1996) with permission.
[b]Percentage of cells with lipid droplets ≥3 μm in diameter; the value in parentheses gives the total number of cells in the injection field.
[c]Data expressed as a percentage of the uninjected control value.

trast, induction of adipose conversion in the cells expressing the dominant-negative mutant vimentin was found to result in a significant reduction in lipid droplet accumulation (Lieber and Evans, 1996). This impaired ability to accumulate lipid droplets was not associated with an effect on triglyceride synthesis. Although there was a failure to form prominent lipid droplets or to accumulate [^{14}C]-palmitate-labeled triglycerides (Figure 2), expression of the dominant-negative vimentin did not affect the induction of the marker enzyme glycerol-3-phosphate dehydrogenase or the incorporation of [^{14}C]-palmitate into triglycerides during a 10-min pulse label. Rather, pulse-chase experiments indicated that the failure of these cells to accumulate triglyceride was associated with an increased rate of turnover (Lieber and Evans, 1996). This effect was specific for triglycerides, as the synthesis and turnover of phospholipids (Lieber and Evans, 1996) and cholesteryl esters (J. G. Lieber and R. M. Evans, unpublished observation) were not altered by perturbation of vimentin IF organization. These studies indicated that vimentin IFs provide a function that influences lipid stability during adipose conversion of 3T3-L1 cells.

The mechanism by which vimentin IFs might affect the stability of accumulated triglycerides has not yet been determined. While the meta-

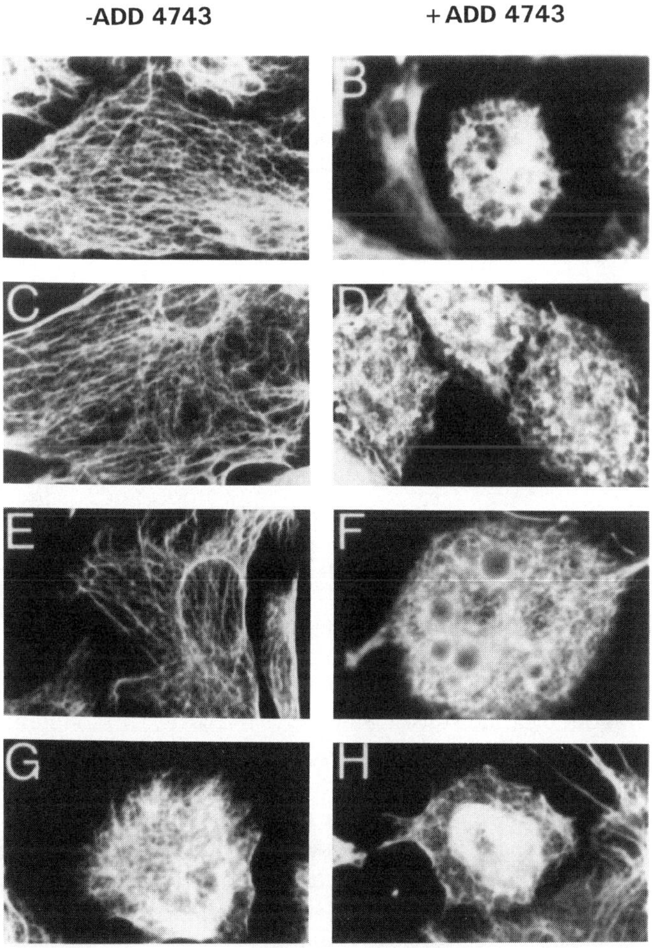

FIGURE 1. Hormone-inducible disruption of vimentin filament organization during adipose conversion. The 83ΔC/9A2 cell line was derived from 3T3-L1 cells by stable transfection with a cDNA encoding a COOH-terminally truncated mutant vimentin under the regulation of the hormone response element of mouse mammary tumor virus. The figure shows the antivimentin immunofluorescence staining pattern of 3T3-L1 cells (A–D) and 83ΔC 9A2 cells (E–H) that were untreated (A, E) or treated with 10^{-7} M dexamethasone for 96 hr (C, G), and 96 hr after induction of adipose conversion with 2 μg/ml ADD 4743 and 10^{-7} M insulin alone (B, F) or in the presence of 10^{-7} M dexamethasone (D, H). Reproduced from Lieber and Evans (1996) with permission.

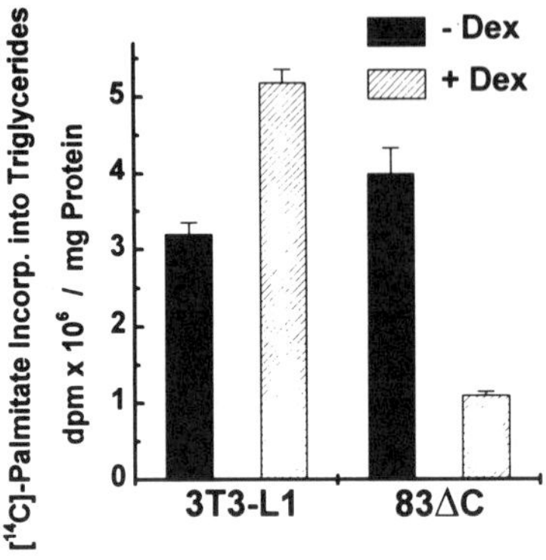

FIGURE 2. The effect of dexamethasone on the accumulation of [^{14}C]-palmitate-labeled triglycerides during adipose conversion in 3T3-L1 cells and 83ΔC/9A2 cells that express the hormone-inducible dominant-negative mutant vimentin. Adipose conversion was induced with 2 μg/ml ADD 4743 and 10^{-7} M insulin in the absence (−Dex) or presence of 10^{-7} M dexamethasone (+Dex). At 96 hr after induction the cells were radiolabeled with [^{14}C]-palmitate for 24 hr and incorporation into triglycerides determined. The values represent the average obtained from triplicate cultures ± S.E.M. Reproduced from Lieber and Evans (1996) with permission.

bolic processes that are required for lipid accumulation in preadipose cell lines have been extensively examined (Ailhaud *et al.*, 1992; Smyth *et al.*, 1993), the molecular basis of the movement of lipid components to and from triglyceride-rich lipid droplets is not understood, and the proteins that are associated with the periphery of lipid droplets have only begun to be characterized (Greenberg *et al.*, 1992). The rate-limiting step in triglyceride lipolysis is catalyzed by hormone-sensitive lipase. It has been reported that the distributions of hormone-sensitive lipase and other lipid droplet components are altered under conditions of increased triglyceride turnover. In isoproterenol-stimulated preadipose cells, Egan *et al.* (1992) described a translocation of hormone-sensitive lipase to the surface of the lipid droplets and speculated that lipid droplet surface proteins such as perilipin, a lipid droplet component that is expressed during adipose conversion, might serve to regulate this interaction. The hypothesis that perturbation of the vimentin IF structure during the formation of lipid droplets interferes with some aspect of this process and results in an inappropriate hydrolysis of triglycerides is testable. It should be possible to determine if the distribution of hormone-sensitive lipase is affected by perturbation of vimentin IFs during adipose conversion in preadipose cells.

5. VIMENTIN EXPRESSION IN IF-FREE ADRENAL CELLS AND THE METABOLISM OF SPECIFIC CELLULAR LIPIDS

5.1. Intracellular Cholesterol Transport

Cholesterol is required for normal membrane biogenesis and steroidogenesis. Cells obtain cholesterol both from endogenous synthesis and from

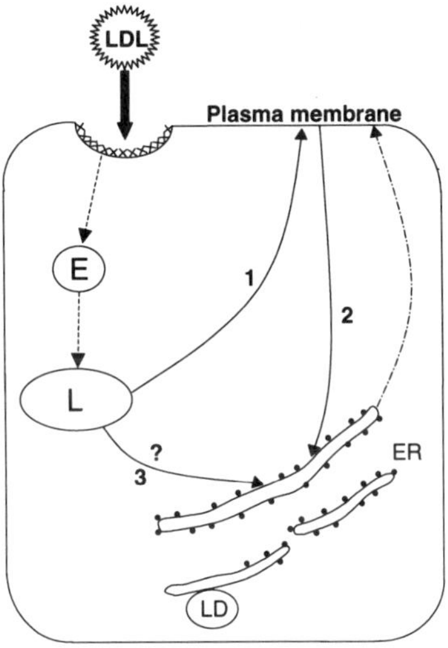

FIGURE 3. Intracellular cholesterol transport in mammalian cells. Endogenous cholesterol synthesized in the ER is trafficked to the plasma membrane (- - →). Exogenous cholesterol, derived primarily from LDL, is taken up by receptor-mediated endocytosis and transported via endosomes (E) to the lysosome (L), where cholesteryl esters are hydrolyzed to free cholesterol. Lysosomal cholesterol can then be trafficked to the plasma membrane (1 →). Excess plasma membrane cholesterol can be trafficked from the plasma membrane to the ER (2 →) for esterification by ACAT, and storage as cholesteryl-ester-rich lipid droplets (LD). Recent observations have also suggested a postlysosomal pathway that bypasses the plasma membrane (3 →) (see Section 5.1).

the uptake of cholesteryl ester-rich lipoproteins, principally low-density lipoproteins (LDL). A summary diagram of the pathways thought to be involved in intracellular cholesterol transport is shown in Figure 3. Cholesterol produced by synthesis in the endoplasmic reticulum (ER) has been shown to be transported to the plasma membrane by a process that is rapid and energy- and temperature-dependent, but distinct from the secretory pathway (Liscum and Underwood, 1995). Trafficking of endogenously synthesized cholesterol is not affected by drugs or agents that disrupt the cytoskeleton or inhibit lysosomal function (Liscum and Underwood, 1995). Exogenous uptake of cholesterol occurs primarily through receptor-mediated endocytosis of LDL (Brown and Goldstein, 1986). After internalization via clathrin-coated pits, the receptor : apolipoprotein complex enters the endosomal system where the receptor is recycled to the plasma membrane and the lipoprotein component is delivered to the lysosome. Once in the lysosome, enzymes degrade the LDL particle, and the lipid core of cholesteryl ester is hydrolyzed to yield free cholesterol (Brown and Goldstein, 1986). This is referred to as “lysosomal” or “LDL-derived” cholesterol.

The subsequent intracellular transport of lysosomal cholesterol has not been precisely defined. Studies of the human disorder Niemann–Pick type C disease (Liscum *et al.*, 1989; Argoff *et al.*, 1991), mutant cells selected for defective cholesterol transport (Cadigan *et al.*, 1988; Dahl *et al.*, 1992), and of the effect of a variety of inhibitors (Liscum, 1990) have indicated that transport of lipoprotein-derived cholesterol occurs independently from the transport of endogenously synthesized cholesterol. Although there is good evidence that the postlysosomal transport of cholesterol involves specific cytoplasmic components, it is not clear whether this involves vesicle-mediated transport or a carrier molecule (Liscum and Underwood, 1995). There are also significant differences in the reported fate or route of intracellular trafficking of cholesterol after it leaves the lysosome. Initial studies with fibroblasts had demonstrated a rapid and unregulated transport of free cholesterol from lysosome to the plasma membrane (Brasaemle and Attie, 1990; Johnson *et al.*, 1990). This transport is reported to be energy independent and not affected by agents that disrupt the cytoskeleton (Liscum and Underwood, 1995). The plasma membrane, therefore, acts as an acceptor of lysosomal cholesterol and also as a donor of free cholesterol both to acyl CoA:cholesterol acyltransferase (ACAT) in the ER for esterification and to suitable acceptors in the extracellular medium. The subsequent transport of plasma membrane cholesterol to the site of esterification is reported to be energy dependent (Skiba *et al.*, 1996) and inhibited by anticytoskeletal drugs (Liscum, 1990; Nagy and Freeman, 1990) and brefeldin A (Stein *et al.*, 1992).

In contrast to the evidence that lysosomal cholesterol is first trafficked through the plasma membrane, some recent studies have suggested that lysosomal cholesterol may be transported from the lysosome to the ER by a pathway that bypasses the plasma membrane (Figure 3). Inhibition of lysosomal cholesterol transport using hydrophobic amines (Underwood *et al.*, 1996) suggested an additional pathway that is more sensitive to inhibition than the trafficking of lysosomal cholesterol through the plasma membrane. Based on genetic complementation of mutant CHO cell lines, Spillane and co-workers demonstrated increased lysosomal cholesterol transport that was independent of the plasma membrane cholesterol pool (Spillane *et al.*, 1995). Similarly, using cyclodextrins to follow the intracellular movement of cholesterol, Neufeld *et al.* (1996) observed that as much as 30% of the lysosomal cholesterol in these experiments moved to the ER by a plasma membrane-independent mechanism. Additional support for transport of lysosomal cholesterol to the ER by a mechanism that bypasses the plasma membrane has come from studies of fibroblasts from individuals with the hereditary lipid storage disease Niemann Pick type C. Blanchette-Mackie *et al.* (1988) have shown that under appropriate conditions, in addition to an abnormal lysosomal storage, lipoprotein-derived cholesterol accumulates in the Golgi complex of these cells, implicating the Golgi compartment in the postlysosomal trafficking of cholesterol. Therefore, there now appears to be significant evidence for postlysosomal trafficking of cholesterol to the ER by some mechanism that does not require transport to the plasma membrane, but the nature of these various pathways of trafficking remains to be characterized.

5.2. Lipoprotein Cholesterol

Cholesterol derived from LDL is an important substrate for steroidogenesis (Faust *et al.*, 1977). Various reports had indicated that drugs that disrupt microfilaments or microtubules affect steroid production, suggesting that some element of the cytoskeleton is involved in the intracellular movement of cholesterol in steroidogenic cells (Temple and Wolff, 1973; Crivello and Jefcoate, 1980). Inhibitors of microtubule assembly had been shown to inhibit intracellular movement of cholesterol in adrenal cells, leading Crivello and Jefcoate (1980) to conclude that microtubules were involved in this process. However, these experiments were carried out before it was widely appreciated that drugs that affect microtubules also result in perinuclear aggregation of IFs (Geuens *et al.*, 1983). The close physical association of vimentin IFs and cholesterol-rich lipid droplets in steroidogenic cells suggested that vimentin might be involved in some aspect of cholesterol transport (Almahbobi *et al.*, 1992a).

In order to examine more directly the role of vimentin in the intracellular metabolism of cholesterol in adrenal cells, Sarria *et al.* (1990) isolated individual cell lines from a human adrenal tumor cell line, SW-13, that spontaneously contained vimentin or lacked any detectable cytoplasmic IF network (IF-free cells) (Hedberg and Chen, 1986). Because it was not known what other differences might exist between SW-13 cell lines that spontaneously contain or lack vimentin, IF-free SW-13 cells were transfected with a mouse vimentin transgene, and stable cell lines with various levels of vimentin expression were established (Sarria *et al.*, 1992). The initial characterization of the metabolism of the major lipid components of these cells indicated that while there was no significant difference in phospholipid or glyceride synthesis, the IF-free cells exhibited higher levels of cholesterol synthesis and lower levels of cholesterol esterification than the cells that contained vimentin. This difference in cholesterol metabolism between cells that contain or lack vimentin was found to depend on the presence of low-density lipoprotein (LDL) in the culture medium, and showed that IF-free cells had an impaired ability to store LDL-cholesterol (Sarria *et al.*, 1992). The differences in the capacity of these cell lines to esterify lipoprotein-derived cholesterol appear to be related specifically to vimentin expression. Table II shows that the incorporation of [^{14}C]-oleate into cholesteryl esters was increased in relation to the level of vimentin expression in adrenal tumor cells, derived from a parental IF-free cell line, that stably express a vimentin cDNA.

Table II
Relative Vimentin Expression and Cholesterol Esterification in SW-13 Cell Lines Stably Transfected with a Mouse Vimentin cDNA[a]

Cell line	Relative vimentin expression[b]	Percent of cells that contain vimentin[c]	[^{14}C]Oleate incorporation into cholesterol esters (cpm $\times$ 10^3/mg protein/hr)[d]
cl.1 Vim$^+$	21.3	97.4	60.6 $\pm$ 7.1
T3M Vim$^+$	3.9	87.1	43.7 $\pm$ 8.3
T4M Vim$^+$	1.2	28.3	26.2 + 4.2
T19M Vim$^-$	0.4	1.5	17.5 $\pm$ 6.8
cl.2 IF-Free	0	0.5	14.5 $\pm$ 7.4

[a]The parental cl.2 IF-free cell line and the cl.1 vim$^+$ line that expresses human vimentin were included as references. The stable lines T3M, T4M, and T19M were derived from cl.2 vim$^-$ cells following transfection with a mouse vimentin cDNA (Sarria *et al.*, 1990). Cells were incubated in medium containing 50 μg ml LDL for 24 hr before labeling with [^{14}C]-oleate. Adapted from Sarria *et al.* (1992) with permission.
[b]An SDS–PAGE of Triton-insoluble proteins from [^{35}S]-methionine-labeled cells was scanned for [^{35}S]-vimentin as described in Sarria *et al.* (1992). Values represent normalized pixel values − background $\times$ 10^5.
[c]Percent of cells that contained vimentin IFs visible by indirect immunofluorescence.
[d]Values represent the mean of 4–6 determinations $\pm$ S.E.M.

5.3. Intermediate Filaments and Lysosomal Cholesterol Transport

As described in Section 5.2, studies with IF-free and vimentin-expressing lines of SW-13 adrenal tumor cells had indicated that IF-free cells exhibited an impaired ability to store lipoprotein-derived cholesterol compared with cells that express vimentin IFs. This effect of vimentin on cholesterol metabolism appears to involve some step in postlysosomal transport. Comparison of IF-free and vimentin-expressing SW-13 cells indicated that there was no measurable difference in the uptake of [^{125}I]-LDL or the subsequent lysosomal hydrolysis of LDL-cholesteryl esters (Sarria *et al.*, 1992), indicating that the impaired capacity of IF-free cells to accumulate cholesteryl esters was not associated with an effect on delivery of cholesterol to the lysosome. Moreover, the presence or absence of vimentin IFs had no effect on esterification of endogenously synthesized cholesterol or 25-hydroxycholesterol stimulation of ACAT activity, indicating that the impaired capacity of the IF-free cells to esterify lysosmal cholesterol was not due to a difference in ACAT function (Sarria *et al.*, 1992). Moreover, anti-ACAT immunofluorescence of IF-free and vimentin-expressing SW-13 cell lines (Figure 4) indicated that vimentin expression did not obviously affect the cytoplasmic distribution of ACAT in these cells.

The uptake and subsequent metabolism of LDL-cholesterol was examined in vimentin-expressing and IF-free cells with time after a 2-hr pulse

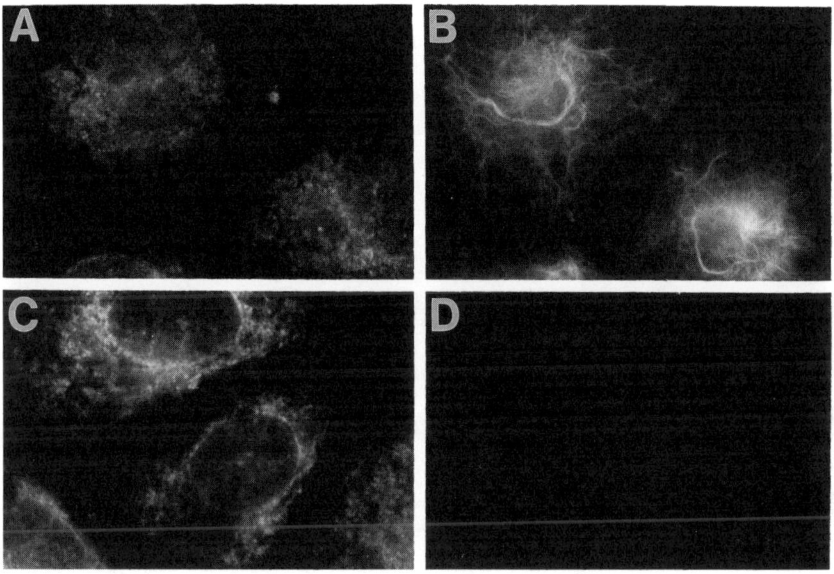

FIGURE 4. The presence or absence of vimentin does not affect the cytoplasmic distribution of ACAT. The figure shows the anti-ACAT (gift of Dr. T. Y. Chang) (A, C) and antivimentin (B, D) immunofluorescence of SW-13 vim$^+$ (A, B) and IF-free (C, D) cell lines.

with [^{3}H]-cholesteryl linoleate-LDL. Both the culture medium and the monolayer were recovered, the lipids extracted, and the specific radioactivity in cholesterol and specific cholesteryl esters determined. As shown in Figures 5A and 5B, the rapid initial loss of [^{3}H]-cholesteryl linoleate and appearance of [^{3}H]-cholesterol in the monolayers was reflective of delivery of the cholesteryl linoleate to the lysosome and subsequent deesterification. This formation of free cholesterol from LDL was not appreciably different in the IF-free and vimentin-expressing cell lines. Over the next 46 hr the monolayer content of [^{3}H]-cholesterol gradually declined in all three cell lines to a similar extent (Figure 5B). The rate of appearance of [^{3}H]-cholesteryl oleate and palmitate in the cells is indicative of the arrival of lysosomally derived cholesterol at the ER, and therefore represents postlysosomal transport. During the initial 12 hr after the pulse, there was a rapid formation of [^{3}H]-cholesteryl oleate and palmitate in the cells that expressed vimentin, whereas reesterification was not detectable in the IF-free cells until 24 hr after the initial pulse (Figure 5C). These results indicate that some aspect of postlysosomal cholesterol transport is affected by vimentin IFs in SW-13 cells. Because the chase was conducted in culture medium containing LDL and serum proteins that can act as cholesterol

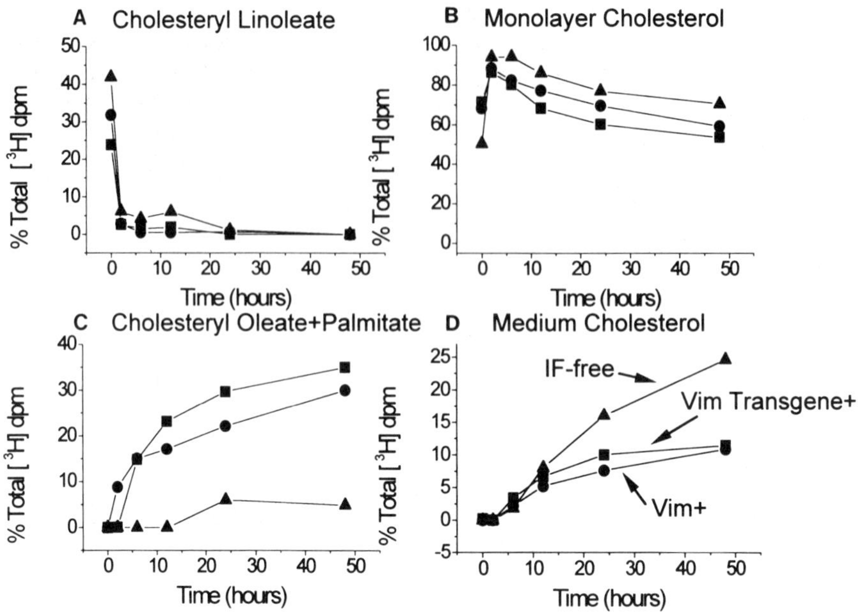

FIGURE 5. Uptake and metabolism of [^{3}H]-cholesteryl linoleate-LDL in vim$^+$ and IF-free SW-13 cell lines. Cells were cultured in medium containing 5% Controlled Process Serum Replacement (CPSR) and 50 μg/ml LDL for 4 days. Cultures were pulse-labeled for 2 hr in the presence of [^{3}H]-cholesteryl linoleate-loaded LDL (CL-LDL, 2 μg/ml, 3×10^7 cpm/mg protein) in medium containing 5% CPSR and then chased in the same medium containing 50 μg/ml unlabeled LDL. At the indicated times after removal of the labeled CL-LDL, the culture medium and cell monolayer content of free and esterified cholesterol was determined. (A) Percent total [^{3}H] present in the monolayer as cholesteryl linoleate. (B) Percent total [^{3}H] present in the monolayer as free cholesterol. (C) Percent total [^{3}H] present in the monolayer as cholesteryl oleate and palmitate. (D) Percent total [^{3}H] recovered as free cholesterol in the culture medium. Each time point represents the average of duplicate determinations for SW-13/cl.1 vim$^+$ that spontaneously express vimentin (●), SW-13/cl.2 IF-free cells (▲), and SW-13/T3M vim$^+$ cells, derived from cl.2, that express a vimentin transgene (■). Adapted with from Sarria *et al.* (1992) with permission.

acceptors, the medium was also examined for desorbed [^{3}H]-cholesterol. As shown in Figure 5D, during the initial 2 hr chase there was no detectable [^{3}H]-cholesterol in the culture medium. During the next 10 hr, [^{3}H]-cholesterol appeared in the medium in all three cell lines at similar rates. Thereafter, the rate of cholesterol loss from the monolayer into the medium was greater in the IF-free cells compared with the cells that expressed vimentin. This difference in the rate of appearance of [^{3}H]-cholesterol in the culture medium appeared to be inversely related to the difference in reesterification in the monolayers.

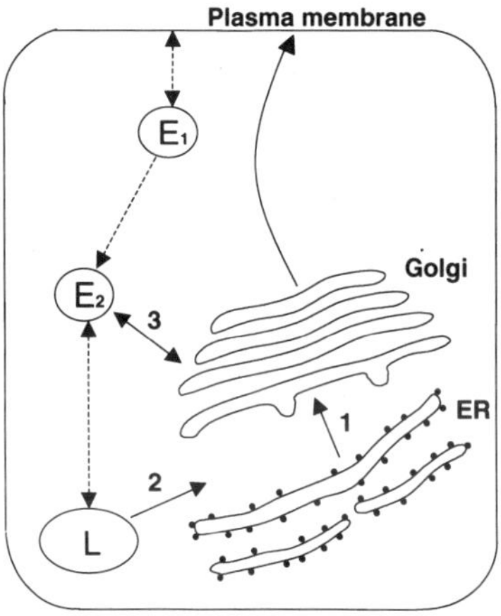

FIGURE 6. Glycosphingolipid metabolism. Following *de novo* synthesis in the ER and Golgi, GSLs are transported to intracellular sites or to the plasma membrane (1 →). Alternatively, GSLs can be recycled (↔) from the plasma membrane or intracellular sites through early (E_1) and late (E_2) endosomes and/or lysosomes (L), and components used for synthesis in recycling pathways (2 → and 3 →) (see Section 5.4).

5.4. Glycosphingolipids

Glycosphingolipids (GSLs) are a family of carbohydrate-containing sphingolipids. They play an important role in the nervous system, and defects in sphingolipid metabolism are associated with several human genetic diseases (Pennock, 1994; Marinetti, 1990). GSLs are synthesized *de novo* in the ER from palmitate and serine to form a sphingoid base. Ceramide results from addition of a fatty acid acyl group to this base. GSLs are then formed by the transfer of sugar residues to ceramide from the appropriate sugar nucleotide, with each step being catalyzed by a specific glycosyltransferase. Both the ER and the Golgi apparatus contain at least some of the unique transferases necessary for the synthesis of GSLs, but the more complex species are assembled in the Golgi. Alternatively, GSLs can be synthesized from recycled components of sphingolipids. Lysosomal hydrolysis can generate sphingosine and/or ceramide intermediates which, when transported to the ER or Golgi, can be reassembled into GSLs. Following synthesis, GSLs are transported to the cell surface and to their specific intracellular locations. Figure 6 illustrates the *de novo* and recycling pathways of GSLs .

A possible association of vimentin IFs with the intracellular distribution of GSLs was first suggested by the observations of Gillard *et al.* (1991, 1992), who found that specific GSLs colocalized with vimentin IFs in a variety of cell lines. To determine if this association might have functional consequences, Gillard *et al.* (1994) independently isolated IF-free SW-13 adrenal tumor cell lines and derived stable transfectant lines that expressed a vimentin transgene. A comparison of the biosynthesis and transport of GSLs in the presence and absence of a vimentin IF network in these cells showed that the IF-free cells had a decreased ability to incorporate radiolabeled sugars into GSLs, and this defect could be corrected by expressing a mouse vimentin cDNA (Gillard *et al.*, 1994). This defect in GSL synthesis did not affect all GSLs equally, but was most pronounced for LacCer and Gb3Cer, both of which are early precursors for more complex GSLs and were found to be synthesized largely by recycling pathways. Additional studies using metabolic inhibitors to specifically inhibit *de novo* biosynthesis and/or recycling pathways showed that the difference in the capacity of IF-free and vimentin-expressing SW-13 cells to accumulate LacCer and Gb3Cer GSLs was primarily associated with the recycling pathway. These observations indicate that vimentin expression in these cells has an effect on lysosome to ER and/or endosome-to-Golgi trafficking of sphingolipid intermediates (Gillard *et al.*, 1996).

5.5. Lipid Droplets

As described in Section 4, perturbation of vimentin IF organization in preadipose cells impaired triglyceride-rich lipid droplet formation during adipose conversion. Although vimentin in adrenal tumor cells does not appear to undergo the sort of dramatic reorganization seen during adipose conversion of 3T3-L1 cells, we have examined the effect of vimentin expression and organization on the capacity of SW-13 adrenal tumor cell lines to form lipid droplets. In these experiments, cells were depleted of endogenous lipid droplets by culture in medium containing a delipidized serum replacement (Sigma CPSR-1). Lipid droplet formation was then induced by addition of 1 mM oleate-BSA to the culture medium. Oleate-induced lipid droplets were examined 48 hr later by Nile red fluorescence microscopy (Figure 7 and Table III). Under these conditions there was no obvious difference in the capacity of an IF-free and a vimentin containing SW-13 cell line to form lipid droplets (Table III). This result demonstrates that oleate-induced lipid droplet formation in SW-13 cells is not influenced by the presence or absence of vimentin IFs. Perturbation of endogenous

FIGURE 7. The effect of dominant-negative mutant vimentin expression on oleate-BSA-induced lipid droplet formation in SW-13 and 3T3-L1 cell lines. The figure shows the Nile red fluorescence of cells in the absence of oleate-BSA (A, C, E, G) or 48 hr after addition of 1 mM oleate-BSA to the culture medium (B, D, F, H). The cell line SW-13 cl.1 vim$^+$ (A, B) has normal vimentin IF organization, while SW-13/83ΔC/1A4 (C, D) expresses a COOH-terminally truncated vimentin transgene which perturbs endogenous vimentin IF organization (Sarria *et al.*, 1994). The 3T3-L1 83ΔC/9A2 cells exhibit a dexamethasone-inducible expression of the COOH-terminally truncated vimentin transgene and have normal vimentin organization (see Figure 1) in the absence of dexamethasone (E, F) and disrupted vimentin organization after treatment with 10^{-6} M dexamethasone (G, H) (see Figure 1).

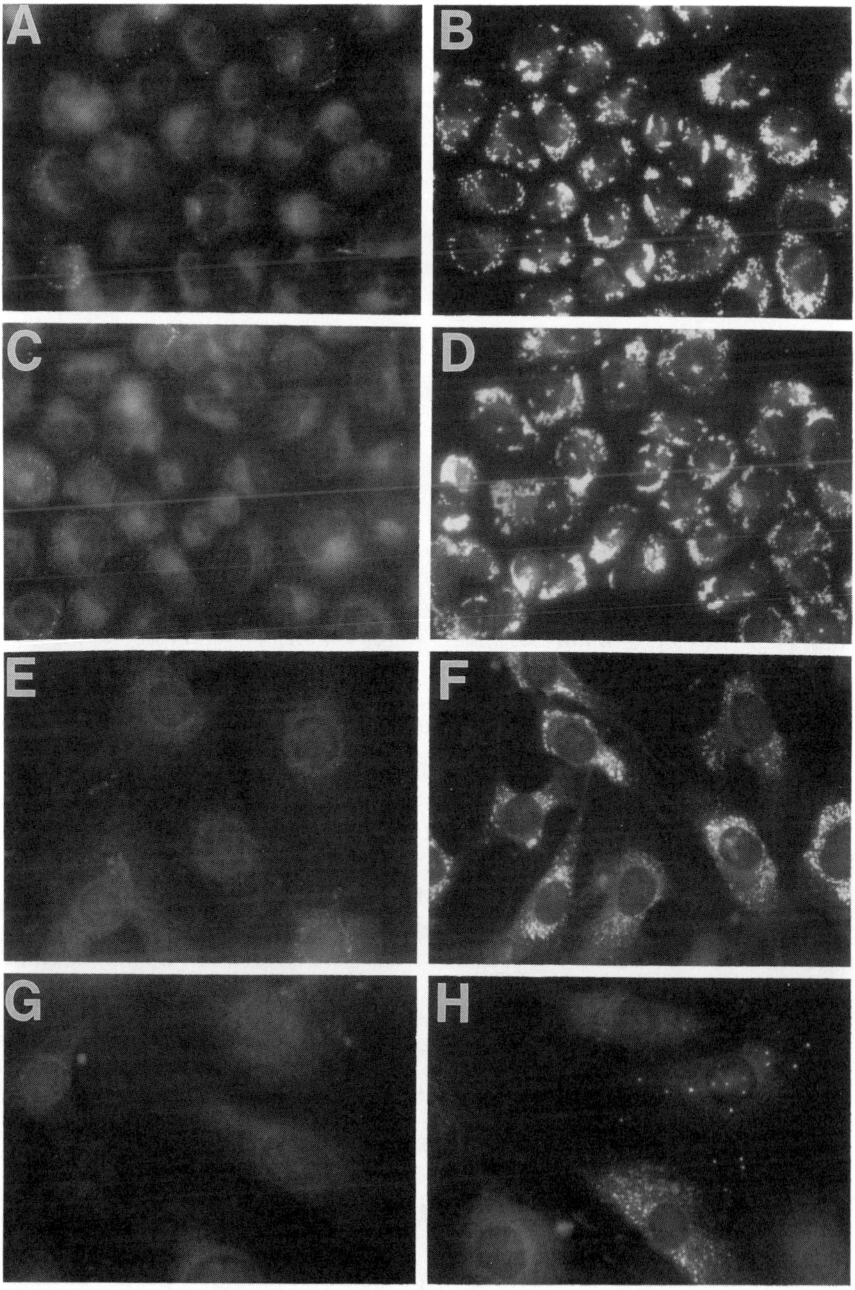
A
B
C
D
E
F
G
H

Table III
Oleate-BSA-Induced Lipid Droplet Formation in SW-13 and 3T3-L1 Cell Lines

Cell line	Additions	Percent of cells with prominent Nile red fluorescence[a]
SW-13/cl.1 Vim$^+$	—	100
SW-13/83ΔC/1A4[b]	—	99
SW-13/cl.2 IF-free	—	99
3T3-L1	—	90
	+ Dexamethasone	89
83ΔC/9A2[c]	—	80
	+ Dexamethasone	33

[a] At 48 hr after addition of 1 mM oleate-BSA to the culture medium, cells cultured in the absence of oleate had essentially no prominent Nile red lipid droplets.
[b] 83ΔC/1A4 was derived from SW-13/cl.1 vim$^+$ and constituitively expresses a COOH-terminally truncated dominant-negative vimentin transgene that disrupts the organization of the endogenous vimentin IFs (Sarria *et al.*, 1994).
[c] 83ΔC/9A2 was derived from 3T3-L1 and expresses the COOH-terminally truncated dominant-negative vimentin transgene that disrupts the organization of the endogenous vimentin IFs in a dexamethasone-inducible manner (Lieber and Evans, 1996).

vimentin IF organization similarly had no apparent effect on lipid droplet formation in SW-13 cells. As shown in Figure 7 and Table III, lipid droplet formation in a stable cell line derived from vimentin-positive SW-13 cells that expresses a dominant-negative mutant vimentin and exhibits disrupted vimentin IF organization (Sarria *et al.*, 1994) was not observably different from the parental SW-13 cell line. As a positive control, expression of the same dominant-negative mutant vimentin was induced in stably transfected 3T3-L1 cells and its affect on oleate-induced lipid droplet formation examined. As shown in Table III, compared with untransfected 3T3-L1 cells, induction of the synthesis of the dominant-negative vimentin resulted in an obvious decrease in oleate-induced formation of lipid droplets in the transfectant cell line. These results indicate that although perturbation of vimentin IF organization affected lipid droplet formation in 3T3-L1 cells, perturbation of vimentin IF organization in SW-13 cells had no obvious effect on lipid droplet formation in response to exogenous oleate. These studies show that vimentin IFs are not required for the formation of lipid droplets in all cells, and the apparent relationship between vimentin IF organization and lipid droplet formation in 3T3-L1 cells may involve some cell-type-specific factor.

6. VIMENTIN EXPRESSION IN IF-FREE FIBROBLASTS

The triglyceride metabolism in preadipose cells and the metabolism of cholesterol and GSLs in adrenal cortical cells may represent cell-type-specific examples of an effect of vimentin on lipid metabolism. To investigate whether vimentin may have similar effects on triglyceride or cholesterol metabolism in other cell types, experiments were performed to compare triglyceride accumulation and cholesterol esterification in primary embryo fibroblasts derived from mice that were homozygous for a vimentin null mutation (−/−) (Colucci-Guyon *et al.*, 1994) with that of fibroblasts similarly derived from vim$^{+/+}$ mice (gift of C. Babinet). Although initial experiments that examined the incorporation of [^{14}C]-oleate into triglycerides and cholesteryl ester showed a substantially greater level of triglyceride synthesis in the vim$^{+/+}$ cells compared to the vim$^{-/-}$ cultures, there was a notable difference in the growth characteristics of these cultures (vim$^{+/+}$ cells grew more rapidly), which made these results impossible to compare on an equivalent basis (data not shown). Therefore, similar experiments were conducted with an immortalized cell line derived from vim$^{-/-}$ fibroblasts. These cells (Mtetvim-8 cells) contain no other detectable cytoplasmic IF proteins, but express a vimentin transgene in a tetracycline-regulatable manner (Holwell *et al.*, 1997). In the presence of low concentrations of tetracycline (0.1 μg/ml), Mtetvim-8 cells contain little detectable vimentin. When tetracycline is removed from the culture medium, the cells rapidly express vimentin, and within 72 hr contain abundant vimentin IF networks (Holwell *et al.*, 1997). To determine the effect of vimentin synthesis on triglyceride metabolism and cholesterol esterification, we determined the rate of incorporation of [^{14}C]-oleate into triglycerides and cholesteryl ester in LDL-loaded cells in the presence or absence of vimentin expression. As shown in Figure 8, vimentin expression had no significant effect on the rate of incorporation of the labeled fatty acid into either triglycerides or cholesteryl oleate. These results indicate that in fibroblasts, vimentin expression and assembly into extended filament networks by itself does not measurably affect the accumulation of triglycerides. However, whether vimentin may affect cholesterol transport in these cells remains unclear. The level of total cholesterol esterification in these experiments was barely detectable, and experiments with radiolabeled cholesteryl linoleate LDL were unable to measure reesterification of lipoprotein-derived cholesterol (not shown), indicating that these fibroblasts express very low levels of ACAT. Because of this low esterification activity, it is not possible to draw any conclusions on the possible role of vimentin in the transport of lipoprotein-derived cholesterol in these cells.

7. THE "DILEMMA" OF THE VIMENTIN KNOCKOUT MOUSE

The targeted inactivation of individual IF protein genes has produced a variety of phenotypes that range from dramatic effects on the capacity of

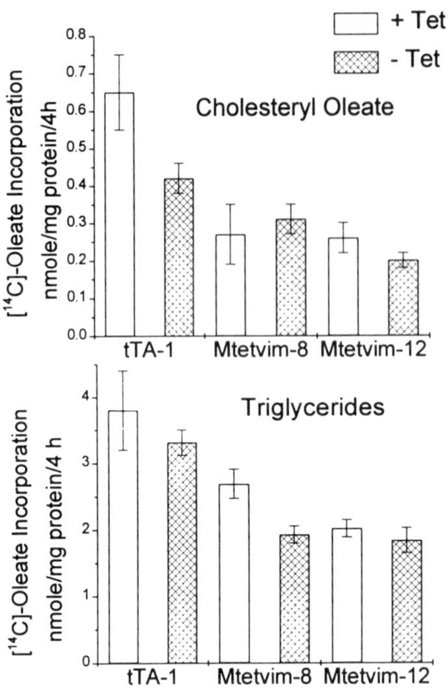

FIGURE 8. The effect of tetracycline-regulated vimentin expression on the incorporation of [^{14}C]-oleate into cholesteryl oleate and triglycerides in vim$^{-/-}$ fibroblasts. The Mtetvim-8 and Mtetvim-12 cell lines were derived from the IF-free vim$^{-/-}$ tTA-1 cell line and contain a mouse vimentin transgene under the regulation of the tetracycline-responsive promoter. In the presence of 0.1 μg/ml tetracycline (+Tet), Mtetvim-8 and Mtetvim-12 cells contain little vimentin, but 72 hr after tetracycline removal from the medium (−Tet), these cells express the vimentin transgene and contain vimentin IFs (Holwell *et al.*, 1997). Values represent the average of triplicate determinations ± S.E.M.

cells in some tissues to resist mechanical injury, to phenotypes that are only slowly being appreciated. Perhaps the most surprising result has been the observation that mice that were homozygous for a vimentin null mutation (−/−) exhibited no obvious abnormality (Colucci-Guyon *et al.*, 1994). Subsequent studies have revealed a phenotype in these mice that is both subtle and unexpected. Galou *et al.* (1996) found that in certain glial cells that normally coexpress vimentin and GFAP, GFAP filaments were also absent in the vim$^{-/-}$ mice. The absence of GFAP IFs in these vim$^{-/-}$ glial cells was not due to an inability to express GFAP. Transfection of cultured vim$^{-/-}$ astrocytes with a vimentin cDNA restored the vimentin–GFAP filament network, suggesting that in these cells vimentin was required for coassembly with GFAP (Galou *et al.*, 1996). This result was unexpected since GFAP assembles into homopolymeric IFs *in vitro* (Ralton *et al.*, 1994) and GFAP filaments form a network in IF-free cells (Chen and Liem, 1994). Moreover, in the vim$^{-/-}$ mice, those glial cells that do not normally coexpress vimentin were found to have normal GFAP IFs.

A role for vimentin in lipid accumulation in adipose tissue or some aspect of intracellular lipid trafficking in steroidogenic tissues might be expected to have obvious consequences in mice that do not express vimentin. However, vim$^{-/-}$ mice do not exhibit any obvious defect in adipose development, and Colucci-Guyon *et al.* (1994) showed that adipocytes from these animals clearly can form lipid droplets. Similarly, there is no evidence yet that the vim$^{-/-}$ mice exhibit any defect in steroid hormone production or cholesterol metabolism. Various explanations have been given to account for this absence of a more profound phenotype in the vimentin knockout mouse, and to rationalize this lack of apparent association of the phenotype of the knockout mouse with the apparent effects of manipulating vimentin organization or expression on lipid metabolism in cultured cells. Most commonly, it has been speculated there is some redundancy in the function provided by vimentin, or that cells can compensate for the absence of these IFs. A knockout mutation of the keratin expressed in simple epithelia (keratin K8) has been shown to exhibit a significant variability in lethality, depending on the genetic background of the mice, suggesting that there can be modifiers of an IF-null phenotype (Baribault *et al.*, 1994). However, no equivalent structure or structural adaptation has yet been identified in the vimentin knockout mouse (Colucci-Guyon *et al.*, 1994) or IF-free cultured cells (Sarria *et al.*, 1994; Holwell *et al.*, 1997). Another, and perhaps more appealing idea is that the phenotype is subtle, and that the appropriate experiment in vimentin-null mice has not yet been done. Finally, it has also been suggested the highly conserved primary structure of vimentin IFs indicates a function that would confer some advantage for animals in a natural environment that is not apparent in a laboratory setting (Herrmann *et al.*, 1996).

8. SUMMARY

Vimentin IFs form close associations with the lipid droplets that are characteristic of adipose and steroidogenic cells. There is good evidence that changes in vimentin expression or organization can alter the metabolism of specific lipid components in cultured preadipose and adrenal cell lines. However, the effect of vimentin on triglyceride stability that is observed in preadipose cells is not obviously reflected in adrenal cells or fibroblasts. Conversely, an effect of vimentin on the metabolism of lipoprotein-derived cholesterol observed in adrenal cells is not apparent in preadipose cells. While the complexity of the phenotypes observed in these cells might be associated with cell-type-specific differences in the metabolism of triglycerides or lipoprotein-derived cholesterol, these studies have not yet revealed a general role of vimentin in lipid metabolism that would indicate a common mechanism in all cell types. A key issue that needs to be addressed is whether the effect of vimentin IFs on the stability of triglycerides or the trafficking of GSLs and lysosomal cholesterol is due to a direct participation of vimentin IFs in some aspect of these processes, or perhaps reflects an indirect response of the lipid metabolism of these cells to an effect of vimentin on some other cellular process.

ACKNOWLEDGMENTS. We are grateful to Tricia Holwell for the experiments with the tetracycline-regulatable expression of vimentin in fibroblasts and her technical expertise. This work was supported by NIH grant HL 51850.

9. REFERENCES

Ailhaud, G., Grimaldi, P., and Negrel, R., 1992, Cellular and molecular aspects of adipose tissue development, *Annu. Rev. Nutr.* **12:**207–233.

Almahbobi, G., 1995, Adhesion of intermediate filaments and lipid droplets in adrenal cells studied by field emission scanning electron microscopy, *Cell Tiss. Res.* **281:**387–390.

Almahbobi, G., and Hall, P. F., 1990, The role of intermediate filaments in adrenal steroidogenesis, *J. Cell Sci.* **97:**679–687.

Almahbobi, G., Williams, L. J., and Hall, P. F., 1992a, Attachment of mitochondria to intermediate filaments in adrenal cells—Relevance to the regulation of steroid synthesis, *Exp. Cell Res.* **200:**361–369.

Almahbobi, G., Williams, L. J., and Hall, P. F., 1992b, Attachment of steroidogenic lipid droplets to intermediate filaments in adrenal cells, *J. Cell Sci.* **101:**383–393.

Almahbobi, G., Williams, L. J., Han, X. G., and Hall, P. F., 1993, Binding of lipid droplets and mitochondria to intermediate filaments in rat Leydig cells, *J. Reprod. Fertil.* **98:**209–217.

Amanuma, K., Kanaseki, T., Ikeuchi, Y., Ohkuma, S., and Takano, T., 1986, Studies on fine structure and location of lipids in quick-freeze replicas of atherosclerotic aorta of WHHL rabbits, *Virchows. Arch. [A]* **410:**231–238.

Argoff, C. E., Comly, M. E., Blanchette-Mackie, E. J., Kruth, H. S., Pye, H. T., Goldin, E., Kaneski, C., Vanier, M. T., Brady, R. O., and Pentchev, P. G., 1991, Type C Niemann–Pick disease: Cellular uncoupling of cholesterol homeostasis is linked to the severity of disruption in the intracellular transport of exogenously derived cholesterol, *Biochim. Biophys. Acta* **1096:**319–327.

Baribault, H., Penner, J., Iozzo, R. V., and Wilsonheiner, M., 1994, Colorectal hyperplasia and inflammation in keratin 8-deficient FVB/n mice, *Gene Dev.* **8:**2964–2973.

Blanchette-Mackie, E. J., Dwyer, N. K., Amende, L. M., Kruth, H. S., Butler, J. D., Sokol, J., Comly, M. E., Vanier, M. T., August, J. T., Brady, R. O., and Pentchev, P. G., 1988, Type-C Niemann–Pick disease: Low density lipoprotein uptake is associated with premature accumulation in the Golgi complex and excessive cholesterol storage in lysosomes, *Proc. Natl. Acad. Sci. USA* **85:**8022–8026.

Blanchette-Mackie, E. J., Dwyer, N. K., Barber, T., Coxey, R. A., Takeda, T., Rondinone, C. M., Theodorakis, J. L., Greenberg, A. S., and Londos, C., 1995, Perilipin is located on the surface layer of intracellular lipid droplets in adipocytes, *J. Lipid Res.* **36:**1211–1226.

Brandes, R., Arad, R., Gaathon, A., and Bartana, J., 1993, Induction of adipose conversion in 3T3-L1 cells is associated with an early phosphorylation of a protein partly homologous with mouse vimentin, *FEBS Lett.* **333:**179–182.

Brasaemle, D. L., and Attie, A. D., 1990, Rapid intracellular transport of LDL-derived cholesterol to the plasma membrane in cultured fibroblasts, *J. Lipid Res.* **31:**103–112.

Brown, M. S., and Goldstein, J. L., 1986, A receptor mediated pathway for cholesterol homeostasis, *Science* **232:**34–47.

Cadigan, K. M., Heider, J. G., and Chang, T. Y., 1988, Isolation and characterization of Chinese hamster ovary cell mutants deficient in acyl-coenzyme A: Cholesterol acyltransferase activity, *J. Biol. Chem.* **263:**274–282.

Chen, W. J., and Liem, R. K. H., 1994, The endless story of the glial fibrillary acidic protein, *J. Cell Sci.* **107:**2299–2311.

Colucci-Guyon, E., Portier, M. M., Dunia, I., Paulin, D., Pournin, S., and Babinet, C., 1994, Mice lacking vimentin develop and reproduce without an obvious phenotype, *Cell* **79:**679–694.

Crivello, J. F., and Jefcoate, C. R., 1980, Intracellular movement of cholesterol in rat adrenal cells, *J. Biol. Chem.* **255:**8144–8151.

Dahl, N. K., Reed, K. L., Daunais, M. A., Faust, J. R., and Liscum, L., 1992, Isolation and characterization of Chinese hamster ovary cells defective in the intracellular metabolism of low density lipoprotein-derived cholesterol, *J. Biol. Chem.* **267:**4889–4896.

Egan, J. J., Greenberg, A. S., Chang, M.-K., Wek, S., Moos Jr., M. C., and Londos, C., 1992, Mechanism of hormone-stimulated lipolysis in adipocytes: Translocation of hormone sensitive lipase to the lipid storage droplet, *Proc. Natl. Acad. Sci. USA* **89:**8537–8541.

Evans, R. M., 1998, Vimentin: The conundrum of the intermediate filament gene family, *Bioessays* **20:**79–86.

Evans, R. M., and Fink, L. M., 1982, An alteration in the phosphorylation of vimentin-type intermediate filaments is associated with mitosis in cultured mammalian cells, *Cell* **29:**43–52.

Eyer, J., and Peterson, A., 1994, Neurofilament-deficient axons and perikaryal aggregates in viable transgenic mice expressing a neurofilament-beta-galactosidase fusion protein, *Neuron* **12:**389–405.

Faust, J. R., Goldstein, J. L., and Brown, M. S., 1977, Receptor-mediated uptake of low density lipoprotein and utilization of its cholesterol for steroid synthesis in cultured mouse adrenal cells, *J. Biol. Chem.* **252:**4861–4871.

Franke, W. W., Hergt, M., and Grund, C., 1987, Rearrangement of the vimentin cytoskeleton during adipose conversion: Formation of an intermediate filament cage around lipid globules, *Cell* **49:**131–141.

Fuchs, E., and Weber, K., 1994, Intermediate filaments: Structure, dynamics, function, and disease, *Annu. Rev. Biochem.* **63:**345–382.

Galou, M., Colucci-Guyon, E., Ensergueix, D., Ridet, J.-L., Gimenez y Ribotta, M., Privat, A., Babinet, C., and Dupouey, P., 1996, Disrupted glial fibrillary acidic protein network in astrocytes from vimentin knockout mice, *J. Cell Biol.* **133:**853–863.

Geuens, G., de Brabander, M., Nuydens, R., and Demay, J., 1983, The interaction between microtubules and intermediate filaments in cultured cells treated with taxol and nocodazole, *Cell Biol. Int. Rep.* **7:**35–47.

Gillard, B. K., Heath, J. P., Thurmon, L. T., and Marcus, D. M., 1991, Association of glycosphingolipids with intermediate filaments of human umbilical vein endothelial cells, *Exp. Cell Res.* **192:**433–444.

Gillard, B. K., Thurmon, L. T., and Marcus, D. M., 1992, Association of glycosphingolipids with intermediate filaments of mesenchymal, epithelial, glial, and muscle cells, *Cell Motil. Cytoskel.* **21:**255–271.

Gillard, B. K., Thurmon, L. T., Harrell, R. G., Capetanaki, Y., Saito, M., Yu, R. K., and Marcus, D. M., 1994, Biosynthesis of glycosphingolipids is reduced in the absence of a vimentin intermediate filament network, *J. Cell Sci.* **107:**3545–3555.

Gillard, B. K., Harrell, R. G., and Marcus, D. M., 1996, Pathways of glycolipid biosynthesis in SW-13 cells in the presence and absence of vimentin intermediate filaments, *Glycobiology* **6:**33–42.

Greenberg, A. S., Egan, J. J., Wek, S. A., Garty, N. G., Blanchette-Mackie, E. J., and Londos, C., 1992, Perilipin, a major hormonally regulated adipocyte-specific phosphoprotein associated with the periphery of lipid storage droplets, *J. Biol. Chem.* **266:**11341–11346.

Hall, P. F., 1992, Cellular organization for steroidogenesis, *Int. Rev. Cytol.* **86:**53–95.

Hedberg, K. K., and Chen, L. B., 1986, Absence of intermediate filaments in a human adrenal cortex carcinoma derived cell line, *Exp. Cell Res.* **163:**509–517.

Herrmann, H., Münick, M. D., Brettel, M., Fouquet, B., and Markl, J., 1996, Vimentin in a cold-water fish, the rainbow trout: Highly conserved primary structure but unique assembly properties, *J. Cell Sci.* **109:**569–578.

Holwell, T. A., Schweitzer, S. C., and Evans, R. M., 1997, Tetracycline regulated expression of vimentin in fibroblasts derived from vimentin null mice, *J. Cell Sci.* **110:**1947–1956.

Horkovics-Kovats, S., and Traub, P., 1990, Specific interaction of the intermediate filament protein vimentin and its isolated N-terminus with negatively charged phospholipids as determined by vesicle aggregation, fusion, and leakage measurements, *Biochemistry* **29:**8652–8657.

Inagaki, M., Matsuoka, Y., Tsujimura, K., Ando, S., Tokui, T., Takahashi, T., and Inagaki, N., 1996, Dynamic property of intermediate filaments: Regulation by phosphorylation, *Bioessays* **18:**481–487.

Johnson, W. J., Chacko, G. K., Phillips, M. C., and Rothblat, G. H., 1990, The efflux of lysosomal cholesterol from cells, *J. Biol. Chem.* **265:**5546–5553.

Klymkowsky, M. W., Bachant, J. B., and Domingo, A., 1989, Functions of intermediate filaments, *Cell Motil. Cytoskel.* **14:**309–331.

Li, Z., Colucci-Guyon, E., Pincon-Raymond, M., Mericskay, M., Pournin, S., Paulin, D., and Babinet, C., 1996, Cardiovascular lesions and skeletal myopathy in mice lacking desmin, *Dev. Biol.* **175:**362–366.

Lieber, J. G., and Evans, R. M., 1996, Disruption of the vimentin intermediate filament system during adipose conversion of 3T3-L1 cells inhibits lipid droplet accumulation, *J. Cell Sci.* **109:**3047–3058.

Liscum, L., 1990, Pharmacological inhibition of the intracellular transport of low-density lipoprotein-derived cholesterol in Chinese hamster overy cells, *Biochim. Biophys. Acta* **1045:**40–48.

Liscum, L., and Underwood, K. W., 1995, Intracellular cholesterol transport and compartmentation, *J. Biol. Chem.* **270:**15443–15446.

Liscum, L., Ruggiero, R. M., and Faust, J. R., 1989, The intracellular transport of low density lipoprotein-derived cholesterol is defective in Niemann–Pick type C fibroblasts, *J. Cell Biol.* **108:**1625–1636.

Luckenbill, L., and Cohen, A. S., 1966, The association of lipid droplets with cytoplasmic filaments in avian subsynovial cells, *J. Cell Biol.* **31:**195–199.

Marinetti, G. V., 1990, Sphingolipidoses, in *Disorders of Lipid Metabolism* (G. V. Marinetti, ed.), pp. 205–214, Plenum Press, New York.

Marszalek, J. R., Williamson, T. L., Lee, M. K., Xu, Z., Hoffman, P. N., Becher, M. W., Crawford, T. O., and Cleveland, D. W., 1996, Neurofilament subunit NF-H modulates axonal diameter by selectively slowing neurofilament transport, *J. Cell Biol.* **135:**711–724.

McGookey, D. J., and Anderson, R. G. W., 1983, Morphological characterization of the cholesterol ester cycle in cultured mouse macrophage foam cells, *J. Cell Biol.* **97:**1156–1168.

Milner, D. J., Weitzer, G., Tran, D., Bradley, A., and Capetanaki, Y., 1996, Disruption of muscle architecture and myocardial degeneration in mice lacking desmin, *J. Cell Biol.* **134:**1255–1270.

Nagy, L., and Freeman, D. A., 1990, Effect of cholesterol transport inhibitors on steroidogenesis and plasma membrane cholesterol transport in cultured MA-10 Leydig tumor cells, *Endocrinology* **126:**2267–2276.

Neufeld, E. B., Cooney, A. M., Pitha, J., Dawidowicz, E. A., Dwyer, N. K., Pentchev, P. G., and Blanchette-Mackie, E. J., 1996, Intracellular trafficking of cholesterol monitored with a cyclodextrin, *J. Biol. Chem.* **271:**21604–21613.

Novikoff, A. B., Novikoff, P. M., Rosen, O. M., and Rubin, C. S., 1980, Organelle relationships in cultured 3T3-L1 preadipocytes, *J. Cell Biol.* **87:**180–196.

Pennock, C. A., 1994, Lysosomal storage disorders, in *The Inherited Metabolic Diseases*, 2nd ed. (J. B. Holton, ed.), pp. 205–242, Churchill Livingstone, Edinburgh.

Perides, G., Scherbarth, A., and Traub, P., 1986, Influence of phospholipids on the formation and stability of vimentin type intermediate filaments, *Eur. J. Cell Biol.* **42:**268–280.

Perides, G., Harter, C., and Traub, P., 1987, Electrostatic and hydrophobic interactions of the intermediate filament protein vimentin and its amino terminus with lipid bilayers, *J. Biol. Chem.* **262:**13742–13749.

Quinlan, R., Hutchison, C., and Lane, B., 1994, Intermediate filaments, in *Protein Profile*, Vol. 1 (P. Sheterline, ed.), pp. 779–802, Academic Press, London.

Ralton, J. E., Lu, X., Hutcheson, A. M., and Quinlan, R. A., 1994, Identification of two N-terminal non-alpha-helical domain motifs important in the assembly of glial fibrillary acidic protein, *J. Cell Sci.* **107:**1935–1948.

Sarria, A. J., Nordeen, S. K., and Evans, R. M., 1990, Regulated expression of vimentin cDNA in cells in the presence and absence of a pre-existing vimentin filament network, *J. Cell Biol.* **111:**553–566.

Sarria, A. J., Panini, S. R., and Evans, R. M., 1992, A functional role for vimentin filaments in the metabolism of lipoprotein-derived cholesterol in human SW-13 cells, *J. Biol. Chem.* **267:**19455–19463.

Sarria, A. J., Lieber, J. G., Nordeen, S. K., and Evans, R. M., 1994, The presence or absence of a vimentin-type intermediate filament network affects the shape of the nucleus in human SW-13 cells, *J. Cell Sci.* **107:**1593–1607.

Seifert, G. J., Lawson, D., and Wiche, G., 1992, Immunolocalization of the intermediate filament-associated protein plectin at focal contacts and actin stress fibers, *Eur. J. Cell Biol.* **59:**138–147.

Skiba, P. J., Zha, X., Maxfield, F. R., Schissel, S. L., and Tabas, I., 1996, The distal pathway of lipoprotein-induced cholesterol esterification, but not sphingomyelinase-induced cholesterol esterification, is energy dependent, *J. Biol. Chem.* **271:**13392–13400.

Smyth, M. J., Sparks, R. L., and Wharton, W., 1993, Proadipocyte cell lines: Models of cellular proliferation and differentiation, *J. Cell Sci.* **106:**1–9.

Spillane, D. M., Reagan, J. W., Kennedy, N. J., Schneider, D. L., and Chang, T. Y., 1995, Translocation of both lysosomal LDL-derived cholesterol and plasma membrane cholesterol to the endoplasmic reticulum for esterification may require common cellular factors involved in cholesterol egress from the acidic compartments (lysosomes, endosomes), *Biochim. Biophys Acta* **1254:**283–294.

Stein, O., Bennaim, M., Dabach, Y., Hollander, G., and Stein, Y., 1992, Modulation of sphingomyelinase-induced cholesterol esterification in fibroblasts, CaCo2 cells, macrophages and smooth muscle cells, *Biochim. Biophys. Acta* **1126:**291–297.

Svitkina, T. M., Verkhovsky, A. B., and Borisy, G. B., 1996, Plectin sidearms mediate interaction of intermediate filaments with microtubules and other components of the cytoskeleton, *J. Cell Biol.* **135:**991–1007.

Temple, R., and Wolff, J., 1973, Stimulation of steroid secretion by antimicrotubular agents, *J. Biol. Chem.* **248:**2691–2698.

Traub, P., Perides, G., Scherbarth, A., and Traub, U., 1985, Tenacious binding of lipids to vimentin during its isolation and purification from Ehrlich ascites tumor cells, *FEBS Lett.* **193:**217–221.

Traub, P., Perides, G., Kuhn, S., and Scherbarth, A., 1987, Efficent interaction of nonpolar lipids with intermediate filaments of the vimentin type, *Eur. J. Cell Biol.* **43:**55–64.

Underwood, K. W., Abdemariam, B., McWilliams, G. L., and Liscum, L., 1996, Quantitative analysis of hydrophobic amine inhibition of intracellular cholesterol transport, *J. Lipid Res.* **37:**1556–1568.

Wood, E. M., 1967, An ordered complex of filaments surrounding the lipid droplets in developing adipose cells, *Anat. Rec.* **157:**437–448.

Chapter 16

Desmin Cytoskeleton in Muscle Integrity and Function

Yassemi Capetanaki and Derek J. Milner

1. INTRODUCTION

Muscle cells contain several structural domains highly organized in a characteristic striated appearance which allows high efficiency in force generation and movement. In addition to the dominant structural feature, the contractile apparatus composed primarily of interdigitating actin and myosin filaments, striated muscle contains at least three cytoskeletal systems: the intrasarcomeric cytoskeleton, located within the myofibrillar lattice itself and composed of titin and nebulin, the extrasarcomeric cytoplasmic filaments composed mainly of intermediate filaments (IFs), and the membrane skeleton (for review see Wang, 1985; Lazarides and Capetanaki, 1986; Price, 1991). The mechanism by which these domains are generated during muscle development and damage repair, the way they assemble together to form the ultimate supramolecular phenotype of the muscle, and the way they are maintained together in adulthood to guarantee proper muscle function are issues of intensive investigation.

Yassemi Capetanaki and Derek J. Milner Department of Cell Biology, Baylor College of Medicine, Houston, Texas 77030.

Subcellular Biochemistry, Volume 31: Intermediate Filaments, edited by Herrmann and Harris. Plenum Press, New York, 1998.

The IF-containing lattices interconnect the myofibrils to each other, through the Z-discs, and the entire contractile apparatus to the sarcolemma cytoskeleton, several cytoplasmic organelles and the nucleus. This allows the formation of a continuous network that could be involved in diverse functions including cell integrity, force transmission, and mechanochemical signaling.

In this review we summarize recent studies on the function of desmin, the major protein of the IF network in muscle, which demonstrate that desmin is indeed essential for maintenance of muscle integrity and function.

2. DESMIN IN DEVELOPING MUSCLE

The IF cytoskeleton of mature muscle is composed predominantly of the muscle-specific IF protein desmin (Lazarides and Hubbard, 1976) and lower amounts of synemin (Granger and Lazarides, 1980) and paranemin (Breckler and Lazarides, 1982), which are not muscle-specific, but, as recent sequence analysis has revealed, are also members of the IF family (Becker *et al.*, 1995; Hemken *et al.*, 1997). During early muscle development, in addition to vimentin, the IF protein nestin, originally identified as brain stem cell-specific (Sejernsen and Lendahl, 1993; Kachinsky *et al.*, 1994), is also transiently coexpessed with desmin both in skeletal and cardiac muscle (Kachinsky *et al.*, 1994, 1995). Desmin is encoded by a single-copy gene (Capetanaki *et al.*, 1984a, b; Quax *et al.*, 1985) which is expressed in all muscle tissues, cardiac, skeletal, and smooth (for review see Lazarides, 1982; Lazarides and Capetanaki, 1986). It is one of the earliest known myogenic markers both in heart and in somites (Kaufman and Foster, 1988; Herrmann *et al.*, 1989; Schaart *et al.*, 1989; Choi *et al.*, 1990; Z. Lin *et al.*, 1994; Kuisk *et al.*, 1996). During mouse development, desmin first appears at 8.25 days postcoitum (dpc) in the neuroectoderm, where it is transiently expressed with vimentin, nestin, and keratins. At 8.5 dpc it appears in the heart rudiment, and at day 9 it can be detected in the myotome (Fürst *et al.*, 1989; Schaart *et al.*, 1989). Recent studies in transgenic mice have shown that the desmin gene is activated at least $^1/_2$–1 day earlier than the time the protein can be detected (Kuisk *et al.*, 1996). In contrast to most muscle-specific genes, desmin is also expressed at low levels in satellite cells (Allen *et al.*, 1991) and replicating myoblasts (Kaufman and Foster, 1988). During development, desmin expression precedes all other muscle-specific structural genes and the mHLH transcription factors myoD, myogenin, and MRF4 (Olson, 1990; Weintraub *et al.*, 1991), with the exception of Myf-5 (for review see Sassoon, 1993; Lyons and Buckingham, 1992;

Li and Capetanaki, 1993), suggesting that it might play some modulating role in myogenic commitment and differentiation. Studies on the regulation of desmin gene expression have demonstrated that all four of the myogenic HLH transcription factors in cooperation with members of the MEF2 family of transcription regulators (for review see Olson *et al.*, 1995) control desmin expression during skeletal muscle development (Li and Capetanaki, 1993, 1994; Kuisk *et al.*, 1996). Furthermore, it has been shown that MEF2 factors are also major regulators of desmin gene expression in cardiac muscle during mouse embryogenesis (Kuisk *et al.*, 1996).

3. THE DESMIN IF SCAFFOLD OF STRIATED MUSCLE

In mature striated muscle, desmin IFs form a three-dimensional scaffold which seems to extend across the entire diameter of the myofiber (Figure 1; for review see Lazarides and Capetanaki, 1986). The IFs surround the Z-discs, extend from one Z-disc to the other, and seem to traverse and potentially associate with other cellular organelles, including the mitochondria, sarcoplasmic reticulum, as well as the T-tubules, in the case of cardiac muscle (Pardo *et al.*, 1983; Thornell *et al.*, 1983; Tokuyasu *et al.*, 1983a). Finally, IFs extend from the Z-disc to the plasma membrane at the level of the costameres (see below) and intercalated discs in the case of cardiac muscle (Craig and Pardo, 1983), and they project from the Z-discs of the perinuclear myofibrils to the nuclear membrane (Figure 1; Behrendt, 1977; Ferrans and Roberts, 1973; Tokuyasu *et al.*, 1983a). A lower proportion of IFs also seem to form a longitudinal lattice which appears to form a cage around the myofibrils (Tokuyasu *et al.*, 1983a, b). On the basis of these associations with the IF lattice it has been proposed that the desmin or desmin and vimentin IFs could mechanically integrate all contractile actions of a muscle fiber by linking individual myofibrils laterally at the periphery of the Z-discs and by linking Z-discs to the plasma membrane, nuclear membrane, and other cytoplasmic membranous organelles (Granger and Lazarides, 1980). Synemin and paranemin are present in both skeletal and cardiac muscle and, at least in mice, they colocalize with desmin. However, in avians, only synemin is found around the Z-discs of mature skeletal muscle, and only paranemin around the Z-discs (and at the intercalated discs) of mature cardiac muscle cells (Price and Lazarides, 1983).

Costameres are periodic riblike regions of membrane skeleton overlying the flanks of the Z-disks of subsarcolemmal myofibrils (Craig and

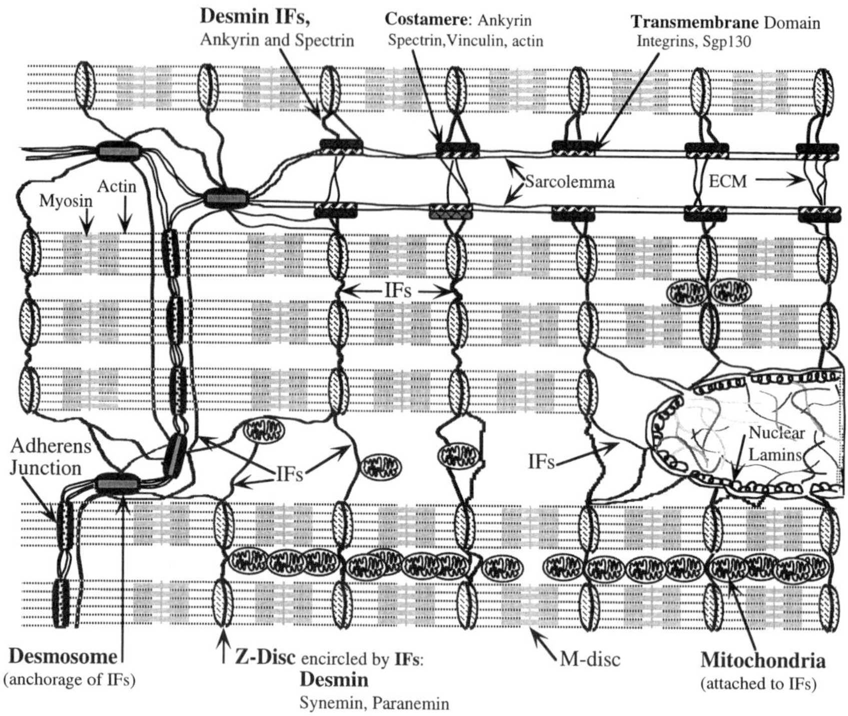

FIGURE 1. Desmin scaffold in striated muscle. Schematic representation of adjacent cardiac myofibers showing the distribution of IFs (mainly desmin) in relationship to the Z-discs of myofibrils and different membranous compartments and organelles including the costameres of the sarcolemma, desmosomes of the intercalated discs, mitochondria, and nucleus. Potential connections with the SR and T systems are not shown here. ECM: Extracellular matrix components, including collagen, fibronectin, laminin, and vitronectin. The titin/connectin endosarcomeric lattice and other sarcomere and sarcomere-associated proteins are not shown. The IF-associated protein plectin, which is also localized at the Z-disc, is also not shown.

Pardo, 1983). It has been shown, mostly in cardiac and in several skeletal muscles, that IFs link the Z-discs to costameres, most possibly through spectrin and ankyrin (for review see Lazarides and Capetanaki, 1986; Nelson and Lazarides, 1984). The major components of costamers in all muscles are α- and β-spectrins, ankyrin, γ-actin, vinculin, and possibly talin and dystrophin (for review see Lazarides and Capetanaki, 1986). The molecular basis of the association of the IF proteins with the costamere membrane skeleton is suggested by biochemical evidence of binding of IFs to spectrin (Frappier *et al.*, 1987; Langley and Cohen, 1987) and to ankyrin (Georgatos and Blobel, 1987a, b; Georgatos *et al.*, 1987). The contribution of desmin to the assembly and stabilization of the structure and composition of costameres as well as the nature of the association of myofibrils with costameres are unknown. Some muscle cells are interconnected in part by extracellular matrix (ECM) filaments localized adjacent to the Z-disks (Borg and Caulfield, 1980; Caufield and Borg, 1979). There is also a strong correspondence between the presence of costameres with attached IFs and extracellular filaments enriched overlying the costameres (Shear and Bloch, 1985). The linkage of the extracellular matrix and the costamere/cytoskeletal complex appears to involve a number of proteins including the integral membrane sialoglycoprotein Sgp130 (Rogalski, 1987) and several integrins, the receptor of a variety of extracellular matrix components including fibronectin, vitronectin, laminin, and different collagen types (for review see Albelda and Buck, 1990). It seems that costameres might require this cell–ECM interaction to be organized. However, it is unknown if desmin IFs influence membrane–ECM organization. Collagen is a major ECM component in muscle. The collagen struts coursing between muscle cells could structurally integrate the cells across the diameter of the entire myocardium. This way the IF cytoskeleton/myofibrils as a unit could act to distribute force throughout a muscle cell, and then through their connection with the ECM via membrane skeleton/transmembrane protein components would distribute force and coordinate movement in the transverse axis across the whole muscle tissue. The longitudinal force transduction, which is the most important element in muscle function, is mediated at the ends of striated cells by specialized junctions, the intercalated disks in cardiac muscle and myotendinous junctions in both skeletal and cardiac muscle. Though these are the attachment sites for actin filaments, intercalated disks are believed to be further strengthened by loops of intermediate filaments that connect desmosomes (for review see Green and Jones, 1990).

Striated muscle was one of the first cell types in which the attachment of IFs to the nuclear membrane was very clearly demonstrated (Ferrans and Roberts, 1973; Behrendt, 1977). Biochemical studies have further shown that vimentin and desmin bind to lamin B of the karyoskeleton through their carboxy termini (Georgatos and Blobel, 1987a, b; Georgatos *et al.*, 1987). The existence of a transcellular desmin–lamin B filament network has also been reported for cardiac myocytes (Lockard and Bloom, 1993). Consequently, the IFs form part of a network that connects the extracellular and the nuclear matrix and thus could participate in mechanochemical signaling to the nucleus both from the outside of the cell as well as from the contractile apparatus in regard to its physiological state. The latter is obviously possible due to the mutual anchorage of the IF lattice and the contractile apparatus at the Z-disc (Figure 1).

4. THE IF CYTOSKELETON OF SMOOTH MUSCLE

As in striated muscle, the IFs in smooth muscle cells are mainly composed of desmin and they seem to associate with Z-disc-like structures called dense bodies (Bagby, 1983). It has been shown that both membrane-associated dense bodies (MADBs) and the cytoplasmic dense bodies (CDBs) are associated with IFs (Bagby, 1983). The MADBs are dense areas on the inner surface of the membrane, between the caveolae, consisting of partially fibrous and partially amorphous material (Gabella, 1979). The less stable CDBs are not obviously associated with the membrane. CDBs are elongated and longitudinally oriented. It has further been shown that IFs are intimately associated with the periphery of CDBs and make a "corridor" which extends to the next CDB (Cooke, 1976). Models of cytoskeletal organization in smooth muscle have been proposed based on these and several other observations (Cooke and Fay, 1972; Small and Sobieszek, 1980; Bagby and Corey, 1981). The common feature in these models include the proposal that the cytoskeleton is comprised mainly of CDBs joined to each other by IFs and perhaps titin and are connected to MADBs either at the terminal portions of the cells (Cooke and Fay, 1972) or equally frequently at the mid portion and terminal portion of the cell (Small and Sobieszek, 1980). Furthermore, it has been suggested that thin filaments are also attached to the CDBs, thus providing an interaction between the cytoskeleton and contractile elements as in striated muscle (Bagby, 1983). This is why they are thought to have a Z-disk-like function. The physiological role of IFs in smooth muscle is largely unknown.

5. HYPOTHESES ON THE FUNCTION OF IF CYTOSKELETON IN MUSCLE

The role of the IF cytoskeleton in muscle morphogenesis and function has been speculated on throughout the years (for earlier reviews see Lazarides, 1982; Lazarides and Capetanaki, 1986). One of the main restrictions on the development of different hypotheses was the fact that desmin is the product of a single gene expressed in all muscle types and throughout development. Consequently, the challenge was to either assume multiple functions in different muscle types and developmental stages or to look for a function common to all muscle tissues throughout development. An additional issue related to potential differences in the function of the different IF proteins in the same muscle cell. Is there a functional difference between desmin and vimentin or are they redundant for security purposes? Why do adult muscle cells need desmin and not vimentin? Why does growing muscle need more IFs than mature muscle and how does this correlate with their function in adult tissue?

Two hypotheses have been developed regarding the function of desmin. The older one was based on several of the morphological observations described above, which suggested that in terminally differentiated myotubes a potential role for desmin is to form a transcytoplasmic integrating matrix, linking individual myofibrils laterally at their Z-disks and to the sarcolemma, thus being responsible for their lateral alignment, their striated appearance, and the normal function of the contractile apparatus (for review see Granger and Lazarides, 1979; Lazarides, 1982; Lazarides and Capetanaki, 1986). According to the second hypothesis, desmin might mediate signal transduction and transport processes from the cell surface to the nucleus. This suggestion was based both on the unique developmental pattern of the expression of desmin, described above, as well as its position in the cell, where it interlinks the sarcolemma with the nuclear envelope by specifically binding to ankyrin, other components of the costameres, and lamin B, respectively. The data discussed in this chapter, obtained by *in vitro* studies, support the latter hypothesis, whereas *in vivo* knockout experiments are in favor of the older suggestions. However, these hypotheses are not mutually exclusive, as desmin could participate in both muscle integrity and strength as well as in mechanochemical signaling within the muscle cell.

6. DESMIN IS CRUCIAL FOR MYOGENESIS *IN VITRO*

During muscle development an embryo must coordinate cell division, cell migration, cell recognition, cell adhesion, cell fusion, and, ultimately, formation and maintenance of a properly functioning muscle. During these developmental processes, signals by extracellular morphogens, extracellular matrix, and direct cell-to-cell interactions must be sensed by the cells through different cell signaling molecules and transmitted to the nucleus where, in coordination with internal signals, induction of muscle-specific gene expression takes place. A variety of independent observations have suggested that changes in cell shape due to interactions with either the extracellular matrix or neighboring cells lead to reorganization of the cytoskeleton that causes changes in gene expression, possibly by directly interacting with the nuclear matrix or by activating and facilitating the transport of regulatory factors to the nucleus (for review see Ingber, 1993; Benya and Shaffer, 1982; Bissell *et al.*, 1982; Collard *et al.*, 1995; Ben-Ze'ev, 1991; Wang *et al.*, 1993; Li *et al.*, 1994; Lin *et al.*, 1995; Rosette and Karin, 1995; Streuli *et al.*, 1995; Tremble *et al.*, 1995).

Studies using antisense RNA strategies (H. Li *et al.*, 1994) in differentiating C2C12 myoblasts have demonstrated that inhibition of desmin expression not only blocks myoblast fusion and myotube formation, but also interferes with the normal induction of muscle-specific gene expression, including the myogenic HLH transcription regulators myoD and myogenin (H. Li *et al.*, 1994). These results suggested that the muscle-specific IF protein desmin, so far considered merely a component of the supporting cell architecture, might be a central molecule regulating myogenesis through direct or indirect signaling that can modulate nuclear events. The possibility that desmin might be involved in signal transduction and transport processes between the cell surface and the nucleus is supported by several pieces of recent information, including its position in the cell, where it interlinks the sarcolemma with the nuclear envelope by specifically binding to ankyrin, other components of the costameres, and lamin B, respectively (for review see Lazarides and Capetanaki, 1986; Price, 1991; Nelson and Lazarides, 1984; Tokuyasu *et al.*, 1983a, b; Georgatos *et al.*, 1987; Georgatos and Blobel, 1987a, b; Lockard and Bloom, 1993). Lamin B binds to the nuclear matrix attachment regions (MARs) (Ludèrus *et al.*, 1992), which can activate gene transcription by binding to the nuclear matrix (Stief *et al.*, 1989). In addition, lamin A and desmin can also specifically bind MARs *in vitro* (Ludèrus *et al.*, 1994). Furthermore, IFs serve both as sites of attachment as well as major substrates of several kinases that seem to be involved in signal transduction (for review see Eriksson *et al.*, 1992). Finally, recent data on mechanotransduction across the cell surface and through the cytoskeleton, including the intermediate filament system (Wang *et al.*, 1993; Feng *et al.*, 1994; for review see Ingber, 1993), favor this hypothesis.

The intriguing results obtained with C2C12 cells were confirmed with desmin null embryonic stem (ES) cells (Weitzer *et al.*, 1995), using a differentiating system based on ES cell-derived embryoid bodies (EBs) which recapitulates the *in vivo* cardiac, skeletal, and smooth muscle myogenesis of mouse embryos (Weitzer *et al.*, 1995). In this system skeletal myogenesis is totally inhibited in desmin-null mutant EBs, as manifested by the absence of myotube formation, contractile activity, and myoD, myogenin, myf5, and myosin heavy-chain expression. Smooth muscle formation is also completely blocked in the absence of desmin. On the other hand, there is no obvious effect on cardiomyocyte differentiation in these desmin-null mutant EBs (Weitzer *et al.*, 1995). These data suggest that in contrast to early cardiomyocyte differentiation, desmin is indispensable for skeletal and smooth muscle formation *in vitro*.

At present, we can only speculate that the loss of desmin either directly blocks some specific step during early myogenic commitment and differentiation or alternatively it interferes with the maintenance of the myogenic phenotype. How could desmin be involved in these processes? Myogenesis can be divided into several phases, including cell commitment, exit from the cell cycle, initiation of differentiation, cell recognition and alignment, cell adhesion, and cell fusion (for review see Wakelam, 1985). Desmin could actively participate in one or more of these processes, during which extensive signal transduction from the cell surface to the nucleus takes place. If desmin somehow modulates signals that lead to a postmitotic state, this could explain the absence of an obvious effect during cardiac cell differentiation, since these cells, in contrast to skeletal myoblasts, can both proliferate and differentiate simultaneously. If desmin is involved in cell recognition or attachment processes rather than fusion itself, this might explain why smooth muscle development is also severely inhibited in desmin-null mutant EBs, since these cells do not fuse. Desmin, through its interaction with the membrane skeleton (for review see Lazarides and Capetanaki, 1986; Nelson and Lazarides, 1984; Georgatos *et al.*, 1987) and particularly components of the costameres (Tokuyasu *et al.*, 1983a), might be involved in integrating and stabilizing several of their potential functions, including cell recognition, adhesion, and signal transduction. Adhesion molecules and receptors have already been reported to play important role(s) in myogenesis (Menko and Boettinger, 1987; Rosen *et al.*, 1992; Holt *et al.*, 1994). Finally, desmin could be directly involved in myoblast fusion. This possibility could be favored by several reported observations. Ectopic expression of desmin in the lenses of transgenic mice causes, among other abnormalities, partial plasma membrane fusion (Krimpenfort *et al.*, 1988). Furthermore, our antisense studies (Li *et al.*, 1994) as well as other investigations (Choi *et al.*, 1990; Z. Lin *et al.*, 1994; Tao and Ip, 1991) have suggested that fusion can take place only between desmin-positive cells. Finally, the correct quantity of desmin seems to be necessary for maintaining the phenotype in newly formed myotubes, because reduced levels of desmin protein in heterozygous EBs caused a collapse of most myotubes (Weitzer *et al.*, 1995).

As with the antisense RNA studies, the EB studies also demonstrated that in the absence of desmin the expression of the mHLH regulators is suppressed. Therefore, it could be speculated that desmin regulates myogenesis by influencing nuclear events, possibly through its potential involvement in myogenic signal transduction and transport processes between the cell surface and the nucleus. This could be a way by which this part of the cytoskeleton can participate in the establishment and maintenance of the myogenic phenotype, in the initiation of differentiation, fusion, and maintenance of the terminally differentiated state. As pointed out above, desmin could be responsible for one or more of these processes.

However, the observation that the absence of desmin also causes suppression of myf5, the only member of the myoD family to be expressed in the mouse embryo as early as desmin and prior to muscle cell differentiation (Smith *et al.*, 1994), suggests that desmin must not participate mainly in terminal differentiation of skeletal muscle, but rather acts upstream in processes crucial for the establishment and maintenance of the myogenic phenotype. The data do not directly address the mechanism of mHLH factor suppression. The desmin antisense studies showed that the inhibition of at least myoD and myogenin was at the RNA level (Li *et al.*, 1994). It is possible that desmin might be involved in the nuclear transport or stabilization of the RNA or protein of these factors and when this process is blocked, even partially, these proteins cannot efficiently transactivate their own genes and their downstream partners, and are quickly degraded. Analysis of myf5 expression in the desmin-null spindle-shaped cells of the EBs showed that a small fraction of them do seem to express low levels of this factor. If stabilization is a level at which desmin might function, this low signal might be due to residual myf5 protein. Alternatively, it might be possible that desmin directly influences nuclear events, including the activation of the mHLH factors, through its connection to the nuclear matrix-associated lamins (Georgatos and Blobel, 1987a, b), which in turn specifically bind to MARs (Ludèrus *et al.*, 1992, 1994). Thus, one could envision that the connection of the cytoplasmic IF with the nuclear lamins could influence nuclear events that might be modulated through lamin–MARs association. Indeed, recent studies with cardiomyocytes have suggested that it is the IF network in mechanically stressed myocytes that transmits the stretch message to chromatin (Bloom *et al.*, 1996). Specifically, it was shown that when cardiomyocytes are stretched there are changes in the spatial arrangement of both the desmin–lamin IF network and the nuclear-envelope-associated chromatin. Studies on mechanotransduction across the cell surface and through the cytoskeleton, including the IF system (Wang *et al.*, 1993; for review see Ingber, 1993), support the above considerations. Furthermore, a variety of independent observations in different cell types have suggested that changes in cell shape due to interactions with either the extracellular matrix or neighboring cells cause changes in gene expression (Benya and Shaffer, 1982; Chen *et al.*, 1997; Clayton *et al.*, 1985; Ben-Ze'ev, 1991; Li *et al.*, 1994; C. Q. Lin *et al.*, 1995; Rosette and Karin, 1995; Streuli *et al.*, 1995; Tremble *et al.*, 1995). Though the exact mechanism(s) of this signal transduction process has not been revealed, it has been speculated that the cytoskeleton is responsible for converting changes in the cytoarchitecture to effects on gene transcription (Bissell *et al.*, 1982; Ingber, 1993; Wang *et al.*, 1993; Forgacs, 1995). The antisense studies (Li *et al.*, 1994) and the desmin knockout experiments in differentiating EBs (Weitzer *et al.*, 1995) provide the first direct evidence of gene regulation by the cytoskeleton and specifically the IF system during myogenesis. The significance of the previously reported sequence homology between desmin and the mHLH regulators (Li *et al.*, 1994) for the presently speculated mechanisms of desmin function remains elusive. Relationship and association of other members of the IF with several transcription factors and their potential involvement in the modulation of nuclear events have previously been reported (Capetanaki *et al.*, 1990; Mancini *et al.*, 1994; for review see Traub and Shoeman, 1994). It was recently demonstrated that disruption of IFs interferes with normal neuron development in *Xenopus laevis* (W. Lin and Szaro, 1996). Finally, it could be speculated that IFs and the cytoskeleton generally could modulate gene activity by different pathways, including direct interaction with the nuclear matrix and/or participation in the control of transport of regulatory factors to the nucleus. The latter possibility has been suggested from recent studies which showed that depolymerization of microtubules controls gene expression by activating NF-kB (Rosette and Karin, 1995). The exact mechanism of the above possibilities remains elusive. The connection of the different cytoskeletal networks through linker proteins, suggested in different systems (Gard *et al.*, 1997; Svitkina *et al.*, 1996; Yang *et al.*, 1996), allows redundancy in these important processes. On the other hand, the elucidation of the mechanism by which each network of the cytoskeleton participates in signaling becomes both challenging and more complicated.

7. PRENATAL MUSCLE DEVELOPMENT IN THE ABSENCE OF DESMIN

Studies with desmin-null mice demonstrated that prenatal muscle development can take place in the absence of desmin (Capetanaki *et al.*, 1997; Li *et al.*, 1996; Milner *et al.*, 1996). The desmin-null mice display no obvious anatomical defects when compared to wild-type and heterozygous littermates, except for slightly reduced size in a small percentage of individuals. Both male and female desmin-null mice are fertile, although, depending on the genetic background, null females produce slightly or much smaller litter sizes when compared to wild-type females. The observed low or high percentage of prenatal and perinatal lethality is consistent with the background dependent litter size decrease. The cause of the lethality has not been completely resolved. However, the severe muscle defects discussed below provide some explanations (Capetanaki *et al.*, 1997; Milner *et al.*, 1996).

Obviously, the available *in vivo* data do not support the *in vitro* data. However, the difference could be explained by the potential use of other IF proteins *in vivo* to compensate for the absence of desmin during development. Such compensatory mechanisms seem to be very efficient *in vivo* not only for members of the IF family (Lloyd *et al.*, 1995; Colucci-Guyon *et al.*, 1994; Gomi *et al.*, 1995; Porter *et al.*, 1996), but also for several other genes, particularly in muscle (Olson and Klein, 1994). In embryoid bodies, however, these compensatory mechanisms seem to take place effectively only in cardiac and not in skeletal and smooth muscle development (Weitzer *et al.*, 1995). In addition to vimentin and desmin, nestin is also expressed during skeletal and cardiac early muscle development (Kachinsky *et al.*, 1994, 1995). It should be noted at this point that vimentin-null mice have not demonstrated any muscle or other abnormal phenotype (Colucci-Guyon *et al.*, 1994). Since vimentin is always coexpressed with another IF that could compensate for each function, these results were not that surprising. Generation of double- or triple-null mutations of desmin, vimentin, and nestin will directly address whether or not one of these IF proteins can compensate for the absence of one or both the others during development *in vivo*. However, an additional complication is the recognition that synemin and paranemin, originally identified as IF-associated proteins (Granger and Lazarides, 1980; Becker and Lazarides, 1982), are also members of the IF family (Becker *et al.*, 1995; Hemken *et al.*, 1997). Like desmin, synemin and paranemin also show similar localization at the Z-line in striated muscle (Granger and Lazarides, 1980; Muguruma *et al.*, 1981; Price and Lazarides, 1983). Finally, compensation by IF proteins that have yet to be identified or by non-IF cytoskeletal proteins is also possible.

During embryonic development of skeletal muscle, levels of desmin expression increase as terminal differentiation proceeds, and desmin filaments reorganize at the periphery of the Z-discs (Granger and Lazarides, 1978; Gard and Lazarides, 1980). Vimentin expression is initially high in replicating myoblasts, but gradually decreases as terminal differentiation occurs (Bennett *et al.*, 1979; Gard and Lazarides, 1980). Like desmin, vimentin can be detected at the Z-discs at low levels, but it gradually disappears and is at very low levels or totally absent from adult skeletal muscle (Tokuyasu *et al.*, 1984). Nestin has also been shown to be present in replicating myoblasts and early myotubes, but, like vimentin, it is also absent in adult skeletal muscle (Sejernsen and Lendahl, 1993; Kachinsky *et al.*, 1994). Recent knockouts of keratins 14 and 10 have suggested that in the absence of these IF subunits, expression of other keratin IFs is able to at least partially compensate for their loss (Lloyd *et al.*, 1995; Porter *et al.*, 1996). Analysis of the desmin-null mice revealed that neither vimentin nor nestin is expressed in tongue and thigh skeletal muscle of adult mice. The same is true for cardiac muscle. Thus, we concluded that in the absence of desmin, neither vimentin nor nestin expression is maintained in adult striated muscle (Milner *et al.*, 1996). Similarly, no induction of nonmuscle IF proteins, including NF-H and GFAP, was detected. However, recent data, although preliminary, show that in the absence of desmin, synemin and paranemin associate with the Z-disc and intercalated discs with much lower affinity (Milner and Capetanaki, unpublished data).

8. MYOFIBRILLOGENESIS CAN TAKE PLACE IN THE ABSENCE OF DESMIN

Analysis of the desmin-null muscle has shown that myofibrils are assembled in these mice (Capetanaki *et al.*, 1997; Li *et al.*, 1996; Milner *et al.*, 1996), thus demonstrating that the presence of desmin is not necessary for the assembly and initial lateral alignment of myofibrils, as was suggested originally by Hill and colleagues (Hill *et al.*, 1986). Disruption of IFs in cultured muscle cells (Schultheiss *et al.*, 1991) or *Xenopus* muscle (Cary and Klymkowsky, 1995), using truncated-desmin dominant mutants in transient transfection experiments agree with these data, in that desmin is not absolutely necessary for myofibril assembly. However, recent studies with desmin null mice demonstrated that the second wave of myofibrillogenesis during regeneration is often abortive (Li *et al.*, 1997). Further analysis of this issue may be crucial for the unraveling of the mechanism of the obtained muscle defects in mice lacking desmin. Furthermore, the absence of desmin does not influence the localization or level of expression of at least some individual myofibrillar proteins. In striated muscle, sarcomeric α-actinin is a major component of the Z-disc. We were unable to detect any discernible difference in localization or intensity of α-actinin staining in wild-type or desmin-null cardiac or skeletal muscle. Similar results were obtained for tropomyosin, nebulin, and myosin heavy chain (Milner *et al.*, 1996). Thus it was concluded that the absence of desmin does not influence the localization and levels of expression of these proteins of the contractile apparatus. However, as mentioned above, the localization of synemin and paranemin to the Z-disc might be influenced. In cardiac muscle, in addition to the Z-disc, desmin is predominantly localized at the intercalated discs, where most possibly it interacts with at least desmoplakin. Our studies have also shown that the absence of desmin does not interfere with the localization of desmoplakin in the desmosomes of the intercalated discs (Milner and Capetanaki, unpublished data). However, defective intercalated discs have been observed in desmin-null cardiac muscle (Capetanaki *et al.*, 1997).

9. MAINTENANCE OF INTEGRATION OF MUSCLE STRUCTURE AND FUNCTION REQUIRES THE PRESENCE OF DESMIN (ITS ABSENCE LEADS TO CARDIOVASCULAR AND SKELETAL MUSCLE DEFECTS)

Recent gene targeting studies demonstrated that mice lacking desmin have defects in all three muscle types (Capetanaki *et al.*, 1997; Li *et al.*, 1996; Milner *et al.*, 1996). Common features in all cases include overt cell and tissue damage with signs of cell detatchment, instability, and degeneration. While most desmin-null mice are anatomically identical to wild-type and heterozygous mice, by the second week after birth it becomes apparent that these mice are somewhat more lethargic than their wild-type and heterozygous littermates. Furthermore, depending on the genetic background, desmin-null mice show a low or high frequency of prenatal and perinatal lethality, possibly due to cardiac arrest. Desmin-null mice are easily identified upon dissection by the presence of extensive calcium deposits in the myocardium. Histological and electron microscopic analyses have demonstrated that lack of desmin leads to a general loss of cardiomyofiber tension and considerable disruption and disorganization of the cardiac myofibers and myofibrils. Extensive interstitial fibrosis and necrosis are hallmarks of the observed defects. Degeneration of the myocardium is initially more prominent throughout the right ventricle, and the reason for this remains elusive. While analysis of hearts from neonates shows little degeneration and no calcification, these defects are visible as early as 1 week after birth in some individuals and gradually increase with age leading to dilated chambers and extremely thin walls. In brief, lack of desmin leads to dilated cardiomyopathy and sudden death (Milner *et al.*, 1996; Milner and Capetanaki, 1997).

Electron microscopic analysis of desmin-null ventricular muscle has revealed that while there are some apparently normal fibers, several myofibers demonstrate severely disrupted architecture (Figures 2 and 3). In contrast to wild-type fibers, which display laterally aligned myofibrils and nicely packed strands of mitochondria, myofibers lacking desmin lose the normal lateral alignment of myofibrils, which are often disrupted (Milner *et al.*, 1996; Li *et al.*, 1996). Loss of myofibril anchorage to the plasma membrane, both at the intercalated discs and costamere sites, is one of the hallmarks of the observed defects (Figure 3). Additionally, the mitochondria lose their positioning, organization, and shape and they often appear to be swollen and disintegrating (Figures 2 and 3). Abnormalities in mitochondria are very prominent in cardiomyocytes lacking desmin (Milner *et al.*, 1996; Capetanaki *et al.*, 1997).

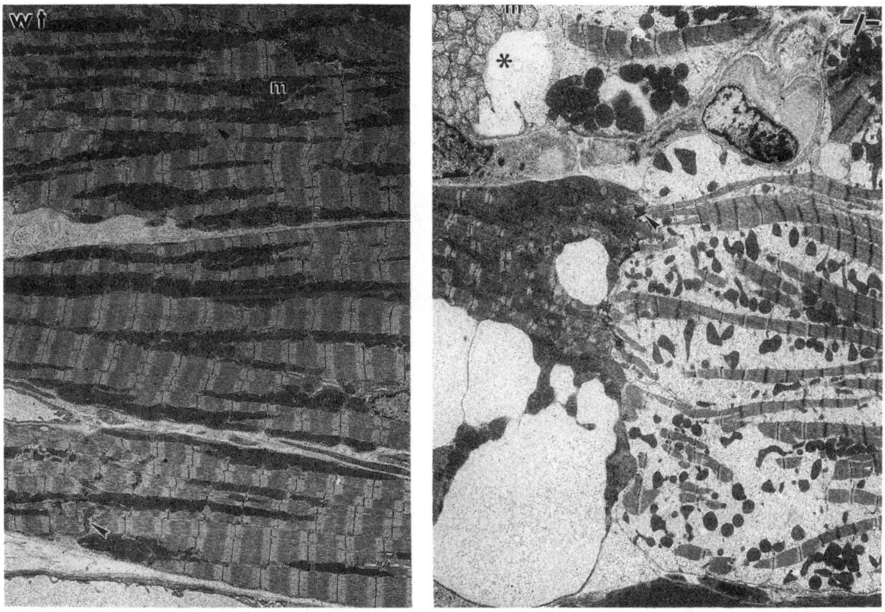

FIGURE 2. Extensive myocardium damage and degeneration in the absence of desmin. Transmission electron microscopy of cardiac muscle from wild-type (wt) and desmin-null (−/−) mice at low magnification (×1500) reveals a variety of severe structural abnormalities. Complete loss of coordination and alignment between neighboring myocytes and extensive degeneration are the main obvious defects in the present section. Myofibril, mitochondrial, and nuclear structural abnormalities and loss of positioning are obvious in all cells. In extreme cases, as in the upper left myocyte (*), myofibrils are replaced by mitochondria which degenerate leaving huge empty spaces surrounded by membranous structures. In contrast, wild-type cardiac muscle shows aligned arrays of myofibrils terminating and inserting at intercalated discs (arrowheads). Mitochondria (m) form nicely packed strands between myofibrils.

Histological examination of skeletal muscle of the tongue, leg, and diaphragm revealed similar defects in the integrity and organization of the myofibers, although to a lesser extent than that seen in cardiac muscle (Milner *et al.*, 1996). Abnormalities included loss of myofiber tension, ragged and disorganized myofibers with fewer, if any, discernible striations, some loss of stable nuclear positioning, and indications of degeneration (Milner *et al.*, 1996). The extent of damage depends on the origin of the muscle examined. In the tongue, a majority of the fibers display the ragged,

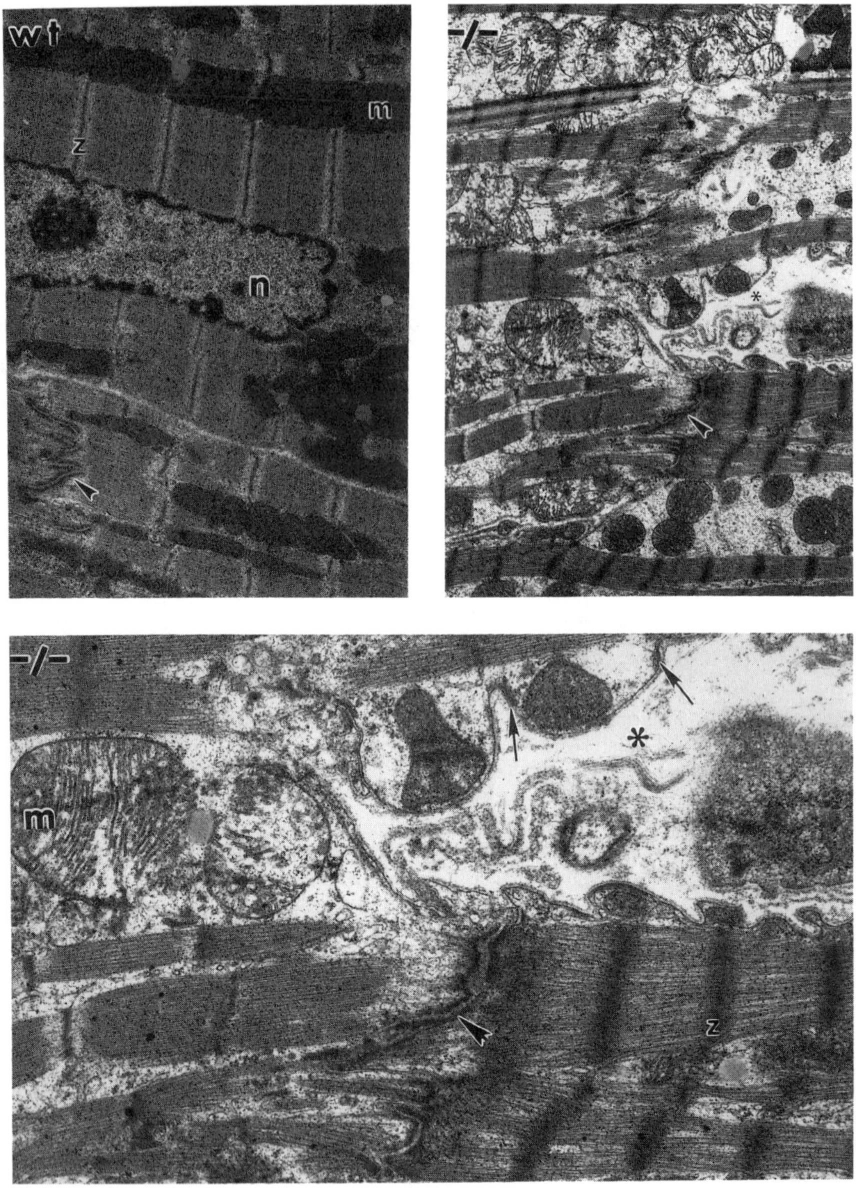

FIGURE 3. Loss of integrity between myofibrils, sarcolemma, and mitochondria in muscle lacking desmin. Transmission electron microscopy of cardiac muscle from wild-type (wt) and mice lacking desmin (−/−). The myofibrils are disorganized, disrupted at the level of the intercalated disk (arrowheads), and separated from each other, from the sarcolemma (arrows indicate the separation of Z-disc from costameres at higher magnification), and mitochonria. Loss of cell adhesion and generation of abnormally enlarged intracellular space is obvious (see area with asterisk). m: Mitochondria; n: nucleus.

disintegrating morphology, while in muscle from the back, relatively few fibers show this morphology. Cross sections of skeletal muscle from desmin-null and wild-type mice demonstrate that in mice lacking desmin, many fibers are also ragged in appearance, are of smaller diameter, and frequently are not closely packed when compared to wild-type muscle. Additionally, there appear to be more nuclei present in cross sections from null mice, including some fibers which appear to have internal clusters of multiple nuclei. Electron microscopy of skeletal muscle from the diaphragm clearly revealed that, as in cardiac muscle, some myofibers from desmin-null animals appear relatively normal, while others show disrupted intracellular architecture, including abnormally registered sarcomeres with missing or disoriented Z-discs and an absence of discernible A and I bands. Similarly, an additional severe effect was the disorganization of mitochondria. While the mitochondria are found in pairs at the Z-disc in wild-type diaphragm, this organization is lost in a number of fibers in desmin-null muscle. Additionally, a number of mitochondria also show the swollen, disintegrating morphology similar to those seen in some desmin-null cardiac myofibers. Indeed, association of mitochondria with intermediate filament networks has been reported (Gard *et al.*, 1997; Tokuyasu *et al.*, 1983a, b). These, together with the defects described above, have suggested that desmin must play a crucial role in mitochondrial integrity and function. Among the different skeletal muscle types studied, tongue demonstrated the most severe myofiber disruption and disintegration both by histological and electron microscopic analysis (Milner *et al.*, 1996). The diaphragm was also severely affected, while skeletal muscle from the thigh is affected to a somewhat lesser degree. The muscles of the back show the least disruption and degeneration. These results suggest a possible correlation between the severity of the damage and the extent of muscle usage.

Similar analyses also revealed parallel defects in smooth muscle lacking desmin (Milner *et al.*, 1996). Abnormalities of the aorta from desmin-null mice include loose organization of smooth muscle cells, loss of nuclear shape, a preponderance of empty space around the myofiber nuclei, hypoplasia, and a low degree of degeneration at both the outtrack and the thoracic levels (Milner *et al.*, 1996). These defects have been confirmed at the electron microscopic level (Figure 4). The effects of the absence of desmin in gastric smooth muscle were also generally milder than those seen in other muscles (Milner *et al.*, 1996). Partial loss of the outer longitudinal smooth muscle layer and some hypoplasia of the circular layer muscle of the intestine have been observed. Loose cellular organization was also very pronounced, thus establishing this defect also as one of the common hallmarks in all muscle types lacking desmin.

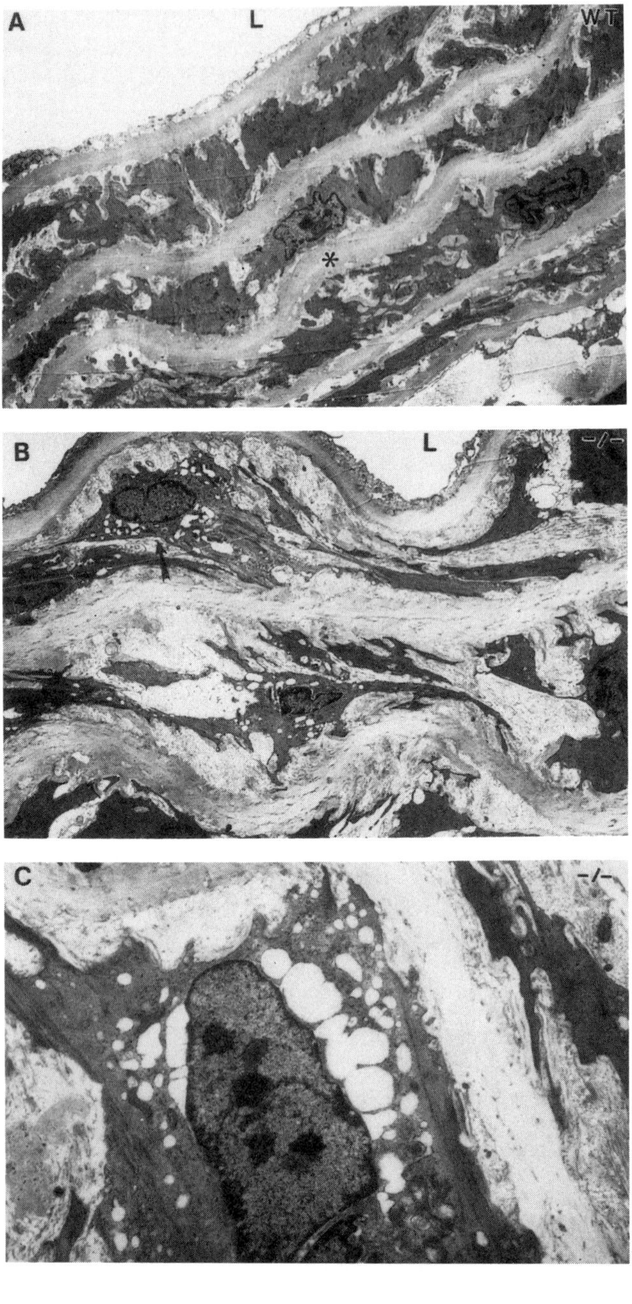
A
L
WT
*
B
L
-/-
C
-/-

10. IS DESMIN CRUCIAL IN MUSCLE INTEGRITY AND STRENGTHENING OR TISSUE REPAIR AND REGENERATION?

In summary, histological and electron microscopic analysis in both heart and skeletal muscle tissues reveals severe disruption of muscle architecture and degeneration. Structural abnormalities include loss of myofibril anchorage to the sarcolemma, perturbation of lateral alignment of myofibrils, abnormal mitochondrial number and organization, and loss of nuclear shape and positioning. Loose cell adhesion and increased intercellular space are prominent defects in all muscle types. The consequences of these abnormalities are most severe in the heart, which exhibits progressive degeneration and necrosis of the myocardium accompanied by extensive calcification. There seems to be a direct correlation between severity of damage and muscle usage. This possibly could be due to increased susceptibility to normal mechanical damage and/or to proper repair deficiency in the absence of desmin. This important issue is still elusive. The former possibility is supported by the fact that damage and degeneration of cardiac muscle, which is constantly contracting, is the most severe that we observe, and that there seems to be a range of severity of damage seen in skeletal muscle. Very active muscles, such as the diaphragm and tongue, show a higher amount of degeneration than muscles which are less active, such as those of the back. However, only a fraction of the cell population in a given muscle tissue is defective. This indicates that though desmin itself is not absolutely necessary for the formation and assembly of myofibrils, it is required for maintenance of their integrity and possibly for their efficient repair or regeneration after mechanical damage. The former possibility is also consistent with the hypothesis that IFs form a transcytoplasmic integrating matrix that might contribute to the integrity and strength of both myofibrils and myofibers (Lazarides, 1980; Lazarides and Capetanaki, 1986). The role of IFs in cell integrity and strength has also been suggested from numerous studies with keratins (Coulombe *et al.*, 1991a, b; Bonifas *et al.*, 1991; Vassar *et al.*, 1991; Cheng *et al.*, 1992; Chipev *et al.*, 1992; Lane *et al.*, 1992; Rothnagel *et al.*, 1992; Reis *et al.*, 1994; Torchard, 1994;

FIGURE 4. Smooth muscle defects in desmin deficient mice. Electron micrographs of the aorta from wild type (A) and desmin null mice (B, C) demonstrate loss of smooth muscle cells, formation of vacuoles throughout the cells, especially around the nuclei (arrow), and disorganized elastic fibers (*), often surrounded by extensive, lower density material. (C) Higher magnification (original mag. 5000) view of a smooth muscle cell from a desmin deficient aorta showing severe perinuclear vacuole formation. Original magnification in A and B is 2000. L: lumen of the aorta.

for review see Fuchs and Weber, 1994; McLean and Lane, 1995). Thus, muscle that is deficient in desmin could be more fragile than wild-type muscle and as such would be more easily damaged during contractile activity. This would lead one to expect that more actively contracting muscle would incur more damage and thus undergo more extensive degeneration.

Recent muscle-repair studies, using the crashed tibialis model with our desmin null mice, suggested that although regeneration does take place, there is loss of proper control of the myogenic program (Anderson and Capetanaki, unpublished data). The mechanism by which desmin could contribute to muscle repair and regeneration might involve among others its participation in signaling to the nucleus through the cytoskeleton, as discussed above. It is well known that growth, differentiation, and function of skeletal muscle takes place in response to mechanical stress generated during normal movement and exercise. It is also well known that external load plays a critical role in determining muscle mass and its phenotype in both cardiac and skeletal muscles *in vivo*. Little is known, however, as to how mechanical load is initially converted into intracellular signals of gene regulation (Vandenburgh, 1987, 1992; Watson, 1991). Recent studies have suggested that this mechanical stress transmits signals from the extracellular matrix through integral membrane proteins to the cytoskeleton and through the cytoskeleton to the nucleus (Wang *et al.*, 1993; Li *et al.*, 1994), where an increase in the transcription of genes coding for metabolic and contractile proteins takes place. It is reasonable to consider that the IF network and specifically desmin could be a potential candidate which could contribute to such a "mechanotranscription" coupling process. Indeed, data from our *in vitro* studies described above support this hypothesis (Li *et al.*, 1994; Weitzer *et al.*, 1995). Furthermore, recent data from others have suggested that, in cardiac hypertrophy, desmin IFs can mediate stretch-induced changes in chromatin (Bloom *et al.*, 1996). If indeed desmin participates in this mechanical signal transduction, most possibly its own expression also must be modulated by the mechanical stimuli. Indeed, recent data have demonstrated that both mechanical stretch and contraction cause robust increase in desmin mRNA content within 6–10 hr and this induction seems to be controlled at the transcriptional level (Watson *et al.*, 1996). It is interesting to note that such induction might be preceded by desmin proteolysis. It was recently demonstrated in a rabbit muscle injury model that desmin can be lost within 5 min of initiation of exercise and it was suggested that secondary proteolysis of the IF network occurs in preparation for muscle repair or remodeling (Lieber *et al.*, 1996). The relationship between these IF proteolytic products and those proposed earlier to potentially enter the nucleus (for review see Traub and Shoeman, 1994) is completely unknown. On the other hand, it has also been suggested that during muscle repair the IF network might serve as a scaffold in which the newly formed sarcomeres are assembled (Eisenberg, 1983). Indeed, recent studies with independently generated desmin null mice showed that myofibrillogenesis during regeneration is often aberrant (Li *et al.*, 1997). Though the exact mechanism has not been revealed, it is obvious that lack of desmin could affect any of these processes. Furthermore, transgenic and knockout mouse models have provided evidence directly linking abnormalities of other IFs to damage repair and degenerative processes. These include neurofilament abnormalities in neurodegenerative diseases (Cote *et al.*, 1993; Xu *et al.*, 1993; Collard *et al.*, 1995; for review see Lee and Cleveland, 1994) and delayed wound healing in BPAG1, or keratin-bundling protein, knockout mice (Guo *et al.*, 1995), possibly due to abnormal keratin filament behavior. Finally, potential defects in signaling and cell movement in all these cases are supported by reports from glial fibrillary acidic protein-null mice, which have suggested involvement of the glial cell-specific IF in communication between Bergmann glial and Purkinje cells (Shibuki *et al.*, 1996). However, the predominance of the cardiac defects in the right ventricle cannot be explained at present by either hypothesis.

As with striated muscle, the abnormalities observed in smooth muscle lacking desmin could be due to tissue instability as a consequence of the absence of desmin and/or an inability to repair mechanical damage resulting from normal activity of the muscle. The fact that organs containing smooth muscle are less affected may be due to the sustained expression of vimentin, which, in contrast to striated muscle, is maintained after differentiation (Schmid *et al.*, 1982). Preliminary data with uterine smooth muscle did not demonstrate extreme differences between nonpregnant wild-type and desmin-null tissue. However, uteri from pregnant desmin-null females show marked hypoplasia of smooth muscle compared to pregnant wild-type uterine tissue, suggesting the potential involvement of desmin in smooth muscle growth (unpublished data).

11. DESMIN IN MITOCHONDRIAL INTEGRITY AND FUNCTION

It is well known that the location of the mitochondria and their conformation are closely related to the Z bands. The abundance of IFs in the areas where mitochondria are densely packed has suggested potential association of desmin with the mitochondrial membrane and consequently an essential role in the maintenance of their positioning (Georgatos and Maison, 1996; Tokuyasu *et al.*, 1983a). Indeed, a prominent feature in desmin-null mice is the alteration of mitochondria in muscle cells. Such alterations mainly include loss of positioning, increase in size and number, aggregation, degeneration, and inclusion in autophagic vacuoles (Milner *et al.*, 1996; Milner and Capetanaki, 1997; Capetanaki *et al.*, 1997). This is particularly prominent in cardiac muscle, where mitochondria are very abundant (Figure 2). After exercise overload these abnormalities become much worse (Milner *et al.*, 1996; Milner and Capetanaki, 1997). These data demonstrated a significant role of desmin IFs in mitochondria positioning, integrity, and normal function. However, the origin of muscle degeneration and its relationship to the obtained calcification and mitochondrial abnormalities in cardiac muscle remains elusive. One possibility is that the absence of desmin could affect not only positioning of mitochondria, but also stabilization of their structure, particularly their surface membrane and consequently their function. Loss of mitochondrial function then would lead to energy deprivation and death. This possibility is favored by the fact that in muscle rich in mitochondria, like heart, the onset of mitochondrial damage is very early (D. J. Milner and Y. Capetanaki, unpublished data). The other possibility is that the alterations in mitochondria are just the consequence of degeneration from other causes. Cell death can be induced by many forms of stress, including conditions that greatly elevate intracellular Ca^{2+}. Indeed the pivotal role of intracellular Ca^{2+} concentrations for cell injury and necrosis has been established (Jackson, 1993). Recent data also suggest the involvement of cellular Ca^{2+} homeostasis in the pathogenesis of chronic myocarditis and cardiomyopathy (Tominaga *et al.*, 1993; Sen and Smith, 1994; Dhalla *et al.*, 1974). Consequently, possible destabilization of the sarcolemma and the other connected membrane systems, particularly sarcoplasmic reticulum, in the absence of desmin could cause increase of intracellular Ca^{2+}, which could lead to death either by inducing Ca^{2+}-dependent proteases, such as calpain, or by impairing electron transport and oxidative phosphorylation, due to excessive mitochondrial Ca^{2+} uptake. Changes in the mitochondrial membrane itself could alter its Ca^{2+} uptake capacity as well. Thus both pathways can contribute to the final observed defects. It must be noted here that the cardiac pathology and calcification in desmin-null mice resembles an age-related cardiomyopathy (dystrophic cardiac calcinosis, DCC) which occurs in mice and was recently localized on proximal chromosome 7 (Ivandic *et al.*, 1996). Mouse desmin is excluded, as it is localized on chromosome 1. The calcification occurring in DCC has been observed ultrastracturally to be associated initially with mitochondrial membranes alone or with involved cytoplasm (VanFleet and Ferrans, 1987). Over time, deposits can become quite extensive and be macroscopically visible in severe cases as with the desmin-null hearts (Milner *et al.*, 1996; Li *et al.*, 1996). It seems plausible that mitochondrial damage and Ca^{2+} imbalance are tightly linked and may both ultimately lead to cell death observed in the desmin-null mice. The unraveling of this issue will be of great importance to our understanding of both IF function as well as muscle structure, function, and disease.

12. DESMIN IN FORCE TRANSMISSION

Physiological and biochemical experiments have demonstrated conclusively that actin–myosin interaction is responsible for force generation. However, there is growing evidence that force transmission from this point outward is more complex. Force transmission from the myosin filaments to the Z-disk is mediated both by the actin filament and by the titin molecule. In addition to sarcomere force transmission in series, evidence exists for lateral force transmission to adjacent sarcomeres, possibly via the IF cytoskeleton, then through the costameres to extracellular matrix and to the neighboring myofiber (for review see Patel and Lieber, 1997) Thus the force generated by a single sarcomere can easily be transmitted radially as well as longitudinally. It is believed that this arrangement provides a great deal of mechanical redundancy within the muscle cell so that when a serial sarcomere is inactive or damaged, longitudinal force can still be transmitted via the IF network either longitudinally around the damaged sarcomere within the same myofibril or laterally to adjacent myofibrils. Previous isometric force measurements in isolated quail skeletal myotubes, expressing truncated desmin, suggested the importance of desmin in force transmission (Feng *et al.*, 1994). Similar experiments with myotubes from our desmin null mice confirmed those data (Sweeney and Capetanaki, unpublished data). Furthermore, work overload studies showed that muscle lacking desmin demonstrates high levels of fatigue and heart failure (Milner and Capetanaki, 1997). Recent studies using echocardiography and doppler methods to measure systolic and diastolic cardiac function in desmin null mice demonstrated the importance of desmin in cardiac muscle contractility (Taffet, Milner, and Capetanaki, unpublished data). In the case of skeletal muscle, similar results have been obtained with independently generated desmin null mice (Li *et al.*, 1997). In both cases, precise estimation of the difference in susceptibility to damage between wild type and desmin null muscle is essential for the proper interpretation of the obtained data, and remains to be done. At what degree does the absence of desmin contribute to the obtained altered mechanical properties of the muscle and how much is due to degeneration and fibrosis? One of the most direct effects of the lack of desmin on muscle is loss of resistance to endurance. The elucidation of the mechanism by which this loss takes place might unravel more precisely the cellular mechanism of desmin function in muscle. There are several ways the observed muscle weakness can also contribute to the muscle defects developed in the absence of desmin along with insufficient transmission of mechanical stress signals responsible for muscle growth and differentiation and metabolic changes to compensate for increased energy demands.

13. DESMIN AND MUSCLE DISEASE

There are several reported myopathies and cardiomyopathies with abnormally accumulated granular and filamentous aggregates of desmin (for review see Goebel, 1995). There is no case, however, where desmin is completely absent. Similarly, there is no evidence demonstrating that these abnormalities in desmin distribution have a causal effect on the disease. Muscle damage and degeneration is the hallmark of most common myopathies, cardiomyopathies, and muscular dystrophies (Grounds and Yablonka-Reuveni, 1993). In addition, several of the lesions developed in all three muscle types of the desmin-null mice are reminiscent of those reported in some familial desminopathies (Vajsar *et al.*, 1993; Thornell *et al.*, 1983; Horowitz and Schmalbruch, 1993; Ariza *et al.*, 1995; for review see Goebel, 1995). These include dilation of the ventricle chamber, hypoplasia of the aorta, thinned intestinal walls, and distended and fragmented Z-lines (Ariza *et al.*, 1995). The data on the desmin-null mice support the idea that the observed abnormalities in desmin distribution in several reported myopathies might have a causal effect. This hypothesis is favored by the extensive similarity of the desmin-abnormal aggregates to those of keratin filament aggregates which relate directly to different skin diseases (Coulombe *et al.*, 1991a, b; Bonifas *et al.*, 1991; Vassar *et al.*, 1991; Cheng *et al.*, 1992; Chipev *et al.*, 1992; Lane *et al.*, 1992; Rothnagel *et al.*, 1992; Reis *et al.*, 1994; Torchard, 1994; for review see Fuchs and Weber, 1994; McLean and Lane, 1995; Fuchs and Cleveland, 1998). Furthermore, natural human mutations in K14 (Chan *et al.*, 1994; Rugg *et al.*, 1995) and the knockout of the K14 gene (Lloyd *et al.*, 1995) provided further support for the notion that cell fragility is a consequence of the absence of the keratin filament network rather than the presence of insoluble aggregates. So far two cases of missense mutations in the desmin gene have been identified, a single one in a family suffering from hypertrophic cardiomyopathy (Brown *et al.*, 1995) and a cluster of missense mutations at the C-terminal end of the desmin rod domain in different families with desmin-related cardiac and skeletal myopathies (Goldfarb *et al.*, 1998). On the other hand, no mutation, at least in the coding region, of the desmin gene was identified in a familial desmin-related myopathy (Vicart *et al.*, 1996). Consequently, it could be suggested that the desmin-null mouse can provide an animal model for several skeletal and cardiac myopathies which closely resemble those with abnormalities in desmin distribution (Thornell *et al.*, 1983; Goebel, 1995; Vajsar *et al.*, 1993; Horowitz and Schmalbruch, 1993; Ariza *et al.*, 1995).

14. CONCLUSIONS

The studies summarized here indicate that, in contrast to vimentin, the desmin IF cytoskeleton is essential for maintaining integration of structure and function in muscle. This could be achieved by generating a continuous scaffold together with extracellular and nuclear matrix through the membrane cytoskeleton that extends throughout the muscle fibers. This cytoskeletal network could serve in coordination of contraction and relaxation, in transmission of mechanical forces longitudinally and laterally, propagation of mechanochemical signals throughout the myofiber, to the nucleus, other cellular organelles, and the entire muscle in a coordinated fashion, as well as in stabilizing the positioning of cellular organelles and strengthening the muscle. Some of these functions could be common to embryonic and adult muscle. During muscle morphogenesis such functions are of great importance and might be guaranteed by redundancy and flexibility in the use of other members of the IF family or even other cytoskeletal proteins. The double or multiple knockouts of desmin, vimentin, nestin, synemin, and paranemin might be necessary to address this issue. Among others, the challenge in the near future will be to unravel which of the above diverse functions comes first, which are IF specific, and which are desmin specific. The knockout studies demonstrated that desmin does deserve its name (Greek *desmos* = "link"), and it is remarkable that one of its major functions is to mechanically integrate muscle structure and function as originally suggested almost two decades ago (Lazarides, 1980).

ACKNOWLEDGMENTS. The research summarized here was supported by NIH grant AR 39617-01 and an MDA grant to Y.C.

15. REFERENCES

Albelda, S. M., and Buck, C. A., 1990, Integrins and other cell adhesion molecules, *FASEB J.* **4:**2868–2880.

Allen, R. E., Rankin, L. L., Greene, E. A., Boxhorn, L. K., Johnson, S. E., Taylor, R. G., and Pierce, P. A., 1991, Desmin is present in proliferating rat muscle satellite cells but not in bovine muscle satellite cells, *J. Cell Physiol.* **149:**525–535.

Ariza, A. J., Coll, M. T., Fernandez-Figueras, M. D., Lopez, J. L., Mate, O., Garcia, A., Fernandez-Vasalo, A., and Navas-Palacios, J. J., 1995, Desmin myopathy: A multisystem disorder involving skeletal, cardiac and smooth muscle, *Hum. Pathol.* **26:**1032–1037.

Bagby, R. M., 1983, Organization of contractile elements, in *The Biochemistry of Smooth Muscle*, (N. L. Stephens, ed.), Vol. 1, pp. 1–84, CRC Press, Boca Raton, Florida.

Bagby, R. M., and Corey, M. D., 1981, Vertebrate smooth muscle contractile elements attach to an axial cytoskeleton, *Physiologist* **24:**89.

Becker, B., Bellin, R. M., Sernett, S. W., Huiatt, T. W., and Robson, R. M., 1995, Synemin contains the rod domain of intermediate filaments, *Biochem. Biophys. Res. Commun.* **213:**796–802.

Behrendt, H., 1977, Effect of anabolic steroids on rat heart muscle cells. I. Intermediate filaments, *Cell Tiss. Res.* **180:**303–315.

Bennett, G. S., Fellini, S. A., Toyama, Y., and Holtzer, H., 1979, Redistribution of intermediate filament subunits during skeletal myogenesis and maturation *in vitro*, *J. Cell Biol.* **82:**577–584.

Benya, P. D., and Shaffer, J. D., 1982, Dedifferentiated chondrocytes reexpress the differentiated collagen phenotype when cultured in agarose gels, *Cell* **30:**215–224.

Ben-Ze'ev, A., 1991, Animal cell shape changes and gene expression, *Bioessays* **13:**207–212.

Bissell, M. J., Hall, H. G., and Parry, G., 1982, How does the extracellular matrix direct gene expression? *J. Theor. Biol.* **99:**31–68.

Bloemendal, H., 1990, Plasma membrane–cytoskeleton damage in eye lenses of transgenic mice expressing desmin, *Eur. J. Cell Biol.* **53:**59–74.

Bloom, S., Lockard, V. G., and Bloom, M., 1996, Intermediate filament-mediated stretch-induced changes in chromatin: A hypothesis for growth initiation in cardiac myocytes, *J. Mol. Cell Cardiol.* **28:**2123–2127.

Bonifas, J. M., Rothman, A. L., and Epstein, E. H., 1991, Epidermolysis bullosa simplex: Evidence in two families for keratin gene abnormalities, *Science* **254:**1202–1205.

Borg, T. K., and Caulfield, J. B., 1980, Morphology of connective tissue in skeletal muscle, *Tiss. Cell* **12:**197–207.

Breckler, J., and Lazarides, E., 1982, Isolation of a new high molecular weight protein associated with desmin and vimentin filaments from avian embryonic skeletal muscle, *J. Cell Biol.* **92:**795–806.

Brown, B. D., Scheffold, T., Rottbauer, W., Zehelein, J., Franz, W., Rodemer, E., and Katus, H. A., 1995, Intermediate filament desmin gene missense mutation found in a family suffering from hypertrophic cardiomyopathy [Abstract], *Circulation* **92:**I233.

Capetanaki, Y., Ngai, J., and Lazarides, E., 1984a, Characterization and regulation in the expression of a gene encoding for the intermediate filament protein desmin, *Proc. Natl. Acad. Sci. USA* **81:**6909–6912.

Capetanaki, Y. G., Ngai, J., and Lazarides, E., 1984b, Regulation of the expression of genes coding for the intermediate filament subunits vimentin, desmin, and glial fibrillary acidic protein, in *Molececular Biology of the Cytoskeleton* (G. G. Borisy, D. W. Cleveland, and D. B. Murphy, eds.), pp. 415–434, Cold Spring Harbor Laboratory, Cold Spring Harbor, New York.

Capetanaki, Y., Kuisk, I., Rothblum, K., and Starnes, S., 1990, Mouse vimentin: Structural relationship to fos, jun, CREB and tpr, *Oncogene* **5:**645–655.

Capetanaki, Y., Milner, D., and Weitzer, G., 1997, Desmin in muscle formation and maintenance: Knock outs and consequenses, *Cell Struct. Funct.* **22:**103–116.

Cary, R. B., and Klymkowsky, M. W., 1995, Disruption of intermediate filament organization leads to structural defects at the intersomite junction in *Xenopus* myotomal muscle, *Development* **121:**1041–1052.

Caulfield, J. B., and Borg, T. K., 1979, The collagen network of the heart, *Lab. Invest.* **40:**364–372.

Chan, Y. M., Anton-Lamprecht, I., Yu, Q. C., Jackel, A., Zabel, B., Ernst, J. B., and Fuchs, E., 1994, A human keratin 14 "knockout": The absence of K14 leads to severe epidermolysis bullosa simplex and a function for an intermediate filament protein, *Genes Dev.* **8:**2574–2587.

Chen, C. S., Mrksick, M., Huang, S., Whitesides, G. M., and Ingber, D. E., 1997, Geometric control of cell life and death, *Science* **276:**1425–1428.

Cheng, J., Snyder, A. J., Yu, Q. C., Letai, A., Paller, A. S., and Fuchs, E., 1992, The genetic basis of epidermolytic hyperkeratosis: A disorder of differentiation-specific epidermal keratin genes, *Cell* **70:**811–819.

Chipev, C. C., Korge, B. P., Markova, N., Bale, S. J., DiGiovanna, J. J., Compton, J. G., and Steinert, P. M., 1992, A leucine → proline mutation in the H1 subdomain of keratin 1 causes epidermolytic hyperkeratosis, *Cell* **70:**821–828.

Choi, J., Costa, M. L., Mermelstein, C. S., Chagas, C., Holtzer, S., and Holtzer, H., 1990, MyoD converts primary dermal fibroblasts, chondroblasts, smooth muscle and retinal pigmented epithelial cells into striated mononucleated and multinucleated myotubes, *Proc. Natl. Acad. Sci. USA* **87:**7988–7992.

Clayton, D. F., Harrelson, A. L., and Darnell, J., Jr., 1985, Dependence of liver-specific transcription on tissue organization, *Mol. Cell Biol.* **5:**2623–2632.

Collard, J. F., Cote, F., and Julien, J. P., 1995, Defective axonal transport in a transgenic mouse model of amyotrophic lateral sclerosis, *Nature* **375:**61–64.

Colucci-Guyon, E., Portier, M.-M., Dunia, I., Paulin, D., Pournin, S., and Babinet, C., 1994, Mice lacking vimentin develop and reproduce without an obvious phenotype, *Cell* **79:**579–694.

Cooke, P., 1976, A filamentous cytoskeleton in vertebrate smooth muscle fibres, *J. Cell Biol.* **68:**539–556.

Cooke, P. H., and Fay, F. S., 1972, Correlation between fiber length, ultrastructure and the length–tension relationship of mammalian smooth muscle, *J. Cell Biol.* **52:**105–116.

Cote, F., Collard, J., and Julien, J., 1993, Progressive neuropathy in transgenic mice expressing the human neurofilament heavy gene: A mouse model for amyotrophic lateral sclerosis, *Cell* **73:**35–46.

Coulombe, P. A., Hutton, M. E., Letai, A., Herbert, A., Paller, S. A., and Fuchs, E., 1991a, Point mutations in human keratin 14 genes of epidermolysis bullosa simplex patients: Genetic and functional analyses, *Cell* **66:**1301–1311.

Coulombe, P. A., Hutton, M. E., Vassar, R., and Fuchs, E., 1991b, A function for keratins and a common thread among different types of epidermolysis bullosa simplex disease, *J. Cell Biol.* **115:**1661–1674.

Craig, S. W., and Pardo, J. V., 1983, Gamma actin, spectrin, and intermediate filament proteins colocalize with vinculin at costameres, myofibril-to-sarcolemma attachment sites, *Cell Motil.* **3:**449–462.

Dhalla, H. S., Sulakhe, P. V., Fedeleson, M., and Yates, J. C., 1974, Molecular abnormalities in cardiomyopathy, *Adv. Cardiol.* **13:**282–300.

Eisenberg, B. R., 1983, Quantitative ultrastructure of mammalian skeletal muscle, in *Skeletal Muscle* (L. D. Peachey, R. H. Adrian, and S. R. Geiger, eds.), pp. 73–112, American Physiological Society, Baltimore.

Eriksson, J. E., Opal, P., and Goldman, R. D., 1992, Intermediate filament dynamics, *Curr. Opin. Cell Biol.* **4:**99–104.

Feng, H. S., Yang, Z. H., Yu, K. R., Hijikata, T., Holzer, H., and Sweeney, H. L., 1994, Truncated desmin expression and isometric force measurement in isolated quail skeletal myotubes [Abstract], *Mol. Biol. Cell* **5:**27a.

Ferrans, V. J., and Roberts, W. C., 1973, Intermyofibrillar and nuclear–myofibrillar connections in human and canine myocardium: An ultrastructural study, *J. Mol. Cell Cardiol.* **5:**247–257.

Forgacs, G., 1995, On the possible role of cytoskeletal filamentous networks in intracellular signalling: An approach based on percolation, *J. Cell Sci.* **108:**2131–2143.

Frappier, T., Regnouf, F., and Pradel, L. A., 1987, Binding of brain spectrin to the 70-kDA neurofilament subunit protein, *Eur. J. Biochem.* **169:**651–657.

Fuchs, E., 1994, Intermediate filaments and disease: Mutations that cripple cell strength, *J. Cell. Biol.* **125:**511–516.

Fuchs, E., and Cleveland, D., 1998, A structural scaffolding of intermediate filaments in health and disease, *Science* **279:**514–519.

Fuchs, E., and Weber, K., 1994, Intermediate filaments: Structure, dynamics, function, and disease, *Annu. Rev. Biochem.* **63:**345–382.

Fürst, D. O., Osborn, M., and Weber, K., 1989, Myogenesis in the mouse embryo: Differential onset of expression of myogenic proteins and the involvement of titin in myofibril assembly, *J. Cell Biol.* **109:**517–527.

Gabella, G., 1979, Smooth muscle cell junctions and structural aspects of contraction, *Br. Med. Bull.* **35:**213.

Gard, D. L., and Lazarides, E., 1980, The synthesis and distribution of desmin and vimentin during myogenesis *in vitro*, *Cell* **19:**263–275.

Gard, D. L., Cha, B. J., and King, E., 1997, The organization and animal–vegetal asymmetry of cytokeratin filaments in stage VI *Xenopous* oocytes is dependent upon F-actin and microtubules, *Dev. Biol.* **184:**95–114.

Georgatos, S. D., and Blobel, G., 1987a, Two distinct attachment sites for vimentin along the plasma membrane and the nuclear envelope in avian erythrocytes: A basis for a vectorial assembly of intermediate filaments, *J. Cell Biol.* **105:**105–115.

Georgatos, S. D., and Blobel, G., 1987b, Lamin B constitutes an intermediate filament attachment site at the nuclear envelope, *J. Cell Biol.* **105:**117–125.

Georgatos, S. D., and Maison, C., 1996, Integration of intermediate filaments into cellular organelles, *Int. Rev. Cytol.* **164:**91–138.

Georgatos, S. D., Weber, K., Geisler, N., and Blobel, G., 1987, Binding of two desmin derivatives to the plasma membrane and the nuclear envelope of avian erythrocytes: Evidence for a conserved site-specificity in intermediate filament–membrane interactions, *Proc. Natl. Acad. Sci. USA* **84:**6780–6784.

Goebel, H. H., 1995, Desmin-related neuromuscular disorders, *Muscle Nerve* **18:**1306–1320.

Goldfarb, L. G., Park, K.-Y., Cervenákova, L., Gorokhova, S., Lee H.-S., Vasconcelos, O., Nagle, J. W., Semino-Mora, C., Kasserwani, H., Sivakumar, K., and Dalakas, M.C., 1998, A cluster of missense mutations at the C-terminal end of the desmin rod domain in familial cardiac and skeletal myopathy, *Nature Genetics*, in press.

Gomi, H., Yokoyama, T., Fujimoto, K., Ikeda, T., Katoh, A., Itoh, T., and Itohara, S., 1995, Mice devoid of the glial fibrillary acidic protein develop normally and are susceptible to scrapie prions, *Neuron* **14:**29–41.

Granger, B. L., and Lazarides, E., 1978, The existence of an insoluble Z-disc scaffold in chicken skeletal muscle, *Cell* **15:**1253–1268.

Granger, B. L., and Lazarides, E., 1979, Desmin and vimentin coexist at the periphery of the myofibril Z disc, *Cell* **18:**1053–1063.

Granger, B. L., and Lazarides, E., 1980, Synemin: A new high molecular weight protein associated with desmin and vimentin filaments in muscle, *Cell* **22:**727–738.

Green, K. J., and Jones, J. C. R., 1990, Interaction of intermediate filaments with the cell surface, in *Cellular and Molecular Biology of Intermediate Filaments* (R. D. Goldman, and

P. M. Steinert, eds.), pp. 147–174, Plenum Press, New York.

Grounds, M. D., and Yablonka-Reuveni, Z., 1993, Molecular and cell biology of skeletal muscle regeneration, in *Molecular and Cell Biology of Muscular Dystrophy* (T. Partridge, ed.), pp. 210–256, Chapman and Hall, London.

Guo, L., Degenstein, L., Dowling, J., Yu, Q.-C., Wollmann, R., Perman, B., and Fuchs, E., 1995, Gene targeting of BPAG1: Abnormalities in mechanical strength and cell migration in stratified epithelia and neurologic degeneration, *Cell* **81:**233–243.

Hemken, P. M., Becker, B., Bellin, R. M., Sernett, S. W., Huiatt, T. W., and Robson, R. M., 1997, Molecular characterization of the novel intermediate filament protein paranemin: Sequence reveals EAP-300 and IFAP-400 are highly homologous to paranemin, *J. Biol. Chem.* **272:**32489–32499.

Herrmann, H., Fouquet, B., and Franke, W. W., 1989, Expression of intermediate filament proteins during development of *Xenopus laevis*, II. Identification and molecular characterization of desmin, *Development* **105:**299–307.

Hill, C. S., Duran, S., and Lin, Z. X., 1986, Titin and myosin, but not desmin, are linked during myofibrillogenesis in postmitotic mononucleated myoblasts, *J. Cell Biol.* **103:**2185–2196.

Holt, C. E., Lemaire, P., and Gurdon, J. B., 1994, Cadherin-mediated cell interactions are necessary for the activation of MyoD in *Xenopus* mesoderm, *Proc. Natl. Acad. Sci. USA* **91:**10844–10848.

Horowitz, S. H., and Schmalbruch, H., 1993, Autosomal dominant distal myopathy with desmin storage: A clinicopathologic and electrophysiologic study of a large kinship, *Muscle Nerve* **17:**151–160.

Ingber, D. E., 1993, The riddle of morphogenesis: A question of solution chemistry or molecular cell engineering? *Cell* **75:**1249–1252.

Ivandic, B. T., Qiao, J. H., Machleder, D., Liao, F., Drake, T. A., and Lusis, A. G., 1996, A locus on chromosome 7 determines myocardial cell necrosis and calcification (dystrophic cardiac calcinosis in mice), *Proc. Natl. Acad. Sci. USA* **93:**5483–5488.

Jackson, M. J., 1993, Molecular mechanisms of muscle damage, in *Molecular and Cell Biology of Muscular Dystrophy* (T. Partridge, ed.), pp. 257–282, Chapman and Hall, London.

Kachinsky, A. M., Dominov, J. A., and Miller, J. B., 1994, Myogenesis and the intermediate filament protein nestin, *Dev. Biol.* **165:**216–228.

Kachinsky, A. M., Dominov, J. A., and Miller, J. B., 1995, Intermediate filaments in cardiac myogenesis: Nestin in the developing mouse heart, *J. Histochem. Cytochem.* **43:**843–847.

Kaufman, S. J., and Foster, R., 1988, Replicating myoblasts express a muscle-specific phenotype, *Proc. Natl. Acad. Sci. USA* **85:**9606–9610.

Krimpenfort, P. J., Schaart, G., Pieper, F. R., Ramaekers, F. C., Cuypers, H. T., van den Heuvel, R. M., Vree-Egberts, W. T., van Eys, G. J., Berns, A., and Bloemendal, H., 1988, Tissue-specific expression of a vimentin–desmin hybrid gene in transgenic mice, *EMBO J.* **7:**941–947.

Kuisk, I. R., Li, H., Tran, D., and Capetanaki, Y., 1996, A single MEF2 site governs desmin transcription in both heart and skeletal muscle during mouse embryogenesis, *Dev. Biol.* **174:**1–13.

Lane, E. B., Rugg, E. L., Navsaria, H. J., Leigh, I. M., Heagerty, A. H. M., Ishida-Yamamoto, A., and Eady, R. A. J., 1992, A mutation in the conserved helix termination peptide of keratin 5 in hereditary skin blistering, *Nature* **356:**244–246.

Langley, R. C., and Cohen, C. M., 1987, Cell type-specific association between two types of spectrin and two types of intermediate filaments, *Cell Motil. Cytoskel.* **8:**165–173.

Lazarides, E., 1980, Intermediate filaments as mechanical integrators of cellular space, *Nature* **238:**249–256.

Lazarides, E, 1982, Intermediate filaments: A chemically heterogeneous, developmentally regulated class of proteins, *Annu. Rev. Biochem.* **51:**219–250.

Lazarides, E., and Capetanaki, Y., 1986, The striated muscle cytoskeleton: Expression and assembly in development, in *Molecular Biology of Muscle Development* (C. Emerson *et al.*, eds.), pp. 749–772, Liss, New York.

Lazarides, E., and Hubbard, B. D., 1976, Immunological characterization of the subunit of the 100 Å filaments from muscle cells, *Proc. Natl. Acad. Sci. USA* **73:**4344–4348.

Lee, M. K., and Cleveland, D. W., 1994, Neurofilament function and dysfunction: Involvement in axonal growth and neuronal disease, *Curr. Opin. Cell Biol.* **6:**34–40.

Li, H., and Capetanaki, Y., 1993, Regulation of the mouse desmin gene: Transactivation by MyoD, myogenin, MRF4 and Myf5, *Nucleic Acids Res.* **21:**335–343.

Li, H., and Capetanaki, Y., 1994, An E box in the desmin promoter cooperates with the E box and MEF2 sites of a distal enhancer to direct muscle-specific transcription, *EMBO J.* **13:**3580–3589.

Li, H., Choudhary, S. K., Milner, D. J., Munir, M. I., Kuisk, I. R., and Capetanaki, Y., 1994, Inhibition of desmin expression blocks myoblast fusion and interferes with the myogenic regulators myoD and myogenin, *J. Cell Biol.* **124:**827–841.

Li, Z., Colucci-Guyon, E., Pincon-Reymond, M., Mericskay, M., Pournin, S., Paulin, D., and Babinet, C., 1996, Cardiovascular lesions and skeletal myopathy on mice lacking desmin, *Dev. Biol.* **175:**362–366.

Li, Z., Mericskay, M., Agbulut, O., Butler-Browne, G., Carlsson, L., Thornell, L.-E., Babinet, C., and Pautin, D., 1997, Desmin is essential for the tensite strength and integrity of myofibrils but not for myogenic commitment, differentiation, and fusion of skeletal muscle, *J. Cell Biol.* **139:**129–144.

Lieber, R. L., Thornell, L.-E., and Friden, J., 1996, Muscle cytoskeletal disruption occurs within the first 15 minutes of cyclic eccentric contraction, *J. Appl. Physiol.* **80:**278–284.

Lin, C. Q., Dempsey, P. J., Coffey, R. J., and Bissell, M. J., 1995, Extracellular matrix regulates whey acidic protein gene expression by suppression of TGF-α in mouse mammary epithelial cells: Studies in culture and in transgenic mice, *J. Cell Biol.* **129:**1115–1126.

Lin, W., and Szaro, B. G., 1996, Effects of intermediate filament disruption on the early development of the peripheral nervous system of *Xenopus laevis*, *Dev. Biol.* **179**:197–211.

Lin, Z., Lu, M.-H., Schultheiss, T., Choi, J., Holtzer, S., DiLullo, C., Fischman, D. A., and Holtzer, H., 1994, Sequential appearance of muscle-specific proteins in myoblasts as a function of time after cell division: Evidence for a conserved myoblast differentiation program in skeletal muscle. *Cell Motil. Cytoskel.* **29:**1–19.

Lloyd, C., Yu, Q.-C., Cheng, J., Turksen, K., Degenstein, L., Hutton, E., and Fuchs, E., 1995, The basal keratin network of stratified squamous epithelia: Defining K15 function in the absence of K14, *J. Cell. Biol.* **129:**1329–1344.

Lockard, V. G., and Bloom, S., 1993, Trans-cellular desmin–lamin B intermediate filament network in cardiac myocytes, *J. Mol. Cell Cardiol.* **25:**303–309.

Ludèrus, M. E. E., de Graaf, A., Mattia, E., Den Blaauwen, J. L., Grande, M. A., de Jong, L., and van Driel, R., 1992, Binding of matrix attachment regions to lamin B1, *Cell* **70:**949–959.

Ludèrus, M. E. E., Den Blaauwen, J. L., De Smit, O. J. B., Compton, D. A., and van Driel. R., 1994, Binding of matrix attachment regions to lamin polymers involves single-stranded regions and the minor groove, *Mol. Cell Biol.* **14:**6297–6305.

Lyons, G. E., and Buckingham, M. E., 1992, Developmental regulation of myogenesis in the mouse, *Semin. Dev. Biol.* **3:**243–253.

Mancini, M. A., Shan, B., Nickerson, J. A., Penman, S., and Lee, W. H., 1994, The retinoblastoma gene product is a cell cycle-dependent, nuclear matrix-associated protein, *Proc. Natl. Acad. Sci. USA* **91:**418–422.

McLean, W. H. I., and Lane, E. B., 1995, Intermediate filaments in disease, *Curr. Opin. Cell Biol.* **7:**118–125.

Menko, A. S., and Boettinger, D., 1987, Occupation of the extracellular matrix receptor, integrin, is a control point for myogenic differentiation, *Cell* **51:**51–57.

Milner, D. J., and Capetanaki, Y., 1997, Dilated cardiomyopathy in mice lacking desmin, in *Molecular Biology of Muscle Development* [Abstract], Keyston Symposia, p. 42.

Milner, D. J., Weitzer, G., Tran, D., Bradley, A., and Capetanaki, Y., 1996, Disruption of muscle architecture and myocardial degeneration in mice lacking desmin, *J. Cell Biol.* **134:**1255–1270.

Muguruma, M., Kobayashi, K., Fukazawa, T., and Ohashi, K. 1981, A new 220,000 dalton protein located in the z lines of vertebrate skeletal muscle, *J. Biochem.* **89:**1981–1984.

Nelson, W. J., and Lazarides, E., 1984, Globin (ankyrin) in striated muscle: Identification of the potential membrane recepter for erythroid spectrin in muscle cells, *Proc. Natl. Acad. Sci. USA* **81:**3292–3296.

Olson, E. N., 1990, MyoD family: A paradigm for development? *Genes Dev.* **9:**1454–1461.

Olson, E. N., and Klein, W. H., 1994, bHLH factors in muscle development: Dead lines and commitments, what to leave in and what to leave out, *Genes Dev.* **8:**1–8.

Olson, E. N., Perry. M., and Schultz, R. A., 1995, Regulation of muscle differentiation by the MEF2 family of MADS box transcription factors, *Dev. Biol.* **172:**2–14.

Pardo, J. V., Siliciano, J. D., and Craig, S. W., 1983, Vinculin is a component of an extensive network of myofibril-sarcolemma attachment regions in cardiac muscle fibers, *J. Cell Biol.* **97:**1081–1088.

Patel, T. J., and Lieber, R. L., 1997, Force transmission in skeletal muscle: From actomyosin to external tendons, *Exerc. Sport Sci. Rev.* **25:**321–363.

Porter, R. M., Leitgeb, S., Melton, D. W., Swensson, O., Eady, R. A. J., and Magin, T. M., 1996, Gene targeting of the mouse cytokeratin 10 locus: Severe skin fragility and changes of cytokeratin expression in the epidermis, *J. Cell Biol.* **132:**925–936.

Price, M. G., 1991, Striated muscle endosarcomeric and exosarcomeric lattices, *Adv. Struct. Biol.* **1:**175–207.

Price, M., and Lazarides, E., 1983, Expression of IF-associated proteins paranemin and synemin in chicken development, *J. Cell Biol.* **97:**1860–1874.

Quax, W., van den Broek, L., Vree Egberts, W., Ramaekers, F., and Bloemendal, H., 1985, Characterization of the hamster desmin gene: Expression and formation of desmin filaments in nonmuscle cells after gene transfer, *Cell* **43:**327–338.

Reis, A., Hennes, H.-C., Langbein, L., Digweed, M., Mischke, D., Drechsler, M., Schrock, E., Royker-Pokora, B., Franke, W. W., Sperling, K., and Kuster, W., 1994, Keratin 9 gene mutations in epidermolytic palmoplantar keratoderma (EPKK), *Nat. Genet.* **6:**174–179.

Rogalski, A. A., 1987, A plasma membrane integral sialoglycoprotein (Sgp 130) molecularly distinguishes nonjunctional dense plaque sites of microfilament attachment, *J. Cell Biol.* **105:**819–831.

Rosen, G. D., Sanes, J. R., LaChance, R., Cunningham, J. M., Roman, J., and Dean, D. C., 1992, Roles for the integrin VLA-4 and its counter receptor VCAM-1 in myogenesis, *Cell* **69:**1107–1119.

Rosette, C., and Karin, M., 1995, Cytoskeletal control of gene expression: Depolarization of microtubules activates NF-kB, *J. Cell Biol.* **128:**1111–1119.

Rothnagel, J. A., Dominey, A. M., Dempsey, L. D., Longley, M. A., Greenhalgh, D. A., Gagne, T. A., Huber, M., Frenk, E., Hohl, D., and Roop, D. R., 1992, Mutations in the rod domins of keratins 1 and 10 in epidermolytic hyperkeratosis, *Science* **257:**1128–1130.

Rugg, E. L., McLean, W. H. I., Lane, E. B., Pitera, R., McMillan, J. R., Dopping-Hepenstal, P.

J. C., Navsaria, H. A., Leigh, I. M., and Eady, R. A. J., 1995, A functional "knock-out" for human keratin 14, *Genes Dev.* **8:**2563–2573.

Sassoon, D. A., 1993, Myogenic regulatory factors: Dissecting their role and regulation during vertebrate embryogenesis, *Dev. Biol.* **156:**11–23.

Schaart, G., Viebahn, C., Langmann, W., and Raemakers, F., 1989, Desmin and titin expression in early post-implantation mouse embryos, *Development* **107:**581–616.

Schmid, E., Osborn, M., Rungger-Brandle, E., Gabbiani, G., Weber, K., and Franke, W. W., 1982, Distribution of vimentin and desmin filaments in smooth muscle tissue of mammalian and avian aorta, *Exp. Cell Res.* **137:**329–340.

Schultheiss, T., Lin, Z., Ishikawa, H., Zamir, I., Stoeckert, C. J., and Holtzer, H., 1991, Desmin/vimentin intermediate filaments are dispensible for many aspects of myogenesis, *J. Cell Biol.* **114:**953–966.

Sejernsen, T., and Lendahl, U., 1993, Transient expression of the intermediate filament nestin during skeletal muscle development, *J. Cell Sci.* **106:**1291–1300.

Sen, L., and Smith, T. W., 1994, T-type Ca^{2+} channels are abnormal in genetically determined cardiomyopathic hamster hearts, *Circ. Res.* **75:**149–155.

Shear, C. R., and Bloch, R. J., 1985, Vinculin in subsarcolemmal densities, *J. Cell Biol.* **101:**240–256.

Shibuki, K., Gomi, H., Chen, L., Bao, S., Kim, J. J., Wakatsuki, H., Fujisaki, T., Fujimoto, K., Katoh, A., Ikeda, T., Chen, C., Thompson, R. F., and Itohara, S., 1996, Deficient cerebellar long-term depression, impaired eyeblink conditioning and normal motor coordination in GFAP mutant mice, *Neuron* **16:**587–599.

Small, J. V., and Sobieszek, A., 1980, The contractile apparatus of smooth muscle, *Int. Rev. Cytol.* **64:**241.

Smith, T. H., Kachinsky, A. M., and Miller, J. B., 1994, Somite subdomains, muscle cell origins and the four muscle regulatory proteins, *J. Cell Biol.* **127:**95–105.

Stief, A., Winter, D. M., Strätling, W. H., and Sippel, A. E., 1989, A nuclear DNA attachment element mediates elevation and position-independent gene activation, *Nature* **341:**343–345.

Streuli, C. H., Schmidhauser, C., Bailey, N., Yurchenco, P., Skubitz, A. P. N., Roskelley, C., and Bissell, M. J., 1995, Laminin mediates tissue-specific gene expression in mammary epithelia, *J. Cell Biol.* **129:**591–603.

Svitkina, T. M., Verkhovsky, A. B., and Borisy, G. G., 1996, Plectin sidearms mediate interaction of intermediate filaments with microtubules and other components of the cytoskeleton, *J. Cell Biol.* **135:**991–1007.

Tao J.-X., and Ip, W., 1991, Site-specific antibodies block kinase A phosphorylation of desmin *in vitro* and inhibit incorporation of myoblasts into myotubes, *Cell Motil. Cytoskel.* **19:**109–120.

Thornell, L. E., Eriksson, A., and Erdstom, L., 1983, Intermediate filaments in human myopathies, in *Cell and Muscle Motility* (R. B. Dowben and J. W. Shay, eds.), 85–135, Plenum Press, New York.

Tokuyasu, K. T., Dutton, A. H., and Singer, S. J., 1983a, Immunoelectron microscopic studies of desmin (skeletin) localization and intermediate filament organization in chicken skeletal muscle, *J. Cell Biol.* **96:**1727–1735.

Tokuyasu, K. T., Dutton, A. H., and Singer, S. J., 1983b, Immunoelectron micropscopic studies of desmin (skeletin) localization and intermediate filament organization in chicken cardiac muscle, *J. Cell Biol.* **96:**1736–1742.

Tokuyasu, K. T., Maher, P. A., and Singer, S. J., 1984, Distributions of vimentin and desmin in developing chick myotubes *in vivo*. I. Immunofluorescence study, *J. Cell Biol.* **98:**1961–1972.

Tominaga, M., Matsumori, A., Horie, M., Yoshida, H., and Okada, Y., 1993, *J. Clin. Invest.* **91:**1231–1234.

Torchard, D., 1994, Epidermolytic palmoplantar keratoderma cosegregates with a keratin 9 mutation in a pedigree with breast and ovarian cancer, *Nat. Genet.* **6:**106–110.

Traub, P., and Shoeman, R. L., 1994, Intermediate filament proteins: Cytoskeletal elements with gene-regulatory function, *Int. Rev. Cytol.* **154:**1–101.

Tremble, P., Damsky, C. H., and Werb, Z., 1995, Components of the nuclear signaling cascade that regulate collagenase gene expression in response to integrin-derived signals, *J. Cell Biol.* **129:**1707–1720.

Vajsar, J., Becker, L. E., Freedom, R. M., and Murphy, E. G., 1993, Familial desminopathy: Myopathy with accumulation of desmin-type intermediate filaments, *J. Neurol. Neurosurg. Psychiatry* **56:**644–648.

Vandenburgh, H. H., 1987, Motion into mass: How does tention stimulate muscle growth? *Med. Sci. Sports Exercise* **19:**S142–S149.

Vandenburgh, H., 1992, Mechanical forces and their second messengers in stimulating cell growth *in vitro*, *Am. J. Physiol.* **262**(*Regul. Integrative Comp. Physiol.* **31**)**:**R350–R355.

Van Fleet, J. F., and Ferrans, V. J., 1987, Ultrastractural changes in inherited cardiac calcinosis of DBA/2 mice, *Am. J. Veterin. Res.* **48:**255–261.

Vassar, R., Coulombe, P. A., Degenstein, L., Albers, K., and Fuchs, E., 1991, Mutant keratin expression in transgenic mice causes marked abnormalities resembling a human genetic skin disease, *Cell* **64:**365–380.

Vicart, P., Dupret, J.-M., Hajan, J., Li, J., Gyapey, G., Krishnamoorthy, R., Weissenbach, J., Fardeau, M., and Paulin, D., 1996, Human desmin gene: cDNA sequence, regional localization and exclusion of the locus in a familial desmin-related myopathy, *Hum. Genet.* **98:**422–429.

Wakelam, M. J. O., 1985, The fusion of myoblasts, *Biochem. J.* **228:**1–12.

Wang, K., 1985, Sarcomere-associated cytoskeletal lattices in striated muscle, *Cell Muscle Motil.* **6:**315–369.

Wang, N., Butler, J. P., and Ingber, D. E., 1993, Mechanotransduction across the cell surface and through the cytoskeleton, *Science* **260:**1124–1127.

Watson, P. A., 1991, Function follows form: Generation of intracellular signals by cell deformation, *FASEB J.* **5:**2013–2019.

Watson, P. A., Hannan, R., Carl, L. L., and Giger, K. E., 1996, Desmin gene expression in cardiac myocytes is responsive to contractile activity and stretch, *Am. J. Physiol. Cell Physiol.* **39:**C1228–C1235.

Weintraub, H., Davis, R., Tapscott, S., Thayer, M., Krause, M., Benezra, R., Blackwell, T. K., Turner, D., Rupp, R., Hollenberg, S., Zhuang, Y., and Lassar, A., 1991, The myoD gene family: Nodal point during specification of the muscle cell lineage, *Science* **251:**761–776.

Weitzer, G., Milner, D. J., Kim, J.-U., Bradley, A., and Capetanaki, Y., 1995, Cytoskeletal control of myogenesis: A desmin null mutation blocks the myogenic pathway during embryonic stem cell differentiation, *Dev. Biol.* **172:**422–439.

Xu, Z., Cork, L., Griffin, J., and Cleveland, D. W., 1993, Increased expression of neurofilament NF-L produces morphological alterations that resemble the pathology of human motor neuron disease, *Cell* **73:**23–33.

Yang, Y., Dowing, J., Yu, Q.-C., Kouklis, P., Cleveland, D. W., and Fuchs, E., 1996, An essential cytoskeletal linker protein connecting actin microfilaments to intermediate filaments, *Cell* **86:**655–665.

Chapter 17

The Complexity of Intermediate Filaments in the Nervous System

Conrad L. Leung, Roberto L. Flores, and Ronald K. H. Liem

1. INTRODUCTION

In the mammalian nervous system, at least eight intermediate filament (IF) proteins are expressed in developing and mature neurons and glial cells. They are the neurofilament triplet proteins (NF-L, NF-M, and NF-H), α-internexin, peripherin, nestin, vimentin, and the glial fibrillary acidic protein (GFAP). The neurofilament triplet proteins (NFTPs) have apparent molecular weights of 68, 145, and 200 kDa on SDS–PAGE and are named NF-L, NF-M, and NF-H, respectively, for low-, middle-, and high-molecular-weight neurofilament subunits. They are expressed in mature neurons of both the central and peripheral nervous systems (CNS, PNS) and were first identified by axonal transport studies (Hoffman and Lasek, 1975) and subsequently by the biochemical isolation of intermediate filaments from ner-

Conrad L. Leung Department of Biochemistry and Molecular Biophysics, Columbia University College of Physicians and Surgeons, New York, New York 10032. **Roberto L. Flores and Ronald K. H. Liem** Departments of Pathology and of Anatomy and Cell Biology, Columbia University College of Physicians and Surgeons, New York, New York 10032.

Subcellular Biochemistry, Volume 31: Intermediate Filaments, edited by Herrmann and Harris. Plenum Press, New York, 1998.

vous tissues (Liem *et al.*, 1978). In addition, the NFTPs are more closely related to each other than to other members of the intermediate filament protein gene family, and are often called type IV intermediate filament proteins based on sequence homology and gene structure (Steinert and Roop, 1988; for review see Fliegner and Liem, 1991). α-Internexin was first isolated from optic nerve and spinal cord and found to be particularly abundant in the developing nervous system (Pachter and Liem, 1985). Although it is present in both the CNS and PNS early in development, it is mainly CNS-specific in the mature mammalian nervous system. α-Internexin is closely related to the neurofilament triplet proteins and is therefore also classified as a type IV intermediate filament protein (Fliegner *et al.*, 1990). Peripherin was first identified as an intermediate filament protein from neuroblastoma (Portier *et al.*, 1982) and its cDNA was cloned from differentiated PC12 cells (Leonard *et al.*, 1988). Peripherin is found mainly in the peripheral nervous system and belongs to a different class of intermediate filament proteins, called type III intermediate filament proteins, which also consists of vimentin, GFAP, and desmin. Vimentin is present in only a few adult neurons. In developing embryos, vimentin can be found in neuroepithelial stem cells and glioblasts, the immature precursor cells that also express another IF protein called nestin. Nestin is expressed in immature muscle cells in addition to the neuronal and glial precursor cells. Nestin has little homology to the other IF proteins, but its genomic structure would put it in the same type as the NFTPs and α-internexin (Lendahl *et al.*, 1990). In glial cells, vimentin (and presumably nestin) is progressively replaced by the glial-specific GFAP, although vimentin continues to be expressed in some astrocytes (Bovolenta *et al.*, 1984). Vimentin appears to be the primary intermediate filament protein in Schwann cells in the PNS (Autilio-Gambetti *et al.*, 1982). In contrast, oligondendrocytes are among the few cells in the nervous system which are devoid of intermediate filaments (Lee *et al.*, 1993).

The variety of intermediate filaments in nonmammalian nervous systems has been most extensively studied in *Xenopus laevis*. In addition to the neurofilament triplet proteins and peripherin, two proteins which share sequence features with α-internexin have been identified. The combined developmental distributions of these two proteins, XNIF and xefiltin, parallel that of α-internexin in mammalian embryos (Charnas *et al.*, 1992; Zhao and Szaro, 1997). Xefiltin is the closest relative of α-internexin and is also closely related to the goldfish intermediate filament protein gefiltin (see Chapter 1 by Markl and Schechter in this book). Another *Xenopus laevis* protein called tanabin is most closely related to nestin, although tanabin has a somewhat different localization, since it is expressed in the growth cones (Hemmati-Brivanlou *et al.*, 1992). Thus, there are proteins related to the various mammalian neuronal IF proteins in *Xenopus laevis*, although their distributions are somewhat different from those of the mammalian proteins. Some functional studies have been performed on *Xenopus laevis* intermediate filaments, which will be described later.

2. ASSEMBLY OF NEURONAL INTERMEDIATE FILAMENTS

Intermediate filaments are made up of a large family of proteins that can either self-assemble to form homopolymeric filaments or coassemble with other family members to form heteropolymeric filaments. A typical intermediate filament protein exhibits a tripartite structure, consisting of a highly conserved central α-helical rod domain of approximately 310 amino acid residues and variable amino- and carboxyl-terminal head and tail domains (Steinert and Roop, 1988). The rod domain contains long stretches of heptad repeats that mediate the formation of coiled-coil dimers, the first step in intermediate filament assembly (see Chapter 10 by Herrmann and Aebi in this book). To form a dimer, the two intermediate filament polypeptide chains are held together by an apolar stripe on the α-helical rod domain. The pair of intermediate filament polypeptides in a dimer are oriented parallel and in-register with one another. The two dimers then interact with each other to form a tetramer, the protofilament. Cross-linking and electron microscopy studies have revealed that the two dimers of a tetramer are aligned in a staggered and antiparallel array (Geisler *et al.*, 1985; Stewart *et al.*, 1989; Quinlan *et al.*, 1989; Albers and Fuchs, 1992). Because of this symmetrical configuration, intermediate filaments do not have polarity, unlike microfilaments and microtubules. Much less is known about the mechanism of intermediate filament assembly beyond the tetrameric stage, but tetramers have long been considered to be the building blocks of intermediate filament: they assemble radially to reach a diameter of 8–12 nm and longitudinally to elongate the filament.

The assembly and coassembly properties of intermediate filaments have been studied extensively both *in vitro* and *in vivo*. *In vitro*, purified intermediate filament proteins have been tested for their ability to self-polymerize or copolymerize with other purified intermediate filament proteins into 10-nm filaments under appropriate buffer conditions. *In vivo*, enhanced expression of intermediate filament proteins in transfected cells has been employed to define the assembly characteristics of the expressed intermediate filament proteins under physiological conditions. Among the mammalian intermediate filament proteins that we have discussed, only the neurofilament triplet proteins (and probably nestin) are incapable of self-assembly into homopolymeric filament in transfected cells that are devoid of an endogenous cytoplasmic intermediate filament network (Ching and Liem, 1993; Lee *et al.*, 1993). However, NF-L can self-assemble into smooth filaments *in vitro*, whereas NF-M and NF-H only form irregular filamentous aggregates (Liem and Hutchison, 1982). Nevertheless, in the presence of NF-L, these two proteins can coassemble into heteropolymeric filaments both *in vitro* (Geisler and Weber, 1981; Liem and Hutchison, 1982) and *in vivo* (Chin and Liem, 1990; Ching and Liem, 1993; Lee *et al.*, 1993). To verify that both NF-L and NF-M are required to form an intermediate filament network, Lee *et al.* (1993) used two sets of transgenic mice that expressed NF-L or NF-M driven by an MSV promoter and examined the oligodendrocytes, which normally do not have any cytoplasmic intermediate filaments. In transgenic mice expressing either NF-L or NF-M, no IF network could be observed by electron microscopy in the oligodendrocytes. However, oligodendrocytes from mice carrying both NF-L and NF-M transgenes showed that expressing both NF-L and NF-M yielded bundles of intermediate filaments that are indistinguishable from those found in neurons.

The carboxyl-terminal regions of NF-M and NF-H polypeptides appear to emerge from the filament backbone when visualized by electron microscopy after purified bovine NF-M or NF-H is reassembled with NF-L *in vitro* (Hisanaga and Hirokawa, 1990). Together with the findings that the protrusions along the filament backbone are reactive only with antibodies directed against the tail domains of NF-M and NF-H (Hirokawa *et al.*, 1984; Mulligan *et al.*, 1991), these results suggest that NF-L appears to form the core structure of neurofilaments, and NF-M and NF-H coassemble onto this backbone with their tail domains projecting out of the filaments. Further evidence comes from photobleaching experiments using microinjected fluorescently labeled neurofilament proteins, which have shown that NF-H is the most dynamic component of the NF array and interchanges more rapidly than NF-L subunits (Takeda *et al.*, 1994).

To characterize further the assembly properties of NFTPs, it is important to identify the first-formed dimers. By using the yeast two-hybrid system, two independent studies have shown that among all possible dimer configurations, NF-L/NF-L homodimers, NF-L/NF-M heterodimers, and NF-L/NF-H heterodimers are the most favored combinations (Leung and Liem, 1996; Carpenter and Ip, 1996). Presumably, these dimers can then form NF-L containing homo- and heterotetramers, which will ultimately form NFTP heteropolymers in the axon. In addition, the two-hybrid results also confirm that an intact rod domain is sufficient to mediate dimer formation. If so, what are the functions of the head and tail domains? To understand the contributions of these domains to neurofilament formation, cultured cells transfected with different combinations of truncated and full-length NFTPs have been examined using immunocytochemistry. In cell lines without an endogenous cytoplasmic intermediate filament network, full-length NF-L or NF-M can compensate for the deletion of the tail domain of its coassembly partner, but not for the removal of the head domain (Ching and Liem, 1993; Lee *et al.*, 1993). These studies indicate that the tail domains of NF-L and NF-M are dispensable for normal coassembly into a filamentous network. However, unlike NF-M, the tail domain of NF-H is critical for its coassembly with NF-L because transfected tailless NF-H fails to form an extensive filament network with full-length NF-L (Sun *et al.*, 1997). Furthermore, the head domain of NF-H is also necessary for its coassembly with NF-L. In general, the head domains of the NFTPs appear to be more important than the tail domains in the assembly of the intermediate filament network. However, whether they exert their influence on the dimer, the tetramer, or high-order structures remains unclear.

α-Internexin, peripherin, and GFAP are all able to self-assemble into homopolymeric filaments in transfected cells (Ching and Liem, 1993; Chen and Liem, 1994a; Ho *et al.*, 1995; Cui *et al.*, 1995). α-Internexin can also coassemble with each of the neurofilament triplet proteins to form heteropolymeric filaments (Ching and Liem, 1993). Transient transfections using peripherin cDNA have shown that deletion of the peripherin head domain abolishes its ability to self-assemble in transfected cells (Ho *et al.*, 1995; Cui *et al.*, 1995). In contrast, filamentous arrays are still maintained in cells transfected with tailless peripherin. In certain axons of rat sciatic nerve, peripherin coexists with the NFTPs. From immunoelectron microscopic studies, it appears that peripherin and NFTPs coassemble to form a unique intermediate filament network in these nerve cells (Parysek *et al.*, 1991). In transfection studies of GFAP, the entire head domain is shown to be required for the *de novo* assembly of GFAP, while one-fifth of the tail domain can be dispensed with without significant effects on assembly (Chen and Liem, 1994a). However, complete deletion of the tail domain of GFAP results in the formation of polymorphic aggregates in transfected cells. *In vitro*, self-assembly studies using bacterially expressed GFAP also showed that tailless GFAP is incompetent in self-assembly (Quinlan *et al.*, 1989). The assembly characteristics of nestin have not been studied in detail since its initial discovery, but so far there is no report of the expression of nestin in the absence of vimentin (Lendahl *et al.*, 1990).

3. DEVELOPMENTAL EXPRESSION OF INTERMEDIATE FILAMENTS IN THE NERVOUS SYSTEM

Prior to differentiation into neurons, rapidly dividing neuroepithelial cells span the thickness of the folding neural plate. Vimentin (Tapscott *et al.*, 1981) and nestin (Hockfield and McKay, 1985) are expressed in nearly all mitotic neuroepithelial cells within the neural tube. After neural tube closure, individual neuroepithelial cells exit the cell cycle, form postmitotic neurons, and migrate out of the germinal zone to the ventrolateral surface of the tube, where differentiation occurs. As neurons withdraw from the cell cycle, neuronal IF expression commences and vimentin expression ceases in the hindbrain and rostro–spinal cord region of the neural tube and spreads in a rostrocaudal fashion (Tapscott *et al.*, 1981; Raju *et al.*, 1981; Fliegner *et al.*, 1994). As individual cells become postmitotic, brief coexpression of neuronal IFs and vimentin can be detected (Bignami *et al.*, 1982; Cochard and Pauling, 1984).

Although α-internexin, NF-L, and NF-M have all been reported to be expressed in the early-developing nervous system (Kaplan *et al.*, 1990; Tapscott *et al.*, 1981), studies from our laboratory and others have shown that α-internexin is generally the earliest neuronal IF protein expressed in differentiated neurons within the CNS. At embryonic day 12 (E12) in the rat, NF-L and NF-M expression is limited to the myelencephalon and spinal cord, whereas α-internexin expression extends anteriorly to the telencephalon (Kaplan *et al.*, 1990; Fliegner *et al.*, 1994). By E16, when α-internexin is at its peak of expression in the telencephalon, *in situ* hybridization studies have shown the presence of α-internexin mRNA in the differentiated neurons and vimentin mRNA in mitotic neuroblasts, with a clear demarcation between the regions where the two mRNAs are expressed (Fliegner *et al.*, 1994). The levels of NF-L and NF-M transcripts are still low at this point, revealing the limitations of these proteins as markers for differentiated neurons. Consistent with this observation, there is a major increase of α-internexin expression compared to a minute increase in NF-L expression in the retinoic acid-induced differentiation of P19 embryonal carcinoma cells (Chiu *et al.*, 1995). Moreover, only α-internexin is expressed during the formation of the lamellipodia and minor processes in cultured hippocampal cells, and not until axon extension do the cells express NF-L or NF-M (Benson *et al.*, 1996). In cultured cerebellar granule neurons, α-internexin is expressed in the absence of the NFTPs (Chien *et al.*, 1996). Indeed, the early expression of α-internexin in the thin processes of young neurons is consistent with its preferential expression in small-caliber neurons (Fliegner *et al.*, 1994; Kaplan *et al.*, 1990), suggesting that the protein may play a role in the stabilization of thin neuronal processes. Additionally, the ability of α-internexin to form homopolymers and heteropolymers with NFTPs (Ching and Liem, 1993) suggests that the protein serves as scaffolding upon which other neurofilament proteins can polymerize to form the filament network. One of the few exceptions to α-internexin primacy in developmental expression exists in the human hippocampus (Arnold and Trojanowski, 1996), where NF-L and NF-M expression precede α-internexin. Furthermore, since no homologue of α-internexin has yet been identified in the chick, NF-L and NF-M may in fact be the earliest neuronal IFs expressed in avian neurons (Tapscott *et al.*, 1981). Throughout development, NF-L and NF-M are generally expressed concurrently in mammals (Shaw and Weber, 1982). However, other investigators have detected NF-M in certain chick neuroepithelial tissue and nonneural tissue such as cardiac myoblasts, indicating a possible, transient role of NF-M in the development of these tissues in the chick (Bennett and DiLullo, 1985). Another exception to NF-L/NF-M coexpression occurs in the mouse adrenal medulla (Troy *et al.*, 1990), which expresses only NF-M. In human T lymphocytes, NF-H has been detected (Murphy *et al.*, 1993), providing another exception to the neuronal specificity of NF proteins. These exceptions to the neural specificity of NF proteins suggest that selected cells may rely on the individual properties of these proteins for their cellular needs.

The last neurofilament to be expressed in the nervous system is NF-H. Initial studies on NF-H expression in the rat revealed that NF-H cannot be detected prior to P5 (Shaw and Weber, 1982). More detailed studies, however, have detected the protein immunohistochemically as early as E15 in the rat (Carden *et al.*, 1987). NF-H is first expressed in the trigeminal ganglia, ventral horn motor neurons, dorsal root ganglia, sympathetic ganglia, and distal peripheral nerve (Carden *et al.*, 1987). Concomitant with the postnatal increase of NF-H, there is a parallel upregulation of NF-L and NF-M expression (Schlaepfer and Bruce, 1990). In the adult rat, NF-H is highly expressed in large neurons, such as the brainstem motor neuron nuclei, the ventral horn of the spine, and the Purkinje cells of the cerebellum, and at lower levels in the eye, cortex, thalamus, and hippocampus (Shaw and Weber, 1983; Roussel *et al.*, 1991; Dautigny *et al.*, 1988). The delayed expression of NF-H and the correlation between NF-H expression and the establishment of an adult pattern of neuronal IF expression provide support for the role of NF-H in the stabilization of mature neurons. Electrophysiological studies performed on E14 rat cortical neurons cocultured with mature type 1 astrocytes indicate the importance of NF-H in the maturation of neuronal cells: expression of nonphosphorylated NF-H occurs just before the neuron becomes excitable, and NF-H is phosphorylated just after neurons make synaptic connections (Steinschneider *et al.*, 1996).

In the cerebellum, α-internexin is the first neuronal IF protein to be expressed in most neurons (Chien *et al.*, 1996). In the mouse, α-internexin can be detected as early as postnatal day 5 (P5) in parallel fibers, Purkinje cells, and basket cell processes. The axons in the white matter express α-internexin as well as the NFTPs from P5 on. By P7, the NFTPs can be detected in the Purkinje cell processes and the basket cell processes show moderate levels of NF-L and NF-M. By P9, the basket cells express all

NFTPs. Transient expression of NF-H has been detected in the Purkinje cell bodies of the rat (Sawant *et al.*, 1994) shortly after initial neuronal IF expression. α-Internexin is the only IF in the parallel fibers throughout development in the mouse (Chien *et al.*, 1996) and rat (Chiu *et al.*, 1989; Kaplan *et al.*, 1990), but there are reports of NF-L transcripts during cerebellar development in the mouse (Kure and Brown, 1995). It must also be noted that in the guinea pig (Vickers and Costa, 1992) and the cat (Riederer *et al.*, 1996), the NFTPs have been detected at moderate levels in parallel fibers, suggesting that there may be variant expression of NFTPs in parallel fibers across species.

Unlike the NFTPs, peripherin is largely localized to the peripheral nervous system (Greene, 1989). Extensive developmental studies on peripherin expression have shown that initial peripherin immunoreactivity occurs at E9 in the mouse (Troy *et al.*, 1990) and E12 in the rat (Escurat *et al.*, 1990) in cervical ventral roots, as well as enteric nervous system, sympathetic, acoustic, facial, and dorsal root ganglia, and trigeminal nerve. In the mouse, peripherin expression is delayed in the dorsal root and sympathetic ganglia relative to the other regions (Troy *et al.*, 1990). In the cervical spine, its expression extends caudally, avoiding the cortical regions (Troy *et al.*, 1990). In the dorsal root, peripherin expression is largely localized to small neurons; in contrast to larger neurons, which mostly express the NFTPs (Goldstein *et al.*, 1991; Goldstein *et al.*, 1996). In the adult, peripherin is expressed along the wall of the small intestine (Parysek and Goldman, 1988), in almost all cranial nerves, dentate nucleus, medial and lateral vestibular nuclei, red nucleus, tegmental nucleus, reticular neurons, and various thin fibers in the brain stem and cerebellum (Leonard *et al.*, 1988, Brody *et al.*, 1989). No peripherin protein has been detected in the adrenal medulla (Troy *et al.*, 1990), although its transcript has been detected at low levels (Leonard *et al.*, 1988; Parysek and Goldman, 1988), suggesting a posttranscriptional control of the mRNA.

In the rat, NF-L, NF-M, α-internexin, and peripherin can be detected by E16 in the retinal ganglion cell layer as well as the optic nerve (Shaw and Weber, 1983; Kaplan *et al.*, 1990; Escurat *et al.*, 1990). However, peripherin has not been detected in the mouse retina at any point during development, although it is expressed in adult optic nerve (Troy *et al.*, 1990). α-Internexin expression is more widespread than the other NF proteins: it is detected in some radial fibers in the outer layers of the retina (Kaplan *et al.*, 1990) and is transiently expressed in amacrine cells during development (Chien and Liem, 1995). α-Internexin, NF-L, and NF-M are all already expressed in the horizontal cell processes of newborn mice and rats (Shaw and Weber, 1983; Chien and Liem, 1995), whereas NF-H is first expressed in the horizontal cell processes by P10 in the rat (Shaw and Weber, 1983) and in both the horizontal cells and ganglion nerve fibers by P5 in the mouse (Chien and Liem, 1995). NF-H expression spreads to the inner plexiform layer and optic nerve by P13 in the mouse.

4. REGULATION OF GENE EXPRESSION

As intermediate filaments have distinct distributions in various cell types, the study of neuronal intermediate filament gene expression provides an excellent approach toward understanding neuronal-specific gene regulation. In the past few years, extensive efforts have been made in identifying the important *cis*- and *trans*-acting elements that together determine the spatial and temporal expression of neuronal intermediate filaments during the development of the nervous system. *In vitro*, DNase I footprinting assays, DNA electrophoretic-mobility-shift analyses, and methylation interference assays have widely been used in identifying regulatory elements. *In vivo*, transfections of neuronal cell lines with dissected neuronal intermediate filament promoters fused to reporter genes are used in testing neuronal specificity. Finally, transgenic mice are established to ultimately define regions of neuronal intermediate filament genes necessary for their unique expression pattern under physiological condition.

The control and coordination of the expression of NFTPs are very complicated. As described, NF-L and NF-M are expressed before NF-H and all are maintained throughout the mature nervous system and are especially abundant in large axons. Therefore, these three genes may share similar control in their cell-type-specific expression, but vary in their temporal control. As one may predict, all three NF gene promoters share similar functional features: a basal promoter, a weak neuronal enhancer, and a distal negative regulatory element (Shneidman *et al.*, 1992; Yazdanbakhsh *et al.*, 1993; Schwartz *et al.*, 1994). However, none of these promoters is able to confer neuronal cell-line-specific expression. Utilizing any of these promoters to drive the expression of reporter genes yields no difference between neuronal and nonneuronal cell lines (Hsu *et al.*, 1995; Nakahira *et al.*, 1990; Shneidman *et al.*, 1992; Reeben *et al.*, 1995; Zopf *et al.*, 1990). *In vitro* transcription assays using extracts from brain and liver tissues also fail to demonstrate any tissue specificity of the mouse NF-H promoter containing 5′ flanking region from −1314 to −115 (Schwartz *et al.*, 1994). The search for the neuronal specificity of the neuronal intermediate filament gene promoters has been clarified by the generation of different transgenic mouse lines: NF-L (Charron *et al.*, 1995a,b; Lee *et al.*, 1994; Leconte *et al.*, 1994; Reeben *et al.*, 1993; Yazdanbakhsh *et al.*, 1993; Beaudet *et al.*, 1992), NF-M (Tu *et al.*, 1995; Wong *et al.*, 1995), and NF-H (Cote *et al.*, 1993, 1994;

Marszalek *et al.*, 1996; Xu *et al.*, 1996). Collectively, these studies show that in addition to a basal promoter, intragenic sequences are required to enhance and direct neuronal-specific expression of the transgene. *In vitro* studies on identifying *trans*-regulatory elements have led to very limited success; in the case of the NF-L promoter, four DNase I hypersensitive sites (HSS) have been reported, but no brain-specific transcription factors have been isolated (Ivanov and Brown, 1992; Kure and Brown, 1997; Kure *et al.*, 1996).

α-Internexin is mainly expressed in granule neurons in the cerebellum, as well as in small neurons in the CNS and PNS. Analysis of the rat α-internexin promoter in transfected cell lines reveals that the region from −77 to +73 relative to the transcription start site is required for basal transcription. The −254 to −78 region, which contains several potential regulatory elements, further enhances the basal expression level (Ching and Liem, 1991). The α-internexin gene promoter is regulated by Brn-3a, Brn-3b, and Brn-3c, members of the POU (Pit-Oct-Unc) family of transcription factors, which are expressed predominantly in neuronal cells; Brn-3a and Brn-3c activate, but Brn-3b represses α-internexin expression (Budhram-Mahadeo *et al.*, 1996; Budhram-Mahadeo, 1995). Similar to the results obtained with the NFTP promoters, a 5-kb α-internexin flanking sequence is unable to drive neuron-specific expression in cultured neuroblastomas (Ching and Liem, 1991), and yet with the presence of intragenic sequences, a 1.2-kb α-internexin promoter is sufficient to direct neuron-specific expression in transgenic mice (G. Y. Ching, R. L. Flores, C. L. Chien, and R. K. Liem, unpublished data).

Peripherin is principally expressed in the peripheral nervous system. In neuronal cell lines, the first 98 bp preceding the transcription start site of the peripherin gene is sufficient to confer cell-type-specific expression (Desmarais *et al.*, 1992). Within this region, three DNase I-protected regions, PER1, PER2, and PER3, have been identified. PER1 contains the TATA motif and is implicated in playing a crucial role in cell-type-specific expression (Desmarais and Royal, 1996). PER2 and PER3 act as positive modulators. In another study, a negative regulatory element (NRE) was identified at −173 (Thompson *et al.*, 1992). This region is important in repressing peripherin expression in undifferentiated and nonneuronal cells. NF1-L was shown to bind to the NRE in a complex with other nuclear proteins (Adams *et al.*, 1995). Even though transfection studies indicate that a short peripherin promoter sequence is sufficient to direct cell-line-specific expression of the reporter gene, a longer promoter sequence and intragenic regions are required to direct the precise cell-type-specific expression of the peripherin transgenes in transgenic mice (Leconte *et al.*, 1996; Belecky-Adams *et al.*, 1993). Without introns, a 5.8-kb peripherin 5′-flanking sequence can only drive the *lacZ* reporter transgene with correct temporal and neural tissue-specific expression, but precise cell-type-specific expression can only be achieved with the presence of introns.

Unlike neuronal intermediate filament proteins, GFAP is expressed exclusively in glial cells. *In vitro*, using a novel retrovirus infection system, a 256-bp GFAP promoter is sufficient to direct astrocyte-specific expression of a *lacZ* reporter gene in primary glial cultures (Baba *et al.*, 1997). However, to obtain faithful glial-specific expression in transgenic mice, 2kb of the 5′-flanking region of the GFAP gene is required. Moreover, this fragment also supports a response to brain injury by upregulation of GFAP expression (Galou *et al.*, 1994; Danks *et al.*, 1995; Brenner *et al.*, 1994; reviewed by Brenner and Messing, 1996). By DNase I footprinting assays, three *trans*-acting factor-binding sites (GFI, GFII, and GFIII) have been identified (Miura *et al.*, 1990). GFI and GFIII are implicated in directing glial-specific expression, while GFII is important in enhancing promoter activity. During early development, nestin is expressed in neuroepithelial cells and muscle progenitor cells. Transgenic studies indicated that the first and second introns of the nestin gene are necessary for directing gene expression in developing muscle and neural precursor cells, respectively (Zimmerman *et al.*, 1994; Lothian and Lendahl, 1997).

5. PHOSPHORYLATION OF INTERMEDIATE FILAMENTS IN THE NERVOUS SYSTEM

Most of the intermediate filament proteins are posttranslationally modified by phosphorylation and glycosylation. Phosphorylation always occurs at the head and tail domains of intermediate filament proteins and plays a crucial role in the modulation of intermediate filament assembly. Phosphorylation at the head domains of vimentin, GFAP, and NF-L by PKN almost completely abolishes their ability to form filaments *in vitro* (Nakamura *et al.*, 1996; Inagaki *et al.*, 1996; Matsuzawa *et al.*, 1997, Mukai *et al.*, 1996; see also Chapter 14 by Matsuzawa *et al.* in this book). Phosphorylation of the NF-L head domain by protein kinase A (PKA) also prevents its self-assembly and induces the disassembly of NF-L filaments *in vitro* (Ando *et al.*, 1989; Hisanaga *et al.*, 1990; Nakamura *et al.*, 1990). This kind of dynamic regulation is required in a number of cellular events. For instance, at the late mitotic phase of dividing glial cells, GFAP is disassembled at the cleavage furrow for successful cell division (Inagaki *et al.*, 1994; Chapter 14 by Matsuzawa *et al.* in this book). Phosphorylation of GFAP at the head domain by a Rho-associated kinase has been proposed to act as the signaling messenger for GFAP disassembly (Sekimata *et al.*, 1996;

Kosako *et al.*, 1997). In another study, phosphorylation on Ser55 of NF-L by PKA has been observed with a rapid turnover rate immediately following NF-L synthesis in neurons (Sihag and Nixon, 1991). This process may regulate neurofilament assembly or initial interactions during axonal transport in neurons (Betts *et al.*, 1997; Nixon *et al.*, 1994). Many kinases, including PKA, PKC, CaMKII, and p34cdc2, have been studied for their ability to phosphorylate neuronal intermediate filament proteins *in vitro* (Schroder *et al.*, 1997; Chou *et al.*, 1996; Tsujimura *et al.*, 1994; Tanaka *et al.*, 1993; Guan *et al.*, 1992; Hisanaga *et al.*, 1991). Obviously, it is difficult to determine the effects on neuronal intermediate filaments of each of these kinases in the nervous system, but they certainly can play a part in regulating the dynamics of neuronal intermediate filaments. Also, it is important to mention that phosphorylation is a dynamic process; the actual phosphorylation state of a protein results from the balance between the activities of the kinases and phosphatases.

As described in the next section, phosphorylation also mediates intermediate filament-specific functions in neurons, at least in the case of the NFTPs, which are well known for their high level of phosphorylation. While NF-L in mature neurons contains just a few phosphate groups, NF-M and NF-H contain as many as 25 and 50 moieties respectively, depending on species (Julien and Mushynski, 1981, 1983; Shecket and Lasek, 1982; Jones and Williams, 1982; Wong *et al.*, 1984; Carden *et al.*, 1985). Recently, most of these phosphates have been mapped to Lys-Ser-Pro (KSP) repeats in the carboxyl-terminal tail domains of NF-M and NF-H (Betts *et al.*, 1997; Bennett and Quintana, 1997; Elhanany *et al.*, 1994). The extensively phosphorylated tail domains of these two proteins have no analogy among other intermediate filaments. The extensive phosphorylation of NF-H and NF-M has led to a search for the kinase(s) that are responsible for phosphorylating the KSP repeats in their tail domains. The cell-cycle kinase cdc2 has been shown to phosphorylate neurofilaments and cause the characteristic gel mobility shift of NF-H upon phosphorylation (Hisanaga *et al.*, 1991). Unfortunately, this result has little physiological relevance since cdc2 kinase is absent in terminally differentiated neurons. Therefore, the search for neurofilament kinases has focused on cdc2 kinase homologues, especially one known as Cdk5. In conjunction with its activator, p35, Cdk5 has been shown to phosphorylate NF-H extensively in transiently transfected cells (Sun *et al.*, 1996; Guidato *et al.*, 1996). Moreover, unlike cdc2, Cdk5 kinase activity is restricted to the nervous system and maintained throughout development in rodents (Ino *et al.*, 1994; Tsai *et al.*, 1993). However, *in vitro* phosphorylation studies have shown that there are two consensus sequences for NF-H phosphorylation: KSPXK and KSPXY. Cdk5/p35 is able to phosphorylate one of these KSP sequences (KSPXK), but not the other

(Sun *et al.*, 1996). The identity of other kinase(s) that can phosphorylate this KSPXY motif is still unknown. Multiple cdc2-like kinases have been identified and some of them are expressed in the nervous system (Okuda *et al.*, 1992, Meyerson *et al.*, 1992; Gao *et al.*, 1996), but so far none of them has been shown to phosphorylate NF-H. Recently, an isoform of p35 designated p39 has been reported (Tang *et al.*, 1995). The precise role of p39 as an activator of Cdk5 in NF-H phosphorylation is under investigation in our laboratory. In addition to phosphorylation, NFTPs are also modified by O-linked N-acetylglucosamine (O-GlcNAc) (Dong *et al.*, 1993, 1996). The function of this modification is not fully understood, but it is interesting to note that some of the KSP repeats at the tail domain of NF-H can also undergo this kind of modification.

6. FUNCTIONS OF INTERMEDIATE FILAMENTS IN THE NERVOUS SYSTEM

The function of neurofilaments as a major determinant of axonal diameter has long been suspected from the close correlation between the number of filaments in axons and the axonal diameter. Independent animal models have been established to support this hypothesis. The Japanese quiver quail (quv) has a mutation in the NF-L gene and as a result develops axons devoid of NFs. In this animal, normal axonal growth is severely inhibited (Yamasaki *et al.*, 1991, 1992; Ohara *et al.*, 1993). In transgenic mice overexpressing an NF-H/β-galactosidase fusion protein, most of the neurofilament proteins are retained in the perikarya of neurons and the resulting neurofilament-free axons are smaller in diameter (Eyer and Peterson, 1994). Among the NFTPs, the importance of NF-H in mediating axonal radial growth has been further elucidated. Moderately enhanced expression of murine NF-H in transgenic mice increases total neurofilament content in neurons and results in larger myelinated axons (Marszalek *et al.*, 1996). However, further increase of NF-H expression in transgenic mice decreases the number of neurofilaments in neurons and strongly inhibits axonal radial growth, probably the result of impaired axonal transport. In another transgenic study, overexpression of human NF-M led to a reduction of fully phosphorylated NF-H and an increase of neurofilament packing density, suggesting that phosphorylation of NF-H is essential for maintaining the interfilament spacing (Tu *et al.*, 1995). Similar results were obtained in studies utilizing the Trembler mutant mouse, which demonstrated that a decrease in NF-H phosphorylation resulted in an increase in neurofilament density and a concomitant decrease in axonal diameter (de Waegh *et al.*, 1992).

Antibody injection experiments conducted on *Xenopus laevis* embryos showed that injection of antibodies to NF-M into a single blastomere of two-cell-stage embryos resulted in abnormal accumulations of neurofilaments in the cell bodies and retardation of peripheral nerve development. In culture, the anti-NF-M-injected blastomeres were shorter than uninjected controls by 21 hr after plating and became morphologically abnormal (Szaro *et al.*, 1991; Lin and Szaro, 1995). Defects included variations in diameter, poorly defined separations between the growth cones and neurites, and more collateral branching. These results support the hypothesis that neurofilaments help maintain the characteristic shapes of axons.

Gene knockout, antisense oligonucleotide, and mRNA experiments and antibody microinjection studies have given rather contradictory results about the potential functions of the other intermediate filament proteins expressed in the nervous system. Both vimentin and GFAP have been knocked out, but surprisingly the resulting mice showed no obvious deleterious phenotype (Colucci-Guyon *et al.*, 1994; Gomi *et al.*, 1995; Pekny *et al.*, 1995). This is in contrast to the results obtained from antisense studies in cultured cells, where the downregulation of vimentin in neuroblastoma cells resulted in the inability of the cells to grow neurites (Shea *et al.*, 1993; see also Chapter 18 by Shea *et al.* in this book). In addition, stable transfections of an antisense GFAP construct in astrocytoma cells resulted in the inability of these cells to grow processes in the presence of neurons (Weinstein *et al.*, 1991). This latter effect could be rescued by the reexpression of GFAP in these astrocytoma cell lines (Chen and Liem, 1994b).

Since astrocytes from the GFAP knockout mice are able to grow processes in the presence of neurons, the effect on process outgrowth appears to be specific for the astrocytoma cells. Antisense studies of peripherin in PC12 cells did not cause a similar inhibition of process outgrowth (Troy *et al.*, 1992). So far there is no report of a peripherin gene knockout in transgenic mice. Overall, these studies show that the correlation between intermediate filament protein expression and process outgrowth is observed only in some culture systems. In contrast, there is a clear indication of the function of neuronal intermediate filament proteins in determining radial growth of the axon. Finally, as discussed below, overexpression of intermediate filament proteins in axons can lead to neuronal degeneration.

7. NEURONAL INTERMEDIATE FILAMENTS AND DISEASE

There has been a recent explosion of studies linking neurofilament protein dysfunction to specific neurological disorders. Through the use of various mouse models and gene characterization, neuronal IF proteins and their associated proteins have been designated as key players in the pathogenesis of motor neuron disease and dystonia musculorum. Motor neuron disease is characterized by selective degeneration of motor neurons and muscle atrophy. Paralysis and death ensue within a few years after the first clinical symptoms manifest themselves. The best-known type of motor neuron disease, amyotrophic lateral sclerosis (ALS) or Lou Gehrig's disease, named after one of its most famous victims, affects roughly 1 in 10,000 people. ALS can be divided into two subcategories: familial ALS (FALS), which accounts for 10% of all ALS cases and is inheritable through a dominant allele, and sporadic ALS (SALS), which is clinically indistinguishable from FALS.

On the cellular level, degenerating motor neurons are characterized by swollen cell bodies of the spinal ventral horn motor neurons (Carpenter, 1968). Immunochemical and electron microscopic studies on neurofilaments (NFs) from both forms of ALS (Hirano *et al.*, 1984a, b) reveal abnormal neurofilamentous accumulation within the cell bodies of motor neurons, forming disorganized IF structures. In FALS, swollen axons with densely packed filaments are a common feature. As previously stated in this chapter, NFs are responsible for maintaining axonal caliber and are highly expressed in larger neurons, such as motor neurons. The structural role of NF proteins and their abundance in motor neurons, combined with their aberrant expression and assembly pattern in pathological ALS cells, led to the creation of a number of transgenic mice in hopes of mimicking the symptoms of ALS.

Transgenic mice that overexpress three to four times wild-type levels of NF-L (Xu *et al.*, 1993) or two times normal levels of human NF-H (Cote *et al.*, 1993) have been generated and studied. Swollen motor neuron cell bodies and axons containing densely packed neurofilaments are common features of both types of transgenic mice, as is atrophy of the skeletal muscle tissue. The NF-L-overexpressing mouse displayed a significant degree of axonal degeneration and most of these mice died prematurely (within 3 weeks after birth). Although both types of transgenic mice mimicked many clinical and pathological conditions of ALS, they did not suffer significant motor neuron loss.

A study linking mutations in the gene encoding Cu/Zn-binding superoxide dismutase (SOD1) to FALS (Rosen *et al.*, 1993) introduced a new potential cause for the disease. SOD1, which is present in a variety of tissues (Crapo *et al.*, 1992), converts the toxic superoxide anion radical O_2^- to hydrogen peroxide. The SOD1 variants, expressed in roughly 20% of FALS patients, are missense mutations (some of which are as conservative as Gly to Ala), mostly occurring in the highly conserved regions of the enzyme

(Rosen *et al.*, 1993). Although one would be tempted to suggest a loss-of-function mutation, this is inconsistent with the dominant phenotype of inheritable ALS. Mice overexpressing wild-type SOD1 do not manifest ALS symptoms (Avraham *et al.*, 1988, 1991; Yarom *et al.*, 1988), suggesting that increased SOD1 activity as a result of the mutation is not the cause of ALS. Instead, results from the studies of transgenic mice carrying an ALS-linked point mutation of the human SOD1 gene suggest that ALS symptoms are caused by an altered catalytic property gained by mutant SOD1 (Gurney *et al.*, 1994; Tu *et al.*, 1996). Only mice expressing the largest amounts of the transgene displayed clinical and pathological ALS symptoms suggesting a gain-of-function mutation.

How does a gain-of-function mutation of a protein which is expressed by many different cell types selectively kill motor neurons and how do the previous ALS models provided by transgenic mice overexpressing neurofilament proteins fit into the pathogenesis of ALS? It must be emphasized that FALS patients with the SOD1 mutations and the SOD1 mutant transgenic mice show abnormal accumulations of NFs in motor neurons, also observed in the NF-overexpressing transgenic mice. This similarity suggests that SOD1 mutations and neurofilamentous accumulations function in the same pathological cascade. Two hypotheses have been proposed for the gain-of-function mutation. The first hypothesis suggests that a decreased scavenging activity by mutant SOD1 promotes a reaction between superoxide and NO, forming a peroxynitrate anion ($ONOO^-$). This anion then reacts with SOD1, creating a nitronium-like intermediate which can nitrate tyrosine residues and thereby damage other proteins, such as the neurofilament proteins (Beckman *et al.*, 1993). The second hypothesis states that mutant SOD1 has hyperactive catalytic activity with hydrogen peroxide and oxidizes substrates at an abnormally elevated rate (Wiedau-Pazos *et al.*, 1996).

Could neurofilaments be a target for the mutated SOD1 protein? A possible answer was provided by the study of a transgenic mouse expressing low levels of NF-L with a leucine-to-proline substitution within the rod domain (Lee *et al.*, 1994). As noted above, the highly conserved rod region of NF proteins is responsible for dimer formation. Disruption of the rod domain in a number of different keratins causes skin blistering diseases due to altered keratin assembly (Fuchs, 1995). Thus, a similar mutation in NF proteins could also lead to an abnormal NF assembly pattern. The transgenic mice carrying the NF-L mutation display the typical ALS symptoms: perikaryal swellings, abnormal neurofilamentous accumulations, and muscle atrophy (Lee *et al.*, 1994). However, unlike the previous neurofilament protein-overexpressing mouse models (Cote *et al.*, 1993; Xu *et al.*, 1993) , the transgenic mice with the NF-L mutation also suffer selective motor neuron death. Since the mutation in the NF-L rod domain leads to aberrant assembly of the neurofilaments, the pathogenesis of FALS with SOD1 mutation can be explained if the mutant SOD1 modifies the structure of the neurofilament proteins and alters their normal assembly pattern leading to neurofilamentous accumulations. This accumulation of filaments could ultimately prove detrimental to the neuron, as slow axonal transport is compromised and the transport of proteins and organelles down the axon is inhibited (Cleveland, 1996). This scenario is consistent with the selective toxicity of mutated SOD1 in FALS, because neurofilaments are predominantly expressed in motor neurons. Moreover, the late onset of ALS implies a slow buildup of toxic proteins in neuronal cells, which requires SOD1 substrates with long half-lives, such as the NFTPS, since the damaged proteins have to accumulate over time.

Since mutations in SOD1 only account for roughly 20% of all FALS cases (2% of all ALS cases), there must be other mechanisms which lead to ALS. One possibility is that there are mutations in NFTPs which lead directly to alterations in neurofilament structures. Mutations in the NF-H gene have been detected in ALS patients, further supporting the role of NF dysfunction in the pathogenesis of ALS. In a study of SALS patients, two genetic variants of NF-H were described in four patients (Figlewicz *et al.*, 1994). Both of these mutations resulted in deletions in the KSP repeat domain of the NF-H gene, which contains the phosphorylation sites of NF-H. As stated previously in this chapter, this domain may be responsible for maintaining the normal axonal architecture. Although two other studies failed to detect mutations in NFTPs in ALS patients (Vechio *et al.*, 1996; Rooke *et al.*, 1996), these studies utilized a smaller patient group with few if any SALS-affected patients. Thus, it is not surprising that NF-H mutations were not found in these two studies. However, the causes of the majority of ALS (familial and sporadic) cases are still unknown. As additional mutations are found, it will also be important to learn how these mutated genes influence neurofilament structure and assembly.

Studies in our laboratory indicate that the abnormal expression of neuronal intermediate filament proteins may result in pathologies which are not limited to motor neurons. By overexpressing wild-type α-internexin in transgenic mice, we detected proximal axonal swellings in the Purkinje cells of the cerebellum, as well as neurofilamentous accumulations in the cell bodies of some neurons in the thalamus and pyramidal neurons in the neocortex. No morphological abnormalities were detected in the spinal cord. Behavioral studies show that these mice display a defect in balance, consistent with the observed morphological abnormalities (G. Y. Ching, R. L. Flores, C. L. Chien, and R. K. Liem, unpublished data).

Initial studies linking neurofilament protein dysfunction to the sensory neuron disease dystonia musculorum came unexpectedly through a knockout mouse created to reproduce the autoimmune skin disorder bullous pemphigoid. Skin cells rely heavily on keratins to preserve their physical integrity. BPAG1, a major antigen characterizing bullous pemphigoid autoimmunity (Stanley, 1993), is part of the hemidesmosome, an intracellular structure responsible for anchoring keratin filaments to the cell membrane. Sequence analysis of the BPAG1 tail domain reveals sequence homology to desmoplakin and plectin, both of which bind to keratins (Stappenbeck *et al.*, 1993; Wiche *et al.*, 1993; Kouklis *et al.*, 1994). Since BPAG1 contains a putative keratin-binding site and is an antigenic target in bullous pemphigoid, a BPAG1 knockout mouse was created in hopes of dissociating keratin filaments from the hemidesmosome and mimicking the skin disorder. This knockout mouse lacked association between keratin filaments and malformed hemidesmosomes, resulting in skin blistering, hair loss, and compromised skin cell migration (Guo *et al.*, 1995). Surprisingly, the mice also suffered sensory nerve degeneration and dystonia, typical of dystonia musculorum (*dt*/*dt*) mice. Although immunohistochemistry revealed that at least one strain of the *dt*/*dt* mice lacked the BPAG1 antigen (Guo *et al.*, 1995), previous immunofluorescence studies (Owaribe *et al.*, 1990) did not reveal BPAG1 expression in nervous tissue. This suggests that BPAG1 may exist in an alternate form in the nervous system (Guo *et al.*, 1995). Evidence in support of this hypothesis came through the characterization of dystonin, a candidate dt gene (Brown *et al.*, 1995). Three dystonin transcripts were identified, all of which contained the same sequence coding for the rod and tail domains of BPAG1. The shortest transcript (BPAG1e) is expressed in keratinocytes; the larger two (to be collectively classified as BPAG1n) are expressed in nervous tissues. Evidence in support of dystonin as the dt effector gene was obtained from genetic analysis of spontaneous dt mutant mice, which showed the presence of truncated dystonin transcripts in all the mutant strains (Brown *et al.*, 1995). Indeed, *in situ* hybridization and immunohistochemistry reveal that BPAG1n is selectively expressed in sensory neurons, the targets of dystonia musculorum (Brown *et al.*, 1995; Yang *et al.*, 1996). A distinct feature of the larger transcripts is the presence of an N-terminal actin-binding domain (Brown *et al.*, 1995; Yang *et al.*, 1996). With a putative binding sequence to actin at the N-terminus and a putative NF-binding sequence at the C-terminus, the BPAG1n protein could serve as an anchor connecting NFs to actin. Thus, the neurodegenerative phenotype of BPAG1 null mice and dt/dt mice could be due to the lack of NF–actin binding, ultimately leading to the collapse of neurons.

A series of *in vitro* and *in vivo* studies (Yang *et al.*, 1996) demonstrated that both binding domains of BPAG1n are indeed functional. The tail region of dystonin, which is expressed in BPAG1e and BPAG1n, showed significant binding to NF proteins, and the actin-binding domain of BPAG1n proved functional with an association constant comparable to that of other actin-binding proteins. Finally, full-length dystonin-coaligned NF proteins and actin in cotransfected cells further demonstrate an ability to link microfilaments with intermediate filaments.

8. CONCLUSION

In this review, we have described the complexity of the intermediate filaments in the nervous system. This complexity is developmentally regulated and different regions of the nervous system can also express distinct IF proteins with varying assembly characteristics and functions. Misregulations of IF proteins have been shown to lead to various pathological conditions. This complexity is also reflected by the generation of animal models. For example, transgenic mice in which either vimentin or GFAP has been knocked out do not show any obvious changes. In contrast, NFTPs are clearly the determinants of axonal caliber, and phosphorylation of the tails of NF-H and NF-M contributes to this function. However, the specific functions of each of the neuronal intermediate filament proteins have not yet been elucidated. These specific functions may involve specific interacting proteins, such as BPAGn, which has recently been shown to interact with neurofilaments in sensory neurons. We expect that additional interacting proteins will be identified in the near future and that the identification of these proteins may help us in elucidating the specific functions of various neuronal intermediate filament proteins.

Acknowledgments. We thank members of the laboratory, especially Dr. Gee Ching and Beth Rosen, for their scientific and editorial input.

9. REFERENCES

Adams, A. D., Choate, D. M., and Thompson, M. A., 1995, NF1-L is the DNA-binding component of the protein complex at the peripherin negative regulatory element, *J. Biol. Chem.* **270:**6975–6983; Erratum, *J. Biol. Chem.* 270:19668 (1995).

Albers, K., and Fuchs, E., 1992, The molecular biology of intermediate filament proteins, *Int. Rev. Cytol.* **134:**243–279.

Ando, S., Tanabe, K., Gonda, Y., Sato, C., and Inagaki, M., 1989, Domain- and sequence-specific phosphorylation of vimentin induces disassembly of the filament structure, *Biochemistry* **28:**2974–2979.

Arnold, S. E., and Trojanowski, J. Q., 1996, Human fetal hippocampal development: II. The neuronal cytoskeleton, *J. Comp. Neurol.* **367:**293–307.

Autilio-Gambetti, L., Sipple, J., Sudilovsky, O., and Gambetti, P., 1982, Intermediate filaments of Schwann cells, *J. Neurochem.* **38:**774–780.

Avraham, K. B., Schickler, M., Sapoznikov, D., Yarom, R., and Groner, Y., 1988, Down's syndrome: Abnormal neuromuscular junction in tongue of transgenic mice with elevated levels of human Cu/Zn-superoxide dismutase, *Cell* **54:**823–829.

Avraham, K. B., Sugarman, H., Rotshenker, S., and Groner, Y., 1991, Down's syndrome: Morphological remodelling and increased complexity in the neuromuscular junction of transgenic CuZn-superoxide dismutase mice, *J. Neurocytol.* **20:**208–215.

Baba, H., Nakahira, K., Morita, N., Tanaka, F., Akita, H., and Ikenaka, K., 1997, GFAP gene expression during development of astrocyte, *Dev. Neurosci.* **19:**49–57.

Beaudet, L., Charron, G., Houle, D., Tretjakoff, I., Peterson, A., and Julien, J. P., 1992, Intragenic regulatory elements contribute to transcriptional control of the neurofilament light gene, *Gene* **116:**205–214.

Beckman, J. S., Carson, M., Smith, C. D., and Koppenol, W. H., 1993, ALS, SOD and peroxynitrite, *Nature* **364:**584.

Belecky-Adams, T., Wight, D. C., Kopchick, J. J., and Parysek, L. M., 1993, Intragenic sequences are required for cell type-specific and injury-induced expression of the rat peripherin gene, *J. Neurosci.* **13:**5056–5065.

Bennett, G. S., and DiLullo, C., 1985, Transient expression of a neurofilament protein by replicating neuroepithelial cells of the embryonic chick brain, *Dev. Biol.* **107:**107–127.

Bennett, G. S., and Quintana, R., 1997, Identification of Ser–Pro and Thr–Pro phosphorylation sites in chicken neurofilament-M tail domain, *J. Neurochem.* **68:**534–543.

Benson, D. L., Mandell, J. W., Shaw, G., and Banker, G., 1996, Compartmentation of alpha-internexin and neurofilament triplet proteins in cultured hippocampal neurons, *J. Neurocytol.* **25:**181–196.

Betts, J. C., Blackstock, W. P., Ward, M. A., and Anderton, B. H., 1997, Identification of phosphorylation sites on neurofilament proteins by nanoelectrospray mass spectrometry, *J. Biol. Chem.* **272:**12922–12927.

Bignami, A., Raju, T., and Dahl, D., 1982, Localization of vimentin, the nonspecific intermediate filament protein, in embryonal glia and in early differentiating neurons. *In vivo* and *in vitro* immunofluorescence study of the rat embryo with vimentin and neurofilament antisera, *Dev. Biol.* **91:**286–295.

Bovolenta, P., Liem, R. K., and Mason, C. A., 1984, Development of cerebellar astroglia: Transitions in form and cytoskeletal content, *Dev. Biol.* **102:**248–259.

Brenner, M., and Messing, A., 1996, GFAP transgenic mice, *Methods* **10:**351–364.

Brenner, M., Kisseberth, W. C., Su, Y., Besnard, F., and Messing, A., 1994, GFAP promoter directs astrocyte-specific expression in transgenic mice, *J. Neurosci.* **14:**1030–1037.

Brody, B. A., Ley, C. A., and Parysek, L. M., 1989, Selective distribution of the 57 kDa neural intermediate filament protein in the rat CNS, *J. Neurosci.* **9:**2391–2401.

Brown, A., Bernier, G., Mathieu, M., Rossant, J., and Kothary, R., 1995, The mouse dystonia musculorum gene is a neural isoform of bullous pemphigoid antigen 1, *Nat. Genet.* **10:**301–306.

Budhram-Mahadeo, V., Morris, P. J., Lakin, N. D., Theil, T., Ching, G. Y., Lillycrop, K. A., Moroy, T., Liem, R. K., and Latchman, D. S., 1995, Activation of the alpha-internexin promoter by the Brn-3a transcription factor is dependent on the N-terminal region of the protein, *J. Biol. Chem.* **270:**2853–2858.

Budhram-Mahadeo, V., Morris, P. J., Lakin, N. D., Dawson, S. J., and Latchman, D. S., 1996, The different activities of the two activation domains of the Brn-3a transcription factor are dependent on the context of the binding site, *J. Biol. Chem.* **271:**9108–9113.

Carden, M. J., Schlaepfer, W. W., and Lee, V. M., 1985, The structure, biochemical properties, and immunogenicity of neurofilament peripheral regions are determined by phosphorylation state, *J. Biol. Chem.* **260:**9805–9817.

Carden, M. J., Trojanowski, J. Q., Schlaepfer, W. W., and Lee, V. M., 1987, Two-stage expression of neurofilament polypeptides during rat neurogenesis with early establishment of adult phosphorylation patterns, *J. Neurosci.* **7:**3489–3504.

Carpenter, S., 1968, Proximal axonal enlargement in motor neuron disease, *Neurology* **18:**841–851.

Carpenter, D. A., and Ip, W., 1996, Neurofilament triplet protein interactions: Evidence for the preferred formation of NF-L-containing dimers and a putative function for the end domains, *J. Cell Sci.* **109:**2493–2498.

Charnas, L. R., Szaro, B. G., and Gainer, H., 1992, Identification and developmental expression of a novel low molecular weight neuronal intermediate filament protein expressed in *Xenopus laevis*, *J. Neurosci.* **12:**3010–3024.

Charron, G., Guy, L. G., Bazinet, M., and Julien, J. P., 1995a, Multiple neuron-specific enhancers in the gene coding for the human neurofilament light chain, *J. Biol. Chem.* **270:**30604–30610.

Charron, G., Julien, J. P., and Bibor-Hardy, V., 1995b, Neuron specificity of the neurofilament light promoter in transgenic mice requires the presence of DNA unwinding elements, *J. Biol. Chem.* **270:**25739–25745.

Chen, W. J., and Liem, R. K., 1994a, The endless story of the glial fibrillary acidic protein, *J. Cell Sci.* **107:**2299–2311.

Chen, W. J., and Liem, R. K., 1994b, Reexpression of glial fibrillary acidic protein rescues the ability of astrocytoma cells to form processes in response to neurons, *J. Cell Biol.* **127:**813–823.

Chien, C. L., and Liem, R. K., 1995, The neuronal intermediate filament, alpha-internexin is transiently expressed in amacrine cells in the developing mouse retina, *Exp. Eye Res.* **61:**749–756.

Chien, C.-L., Mason, C. A., and Liem, R. K. H., 1996, α-Internexin is the only neuronal intermediate filament expressed in developing cerebellar granule neurons, *J. Neurobiol.* **29:**304–318.

Chin, S. S., and Liem, R. K., 1990, Transfected rat high-molecular-weight neurofilament (NF-H) coassembles with vimentin in a predominantly nonphosphorylated form, *J. Neurosci.* **10:**3714–3726.

Ching, G. Y., and Liem, R. K., 1991, Structure of the gene for the neuronal intermediate filament protein alpha-internexin and functional analysis of its promoter, *J. Biol. Chem.* **266:**19459–19468.

Ching, G. Y., and Liem, R. K., 1993, Assembly of type IV neuronal intermediate filaments in nonneuronal cells in the absence of preexisting cytoplasmic intermediate filaments, *J. Cell Biol.* **122:**1323–1335.

Chiu, F. C., Barnes, E. A., Das, K., Haley, J., Socolow, P., Macaluso, F. P., and Fant, J., 1989, Characterization of a novel 66 kd subunit of mammalian neurofilaments, *Neuron* **2:**1435–1445.

Chiu, F. C., Feng, L., Chan, S. O., Padin, C., and Federoff, J. H., 1995, Expression of neurofilament proteins during retinoic acid-induced differentiation of P19 embryonal carcinoma cells, *Brain Res. Mol. Brain Res.* **30:**77–86.

Chou, Y. H., Opal, P., Quinlan, R. A., and Goldman, R. D., 1996, The relative roles of specific N- and C-terminal phosphorylation sites in the disassembly of intermediate filament in mitotic BHK-21 cells, *J. Cell Sci.* **109:**817–826.

Cleveland, D. W., 1996, Neuronal growth and death: Order and disorder in the axoplasm, *Cell* **84:**663–666.

Cochard, P., and Paulin, D., 1984, Initial expression of neurofilaments and vimentin in the central and peripheral nervous system of the mouse embryo *in vivo*, *J. Neurosci.* **4:**2080–2094.

Colucci-Guyon, E., Portier, M. M., Dunia, I., Paulin, D., Pournin, S., and Babinet, C., 1994, Mice lacking vimentin develop and reproduce without an obvious phenotype, *Cell* **79:**679–694.

Cote, F., Collard, J. F., and Julien, J. P., 1993, Progressive neuronopathy in transgenic mice expressing the human neurofilament heavy gene: A mouse model of amyotrophic lateral sclerosis, *Cell* **73:**35–46.

Cote, F., Collard, J. F., Houle, D., and Julien, J. P., 1994, Copy-dependent and correct developmental expression of the human neurofilament heavy gene in transgenic mice, *Brain Res. Mol. Brain Res.* **26:**99–105.

Crapo, J. D., Oury, T., Rabouille, C., Slot, J. W., and Chang, L. Y., 1992, Copper,zinc superoxide dismutase is primarily a cytosolic protein in human cells, *Proc. Natl. Acad. Sci. USA* **89:**10405–10409.

Cui, C., Stambrook, P. J., and Parysek, L. M., 1995, Peripherin assembles into homopolymers in SW13 cells, *J. Cell Sci.* **108:**3279–3284.

Danks, R. A., Orian, J. M., Gonzales, M. F., Tan, S. S., Alexander, B., Mikoshiba, K., and Kaye, A. H., 1995, Transformation of astrocytes in transgenic mice expressing SV40 T antigen under the transcriptional control of the glial fibrillary acidic protein promoter, *Cancer Res.* **55:**4302–4310.

Dautigny, A., Pham-Dinh, D., Roussel, C., Felix, J. M., Nussbaum, J. L., and Jolles, P., 1988, The large neurofilament subunit (NF-H) of the rat: cDNA cloning and *in situ* detection, *Biochem. Biophys. Res. Commun.* **154:**1099–1106.

Desmarais, D., and Royal, A., 1996, The TATA motif is a target for efficient transcriptional activation and nerve growth factor induction of the peripherin gene, *J. Biol. Chem.* **271:**24976–24981.

Desmarais, D., Filion, M., Lapointe, L., and Royal, A., 1992, Cell-specific transcription of the peripherin gene in neuronal cell lines involves a *cis*-acting element surrounding the TATA box, *EMBO J.* **11:**2971–2980.

de Waegh, S. M., Lee, V. M., and Brady, S. T., 1992, Local modulation of neurofilament phosphorylation, axonal caliber, and slow axonal transport by myelinating Schwann cells, *Cell* **68:**451–463.

Dong, D. L., Xu, Z. S., Chevrier, M. R., Cotter, R. J., Cleveland, D. W., and Hart, G. W., 1993, Glycosylation of mammalian neurofilaments. Localization of multiple O-inked N-acetylglucosamine moieties on neurofilament polypeptides L and M, *J. Biol. Chem.* **268:**16679–16687.

Dong, D. L., Xu, Z. S., Hart, G. W., and Cleveland, D. W., 1996, Cytoplasmic O-GlcNAc modification of the head domain and the KSP repeat motif of the neurofilament protein neurofilament-H, *J. Biol. Chem.* **271:**20845–20852.

Elhanany, E., Jaffe, H., Link, W. T., Sheeley, D. M., Gainer, H., and Pant, H. C., 1994, Identification of endogenously phosphorylated KSP sites in the high-molecular-weight rat neurofilament protein, *J. Neurochem.* **63:**2324–2335.

Escurat, M., Djabali, K., Gumpel, M., Gros, F., and Portier, M. M., 1990, Differential expression of two neuronal intermediate-filament proteins, peripherin and the low-molecular-mass neurofilament protein (NF-L), during the development of the rat, *J. Neurosci.* **10:**764–784.

Eyer, J., and Peterson, A., 1994, Neurofilament-deficient axons and perikaryal aggregates in viable transgenic mice expressing a neurofilament-beta-galactosidase fusion protein, *Neuron* **12:**389–405.

Figlewicz, D. A., Krizus, A., Martinoli, M. G., Meininger, V., Dib, M., Rouleau, G. A., and Julien, J. P., 1994, Variants of the heavy neurofilament subunit are associated with the development of amyotrophic lateral sclerosis, *Hum. Mol. Genet.* **3:**1757–1761.

Fliegner, K. H., and Liem, R. K., 1991, Cellular and molecular biology of neuronal intermediate filaments, *Int. Rev. Cytol.* **131:**109–167.

Fliegner, K. H., Ching, G. Y., and Liem, R. K., 1990, The predicted amino acid sequence of alpha-internexin is that of a novel neuronal intermediate filament protein, *EMBO J.* **9:**749–755.

Fliegner, K. H., Kaplan, M. P., Wood, T. L., Pintar, J. E., and Liem, R. K., 1994, Expression of the gene for the neuronal intermediate filament protein alpha-internexin coincides with the onset of neuronal differentiation in the developing rat nervous system, *J. Comp. Neurol.* **342:**161–173.

Fuchs, E., 1995, Keratins and the skin, *Annu. Rev. Cell Dev. Biol.* **11:**123–153.

Galou, M., Pournin, S., Ensergueix, D., Ridet, J. L., Tchelingerian, J. L., Lossouarn, L., Privat, A., Babinet, C., and Dupouey, P., 1994, Normal and pathological expression of GFAP promoter elements in transgenic mice, *Glia* **12:**281–293.

Gao, C. Y., Chauthaiwale, V. M., Rampalli, A. M., and Zelenka, P. S., 1996, Expression of alternatively spliced PCTAIRE-1 mRNA in PC12 cells and neonatal rat brain, *Gene* **176:**243–247.

Geisler, N., and Weber, K., 1981, Self-assembly *in vitro* of the 68,000 molecular weight component of the mammalian neurofilament triplet proteins into intermediate-sized filaments, *J. Mol. Biol.* **151:**565–571.

Geisler, N., Kaufmann, E., and Weber, K., 1985, Antiparallel orientation of the two double-stranded coiled-coils in the tetrameric protofilament unit of intermediate filaments, *J. Mol. Biol.* **182:**173–177.

Goldstein, M. E., House, S. B., and Gainer, H., 1991, NF-L and peripherin immunoreactivities define distinct classes of rat sensory ganglion cells, *J. Neurosci. Res.* **30:**92–104.

Goldstein, M. E., Grant, P., House, S. B., Henken, D. B., and Gainer, H., 1996, Developmental regulation of two distinct neuronal phenotypes in rat dorsal root ganglia, *Neuroscience* **71:**243–258.

Gomi, H., Yokoyama, T., Fujimoto, K., Ikeda, T., Katoh, A., Itoh, T., and Itohara, S., 1995, Mice devoid of the glial fibrillary acidic protein develop normally and are susceptible to scrapie prions, *Neuron* **14:**29–41.

Greene, L. A., 1989, A new neuronal intermediate filament protein, *Trends Neurosci.* **12:**228–230.

Guan, R. J., Hall, F. L., and Cohlberg, J. A., 1992, Proline-directed protein kinase (p34cdc2/p58cyclin A) phosphorylates bovine neurofilaments, *J. Neurochem.* **58:**1365–1371.

Guidato, S., Tsai, L. H., Woodgett, J., and Miller, C. C., 1996, Differential cellular phosphorylation of neurofilament heavy side-arms by glycogen synthase kinase-3 and cyclin-dependent kinase-5, *J. Neurochem.* **66:**1698–1706.

Guo, L., Degenstein, L., Dowling, J., Yu, Q. C., Wollmann, R., Perman, B., and Fuchs, E., 1995, Gene targeting of BPAG1: Abnormalities in mechanical strength and cell migration in stratified epithelia and neurologic degeneration, *Cell* **81:**233–243.

Gurney, M. E., Pu, H., Chiu, A. Y., Dal Canto, M. C., Polchow, C. Y., Alexander, D. D., Caliendo, J., Hentati, A., Kwon, Y. W., Deng, H. X., *et al.*, 1994, Motor neuron degeneration in mice that express a human Cu,Zn superoxide dismutase mutation, *Science* **264:**1772–1775; Erratum, *Science* **269:**149 (1995).

Hemmati-Brivanlou, A., Mann, R. W., and Harland, R. M., 1992, A protein expressed in the growth cones of embryonic vertebrate neurons defines a new class of intermediate filament protein, *Neuron* **9:**417–428.

Hirano, A., Nakano, I., Kurland, L. T., Mulder, D. W., Holley, P. W., and Saccomanno, G., 1984a, Fine structural study of neurofibrillary changes in a family with amyotrophic lateral sclerosis, *J. Neuropathol. Exp. Neurol.* **43:**471–480.

Hirano, A., Donnenfeld, H., Sasaki, S., and Nakano, I., 1984b, Fine structural observations of neurofilamentous changes in amyotrophic lateral sclerosis, *J. Neuropathol. Exp. Neurol.* **43:**461–470.

Hirokawa, N., Glicksman, M. A., and Willard, M. B., 1984, Organization of mammalian neurofilament polypeptides within the neuronal cytoskeleton, *J. Cell Biol.* **98:**1523–1536.

Hisanaga, S., and Hirokawa, N., 1990, Molecular architecture of the neurofilament. II. Reassembly process of neurofilament L protein *in vitro*, *J. Mol. Biol.* **211:**871–882.

Hisanaga, S., Gonda, Y., Inagaki, M., Ikai, A., and Hirokawa, N., 1990, Effects of phosphorylation of the neurofilament L protein on filamentous structures, *Cell. Regul.* **1:**237–248.

Hisanaga, S., Kusubata, M., Okumura, E., and Kishimoto, T., 1991, Phosphorylation of neurofilament H subunit at the tail domain by CDC2 kinase dissociates the association to microtubules, *J. Biol. Chem.* **266:**21798–21803.

Ho, C. L., Chin, S. S., Carnevale, K., and Liem, R. K., 1995, Translation initiation and assembly of peripherin in cultured cells, *Eur. J. Cell Biol.* **68:**103–112.

Hockfield, S., and McKay, R. D., 1985, Identification of major cell classes in the developing mammalian nervous system, *J. Neurosci.* **5:**3310–3328.

Hoffman, P. N., and Lasek, R. J., 1975, The slow component of axonal transport. Identification of major structural polypeptides of the axon and their generality among mammalian neurons, *J. Cell Biol.* **66:**351–366.

Hsu, C., Janicki, S., and Monteiro, M. J., 1995, The first intron of the mouse neurofilament light gene (NF-L) increases gene expression, *Brain Res. Mol. Brain Res.* **32:**241–251.

Inagaki, M., Nakamura, Y., Takeda, M., Nishimura, T., and Inagaki, N., 1994, Glial fibrillary acidic protein: Dynamic property and regulation by phosphorylation, *Brain Pathol.* **4:**239–243.

Inagaki, N., Tsujimura, K., Tanaka, J., Sekimata, M., Kamei, Y., and Inagaki, M., 1996, Visualization of protein kinase activities in single cells by antibodies against phosphorylated vimentin and GFAP, *Neurochem. Res.* **21:**795–800.

Ino, H., Ishizuka, T., Chiba, T., and Tatibana, M., 1994, Expression of CDK5 (PSSALRE kinase), a neural cdc2-related protein kinase, in the mature and developing mouse central and peripheral nervous systems, *Brain Res.* **661:**196–206.

Ivanov, T. R., and Brown, I. R., 1992, Interaction of multiple nuclear proteins with the promoter region of the mouse 68-kDa neurofilament gene, *J. Neurosci. Res.* **32:**149–158.

Jones, S. M., and Williams, R. C., Jr., 1982, Phosphate content of mammalian neurofilaments, *J. Biol. Chem.* **257:**9902–9905.

Julien, J. P., and Mushynski, W. E., 1981, A comparison of *in vitro-* and *in vivo-*phosphorylated neurofilament polypeptides, *J. Neurochem.* **37:**1579–1585.

Julien, J. P., and Mushynski, W. E., 1983, The distribution of phosphorylation sites among identified proteolytic fragments of mammalian neurofilaments, *J. Biol. Chem.* **258:**4019–4025.

Kaplan, M. P., Chin, S. S., Fliegner, K. H., and Liem, R. K., 1990, Alpha-internexin, a novel neuronal intermediate filament protein, precedes the low molecular weight neurofilament protein (NF-L) in the developing rat brain, *J. Neurosci.* **10:**2735–2748.

Kosako, H., Amano, M., Yanagida, M., Tanabe, K., Nishi, Y., Kaibuchi, K., and Inagaki, M., 1997, Phosphorylation of glial fibrillary acidic protein at the same sites by cleavage furrow kinase and Rho-associated kinase, *J. Biol. Chem.* **272:**10333–10336.

Kouklis, P. D., Hutton, E., and Fuchs, E., 1994, Making a connection: Direct binding between keratin intermediate filaments and desmosomal proteins, *J. Cell Biol.* **127:**1049–1060.

Kure, R., and Brown, I. R., 1995, Expression of low-molecular-weight neurofilament (NF-L) mRNA during postnatal development of the mouse brain, *Neurochem. Res.* **20:**833–846.

Kure, R., and Brown, I. R., 1997, Developmental analysis of factors binding to the mouse 68-kDa neurofilament promoter, *Neurochem. Res.* **22:**555–562.

Kure, R., Ivanov, T. R., and Brown, I. R., 1996, Characterization of DNase I hypersensitive sites in the mouse 68-kDa neurofilament gene, *Neurochem. Res.* **21:**713–722.

Leconte, L., Semonin, O., Zvara, A., Boisseau, S., Poujeol, C., Julien, J. P., and Simonneau, M., 1994, Both upstream and intragenic sequences of the human neurofilament light gene direct expression of *lacZ* in neurons of transgenic mouse embryos, *J. Mol. Neurosci.* **5:**273–295.

Leconte, L., Santha, M., Fort, C., Poujeol, C., Portier, M. M., and Simonneau, M., 1996, Cell type-specific expression of the mouse peripherin gene requires both upstream and intragenic sequences in transgenic mouse embryos, *Brain Res. Dev. Brain Res.* **92:**1–9.

Lee, M. K., Xu, Z., Wong, P. C., and Cleveland, D. W., 1993, Neurofilaments are obligate heteropolymers *in vivo*, *J. Cell Biol.* **122:**1337–1350.

Lee, M. K., Marszalek, J. R., and Cleveland, D. W., 1994, A mutant neurofilament subunit causes massive, selective motor neuron death: implications for the pathogenesis of human motor neuron disease, *Neuron* **13:**975–988.

Lendahl, U., Zimmerman, L. B., and McKay, R. D., 1990, CNS stem cells express a new class of intermediate filament protein, *Cell* **60:**585–595.

Leonard, D. G., Gorham, J. D., Cole, P., Greene, L. A., and Ziff, E. B., 1988, A nerve growth factor-regulated messenger RNA encodes a new intermediate filament protein, *J. Cell Biol.* **106:**181–193.

Leung, C. L., and Liem, R. K. H., 1996, Characterization of interactions between the neurofilament triplet proteins by the yeast two-hybrid system, *J. Biol. Chem.* **271:**14041–14044.

Liem, R. K., and Hutchison, S. B., 1982, Purification of individual components of the neurofilament triplet: Filament assembly from the 70,000-dalton subunit, *Biochemistry* **21:**3221–3226.

Liem, R. K., Yen, S. H., Salomon, G. D., and Shelanski, M. L., 1978, Intermediate filaments in nervous tissues, *J. Cell Biol.* **79:**637–645.

Lin, W., and Szaro, B. G., 1995, Neurofilaments help maintain normal morphologies and support elongation of neurites in *Xenopus laevis* cultured embryonic spinal cord neurons, *J. Neurosci.* **15:**8331–8344.

Lothian, C., and Lendahl, U., 1997, An evolutionarily conserved region in the second intron of the human nestin gene directs gene expression to CNS progenitor cells and to early neural crest cells, *Eur. J. Neurosci.* **9:**452–462.

Marszalek, J. R., Williamson, T. L., Lee, M. K., Xu, Z., Hoffman, P. N., Becher, M. W., Crawford, T. O., and Cleveland, D. W., 1996, Neurofilament subunit NF-H modulates axonal diameter by selectively slowing neurofilament transport, *J. Cell Biol.* **135:**711–724.

Matsuzawa, K., Kosako, H., Inagaki, N., Shibata, H., Mukai, H., Ono, Y., Amano, M., Kaibuchi, K., Matsuura, Y., Azuma, I., and Inagaki, M., 1997, Domain-specific phosphorylation of vimentin and glial fibrillary acidic protein by PKN, *Biochem. Biophys. Res. Commun.* **234:**621–625.

Meyerson, M., Enders, G. H., Wu, C. L., Su, L. K., Gorka, C., Nelson, C., Harlow, E., and Tsai, L. H., 1992, A family of human cdc2-related protein kinases, *EMBO J.* **11:**2909–2917.

Miura, M., Tamura, T., and Mikoshiba, K., 1990, Cell-specific expression of the mouse glial fibrillary acidic protein gene: Identification of the *cis*- and *trans*-acting promoter elements for astrocyte-specific expression, *J. Neurochem.* **55:**1180–1188.

Mukai, H., Toshimori, M., Shibata, H., Kitagawa, M., Shimakawa, M., Miyahara, M., Sunakawa, H., and Ono, Y., 1996, PKN associates and phosphorylates the head-rod domain of neurofilament protein, *J. Biol. Chem.* **271:**9816–9822.

Mulligan, L., Balin, B. J., Lee, V. M., and Ip, W., 1991, Antibody labeling of bovine neurofilaments: Implications on the structure of neurofilament sidearms, *J. Struct. Biol.* **106:**145–160.

Murphy, A., Breen, K. C., Long, A., Feighery, C., Casey, E. B., and Kelleher, D., 1993, Neurofilament expression in human T lymphocytes, *Immunology* **79:**167–170.

Nakahira, K., Ikenaka, K., Wada, K., Tamura, T., Furuichi, T., and Mikoshiba, K., 1990, Structure of the 68-kDa neurofilament gene and regulation of its expression, *J. Biol. Chem.* **265:**19786–19791.

Nakamura, Y., Takeda, M., Angelides, K. J., Tanaka, T., Tada, K., and Nishimura, T., 1990, Effect of phosphorylation on 68 KDa neurofilament subunit protein assembly by the cyclic AMP dependent protein kinase *in vitro*, *Biochem. Biophys. Res. Commun.* **169:**744–750.

Nakamura, Y., Takeda, M., and Nishimura, T., 1996, Dynamics of bovine glial fibrillary acidic protein phosphorylation, *Neurosci. Lett.* **205:**91–94.

Nixon, R. A., Lewis, S. E., Mercken, M., and Sihag, R. K., 1994, [32P]Orthophosphate and [35S]methionine label separate pools of neurofilaments with markedly different axonal transport kinetics in mouse retinal ganglion cells *in vivo*, *Neurochem. Res.* **19:**1445–1453.

Ohara, O., Gahara, Y., Miyake, T., Teraoka, H., and Kitamura, T., 1993, Neurofilament deficiency in quail caused by nonsense mutation in neurofilament-L gene, *J. Cell Biol.* **121:**387–395.

Okuda, T., Cleveland, J. L., and Downing, J. R., 1992, PCTAIRE-1 and PCTAIRE-3, two members of a novel cdc2/CDC28-related protein kinase gene family, *Oncogene* **7:**2249–2258.

Owaribe, K., Kartenbeck, J., Stumpp, S., Magin, T. M., Krieg, T., Diaz, L. A., and Franke, W. W., 1990, The hemidesmosomal plaque. I. Characterization of a major constituent protein as a differentiation marker for certain forms of epithelia, *Differentiation* **45:**207–220.

Pachter, J. S., and Liem, R. K., 1985, alpha-Internexin, a 66-kD intermediate filament-binding protein from mammalian central nervous tissues, *J. Cell Biol.* **101:**1316–1322.

Parysek, L. M., and Goldman, R. D., 1988, Distribution of a novel 57 kDa intermediate filament (IF) protein in the nervous system, *J. Neurosci.* **8:**555–563.

Parysek, L. M., McReynolds, M. A., Goldman, R. D., and Ley, C. A., 1991, Some neural intermediate filaments contain both peripherin and the neurofilament proteins, *J. Neurosci. Res.* **30:**80–91.

Pekny, M., Leveen, P., Pekna, M., Eliasson, C., Berthold, C. H., Westermark, B., and Betsholtz, C., 1995, Mice lacking glial fibrillary acidic protein display astrocytes devoid of intermediate filaments but develop and reproduce normally, *EMBO J.* **14:**1590–1598.

Portier, M. M., Croizat, B., and Gros, F., 1982, A sequence of changes in cytoskeletal components during neuroblastoma differentiation, *FEBS Lett.* **146:**283–288.

Quinlan, R. A., Moir, R. D., and Stewart, M., 1989, Expression in *Escherichia coli* of fragments of glial fibrillary acidic protein: Characterization, assembly properties and paracrystal formation, *J. Cell Sci.* **93:**71–83.

Raju, T., Bignami, A., and Dahl, D., 1981, *In vivo* and *in vitro* differentiation of neurons and astrocytes in the rat embryo. Immunofluorescence study with neurofilament and glial filament antisera, *Dev. Biol.* **85:**344–357.

Reeben, M., Halmekyto, M., Alhonen, L., Sinervirta, R., Saarma, M., and Janne, J., 1993, Tissue-specific expression of rat light neurofilament promoter-driven reporter gene in transgenic mice, *Biochem. Biophys. Res. Commun.* **192:**465–470.

Reeben, M., Neuman, T., Palgi, J., Palm, K., Paalme, V., and Saarma, M., 1995, Characterization of the rat light neurofilament (NF-L) gene promoter and identification of NGF and cAMP responsive regions, *J. Neurosci. Res.* **40:**177–188.

Riederer, B. M., Porchet, R., and Marugg, R. A., 1996, Differential expression and modification of neurofilament triplet proteins during cat cerebellar development, *J. Comp. Neurol.* **364:**704–717.

Rooke, K., Figlewicz, D. A., Han, F. Y., and Rouleau, G. A., 1996, Analysis of the KSP repeat of the neurofilament heavy subunit in familiar amyotrophic lateral sclerosis, *Neurology* **46:**789–790.

Rosen, D. R., Siddique, T., Patterson, D., Figlewicz, D. A., Sapp, P., Hentati, A., Donaldson, D., Goto, J., O'Regan, J. P., Deng, H. X., *et al.*, 1993, Mutations in Cu/Zn superoxide dismutase gene are associated with familial amyotrophic lateral sclerosis, *Nature* **362:**59–62; Erratum, *Nature* **364:**362 (1993).

Roussel, G., Felix, J. M., Dautigny, A., Pham-Dinh, D., Hindelang, C., Jolles, P., and Nussbaum, J. L., 1991, *In situ* localization of NF-H neurofilament subunit mRNAs in rat brain, *Dev. Neurosci.* **13:**98–103.

Sawant, L. A., Hasgekar, N. N., and Vyasarayani, L. S., 1994, Developmental expression of neurofilament and glial filament proteins in rat cerebellum, *Int. J. Dev. Biol.* **38:**429–437.

Schlaepfer, W. W., and Bruce, J., 1990, Simultaneous up-regulation of neurofilament proteins during the postnatal development of the rat nervous system, *J. Neurosci. Res.* **25:**39–49.

Schroder, N., de Mattos-Dutra, A., Sampaio de Freitas, M., Fogaca Lisboa, C. S., Zilles, A. C., Pessoa-Pureur, R., and Izquierdo, I., 1997, The effects of behavioral tasks on the *in vitro* phosphorylation of intermediate filament subunits of rat hippocampus are mediated by CaMKII and PKA, *Brain Res.* **749:**275–282.

Schwartz, M. L., Katagi, C., Bruce, J., and Schlaepfer, W. W., 1994, Brain-specific enhancement of the mouse neurofilament heavy gene promoter *in vitro*, *J. Biol. Chem.* **269:**13444–13450.

Sekimata, M., Tsujimura, K., Tanaka, J., Takeuchi, Y., Inagaki, N., and Inagaki, M., 1996, Detection of protein kinase activity specifically activated at metaphase-anaphase transition, *J. Cell Biol.* **132:**635–641.

Shaw, G., and Weber, K., 1982, Differential expression of neurofilament triplet proteins in brain development, *Nature* **298:**277–279.

Shaw, G., and Weber, K., 1983, The structure and development of the rat retina: An immunofluorescence microscopical study using antibodies specific for intermediate filament proteins, *Eur. J. Cell Biol.* **30:**219–232.

Shea, T. B., Beermann, M. L., and Fischer, I., 1993, Transient requirement for vimentin in neuritogenesis: Intracellular delivery of anti-vimentin antibodies and antisense oligonucleotides inhibit neurite initiation but not elongation of existing neurites in neuroblastoma, *J. Neurosci. Res.* **36:**66–76.

Shecket, G., and Lasek, R. J., 1982, Neurofilament protein phosphorylation. Species generality and reaction characteristics, *J. Biol. Chem.* **257:**4788–4795.

Shneidman, P. S., Bruce, J., Schwartz, M. L., and Schlaepfer, W. W., 1992, Negative regulatory regions are present upstream in the three mouse neurofilament genes, *Brain Res. Mol. Brain Res.* **13:**127–138.

Sihag, R. K., and Nixon, R. A., 1991, Identification of Ser-55 as a major protein kinase A phosphorylation site on the 70-kDa subunit of neurofilaments. Early turnover during axonal transport, *J. Biol. Chem.* **266:**18861–18867.

Stanley, J. R., 1993, Cell adhesion molecules as targets of autoantibodies in pemphigus and pemphigoid, bullous diseases due to defective epidermal cell adhesion, *Adv. Immunol.* **53:**291–325.

Stappenbeck, T. S., Bornslaeger, E. A., Corcoran, C. M., Luu, H. H., Virata, M. L., and Green, K. J., 1993, Functional analysis of desmoplakin domains: Specification of the interaction with keratin versus vimentin intermediate filament networks, *J. Cell Biol.* **123:**691–705.

Steinert, P. M., and Roop, D. R., 1988, Molecular and cellular biology of intermediate filaments, *Annu. Rev. Biochem.* **57:**593–625.

Steinschneider, R., Delmas, P., Nedelec, J., Gola, M., Bernard, D., and Boucraut, J., 1996, Appearance of neurofilament subunit epitopes correlates with electrophysiological maturation in cortical embryonic neurons cocultured with mature astrocytes, *Brain Res. Dev. Brain Res.* **95:**15–27.

Stewart, M., Quinlan, R. A., and Moir, R. D., 1989, Molecular interactions in paracrystals of a fragment corresponding to the alpha-helical coiled-coil rod portion of glial fibrillary acidic protein: Evidence for an antiparallel packing of molecules and polymorphism related to intermediate filament structure, *J. Cell Biol.* **109:**225–234.

Sun, D., Leung, C. L., and Liem, R. K. H., 1996, Phosphorylation of the high molecular weight neurofilament protein (NF- H) by Cdk5 and p35, *J. Biol. Chem.* **271:**14245–14251.

Sun, D., Macioce, P., Chin, S. S., and Liem, R. K., 1997, Assembly properties of amino- and carboxyl-terminally truncated neurofilament NF-H proteins with NF-L and NF-M in the presence and absence of vimentin, *J. Neurochem.* **68:**917–926.

Szaro, B. G., Grant, P., Lee, V. M., and Gainer, H., 1991, Inhibition of axonal development after injection of neurofilament antibodies into a *Xenopus laevis* embryo, *J. Comp. Neurol.* **308:**576–585.

Takeda, S., Okabe, S., Funakoshi, T., and Hirokawa, N., 1994, Differential dynamics of neurofilament-H protein and neurofilament-L protein in neurons, *J. Cell Biol.* **127:**173–185.

Tanaka, J., Ogawara, M., Ando, S., Shibata, M., Yatani, R., Kusagawa, M., and Inagaki, M., 1993, Phosphorylation of a 62 kd porcine alpha-internexin, a newly identified intermediate filament protein, *Biochem. Biophys. Res. Commun.* **196:**115–123.

Tang, D., Yeung, J., Lee, K. Y., Matsushita, M., Matsui, H., Tomizawa, K., Hatase, O., and Wang, J. H., 1995, An isoform of the neuronal cyclin-dependent kinase 5 (Cdk5) activator, *J. Biol. Chem.* **270:**26897–26903.

Tapscott, S. J., Bennett, G. S., and Holtzer, H., 1981, Neuronal precursor cells in the chick neural tube express neurofilament proteins, *Nature* **292:**836–838.

Thompson, M. A., Lee, E., Lawe, D., Gizang-Ginsberg, E., and Ziff, E. B., 1992, Nerve growth factor-induced derepression of peripherin gene expression is associated with alterations in proteins binding to a negative regulatory element, *Mol. Cell Biol.* **12:**2501–2513.

Troy, C. M., Brown, K., Greene, L. A., and Shelanski, M. L., 1990, Ontogeny of the neuronal intermediate filament protein, peripherin, in the mouse embryo, *Neuroscience* **36:**217–237.

Troy, C. M., Greene, L. A., and Shelanski, M. L., 1992, Neurite outgrowth in peripherin-depleted PC12 cells, *J. Cell Biol.* **117:**1085–1092.

Tsai, L. H., Takahashi, T., Caviness, V. S., Jr., and Harlow, E., 1993, Activity and expression pattern of cyclin-dependent kinase 5 in the embryonic mouse nervous system, *Development* **119:**1029–1040.

Tsujimura, K., Tanaka, J., Ando, S., Matsuoka, Y., Kusubata, M., Sugiura, H., Yamauchi, T., and Inagaki, M., 1994, Identification of phosphorylation sites on glial fibrillary acidic protein for cdc2 kinase and Ca(2+)-calmodulin-dependent protein kinase II, *J. Biochem.* (Tokyo) **116:**426–434.

Tu, P. H., Elder, G., Lazzarini, R. A., Nelson, D., Trojanowski, J. Q., and Lee, V. M., 1995, Overexpression of the human NFM subunit in transgenic mice modifies the level of endogenous NFL and the phosphorylation state of NFH subunits, *J. Cell Biol.* **129:**1629–1640.

Tu, P. H., Raju, P., Robinson, K. A., Gurney, M. E., Trojanowski, J. Q., and Lee, V. M., 1996, Transgenic mice carrying a human mutant superoxide dismutase transgene develop neuronal cytoskeletal pathology resembling human amyotrophic lateral sclerosis lesions, *Proc. Natl. Acad. Sci. USA* **93:**3155–3160.

Vechio, J. D., Bruijn, L. I., Xu, Z., Brown, R. H., Jr., and Cleveland, D. W., 1996, Sequence variants in human neurofilament proteins: Absence of linkage to familial amyotrophic lateral sclerosis, *Ann. Neurol.* **40:**603–610.

Vickers, J. C., and Costa, M., 1992, Subpopulations of neurons in the guinea-pig inferior mesenteric ganglia distinguished by the differential distribution of neurofilament triplet epitopes, *J. Chem. Neuroanat.* **5:**417–426.

Weinstein, D. E., Shelanski, M. L., and Liem, R. K., 1991, Suppression by antisense mRNA demonstrates a requirement for the glial fibrillary acidic protein in the formation of stable astrocytic processes in response to neurons, *J. Cell Biol.* **112:**1205–1213.

Wiche, G., Gromov, D., Donovan, A., Castanon, M. J., and Fuchs, E., 1993, Expression of plectin mutant cDNA in cultured cells indicates a role of COOH-terminal domain in intermediate filament association, *J. Cell Biol.* **121:**607–619.

Wiedau-Pazos, M., Goto, J. J., Rabizadeh, S., Gralla, E. B., Roe, J. A., Lee, M. K., Valentine, J. S., and Bredesen, D. E., 1996, Altered reactivity of superoxide dismutase in familial amyotrophic lateral sclerosis, *Science* **271:**515–518.

Wong, J., Hutchison, S. B., and Liem, R. K., 1984, An isoelectric variant of the 150,000-dalton neurofilament polypeptide. Evidence that phosphorylation state affects its association with the filament, *J. Biol. Chem.* **259:**10867–10874.

Wong, P. C., Marszalek, J., Crawford, T. O., Xu, Z., Hsieh, S. T., Griffin, J. W., and Cleveland, D. W., 1995, Increasing neurofilament subunit NF-M expression reduces axonal NF-H, inhibits radial growth, and results in neurofilamentous accumulation in motor neurons, *J. Cell Biol.* **130:**1413–1422.

Xu, Z., Cork, L. C., Griffin, J. W., and Cleveland, D. W., 1993, Increased expression of neurofilament subunit NF-L produces morphological alterations that resemble the pathology of human motor neuron disease, *Cell* **73:**23–33.

Xu, Z., Marszalek, J. R., Lee, M. K., Wong, P. C., Folmer, J., Crawford, T. O., Hsieh, S. T., Griffin, J. W., and Cleveland, D. W., 1996, Subunit composition of neurofilaments specifies axonal diameter, *J. Cell Biol.* **133:**1061–1069.

Yamasaki, H., Itakura, C., and Mizutani, M., 1991, Hereditary hypotrophic axonopathy with neurofilament deficiency in a mutant strain of the Japanese quail, *Acta Neuropathol.* (Berlin) **82:**427–434.

Yamasaki, H., Bennett, G. S., Itakura, C., and Mizutani, M., 1992, Defective expression of neurofilament protein subunits in hereditary hypotrophic axonopathy of quail, *Lab. Invest.* **66:**734–743.

Yang, Y., Dowling, J., Yu, Q. C., Kouklis, P., Cleveland, D. W., and Fuchs, E., 1996, An essential cytoskeletal linker protein connecting actin microfilaments to intermediate filaments, *Cell* **86:**655–665.

Yarom, R., Sapoznikov, D., Havivi, Y., Avraham, K. B., Schickler, M., and Groner, Y., 1988, Premature aging changes in neuromuscular junctions of transgenic mice with an extra human CuZnSOD gene: A model for tongue pathology in Down's syndrome, *J. Neurol. Sci.* **88:**41–53.

Yazdanbakhsh, K., Fraser, P., Kioussis, D., Vidal, M., Grosveld, F., and Lindenbaum, M., 1993, Functional analysis of the human neurofilament light chain gene promoter, *Nucleic Acids Res.* **21:**455–461.

Zhao, Y., and Szaro, B. G., 1997, Xefiltin, a new low molecular weight neuronal intermediate filament protein of *Xenopus laevis*, shares sequence features with goldfish gefiltin and mammalian alpha-internexin and differs in expression from XNIF and NF-L, *J. Comp. Neurol.* **377:**351–364.

Zimmerman, L., Parr, B., Lendahl, U., Cunningham, M., McKay, R., Gavin, B., Mann, J., Vassileva, G., and McMahon, A., 1994, Independent regulatory elements in the nestin gene direct transgene expression to neural stem cells or muscle precursors, *Neuron* **12:**11–24; Erratum, *Neuron* **12:**1388 (1994).

Zopf, D., Dineva, B., Betz, H., and Gundelfinger, E. D., 1990, Isolation of the chicken middle-molecular weight neurofilament (NF-M) gene and characterization of its promoter, *Nucleic Acids Res.* **18:**521–529.

Chapter 18

Extensive Phosphorylation and Axonal Transport of Triton-Soluble Neurofilament Subunits

Thomas B. Shea, Cheolwha Jung, Jason Yabe, DongLing Ma, and Itzhak Fischer

1. INTRODUCTION

1.1. Extensive Phosphorylation of the High-Molecular-Weight Neurofilament Protein (NF-H) Prior to Incorporation into Triton-Soluble Structures

The extensively cross-linked neurofilament (NF) network is thought to confer structural support for the establishment and maintenance of the axonal cytoskeleton by interaction of NFs with each other and with axonal microtubules (MTs) (for reviews, see Brady, 1993; Matus, 1987; Nixon, 1993; Shea and Beermann, 1994). Consistent with this role are the observa-

---

Thomas B. Shea, Cheolwha Jung, and Jason Yabe Center for Cellular Neurobiology and Neurodegeneration Research, Department of Biological Sciences, University of Massachusetts at Lowell, Lowell, Massachusetts 01854. **DongLing Ma and Itzhak Fischer** Department of Neurobiology and Anatomy, Allegheny Medical College, Philadelphia, Pennsylvania, 19129.

Subcellular Biochemistry, Volume 31: Intermediate Filaments, edited by Herrmann and Harris. Plenum Press, New York, 1998.

tions that, once assembled, NFs are relatively resistant to depolymerization and that the polymerized state of NFs is highly stable in mature axons, where NFs are most abundant (Georges *et al.*, 1986). While NFs exhibit more stability than the other major axonal cytoskeletal polymers (MTs and actin filaments), a growing number of studies provide evidence that NFs, or at least certain populations of NFs, exhibit dynamics reminiscent of other cytoskeletal strucutures. Phosphorylation sites on the amino-terminal head domain of NF-L and the medium-molecular-weight NF protein (NF-M) (Sihag and Nixon, 1989, 1990; Nixon and Sihag, 1991) influence the ability of NFs to polymerize or depolymerize *in vitro* (Hisanaga *et al.*, 1990; Nakamura *et al.*, 1990). Exchange of soluble low-molecular-weight NF protein (NF-L) subunits and/or subunit addition with those in NF polymers has been demonstrated *in vitro* (Angelides *et al.*, 1989) and *in vivo* (Okabe *et al.*, 1993; Takeda *et al.*, 1994). Such striking behaviors of injected NF subunits suggest that a dynamic equilibrium between soluble and assembled subunits may be a significant feature of the NF network in growing, if not also mature neurons, and/or in newly assembled, if not in all NF populations. In this regard, several observations suggest that NFs within neuronal perikarya might be more dynamic than those in axons. For example, NF subunits are abundant immunocytochemically at both the light and electron microscopic level in Purkinje cell bodies despite the absence of 10-nm filaments (Langley *et al.*, 1988). In NB2a/d1 neuroblastoma cells, where cytoskeletal dynamics within perikarya and events of neurite outgrowth are highlighted, a substantial perikaryal pool of assembly-competent, Triton-soluble, 200-kDa NF-H subunits that represent cytoskeletal precursors coexists with cytoskeleton-associated NF triplet proteins (Shea *et al.*, 1989, 1990, 1995a, b; Shea, 1994).

In the first section of this chapter, we demonstrate a similar Triton-soluble pool of extensively phosphorylated (200-kDa) NF-H subunits in developing mouse brain. We also demonstrate that at least some of these subunits undergo extensive phosphorylation prior to their incorporation into Triton-insoluble structures.

1.2. Axonal Transport of Triton-Soluble Neurofilament (NF) Subunits

The constituents of the axonal cytoskeleton are synthesized within the neuronal perikarya and subsequently delivered into the axon by a process referred to as axonal transport (Okabe and Hirokawa, 1989; Nixon, 1993). Long-unresolved questions include (1) the form(s) in which NF subunits are transported, (2) how they take up residence in the axonal cytoskeleton, and (3) how localized repair/remodeling of the axonal cytoskeleton is effected.

In addition to the above studies highlighting dynamic aspects of NFs, additional studies indicate that even axonal NFs exhibit some degree of dynamic exchange, including the exchange of microinjected NF subunits with axonal NFs in cultured neurons (Okabe *et al.*, 1993; Takeda *et al.*, 1994), axonal transport of transfected NF-M in NF-deficient mice (Terada *et al.*, 1996), unique transport dynamics for NF-H (Tsuda *et al.*, 1997), bidirectional transport of a population of axonal NFs *in situ* (Glass and Griffin, 1994; Watson *et al.*, 1993) and in neuroblastoma cells (Wang and Shea, 1994), and the isolation from axons of nonfilamentous aggregates containing NF along with other axonal cytoskeletal proteins (Weisenberg *et al.*, 1987; Bray and Mills, 1991).

While certain morphological and biochemical evidence has been interpreted to indicate that NF subunits enter axons as polymers (Nixon and Lewis, 1986, 1987; Hoffman and Lasek, 1975), even these studies do not fully address whether or not such polymers undergo axonal transport in Triton-soluble form and subsequently incorporate into the Triton-insoluble cytoskeleton. Should NF subunits and/or newly assembled NFs indeed undergo axonal transport in Triton-soluble form, Triton-soluble, radiolabeled NF subunits should be recovered from axons before the accumulation of substantial radiolabeled NFs within the axonal cytoskeleton, or, at the very least, Triton-soluble populations of NF subunits within axons should exhibit relatively more radioactivity per unit protein than cytoskeleton-associated subunits. Similarly, microinjected "tagged" subunits should initially be detected within Triton-soluble NFs/NF subunits within axons before such subunits accumulate with the Triton-insoluble axonal cytoskeleton. In the second section of this chapter, we examine the form(s) in which NF subunits undergo axonal transport and incorporation into the Triton-insoluble cytoskeleton using three systems (optic pathway *in situ*, differentiated neuroblastoma cells, cultured dorsal root ganglion neurons) and pulse-chase radiolabeling and microinjection analyses.

2. EXPERIMENTAL PROCEDURES

2.1. Harvesting of Total Brains

Total brains were harvested from 3–10 mice of embryonic day 17 (E17) and postnatal days 7–120 (P7–P120) from the B6C3 strain maintained at the E. K. Shriver Center animal facility (Fischer and Shea, 1991). Following cervical dislocation, brains were rapidly removed, homogenized in cold 1% Triton X-100 in 50 mM Tris-HCl (pH 6.8), 5 mM ethylene diamine tetraacetic acid (EDTA), 2 mM phenylmethylsulfonyl fluoride (PMSF), and

50 μg/ml leupeptin. Equivalent amounts (200 μg) of each fraction from all samples were loaded onto gels. Homogenized samples were fractionated into Triton-insoluble cytoskeletons and Triton-soluble fractions as described by Chiu and Norton (1982). Samples were centrifuged for 10 min at 13,000 × *g* at 4°C. The material sedimented by this low-speed centrifugation (the Triton-insoluble cytoskeleton) was resuspended in 1% sodium dodecyl sulfate (SDS) and stored at −70°C. In some experiments, the above Triton-soluble fractions were further centrifuged at 100,000 × *g* for 1 hr (Shea *et al.*, 1990) and the resulting high-speed pellet was resuspended in 1% SDS.

In additional experiments, mouse total brain, rat sciatic nerve, and rat dosal root ganglia were homogenized in a buffer that preserves assembled microtubules [0.1 M PIPES buffer, pH 6.7, containing 1 mM $MgCl_2$, 2 mM ethyleneglycol-aminoethyl tetraacetic acid (EGTA), 0.1 mM EDTA, 2 M glycerol at 30°C; Shelanski *et al.*, 1973). This latter homogenate was centrifuged at 37,000 × *g* for 15 min at 30°C, and the resulting pellet was rinsed in the same buffer and resuspended in 1% SDS. In developmental analyses, equivalent amounts of total protein were loaded onto gels.

2.2. Cell Culture

Mouse NB2a/d1 neuroblastoma cells were considered ideal for these analyses, since sufficient numbers of these cells can be generated to accommodate both radiolabeling and biochemical analyses, as well as microinjection and immunocytochemical analyses (e.g., Burrell *et al.*, 1996; Carter *et al.*, 1997; Cressman and Shea, 1995; Shea, 1995b; Shea *et al.*, 1990, 1995a). Cells were cultured in Dulbecco's modified Eagle's medium (DMEM) containing 10% normal horse serum and differentiated by 3 days of treatment with 1 mM dibutyryl cyclic AMP (dbcAMP) as described previously (Shea *et al.*, 1988a, 1990). Dorsal root ganglion (DRG) neurons were cultured from day 12 chick embryos in Ham's F12 medium containing 10% fetal calf serum and 25 ng/ml nerve growth factor according to Straube-West *et al.* (1996).

2.3. Pulse and Pulse-Chase Radiolabeling

NB2a/d1 cells were pulse- and pulse-chase radiolabeled with ^{35}S-methionine as described previously (Shea *et al.*, 1990, 1997). Briefly, NB2a/d1 cells were incubated in methionine-free medium containing 1% serum for 15 min and then received 500 μCi per plate in 3 ml of total methionine-free medium containing 1% serum and incubated at 37°C for 15 min. Cells were then either harvested (pulse) as described above, or their medium was replaced with complete DMEM with 10% serum and 20× the normal methionine concentration, and incubation was continued for various intervals (pulsechase). Murine retinal ganglion cells were radiolabeled *in situ* by injection of 70 mCi of ^{35}S-methionine via a pulled glass capillary pipette into the vitreous of anesthetized mice as described (Nixon and Logvinenko, 1986; Shea *et al.*, 1997). Mice were sacrificed by cervical dislocation at intervals of 1 and 2 hr, and 1–28 days following injection.

2.4. Fractionation of NB2a/d1 Cells and Optic Pathway

NB2a/d1 cultures were rinsed with Tris-buffered saline (TBS), pH 7.4, and cells were separated into fractions enriched in axonal neurites and separately enriched in perikarya as described previously (Shea *et al.*, 1993). Briefly, cells were scraped from the plate in 50 mM Tris-HCl (pH 6.9) containing 2 mM EDTA, 1 mM PMSF, and 50 μg/ml leupeptin at 4°C and gently homogenized (25 strokes) in a loose-fitting glass-Teflon homogenizer, which disrupted cell bodies and fragmented, but did not fully disrupt, axonal neurites. Homogenized samples were then centrifuged at 5000 × *g* for 5 min, which selectively sedimented axonal neurite fragments (Shea *et al.*, 1993). These fractions were then further separated into Triton-insoluble cytoskeletons and Triton-soluble high-speed pellets and supernatants as described (see also Shea *et al.*, 1993). Briefly, the neurite-enriched pellet was resuspended in the above buffer containing 1% Triton, and to the perikaryal-enriched supernatant fraction Triton was added to 1%. Neurite- and perikaryal-enriched fractions were then separately homogenized by 50 strokes in a tight-fitting glass-Teflon homogenizer and the Triton-insoluble cytoskeleton (referred to as “P1” or “cytoskeleton”) was sedimented by centrifugation at 15,000 × *g* for 15 min as described (Chiu and Norton, 1982). The resultant Triton-soluble supernatant was subjected to an additional high-speed centrifugation (100,000 × *g*, 1 hr, 4°C) to yield a high-speed pellet (“P2” or “soluble polymer,” with specific reference to solubility in Triton) and a high-speed supernatant (“S2” or “monomer”; Shea *et al.*, 1990). We stress that “monomer” is utilized for simplicity only: it remains likely that the Triton-soluble high-speed supernatant contains relatively small, nonsedimentable oligomers (Cohlberg *et al.*, 1995). In additional experiments, cells were homogenized in 1% Triton without prior separation of axonal neurites and perikarya.

For harvesting of retinas and optic axons, retinas were dissected away from the rest of the eye and optic axons dissected into 9 × 1.1 mm segments on a glass slide on dry ice as described (Nixon and Logvinenko, 1986). Retinas and segments from 5–11 mice were pooled and homogenized in 1% Triton X-100 in 50 mM Tris-HCl (pH 6.9) containing 2 mM EDTA, 1 mM PMSF, and 50 μg/ml leupeptin at 4°C by 50 strokes in a tight-fitting glass-Teflon homogenizer and separated into cytoskeleton, soluble polymer, and monomer fractions as described above. Triton-soluble material obtained from total optic pathway and from NB2a/d1 axonal neurites and perikarya was also subjected to gel filtration chromatography; material recovered from the column void volume (where NF polymers are recovered; e.g., Hisanaga and Hirokawa, 1988) was subjected to ultrastructural analyses under negative staining, immunoblot analyses, and autoradiography (below).

2.5. Immunoprecipitation

Immunoprecipitation of fractions from radiolabeled retina and optic pathways was carried out according to Shea *et al.* (1990). Cytoskeleton and soluble polymer fractions were resuspended in 100 μl of 1% SDS and the monomer fraction was substituted with 1% SDS. The fractions were diluted with four volumes of 60 mM Tris-HCl (pH 7.6), 190 mM NaCl, 6 mM EDTA, and 1.25% Triton X-100 and incubated overnight at 4°C with reciprocal shaking with 1:100 dilutions of a polyclonal antisera (H34) that reacts with 200-kDa NF-H, monoclonal antibodies directed against phosphorylated (SMI-31), and nonphosphorylated (SMI-32) NF epitopes (Sternberger and Sternberger, 1983), or a polyclonal antiserum (R39) that precipitates all three NF subunits (Nixon *et al.*, 1988; Shea *et al.*, 1997). Samples were then incubated at room temperature with 10 mg of Protein A-Sepharose complex (Sigma Chemical Co., St. Louis, MO) for 2 hr with reciprocal shaking, followed by extensive washing with 50 mM Tris-HCl (pH 7.5), 150 mM NaCl, 5 mM EDTA, 0.1% Triton X-100, and 0.02% SDS. Elution of immunoprecipitated proteins was carried out by heating at 100°C for 1 min with 100 μl of 125 mM Tris-HCl (pH 6.8), 10 mM dithiothreitol, 2% SDS, 5% β-mercaptoethanol, and 10% glycerol.

2.6. Gel Electrophoresis, Autoradiography, and Immunoblot Analyses

Entire fractions derived from radiolabeled cells and optic pathway were subjected to SDS–gel electrophoresis and transferred to nitrocellulose (e.g., Shea *et al.*, 1990, 1997). The proteins were visualized by sequential reaction of the nitrocellulose replica in TBS overnight at room temperature with the above antibodies and the following additional antibodies: a phosphorylation-independent polyclonal antibody directed against NF-L (L33; Shea *et al.*, 1992); a phosphorylation-independent polyclonal antibody directed against NF-M (M25; Shea *et al.*, 1992); polyclonal antibodies directed against extensively phosphorylated forms of NF-H (anti-200; Shea *et al.*, 1988b), and additional monoclonal antibodies directed against phosphorylated NF epitopes (SMI-34, RT97; Sternberger and Sternberger, 1983;

Anderton *et al.*, 1982). SMI antibodies were obtained from Sternberger-Meyer Immunochemicals (Jarretsville, MD). RT97 was a generous gift of Dr. U. Drager, (Harvard Medical School, Boston, MA). Anti-H-200 was a generous gift of Dr. Ronald Majocha (Brown University, RI). R39 was generously provided by Dr. Doris Dahl (West Roxbury Veteran's Administration Medical Center, Roxbury, MA). Incubation of replicas with the above antibodies was followed by incubation for 2 hr with 1:1000 dilutions of goat anti-rabbit or anti-mouse immunoglobulin G (IgG) conjugated with horseradish peroxidase and development for 5 min with diaminobenzidine in the presence of H_2O_2 (Fischer *et al.*, 1986). Following immunostaining, some nitrocellulose replicas were rinsed 3× in TBS and reprobed with a second antibody using the above protocol. In some experiments, the nitrocellulose replica was incubated at 37°C for 6 hr with 50 units/ml alkaline phosphatase (Sigma) in TBS (pH 8.0) to cleave NF-associated phosphate groups prior to immunostaining as described above. Autoradiographs were generated either from dried SDS gels or from nitrocellulose replicas (Shea *et al.*, 1990, 1997).

2.7. Quantitation of NF Subunit Immunoreactivity

Immunoreactive bands on nitrocellulose replicas were scanned, digitized, and quantified using the Visage 110 Bio Image System (Ann Arbor, MI) consisting of a digitizing camera and digital processing software or on a Macintosh Power PC via a Scion LG-3 frame grabber and NIH Image software as described previously (Fischer and Shea, 1991; Cressman and Shea, 1995). The immunostained replicas were digitized as 1024 × 1024 pixel images with a densitometry calibration wedge. Densitometric analyses were performed on entire bands, as defined by boundaries circumscribed by the operator on the digitized image as displayed on the monitor. The values obtained for the quantitative analyses represent the total integrated optical density (i.e., area × optical density; Crandall and Fischer, 1989; Fischer and Shea, 1991). The range of linear response for staining of immunoblots with NF antibodies was determined by analyzing a concentration curve of Triton-insoluble cytoskeletons from adult mouse brain, and the amount of total protein loaded onto gels was within the linear response range (Fischer and Shea, 1991). The value obtained for Triton-soluble fractions at P60 was divided by that obtained for Triton-insoluble cytoskeletons at P60 for each antibody and the resulting ratio presented; values for each antibody were calculated separately. Scans from three separate experiments demonstrated a variance of no more than 15%, and repeated scans of the same immunoblot were identical. Immunoblots presented are representative of all experimental trials.

2.8. Biotinylation and Microinjection of NF Subunits

Bovine spinal cord was homogenized in 100 mM PIPES (pH 6.6) containing 1 mM EGTA, 1 mM $MgCl_2$, 1 mM PMSF, 1 mM dithiothreitol (DTT), and 1 μg/ml leupeptin. NFs were recovered in the void volume following Sepharose CL-4B gel chromatography (Hisanaga and Hirokawa, 1988; Takeda *et al.*, 1994). As described (Takeda *et al.*, 1994), a single assembly/disassembly cycle was sufficient to eliminate tubulin and other contaminating proteins. Isolated NFs were disassembled in *urea buffer* (100 mM phosphate buffer, pH 7.5, containing 6 M urea, 1 mM EDTA, 1 mM EGTA, 1 mM DTT, 0.5 M PMSF, and 1 μg/ml leupeptin), dialyzed overnight at 4°C followed by incubation at 37°C for 1 hr in *assembly buffer* (20 mM PIPES, pH 6.6, 1 mM EGTA, 1 mM $MgCl_2$, 1 mM EDTA, 1 mM DTT, 0.17 NaCl; e.g., Takeda *et al.*, 1994) and recovered by centrifugation at 100,000 × *g* for 1 hr. The resultant once-cycled NFs were conjugated with 3-(N-maleimidylproprionyl) biocytin (25 ng/ml; Molecular Probes, Eugene, OR) while in assembled form essentially as described previously (Takeda *et al.*, 1994). NF-H and NF-L were subsequently purified by DE-52 chromatography in urea buffer; pure fractions (ascertained by SDS–gel electrophoresis and immunoblot analysis; e.g., Fig. 8) were combined and dialyzed overnight at 4°C against *assembly buffer* (e.g., Takeda *et al.*, 1994) to remove urea, then dialyzed for 6 hr against 10,000 volumes of *injection buffer* (5 mM Hepes, pH 8.5, 1 mM DTT, 1 mM EDTA, 1 mM EGTA), concentrated by centrifugation onto Centricon filters, and stored at −80°C (Takeda *et al.*, 1994). Subunits were mixed 1 : 1 with fluorescein-conjugated dextran (Molecular Probes) just prior to injection and clarified by centrifugation for 10 min in a Beckman airfuge. The resulting supernatant (containing approximately 1 mg/ml NF protein) was microinjected into NB2a/d1 cells differentiated for 3 days with dbcAMP or into DRG neurons cultured for 2–3 days; both cell types were cultured on poly-L-lysine-coated plates (to facilitate adherence during injection) as described (Cressman and Shea, 1995; Straube-West *et al.*, 1996; Wang and Shea, 1994). Cells exhibiting any obvious trauma resulting from microinjection were excluded from further analyses. Additional cultures received 330 nM nocodazole immediately following injection.

2.9. Immunofluorescence and Immunoelectron Microscopy

Cells were fixed with 4% paraformaldehyde in TBS (pH 7.4) for 10 min at room temperature with or without prior extraction for 10 min with 1% Triton X-100 in 50 mM Tris-HCl (pH 6.9) containing 2 mM EDTA and 1 mM PMSF at 4°C essentially as described (Shea *et al.*, 1988a). Microinjected bNF-H and bNF-L were visualized with mouse anti-biotin (1 : 20–

100) followed by a 1:150 dilution of Texas-Red-conjugated goat anti-mouse IgG. Images were captured via a DAGE CCL-72 camera with frame averager and stored in a Macintosh PowerPC 7100AV via NIH Image software and a Scion LG-3 frame grabber as described (Cressman and Shea, 1995). For immunoelectron microscopy, microinjected NB2a/d1 cells were fixed in glutaraldehyde, dehydrated, embedded in resin, and sectioned by conventional methods (e.g., Shea *et al.*, 1993). Sections were then incubated for 1 hr with a 1:100 dilution of anti-biotin, rinsed three times, and incubated with goat anti-mouse or anti-rabbit IgG conjugated to 5-nm colloidal gold particles, respectively, and examined by electron microscopy (EM) (Shea *et al.*, 1993).

All experiments were repeated two to five times; only minor variations were detected among any experiments, and data presented in individual figures are representative of all experiments; densitometric values correspond to the signals obtained from the stained gel, immunoblot, or autoradiograph presented in that figure.

3. RESULTS

3.1. Extensive Phosphorylation of NF-H prior to Incorporation into Triton-Insoluble Structures

3.1.1. NF Subunits Are Recovered in Triton-Insoluble and -Soluble Fractions of Total Brain during Development

Immunoblot analysis of murine brains demonstrated near-linear increases in cytoskeleton-associated NF triplet proteins from postnatal day 14 until 60. Steady-state levels of all three NF subunits remained constant until postnatal day 120 (the last point examined; Figure 1). Despite the clear presence of 200-kDa NF-H as early as postnatal day 14 following immunostaining with "nonrestrictive" antisera (which indicating that NF-H subunits have already undergone extensive phosphorylation; Julien and Mushynski, 1982), immunostaining with a monoclonal antibody (RT97) directed against a late-appearing phosphoepitope (Anderton *et al.*, 1982; Shea *et al.*, 1989) demonstrated that continued development is accompanied by additional NF phosphorylation events (Figure 1).

In addition to cytoskeleton-associated subunits, NF-H migrating with an apparent M_r of 160,000–200,000 was recovered in Triton-soluble fractions from postnatal days 14–60 (Figure 1). The extent of reactivity of soluble NF-H subunits varied with each antibody; soluble subunits were strongly reactive with "nonrestrictive" polyclonal antibodies (e.g., anti-NF-H, anti-H-200, and H34), weakly reactive with monoclonal antibody SMI-

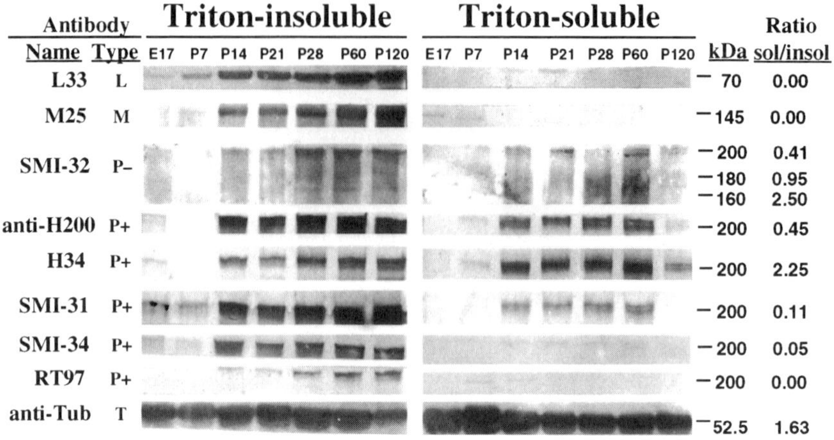

FIGURE. 1. Appearance and posttranslational modification of neurofilament subunits during mouse brain postnatal development. Composite of nitrocellulose replicas of Triton-insoluble and Triton-soluble fractions from mouse brain from embryonic day 17 (E17) and postnatal days 7–120 (P7–P120). All lanes received an equivalent amount (200 μg) of total protein from Triton-soluble or Triton-insoluble fractions. Replicas were reacted with anti-NF-L antibody L33, anti-NF-M antibody M25, antibodies directed against hypophosphorylated (SMI-32) or extensively phosphorylated forms of NF-H (anti-200, H34, SMI-31, SMI-34, RT97), and an antibody against tubulin. Replicas containing respective samples from Triton-insoluble and -soluble fractions were coprocessed using the identical antibody and staining solutions and incubation periods. Only relevant portions of replicas are presented; the respective migratory positions of immunoreactive polypeptides are indicated along the right. Designations regarding antibody specificity are included along the left of the nitrocellulose replicas. L, M, and T indicate specificity of antibodies for NF-L, NF-M, and tubulin, respectively. P− and P+ refer to antibodies reactive toward nonphosphorylated or phosphorylated NF − H isoforms, respectively. Whole-band densitometric analysis was conducted on Triton-soluble and -insoluble fractions in at least three separate immunoblots, one of which is presented here for each antibody, using the Visage computer-assisted densitometric system as described previously (Fischer and Shea, 1991). The value obtained for Triton-soluble fractions at P60 was divided by that obtained for Triton-insoluble cytoskeletons at P60 for each antibody and the resulting ratio plotted; values for each respective antibody were calculated separately. These values are listed on the right of the figure. Reprinted from Shea *et al.* (1997) with permission.

31, and nonreactive with monoclonal antibodies SMI-34 or RT97 (Figure 1). At postnatal day 60, Triton-soluble fractions actually contained a greater percentage of immunoreactivity toward the phospho-dependent polyclonal H34 antibody than did cytoskeletons. Triton-soluble NF-H as recognized by anti-200 (H-200) often migrated as a broad doublet, consistent with the previous demonstration of multiple phosphorylated NF-H variants (Lewis and Nixon, 1988; Goldstein *et al.*, 1987a). Triton-soluble H-200 declined by

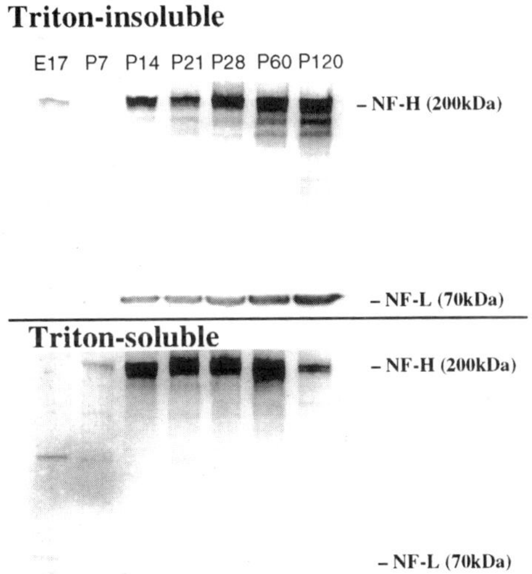

FIGURE 2. Recovery of H-200 in the Triton-soluble fraction is not the result of neurofilament breakage. A nitrocellulose replica of mouse brain from E17-P120 was immunostained with a polyclonal antibody directed against 200-kDa NF-H (H34), which demonstrated the presence of H-200 in both Triton-soluble and -insoluble fractions. This replica was then reprobed with a polyclonal anti-NF-L antibody L33. Note the quantitative recovery of NF-L in the Triton-insoluble fraction. Reprinted from Shea *et al.* (1997) with permission.

P120 as visualized by all antibodies (Figure 1). Immuoreactivity of all samples with an antitubulin antibody confirmed that the decline in NF-H isoforms in P120 Triton-soluble fractions was not a general effect on cytoskeletal proteins (Figure 1).

Unlike NF-H in these analyses of total brain, NF-L and NF-M were at all times quantitatively recovered in the Triton-insoluble fraction (Figures 1 and 2), confirming that the presence of H-200 within Triton-soluble fractions was not the result of NF disruption. It is noteworthy that, while the failure to detect NF-L or NF-M in these analyses could theoretically be attributed to a relatively poorer affinity of antisera for these subunits versus those utilized in analyses of NF-H, the maximum amount of NF-L or NF-M that could be present in soluble fractions yet remain undetected must be less than the trace levels observed in cytoskeletons derived from these brains at E17. Additional polyclonal antisera directed against NF-M and NF-L did not detect either subunit in Triton-soluble fractions (not shown).

To define further the nature of soluble H-200, the Triton-soluble fraction was centrifuged at 100,000 × *g* for 1 hr. This procedure sedimented soluble H-200 as detected by immunoblot analysis (Figure 3). Moreover, H-200 was quantitatively recovered in the pellet when cells were harvested and fractionated under conditions typically utilized to preserve assembled microtubules (e.g., Shelanski *et al.*, 1973) (Figure 3). These findings indicate that most to all of the Triton-soluble H-200 is either associated with other

(Triton-soluble) cytoskeletal polymers such as microtubules (see also Delacourte *et al.*, 1980), or is present in some other nonfilamentous, oligomeric form (e.g., see Weisenberg *et al.*, 1987).

Alkaline phosphatase treatment of immunoblots of Triton-soluble fractions substantially reduced H-200 immunoreactivity with phospho-dependent antibodies (not shown), confirming, as has previously been demonstrated for cytoskeleton-associated H-200 (Julien and Mushynski, 1982) and for Triton-soluble H-200 in neuroblastoma (Shea *et al.*, 1990), that reactivity of these antibodies with Triton-soluble H-200 was derived from phosphorylation. Monoclonal antibodies directed against nonphosphorylated (SMI-32) NF epitopes also demonstrated the developmental appearance of Triton-soluble NF-H isoforms ranging in apparent molecular weight from 160 to 200 kDa (Figure 1), indicating that the decline in H-200 observed at P120 was due at least in part to depletion of the NF-H polypeptide from this fraction, rather than exclusively to dephosphorylation of Triton-soluble NF-H subunits.

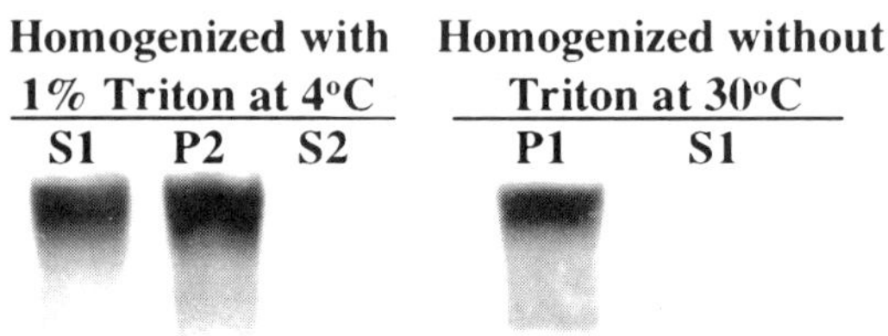

FIGURE 3. Partitioning of H-200 among cellular fractions is dependent upon speed of centrifugation and conditions of homogenization. Left: Following sedimentation of the Triton-insoluble cytoskeleton by low-speed centrifugation (13,000 × *g*, 10 min; see Section 2), the resulting Triton-soluble supernatant (S1) was subjected to a second, high-speed centrifugation (100,000 × *g*, 1 hr), which generated a second pellet (P2) and resulting supernatant (S2). The resulting S1, P2, and S2 fractions from postnatal day 60 were subjected to SDS–gel electrophoresis and immunoblot analysis using anti-NF-H antibody H34. Note that Triton-soluble H-200 immunoreactivity was quantitatively recovered in the fraction sedimented following high-speed centrifugation. Right: Nitrocellulose replicas of mouse brain from postnatal day 60 (P14–P60) were homogenized at 30°C in a microtubule-stabilizing buffer (0.1 M PIPES buffer, pH 6.7, containing 1 mM $MgCl_2$, 2 mM EGTA, 0.1 mM EDTA, 2 M glycerol at 30°C; Shelanski *et al.*, 1973) and fractionated by centrifugation at 37,000 × *g* for 15 min. The resulting pellet and supernatant fractions were subjected to SDS–gel electrophoresis and transferred to nitrocellulose and immunostained with H34. Unlike the situation observed following harvesting in the presence of 1% Triton at 4°C (e.g., Figure 1), harvesting of brain at 30°C in microtubule-stabilizing buffer resulted in the quantitative recovery of H-200 in the pellet. H-200 distribution patterns as presented in the presence and absence of Triton were identical for all times at which soluble H-200 was observed (not shown). Reprinted from Shea *et al.* (1997) with permission.

3.1.2. Newly Synthesized NF-H Subunits Can Be Phosphorylated within Neuronal Perikarya before Incorporation into Triton-Insoluble Structures

It remained to be determined whether or not the appearance of H-200 subunits within Triton-soluble fractions resulted from NF catabolism, or instead whether Triton-soluble H-200 was destined for incorporation into NFs. To accomplish this, we examined NF steady-state levels and synthesis as well as assembly dynamics in fractions of retinal cells (e.g., Nixon *et al.*, 1987).

NF-H subunits of heterogeneous phosphorylation states were immunoprecipitated from Triton-soluble and -insoluble fractions of retinal ganglion cells after mice were injected intravitreously with ^{35}S-methionine. To insure that Triton-soluble radiolabeled NF subunits were indeed monomeric, the Triton-soluble fraction was clarified by high-speed centrifugation at 100,000 × *g* for 1 hr. Within 2 hr after injection of isotope, radioactivity was associated with H-200 and, to a lesser extent, H-160, in the high-speed, Triton-soluble supernatant. By contrast, radiolabel was associated only with H-160 and not H-200 in the Triton-insoluble cytoskeleton at this time (Figure 4). Over the following 48 hr, cytoskeleton-associated

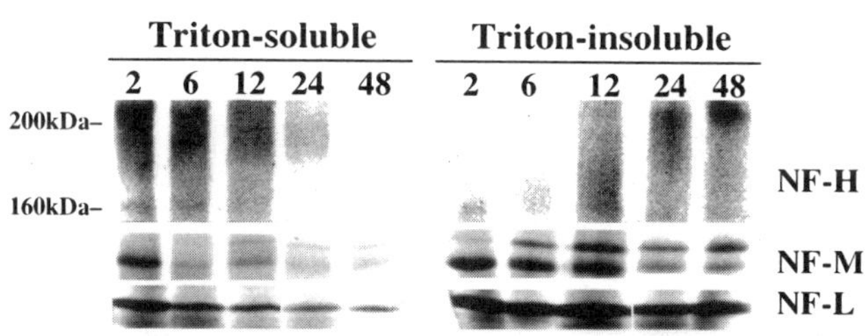

FIGURE 4. Immunoprecipitation of metabollically labeled NF polypeptides from mouse retinal ganglion cells. Adult mice were injected intravitreally with ^{35}S-methionine and sacrificed after postinjection intervals of 2–48 hr. Retinas were fractionated into Triton-insoluble cytoskeletons and high-speed Triton-soluble fractions by sequential centrifugation, and NF polypeptides were immunoprecipitated with an antineurofilament polyclonal antiserum ("R39") which immunoprecipitates all three NF subunits regardless of phosphorylation states (Nixon *et al.*, 1989). Immunoprecipitated material was subjected to gel electrophoresis and autoradiography. Relevant portions of the autoradiogram are presented as indicated. Note that, within 2 hr after injection of isotope, radioactivity was associated with both H-160 and H-200 in the Triton-soluble supernatant, but was associated only with H-160 in the Triton-insoluble cytoskeleton. Note also the progressive depletion over 48 hr of cytoskeleton-associated radiolabeled H-160 and the appearance of a range of isoforms migrating from 160 to 200 kDa. Radiolabeled Triton-soluble H-200 was progressively lost over 48 hr. The majority of NF-M and NF-L isoforms were incorporated into Triton-insoluble structures between 2 and 6 hr. Posttranslationally modified NF-M was first detected within the Triton-insoluble fraction at 6 hr. Reprinted from Shea *et al.* (1997) with permission.

H-160 in retina (Figure 4) was depleted while H-200 appeared. A progressive loss of radiolabeled H-200 from the Triton-soluble fraction was observed over the same 48-hr interval. These data indicate that a percentage of NF-H subunits in mouse CNS are extensively phosphorylated without undergoing prior incorporation into the Triton-insoluble cytoskeleton. The appearance of newly synthesized H-200 subunits within Triton-soluble fractions prior to their appearance in Triton-insoluble fractions was also confirmed from the abilities of three additional NF-H antibodies to immunoprecipitate ^{35}S-methionine-radiolabeled subunits from the Triton-soluble, but not Triton-insoluble fractions 2 hr after the injection of radiolabel (Figure 5). A partially phosphorylated (180-kDa) radiolabeled NF-H isoform was immunoprecipitated from both fractions at this time by SMI-32 (Figure 5).

Both NF-M and NF-L were rapidly incorporated into the retinal Triton-insoluble cytoskeleton. Unlike NF-H, the majority of NF-M had apparently incorporated into Triton-insoluble structures prior to becoming sufficiently phosphorylated to increase their electrophoretic mobility from approximately 139 to 145 kDa (see Nixon *et al.*, 1988).

3.2. Axonal Transport of Triton-Soluble NF Subunits

3.2.1. Newly Synthesized NF Subunits Are Present within NB2a/d1 Axonal Neurites in Triton-Soluble Form

Immunoblot and autoradiographic analyses have previously demonstrated the presence of NF subunits within P1, P2, and S2 fractions of

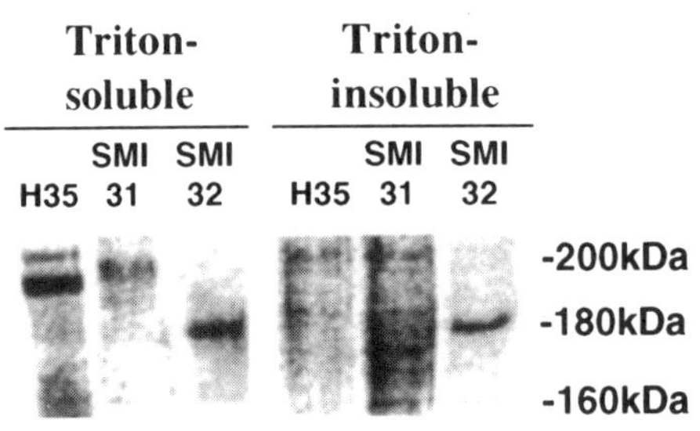

FIGURE 5. Newly synthesized, extensively phosphorylated NF-H is recovered primarily from the high-speed Triton-soluble supernatant of mouse retinal ganglion cells. Retinas of 10 adult mice were fractionated into Triton-insoluble cytoskeletons and high-speed Triton-soluble fractions by sequential centrifugation 2 hr after intravitreal injection of ^{35}S-methionine. NF-H was immunoprecipitated with H34, SMI-31, and SMI-32 and subjected to SDS–gel electrophoresis and autoradiography. Note the recovery of significant radiolabeled H-200 from the Triton-soluble fraction, and substantially less H-200 from Triton-insoluble fractions, following immunoprecipitation with H35 and SMI-31. A radiolabeled isoform migrating at 180 kDa, indicative of partial phosphorylation, was immunoprecipitated from both fractions with SMI-32, but was most prevalent within the Triton-soluble fraction. These findings indicate that NF-H can be phosphorylated prior to incorporation into Triton-insoluble structures and undergoing transport into axons. Reprinted from Shea *et al.* (1997) with permission.

dbcAMP-differentiated NB2a/d1 cells, and have confirmed that subunits within S2 and P2 fractions represent cytoskeletal precursors (Shea, 1995a; Shea *et al.*, 1990, and references therein). We explored in the present study the relative distribution of radiolabel among these NF subunit populations within perikarya and axonal neurites in pulse- and pulse-chase ^{35}S-methionine radiolabeling of dbcAMP-differentiated cells.

Radiolabeled NF subunits were detected in various fractions following the 15-min pulse. Label was associated with NF subunits immunoprecipitated from the perikaryal S2 fraction concordant with *de novo* NF synthesis. Relatively minor levels of radiolabel were associated with P2 and P1 from perikarya (Figure 6). Radiolabel was also associated with axonal NF subunits following this 15-min pulse, indicating that some newly synthesized NF subunits had already undergone translocation into axonal neurites. Comparison of radiolabel within the P2 and P1 of axonal neurites demon-

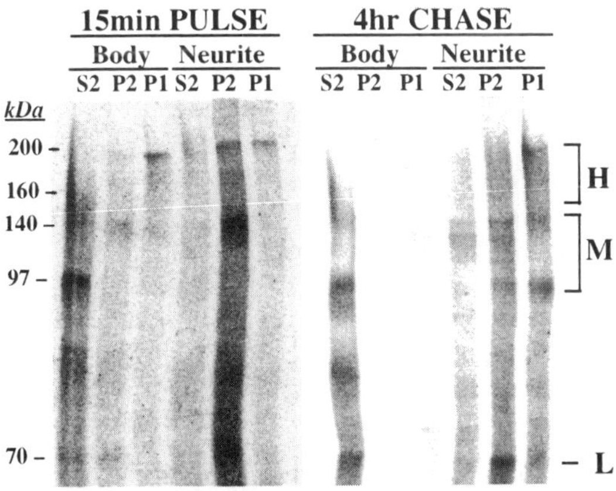

FIGURE 6. Some NF subunits are initially detected within NB2a/d1 axonal neurites in a high-speed sedimentable, but Triton-soluble, form Panels present autoradiographs derived from SDS gels of material immunoprecipitated from the high-speed Triton-soluble and -insoluble fractions derived from NB2a/d1 perikarya or axonal neurites following pulse and pulse-chase ^{35}S-methionine radiolabeling. Following pulse labeling, radiolabel was associated with NF subunits immunoprecipitated from the perikaryal S2 (monomer), and minor levels of radiolabeled NF subunits were detected in the perikaryal P2 (soluble polymer) and P1 (cytoskeleton). Substantial radiolabel was associated with NF subunits in the P2 of axonal neurites; densitometric analysis of radiolabeled axonal NF demonstrated that 80% of radiolabeled NF-H axonal reactivity was recovered within the P2 fraction, while only 20% was associated with the axonal neurite P1 fraction. Following a 4-hr chase in the absence of radiolabel, increased levels of radiolabeled NF subunits were recovered from the axonal neurite P1, while those in the axonal P2 had diminished. Only minor radiolabel was associated with the axonal neurite S2 at either time. Perikaryal radiolabeled subunits declined during chase analyses.

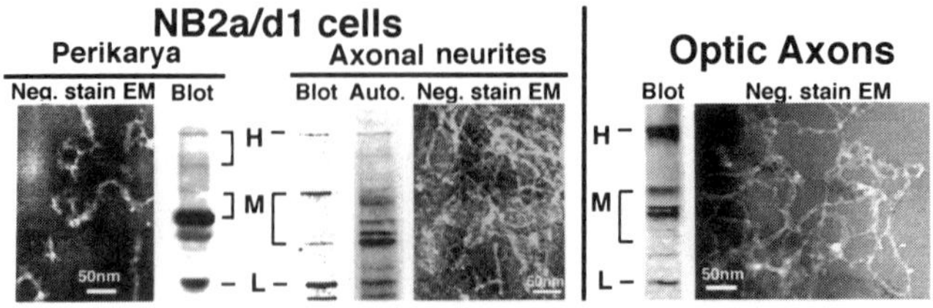

FIGURE 7. Triton-soluble NFs exist within NB2a/d1 perikarya, axonal neurites, and optic axons. Panels present void volume fractions obtained from NB2a/d1 perikarya and neurites and from optic axons as indicated. EM and immunoblot analysis (immunostained with a cocktail of H3, M2, and L3 antibodies) of these fractions confirmed the presence of Triton-soluble NFs that contained all three NF subunits. Perikaryal NFs contained higher ratios of hypophosphorylated NF-H (160kDa) and NF-M (97–125kDa) isoforms. Autoradiographic analyses of identical void volume fractions from axonal neurites following pulse-labeling with ^{35}S-methionine demonstrated that newly synthesized (i.e., radiolabeled) Nf subunits are present within axonal neurites within 15min in the form of NFs, at which time radiolabel is only marginally detectable within the axonal cytoskeleton.

strated that the majority of radiolabeled subunits within axonal neurites were detected within the soluble polymer fraction. Close inspection of the distribution of radiolabeled NF-H compared to that of NF-M and NF-L in fractions obtained after the 15-min chase period suggested that some newly synthesized NF-H was incorporated into perikaryal and neurite cytoskeletons prior to NF-M and NF-L; this phenomenon is independent of *de novo* NF assembly and instead derives from unique exchange of NF-H with existing Triton-insoluble structures (T. B. Shea *et al.*, to be published).

Following a 4-hr chase in the absence of radiolabel, increased levels of radiolabeled NF subunits were recovered from the axonal P1, while those in the P2 had decreased (Figure 6). By contrast, minor radiolabel was associated with the S2 fraction from axonal neurites (Figure 6). Perikaryal radiolabeled subunits underwent a corresponding decline during chase analyses, likely to be a consequence of both axonal transport and selective NF degradation within perikarya (Shea, 1995a). To probe further whether or not the radiolabeled subunits recovered from the P2 fraction of axonal neurites 15min after radiolabeling of cells were associated with NFs, we subjected this material to gel chromatography and analyzed the resultant void volume fractions (where NF polymers are recovered; e.g., Hisanaga and Hirokawa, 1988) by EM, immunoblot analysis, and autoradiography. Radiolabeled NF triplet subunits, including extensively phosphorylated (200-kDa) NF-H, codistributed with NFs within the void volume (Figure 7). This confirms that at least a portion of the Triton-soluble radiolabeled subunits detected

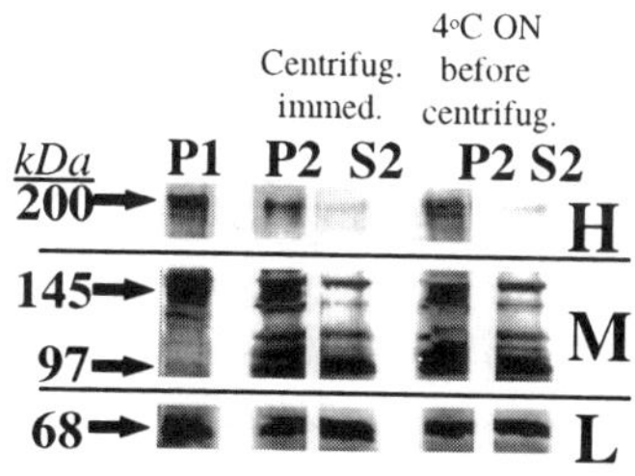

FIGURE 8. Triton-soluble NFs are not generated by artifactual assembly during isolation and incubation. Panels present nitrocellulose replicas immunostained with R39 derived from the S2, P2, and P1 fractions from dbcAMP-treated NB2a/d1 cells as indicated. P2 and S2 fractions were generated from S1 fractions either immediately after separation of cytoskeletons as described in Section 2 ("Centrifug. immed."), or were instead derived from an S1 fraction that was incubated overnight at 4°C with shaking ("4°C ON before centrifug."). Note that no appreciable shift of immunoreactivity from S2 into P2 fractions is observed following overnight incubation. An aliquot of the corresponding P1 fraction is included for comparative purposes only. Note further that the high-speed centrifugation used in these analyses far exceeds the centrifugation procedure used in immunoprecipitation (i.e., approximately 5 min at 5000 RPM in a bench-top centrifuge). These analyses indicate that material immunoprecipitated from P2 fractions from radiolabeled cells (Figure 5) did not artifactually assemble during harvesting and/or the overnight incubation utilized in immunoprecipitation methodologies.

within axonal neurites during pulse-chase analyses (Figure 1) is present in the form of NFs as early as 15 min after labeling. Similar analyses of enriched perikaryal fractions indicated that at least some Triton-soluble NF subunits had already assembled into NFs (Figure 7).

The possibility remained that axonal radiolabeled NFs were derived from assembly of monomeric subunits during the overnight incubation used in immunoprecipitation (e.g., see Section 2). However, this possibility was excluded by overnight incubation of total Triton-soluble fractions under these conditions in the absence of antibody: subsequent immunoblot analyses of P2 and S2 fractions derived from these total Triton-soluble fractions immediately after harvesting of cells or following and overnight incubation revealed no detectable shift of NF subunits from S2 to P2 fractions, indicating that additional NF assembly had not occurred during these incubations (Figure 8).

The above results in NB2a/d1 cells collectively demonstrate that some newly synthesized NF subunits undergo axonal transport in a Triton-soluble form.

3.2.2. Newly Synthesized NF Subunits Are Also Present in a Triton-Soluble Form in Mouse Optic Axons *in Situ*

Immunoblot analyses of retina and optic axons revealed all three subunits in S2, P2, and P1 fractions derived from retina and the entire length of optic axons (Figure 9). Although phospho-NF-H isoforms (as evidenced by

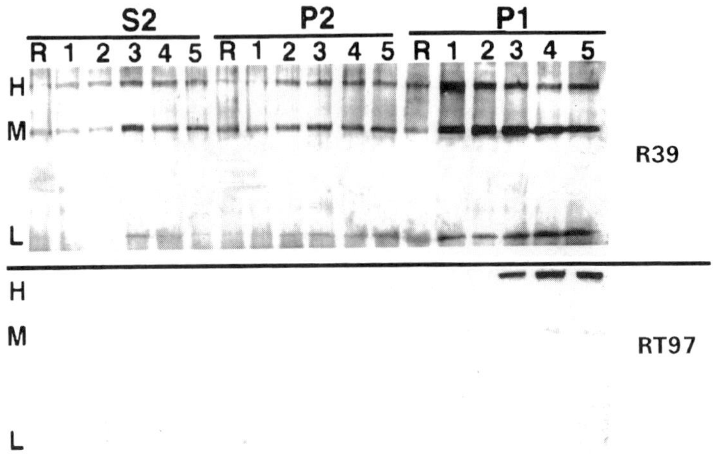

FIGURE 9. Triton-soluble NF subunits are present in retina and optic axons. Panels present nitrocellulose replicas of retinas and 2-mm segments of optic axons following immunostaining with polyclonal antisera R39, which reacts with all three NF subunits (Shea *et al.*, 1997) or monoclonal antibody RT97, which specifically recognizes a developmentally delayed phosphoepitope of NF-H (Anderton *et al.*, 1982). Note that, while R39 strongly labels 200-kDa NF-H in all fractions, RT97 only labels Triton-insoluble 200-kDa NF-H in the latter segments of optic axons. These data indicate that Triton-soluble and -insoluble 200-kDa NF-H are not identically phosphorylated.

migration at 200 kDa) were detected in all fractions from all segments by relatively unrestrictive antibodies such as R39 (Figure 9), and SMI-31 (not shown; see Shea *et al.*, 1997), NF-H exhibited differential phosphorylation among these fractions as indicated by the restriction of the RT97 phosphoepitope to the P1 fraction (Figure 9).

We next conducted comparative analyses of S2, P2, and P1 fractions derived from retina and optic axon segments at various intervals following injection of radiolabel. For radiolabeling analyses, optic axons were cut into 9×1.1 mm segments. Immunoprecipitation followed by autoradiography (Figure 10) demonstrated a broadening peak of radiolabeled subunits within Triton-insoluble cytoskeletons, consistent with previous studies (Lasek *et al.*, 1992, 1994; Nixon and Logvinenko, 1986). In addition to cytoskeleton-associated subunits, however, multiple populations of NF subunits were observed that differed both in relative solubility and in their degree of comigration with other NF subunits. Radiolabel was associated with all three NF subunits in S2 and P2 fractions of optic axon segments during continued analyses from 1 to 28 days after radiolabeling (Figure 11). In order to determine whether or not NF subunits within Triton-soluble

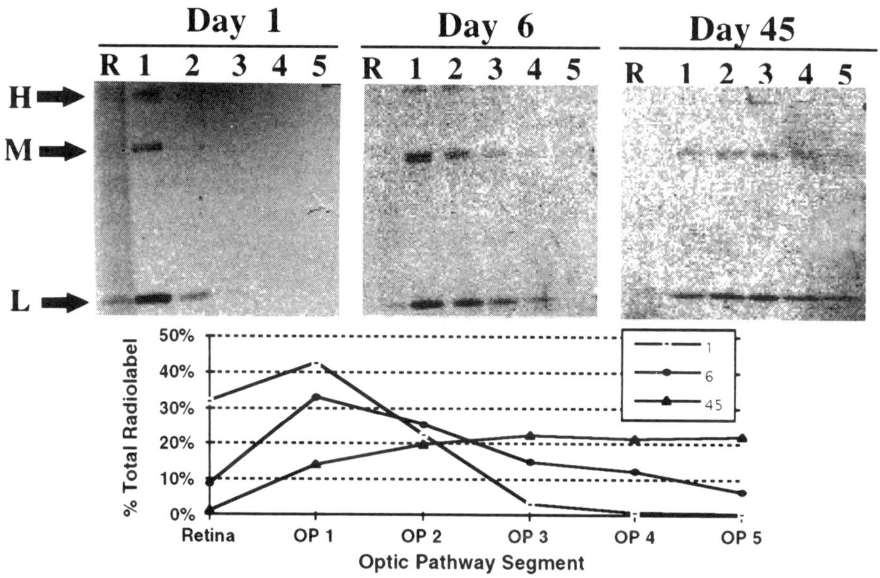

FIGURE 10. Axonal transport of cytoskeleton-associated NF subunits in optic axons. Panels present autoradiographs of immunoprecipitated NFs at days 1, 6, and 45 following radiolabeling. In these analyses, optic axons were dissected into 5×2 mm segments. The accompanying graph presents the relative distribution of immunoprecipitated radiolabeled NF-L calculated as (densitometric value of each segment/densitometric value of all segments) $\times$ 100. Note that, as described in previous studies, the distribution of radiolabeled subunits broadens during continued transport.

fractions displayed distinct levels of newly synthesized subunits versus those within the Triton-insoluble fraction, we calculated the ratios of radiolabeled versus total (Coomassie-stained) NF-L subunits within each fraction of central segments from optic axons (Figure 12) as described above for NB2a/d1 cells. Observation of a relatively higher labeling ratio of subunits versus total protein within Triton-soluble fractions confirms that these subunits represent a unique population(s), and are not derived from artifactual "breakaway" of Triton-insoluble NFs/NF subunits during harvesting. As shown for NB2a/d1 cells, gel chromatography of Triton-soluble material from optic axons followed by immunoblot and ultrastructural analyses indicated that at least some of these Triton-soluble NF subunits are present in the form of NFs (Figure 7).

The above results in mouse optic axons collectively demonstrate that some newly synthesized NF subunits undergo axonal transport in a Triton-soluble form.

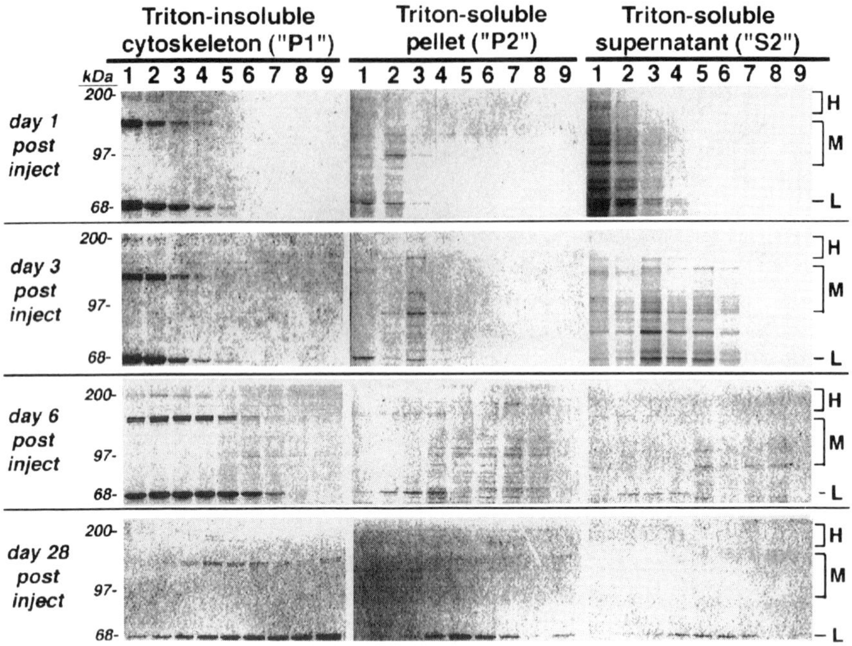

FIGURE 11. Triton-soluble NF subunits undergo transport in optic axons. Panels present autoradiographs of NF subunits immunoprecipitated from S2, P2, and P1 fractions of 9 × 1.1 mm segments of optic axons harvested at 1, 3, 6, and 28 days after intravitreal injection of ^{35}S-methionine. All panels were processed at the same intensity. Visualization of NF-H and NF-M are problematic in Triton-soluble fractions and also in Triton-insoluble fractions at later intervals; longer exposures (not shown) clearly indicate their presence.

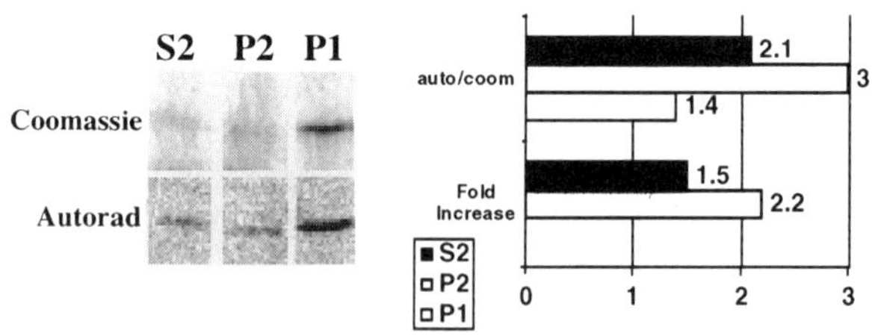

FIGURE 12. Triton-soluble subunits in optic axons represent populations distinct from Triton-insoluble subunits. Panel on left presents a representative 70-kDa region of Coomassie-stained gels and corresponding autoradiographs of material immunoprecipitated S2, P2, and P1 fractions of optic axons (segment 5, day 28 postinjection). Panel on the right presents the ratio (in arbitrary densitometric units) of autoradiographic density/Coomassie brilliant blue (CBB)-stained density for the corresponding images on the left. Note that both S2 and P2 fractions exhibit a relatively higher ratio of radiolabeled to total subunits than do P1 fractions (see text for detailed description). The fold-increase was derived by dividing this ratio for the S2 and P2 fractions by that obtained for the P1 fraction.

3.2.3. Microinjected NF Subunits Undergo Axonal Transport in Triton-Soluble Form in NB2a/d1 Cells and Dorsal Root Ganglion Neurons

The mode of NF subunit axonal transport was also probed by microinjection of "tagged" NF subunits. Bovine spinal cord NFs were conjugated to biotin or rhodamine while assembled essentially according to Hisanaga and Hirokawa (1988) with minor modifications (see Section 2). Biotinylated (bNF-H and bNF-L, respectively) and rhodamine-labeled NF-H and NF-L (rNF-H and rNF-L, respectively) were then purified from these NFs (Figure 13). Labeling was carried out on assembled NFs in an effort to preserve subunit assembly competence; this was confirmed by a subsequent round of assembly–disassembly for biotinylated subunits (not shown; see also Hisanaga and Hirokawa, 1988).

Previous studies demonstrated that biotinylated NF (bNF) subunits were observed throughout the perikaryon and at minor levels within axonal neurites within 15 min following microinjection (Shea *et al.*, 1995b; Wang and Shea, 1994). Similarly, in the present study, bNF-H and bNF-L was detectable within axonal neurites within 15 min and persisted for at least

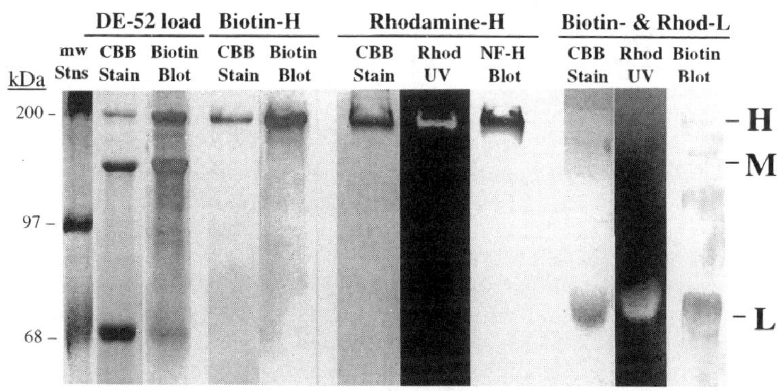

FIGURE 13. Biotin and rhodamine labeling of bovine NF-H. Bovine spinal cord NFs recovered from the void-volume fraction following gel filtration were conjugated to biotin or rhodamine while still assembled as described in Section 2. NF-H and NF-L were then isolated by DE-52 chromatography. Panels present the DE-52 column load (which contains all three NF subunits as indicated on the right of the figure) following biotinylation, and isolated biotinylated NF-H and NF-L (bNF-H and bNF-L, respectively) as visualized by Coomassie staining (CBB stain) and immunoblot analysis with antibiotin antibody (biotin blot). Isolated NF-H is also presented following Coomassie staining, UV illumination at the rhodamine excitation wavelength (Rhod UV), and immunoblot analysis with H3 (NF-H blot). NF-L recovered after chromatography is also presented following Coomassie staining, UV illumination at the rhodamine excitation wavelength (Rhod UV), and immunoblot analysis with antibiotin antibody (biotin blot).

18 hr (Figure 14). We examined the fate of these subunits at intermediate times. Prominent axonal bNF-H and bNF-L immunoreactivity was observed within 2 hr following microinjection (Figure 14). However, axonal immunoreactivity was depleted following Triton-extraction until 4 hr following microinjection (Figure 14), indicating that bNF-H and bNF-L could undergo translocation into axonal neurites substantially prior to incorporation into the Triton-insoluble cytoskeleton. Identical results were observed for rHF-H and rNF-L (not shown).

To examine further the nature of rNF-H within axonal neurites prior to its incorporation into the Triton-insoluble cytoskeleton, cells were extracted with saponin under conditions known to preserve assembled MTs, yet extract unassembled tubulin (e.g., Brown *et al.*, 1992). The rationale for

FIGURE 14. Transport of biotinylated NF-H and NF-L (bNF-H and bNF-L) into axonal neurites of dbcAMP-treated NB2a/d1 cells. Phase-contrast and immunofluorescent images of dbcAMP-treated NB2a/d1 cells following microinjection with bNF-H at the indicated times. Noninjected cells (A) were probed for the distribution of endogenous phosphorylated NFs using SMI-31. Note that endogenous phosphorylated NF-H is present throughout perikarya and neurites. Microinjected cells, located via coinjection of fluorescein-conjugated dextran ("tracer"), were probed for bNF-H and bNF-L using antibiotin antibody with and without prior Triton extraction as indicated (see Section 2). For images at 15 min and 18 hr after microinjection (B), cells are presented under phase-contrast illumination and UV illumination to visualize tracer and biotin (bNF-H). The portion of the axonal neurite of each of these fields annotated by a dotted white line is also presented under UV illumination at higher magnification ("Axon"), with exposure conditions optimized for the axonal neurite. All cells and axon segments for bNF-H and bNF-L were captured, processed, and printed under exposure and magnification settings. At 15 min after injection, note the colocalization of nearly all bNF-H and bNF-L along with coinjected tracer within the perikaryon, and only minor translocation of bNF-H, bNF-L, and tracer along the proximal portion of the axonal neurite. Following 18 hr of incubation, however, perikaryal bNF-H was depleted, and bNF-H was instead localized along the entire axonal neurite length and within the axon hillock; by contrast, at this time, the bulk of the tracer remained within the cell body. Consistent with the distribution of endogenous NF-L (e.g., Shea *et al.*, 1988a), bNF-L was distributed throughout perikarya and axonal neurites at 12 hr. Microinjected cells were examined at intermediate time points as indicated (C). NB2a/d1 cells, and especially their neurites (arrows), are difficult to visualize under phase-contrast optics following Triton extraction prior to fixation; retention of biotin-labeling following 4 hr of incubation facilitates their localization. Following 2 hr of incubation, both bNF-H and bNF-L had undergone translocation along the length of axonal neurites (arrows), but were depleted from neurites, but not perikarya following Triton extraction. By contrast, degrees of bNF-H and bNF-L were retained within the proximal portion of axonal neurites by 4 hr following microinjection. Coinjected fluorescein-conjugated tracer was not retained following extraction; note the lack of fluorescent signal in the representative field of Triton-extracted, microinjected cells (bNF-L-injected cells extracted with Triton 4 hr after microinjection). Image is also presented under fluorescein optics (to visualize tracer; bottom right panel). As described previously (e.g., Miller and Joshi, 1996), during extraction some injected subunits remain adjacent to the cell and produce a fluorescent halo.

this extraction was that such conditions would presumably also extract monomeric NF subunits, but may retain polymerized NF subunits and/or subunits associated with MTs (e.g., Shea and Beermann, 1994; Terada *et al.*, 1996). NB2a/d1 cells extracted under these conditions appear more substantive under phase contrast microscopy than do those extracted with 1% Triton (Figures 14 and 15). As with biotinylated subunits, rNF-H and rNF-

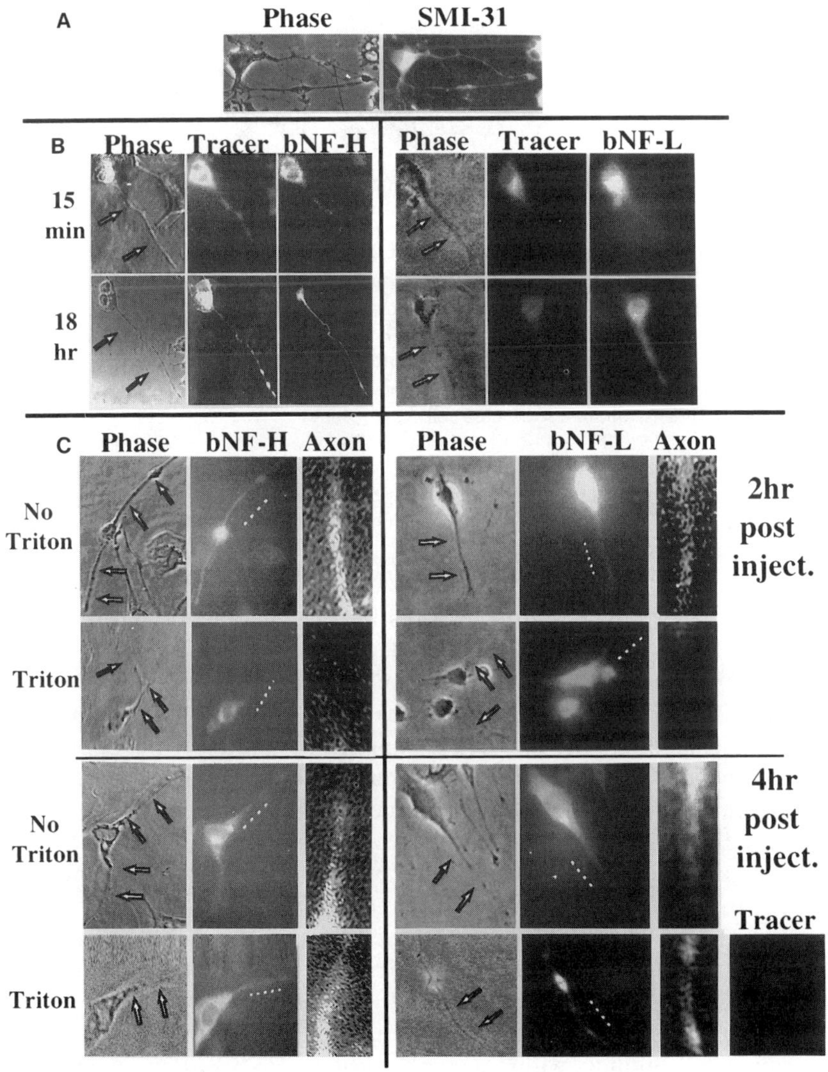

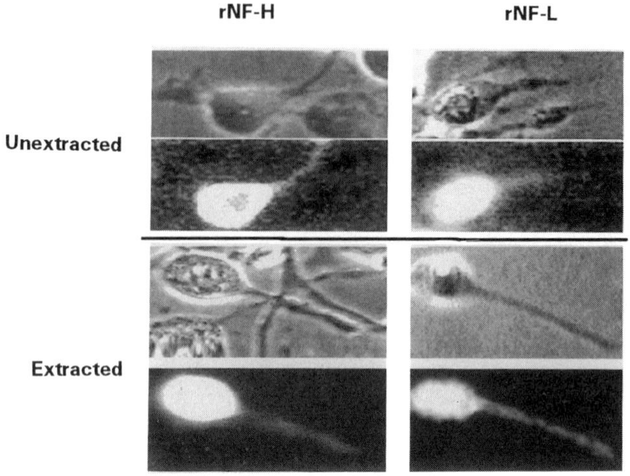

FIGURE 15. Differential extraction of dbcAMP-treated NB2a/d1 cells. Panels present phase-contrast, ultrastructural, and immunocytochemical images of paraformaldehyde-fixed unextracted NB2a/d1 cells and cells extracted before fixation with saponin or Triton as indicated. Cells extracted with saponin in the presence of taxol appear more substantive under phase contrast microscopy and retain significantly more axonal cytoskeletal polymers (both MTs and intermediate-sized filaments) as observed in ultrastructural analyses than do those extracted with 1% Triton. In unextracted cells, SMI-31 immunoreactivity, indicative of endogenous phosphorylated NFs and/or NF subunits, is prominent within perikarya and neurites. Extraction with saponin in the presence of taxol (see Section 2) retains significant perikaryal immunoreactivity and the majority of axonal neuritic immunoreactivity. By contrast, following extraction with 1% Triton, most perikaryal immunoreactivity is lost, while substantial axonal neuritic immunoreactivity is retained.

L were both observed within neurites by 15 min following injection into perikarya. However, in contrast to depletion of microinjected subunits following Triton extraction at this time (above), rNF-H and rNF-L were retained in axonal neurites following extraction with saponin (Figure 15). We next examined whether or not soluble NFs/NF subunit populations were detectable in peripheral nerve. Immunoblot of both sciatic nerve and DRG demonstrated a developmental increase in soluble NF subunits, with a delayed appearance of phospho-NF-H (Figure 16).

We also microinjected bNF-H and bNF-L into cultured DRG neurons. Similar to results obtained in NB2a/d1 cells, bNF-L was readily detectable within axons by 15 min after injection, yet did not survive Triton extraction at 30 min after injection. However, microinjected subunits were retained when DRG neurons were extracted 1 hr after injection (Figure 17). A remaining concern was that the initial appearance of (Triton-soluble) microinjected subunits within axonal neurites of NB2a/d1 cells and DRG neurons was due to diffusion and/or pressure from injection itself.

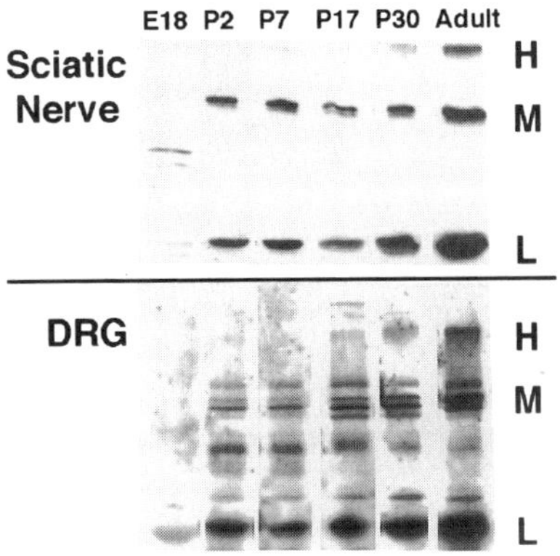

FIGURE 16. Soluble NF subunits are recovered from peripheral nerve tissue. Rat DRG and sciatic nerve at different ages were dissected, homogenized in five volumes of buffer containing 50 mM Tris, pH 7.5, 2 mM EDTA and protease inhibitors (1 mM PMSF, 1% aprotinin, 25 μM leupeptin, and 25 μM pepstatin), and spun at 15,000 × *g* for 10 min. Equal amounts of the resultant supernatant (40 μg) were separated on a 4–10% SDS–PAGE, transferred onto nitrocellulose membrane, and visualized with the R39 antibody.

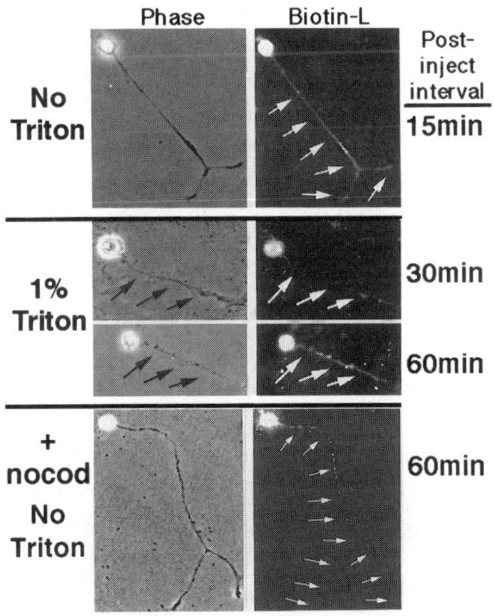

FIGURE 17. Microinjected subunits undergo transport within DRG axons in a Triton-soluble, nocodazole-sensitive form. Panels present phase-contrast and immunofluorescent images of DRG neurons following microinjection with bNF-L at the indicated times. At 15, 30, or 60 min following microinjection, cells were fixed and processed for immunofluorescence with or without prior extraction for 15 min with 1% Triton (see Section 2). Following 15 min of incubation after injection, bNF-L had undergone translocation along the length of axons (arrows), but was depleted from axons (but not perikarya) following Triton extraction for as long as 30 min after injection. By contrast, some bNF-L was retained within axons by 60 min following microinjection. Alternate cultures received 330 nM nocodazole immediately following injection, and were processed for immunofluorescence after 60 min of incubation without extraction. Note that nocodazole substantially diminished the extent of axonal transport of microinjected bNF-L.

However, during ongoing studies of axonal transport motor systems, we observed that treatment with the MT-depolymerizing drug nocodazole (e.g., Baas *et al.*, 1993) immediately following microinjection markedly reduced the transport of microinjected NF subunits into DRG axons

(Figure 17). Prevention of appearance of microinjected subunits within neurites following application of a drug demonstrates that diffusion or pressure from injection itself cannot be responsible for the level or extent of axonal transport of microinjected subunits demonstrated in these cell systems.

The above results in NB2a/d1 cells and DRG neurons demonstrate that some NF subunits undergo axonal transport in Triton-soluble form.

4. DISCUSSION

4.1. Extensive Phosphorylation of NF-H prior to Incorporation into Triton-Soluble Structures

Previous studies of NF proteins during development focused on large myelinated axons enriched in NFs (e.g., Pachter and Liem, 1984; Carden *et al.*, 1985), which of necessity emphasizes the properties of the axonal cytoskeleton, which is the most stable portion of the neuronal cytoskeleton. Our observation of unique soluble forms of NF-H within total brain may reflect the presence of perikarya. While soluble H-200 reacted strongly with multiple phospho-dependent polyclonal antibodies, it reacted poorly or not at all with monoclonal antibodies directed against developmentally delayed phosphoepitopes (e.g., those recognized by RT97 and SMI-34; Anderton *et al.*, 1982; Shea *et al.*, 1989), suggesting that certain phosphorylation events may be restricted to cytoskeleton-associated NF-H. The confinement of abundant Triton-soluble H-200 immunoreactivity to the first few months of postnatal development, coupled with minimal to no cross-reactivity with commonly utilized phosphate-dependent monoclonal antibodies (e.g., SMI-31, SMI-34, and RT97), represent likely reasons why this Triton-soluble subunit pool has not been well appreciated.

Our finding that NF-H solubility behavior differs from that of NF-M and NF-L leaves open the possibility that the initial interaction of NF-H with NFs may be more dynamic than previously recognized. That NF-H differs in dynamics from NF-M and NF-L is consistent with previous studies demonstrating (1) delayed gene expression of NF-H during development (e.g., Shaw and Weber, 1982), (2) variable stoichiometry of NF-H relative to the other two subunits (e.g., Hirokawa *et al.*, 1984; Scott *et al.*, 1985), and (3) the considerably higher rate of association of NF-H than that of NF-L with NFs in DRG cell axons (Takeda *et al.*, 1994).

NFs composed of only NF-L and NF-M are observed during initial neuritogenesis in rat spinal cord (Carden *et al.*, 1985). These doublet NFs gradually decline during embryonic development as triplet NFs increase.

Because NF-H expression coincides temporally with the stabilization of neuronal circuitries, it was suggested that the presence of NF-H may be required to mediate the slowing of axonal cytoskeletal transport and growth of axonal caliber (Carden *et al.*, 1985). Although the developmental conversion of doublet NFs to triplet NFs is fairly generalized (Trojanowski *et al.*, 1986), it remains an open question whether or not NF-H subunits are recruited into preexisting doublet NFs. However, this possibility is consistent with the findings of the present study.

Metabolically radiolabeled H-200 in the retina was detected 10 hr earlier in the high-speed Triton-soluble fraction than in corresponding Triton-insoluble cytoskeletons, indicating that, as previously demonstrated in NB2a/d1 cells (Shea *et al.*, 1990), a certain part of soluble NF-H subunits in mouse central nervous system undergoes extensive phosphorylation. Radiolabeled Triton-soluble H-200 disappeared concomitantly with the appearance of radiolabeled H-200 in Triton-insoluble cytoskeletons. In addition, however, newly synthesized NF-H subunits within the Triton-insoluble cytoskeleton, which displayed an apparent molecular mass of 160 kDa, were progressively converted to H-185 and H-200 isoforms. Accordingly, whether Triton-soluble H-200 incorporates into cytoskeletons or is instead depleted by proteolysis remains unclear. Incorporation of microinjected 200-kDa NF-H subunits into the cytoskeleton of NB2a/d1 cells and cultured DRG neurons here and in previous studies (Shea *et al.*, 1995a, b; Takeda *et al.*, 1994; Wang and Shea, 1994) is consistent with the former possibility. By contrast, the selective degradation of perikaryal NFs (e.g., Shea, 1995a, b) is consistent with the latter possibility.

Previous studies in several systems indicate that NF-H phosphorylation initiates within perikarya (Black and Lee, 1988; Koliatsos *et al.*, 1989; Shea *et al.*, 1990), and that the extent of perikaryal NF-H phosphorylation is a function of its residence time within perikarya (Black and Lee, 1988). Furthermore, phosphorylated NF epitopes accumulate within perikarya following treatments that disrupt axonal transport, such as axotomy (Drager and Hofbauer, 1984; Cork *et al.*, 1986; Goldstein *et al.*, 1987b) or exposure to certain neurotoxins (Hoffman *et al.*, 1978; Bizzi *et al.*, 1984; Parhad *et al.*, 1986; Gold *et al.*, 1988; Lapadula *et al.*, 1988). One interpretation of these findings is that the extensive phosphorylation and assembly of NF-H subunits are largely independent mechanisms and that, in adult neurons under normal conditions, most subunits undergo assembly and transport at faster rates than the rate of phosphorylation (Shea *et al.*, 1990). This line of reasoning is consistent with the observation of extensively phosphorylated NF-H within perikarya only following substantially increased levels accompanying trauma (above) or at particular stages in development.

4.2. Axonal Transport of Triton-Soluble NF Subunits

Our findings demonstrate that Triton-soluble NF subunits undergo axonal transport. Analyses of high-speed supernatant and pellets derived from Triton-soluble material was key to this demonstration. By contrast, previous studies have either discarded the Triton-soluble fraction or instead analyzed total homogenates (since immunoblot analyses demonstrate very little total protein), or derived a "Triton-insoluble cytoskeleton" by subjecting initial homogenates to a high-speed centrifugation (which effectively combines the P1 and P2 fractions detailed here).

As previously demonstrated in neuroblastoma cells (Shea *et al.*, 1990), triplet NFs were recovered from the soluble polymer fraction and migrated through this fraction in pulse-chase radiolabeling analyses, indicating that assembly of NF subunits is distinct from their incorporation into the Triton-insoluble cytoskeleton. Furthermore, recovery of radiolabeled NF subunits from perikaryal S1, but not the neuritic S1, yet from both perikaryal and axonal neuritic P2 fractions, suggests that at least some NF subunits undergo axonal transport in the form of assembled NFs. This possibility is further supported by (1) the recovery of NFs from axonal neuritic P2 fraction and (2) the observation of NF subunits within this fraction from undifferentiated cells, which lack axonal neurites (e.g., Shea *et al.*, 1988a) and could therefore only form NFs within their perikarya or at the base of putative, filopodia-like neurites (Shea *et al.*, 1990). While the analyses of radiolabeled subunits were more complex in the retinal ganglion cell system, our data obtained in this system also demonstrated the presence of unique Triton-soluble populations of NFs and NF subunits within axons *in situ*. Analyses of the fate of biotinylated NF subunits following microinjection into NB2a/d1 cells and DRG neurons supports this conclusion: both NF-H and NF-L were observed to translocate into axons significantly before these subunits survived extraction with Triton, clearly indicating that axonal transport of NF subunits is not necessarily equivalent to incorporation into the Triton-insoluble cytoskeleton.

The data of this study demonstrate that Triton-soluble NF subunits include precursors for the axonal cytoskeleton. While our radiolabeling studies in NB2a/d1 cells and retinal ganglion neurons strongly suggest that at least some subunits undergo entry into axons in the form of NFs, our data are also fully consistent with the additional possibilities that (1) unassembled subunits may enter axonal neurites and undergo initial assembly within axons, and/or (2) newly synthesized subunits undergo continued exchange among Triton-soluble and/or Triton-insoluble NFs. In this regard, recent studies have indicated that isolated NF-M subunits may translocate into axons along with MTs (Terada *et al.*, 1996). Accordingly, although biotinylated subunits were observed within NFs in axonal neurites in immuno-EM, some of the biotin signal observed within axonal neurites by immunofluorescent analyses may therefore be derived from the independent translocation of individual subunits, perhaps in association with translocating MTs (e.g., Terada *et al.*, 1996). Similarly, in our radiolabeling analyses, it is possible that some NF subunits sedimenting during high-speed centrifugation did so via cosedimentation with Triton-stable, cold-stable MTs. This possibility is substantiated by detection in immunoblot analysis of fodrin, tubulin, MAP1, and MAP2 in P2 fractions and void volume fractions from gel filtration (not shown; to be published). Our demonstration of NFs within these fractions does not exclude the possibility of additional cytoskeletal precursor structures such as the 25-nm globular structures "scaps" (slow-component associated particles) described by Weisenberg and colleagues, which are apparently comprised of the above cytoskeletal proteins and NFs (Weisenberg *et al.*, 1987). In this regard, 20- to 25-nm globular structures were occasionally observed along with NFs within the void volume following gel filtration chromatography; analyses of these structures and their potential role as cytoskeletal precursors will be the focus of subsequent studies. In addition, the distribution of NF subunit heterotetramers, shown to represent soluble assembly intermediates for NFs (Cohlberg *et al.*, 1995), has not been investigated. However, it is reasonable to consider that any nonfilamentous oligomeric structures lie upstream of the Triton-soluble NFs shown here to represent cytoskeletal precursors.

During chase analyses, significant radiolabeled NF-H migrating at approximately 180 kDa incorporated into, and was subsequently depleted from, the perikaryal Triton-insoluble cytoskeleton of NB2a/d1 cells. This finding is consistent with the routine detection of hypo- and partially phosphorylated NF-H within the perikaryal cytoskeleton (Shea *et al.*, 1988a, b, 1990). Moreover, our previous analyses have demonstrated that selective proteolysis of perikaryal subunits represents a major mechanism for their depletion (Shea, 1995a, b). The possibility should not be overlooked, however, that some of perikaryal cytoskeleton-associated NF-H is subsequently transported into axonal neurites. Accordingly, while our data clearly demonstrate that some subunits translocate into axonal neurites prior to incorporation into Triton-insoluble structures, an additional population of subunits may incorporate into Triton-insoluble structures within perikarya and subsequently translocate into axonal neurites in a Triton-insoluble form (e.g., see Lasek *et al.*, 1992 and references therein).

The extensively cross-linked NF network is thought to confer structural support for the establishment and maintenance of the axonal cytoskeleton (for reviews, see Brady, 1993; Matus, 1987; Nixon, 1993; Shea and Beermann, 1994). However, such a function does not exclude the likelihood that some regional repair/replacement-oriented exchange of NF subunits and/or polymers does occur, and, furthermore, that any such dynamics is likely to be even more pronounced among Triton-soluble NF polymers that represent precursors for the NF cytoskeleton. Our findings in the present and previous studies indicate that the phosphorylation, assembly, axonal transport, and incorporation of NFs into Triton-insoluble structures represent distinct events. Some of these conclusions were also derived following stretching of nerve fibers emanating from cat dorsal root ganglia, where radiolabeled NFs were transported as soluble components that subsequently deposited into the cytoskeleton (Ochs *et al.*, 1989), and by photobleaching analyses, which demonstrated apparent regional exchange of NF-H and NF-L with nontranslocating NF polymers (Okabe *et al.*, 1993; Takeda *et al.*, 1994). Notably, our observation of differential phosphorylation of Triton-insoluble and -soluble axonal NF-H as revealed by differential RT97 immunoreactivity is consistent with the notion that key phosphorylation events may regulate the integrity of association of NFs with the cytoskeletal matrix (e.g., Nixon and Lewis, 1987; Tsuda *et al.*, 1997). The findings of the present study also provide support for the suggestion (Nixon and Lewis, 1986, 1987) that some NF subunits enter and move along axons as NFs; this earlier conclusion was based on the observation that newly synthesized NFs enter optic axons in the same ratio as found in resident axonal NF subunits.

4.3. Concluding Remarks

The present findings underscore that posttranslational modification, assembly, and axonal transport of NF subunits represent distinct, nonobligate, experimentally separable events. They further indicate that multiple populations of axonal NF subunits exist within perikarya and axons *in situ* and in culture, and moreover suggest that axonal transport of Triton-soluble NFs/NF subunits contributes to the development and maintenance of the axonal cytoskeleton. Remaining to be determined is whether or not NF subunits translocate into axons exclusively in the form of NFs, or whether soluble subunits undergo translocation and subsequent polymerization.

5. SUMMARY

The low abundance of soluble neurofilament (NF) subunits in mature axons has suggested that newly synthesized NF proteins rapidly assemble into highly stable polymers and associate with the Triton X-100-insoluble cytoskeleton. Here we present evidence for multiple populations of NFs and NF subunits, distinguished by differential solubility in Triton, within perikarya and axons of neurons *in situ* and in culture. We further demonstrate, using microinjection of "tagged" NF subunits and by pulse-chase radiolabeling of endogenous NF subunits, that these soluble NF populations represent precursors for incorporation into the axonal cytoskeleton.

6. REFERENCES

Anderton, B. H., Breinburg, D., Downes, M. J., Gree, P. J., Tomilson, B. E., Ulrich, J., Wood, J. N., and Kahn, J., 1982, Monoclonal antibodies show that neurofibrillary tangles and neurofilaments share antigenic determinants, *Nature* **298:**84–86.

Angelides, K. J., Smith, T. E., and Takeda, M., 1989, Assembly and exchange of intermediate filament proteins of neurons: NFs are dynamic structures, *J. Cell. Biol.* **108:**1495–1506.

Baas, P. W., Ahmad, F. J., Pienlowski, T. P., Brown, A., and Black, M. M., 1993, Sites of microtubule stabilization for the axon, *J. Neurosci.* **13:**2177–2185.

Bizzi, A., Crane, R. C., Autilio-Gambetti, L., and Gambetti, P., 1984, Aluminum effect on slow axonal transport, *J. Neurosci.* **4:**722–731.

Black, M. M., and Lee, M. M.-Y., 1988, Phosphorylation of neurofilament proteins in intact neurons: Demonstration of phosphorylation in cell bodies and axons, *J. Neurosci.* **8:**3296–3305.

Brady, S. T., 1993, Motor neurons and neurofilaments in sickness and health, *Cell* **73:**1–3.

Bray, J. J., and Mills, R. G., 1991, Transport complexes associated with slow axonal flow, *Neurochem. Res.* **16:**645–649.

Brown, A., Slaugther, T., and Black, M. M., 1992, Newly-assembled microtubules are concentrated in the proximal and distal regions of growing axons, *J. Cell Biol.* **119:**867–882.

Burrell, K. A., Enayate, Z. E., Powell, J. F., Leigh, P. N., and Gallo, J.-M., 1996, Overexpression of wildtype and mutant Cu/Zn superoxide dismutase in a differentiated mouse neuroblastoma cell line, *Mol. Biol. Cell* **7:**220a.

Carden, M. J., Schlaepfer, W. W., and Lee, V. M., 1985, Two stage expression of neurofilament polypeptides during rat neurogenesis with early establishment of adult phosphorylation patterns, *J. Neurosci.* **7:**3489–3504.

Carter, J. E., Gallo, J.-M., Anderson, V. E. R., Anderton, B. H., and Robertson, J., 1996, Aggregation of neurofilaments in NF-L transfected neuronal cells: Regeneration of the filamentous network by a protein kinase C inhibitor, *J. Neurochem.* **67:**1997–2004.

Chiu, F.-C., and Norton, W. T., 1982, Bulk preparations of mammalian CNS cytoskeletons, *J. Neurochem.* **39:**1252–1260.

Cohlberg, J. A., Hajarian, H., Tran, T., Alipourjeddi, A., and Noveen, A., 1995, Neurofilament protein heterotetramers as assembly intermediates, *J. Biol. Chem.* **270:**9334–9339.

Cork, L. C., Sternberger, N. H., Sternberger, L. A., Casanova, M. F., Struble, R. G., and Price, D. L., 1986, Phosphorylated neurofilament antigens in neurofibrillary tangles in Alzheimer's disease, *J. Neuropathol. Exp. Neurol.* **45:**56–64.

Crandall, J. E., and Fischer, I., 1989, Developmental expression of microtubule-associated protein 2 in expression in regions of mouse brain, *J. Neurochem.* **53:**1910–1917.

Cressman, C. M., and Shea, T., 1995, Hyperphosphorylation of tau and filopodial retraction following microinjection of protein kinase C catalytic subunits, *J. Neurosci. Res.* **42:**648–656.

Delacourte, A., Filleatreau, G., Boutteau, F., Biserte, G., and Schrevel, J., 1980, Study of the 10nm-filament fraction isolated during the standard microtubule preparation, *Biochem. J.* **191:**543–546.

Drager, U. C., and Hofbauer, A., 1984, Antibodies to heavy neurofilament subunit detect a subpopulation of damaged ganglion cells in retina, *Nature* **309:**624–626.

Fischer, I., and Shea, T. B., 1991, Differential appearance of extensively phosphorylated forms of the high molecular weight neurofilament protein in regions of mouse brain during post-natal development, *J. Neuroimmunol.* **31:**73–81.

Fischer, I., Shea, T. B., Sapirstein, V. S., and Kosik, K. S., 1986, Expression and distribution of microtubule-associated protein 2 (MAP2) in neuroblastoma and primary neuronal cells, *Dev. Brain Res.* **25:**99–109.

Georges, E., Lefebvre, S., and Mushynski, W. E., 1986, Dephosphorylation of neurofilaments by exogenous phosphatases has no effect on reassembly of subunits, *J. Neurochem.* **47:**477–483.

Glass, J. D., and Griffin, J. W., 1994, Retrograde transport of radiolabeled cytoskeletal proteins in transected nerves, *J. Neurosci.* **14:**3915–3921.

Gold, B. G., Price, D. L., Griffin, J. W., Rosenfield, J., Hoffman, P. N., Sternberger, N. H., and Sternberger, L. A., 1988, Neurofilament antigens in acrylamide neuropathy, *J. Neuropathol. Exp. Neurol.* **47:**145–157.

Goldstein, M. E., Sternberger, L. A., and Sternberger, N. H., 1987a, Varying degrees of phosphorylation determine microheterogeneity of the heavy neurofilament polypeptide (NF-H), *J. Neuroimmunol.* **14:**135–148.

Goldstein, M. E., Cooper, H. S., Bruce, J., Carden, M. J., Lee, V. M.-Y., and Schlaepfer, W. W., 1987b, Phosphorylation of neurofilament proteins and chromatolysis following transection of rat sciatic nerve, *J. Neurosci.* **7:**1586–1594.

Hirokawa, N., 1993, Axonal transport and the cytoskeleton, *Curr. Opin. Neurobiol.* **3:**724–731.

Hirokawa, N., Glicksman, M. A., and Willard, M. B., 1984, Organization of mammalian neurofilament polypeptides within the neuronal cytoskeleton, *J. Cell Biol.* **98:**1523–1536.

Hisanaga, S., and Hirowawa, N., 1988, Structure of the peripheral domains of neurofilaments revealed by low angle rotary shadowing, *J. Mol. Biol.* **202:**297–305.

Hoffman, P. N., and Lasek, R. J., 1975, The slow component of axonal transport. Identification of major structural polypeptides of the axon and their generality among mammalian axons, *J. Cell Biol.* **66:**351–366.

Hoffman, P. N., Clark, A., Carroll, P. T., and Price, D. L., 1978, Slow axonal transport of neurofilament proteins: Impairment by B,B′-iminopropionitrile administration, *Science* **202:**634–635.

Julien, J.-P., and Mushynski, W. E., 1982, Multiple phosphorylation sites in mammalian neurofilament polypeptides, *J. Biol. Chem.* **257:**10467–10470.

Koliatsos, V. E., Applegate, M. D., Kitt, C. A., Walker, L. C., DeLong, M. R., and Price, D. L., 1989, Aberrant phosphorylation of neurofilaments accompanies transmitter-related changes in rat septal neurons following transection of the fimbria-fornix, *Brain Res.* **482:**205–218.

Langley, O. K., Sternberger, N. H., and Sternberger, L. A., 1988, Expression of neurofilament proteins by Purkinje cells: Ultrastructural immunolocalization with monoclonal antibodies, *Brain Res.* **457:**12–20.

Lapadula, D. M., Suwita, E., and Abou-Donia, M. B., 1988, Evidence for multiple mechanisms responsible for 2,5-hexanedione-induced neuropathy, *Brain Res.* **458:**123–131.

Lasek, R. J., Paggi, P., and Katz, M. J., 1992, The maximum rate of neurofilament transport in axons: A view of molecular transport mechanisms continuously engaged, *Brain Res.* **616:**58–64.

Lasek, R. J., Paggi, P., and Katz, M. J., 1994, Slow axonal transport mechanisms move neurofilaments relentlessly in mouse optic axons, *J. Cell Biol.* **117:**607–616.

Lewis, S. E., and Nixon, R. A., 1988, Multiple phosphorylated variants of the high molecular weight NF subunit of NFs in axons of retinal ganglion cell neurons: Characterization and evidence for their differential association with stationary and moving NFs, *J. Cell Biol.* **107:**2689–2701.

Matus, A., 1987, Putting together the neuronal cytoskeleton, *Trends Neurosci.* **10:**186–188.

Miller, K. E., and Joshi, H. C., 1996, Tubulin transport in neurons, *J. Cell Biol.* **183:**1355–1366.

Nakamura, Y., Takeda, M., Angelides, K. J., Tanaka, T., Tada, K., and Nishimura, T., 1990, Effect of phosphorylation on 68 kDa neurofilament subunit protein assembly by the cyclic AMP-dependent protein kinase *in vitro*, *Biochem. Biophys. Res. Commun.* **169:**744–750.

Nixon, R. A., 1993, The regulation of neurofilament protein dynamics by phosphorylation: Clues to neurofibrillary pathology? *Brain Pathol.* **3:**29–38.

Nixon, R. A., and Lewis, S. E., 1986, Differential turnover of phosphate groups on neurofilament subunits in mammalian neurons *in vivo*, *J. Biol. Chem.* **261:**16298–16301.

Nixon, R. A., and Lewis, S. E., 1987, Phosphorylation and dephosphorylation of neurofilament proteins in retinal ganglion cell neurons *in vivo*, in: *Molecular Mechanisms of Neuronal Responsiveness* (Y. H. Ehrlich, R. H. Lenox, E. Kornecki, and W. O. Berry, eds.), pp. 167–186, Plenum Press, New York.

Nixon, R. A., and Logvinenko, K. B., 1986, Multiple fates of newly-synthesized neurofilament proteins: Evidence for a stationary neurofilament network distributed nonuniformly along axons of retinal ganglion cell neurons, *J. Cell Biol.* **102:**647–659.

Nixon, R. A., and Sihag, R. K., 1991, Neurofilament phosphorylation: A new look at regulation and function, *Trends Neurosci.* **14:**501–506.

Nixon, R. A., Lewis, S. E., and Marotta, C. A., 1987, Posttranslational modification of neurofilament proteins by phosphate during axoplasmic transport in retinal ganglion cell neurons, *J. Neurosci.* **7:**1145–1158.

Nixon, R. A., Lewis, S. E., Dahl, D., Marotta, C. A., and Drager, U. C., 1988, Early postranslational modifications of the three neurofilament subunits in mouse retinal ganglion cells: Neuronal sites and time course in relation to subunit polymerizaion and axonal transport, *Mol. Brain Res.* **5:**93–108.

Ochs, S., Jersild, J. K., and Li, J.-M., 1989, Slow transport of freely movable cytoskeletal components shown by beading partition of nerve fibers in the cat, *Neuroscience* **33:**421–430.

Okabe, S., and Hirokawa, H., 1989, Incorporation and turnover of biotin-labeled actin microinjected into fibroblastic cells: An immunoelectron microscopic study, *J. Cell Biol.* **109:**1581–1595.

Okabe, S., Miyasaka, H., and Hirokawa, N., 1993, Dynamics of the neuronal intermediate filaments, *J. Cell Biol.* **121:**375–386.

Pachter, J. S., and Liem, R. K. H., 1984, The differential appearance of neurofilament triplet polypeptides in the developing rat optic nerve, *Dev. Biol.* **103:**200–210.

Parhad, I. M., Griffin, J. W., Hoffman, P. N., and Koves, J. F., 1986, Selective interruption of axonal transport of neurofilament proteins in the visual system by B′,B′-iminodipropionitrile (IDPN) intoxication, *Brain Res.* **363:**315–324.

Scott, D., Smith, K. E., O'Brien, B., and Angelides, K. J., 1985, Characterization of mammalian neurofilament triplet proteins: Subunit stochiometry and morphology of native and reconstituted filaments, *J. Biol. Chem.* **260:**10736–10797.

Shaw, G., and Weber, K., 1982, Differential expression of neurofilament triplet proteins in brain development, *Nature* **298:**276–299.

Shea, T. B., 1994, Triton-soluble phosphovariants of the high molecular weight neurofilament subunit from NB2a/d1 cells are assembly-competent: Implications for normal and abnormal neurofilament assembly, *FEBS Lett.* **343:**131–136.

Shea, T. B., 1995a, Differential synthesis and cytoskeletal deposition of neurofilament subunits before and during axonal outgrowth in NB2a/d1 cells: Evidence that segregation of phosphorylated subunits within the axonal cytoskeleton involves selective deposition, *J. Neurosci. Res.* **40:**225–232.

Shea, T. B., 1995b, Leupeptin enhances phosphorylated neurofilament immunoreactivity in Triton-insoluble cytoskeletons of mouse NB2a/d1 cells, *Neurosci. Res. Commun.* **15:**143–148.

Shea, T. B., and Beermann M. L., 1994, Sequential requirements for NFs, MTs, and the MT-associated proteins MAP1B and tau in axonal neurite initiation, elongation and stabilization, *Mol. Cell Biol.* **5:**863–875.

Shea, T. B., Sihag, R. K., and Nixon, R. A., 1988a, Neurofilament triplet proteins of NB2a/d1 neuroblastoma: Posttranslational modification and incorporation into the cytoskeleton during differentiation, *Dev. Brain Res.* **43:**97–109.

Shea, T. B., Majocha, R. E., Marotta, C. A., and Nixon, R. A., 1988b, Soluble, phosphorylated forms of the high molecular weight neurofilament protein in perikarya of cultured neuronal cells, *Neurosci. Lett.* **97:**291–297.

Shea, T. B., Beermann, M. L., and Nixon, R. A., 1989, Appearance and localization of phosphorylated variants of the high molecular weight neurofilament protein in NB2a/d1 cytoskeletons during differentiation, *Dev. Brain Res.* **50:**142–146.

Shea, T. B., Sihag, R. K., and Nixon, R. A., 1990, Dynamics of phosphorylation and assembly of the high molecular weight NF subunit in NB2a/d1 neuroblastoma, *J. Neurochem.* **55:**1784–1792.

Shea, T. B., Fischer, I., Paskevich, P. A., and Beermann, M. L., 1993, The protein phosphatase inhibitor okadaic acid increases axonal NFs and neurite caliber, and decreases axonal MTs in NB2a/d1 cells, *J. Neurosci. Res.* **35:**507–521.

Shea, T. B., Beermann, M. L., and Nixon, R. A., 1995a, Aluminum treatment of intact neuroblastoma cells alters neurofilament subunit phosphorylation, solubility and proteolysis, *Mol. Chem. Neuropathol.* **26:**1–11.

Shea, T. B., Wheeler, E., and Wang, F.-S., 1995b, Neurofilament translocation into axonal neurites is distinct from incorporation into the cytoskeleton, *Mol. Biol. Cell* **6:**375a.

Shea, T. B., Dahl, D., Nixon, R. A., and Fischer, I., 1997, Triton-soluble phosphovariants of the heavy neurofilament subunit in developing and mature mouse central nervous system, *J. Neurosci. Res.* **48:**515–523.

Shelanski, M. L., Gaskin, F., and Cantor, C. R., 1973, Assembly of microtubules in the absence of added nucleotide, *Proc. Natl. Acad. Sci. USA* **70:**765–768.

Sihag, R. K., and Nixon, R. A., 1989, *In vivo* phosphorylation of distinct domains of the 70-kDa neurofilament subunit involves different protein kinases, *J. Biol. Chem.* **264:**457–464.

Sihag, R. K., and Nixon, R. A., 1990, Phosphorylation of the amino terminal head domain of the middle molecular mass 145 kDa subunit of neurofilaments: Evidence for regulation by second-messenger dependent protein kinases, *J. Biol. Chem.* **265:**4166–4171.

Sternberger, L. A., and Sternberger, N. H., 1983, Monoclonal antibodies distinguish phosphorylated and nonphosphorylated forms of neurofilaments *in situ*, *Proc. Natl. Acad. Sci. USA* **80:**6126–6130.

Straube-West, K., Loomis, P. A., Opal, P., and Goldman, R. D., 1996, Alterations in neuronal intermediate filament organization: Functional implications and the induction of pathological changes related to motor neuron disease, *J. Cell Sci.* **109:**2319–2329.

Takeda, S., Okabe, S., Funakoshi, T., and Hirokawa, N., 1994, Differential dynamics of neurofilament-H protein and neurofilament-L protein in neurons, *J. Cell Biol.* **127:**173–185.

Terada, S., Nakata, T., Peterson, A. C., and Hirokawa, N., 1996, Visualization of slow axonal transport *in vivo*, *Science* **273:**784–788.

Trojanowski, J. Q., Walkenstein, N., and Lee, V. M.-Y., 1986, Expression of neurofilament subunits in neurons of the central and peripheral nervous system: An immunohistochemical study with monoclonal antibodies, *J. Neurosci.* **6:**650–660.

Tsuda, M., Tashiro, T., and Komiya, Y., 1997, Increased solubility of high-molecular-mass neurofilament subunit by suppression of dephosphorylation: Its relation to axonal transport, *J. Neurochem.* **68:**2558–2565.

Wang, F. S., and Shea, T. B., 1994, Transport and incorporation of biotinylated high mw NF protein into the axonal cytoskeleton in NB2a/d1 cells, *Mol. Biol. Cell* **5:**550a.

Watson, D. W., Glass, J. D., and Griffin, J. W., 1993, Redistribution of cytoskeletal proteins in mammalian axons disconnected from their cell bodies, *J. Neurosci.* **13:**4354–4360.

Weisenberg, R. C., Flynn, J., Gao, B., Awodi, S., Skee, F., Goodman, S. R., and Riederer, B. M., 1987, Microtubule gelation-contraction: Essential components and relation to slow axonal transport, *Cell Motil. Cytoskel.* **10:**331–340.

Chapter 19

Lamin Assembly *in Vivo*

Georg Krohne

1. THE NUCLEAR LAMINA: STRUCTURE AND BIOCHEMICAL PROPERTIES

This review focuses on the structure of the nuclear lamina and the comparison of the assembly of lamins *in vivo* and *in vitro*. Other aspects of lamin biology have been discussed in detail in recent reviews (Hutchison *et al.*, 1994; Lourim and Krohne, 1994; Moir *et al.*, 1995) and by Ye *et al.* in Chapter 20 of this book.

The nuclear envelope of eukaryotic cells is a membrane system composed of three separate membrane domains: (i) The outer nuclear membrane is in continuity with the endoplasmic reticulum; (ii) the membraneous wall of the nuclear pore complexes is located where the outer and the inner nuclear membranes fuse to form the pore channel; (iii) the inner nuclear membrane is associated between the pores with a meshwork of proteins, the nuclear lamina. Integral membrane proteins of the inner nuclear membrane interact with the nuclear lamina (for review see Chapter 20 by Ye *et al.* in this book). The nuclear lamina can be seen in certain cells of invertebrates and vertebrates as an electron-dense layer of more than 50 nm thickness interposed between the inner nuclear membrane

Georg Krohne Division of Electron Microscopy, Biocenter of the University of Würzburg, D-97074 Würzburg, Germany.

Subcellular Biochemistry, Volume 31: Intermediate Filaments, edited by Herrmann and Harris. Plenum Press, New York, 1998.

and the peripheral chromatin (Figure 1A; Höger *et al.*, 1991a). This thick nuclear lamina has been described as "fibrous lamina" (Fawcett, 1966; Ghadially *et al.*, 1972), "zonula nucleum limitans" (Mazanec, 1967; Patrizi and Poger, 1967), or "dense lamina" (Kalifat *et al.*, 1967) [for a review on the thick nuclear lamina of unicellular organism see Schmidt *et al.* (1995)]. However, in the majority of cells the nuclear lamina is so thin that it is only visible as a distinct structure at the level of the inner nuclear membrane when lipids have been removed by extraction with nonionic detergents (Figures 1B and 1C). Alternatively, the thin nuclear lamina can be visualized in nonextracted cells and tissues when a special fixation and staining protocol is applied that enhances the contrast of the lamina (Stick and Schwarz, 1982).

Whole-mount preparation of detergent-extracted and nonextracted nuclear envelopes has revealed that the lamina is a loose meshwork of protein fibrils interconnecting nuclear pores (Scheer *et al.*, 1976; Dwyer and Blobel, 1976; Stewart and Whytock, 1988; Hutchison *et al.*, 1994). In well-preserved areas of amphibian oocyte nuclear envelopes, lamina filaments form an orthogonal network with a crossover spacing of 52-nm (Aebi *et al.*, 1986).

The association of the nuclear lamina with the nucleoplasmic side of pore complexes is stabilized by hydrophobic interactions. This is why the nuclear lamina and pore complexes of invertebrates and vertebrates are always coisolated as a "pore complex–lamina fraction" when nuclear envelopes are sequentially extracted with nonionic detergents and buffers

FIGURE 1. Electron micrographs of ultrathin sections showing details of the thick (A) and the thin (B–D) nuclear lamina. (A) Human synovial cell with a 60- to 70-nm-thick nuclear lamina (marked by white brackets). Its fibrillar texture can be directly compared with the size and ultrastructure of cytoplasmic intermediate filaments (small arrows) and microtubules (arrowheads). The lamina is in direct contact with the inner nuclear membrane and the peripheral chromatin (ch). It is interrupted in areas of the nuclear envelope containing pore complexes (one pore is marked by a large arrow). A- and B-type lamins have been localized in the entire thick lamina of this cell type (Höger *et al.*, 1991a). (B) The nuclear envelope of a *Xenopus* oocyte that was manually isolated within 1 min (Krohne and Franke, 1983) and immediately fixed for electron microscopy. No nuclear lamina is visible. (C) Manually isolated *Xenopus* oocyte nuclear envelope after extraction with Triton X-100. The lamina is now visible at the level of the inner nuclear membrane as a proteinaecous layer of approximately 6 nm thickness that is in direct contact with the nucleoplasmic side of pore complexes. (D) Electron microscopic immunolocalization of lamins in the nuclear envelope of *Xenopus* oocytes. Lamin antibodies bound to the nuclear envelope were visualized with secondary antibodies coupled to 5-nm collodial gold (antibodies against the *Xenopus* B-type lamin LII; Lourim and Krohne, 1993). Gold particles are exclusively localized between pore complexes at the inner nuclear membrane. cy, Cytoplasmic side of the nuclear envelope; ch, chromatin. Bars, 100 nm.

containing high-salt solutions (Aaronson and Blobel, 1975; Dwyer and Blobel, 1976; P. A. Fisher *et al.*, 1982; Krohne *et al.*, 1978a; Maul *et al.*, 1984; for review see Krohne and Benavente, 1986; Krohne and Franke, 1983).

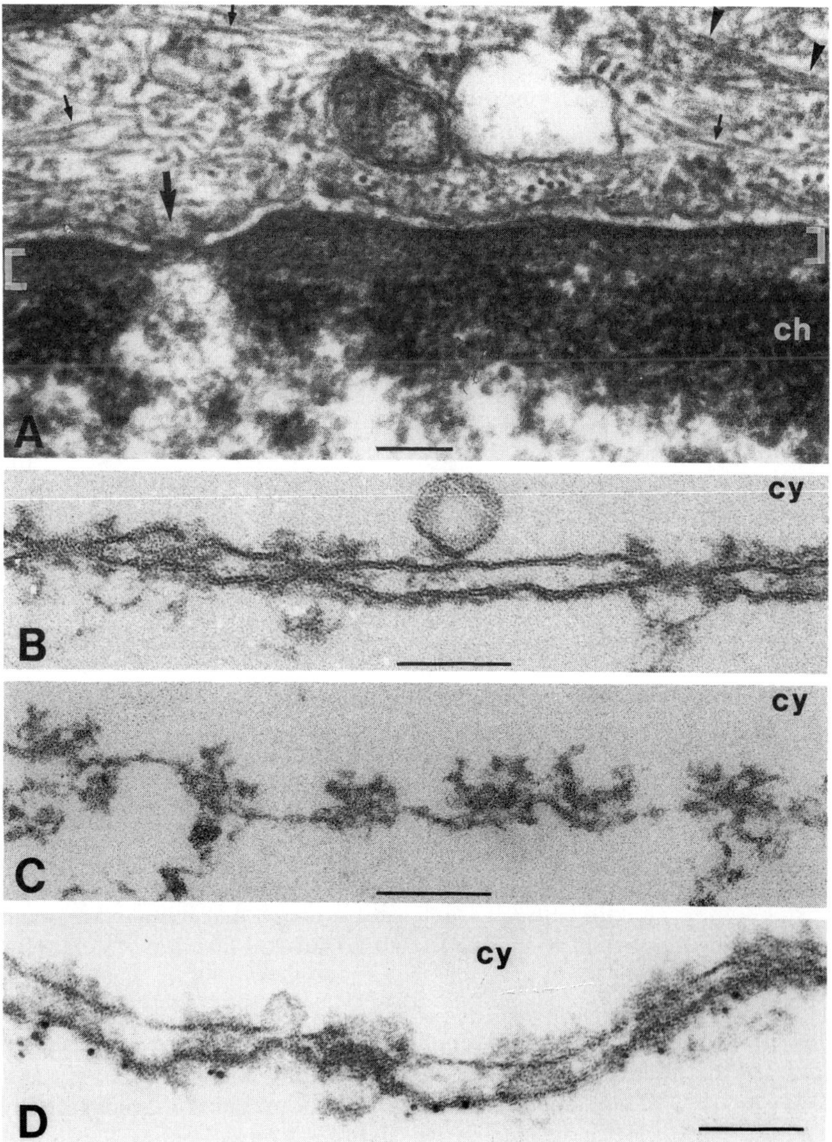

2. LAMINS: DISCOVERY, MOLECULAR STRUCTURE, AND ISOTYPES

2.1. Lamin Discovery

Lamins have apparent molecular masses in sodium dodecyl sulfate–polyacrylamide gel electrophoresis (SDS–PAGE) of 60,000–80,000 Da (for exceptions see Figure 2, n4, n5) and were first identified as major karyoskeletal proteins of the pore complex–lamina fractions in 1975 (Aaronson and Blobel, 1975; for review see Krohne and Benavente, 1986). In 1978 electron microscopic immunolocalizations demonstrated that these proteins are exclusively localized in the nuclear lamina. These proteins have therefore been named "nuclear lamins" or "lamins" (Figure 1D; Gerace *et al.*, 1978; Krohne *et al.*, 1978b) [for comparative electron microscopic immunolocalization of lamins with other nuclear envelope proteins see Gerace *et al.* (1982), Snow *et al.* (1987), and Gajewski *et al.* (1996)]. The cDNA cloning of the first vertebrate lamin in 1986 revealed that lamins are members of the multigene family of intermediate filament (IF) proteins (McKeon *et al.*, 1986; D. Z. Fisher *et al.*, 1986). This discovery has enabled structural and functional analysis of lamins at the molecular level.

2.2. Molecular Structure of the Lamin Molecule

The lamin molecule is composed of three main structural domains (Figure 2, n1–n3). These are the central helical rod domain [354 amino acids; for exceptions see Riemer *et al.* (1993)], a short nonhelical aminoterminus (head domain; 26–55 amino acids), and a long carboxy-terminal tail (160–260 amino acids). The α-helical segments (1A, 1B, 2) of the rod domain contain heptad repeats with hydrophobic amino acids at the first and fourth positions characteristic for coiled-coil-forming proteins (for review see McKeon, 1987).

The carboxy-terminal tail contains two major sequence elements that allow one to distinguish cytoplasmic and nuclear IF proteins. On lamins, these are the nuclear localization signal (NL) and a tetrapeptide (CxxM; C = cysteine, x = any amino acid, M = methionine) at the direct carboxy terminus.

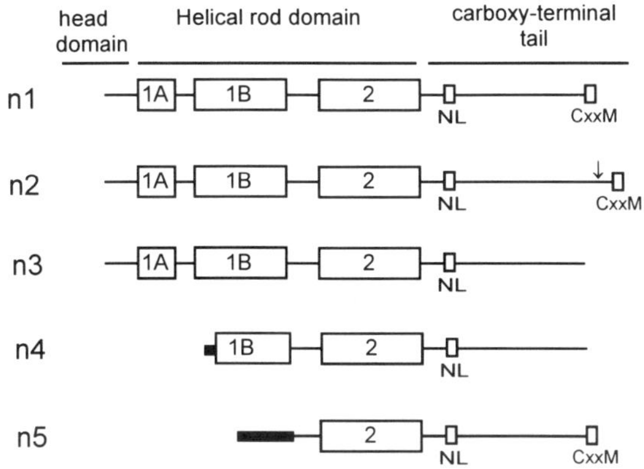

FIGURE 2. Schematic drawing of lamin isotypes. The three major domains (head domain, helical rod domain, carboxy-terminal tail) of this class of intermediate filament proteins are marked. The helical rod domain contains three α-helical segments with heptad repeats (1A, 1B, 2). NL, Nuclear localization sequence; CxxM, CxxM motif at the direct carboxy terminus (C = cysteine; x = any amino acid; M = methionine). (n1) B-type lamins of vertebrates and invertebrates including isotypes of the B1- and the B2-type lamins [for sequence comparisons of vertebrate B-type lamins see Höger *et al.* (1990)]. (n2) A-type lamins of vertebrates containing a CxxM motif in their primary amino acid sequences. During processing this motif including the carboxy-terminal 18 (mammals; Weber *et al.*, 1989) or 20 (chicken; Hennekes and Nigg, 1994) amino acids are proteolytically removed (arrow). (n3) Lamins that do not contain a CxxM motif in their amino acid sequences and are comparable in their properties with processed lamin A. These are lamins C of mammals (Lin and Worman, 1993) and *Drosophila* (Bossie and Sanders, 1993). (n4) Mammalian lamin C2, a splice variant of the lamin A gene (Nakajima and Abe, 1995). The filled box marks the unique amino-terminal sequence comprising six amino acids. (n5) Mammalian lamin B3, a splice variant of the lamin B2 gene (Furukawa and Hotta, 1993). The filled box marks the unique amino-terminal sequence comprising 84 amino acids.

2.3. Lamin Isotypes

Two major lamin isotypes can be distinguished according to their behavior during the cell cycle (Gerace and Blobel, 1980) and the posttranslational modification of the CxxM motif. During mitosis B-type lamins remain associated with membranes, whereas A-type lamins form soluble oligomers. Membrane attachment is mediated by the farnesylated and carboxyl methylated cysteine of the CxxM motif that remains permanently attached to B-type lamins (Figure 2, n1). In contrast, the carboxy-terminal 18 (in mammals; Weber *et al.*, 1989) or 20 (in chicken; Hennekes and Nigg, 1994) amino acids including this modified cysteine are proteolytically removed from A-type lamins in two processing steps (Figure 2, n2; for review see Hennekes and Nigg, 1994; Nigg and Hennekes, 1995; Moir *et al.*, 1995).

The CxxM motives are absent (Figure 2, n3) from the primary sequences of the mammalian lamin C, a splice variant of the mammalian lamin A (Lin and Worman, 1993), and lamin C of *Drosophila* (Bossie and Sanders, 1993). Two further observations favor the classification of the *Drosophila* lamin C as an A-type lamin. These are its developmentally regulated expression (Riemer *et al.*, 1995) and its assembly behavior *in vivo* (see below; Klapper *et al.*, 1997).

Two mammalian lamins with shorter helical rod domains are exclusively expressed in the male germ line (Furukawa and Hotta, 1993; Furukawa *et al.*, 1994; Alsheimer and Benavente, 1996). These shorter proteins are splice variants of lamin A (Figure 2, n4; Furukawa *et al.*, 1994; Nakajima and Abe, 1995) and the B-type lamin B2 (Figure 2, n5; Furukawa and Hotta, 1993), respectively. Reconstitution experiments with full-length lamins (Figure 2, n1–n3) and mutants derived therefrom (see below) indicate that the assembly properties of the two germ-line lamins should be drastically altered.

3. LAMIN TARGETING, ASSEMBLY STEPS, AND DISASSEMBLY

3.1. Lamin Targeting

Mutational analysis of lamins has allowed the identification of domains (see Figure 3) that are required for nuclear transport, membrane association, *in vitro* and *in vivo* assembly, mitotic disassembly, *in vitro* chromatin

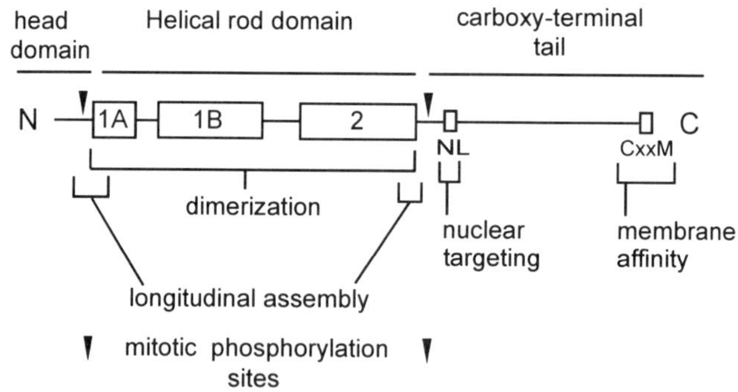

FIGURE 3. Domains of the lamin molecule involved in nuclear targeting (NL), in the interaction with the inner nuclear membrane (CxxM motif; membrane affinity), in individual steps of polymer formation (dimerization, longitudinal assembly), and in the disassembly of the polymer (mitotic phoshorylation sites marked by arrowheads).

binding (Glass *et al.*, 1993; Höger *et al.*, 1991b; Taniura *et al.*, 1995), and lamin degradation during apoptosis (Rao *et al.*, 1996).

The data make it possible to propose the following sequence of steps for assembly of newly synthesized lamins into the lamina of interphase cells.

1. During or immediately after translation, lamins form parallel aligned homodimers (Krohne *et al.*, 1987). For dimerization *in vitro* the head domain and the carboxy-terminal tail are dispensable (Heitlinger *et al.*, 1992; Moir *et al.*, 1991; Krohne *et al.*, 1987).

2. The cysteine of the CxxM motif is modified by farnesylation in the cytoplasm (Firmbach-Kraft and Stick, 1995; Kitten and Nigg, 1991). The carboxy-terminal three amino acids are proteolytically removed from A- and B-type lamins (Figure 2; n1, n2) and the cysteine is carboxyl methylated (for review see Nigg and Hennekes, 1995; Moir *et al.*, 1995).

3. The nuclear localization signal (NL; Figure 3) is required for the nuclear transport of the dimers (Chelsky *et al.*, 1989; Krohne *et al.*, 1989; Loewinger and McKeon, 1988), whereas the nuclear retention of lamins is mediated by domains involved in the assembly of lamin polymers (Schmidt-Zachmann *et al.*, 1993). The activity of the NL can be modulated *in vivo*. Nuclear targeting of lamins is inhibited by the phoshorylation of serines flanking the NL (Haas and Jost, 1993; Hennekes *et al.*, 1993). In the absence of a functional NL, lamin molecules accumulate in the cytoplasm (Figure 4C). Pulse-chase experiments with [^{35}S]methionine in cultured cells (Gerace *et al.*, 1984; Lehner *et al.*, 1986) and the cytoplasmatic microinjection of fluorescently labeled lamins (Benavente and Krohne, 1998; Schmidt and Krohne, 1995; Schmidt *et al.*, 1994; Pugh *et al.*, 1997) demonstrated that these proteins are transported within minutes into the nuclear compartment and are immediately assembled into the nuclear envelope (Figure 4A).

4. The farnesylated cysteine mediates the binding of lamins to the inner nuclear membrane (Kitten and Nigg, 1991; Krohne *et al.*, 1989; Holtz *et al.*, 1989). However, this domain in itself is not sufficient for retention of lamins in the nuclear periphery if one of the domains involved in the assembly of lamin polymers is nonfunctional (see Figure 4B; Schmidt-Zachmann *et al.*, 1993). Lamin molecules lacking the CxxM motif (Figure 2, n3) are efficiently transported into the nucleus; however, their incorporation into the lamina is drastically delayed (Gerace *et al.*, 1984; Pugh *et al.*, 1997; Horton *et al.*, 1992). Fluorescently labeled mammalian lamin A (Figure 2, n2) microinjected into the cytoplasm of cultured cells requires less than 20–30 min to assemble into the lamina (Figure 4A), whereas lamin C (Figure 2, n3) needs more than 20 h (Pugh *et al.*, 1997).

5. Finally, lamin dimers assemble into the preexisting lamina. The highly conserved end segments of the α-helical rod domain, and part of the head domain, are required for the formation of lamin polymers *in vitro* and

in vivo (Heitlinger *et al.*, 1992; Klapper *et al.*, 1997; Stuurman *et al.*, 1996; Moir *et al.*, 1991; Schmidt and Krohne, 1995). Subsequently, in A-type lamins the farnesylated cysteine of the CxxM motif is removed by proteolytic processing.

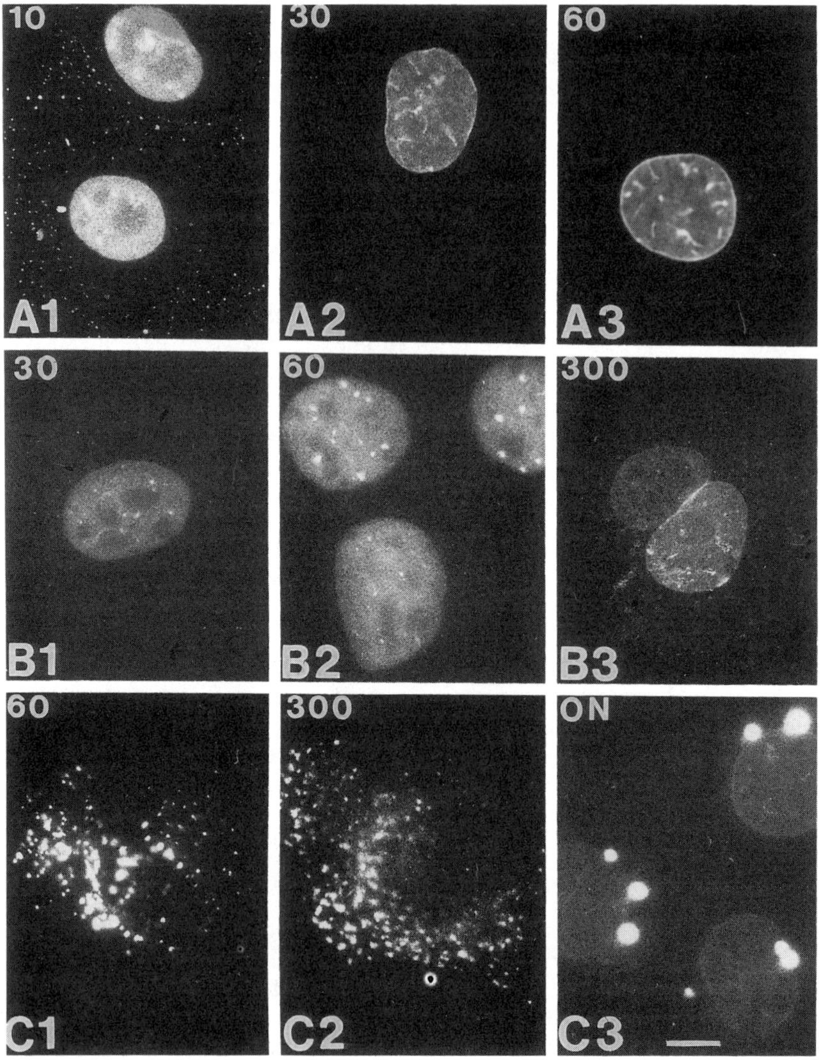

3.2. Steps in the Formation of Lamin Polymers

In vitro reconstitution experiments with bacterially expressed lamins have elucidated individual steps of polymer formation (Figure 5). Reconstituted dimers (Figure 5A) consist of an approximately 52-nm-long rod with two globular particles attached to one end (Aebi *et al.*, 1986). The rod represents the amino-terminal heads and the alpha-helical rod domains, and each globular particle contains one carboxy-terminal tail (Gieffers and Krohne, 1991; Heitlinger *et al.*, 1992; Moir *et al.*, 1991).

Following dimerization, in the first step (Figure 5B) the highly conserved amino-terminal end domain of the rod of one dimer slightly overlaps with the carboxy-terminal end domain of the rod of a second dimer. By this type of interaction longitudinal polymers also named "head-to-tail" aggregates are formed (Aebi *et al.*, 1986; Heitlinger *et al.*, 1991; Stuurman *et al.*, 1996; Moir *et al.*, 1991; Gieffers and Krohne, 1991). The B-type lamin from *Drosophila* (lamin Dmo) needs a few amino acids of its carboxy-terminal tail domain for longitudinal association of lamin dimers (Stuurman *et al.*, 1996).

Under optimal buffer conditions "head-to-tail" polymers of A- as well as B-type lamins aggregate laterally in half-staggered, probably antiparallel fashion, and form paracrystals with alternating light- and dark-stained bands (Figure 5C). The axial repeat of a paracrystal comprises one light- and one dark-stained band and measures 22–25 nm (polymers that have been analyzed by negative staining).

FIGURE 4. Properties of (A) fluorescently labeled (5-iodoacetamidoflourescein) *Xenopus* lamin A and (B, C) two lamin A mutants, analyzed *in vivo* by their microinjection into the cytoplasm of mouse 3T3 cells. Cells were fixed at the indicated time points and the intracellular distribution of lamins was monitored by fluorescence microscopy. (A) Wild-type lamin A. Cells were analyzed 10, 30, and 60 min after microinjection. Lamin molecules are transported within 10 min into the nucleus, and after 30 min the majority of the molecules have been incorporated into the nuclear lamina. The apparent streak- and dotlike intranuclear staining results from numerous invaginations of the nuclear envelope (Schmidt *et al.*, 1994). Reproduced from Schmidt and Krohne (1995) with permission. (B) A *Xenopus* lamin A mutant lacking the complete head domain (see Figure 3). Cells were analyzed 30, 60, and 300 min after microinjection. The mutant molecules are efficiently transported into the nucleus. The intense staining of the nucleoplasm indicates that even after prolonged incubation times (B3) only minimal amounts of the protein have been assembled in the nuclear periphery. (C) A *Xenopus* lamin A mutant lacking the complete carboxy-terminal tail including the nuclear localization signal (NL). Microinjected cells were analyzed after 60 and 300 min and after an overnight incubation (C3, ON). These lamin molecules are not transported into the nucleus, but they can still assemble resulting in the dotlike staining of the cytoplasm. Bar in panel C3 is 10 μm (relevant for all panels). Reproduced from Schmidt and Krohne (1995) with permission.

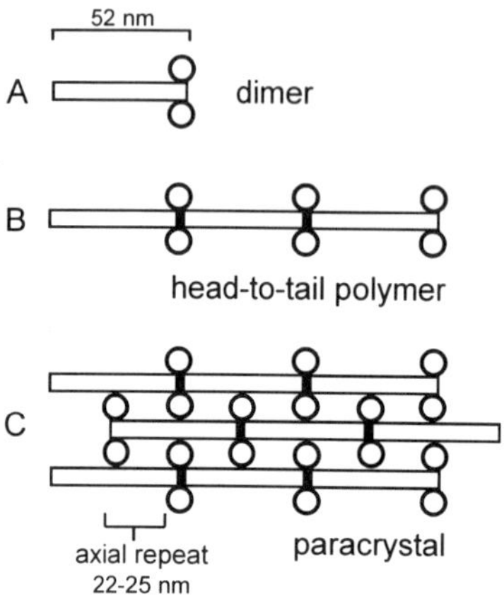

FIGURE 5. Individual steps of the assembly of dimers into longitudinal aggregates (head-to-tail polymers) and paracrystals. (A) The lamin dimer consists of two electron microscopically distinguishable domains. These are the rod domain (shown as a rectangle) and two globular particles (shown as spherical structures) attached to one end. The rod domain contains the amino-terminal head domains and the helical rod domains of two lamin molecules, and each globular particle represents one carboxy-terminal tail. (B) In the "head-to-tail" polymer the putative overlap of the amino-terminal end of the rod domain of one dimer with the carboxy-terminal end of the rod domain of a second dimer is shown as a vertical black bar. (C) Paracrystals are formed by the approximately half-staggered lateral association of "head-to-tail" polymers. The axial repeat of paracrystals comprises one dark- and one light-stained band.

The comparison of the dimer length (52 nm) with the shorter repeat length in polymers (the axial repeat in paracrystals is less than the half-length of a dimer) favors the model that the end segments of the rod domains overlap in "head-to-tail" polymers (Heitlinger *et al.*, 1992; for review see Heins and Aebi, 1994).

In vitro reconstitution experiments performed under identical conditions with *Xenopus* lamin A and a lamin A mutant lacking the complete carboxy terminal have shown that the carboxy-terminal tail does contribute to the alternating light- and dark-stained bands in lamin A paracrystals (Gieffers and Krohne, 1991). However, it is important to note that paracrystals with alternating light- and dark-stained bands can also be formed *in vitro* by mutants of A- and B-type lamins lacking the carboxy-terminal tail if reconstitution buffers are used that contain at least 10 mM of divalent cations (Heitlinger *et al.*, 1992; Stuurman *et al.*, 1996; Sasse *et al.*, 1997). These observations indicate that overlaps of dimers and gaps in the lamin polymer can also contribute to the characteristic morphology of lamin paracrystals (Moir *et al.*, 1991).

3.3. Mitotic Disassembly of Lamins

The mitotic disassembly of lamins is achieved by the phosphorylation of SP/TP motives (S = serine; T = threonine; P = proline) flanking the helical rod domain (see Figure 3, arrowheads; for reviews see Nigg, 1992; Nigg and Hennekes, 1995; Moir *et al.*, 1995). The mutation of these phosphoacceptor sites prevents the disassembly of lamin polymers *in vitro* (Peter *et al.*, 1991) and *in vivo* during mitosis (Heald and McKeon, 1990). *In vitro* phosphorylation of lamin molecules (Figure 2, n1–n3) by cdc2 kinase at these sites inhibits the assembly of dimers into "head-to-tail" polymers (Peter *et al.*, 1991).

4. THE STRUCTURE OF *IN VIVO*-FORMED LAMIN POLYMERS

4.1. The Amphibian Oocyte Nuclear Envelope

In comparison to the detailed molecular characterization of lamins, little is known about the structural organization of lamin polymers *in vivo*. The most direct visualization of lamin polymers was obtained from manually isolated *Xenopus* oocyte nuclear envelopes (Aebi *et al.*, 1986). In the lamina filaments Aebi and co-workers noticed a periodicity of 25 nm, a value identical with the length of the axial repeats in lamin paracrystals. It is most likely that the oocyte lamina filaments consist of few "head-to-tail" polymers of lamins that associated laterally.

4.2. Lamin Polymers Formed in Insect Cells

4.2.1. Polymer Structure of B-Type Lamins

The expression of A- and B-type lamins from vertebrates and invertebrates at high levels in insect cells (Sf9 cells) with the aid of the baculovirus system has provided new insights into lamin polymer formation *in vivo* (Klapper *et al.*, 1997). The aggregates formed by polymerization-competent lamins could be directly seen on ultrathin sections through *in situ* fixed insect cells 48–60 h postinfection.

The B-type lamins LI of *Xenopus* (Krohne *et al.*, 1987) and Dmo of *Drosophila* (Gruenbaum *et al.*, 1988) assembled into polymers with a fibrillar substructure in the cytoplasm (LI, Dmo; Figures 6A and 6C), in the nucleoplasm (LI), and at the inner nuclear membrane (Dmo; Figures 6A and 6B). The polymerization of lamin Dmo at the inner nuclear membrane resulted in the formation of a prominent nuclear lamina that was up to 2 μm thick. The thinnest fibrils in the nuclear and cytoplasmatic polymers of both B-type lamins had diameters of approximately 5 nm (Figure 6C). It can be speculated that the thin, irregularly arranged fibrils in the polymers of

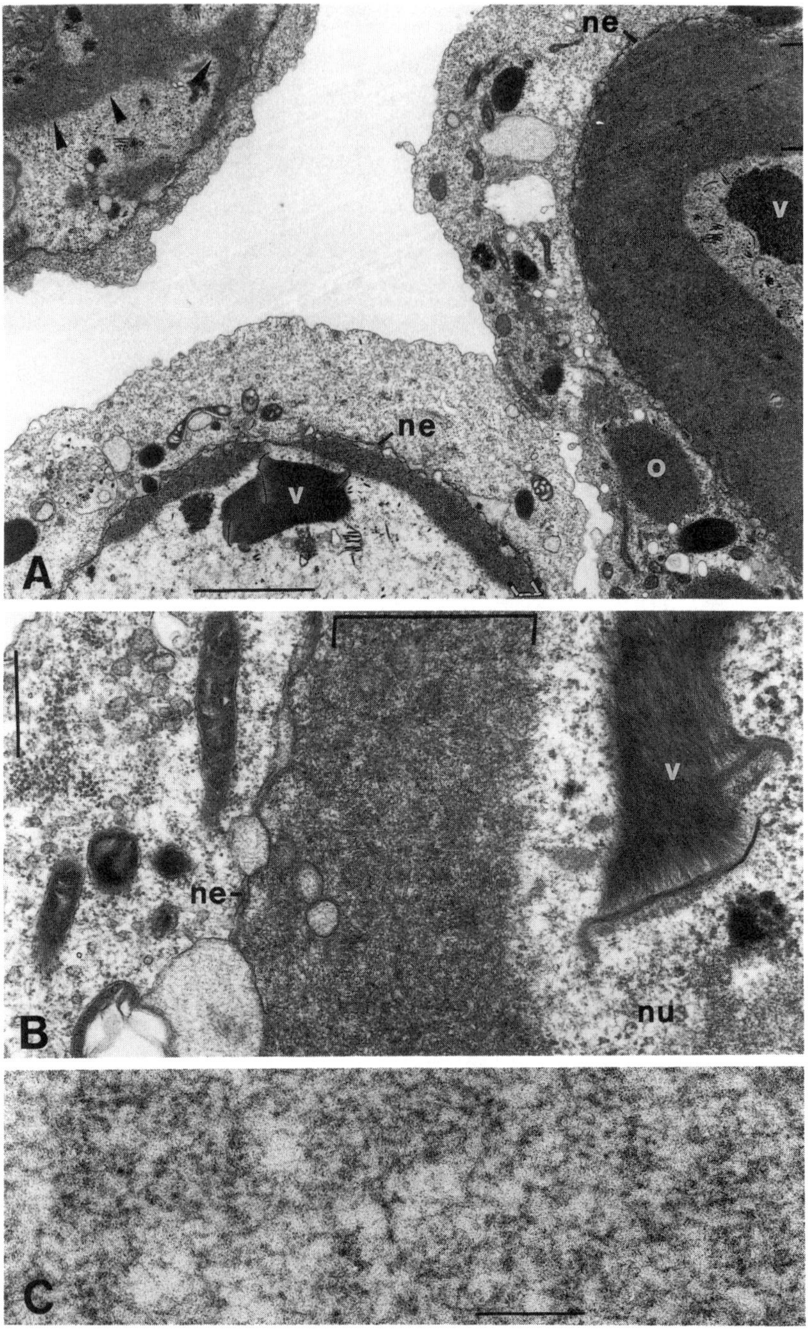
ne
v
ne
v
o
A
ne
v
nu
B
C

lamins LI and Dmo may represent single longitudinal "head-to-tail" aggregates of lamin dimers, or lateral assemblies of a few longitudinal aggregates. The absence of any obvious periodicity in these polymers indicates that the ordered lateral aggregation of B-type lamins is prevented *in vivo*, whereas B-type lamins including lamin Dmo have been assembled *in vitro* into paracrystals (Heitlinger *et al.*, 1991, 1992; Stuurman *et al.*, 1996). At least in the baculovirus system it is unlikely that components of the inner nuclear membrane and/or the lamina of Sf9 cells modulate the morphology of polymers formed by B-type lamins, because all polymers had a very similar ultrastructure, independent of their intracellular localization.

We expressed in Sf9 cells point and deletion mutants of lamin Dmo (Krohne *et al.*, 1998). It has been demonstrated that these mutations influenced the formation of "head-to-tail" polymers *in vitro* (Stuurman *et al.*, 1996; Sasse *et al.*, 1997). All mutants assembled *in vivo* into aggregates of thin fibrils indistinguishable in morphology from the wild type polymers formed in insect cells (Figure 6; Klapper *et al.*, 1997). However, in contrast to the wild type lamin, proteins of all Dmo mutants were easily extractable with physiological buffers (Klapper *et al.*, 1997; Krohne *et al.*, 1998) suggesting that "head-to-tail" interactions of dimers are also *in vivo* important for the stability of the B-type lamin polymer.

4.2.2. Polymer Structure of A-Type Lamins

Xenopus lamin A and *Drosophila* lamin C assembled in the cytoplasm and the nucleoplasm of Sf9 cells into dense aggregates of variable form (Figure 7A) which in several areas exhibit a paracrystalline morphology with an axial repeat of 18.5–20 nm (Figures 7B [inset] and 7C). Nuclear localized Drosophila lamin C paracrystals were often associated with the lamina (paracrystal on the left side in Figures 7A and 7B). Other areas with

FIGURE 6. Lamin polymer formation *in vivo* analyzed with the aid of the baculovirus system. Expression of the B-type lamins Dmo of *Drosophila* (A, B) and LI of *Xenopus* (C) in Sf9 cells. Electron micrographs of ultrathin-sectioned cells are shown. (A, B) Lamin Dmo can form a thick nuclear lamina (marked by brackets) that could extend locally deep into the nucleoplasm (A, arrowheads). Lamin Dmo that has polymerized in the cytoplasm is marked by a white circle (A, cell on the right side). In the cells shown the lamina has a thickness of approximately 2 μm (A, cell on the right side), 0.5 μm (A, cell in the lower half), and 1 μm (B). (C) High magnification of a small area of a lamin LI polymer localized in the cytoplasm. The fibrillar ultrastructure of the polymers is clearly visible and very similar to that of lamin Dmo polymers. The cell containing this polymer was briefly fixed with formaldehyde to stabilize protein aggregates and than incubated with Triton X-100 to extract soluble proteins that could mask the substructure of protein polymers (Klapper *et al.*, 1997). ne, Nuclear envelope; nu, nucleoplasm; v, aggregate of viral proteins. Bars: (A) 2 μm, (B) 500 nm, (C) 100 nm.

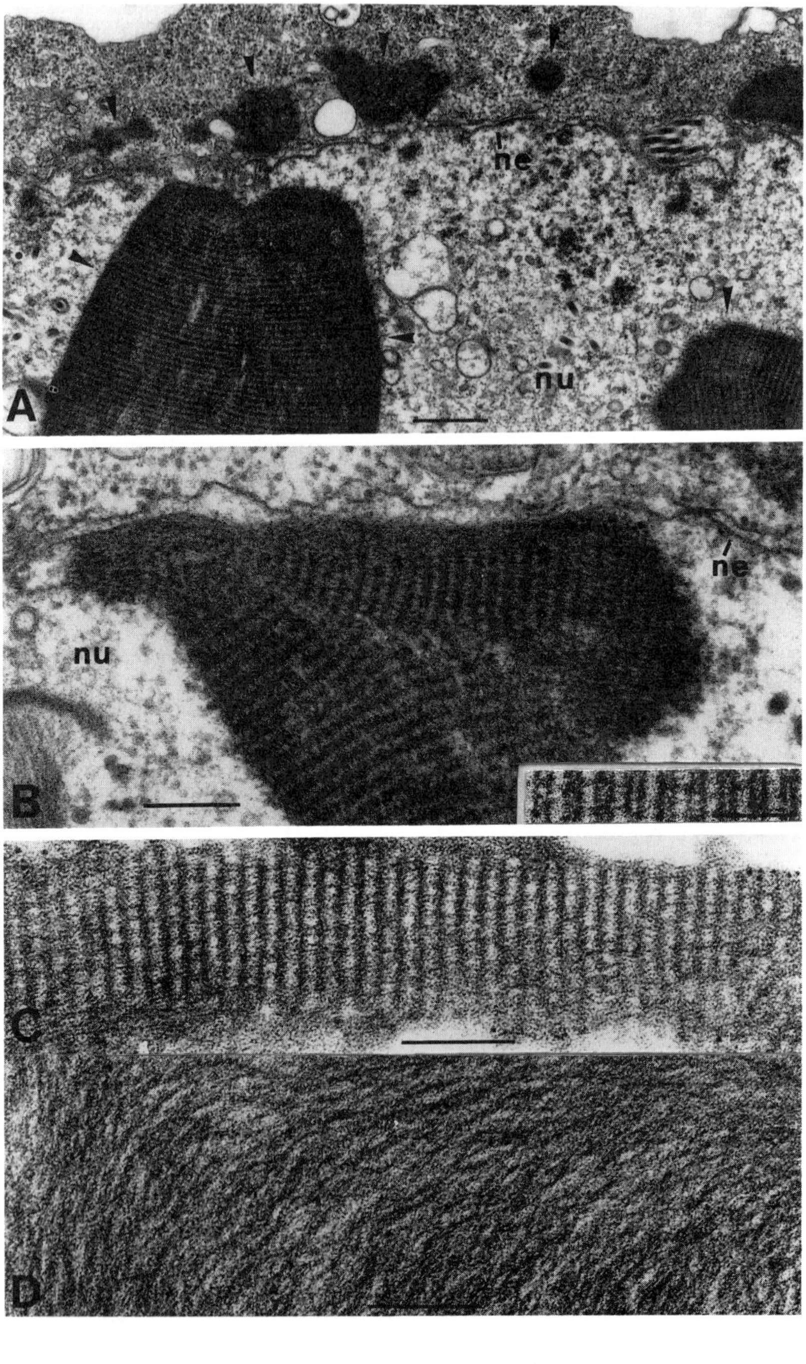
ne
nu
A
ne
nu
B
C
D

more amorphous morphology also contained lamin polymers, as has been verified by electron microscopic immunolocalizations (Klapper *et al.*, 1997). It is important to note that large and small paracrystals could only be identified when they were sectioned parallel to their longitudinal axes. Therefore it is likely that several lamin polymers with periodic substructure are hidden in areas which appear as amorphous aggregates, and that the majority of dimers of lamins A and C have assembled at least into thin paracrystals.

The carboxy-terminal tail of *Xenopus* lamin A contributes to the alternating light- and dark-stained bands observed in paracrystals assembled *in vivo* (Figures 7C and 7D). *Xenopus* lamin A lacking the complete carboxy-terminal tail assembles in the cytoplasm of Sf9 cells into filament bundles composed of fibrils less than 6 nm thick (Figure 7D).

The morphology of polymeric structures formed by *Xenopus* lamin A and the tailless lamin A mutant *in vivo* (Figure 7C, D; Klapper *et al.*, 1997) and *in vitro* (Gieffers and Krohne, 1991) are virtually identical, indicating that the packing of lamin dimers within these structures is the same. Therefore it can be assumed that, similar to *in vitro* assembly, A-type lamins assemble *in vivo* in a longitudinal fashion to yield "head-to-tail" polymers which subsequently associate laterally into paracrystalline structures. The observation that a lamin A molecule lacking the complete amino-terminal head domain (see Figures 3 and 4B; Klapper *et al.*, 1997) can be extracted with physiological salt concentrations from Sf9 cells whereas wild-type molecules remain insoluble under identical experimental conditions suggests that the "head-to-tail" interaction between lamin dimers indeed occurs *in vivo*.

The 18.5- to 20-nm axial repeat of *in vivo* assembled paracrystals of *Xenopus* lamin A and *Drosophila* lamin C measured on ultrathin-sectioned

FIGURE 7. Lamin polymer formation *in vivo* analyzed with the aid of the baculovirus system. Expression of the lamins C of *Drosophila* (A, B, inset in B) and A of *Xenopus* (C, D) in Sf9 cells. Both lamins form polymers in the cytoplasm and the nucleoplasm (marked by arrowheads in A). Nuclear localized lamin C is often found in association with the lamina of the infected insect cells (A, paracrystal on left; B). The electron-dense lamin polymers of variable size (A, B: arrowheads) exhibit in several areas a paracrystalline morphology (A–C) with an axial repeat of 18.5–20 nm. (A) Overview of an Sf9 cell expressing *Drosophila* lamin C; (B, inset in B) higher magnifications of a lamin C paracrystal associated with the inner nuclear membrane. The cells shown in panels A and B have been fixed *in situ*. (C, D) Cytoplasmatic localized polymers formed by (C) wild-type *Xenopus* lamin A and (D) a lamin A mutant lacking the complete carboxy-terminal tail. Both lamin A polymers are shown at the same magnification. Lamin A-expressing cells were extracted with Triton X-100 prior to fixation (C) or processed (D) as described in the legend to Fig. 6C. ne, nuclear envelope; nu, nuclear content. Bars: (A) 4200 nm, (B) 200 nm, and (inset in B–D) 100 nm.

material is similar to the 20- to 22-nm repeat of *in vitro*-formed *Xenopus* lamin A paracrystals that had also been embedded in Epon. The shorter repeat lengths in comparison to negatively stained lamin polymers (22–25 nm; Moir *et al.*, 1991; Aebi *et al.*, 1986; Heitlinger *et al.*, 1991, 1992; Gieffers and Krohne, 1991; Stuurman *et al.*, 1996) result from shrinkage of biological structures during the embedding procedure.

When A-type lamins are expressed at a more physiological level and are normally localized at the nuclear envelope, a paracrystalline morphology is not readily apparent even in the thick nuclear lamina of mammalian cells (Figure 1A; Höger *et al.*, 1991a). At least for the baculovirus system it can be concluded that the assembly of the Drosophila A-type lamin into paracrystals is not inhibited significantly by the inner nuclear membrane and/or components of the lamina (see Figure 7B; Krohne *et al.*, 1998).

When we made mutants of the Drosophila lamin C (Krohne *et al.*, 1998) we realized that we had used for our first study (Klapper *et al.*, 1997) a lamin C cDNA clone that was missing nucleotide G1781 of the published sequence (Bossie and Sanders, 1993). This mutation resulted in a frame shift which completely changed the sequence of the carboxy-terminal 69 amino acids. As a further consequence, the mutant protein was 9 amino acids shorter than the wild type protein. The major difference between the protein polymers was that only the wild type lamin C was found in association with the nuclear lamina of Sf9 cells (Figure 7B), whereas the paracrystals formed by both proteins had very similar axial repeats (compare Figures 7A and 7B with Figures 6 and 7 in Klapper *et al.*, 1997).

4.2.3. Targeting of Overexpressed Lamins

Why do the four wild-type lamins assemble in the cytoplasm of the insect cells? It is most likely that newly synthesized lamin dimers are present in high local concentrations in the cytoplasm, resulting in their immediate assembly into oligomeric structures that are too large to allow their free diffusion and transport into the nucleus. *In vitro* experiments have shown that lamin concentrations of 5 μg/ml are sufficient to enable their rapid polymerization (Glass and Gerace, 1990).

Of the four vertebrate and invertebrate lamins analyzed, only nuclear localized *Drosophila* lamin Dmo assembles into a thick nuclear lamina (compare Figures 1A and 6). The B-type lamin of Sf9 cells is very similar in its biochemical properties to lamin Dmo (Klapper *et al.*, 1997), and therefore most likely mediates the association of lamin Dmo with the nuclear lamina of the butterfly cells. I speculate that the local association of Drosophila lamin C with the lamina of baculovirus infected cells is also facilitated by interactions with the endogenous lamins of Sf9 cells.

The lack of interaction of *Xenopus* lamins with the nuclear lamina can be explained by the low degree of homology between insect and vertebrate lamins. The *Drosophila* lamin Dmo possesses less than 30% sequence identity with *Xenopus* B-type lamins [see sequence comparisons in Bossie and Sanders (1993) and Riemer *et al.* (1993)]. In addition, these data demonstrate that the evolutionary conserved CxxM motif alone is not sufficient for the assembly of vertebrate lamins into the lamina of invertebrate cells.

5. LAMIN ASSEMBLY: INFLUENCE ON NUCLEAR ENVELOPE STRUCTURE AND STABILITY

5.1. The Mammalian Spermatocyte

Investigations concerned with germ-line cells and a *Drosophila* lamin mutant indicate that the lamins can confer mechanical stability and flexibility to the nuclear envelope.

Primary spermatocytes of mammalian cells express in addition to the B-type lamin B1 (Vester *et al.*, 1993) two germ-line-specific lamins (Figure 2, n4, n5) with shorter amino termini than the typical A- and B-type lamins (Alsheimer and Benavente, 1996; Furukawa and Hotta, 1993; Furukawa *et al.*, 1994; Smith and Benavente, 1992). Both shorter lamins lack sequences known to be required for lamin polymer formation (Figure 3). According to the present knowledge about lamin polymerization it can be expected that a lamina formed by such a lamin complement will be physically less stable than the corresponding structure in somatic cells. This notion is in agreement with observations by Stick and Schwarz (1982), who have shown that in contrast to somatic cells, the nuclear envelope of primary spermatocytes is completely disrupted after incubation of the cells in physiological buffers containing nonionic detergents.

The structural differences between the nuclear envelopes of spermatocytes and other cell types have also been demonstrated with a special electron microscopic staining technique that enhances the contrast of the nuclear lamina on ultrathin-sectioned cells (Stick and Schwarz, 1982). In spermatocytes no stainable material was detectable subjacent to the inner nuclear membrane, whereas in all other cells analyzed the lamina was clearly visible as an electron-dense sheet.

These features of the spermatocyte nuclear lamina may be an adaptation to meiosis-specific nuclear functions, such as the attachment of the chromosome ends to the nuclear envelope and the movement of meiotic chromosomes despite their anchorage to the nuclear envelope. These

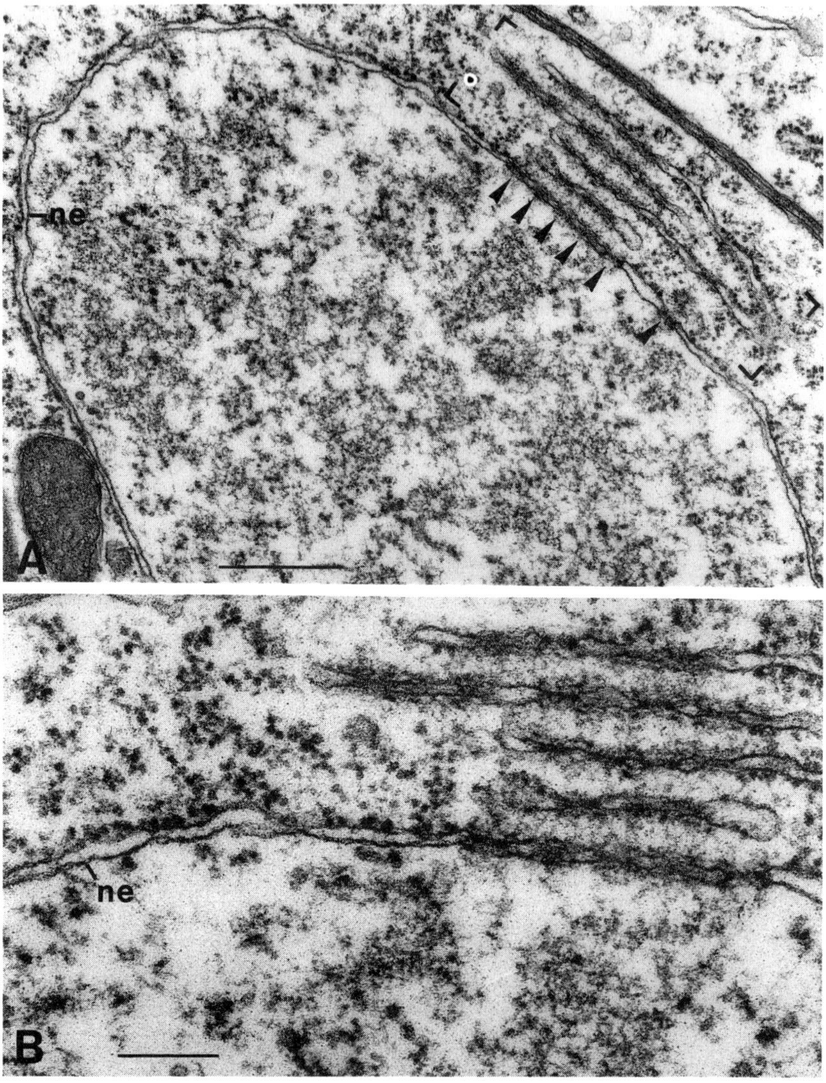

FIGURE 8. Neuronal cell of a *Drosophila melanogaster* fly homozygous for an insertional mutation of the lamin Dmo gene (Lenz-Böhme *et al.*, 1997). (A) Electron micrograph of an ultrathin section showing a cell with clustered nuclear pore complexes (marked by arrowheads) in the nuclear envelope and cytoplasmic annulate lamellae (marked by brackets). Note the absence of pore complexes in the nuclear envelope outside the pore complex cluster. (B) Higher magnification of the area containing the clustered pore complexes. ne, Nuclear envelope. Bars: (A) 500 nm and (B) 200 nm.

functions require a more flexible nuclear periphery with properties that differ from those of somatic cells. The flexibility of the nuclear envelope could be provided by lamins with altered assembly properties.

5.2. A Mutation of the *Drosophila* Lamin Dmo Gene

The description of an insertional mutation of the *Drosophila* lamin Dmo gene clearly demonstrates that lamins are required for the structural integrity of the nuclear envelope in multicellular organisms (Lenz-Böhme *et al.*, 1997). Flies homozygous for this mutation express less than 20% of the amount of lamin Dmo present in wild-type animals, resulting in impaired viability, sterility, and locomotion. Only a small percentage of the homozygous mutants reach the adult stage. In 67% of the neuronal cells of homozygous mutant flies at least one of the following three structural alterations were observed.

These are the clustering of pore complexes in the nuclear envelope (Figure 8), the abundance of stacks of cytoplasmic membranes containing pore complexes (annulate lamellae) that were found as parallel cisternae apposed to the clustered pore complexes in the nuclear envelope (Figure 8), and the presence of incomplete or fragmented nuclear envelopes. The structural alterations at the cellular level in this fly mutant convincingly demonstrate that lamins are essential for the proper spatial organization of pore complexes in the nuclear envelope and this is consistent with a direct interaction between pore complex proteins and the lamins.

The observation of incomplete nuclear envelopes and the high incidence of annulate lamellae in mutant animals can be interpretated as impaired assembly of the nuclear envelope. These observations are consistent with a role of lamins in targeting nuclear envelope precursor vesicles to chromatin (for review see Benavente, 1991; Lourim and Krohne, 1994) and extend the evidence for an important role of lamins in nuclear envelope reformation in multicellular organisms.

ACKNOWLEDGMENTS. I thank Drs. Ricardo Benavente (Biocenter of the University of Würzburg), David Lourim (Biocenter of the University of Würzburg), and Nico Stuurman (Maurice Müller Institute for Microscopy, Biozentrum, University of Basel, Switzerland) for critical reading of the manuscript. The micrographs in Figure 8 were kindly provided by Dr. Bertram Schmitt (Max-Planck-Institut für Hirnforschung, Frankfurt am Main, Germany). Financial support of the Deutsche Forschungsgesellschaft is greatly acknowledged (Kr 758/7-1).

6. REFERENCES

Aaronson, R. P., and Blobel, G., 1975, Isolation of nuclear pore complexes in association with a lamina, *Proc. Natl. Acad. Sci. USA* **72:**1007–1011.

Aebi, U., Cohn, J., Buhle, L., and Gerace, L., 1986, The nuclear lamina is a meshwork of intermediate-type filaments, *Nature* **323:**560–564.

Alsheimer, M., and Benavente, R., 1996, Change of karyoskeleton during mammalian spermatogenesis: Expression pattern of nuclear lamin C2 and its regulation, *Exp. Cell Res.* **228:**181–188.

Benavente, R., 1991, Postmitotic nuclear reorganization events analyzed in living cells, *Chromosoma* **100:**215–220.

Benavente, R., and Krohne, G., 1998, *In vivo* systems to study the dynamics of nuclear lamins, in *Nuclear Structure and Function* (M. Berrios, ed.), pp. 591–602, Academic Press, San Diego, California.

Bossie, C. A., and Sanders, M. M., 1993, A cDNA from *Drosophila melanogaster* encodes a lamin C-like intermediate filament protein, *J. Cell Sci.* **104:**1263–1272.

Chelsky, D., Ralph, R., and Jonak, G., 1989, Sequence requirements for synthetic peptide-mediated translocation to the nucleus, *Mol. Cell. Biol.* **9:**2487–2492.

Dwyer, N., and Blobel, G., 1976, A modified procedure for the isolation of a pore complex-lamina fraction from rat liver nuclei, *J. Cell Biol.* **70:**581–591.

Fisher, D. Z., Chaudhary, N., and Blobel, G., 1986, cDNA sequencing of nuclear lamins A and C reveals primary and secondary structural homology to intermediate filament proteins, *Proc. Natl. Acad. Sci. USA* **83:**6450–6454.

Fisher, P. A., Berrios, M., and Blobel, G., 1982, Isolation and characterization of a proteinaceous subnuclear fraction composed of nuclear matrix, peripheral lamina, and nuclear pore complexes from embryos of *Drosophila melanogaster*, *J. Cell Biol.* **92:**674–686.

Fawcett, D. W., 1966, On the occurrence of a fibrous lamina on the inner aspect of the nuclear envelope in certain cells of vertebrates, *Am. J. Anat.* **119:**129–146.

Firmbach-Kraft, I., and Stick, R., 1995, Analysis of nuclear lamin isoprenylation in *Xenopus* oocytes: Isoprenylation of lamin B3 precedes its uptake into the nucleus, *J. Cell Biol.* **129:**17–24.

Furukawa, K., and Hotta, Y., 1993, cDNA cloning of a germ line specific lamin B3 from mouse spermatocytes and analysis of its function by ectopic expression in somatic cells, *EMBO J.* **12:**97–106.

Furukawa, K., Inagaki, H., and Hotta, Y., 1994, Identification and cloning of an mRNA coding for a germ cell-specific A-type lamin in mice, *Exp. Cell Res.* **212:**426–430.

Gerace, L., and Blobel, G., 1980, The nuclear envelope lamina is reversibly depolymerized during mitosis, *Cell* **19:**277–287.

Gerace, L., Blum, A., and Blobel, G., 1978, Immunocytochemical localization of the major polypeptides of the nuclear pore complex-lamina fraction. Interphase and mitotic distribution, *J. Cell Biol.* **79:**546–566.

Gerace L., Ottaviano, Y., and Kondor-Koch, C., 1982, Identification of a major polypeptide of the nuclear pore complex, *J. Cell Biol.* **95:**826–837.

Gerace, L., Comeau, C., and Benson, M., 1984, Organization and modulation of nuclear structure, *J. Cell Sci. Suppl.* **1:**137–160.

Gajewski, A., Lourim, D., and Krohne, G., 1996, An antibody against a glycosylated integral membrane protein of the *Xenopus laevis* nuclear pore complex: A tool for the study of pore complex membranes, *Eur. J. Cell Biol.* **71:**14–21.

Ghadially, F. N., Bhatnagar, R., and Fuller, J. A., 1972, Waxing and waning of the nuclear fibrous lamina, *Arch. Pathol.* **94:**303–307.

Gieffers, C., and Krohne, G., 1991, *In vitro* reconstitution of recombinant lamin A and a lamin A mutant lacking the carboxy-terminal tail, *Eur. J. Cell Biol.* **55:**191–199.

Glass, J., and Gerace, L., 1990, Lamins A and C bind and assemble at the surface of mitotic chromosomes, *J. Cell Biol.* **111:**1047–1057.

Glass, C. A., Glass, J. R., Taniura, H., Hasel, K. W., Blevitt, J. M., and Gerace, L., 1993, The α-helical rod domain of human lamins A and C contains a chromatin binding site, *EMBO J.* **12:**4413–4424.

Gruenbaum, Y., Landesman, Y., Drees, B., Bare, J. W., Saumweber, H., Paddy, M. R., Sedat, J. W., Smith, D. E., Benton, B. M., and Fisher, P. A., 1988, *Drosophila* nuclear lamin precursor Dm0 is translated from either of two developmentally regulated mRNA species apparently encoded by a single gene, *J. Cell Biol.* **106:**585–596.

Haas, M., and Jost, E., 1993, Functional analysis of phosphorylation sites in human lamin A controlling lamin disassembly, nuclear transport and assembly, *Eur. J. Cell Biol.* **62:**237–247.

Heald, R., and McKeon, F., 1990, Mutations of phosphorylation sites in lamin A that prevent nuclear lamina disassembly in mitosis, *Cell* **61:**579–589.

Heins, S., and Aebi, U., 1994, Making heads and tails of intermediate filament assembly, dynamics and networks, *Curr. Opin. Cell Biol.* **6:**25–33.

Heitlinger, E., Peter, M., Häner, M., Lustig, A., Aebi, U., and Nigg, E. A., 1991, Expression of chicken lamin B2 in *Escherichia coli*: Characterization of its structure, assembly and molecular interactions, *J. Cell. Biol.* **113:**485–495.

Heitlinger, E., Peter, M., Lustig, A., Villiger, W., Nigg, E. A., and Aebi, U., 1992, The role of the head and tail domain in lamin structure and assembly: Analysis of bacterially expressed chicken lamin A and truncated B_2 lamins, *J. Struct. Biol.* **108:**74–91.

Hennekes, H., and Nigg, E., 1994, The role of isoprenylation in membrane attachment of nuclear lamins, *J. Cell Sci.* **107:**1019–1029.

Hennekes, H., Peter, M., Weber, K., and Nigg, E. A., 1993, Phosphorylation by protein kinase C inhibits nuclear import of lamin B2, *J. Cell Biol.* **120:**1293–1304.

Höger, T. H., Zatloukal, K., Waizenegger, I., and Krohne, G., 1990, Characterization of a second highly conserved B-type lamin present in cells previously thought to contain only a single B-type lamin, *Chromosoma* **99:**379–390.

Höger, T., Grund, C., Franke, W., and Krohne, G., 1991a, Immunolocalization of lamins in the thick nuclear lamina of human synovial cells, *Eur. J. Cell Biol.* **54:**150–156.

Höger, T., Krohne, G., and Kleinschmidt, J., 1991b, Interaction of *Xenopus* lamins A and LII with chromatin *in vitro* mediated by a sequence element in the carboxyterminal domain, *Exp. Cell Res.* **197:**280–289.

Holtz, D., Tanaka, R. A., Hartwig, J., and McKeon, F., 1989, The CaaX motif of lamin A functions in conjunction with the nuclear localization signal to target assembly to the nuclear envelope, *Cell* **59:**969–977.

Hutchison, C. J., Bridger, J. M., Cox, L. S., and Kill, I. R., 1994, Weaving a pattern from disparate threads: Lamin function in nuclear assembly and DNA replication, *J. Cell Sci.* **107:**3259–3269.

Kalifat, S. R., Bouteille, M., and DeLarue, J., 1967, Etude ultrastructurale de la lamelle dense observée au contact de la membrane nucleaire interne, *J. Microsc.* (Paris) **6:**1019–1026.

Kitten, G., and Nigg, E., 1991, The CaaX motif is required for isoprenylation, carboxyl methylation, and nuclear membrane association of lamin B2, *J. Cell Biol.* **113:**13–23.

Klapper, M., Exner, K., Kempf, A., Gehrig, C., Stuurman, N., Fisher, P. A., and Krohne, G., 1997, Assembly of A- and B-type lamins studied *in vivo* with the baculovirus system, *J. Cell Sci.* **110:**2519–2532.

Krohne, G., and Benavente, R., 1986, The nuclear lamins. A multigene family of proteins in evolution and differentiation, *Exp. Cell Res.* **162:**1–10.
Krohne, G., and Franke, W. W., 1983, Proteins of pore complex-lamina structures from nuclei and membranes, *Meth. Enzymol.* **96J:**597–608.
Krohne, G., Franke, W. W., and Scheer, U., 1978a, The major polypeptides of the nuclear pore complex, *Exp. Cell Res.* **116:**85–102.
Krohne, G., Franke, W. W., Ely, S., D'Arcy, A., and Jost, E., 1978b, Localization of a nuclear envelope-associated protein by indirect immunofluorescence microscopy using antibodies against a major polypeptide from rat liver fractions enriched in nuclear envelope associated material, *Cytobiologie* **18:**22–38.
Krohne, G., Wolin, S., McKeon, F., Franke, W., and Kirschner, M., 1987, Nuclear lamin L_I of *Xenopus laevis*: cDNA cloning, amino acid sequence and binding specificity of a member of the lamin B subfamily, *EMBO J.* **6:**3801–3808.
Krohne, G., Waizenegger, I., and Höger, T. H., 1989, The conserved carboxy-terminal cysteine of nuclear lamins is essential for lamin association vith the nuclear envelope, *J. Cell Biol.* **109:**2003–2011.
Krohne, G., Stuurman, N., and Kempf, A., 1998, Assembly of Drosophila lamin Dmo and C mutants studied with the baculovirus system, *Eur. J. Cell Biol.*, submitted.
Lehner, C. F., Fürstenberger, G., Eppenberger, H. M., and Nigg, E. A., 1986, Biogenesis of the nuclear lamina: *In vivo* synthesis and processing of nuclear protein precursors, *Proc. Natl. Acad. Sci. USA* **83:**2096–2099.
Lenz-Böhme, B., Wismar, J., Fuchs, S., Reifegerste, R., Buchner, E., Betz, H., and Schmitt, B., 1997, Insertional mutation of the *Drosphila* nuclear lamin Dmo gene results in defective nuclear envelopes, clustering of nuclear porecomplexes, and accumulation of annulate lamellae, *J. Cell Biol.* **137:**1001–1016.
Lin, F., and Worman, H. J., 1993, Structural organization of the human gene encoding nuclear lamin A and nuclear lamin C, *J. Biol. Chem.* **268:**16321–16326.
Loewinger, L., and McKeon, F., 1988, Mutations in the nuclear lamin proteins resulting in their aberrant assembly in the cytoplasm, *EMBO J.* **7:**2301–2309.
Lourim, D., and Krohne, G., 1993, Membrane-associated lamins in *Xenopus* egg extracts: Identification of two vesicle populations, *J. Cell Biol.* **123:**501–512.
Lourim, D., and Krohne, G., 1994, Lamin-dependent nuclear envelope reassembly following mitosis: An argument, *Trends Cell Biol.* **4:**314–318.
Maul, G. G., Baglia, F. A., Newmeyer, D. D., and Ohlsson-Wilhelm, B. H., 1984, The major 67,000 molecular weight protein of the clam oocyte nuclear envolope is lamin-like, *J. Cell Sci.* **67:**69–85.
Mazanec, K., 1967, Presence de la "zonula nucleum limitans" dans quelques cellules humaines, *J. Microsc.* (Paris) **6:**1027–1032.
McKeon, F., 1987, Nuclear lamin proteins and the structure of the nuclear envelope: Where is the function? *BioEssays* **7:**169–173.
McKeon, F., Kirschner, M. W., and Caput, D., 1986, Homologies both in primary and secondary structure between nuclear envelope and intermediate filament proteins, *Nature* **319:**463–468.
Moir, R. D., Donaldson, A. D., and Stewart, M., 1991, Expression in *Escherichia coli* of human lamins A and C: Influence of head and tail domains on assembly properties and pracrystal formation, *J. Cell Sci.* **99:**363–372.
Moir, R. D., Spann, T. P., and Goldmann, R. D., 1995, The dynamic properties and possible functions of nuclear lamins, *Int. Rev. Cytol.* **162B:**141–182.
Nakajima, N., and Abe, K., 1995, Genomic structure of the mouse A-type lamin gene locus encoding somatic and germ cell-specific lamins, *FEBS Lett.* **365:**108–114.

Nigg, E. A., 1992, Assembly–disassembly of the nuclear envelope, *Curr. Opin. Cell Biol.* **4:**105–109.

Nigg, E. A., and Hennekes, H., 1995, The nuclear lamina: Regulation of assembly by posttranslational modifications, in *The Cytoskeleton 45. Colloquium Mosbach,* (B. M. Jockusch, E. Mandelkow, and K. Weber, eds.), pp. 77–87, Springer, Berlin.

Patrizi, G., and Poger, M., 1967, The ultrastructure of the nuclear periphery. The zonula nucleum limitans, *J. Ultrastruct. Res.* **17:**127–136.

Peter, M., Heitlinger, E., Häner, M., Aebi, U., and Nigg, E. A., 1991, Disassembly of *in vitro* formed lamin head-to-tail polymers by cdc2 kinase, *EMBO J.* **10:**1535–1544.

Pugh, G. E., Coates, P. J., Lane, E. B., Raymond, Y., and Quinlan, R., 1997, Distinct nuclear assembly pathways for lamins A and C lead to their increase during quiescence in Swiss 3T3 cells, *J. Cell Sci.* **110:**2483–2493.

Rao, L., Perez, D., and White, E., 1996, Lamin proteolysis facilitates nuclear events during apoptosis, *J. Cell Biol.* **135:**1441–1455.

Riemer, D., Dodemont, H., and Weber, K., 1993, A nuclear lamin of the nematode *Caenorhabditis elegans* with unusual structural features; cDNA cloning and gene organization, *Eur. J. Cell Biol.* **62:**214–223.

Riemer, D., Stuurman, N., Berrios, M., Hunter, C., Fisher, P. A., and Weber, K., 1995, Expression of *Drosophila* lamin C is developmentally regulated: Analogies with vertebrate A-type lamins, *J. Cell Sci.* **108:**3189–3198.

Sasse, B., Lustig, A., Aebi, U., and Stuurman, N., 1997, *In vitro* assembly of *Drosophila* lamin Dmo. Lamin polymerization properties are conserved, *Eur. J. Biochem.* **250:**30–38.

Scheer, U., Kartenbeck, J., Trendelenburg, M. F., Stadler, J., and Franke, W. W., 1976, Experimental disintegration of the nuclear envelope, *J. Cell Biol.* **69:**1–18.

Schmidt, M., and Krohne, G., 1995, *In vivo* assembly kinetics of fluorescently labeled *Xenopus* lamin A mutants, *Eur. J. Cell Biol.* **68:**345–354.

Schmidt, M., Tschödrich-Rotter, M., Peters, R., and Krohne, G., 1994, Properties of fluorescently labeled *Xenopus* lamin A *in vivo*, *Eur. J. Cell Biol.* **65:**70–81.

Schmidt, M., Grossmann, U., and Krohne, G., 1995, The nuclear membrane associated honeycomb structure of the unicellular organism *Amoeba proteus*: On the search for homologies with the nuclear lamina of metazoa, *Eur. J. Cell Biol.* **67:**199–208.

Schmidt-Zachmann, M. S., Dargemont, C., Kühn, L. C., and Nigg, E. A., 1993, Nuclear export of proteins: The role of nuclear retention, *Cell* **74:**493–504.

Smith, A., and Benavente, R., 1992, Identification of a short nuclear lamin protein selectively expressed during meiotic stages of rat spermatogenesis, *Differentiation* **52:**55–60.

Snow, C. M., Senior, A., and Gerace, L., 1987, Monoclonal antibodies identify a group of nuclear pore complex glycoproteins, *J. Cell Biol.* **104:**1143–1156.

Stewart, M., and Whytock, S., 1988, The structure and the interactions of components of nuclear envelopes from *Xenopus* oocyte germinal vesicles observed by heavy metal shadowing, *J. Cell Sci.* **90:**409–423.

Stick, R., and Schwarz, H., 1982, The disappearance of the nuclear lamina during spermatogenesis: An electron microscopic and immunofluorescence study, *Cell Diff.* **11:**235–243.

Stuurman, N., Sasse, B., and Fisher, P. A., 1996, Intermediate filament protein polymerization: Molecular analysis of *Drosophila* nuclear lamin head-to-tail binding, *J. Struct. Biol.* **117:**1–15.

Taniura, H., Glass, C., and Gerace, L., 1995, A chromatin binding site in the tail domain of nuclear lamins that interacts with core histones, *J. Cell Biol.* **131:**33–44.

Vester, B., Smith, A., Krohne, G., and Benavente, R., 1993, Presence of a nuclear lamina in pachytene spermatocytes of the rat, *J. Cell Sci.* **104:**557–563.

Weber, K., Plessmann, U., and Traub, P., 1989, Maturation of nuclear lamin A involves a specific carboxy-terminal trimming, which removes the polyisoprenylation site from the precursor; implications for the structure of the nuclear lamina, *FEBS Lett.* **257**:411–414.

Chapter 20

Nuclear Lamin-Binding Proteins

Qian Ye, Racine M. Barton, and Howard J. Worman

1. NUCLEAR ENVELOPE

The nuclear envelope separates the contents of the nucleus from the cytoplasm in eukaryotic cells. In higher eukaryotic cells, the nuclear envelope is composed of the nuclear lamina, nuclear pore complexes, and nuclear membranes (for reviews see Courvalin and Worman, 1997; Goldberg and Allen, 1995; Wiese and Wilson, 1993; Worman and Courvalin, 1991). The nuclear membranes can be separated into three morphologically distinct, but interconnected domains (Figure 1). Macromolecular structures are associated with each membrane domain. The outer nuclear membrane is continuous at points with the rough endoplasmic reticulum and contains ribosomes on its outer surface. The pore membrane is associated with the nuclear pore complexes, macromolecular structures that provide passageways between the nucleus and cytoplasm (for reviews see Panté and Aebi, 1996; Stewart and Clarkson, 1996; Bastos *et al.*, 1995; Davis, 1995; Moore, 1995). The inner nuclear membrane is associated with the nuclear lamina, a meshwork of intermediate filaments (Aebi *et al.*,

---

Qian Ye, Racine M. Barton, and Howard J. Worman Departments of Medicine and of Anatomy and Cell Biology, College of Physicians and Surgeons, Columbia University, New York, New York 10032. *Present address for QY:* Laboratory of Developmental Hematopoiesis, Sloan-Kettering Institute for Cancer Research, New York, New York 10021.

Subcellular Biochemistry, Volume 31: Intermediate Filaments, edited by Herrmann and Harris. Plenum Press, New York, 1998.

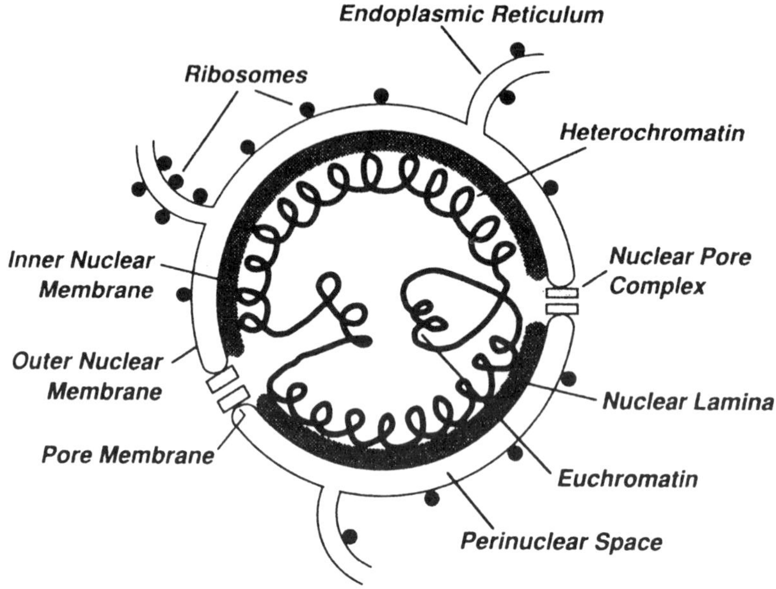

FIGURE 1. Schematic diagram of the nuclear envelope and its associated structures. Reproduced from Worman and Courvalin (1991) by permission.

1986). The lamina is a complex and discontinuous structure in interphase (Paddy *et al.*, 1990; Belmont *et al.*, 1993), and at sites where the lamina is disrupted, the inner nuclear membrane likely interacts directly with chromatin. Lamin filaments may also extend into the interior of the nucleus (Hozak *et al.*, 1995).

The lamina, pore complexes, and chromatin are associated with membranes via integral proteins specific to each membrane domain. Representative integral proteins in the pore and inner nuclear membranes are shown in Figure 2. Lamin B receptors (LBRs) and lamina-associated polypeptides (LAPs) of the inner nuclear membrane interact with chromatin and the lamina. Gp210, an integral protein of the pore membrane domain, may interact with pore complex proteins. Protein components of the pore complexes, lamina, and chromatin that interact with nuclear membrane proteins may in turn interact with intranuclear proteins or nucleic acids, giving rise to the higher ordered structure of the nucleus. As the nuclear lamina interacts with both the inner nuclear membrane and intranuclear proteins, it plays a central role in establishing and maintaining overall nuclear architecture.

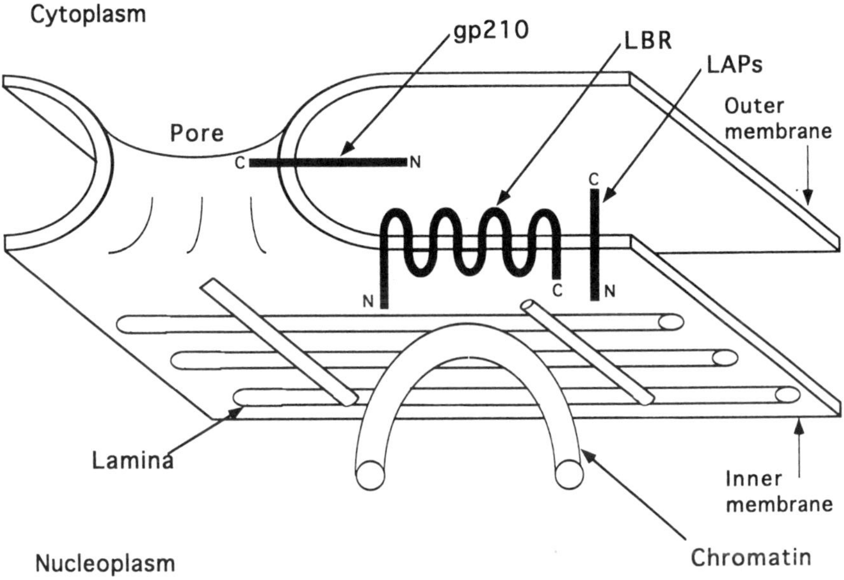

FIGURE 2. Schematic diagram of the nuclear membranes showing the topologies and localizations of representative integral membrane proteins gp210, LBR, and LAPs. The chromatin and lamina are shown associated with the inner nuclear membrane and LBR and LAPs. The pore proteins are not shown. Reproduced from Courvalin and Worman (1997), by permission.

2. NUCLEAR LAMINA

2.1. Structure and Biochemistry

The nuclear lamina is a meshwork of intermediate filaments on the inner aspect of the inner nuclear membrane (Aebi *et al.*, 1986). The lamina forms from the polymerization of lamins, proteins shown by McKeon *et al.* (1986) and Fisher *et al.* (1986) to be members of the intermediate filament protein family. Different lamin proteins interact with themselves and each other to form 10-nm-diameter filaments (Aebi *et al.*, 1986; Georgatos *et al.*, 1988; Ward and Kirschner, 1990; Gieffers and Krohne, 1991; Heitlinger *et al.*, 1991; Moir *et al.*, 1991; Ye and Worman, 1995). Lamins have been identified in a wide range of invertebrate and vertebrate species. Early reports suggested that lamin-like proteins may be present in yeast (Georgatos *et al.*, 1989), however, genes encoding proteins structurally similar to lamins have not been found in the *Saccharomyces cerevisiae* genome, now completely sequenced.

Lamins differ from cytoplasmic intermediate filament proteins in several ways. In vertebrates, the lamins contain a stretch of 42 additional amino acids in their central rod domains (McKeon *et al.*, 1986; Fisher *et al.*, 1986). These additional amino acids are also present in an invertebrate cytoplasmic intermediate filament protein, suggesting that cytoplasmic intermediate filaments may have evolved from a lamin-like ancestral precursor (Dodemont *et al.*, 1990). Lamins contain nuclear localization signals in their tail domains that direct their translocation into the nucleus after synthesis (Holtz *et al.*, 1989). Most lamins also contain a CAAX box at their carboxyl termini and are posttranslationally modified by isoprenylation (Beck *et al.*, 1988; Wolda and Glomset, 1988), specifically farnesylation (Farnsworth *et al.*, 1989). Nuclear lamins are generally divided into two types: A type and B type (Gerace and Blobel, 1980; Krohne and Benavente, 1986). The A-type and B-type lamins differ in their isoelectric points (Gerace and Blobel, 1980). While B-type lamins are permanently farnesylated in cells, A-type lamins either have their farnesyl groups removed by proteolytic processing (Beck *et al.*, 1990; Lutz *et al.*, 1992) or are never farnesylated.

2.2. Human Nuclear Lamins

In mammals, at least four different lamin proteins are expressed in somatic cells, and different isoforms arising from alternative RNA splicing are expressed in germ cells (Table I). One A-type and two B-type lamin genes have been identified in humans. The gene that encodes the A-type lamins A and C, isoforms that arise by alternative RNA splicing, is located on chromosome 1q21.2–21.3 (Lin and Worman, 1993; Wydner *et al.*, 1996). Lamins A and C are identical for the first 566 amino acids. Alternative

Table I
The Human Nuclear Lamin Gene Family

Gene name	Chromosomal localization	Encoded protein	Cell type expressed
LMNA	1q21.2–21.3	Lamin A	Differentiated somatic
		Lamin C	Differentiated somatic
		Lamin C2?	Germ
LMNB1	5q23.2–31.1	Lamin B1	All somatic
LMNB2	19p13.3	Lamin B2	All somatic
		Lamin B3?	Germ

splicing in exon 10 of the lamin A/C gene yields two transcripts, one for lamin C that has 6 unique amino acids and another for lamin A that has 98 unique amino acids (Lin and Worman, 1993). Lamin A is synthesized as a precursor called prelamin A that contains a CAAX box at its carboxyl terminus. Prelamin A is cleaved after nuclear translocation by an endoprotease that removes the last 18 amino acids containing the farnesyl group (Beck *et al.*, 1990; Kilic *et al.*, 1997). Lamin C lacks the carboxyl-terminal CAAX box and is never farnesylated. Lamins A and C are present in equal amounts in the nuclei of most terminally differentiated somatic cells, but both proteins are absent from some undifferentiated and cancer cells (Guilly *et al.*, 1987, 1990; Lebel *et al.*, 1987; Stewart and Burke, 1987; Paulin-Levasseur *et al.*, 1988, 1989; Worman *et al.*, 1988a; Röber *et al.*, 1989, 1990; Kaufmann *et al.*, 1991; Cance *et al.*, 1992; Broers *et al.*, 1993; Hytiroglou *et al.*, 1993). A third isoform similar to lamin C, but with a different head domain, expressed in germ cells, may also arise from the same gene by alternative RNA splicing (Furukawa *et al.*, 1994).

Lamin B1 is encoded by a gene on chromosome 5q23.2–31.3 (Lin and Worman, 1995; Wydner *et al.*, 1996). A second B-type lamin gene encoding lamin B2 is located on chromosome 19p13.3 (Biamanti *et al.*, 1992). Lamin B1 and B2 appear to be expressed in all somatic cells, albeit at different levels. A germ-cell-specific B-type lamin is encoded from the lamin B2 gene and arises from alternative RNA splicing (Furukawa and Hotta, 1993). The findings that B-type lamins are constitutively expressed in somatic cells, that various lamin isoforms arise by alternative RNA splicing, and that A-type lamins are differentially expressed have implications for their cell-type-specific protein–protein interactions.

3. LAMIN B RECEPTOR

3.1. Identification as a Lamin-Binding Protein

Lamins are tightly associated with nuclear membranes although they do not have transmembrane segments. Like cytoplasmic intermediate filaments, nuclear lamins are resistant to extraction with high salt and most detergents, and strong denaturing agents are needed to solubilize them (Gerace *et al.*, 1978). In metaphase, A-type lamins are soluble and distributed randomly in the cytosol, whereas about 50% of B-type lamins remain associated with intracellular membrane vesicles (Gerace and Blobel, 1980; Burke and Gerace, 1986; Chaudhary and Courvalin, 1993). In addition, B-type lamins demonstrate greater resistance to chemical extraction from nuclear membranes than A-type lamins (Gerace *et al.*, 1984). These findings led to the hypothesis that B-type lamins are associated with the inner nuclear membrane via a transmembrane "receptor" which functions as an attachment site for the nuclear lamina.

In the initial experiments to identify B-type-lamin-binding proteins in the inner nuclear membrane, ^{125}I-labeled lamins were incubated with turkey erythrocyte nuclear envelopes that had been extracted with 8M urea to remove endogenous lamins (Worman *et al.*, 1988b). It was found that B-type, but not A-type lamins bound to nuclear envelope membranes in a saturable fashion. An overlay binding assay demonstrated that B-type lamins bound to an abundant protein in turkey erythrocyte nuclear envelopes. This protein had an apparent molecular mass of 58kDa as determined by sodium dodecyl sulfate–polyacrylamide gel electrophoresis (SDS–PAGE), was resistant to alkali extraction from membranes, and was soluble only at relatively high detergent concentration such as 2% Triton X-100 and 2M KCl, indicating that it was an integral membrane protein. Antibodies against the protein partially inhibited the binding of B-type lamins to lamin-depleted nuclear membranes. These data strongly suggested that the protein was a lamin-binding protein in the nuclear envelope and it was termed lamin B receptor (LBR; Worman *et al.*, 1988b). LBR was subsequently localized to the inner nuclear membrane by immunoelectron microscopy using a specific antibody from a battery of monoclonal antibodies raised against abundant proteins of chicken cell nuclei (Bailer *et al.*, 1991). Homologues of avian LBR have been identified in humans and rats (Courvalin *et al.*, 1990; Ye and Worman, 1994). An LBR-like inner nuclear membrane protein has also been identified in an invertebrate (Collas *et al.*, 1996).

Binding of LBR to B-type lamins has been confirmed by several studies. LBR autoantibodies from some patients with primary biliary cirrhosis are antiidiotypic to some B-type lamin autoantibodies, suggesting that the recognized antigens bind to each other (Lassoued *et al.*, 1991). An immunoprecipitation complex from chicken cell nuclear extracts using anti-LBR antibodies contained LBR, B-type lamins, and several other polypeptides, suggesting that LBR and B-type lamins are present in protein complexes at the nuclear envelope (Simos and Georgatos, 1992). When coexpressed in yeast, chicken LBR and lamin B1 colocalize in the nuclear envelope and in adjacent membrane stacks, providing evidence for an LBR–lamin B1 interaction in living cells (Smith and Blobel, 1994). In addition, a glutathione-S-transferase fusion protein of human LBR precipitates B-type lamins from a dilute solution of rat lamins in a pulldown assay (Ye and Worman, 1994). These data demonstrate that LBR binds to B-type lamins and that this interaction may be involved in the attachment of the lamina to the inner nuclear membrane *in vivo*.

3.2. Protein and Gene Structure

Sequencing of cDNA clones for chicken LBR revealed that it is a protein of 637 amino acids (calculated molecular mass approximately 73 kDa) with an amino-terminal domain of 204 amino acids followed by a hydrophobic domain with eight putative transmembrane segments (Worman *et al.*, 1990). Using chicken LBR as a probe, the cDNA for human LBR was isolated and its protein sequence was deduced (Ye and Worman, 1994). Human LBR has 615 amino acids, displays 68% identity in amino acid sequence to chicken LBR, and has the same topological organization. LBR is synthesized without a cleaved amino-terminal signal sequence and the amino-terminal domain that precedes the first transmembrane segment faces the nucleoplasm. The amino-terminal domain is basic with an isoelectric point of approximately 10. Secondary structure analysis of the amino-terminal domain reveals that it has three distinct subdomains: a globular domain between amino acids 1 and 60, a highly charged hinge region between amino acids 60 and 100, and a second globular domain between amino acids 100 and 210 (Ye *et al.*, 1997). The amino-terminal domain of LBR is phosphorylated by protein kinase A and this phosphorylation affects its binding affinity for B-type lamins (Appelbaum *et al.*, 1990). In addition, a threonine residue in the nucleoplasmic domain of LBR is phosphorylated during mitosis by $p34^{cdc2}$ kinase (Courvalin *et al.*, 1992).

The hydrophobic domain of LBR is 30–40% identical in primary structure to yeast and plant proteins that function as sterol reductases (Lorenz and Parks, 1992; Lai *et al.*, 1994; Smith, 1995; Lecain *et al.*, 1996). These proteins contain about 200 amino acids less than LBR and do not have homologous amino-terminal domains. Yeast and plant sterol reductases are probably localized to the endoplasmic reticulum and not the inner nuclear membrane. It is unlikely that they can bind lamins, DNA, or chromatin, as these binding functions reside primarily in the nucleoplasmic, amino-terminal domain of LBR. Whether the transmembrane segments of LBR have enzymatic activities similar to any of these yeast or plant proteins is not entirely clear.

The gene for human LBR has been isolated (Schuler *et al.*, 1994) and localized to chromosome 1q42.1 (Wydner *et al.*, 1996). The human LBR gene is approximately 35 kb long and contains 13 protein-coding exons. The first four exons encode the nucleoplasmic amino-terminal domain and exons 5–13 encode the remainder of the protein. Exons 4 and 5 are separated by a relatively large intron of over 10 kb. The organization of the exons suggests that the LBR gene may have evolved from recombination between a gene encoding a nuclear protein, which became the first four coding exons, and a gene encoding an integral membrane protein similar to yeast sterol reductases, which became the last nine exons (Schuler *et al.*, 1994).

3.3. Targeting to the Inner Nuclear Membrane

LBR is synthesized on endoplasmic reticulum-bound ribosomes, inserted into the membranes, translocated by lateral diffusion to the inner nuclear membrane, and retained there by binding to nuclear proteins and possibly DNA (Soullam and Worman, 1993, 1995; Ellenberg *et al.*, 1997). The nucleoplasmic, amino-terminal domain of LBR can target a heterologous transmembrane segment from a type II integral membrane protein to the inner nuclear membrane (Soullam and Worman, 1993). By itself, the amino-terminal domain accumulates in the nucleoplasm in transfected cells (Soullam and Worman, 1993). Transmembrane segments of LBR devoid of the amino-terminal domain also concentrate in the inner nuclear membrane (Smith and Blobel, 1993; Soullam and Worman, 1995). Hence, LBR has two nonoverlapping inner nuclear membrane targeting signals. The finding that the amino-terminal domain of LBR can target to the inner nuclear membrane suggests that proteins can be retained in the inner nuclear membrane by binding to nuclear ligands such as B-type lamins, heterochromatin proteins, and possibly DNA (see below). The transmembrane segments of LBR can probably oligomerize in the plane of the membrane enabling them to function as a second inner nuclear membrane retention signal. Supporting this hypothesis for inner nuclear membrane targeting and retention is the finding that LBR has a high diffusional mobility in the endoplasmic reticulum membrane, but, upon reaching the inner nuclear membrane, its diffusional mobility is drastically reduced to near zero (Ellenberg *et al.*, 1997).

As inner nuclear membrane proteins must diffuse around the nuclear pore membranes and through the lateral channels of the pore complexes, size constraints imposed by these channels should influence a protein's ability to reach the inner nuclear membrane. Because the lateral channels of the nuclear pore complexes are only 10 nm in diameter (Hinshaw *et al.*, 1992), proteins with cytoplasmically exposed domains of greater than 60 kDa should not be able to reach the inner nuclear membrane. Indeed, if the amino-terminal domain of LBR is enlarged to more than 60 kDa by the insertion of an irrelevant polypeptide or tandem copies of the amino-terminal domain, the constructs can no longer reach the inner nuclear membrane (Soullam and Worman, 1995).

Based on the identified targeting domains of LBR (Smith and Blobel, 1993; Soullam and Worman, 1993, 1995), its diffusional mobility in the endoplasmic reticulum (ER), and inner nuclear membranes (Ellenberg *et al.*, 1997), and the identification of nuclear proteins that bind to LBR (Worman *et al.*, 1988b; Ye and Worman, 1994, 1996; Ye *et al.,* 1997), a general model for the targeting of integral proteins to the nuclear envelope membranes can be proposed. In this model, integral proteins synthesized on endoplasmic reticulum-bound ribosomes with cytoplasmic domains less than approximately 60kDa can rapidly diffuse between the interconnected membranes of the ER and outer, pore, and inner nuclear membranes. If they contain no specific retention sequences, these proteins may transienlty be found in the inner nuclear membrane, but are not retained there, and enter the default secretory pathway as has been demonstrated for some viral integral membrane proteins (Bergman and Singer, 1983; Torrisi and Bonnatti, 1985). Such proteins will not accumulate in the inner nuclear membrane to any significant degree, as mass action forces them back to the endoplasmic reticulum, where their concentration is constantly decreased by vesicle budding to the Golgi. In contrast, integral proteins that reside in the inner nuclear membrane are retained there by binding to nuclear ligands and possibly by oligomerization of transmembrane domains. Inner nuclear membrane proteins must have cytoplasmically exposed domains of less than 60kDa, which is the case for all such proteins that have been characterized by cDNA cloning to date (Worman *et al.*, 1990; Ye and Worman, 1994; Martin *et al.*, 1995; Furukawa *et al.*, 1995; Bione *et al.*, 1994). Binding to nuclear lamins may be a key factor involved in the retention of integral proteins in the inner nuclear membrane, with an example other than LBR being LAP2 (Furukawa *et al.*, 1995).

3.4. Association with Chromatin

Numerous morphological studies have shown that heterochromatin is associated with the inner nuclear membrane and nuclear lamina (Blobel, 1985). Lamins can associate with chromosomes in mitosis (Glass *et al.*, 1993), probably in part by binding to DNA (Luderus *et al.*, 1992) and histones (Yuan *et al.*, 1991; Taniura *et al.*, 1995). Three-dimensional optical sectioning microscopy studies have demonstrated that the lamina is a discontinuous network with spacing between network components (Paddy *et al.*, 1990; Belmont *et al.*, 1993). In the spacing regions, inner nuclear membrane proteins can potentially interact directly with chromatin. The initial primary structure determination of LBR showed that it contained potential DNA-binding motifs (Worman *et al.*, 1990) and several studies have shown that LBR can indeed associate with chromatin and DNA. *In vivo*, LBR-containing vesicles are targeted to decondensing chromosomes early during mitotic nuclear envelope reassembly (Chaudhary and Courvalin, 1993; Buendia and Courvalin, 1997). An LBR-like protein in sea urchin nuclear membranes has also been shown to be responsible for the targeting of membranes to chromatin in an *in vitro* nuclear envelope reassembly assay (Collas *et al.*, 1996). LBR-containing nuclear vesicles and immunopurified LBR on beads associate with chromosomes (Pyrpasopoulou *et al.*, 1996). The amino-terminal domain of LBR also binds directly to double-stranded DNA *in vitro* (Ye and Worman, 1994).

In a yeast two-hybrid screen, two human proteins, $HP1^{Hs\alpha}$ and $HP1^{Hs\gamma}$, homologous to *Drosophila* HP1 have been found to bind to the amino-terminal domain of LBR (Ye and Worman, 1996). The interactions have also been confirmed in several different *in vitro* binding and coimmuno-precipitation assays (Ye and Worman, 1996). *Drosophila* and mammalian HP1-type proteins have been localized to heterochromatin (James and Elgin, 1986; James *et al.*, 1989; Wreggett *et al.*, 1994). The interactions between HP1-type proteins and LBR are mediated by the so-called chromoshadow domain of HP1-type proteins and the second globular domain of LBR (Ye *et al.*, 1997). LBR may therefore preferentially bind to heterochromatin at the inner nuclear membrane.

Thus, several studies demonstrate that LBR interacts with chromatin and heterochromatin through both protein–protein and possibly protein–DNA interactions. In addition to attaching the nuclear lamina, LBR may also be directly involved in the association of chromatin with the inner nuclear membrane. Binding of LBR to chromatin may also be responsible for the targeting of membranes to decondensing chromosomes in anaphase and initiating nuclear envelope reassembly during mitosis (Collas *et al.*, 1996).

4. LAMINA-ASSOCIATED PROTEINS

4.1. Identification and Lamin Binding

LAPs were first discovered using monoclonal antibodies raised against rat liver nuclear envelopes. One single monoclonal antibody recognized 75-kDa (LAP1A), 68-kDa (LAP1B), and 55-kDa (LAP1C) proteins in rat liver nuclear envelopes (Senior and Gerace, 1988). A second antibody recognized a different 63-kDa protein which was termed LAP2 (Foisner and Gerace, 1993). LAP1A, LAP1B, and LAP2 remained associated with the lamina and LAP1C was soluble when nuclear envelopes were extracted with Triton X-100 and increasing concentrations of salt. *In vitro* binding experiments showed that immunoaffinity-purified LAP1A and LAP1B could bind to lamins A/C and B1, while LAP2 only bound lamin B1 (Senior and Gerace, 1988; Foisner and Gerace, 1993). When LAP1C was examined in a cell fusion experiment where rat and mouse or rat and hamster tissue culture cells were fused to form heterokaryons, LAP1C was found to migrate from the rat nucleus to the mouse or hamster cell nucleus, but only if the fused cells expressed lamins A/C (Powell and Burke, 1990). This *in vivo* nuclear accumulation and apparent association with lamins A/C occurred even though the heterokaryon nuclei remained unfused, the mouse or hamster cell lamins remained in their original nuclei, and new protein synthesis was blocked by cyclohexamide.

4.2. Protein Structure

The LAP2 (Furukawa *et al.*, 1995) and LAP1C (Martin *et al.*, 1995) cDNAs have been cloned and sequenced. Sequencing of rodent LAP2 showed it to be the homologue of human thymopoietin-β, a protein from a gene on chromosome 12 encoding several alternatively spliced isoforms proposed to be regulators of gene expression in immunologic cells (Harris *et al.*, 1994, 1995) [for new splice forms characterized in mouse, see Berger *et al.* (1996)]. The exact function of LAP2/thymopoietin-β in immunologic cells, if any, remains to be established. LAPs 1A, 1B, and 1C appear to be alternatively spliced variants from the same gene, although the amino acid sequence of only LAP1C is currently known (Martin *et al.*, 1995). Sequencing of the cDNAs encoding LAP1C and LAP2 predicts that they are both type II integral membrane proteins with single transmembrane segments spanning the inner nuclear membrane, hydrophilic amino-terminal domains of 40–50 kDa that face the nucleoplasm, and short carboxyl-terminal domains in perinuclear space (Martin *et al.*, 1995; Furukawa *et al.*, 1995).

4.3. Association with Chromatin

Immunoaffinity-purified LAP2 has been shown to bind mitotic chromosomes saturably and with high affinity (Foisner and Gerace, 1993). LAP1A and LAP1B also associate with mitotic chromosomes, but their interactions appear to be indirect since purified LAP1A and LAP1B cannot bind to chromosomes in an *in vitro* binding assay. LAP1C does not appear to bind to DNA, but this may be misleading since LAP1C's binding characteristics are changed upon purification and membrane extraction as seen by its *in vivo* association with lamins and its lack of binding *in vitro* (Powell and Burke, 1990; Senior and Gerace, 1988).

LAP association with mitotic chromosomes does not require nuclear lamins. When mitotic chromosomes were incubated with urea-extracted nuclear envelopes depleted of approximately 99% of their lamins, antibodies to LAPs 1 and 2 immunostained the surfaces of the chromosomes. This association was lost, however, when the LAPs were preincubated in mitotic extracts and phosphorylated, suggesting that mitotic phosphorylation inhibited chromosome binding. LAPs incubated in interphase extracts, even with ATP, or LAP2 incubated in mitotic extracts, but treated with alkaline phosphatase, regained chromatin binding (Foisner and Gerace, 1993). The proteins in chromatin with which LAPs may associate have not been identified.

5. OTHER LAMIN-BINDING PROTEINS

5.1. Lamin-Modifying Enzymes

Lamins undergo posttranslational modifications and the enzymes that catalyze these events can be considered lamin-binding proteins. Both A-type and B-type lamins are phosphorylated during mitosis at sites which have been shown to be recognized by $p34^{cdc2}$. Protein kinase C has also been shown to phosphorylate A-type (Haas and Jost, 1993) and B-type (Hocevar et al., 1993; Hennekes *et al.*, 1993; Collas *et al.*, 1997) lamins at sites near their nuclear localization signals. Some A-type and all B-type lamins are farnesylated at a CAAX-box sequence that must be recognized by cellular farnesyltransferase or other polypeptides in an enzymatic complex (for a review see Zhang and Casey, 1996). Lamin A has one further processing event which requires a specific endoprotease. The endoprotease that cleaves farnesylated prelamin A must also transiently recognize sequences in this protein (Kilic *et al.*, 1997).

During apoptosis, the lamina is proteolytically degraded before or during the characteristic fragmentation of DNA into an apoptotic ladder. Lamin cleavage is mediated by one or more interleukin 1β-converting enzyme (ICE)-like proteases and may also involve another serine protease termed the nuclear scaffold (NS) protease (Lazebnik *et al.*, 1995; McConkey, 1996; Takahashi *et al.*, 1996). If cleavage of the nuclear lamina is specifically inhibited using a peptide mimic of the lamin ICE-like protease, chromatin is still fragmented, but the cell is prevented from undergoing complete nuclear fragmentation and apoptosis (Lazebnik *et al.*, 1995). Lamin A was shown to be cleaved *in vitro* by Mch2α, an ICE-like protease, in the middle of the rod domain, producing the same fragments as those seen in apoptotic cells (Takahashi *et al.*, 1996). In thymocytes, peptide inhibitors specific to ICE-like proteases blocked lamin B1 degradation, but inhibitors of Ca^{2+}-dependent serine proteases were even more effective, blocking all apoptotic events, including DNA fragmentation and lamin degradation, events which normally occurred in cells subjected to high Ca^{2+}

levels (McConkey, 1996). These data indicate that there are multiple pathways for apoptosis, but that blocking the degradation of the nuclear lamina is a key for preventing other apoptotic degradations, likely by preventing the fragmentation of the nuclear envelope and preventing the release of lamin-associated heterochromatin, events which both allow other proteases access to nuclear proteins and DNA. Along these lines, expression of an uncleavable lamin mutant attenuates apoptosis and delays cell death and associated DNA fragmentation (Rao *et al.*, 1996).

5.2. Cytoplasmic Intermediate Filament Binding to Lamins

Most interactions between cytoplasmic intermediate filament proteins and lamins have been shown by *in vitro* binding assays. Peripherin, a neuronal cytoplasmic intermediate filament protein, has been shown to bind to B-type lamins (Djabali *et al.*, 1991). Gel overlay assays have also demonstrated that B-type lamins can bind to desmin, but only if desmin's central rod domain was intact (Georgatos *et al.*, 1987). When affinity columns were made with the carboxyl-terminal tail domains of vimentin, B-type lamins bound to the column (Georgatos and Blobel, 1987). Mitotic membrane vesicles that contained B-type lamins were also found to bind to vimentin *in vitro* in a saturable manner (Maison *et al.*, 1995). These findings lead to the speculation that lamins and cytoplasmic intermediate filaments might bind *in vivo*, coming in contact possibly through nuclear pores, and that such interactions stabilize nuclear structure in interphase and possibly organize membrane vesicles that contain lamins in mitosis. Since this binding is difficult to confirm *in vivo* and since no cytoplasmic intermediate filaments have been seen in electron micrographs to penetrate the nuclear pore complex, interactions between lamins and cytoplasmic intermediate filament proteins might just reflect the evolutionary closeness and similar binding properties of members of this protein family.

5.3. *In Vitro* Lamin-Binding Proteins

In vitro binding assays have been used to demonstrate interactions between lamins and other proteins. For most of these protein–protein interactions, the biological significance is not yet clear. As mentioned above, peripherin, vimentin, and desmin bind to lamins *in vitro*. *In vitro* binding assays also suggest that A-type lamins interact with histones (Yuan *et al.*, 1991) and that this interaction is mediated primarily by the tail domains of lamins A and C (Taniura *et al.*, 1995).

The retinoblastoma gene product pRb has also been shown to bind to lamins A/C *in vitro*, the interaction being specific to a part of the rod domain of lamins A/C and to a known protein–protein binding pocket of pRb (Ozaki *et al.*, 1994). This interaction is only seen during the early G_1 phase of the cell cycle when pRb is hypophosphorylated, also the time when pRb associates with viral oncoproteins and functions as a transcription factor. Immunoelectron microscopy staining shows that pRb is inside the nucleus and associated with the lamina and inner nucleoskeleton. Immunostaining also showed colocalization of lamins A/C and pRb at the nuclear periphery in tissue culture cells (Mancini *et al.*, 1994). The significance of this association is unclear, but when taken together with studies showing that lamins extend into the interior of the nucleus (Hozak *et al.*, 1995) and that several transcription factors such as SP-1 and AP-1 (van Wijnen *et al.*, 1993) and c-myc (Waitz and Loidl, 1991) become attached to the nuclear matrix during G_1 or S phases of the cell cycle, it is possible that lamins may play a role in stabilizing transcriptional complexes or in targeting and retaining transcription factors within the nucleus.

5.4. Proteins Colocalizing with Lamins

Several proteins have been identified which colocalize with nuclear lamins in immunofluorescence or immunoelectron microscopy experiments. Their peripheral nuclear envelope localization make it likely that future studies will confirm direct or indirect binding of these proteins to nuclear lamins. In *Drosophila*, the fs(1)Ya protein has been localized to the nuclear envelopes in early embryos. The fs(1)Ya protein is essential for the first nuclear division after the fusion of egg and sperm pronuclei (Lin and Wolfner, 1991). Double-label immunofluorescence of early embryos with antilamin and anti-fs(1)Ya antibodies shows that the proteins colocalize (Lopez *et al.*, 1994). During mitosis, both lamin and fs(1)Ya proteins are released into the cytoplasm, but fs(1)Ya is relocalized to the nuclear envelope in late telophase after the lamina reassembles. The cDNA of fs(1)Ya indicates that it is a nonmembrane protein with two zinc fingers and a $p34^{cdc2}$ consensus phosphorylation site (Lin and Wolfner, 1991). Since the zinc finger motif is known to bind DNA, it is possible that fs(1)Ya is a protein which links the lamina to DNA.

Another *Drosophila* protein which colocalizes with the nuclear lamina is otefin. Antibodies directed against otefin show by immunoelectron microscopy that otefin is on the inner nuclear face of the nuclear envelope, and immunofluorescence labeling shows that otefin colocalizes with the lamina throughout interphase and mitosis (Harel *et al.*, 1989). The cDNA sequence predicts a 45-kDa protein with a hydrophilic amino-terminal domain and a 17-amino acid carboxyl-terminal hydrophobic domain that probably mediates the association of otefin with the inner nuclear membrane (Padan *et al.*, 1990). Otefin is a peripheral protein that can be extracted from nuclear envelopes with detergents (Ashery-Padan *et al.*, 1997). Direct interaction between otefin and other nuclear proteins remains to be shown, but indirect evidence points to otefin as a nonintegral membrane type of lamin-associated protein.

In avian erythrocytes several proteins appear to form a complex with LBR in the inner nuclear membrane (Simos and Georgatos, 1992). One of these is an 18-kDa protein, p18, which appears to be an integral membrane protein (Simos *et al.*, 1996). Antibodies raised against this protein coprecipitate LBR and B-type lamins, indicating that they bind each other *in vitro*. Further indication that p18 binds to B-type lamins *in vivo* is the colocalization of the two proteins in turkey erythrocytes as seen by double immunofluorescent labeling. The function of p18 and whether or not it exists in cell types other than turkey erythrocytes remain to be determined.

Emerin is encoded by the STA gene, which is responsible for the X-linked recessive disease Emery–Dreifuss muscular dystrophy (Bione *et al.*, 1994). Emerin is a 254-amino acid protein with a serine-rich hydrophilic amino-terminal domain and a 34-amino acid carboxyl-terminal hydrophobic transmembrane segment. Thirty-nine amino acids in the amino-terminal domain share 41% homology with the same region in LAP2 and the putative transmembrane domain also has 41% identity to the transmembrane domain of LAP2 (Bione *et al.*, 1994). Emerin has been localized to the inner nuclear membrane of muscle cells, but is absent from this location in the cells of patients with Emery–Dreifuss muscular dystrophy (Manilal *et al.*, 1996; Nagano *et al.*, 1996). Emerin's inner nuclear membrane localization and similarity to LAPs suggest that it may also interact with lamins. How mutations in emerin cause muscular dystrophy is not known.

The MAN antigens were discovered by screening tissue culture cells with serum from a patient with an autoimmune disease (Paulin-Levasseur *et al.*, 1996). Three proteins of 78, 58, and 40 kDa were detected with these antibodies, all of which were extractable from HeLa cells using conditions that also extracted nuclear lamins. Purified murine lamins did not cross-react with the anti-MAN antibodies, indicating that the proteins were nonlamin components associated with the lamina. Immunofluorescence microscopy using MAN autoantibodies labeled various interphase tissue culture cells with the same peripheral nuclear envelope pattern as antilamin antibodies. Mitotic HeLa cells were also labeled with anti-MAN and antilamin antibodies, which showed that the MAN antigens became cytoplasmic during prophase, but became associated with the reassembling nuclear envelope at the same time as the B-type lamins. Complementary DNA cloning and sequencing of these proteins have not yet been reported.

The B-type nuclear lamins are associated with origins of DNA replication in cells and play a role in the assembly of replication-competent nuclei in an *in vitro Xenopus* nuclear assembly assay (Moir *et al.*, 1994; Goldberg *et al.*, 1995). Expression of a dominant-negative lamin A mutant which causes disruption of the nuclear lamina results in redistribution of proteins required for the elongation phase of DNA replication, but not of proteins involved in the initiation phase of DNA replication (Spann *et al.*, 1997). These findings suggest that nuclear lamins may interact with proteins that function in DNA replication.

6. LAMIN-BINDING PROTEINS IN MITOSIS

During mitosis, the nuclear envelope disassembles and two nuclear envelopes subsequently reassemble in the daughter cells. The nuclear membranes either break down into vesicles (Lohka and Masui, 1983; Newport and Spann, 1987; Vigers and Lohka, 1991; Chaudhary and Courvalin, 1993; Buendia and Courvalin, 1997) or are dispersed into an endoplasmic reticulum-like membrane system (Ellenberg *et al.*, 1997; Yang *et al.*, 1997). The nuclear lamina is reversibly depolymerized during mitosis, with A-type lamins becoming soluble in the cytoplasm and B-type lamins becoming partially soluble and partially associated with membranes (Gerace and Blobel, 1980; Burke and Gerace, 1986; Chaudhary and Courvalin, 1993). The depolymerization of the lamina at the start of mitosis results from the phosphorylation of the lamins by $p34^{cdc2}$ and possibly other protein kinases (Gerace and Blobel, 1980; Ward and Kirschner, 1990; Heald and McKeon, 1990; Peter *et al.*, 1990; Dessev *et al.*, 1991; Goss *et al.*, 1994).

The breakdown of the nuclear membranes or their loss of differentiation from the endoplasmic reticulum likely results from the loss of association of inner nuclear membrane proteins with the lamins or other proteins. LBR, which is virtually immobile in the inner nuclear membrane in interphase, has a much higher lateral diffusional mobility in metaphase (Ellenberg *et al.*, 1997). This likely results from dissociation from B-type lamins and HP1-type polypeptides. LBR undergoes site-specific phosphorylation by $p34^{cdc2}$ protein kinase during mitosis and is phosphorylated at different sites during interphase (Courvalin *et al.*, 1992). LAP2 is also phosphorylated during mitosis and this phosphorylation may decrease its binding affinities for lamins and chromatin *in vivo* (Foisner and Gerace, 1993). The cell-cycle-dependent phosphorylation of lamins and their binding proteins LBR and LAPs may result in disruption of protein–protein interactions that occur in interphase. Loss of these interactions early in mitosis may lead to disassembly of the nuclear envelope, and subsequent reestablishment may result in reassembly. Hence, cell-cycle-dependent changes in the interactions of lamin-binding proteins with the lamina and chromatin may be responsible, at least in part, for the disassembly and reassembly of the nucleus during mitosis.

7. REFERENCES

Aebi, U., Cohn, J., Buhle, L., and Gerace, L., 1986, The nuclear lamina is a meshwork of intermediate-type filaments, *Nature* **323:**560–564.

Appelbaum, J., Blobel G., and Georgatos, S. D., 1990, *In vivo* phosphorylation of the lamin B receptor. Binding of lamin B to its nuclear membrane receptor is affected by phosphorylation, *J. Biol. Chem.* **265:**4181–4184.

Ashery-Padan, R., Weiss, A. M., Feinstein, N., and Gruenbaum, Y., 1997, Distinct regions specify the targeting of otefin to the nucleoplasmic side of the nuclear envelope, *J. Biol. Chem.* **272:**2493–2499.

Bailer, S. M., Eppenberger, H. M., Griffiths, G., and Nigg, E. A., 1991, Characterization of A 54-kD protein of the inner nuclear membrane: Evidence for cell cycle-dependent interaction with the nuclear lamina, *J. Cell Biol.* **114:**389–400.

Bastos, R., Panté, N., and Burke, B, 1995, Nuclear pore complex proteins, *Int. Rev. Cytol.* **162B:**257–302.

Beck, L. A., Hosick, T. J., and Sinensky, M., 1988, Incorporation of a product of mevalonic acid metabolism into proteins of Chinese hamster ovary cell nuclei, *J. Cell Biol.* **107:**1307–1316.

Beck, L. A., Hosick, T. J., and Sinensky, M., 1990, Isoprenylation is required for the processing of the lamin A precursor, *J. Cell Biol.* **110:**1489–1499.

Belmont, A. S., Zhai, Y., and Thilenius, A., 1993, Lamin B distribution and association with peripheral chromatin revealed by optical sectioning and electron microscopy tomography, *J. Cell Biol.* **123:**1671–1685.

Berger, R., Theodor, L., Shoham, J., Gokkel, E., Brok-Simoni, F., Avraham, K. B., Copeland, N. G., Jenkins, N. A., Rechavi, G., and Simon, A. J., 1996, The characterization and localization of the mouse thymopoietin/lamina-associated polypeptide 2 gene and its alternatively spliced products, *Genome Res.* **6:**361–370.

Bergman, J., and Singer, J., 1983, Immunoelectron microscopic studies of the intracellular transport of the membrane glycoprotein (G) of vesicular stomatitis virus in infected Chinese hamster ovary cells, *J. Cell Biol.* **97:**1777–1787.

Biamanti, G., Giacca, M., Perini, G., Contreas, G., Zentilin, L., Weighardt, F., Guerra, M., Della Valle, G., Saccone, S., Riva, S., and Falaschi, A., 1992, The gene for a novel human nuclear lamin maps at a highly transcribed locus of chromosome 19p which replicates at the onset of S-phase, *Mol. Cell. Biol.* **12:**3499–3506.

Bione, S., Maestrini, E., Rivella, S., Mancini, M., Regis, S., Romeao, G., and Toniolo, D., 1994, Identification of a novel X-linked gene responsible for Emery–Dreifuss muscular dystrophy, *Nature Genet.* **8:**323–327.

Blobel, G., 1985, Gene gating: A hypothesis, *Proc. Natl. Acad. Sci. USA* **82:**8527–8529.

Broers, J. L. V., Raymond, Y., Rot, M. K., Kuijpers, H., Wagenaar, S. S., and Ramaekers, F. C. S., 1993, Nuclear A-type lamins are differentially expressed in human lung cancer subtypes, *Am. J. Pathol.* **143:**211–220.

Buendia, B., and Courvalin, J.-C., 1997, Domain-specific disassembly and reassembly of nuclear membranes during mitosis, *Exp. Cell Res.* **230:**133–144.

Burke, B., and Gerace L., 1986, A cell free system to study reassembly of the nuclear envelope at the end of mitosis, *Cell* **44:**639–652.

Cance, W. G., Chaudhary, N., Worman, H. J., Blobel, G., and Cordon-Cardo, C., 1992, Expression of the nuclear lamins in normal and neoplastic human tissues, *J. Exp. Clin. Cancer Res.* **11:**233–246.

Chaudhary, N., and Courvalin, J.-C., 1993, Stepwise reassembly of the nuclear envelope at the end of mitosis, *J. Cell Biol.* **122:**295–306.

Collas, P., Courvalin, J.-C., and Poccia, D., 1996, Targeting of membranes to sea urchin sperm chromatin is mediated by a lamin B receptor-like integral membrane protein, *J. Cell Biol.* **135:**1715–1725.

Collas, P., Thompson, L., Fields, A. P., Poccia, D. L., and Courvalin, J.-C., 1997, Protein kinase C-mediated interphase lamin B phosphorylation and solubilization, *J. Biol. Chem.* **272:**21274–21280.

Courvalin, J.-C., and Worman, H. J., 1997, Nuclear envelope protein autoantibodies in primary biliary cirrhosis, *Semin. Liver Dis.* **17:**79–90.

Courvalin, J.-C., Lassoued, K., Worman, H. J., and Blobel, G., 1990, Identification and characterization of autoantibodies against the nuclear envelope lamin B receptor from patients with primary biliary cirrhosis, *J. Exp. Med.* **172:**961–967.

Courvalin, J.-C., Segil, N., Blobel, G., and Worman, H. J., 1992, The lamin B receptor of the inner nuclear membrane undergoes mitosis-specific phosphorylation and is a substrate for $p34^{cdc2}$-type protein kinase, *J. Biol. Chem.* **267:**19035–19038.

Davis, L. I., 1995, The nuclear pore complex, *Annu. Rev. Biochem.* **64:**865–896.

Dessev, G., Iovcheva-Dessev, C., Bischoff, J. R., Beach, D., and Goldman, R. D., 1991, A complex containing $p34^{cdc2}$ and cyclin B phosphorylates the nuclear lamin and disassembles nuclei of clam oocytes *in vitro*, *J. Cell Biol.* **112:**523–533.

Djabali, K., Portier, M.-M., Gros, F., Blobel, G., and Georgatos, S. D., 1991, Network antibodies identify nuclear lamin B as a physiological attachment site for peripherin intermediate filaments, *Cell* **64:**109–121.

Dodemont, H., Riemer, D., and Weber, K., 1990, Structure of an invertebrate gene encoding cytoplasmic intermediate filament (IF) proteins: Implications for the origin and the diversification of IF proteins, *EMBO J.* **9:**4073–4081.

Ellenberg, J., Siggia, E. D., Moreira, J., Smith, C. L., Presley, J. F., Worman, H. J., and Lippincott-Schwartz, J., 1997, Nuclear envelope dynamics and reassembly in living cells: Targeting of an inner nuclear membrane protein in interphase and mitosis, *J. Cell Biol.* **138:**1193–1206.

Farnsworth, C. C., Wolda, S. L., Gelb, M. H., and Glomset, J. A., 1989, Human lamin B contains a farnesylated cysteine residue, *J. Biol. Chem.* **264:**20422–20429.

Fisher, D. Z., Chaudhary, N., and Blobel, G., 1986, cDNA sequencing of nuclear lamins A and C reveals primary and secondary structural homology to intermediate filament proteins, *Proc. Natl. Acad. Sci. USA* **83:**6450–6454.

Foisner, R., and Gerace, L., 1993, Integral membrane proteins of the nuclear envelope interact with lamins and chromosomes, and binding is modulated by mitotic phosphorylation, *Cell* **73:**1267–1279.

Furukawa, K., and Hotta, Y., 1993, cDNA cloning of a germ line specific lamin B3 from mouse spermatocytes and analysis of its function by ectopic expression in somatic cells, *EMBO J.* **12:**97–106.

Furukawa, K., Inagaki, H., and Hotta, Y., 1994, Identification and cloning of an mRNA coding for a germ cell-specific A-type lamin in mice, *Exp. Cell Res.* **212:**426–430.

Furukawa, K., Panté, N., Aebi, U., and Gerace, L., 1995, Cloning of a cDNA for lamina-associated polypeptide 2 (LAP2) and identification of regions that specify targeting to the nuclear envelope, *EMBO J.* **14:**1626–1636.

Georgatos, S. D., and Blobel, G., 1987, Lamin B constitutes an intermediate filament attachment site at the nuclear envelope, *J. Cell Biol.* **105:**117–125.

Georgatos, S. D., Weber, K., Geisler, N., and Blobel, G., 1987, Binding of two desmin derivatives to the plasma membrane and the nuclear envelope of avian erythrocytes: Evidence for a conserved site-specificity in intermediate filament-membrane interactions, *Proc. Natl. Acad. Sci. USA* **84:**6780–6784.

Georgatos, S. D., Stournaras, C., and Blobel, G., 1988, Heterotypic and homotypic association between the nuclear lamins, *Proc. Natl. Acad. Sci. USA* **85:**4325–4329.

Georgatos, S. D., Maroulakou, I., and Blobel, G., 1989, Lamin A, lamin B, and lamin B receptor analogues in yeast, *J. Cell. Biol.* **108:**2069–2082.

Gerace, L., and Blobel, G., 1980, The nuclear envelope lamina is reversibly depolymerized during mitosis, *Cell* **19:**277–287.

Gerace, L., Blum, A., and Blobel, G., 1978, Immunocytochemical localization of the major polypeptides of the nuclear pore complex-lamina fraction. Interphase and mitotic distribution, *J. Cell Biol.* **79:**546–566.

Gerace, L., Comeau, C., and Benson, M., 1984, Organization and modulation of nuclear lamina structure, *J. Cell Sci. Suppl.* **1:**137–160.

Gieffers, C., and Krohne, G., 1991, *In vitro* reconstitution of recombinant lamin A and a lamin A mutant lacking the carboxy-terminal tail, *Eur. J. Cell Biol.* **55:**191–199.

Glass, C. A., Glass, J. R., Taniura, H., Hasel, K. W., Blevitt, J. M., and Gerace, L., 1993, The alpha-helical rod domain of human lamins A and C contains a chromatin binding site, *EMBO J.* **12:**4413–4424.

Goldberg, M. W., and Allen, T. D., 1995, Structural and functional organization of the nuclear envelope, *Curr. Opin. Cell Biol.* **7:**301–309.

Goldberg, M., Jenkins, H., Allen, T., Whitfield, W. G., and Hutchison, C. J., 1995, *Xenopus* lamin B3 has a direct role in the assembly of a replication competent nucleus: Evidence from cell-free egg extracts, *J. Cell Sci.* **108:**3451–3461.

Goss, V. L., Hocevar, B. A., Thompson, L. J., Stratton, C. A., Burns, D. J., and Fields, A. P., 1994, Identification of nuclear beta II protein kinase C as a mitotic lamin kinase, *J. Biol. Chem.* **269:**19074–19080.

Guilly, M.-N., Bensussan, A., Bourge, J. F, Bornens, M., and Courvalin, J.-C., 1987, A human lymphoblastic cell line lacks lamins A and C, *EMBO J.* **6:**3795–3799.

Guilly, M.-N., Kolb, J.-P., Gosti, F., Godeau, F., and Courvalin, J.-C., 1990, Lamins A and C are not expressed at early stages of human lymphocyte development, *Exp. Cell Res.* **189:**145–148.

Haas, M., and Jost, E., 1993, Functional analysis of phosphorylation sites in human lamin A controlling lamin disassembly, nuclear transport and assembly, *Eur. J. Cell Biol.* **62:**237–247.

Harel, A., Zlotkin, E., Nainudel-Epszteyn, S., Feinstein, N., Fisher, P. A., and Gruenbaum, Y., 1989, Persistence of major nuclear envelope antigens in an envelope-like structure during mitosis in *Drosophila melanogaster* embryos, *J. Cell Sci.* **94:**463–470.

Harris, C. A., Andryuk, P. J., Cline, S., Chan, H. K., Natarajan, A., Siekierka, J. J., and Goldstein G., 1994, Three distinct human thymopoietins are derived from alternatively spliced mRNAs, *Proc. Natl. Acad. Sci USA* **91:**6283–6287.

Harris, C. A., Andryuk, P. J., Cline, S., Sierkierka, J. J., and Goldstein, G., 1995, Structure and mapping of the human thymopoietin (TMPO) gene and relationship of human TMPO β to the rat lamin-associated polypeptide 2, *Genomics* **28:**198–205.

Heald, R., and McKeon, F. D., 1990, Mutations of phosphorylation sites in lamin A that prevent nuclear lamina disassembly in mitosis, *Cell* **61:**579–590.

Heitlinger, E., Peter, M., Häner, M., Lustig, A., Aebi, U., and Nigg, E. A., 1991, Expression of chicken lamin B2 in *Escherichia coli:* Characterization of its structure, assembly and molecular interactions, *J. Cell Biol.* **113:**485–495.

Hennekes, H., Peter, M., Weber, K., and Nigg, E. A., 1993, Phosphorylation on protein kinase C sites inhibits nuclear import of lamin B_2, *J. Cell Biol.* **120:**1293–1304.

Hinshaw, J. E., Carragher, B. O., and Milligan, R. A., 1992, Architecture and design of the nuclear pore complex, *Cell* **69:**1133–1141.

Hocevar, B. A., Burns, D. J., and Fields, A. P., 1993, Identification of protein kinase C (PKC) phosphorylation sites on human lamin B. Potential role of PKC in nuclear lamina structural dynamics, *J. Biol. Chem.* **268:**7545–7552.

Holtz, D., Tanaka, R. A., Hartwig, J., and McKeon, F., 1989, The CaaX motif of lamin A functions in conjunction with the nuclear localization signal to target assembly to the nuclear envelope, *Cell* **59:**969–977.

Hozak, P., Sasseville, A. M., Raymond, Y., and Cook, P. R., 1995, Lamin proteins form an internal nucleoskeleton as well as a peripheral lamina in human cells, *J. Cell Sci.* **108:**635–644.

Hytiroglou, P., Choi, S. W., Theise, N. D., Chaudhary, N., Worman, H. J., and Thung, S. N., 1993, The expression of nuclear lamins in human liver: An immunohistochemical study, *Hum. Pathol.* **24:**169–172.

James, T. C., and Elgin, S. C. R., 1986, Identification of a nonhistone chromosomal protein associated with heterochromatin in *Drosophila melanogaster* and its gene, *Mol. Cell. Biol.* **6:**3862–3872.

James, T. C., Eissenberg, J. C., Craig, C., Dietrich, V., Hobson, A., and Elgin, S. C. R., 1989, Distribution patterns of HP1, a heterochromatin-associated nonhistone chromosomal protein of *Drosophila*, *Eur. J. Cell Biol.* **50:**170–180.

Kaufmann, S. H., Mabry, M., Jasti, R., and Shaper, J. H., 1991, Differential expression of nuclear envelope lamins A and C in human lung cancer cell lines, *Cancer Res.* **51:**581–586.

Kilic, F., Dalton, M. B., Burrell, S. K., Mayer, J. P., Patterson, S. D., and Sinensky, M., 1997, *In vitro* assay and characterization of the farnesylation-dependent prelamin A endoprotease, *J. Biol. Chem.* **272:**5298–5304.

Krohne, G., and Benavente, R., 1986, The nuclear lamins. A multigene family of proteins in evolution and differentiation, *Exp Cell Res.* **162:**1–10.

Lai, M. H., Bard, M., Pierson, C. A., Alexander, J. F., Goebl, M., Carter, G. T., and Kirsch, D. R., 1994, The identification of a gene family in the *Saccharomyces cerevisiae* ergosterol biosynthesis pathway, *Gene* **140:**41–49.

Lassoued, K., Danon, F., and Brouet, J.-C., 1991, Human autoantibodies to lamin B receptor are also anti-idiotypic to certain anti-lamin B antibodies, *Eur. J. Immunol.* **21:**1959–1962.

Lazebnik, Y. A., Tkahashi, A., Moir, R. D., Goldman, R. D., Poirier, G. G., Kaufmann, S. H., and Earnshaw, W. C., 1995, Studies of the lamin proteinase reveal multiple parallel biochemical pathways during apoptotic execution, *Proc. Natl. Acad. Sci. USA* **92:**9042–9046.

Lebel, S., Lampron, C., Royal, A., and Raymond, Y., 1987, Lamins A and C appear during retinoic acid-induced differentiation of mouse embryonal teratocarcinoma cells, *J. Cell Biol.* **105:**1099–1104.

Lecain, E., Chenivesse, X., Spagnoli, R., and Pompon, D., 1996, Cloning by metabolic interference in yeast and enzymatic characterization of *Arabidopsis thaliana* sterol delta 7-reductase, *J. Biol. Chem.* **271:**10866–10873.

Lin, H., and Wolfner, M. F., 1991, The *Drosophila* maternal-effect gene *fs(1)Ya* encodes a cell cycle-dependent nuclear envelope component required for embryonic mitosis, *Cell* **64:**49–62.

Lin, F., and Worman, H. J., 1993, Structural organization of the human gene encoding nuclear lamin A and nuclear lamin C, *J. Biol. Chem.* **268:**16321–16326.

Lin, F., and Worman, H. J., 1995, Structural organization of the human gene (LMNB1) encoding nuclear lamin B1, *Genomics* **27:**230–236.

Lohka, M. J., and Masui, Y., 1983, Formation *in vitro* of sperm pronuclei and mitotic chromosomes induced by amphibian ooplasmic components, *Science* **220:**719–721.

Lopez, J. M., Song, K., Hirshfeld, A. B., Lin, H., and Wolfner, M. F., 1994, The *Drosophila* fs(1)Ya protein, which is needed for the first mitotic division, is in the nuclear lamina and in the envelopes of cleavage nuclei, pronuclei, and nonmitotic nuclei, *Dev. Biol.* **163:**202–211.

Lorenz, R. T., and Parks, L. W., 1992, Cloning, sequencing, and disruption of the gene encoding sterol C-14 reductase in *Saccharomyces cerevisiae*, *DNA Cell Biol.* **11:**685–692.

Luderus, M. E., de Graaf, A., Mattia, E., den Blaauwen, J. L., Grande, M. A., de Jong, L., and van Driel, R., 1992, Binding of matrix attachment regions to lamin B1, *Cell* **70:**949–959.

Lutz, R. J., Trujillo, M. A., Denham, K. S., Wenger, L., and Sinensky, M., 1992, Nucleoplasmic localization of prelamin A: Implications for prenylation-dependent lamin A assembly into the nuclear lamina, *Proc. Natl. Acad. Sci. USA* **89:**3000–3004.

Maison, C., Pyrpasopoulou, A., and Georgatos, S. D., 1995, Vimentin-associated mitotic vesicles interact with chromosomes in a lamin B- and phosphorylation-dependent manner, *EMBO J.* **14:**3311–3324.

Mancini, M. A., Shan, B., Nickerson, J. A., Penman, S., and Lee, W.-H., 1994, The retinoblastoma gene product is a cell cycle-dependent, nuclear matrix-associated protein, *Proc. Natl. Acad. Sci. USA* **91:**418–422.

Manilal, S., Man, N. T., Sewry, C. A., and Morris, G. E., 1996, The Emery–Dreifuss muscular dystrophy protein, emerin, is a nuclear membrane protein, *Hum. Mol. Genet.* **5:**801–808.

Martin, L., Crimaudo, C., and Gerace, L., 1995, cDNA cloning and characterization of lamina-associated polypeptide 1C (LAP1C), an integral protein of the inner nuclear membrane, *J. Biol. Chem.* **270:**8822–8828.

McConkey, D. J., 1996, Calcium-dependent, interleukin 1β-converting enzyme inhibitor-insensitive degradation of lamin B1 and DNA fragmentation in isolated thymocyte nuclei, *J. Biol. Chem.* **271:**22398–22406.

McKeon, F. D., Kirschner, M. W., and Caput, D., 1986, Homologies in both primary and secondary structure between nuclear envelope and intermediate filament proteins, *Nature* **319:**463–468.

Moir, R. D., Donaldson, A. D., and Stewart, M., 1991, Expression in *Escherichia coli* of human lamins A and C: Influence of head and tail domains on assembly properties and paracrystal formation, *J. Cell Sci.* **99:**363–372.

Moir, R. D., Montag-Lowy, M., and Goldman, R. D., 1994, Dynamic properties of nuclear lamins: Lamin B is associated with sites of DNA replication, *J. Cell Biol.* **125:**1201–1212.

Moore, M. S., 1995, Nuclear pores: David and Goliath in nuclear transport, *Curr. Biol.* **5:**1339–1341.

Nagano, A., Koga, R., Ogawa, M., Kurano, Y., Kawada, J., Okado, R., Hayashi, Y. K., Tsukahara, T., and Arahata, K., 1996, Emerin deficiency at the nuclear membrane in patients with Emery–Dreifuss muscular dystrophy, *Nat. Genet.* **12:**254–259.

Newport, J., and Spann, T., 1987, Disassembly of the nucleus in mitotic extracts: Membrane vesicularization, lamin disassembly, and chromosome condensation are independent processes, *Cell* **48:**219–230.

Ozaki, T., Saijo, M., Murakami, K., Enomoto, H., Taya, Y., and Sakiyama, S., 1994, Complex formation between lamin A and the retinoblastoma gene product: Identification of the domain on lamin A required for its interaction, *Oncogene* **9:**2649–2653.

Padan, R., Nainudel-Epszteyn, S., Goitein, R., Fainsod, A., and Gruenbaum, Y., 1990, Isolation and characterization of the Drosophila nuclear envelope otefin cDNA, *J. Biol. Chem.* **265:**7808–7813.
Paddy, M. R., Belmont, A. S., Saumweber, H., Agard, D. A., and Sedat, J. W., 1990, Interphase nuclear envelope lamins form a discontinuous network that interacts with only a fraction of the chromatin in the nuclear periphery, *Cell* **62:**89–106.
Panté, N., and Aebi, U., 1996, Molecular dissection of the nuclear pore complex, *Crit. Rev. Biochem. Mol. Biol.* **31:**153–199.
Paulin-Levasseur, M., Scherbarth, A., Traub, U., and Traub, P., 1988, Lack of lamins A and C in mammalian hematopoietic cell lines devoid of intermediate filament proteins, *Eur. J. Cell Biol.* **47:**121–131.
Paulin-Levasseur, M., Scherbarth, A., Giese, G., Röser, K., Bohn, W., and Traub, P., 1989, Expression of nuclear lamins in mammalian somatic cells lacking cytoplasmic intermediate filament proteins, *J. Cell Sci.* **92:**361–370.
Paulin-Levasseur, M., Blake, D. L., Julien, M., and Rouleau, L., 1996, The MAN antigens are non-lamin constituents of the nuclear lamina in vertebrate cells, *Chromosoma* **104:**367–379.
Peter, M., Nakagawa, J., Dorée, M., Labbé, J.-C., and Nigg, E. A., 1990, *In vitro* disassembly of the nuclear lamina and M phase-specific phosphorylation of lamins by cdc2 kinase, *Cell* **61:**591–602.
Powell, L., and Burke, B., 1990, Internuclear exchange of an inner nuclear membrane protein (p55) in heterokaryons: *In vivo* evidence for the interaction of p55 with the nuclear lamina, *J. Cell Biol.* **111:**2225–2234.
Pyrpasopoulou, A., Meier, J., Maison, C., Simos, G., and Georgatos, S. D., 1996, The lamin B receptor (LBR) provides essential docking sites at the nuclear envelope, *EMBO J.* **15:**7108–7119.
Rao, L., Perez, D., and White, E., 1996, Lamin proteolysis facilitates nuclear events during apoptosis, *J. Cell Biol.* **135:**1441–1455.
Röber, R.-A., Weber, K., and Osborn, M., 1989, Differential timing of nuclear lamin A/C expression in the various organs of the mouse embryo and young animal: A developmental study, *Development* **105:**365–378.
Röber, R.-A., Stauter, H., Weber, K., and Osborn, M., 1990, Cells of the cellular immune and hemopoietic system of the mouse lack lamins A/C: Distinction versus other somatic cells, *J. Cell Sci.* **95:**587–598.
Schuler, E., Lin, F., and Worman, H. J., 1994, Characterization of the human gene encoding LBR, an integral protein of the nuclear envelope inner membrane, *J. Biol. Chem.* **269:**11312–11317.
Senior, A., and Gerace, L., 1988, Integral membrane proteins specific to the inner nuclear membrane and associated with the nuclear lamina, *J. Cell Biol.* **107:**2029–2036.
Simos, G., and Georgatos, S. D., 1992, The inner nuclear membrane protein p58 associates *in vivo* with a p58 kinase and the nuclear lamins, *EMBO J.* **11:**4027–4036.
Simos, G., Maison, C., and Georgatos, S. D., 1996, Characterization of p18, a component of the lamin B receptor complex and a new integral membrane protein of the avian erythrocyte nuclear envelope, *J. Biol. Chem.* **271:**12617–12625.
Smith, S., 1995, Cloning and sequence analysis of an ERG24 homolog from *Schizosaccharomyces pombe*, *Gene* **155:**139–140.
Smith, S., and Blobel, G., 1993, The first membrane spanning region of the lamin B receptor is sufficient for sorting to the inner nuclear membrane, *J. Cell Biol.* **120:**631–637.
Smith, S., and Blobel, G., 1994, Colocalization of vertebrate lamin B and lamin B receptor (LBR) in nuclear envelopes and in LBR-induced membrane stacks of the yeast *Saccharomyce cerevisiae*, *Proc. Natl. Acad. Sci. USA* **91:**10124–10148.

Soullam, B., and Worman, H. J., 1993, The amino-terminal domain of the lamin B receptor is a nuclear envelope targeting signal, *J. Cell Biol.* **120:**1093–1100.

Soullam, B., and Worman, H. J., 1995, Signals and structural features involved in integral membrane protein targeting to the inner nuclear membrane, *J. Cell Biol.* **130:**15–27.

Spann, T. P., Moir, R. D., Goldman, A. E., Stick, R., and Goldman, R. D., 1997, Disruption of nuclear lamin organization alters the distribution of replication factors and inhibits DNA synthesis, *J. Cell Biol.* **136:**1201–1212.

Stewart, C., and Burke, B., 1987, Teratocarcinoma stem cells and early mouse embryos contain only a single major lamin polypeptide closely resembling lamin, *B. Cell* **51:**383–392.

Stewart, M., and Clarkson, W. D., 1996, Nuclear pores and macromolecular assemblies involved in nucleocytoplasmic transport, *Curr. Opin. Struct. Biol.* **6:**162–165.

Takahashi, A., Alnemri, E. S., Lazebnik, Y. A., Fernandes-Alnemri, T., Litwack, G., Moir, R. D., Goldman, R. D., Poirier, G. G., Kaufmann, S. H., and Earnshaw, W. C., 1996, Cleavage of lamin A by Mch2α but not CPP32: Multiple interleukin 1β-converting enzyme-related proteases with distinct substrate recognition properties are active in apoptosis, *Proc. Natl. Acad. Sci. USA* **93:**8395–8400.

Taniura, H., Glass, C., and Gerace, L., 1995, A chromatin binding site in the tail domain of nuclear lamins that interacts with core histones, *J. Cell Biol.* **131:**33–44.

Torrisi, M. R., and Bonnatti, S., 1985, Immunocytochemical study of the partition and distribution of Sindbis virus glycoproteins in freeze-fractured membranes of infected baby hamster kidney cells, *J. Cell Biol.* **101:**1300–1306.

van Wijnen, A. J., Bidwell, J. P., Fey, E. G., Penman, S., Lian, J. B., Stein, J. L., and Stein, G. S., 1993, Nuclear matrix association of multiple sequence-specific DNA binding activities related to SP-1, ATF, CCAAT, C/EBP, OCT-1, and AP-1, *Biochemistry* **32:**8397–8402.

Vigers, G. P., and Lohka, M. J., 1991, A distinct vesicle population targets membranes and pore complexes to the nuclear envelope in *Xenopus* eggs, *J. Cell Biol.* **112:**545–556.

Waitz, W. and Loidl, P., 1991, Cell cycle dependent association of *c-myc* protein with the nuclear matrix, *Oncogene* **6:**29–35.

Ward, G. E., and Kirschner, M. W., 1990, Identification of cell cycle-regulated phosphorylation sites in lamin C, *Cell* **61:**561–577.

Wiese, C., and Wilson, K. L., 1993, Nuclear membrane dynamics, *Curr. Opin. Cell Biol.* **5:**387–394.

Wolda, S. L., and Glomset, J. A., 1988, Evidence for modification of lamin B by a product of mevalonic acid, *J. Biol. Chem.* **263:**5997–6000.

Worman, H. J., and Courvalin, J.-C., 1991, Autoantibodies against nuclear envelope proteins in liver disease, *Hepatology* **14:**1269–1279.

Worman, H. J., Lazaridis, I., and Georgatos, S. D., 1988a, Nuclear lamin heterogeneity in mammalian cells. Differential expression of the major lamins and variation in lamin B phosphorylation, *J. Biol. Chem.* **263:**12135–12141.

Worman, H. J., Yuan, J., Blobel, G., and Georgatos, S. D., 1988b, A lamin B receptor in the nuclear envelope, *Proc. Natl. Acad. Sci. USA* **85:**8531–8534.

Worman, H. J., Evans, C. D., and Blobel, G., 1990, The lamin B receptor of the nuclear envelope inner membrane: A polytopic protein with eight potential transmembrane domains, *J. Cell Biol.* **111:**1535–1542.

Wreggett, K. A., Hill, F., James, P. S., Hutchings, A., Butcher, G. W., and Singh, P. B., 1994, A mammalian homologue of *Drosophila* heterochromatin protein 1 (HP1) is a component of constitutive heterochromatin, *Cytogenet. Cell Genet.* **66:**99–103.

Wydner, K. L., McNeil, J. A., Lin, F., Worman, H. J., and Lawrence, J. B., 1996, Chromosomal assignment of human nuclear envelope protein genes LMNA, LMNB1, and LBR by fluorescence *in situ* hybridization, *Genomics* **32:**474–478.

Yang, L., Guan, T., and Gerace, L., 1997, Integral membrane proteins of the nuclear envelope are dispersed throughout the endoplasmic reticulum during mitosis, *J. Cell Biol.* **137:**1199–1210.

Ye, Q., and Worman, H. J., 1994, Primary structure analysis and lamin B and DNA binding of human LBR, an integral protein of the nuclear envelope inner membrane, *J. Biol. Chem.* **269:**11306–11311.

Ye, Q., and Worman, H. J., 1995, Protein–protein interactions between human nuclear lamins expressed in yeast, *Exp. Cell Res.* **219:**292–298.

Ye, Q., and Worman, H. J., 1996, Interaction between an integral protein of the nuclear envelope inner membrane and human chromodomain proteins homologous to *Drosophila* HP1, *J. Biol. Chem.* **271:**14653–14656.

Ye, Q., Callebaut, I., Pezhman, A., Courvalin, J.-C., and Worman, H. J., 1997, Domain-specific interactions of human HP1-type chromodomain proteins and inner nuclear membrane protein LBR, *J. Biol. Chem.* **272:**14983–14989.

Yuan, J., Simos, G., Blobel, G., and Georgatos, S. D., 1991, Binding of lamin A to polynucleosomes, *J. Biol. Chem.* **266:**9211–9215.

Zhang, F. L., and Casey, P. J., 1996, Protein prenylation: Molecular mechanisms and functional consequences, *Annu. Rev. Biochem.* **65:**241–269.

Index